The Many-Body Problem

Advanced Book Classics
David Pines, Series Editor

Anderson, P. W., *Basic Notions of Condensed Matter Physics*
Bethe, H. and Jackiw, R., *Intermediate Quantum Mechanics, Third Edition*
Nozières, P., *Theory of Interacting Fermi Systems*
Pines, D., *The Many-Body Problem*
Quigg, C., *Gauge Theories of the Strong, Weak, and Electromagnetic Interactions*

The Many-Body Problem

David Pines
University of Illinois at Urbana-Champaign
Urbana, Illinois

The Advanced Book Program

Addison-Wesley
Reading, Massachusetts

ISBN 0-201-32834-8

Addison-Wesley is an imprint of Addison Wesley Longman, Inc.

Cover design by Suzanne Heiser

1 2 3 4 5 6 7 8 9 -MA-0100999897
First printing, November 1997

Find us on the World Wide Web at
http://www.aw.com/gb/

Editor's Foreword

Addison-Wesley's *Frontiers in Physics* series has, since 1961, made it possible for leading physicists to communicate in coherent fashion their views of recent developments in the most exciting and active fields of physics—without having to devote the time and energy required to prepare a formal review or monograph. Indeed, throughout its nearly forty-year existence, the series has emphasized informality in both style and content, as well as pedagogical clarity. Over time, it was expected that these informal accounts would be replaced by more formal counterparts—textbooks or monographs—as the cutting-edge topics they treated gradually became integrated into the body of physics knowledge and reader interest dwindled. However, this has not proven to be the case for a number of the volumes in the series: Many works have remained in print on an on-demand basis, while others have such intrinsic value that the physics community has urged us to extend their life span.

The *Advanced Book Classics* series has been designed to meet this demand. It will keep in print those volumes in *Frontiers in Physics* or its sister series, *Lecture Notes and Supplements in Physics*, that continue to provide a unique account of a topic of lasting interest. And through a sizable printing, these classics will be made available at a comparatively modest cost to the reader.

The lectures that introduce the reprints collected in the lecture-note reprint volume, *The Many-Body Problem*, were given over thirty-five years ago, at a time when many-body theory—an integrative approach to the study of the consequences of particle interactions in condensed matter physics, nuclear physics, and atomic physics—was just beginning to emerge as a distinct sub-field of theoretical physics. In the lecture note portion of the volume, I made every effort to bring out the underlying physics of the problems under consideration, while introducing the reader to the field-theoretic techniques whose application to these systems was in its infancy. Many of the papers reprinted in the volume are the primary source for the further developments in this field, and, as such, are required reading for graduate students and experienced researchers alike. Somewhat to my surprise, the lecture notes it contains continue to provide a useful, elementary, and in some ways, unique introduction to the basic ideas and mathematical formalism of many-body theory, and I am pleased that the *Advanced Book Classics* series will make the volume readily accessible to new generations of readers.

David Pines
Urbana, Illinois
October 1997

CONTENTS

PREFACE

These lecture notes represent a revised version of the author's notes on lectures that have been delivered during the past two years at Stanford University (January 1960), IBM Laboratories (April 1960), General Atomic Division of General Dynamics Corporation (July-August 1960), the University of Illinois (September-January 1960-1961), and the Summer School on Liquid Helium of the Italian Physical Society (July 1961). There is appended to the notes a collection of reprints as supplementary reading; therefore I have eliminated those parts of the notes which in large measure duplicate material that is readily available in the reprints. I should like to thank Angelo Bardasis, Daniel Hone, and Seth Silverstein for making their notes on the University of Illinois lectures available to me. The support of the U. S. Army Research Office, Durham, North Carolina, during the preparation of the present version of the notes is gratefully acknowledged.

DAVID PINES

Urbana, Illinois
August 1961

ACKNOWLEDGMENTS

The publisher wishes to acknowledge the assistance of the following organizations in the preparation of this volume:

The American Institute of Physics, for permission to reprint the articles from the Physical Review and Soviet Physics JETP.

The Royal Society, for permission to reprint the articles from their Proceedings.

The Italian Physical Society, for permission to reprint the articles from Il Nuovo cimento.

The Many-Body Problem

1 A SURVEY, WITH EMPHASIS ON PHYSICAL CONSIDERATIONS

INTRODUCTION

The many-body problem is the study of the way that the interaction between particles in a dense system alters the behavior of the isolated, noninteracting particles. These system properties are of interest:

1. Ground-state energy.

2. Elementary excitation spectra: In general, there are two kinds of low-lying excited states of a many-body system; quasi particles, which are modified single-particle excitations, and collective modes, which arise from the correlations in the particle motions brought about by particle interaction.

3. Various temperature-dependent phenomena: The specific heat, phase transitions, P-V diagrams, etc., for the equilibrium state; the study of irreversible processes, such as conductivity, viscosity, etc., for nonequilibrium states.

In these lectures I shall focus my attention almost entirely on the ground-state energy and elementary excitation spectrum of systems at $T = 0$, so that the properties listed under (3) will be discussed only very briefly, if at all. Many of these properties can be obtained by straightforward extension of the zero-temperature techniques.

The recent developments in the many-body problem, which have tended to change it from a quiet corner of theoretical physics to a major crossroad, are perhaps threefold. First, and most important, has been the Bardeen, Cooper, and Schrieffer theory of superconductivity.[28] † Not only

† *Note:* Numbered references refer to the list of papers for supplementary reading; other references may be found at the end of the notes, where they are listed by topic.

has their theory provided an explanation of superconductivity, but it has also stimulated a number of interesting developments in the theory of nuclear structure, He^3, and elementary particles.

Second has been the development of a unified point of view toward many-body problems. By that I mean a qualitative and often quantitative understanding of the interrelationship between quasi-particle and collective behavior in complicated systems, and how both derive from the basic interactions between the particles making up the system. For example, 10 years ago one had the shell model of the nucleus, in which the nucleons were regarded as moving independently in a potential well (Mayer and Jensen). One also had the collective states, associated with nuclear deformations (Bohr and Mottelson) and dipole oscillations (Goldhaber and Teller); there was little understanding of the relationship between the excitations, or how either could be understood in terms of a collection of nucleons interacting via a combination of a strong short-range repulsion and a longer-range attraction. The first breakthrough came when Brueckner and his collaborators showed by the use of multiple-scattering techniques that the strong short-range interactions could be handled mathematically and would not produce an appreciable effect on the single-particle motion, thus justifying the notion of the quasi particles distributed according to the rules of the shell model. Next, and rather more recently, a number of people (Mottelson, Brown, Baranger, Beliaev, and others), using techniques borrowed from superconductivity and the theory of the electron gas, have shown how the collective modes arise from the weak residual interactions between the shell-model quasi particles. Thus we now understand how one passes from the basic nuclear interactions to a gas of quasi particles with rather weak residual interactions which give rise to the collective excitations.

Third has been the development of model solutions for the many-body problem. By model I mean a many-body system whose properties are calculated in a well-defined approximation which can be shown to be valid for a given class of particle interactions and densities. For instance, Hugenholtz and Galitskii showed that the low-density limit represents the limiting case in which the Brueckner theory was correct. As another example, consider the electron gas at densities comparable to those found in metals. Some years ago Bohm and I developed a theory based on the random-phase approximation, in which we showed how the Coulomb interactions gave rise to collective modes, the plasmons, and were also rather strongly screened by particle correlations.[4,5] Thus, one obtained an understanding of the coexistence of quasi-particle and collective modes and how both arose from the electron interaction. However, the theory was somewhat inelegant in that the collective modes were treated on a quite different basis from the quasi-particle modes by introducing collective coordinates; a number of people questioned both the introduction of extra coordinates for this purpose and the validity of the random-phase approximation. The further developments on this problem came when Gell-Mann and Brueckner[6] showed that the random-phase approximation was valid in the high-density limit by an explicit summation of the perturbation series, and Hubbard showed how that series gave rise to the plasmons as well as the single-particle modes, and the relationship between the two.

The solution of the model problems has been aided considerably by the application of the field-theoretic techniques developed by Feynman, Schwinger, Dyson, et al. for quantum electrodynamics. These techniques enable one to sum easily a large class of terms in the perturbation-series expansion for a given quantity; they provide a simple way of estimating the region of validity of a given model calculation. Moreover, the propagator techniques offer a way of formulating rigorously certain questions of fundamental interest (i.e., what we mean by an elementary excitation). The field-theory techniques have been available for some time. What presumably slowed their application was the fact that almost all many-body problems of interest contain divergences if one looks at the first few terms in the perturbation series; it is only when the series is summed that one obtains sensible expressions.

In this section I wish to concentrate on the physical considerations of importance in the many-body problem, with only a rather cursory glance at the mathematical techniques that have been developed. I shall then give a brief survey of a number of specific problems, in order to indicate something of the present status of the field. In succeeding sections I plan to discuss in further detail some of the new mathematical principles and techniques, and to apply them to the following problems: the electron gas, systems of fermions and bosons interacting with repulsive interactions, the electron-phonon system, and superconductivity.

ELEMENTARY EXCITATIONS IN SYSTEMS OF INTERACTING PARTICLES

We shall be particularly concerned with two kinds of elementary excitations in a system of interacting particles. The first kind, which derive from the single-particle excitations, we shall call quasi particles. The second kind, which are of importance in determining the response of the system to an external probe, we shall call density-fluctuation excitations. The latter may be composed of quasi-particle excitations, or represent a new kind of excitation, a collective mode of the system, which would be absent if there were no interaction between the particles.

For a noninteracting system, the energy of a single particle with momentum p and mass m is

$$\epsilon(p) = p^2/2m$$

As a consequence of the interaction between the particles, the single-particle motion is considerably modified. As a particle moves along, it pushes other particles out of its way, drags particles along with it, etc. We may speak of the modified particle as a quasi particle; it may have a very different energy versus momentum dependence from that of a free particle. At sufficiently low temperatures, the system may be regarded as made up of independent quasi particles, and the specific heat may be determined from the quasi-particle spectrum. Moreover, as first shown by Landau, the form of the quasi-particle spectrum will determine the flow properties of the system at very low temperatures. Thus, one may distinguish between normal systems, in which the energy versus momentum curve of the quasi particles is quadratic in the particle momenta, and superfluids, in which it is either linear (liquid helium) or possesses a gap (superconductors). The existence of superfluid behavior in the latter system follows from considerations of Galilean invariance and energy and momentum conservation applied to the creation of an elementary excitation by, say, an impurity, in a slowly moving system.

The notion of a quasi particle can be put on a reasonably rigorous mathematical basis by the introduction of a concept familiar in quantum field theory, the single-particle propagator, defined by

$$G(p,\tau) = -i\langle 0 \,|\, T[C_p(\tau)\, C_p^+(0)] \,|\, 0\rangle$$

Here C_p^+ and C_p are the second quantized creation and annihilation operators for fermions, expressed in the Heisenberg representation, T is the Dyson chronological operator, which orders earlier times to the right, and the expectation value is taken with respect to the exact ground-state wave functions in the Heisenberg representation. $G(p,\tau)$ possesses the following physical significance; if at time $t = 0$, one adds a particle of momentum p to the system (the system wave function is then $\psi = C_p^+ \,|\, 0\rangle$), the probability amplitude that the system will be found in that state at a later time τ is $G(p,\tau)$. Galitskii and Migdal[3] have shown that the analytic behavior near the real ϵ axis of $G(p,\epsilon)$, the Fourier transform of $G(p,\tau)$, determines the energy and lifetime of a quasi particle of momentum p.

Let us look at the quasi-particle spectrum of three systems: normal fermions, superconductors, and bosons. In a fermion system, in the absence of interaction, the system in its ground state will possess a momentum distribution that is a filled Fermi sphere. The probability of finding a particle in a state of momentum p, N_p, is given by

$$\begin{aligned} N_p &= 1 \qquad p < p_F \\ &= 0 \qquad p > p_F \end{aligned}$$

as shown in Fig. 1-1. We are interested in the excitations above this ground state. Excited states of the system may be characterized by the distribution of particles (particles with $p > p_F$ present outside the Fermi sphere) and holes (particles with $p < p_F$ absent inside the Fermi sphere). Measured with respect to the ground state, the energy of a particle of momentum p is $\epsilon(p)$; that of a hole of momentum p′ is $-\epsilon(p')$. At finite temperatures one easily derives the usual free-electron linear specific heat from such considerations.

In the presence of particle interaction, one might hope that vestiges of this noninteracting system behavior would remain, in the following sense:

1. There continues to be discontinuity in N_p at $p = p_F$, the Fermi surface, although the magnitude of the discontinuity may be much altered, as seen in Fig. 1-1.

2. The low-lying excited states are well described by a set of independent quasi particles (and quasi holes); the energy $\epsilon(p)$ of a quasi particle will, in general, be complex, corresponding to the fact that single-particle states are no longer well defined. It is always possible for a particle with $p > p_F$ to scatter against the particles present inside the Fermi surface into another state. Thus, the quasi particles will possess a lifetime. For particles close to the Fermi surface, that lifetime will be very long, since Pauli-principle restrictions cause the scattering probability (and hence the imaginary part of the energy) to go as $(p - p_F)^2$. One further expects a linear term in the specific heat, with an altered coefficient, corresponding to a change in the effective mass of a particle at the Fermi surface. Systems that display such behavior (and we possess no definite proof that such systems exist in nature) are called normal fermion systems.

A more-mathematical way of describing normal fermion systems is to say that, for example, one can calculate $G(p, \epsilon)$ by systematic application of the Feynman diagrammatic techniques, in which the basic elements of the series are the propagators $G_0(p, \epsilon)$ for a noninteracting fermion

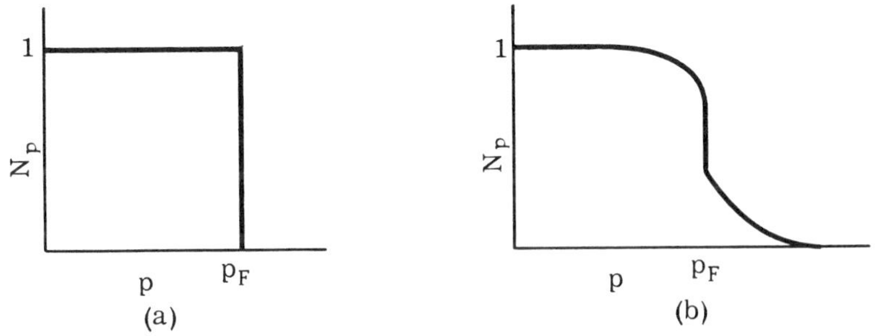

FIG. 1-1. (a) Noninteracting system. (b) Interacting normal fermion system.

system. In other words, normal fermion systems are those for which perturbation theory works. Many elegant theorems for such systems have been recently established by Kohn, Luttinger, and Ward, following the pioneer work of Galitskii and Migdal. Fermions interacting via repulsive forces, either short range or long range, are the best bet for a normal fermion system.

A quite different approach to the study of normal fermion systems is the Fermi liquid theory of Landau.[15,16] Landau assumes at the outset the existence of quasi particles obeying Fermi-Dirac statistics and then shows how the various properties of the system may be expressed in terms of a relatively small number of parameters. Landau's theory is phenomenological; the parameters are to be determined from experiment. In a sense his theory is complementary to the perturbation-theory approach mentioned above, in that it is probably most useful for systems for which systematic application of perturbation theory is not promising. Landau's theory assumes that the lifetimes of the quasi particles are sufficiently long to permit the neglect of any sort of lifetime effects, which means that it is essentially a very low temperature theory. The recent experiments of Wheatley and his collaborators indicate that He^3 behaves like a Fermi liquid at temperatures below about 0.04°K.

The quasi particles in a normal fermion system possess a continuous excitation spectrum; those in a superconductor display an energy gap. In the BCS theory, the quasi-particle energy is

$$E(p) = [\tilde{\epsilon}^2(p) + \Delta^2]^{1/2}$$

where $\tilde{\epsilon}(p)$ is the energy of the "normal-system" quasi particle measured relative to the Fermi surface and Δ is the energy gap. Moreover, there is no discontinuity in N_p at the Fermi surface. The alteration in the excitation spectrum is a consequence of an effective attractive interaction between the electrons in a superconductor; it gives rise directly to many of the phenomena observed in superconductivity (Meissner effect, exponential specific heat, infinite conductivity, etc.). We further remark that on the basis of the BCS theory one expects an energy gap in the excitation spectrum (and accompanying superfluid behavior) for any fermion system with attractive interactions. Therefore it may well be that He^3 at sufficiently low temperatures will display a transition from its Fermi-liquid behavior (viscosity proportional to T^{-2}) to superfluid behavior (a complete absence of viscosity).

In a noninteracting boson system at absolute zero, all particles will be in the state of zero momentum. An excitation of momentum p will possess the free-particle energy $\epsilon(p) = p^2/2m$. However, when interactions between the bosons are taken into account, the quasi-particle excitation spectrum is drastically altered and should, apart from possible pathological cases, be of the form

$$\epsilon(p) = sp$$

in the low-momentum region. The alteration in the excitation spectrum is a natural consequence of the assumption of continued macroscopic occupation of the zero-momentum state in the interacting boson system. It follows provided perturbation theory may be used to determine the propagator for the interacting boson system from that for the free boson system, i.e., provided the boson system is "normal." As shown in specific-heat measurements (a T^3 variation of low temperatures), the quasi-particle spectrum of liquid He^4 takes this form in the low-momentum region.

COLLECTIVE MODES

One does not measure the quasi-particle spectrum directly in an experimental investigation of a many-body system by means of external probes. Such a probe experiment might involve the measurement of the energy and angular distribution of inelastically scattered particles, under circumstances that the Born approximation applies to the description of the scattering act. Examples are the scattering of fast electrons in solids, or slow neutrons in solids and liquid He. The interaction between the particle used as a probe and the particles in the system is of the form

$$\sum_i v(\mathbf{R} - \mathbf{r}_i) = \sum_{ki} v_k e^{i\mathbf{k}\cdot(\mathbf{R}-\mathbf{r}_i)} = \sum_k v_k \rho_k e^{i\mathbf{k}\cdot\mathbf{R}}$$

where v(r) is the effective potential of interaction and v_k and ρ_k are the Fourier transforms of the potential and the density, respectively, defined according to

$$v(r) = \sum_k v_k e^{i\mathbf{k}\cdot\mathbf{r}}$$

$$\rho(r) = \sum_i \delta(\mathbf{r} - \mathbf{r}_i) = \sum_k \rho_k e^{i\mathbf{k}\cdot\mathbf{r}} = \sum_i e^{i\mathbf{k}\cdot(\mathbf{r}-\mathbf{r}_i)}$$

The density fluctuation

$$\rho_k = \sum_i e^{-i\mathbf{k}\cdot\mathbf{r}_i}$$

describes the fluctuations in partial density about the average value N. If the Born approximation applies, the probability per unit time that the probe transfer momentum k and energy ω to the many-body system is given by

$$W = (2\pi) v_k^2 \sum_n (\rho_k^+)_{n0}^2 \delta(\omega - \omega_{n0})$$

where the ω_{n0} are the exact excitation frequencies of the many-body system induced by the density fluctuations ρ_k^+. (We shall take $\hbar = 1$ throughout these lectures.) Thus, the maximum information one gleans about the many-body system is contained in the quantity

$$S(k\omega) = \sum_n |(\rho_k^+)_{n0}|^2 \delta(\omega - \omega_{n0})$$

As Van Hove has first shown (and as we shall see in a subsequent lecture), $S(k\omega)$ is the Fourier transform in space and time of the time-dependent pair-distribution function for the many-particle system. What one measures, therefore, is the spectrum of elementary excitations available to the density fluctuations ρ_k^+ of the system. Since we assume the system initially in its ground state, the measured energies are those of states of momentum k, which are coupled to the ground state by the density fluctuation ρ_k^+.

For a noninteracting system, a typical energy difference ω will be

$$\omega(\mathbf{k}) = (\mathbf{p} + \mathbf{k})^2/2m - p^2/2m = \mathbf{k} \cdot \mathbf{p}/m + k^2/2m$$

corresponding to a single particle making a transition from a state of momentum $\mathbf{p}$ to a state $\mathbf{p} + \mathbf{k}$. For the noninteracting fermion system, the process may be regarded as one in which a particle of momentum $\mathbf{p} + \mathbf{k}$ and a hole of momentum $\mathbf{p}$ are created. Because of the Pauli principle, there is the further restriction that $p \leq p_F$; $|\mathbf{p} + \mathbf{k}| \geq p_F$. Thus particle-hole pairs are created, with a continuous energy spectrum lying between the curves specified by

$$\omega_{min} = 0 \qquad k \leq 2p_F$$

$$= k^2/2m - kp_F/m \qquad k \geq 2p_F$$

and

$$\omega_{max} = k^2/2m + kp_F/m$$

as shown in Fig. 1-2.

When the interaction between the fermions is taken into account, the density-fluctuation-excitation spectrum will change in two ways. First, the spectrum of particle-hole pairs will be changed to one of pairs of quasi particles (a quasi particle and a quasi hole). Second, there will appear a new kind of long wavelength excitation, a collective mode, which in lowest approximation (the random-phase approximation) lies above and is distinct from the quasi-particle continuum. For the electron gas, the collective mode is the plasmon, a quantized plasma oscillation, which for long wavelengths has an energy

$$E(k) \cong \omega_p = (4\pi ne^2/m)^{1/2}$$

For a fermion system with repulsive interactions, the collective mode is called zero sound; for weak interactions it lies only slightly above the particle-hole continuum. These collective modes may be regarded as coherent superpositions of particle-hole pairs. In both cases, as the wavelength is decreased, the collective modes merge into the particle-hole continuum, as shown in Fig. 1-2.

Collective modes resemble sound waves, in that they correspond to oscillations in the particle density. However, they differ from sound waves in the mechanism responsible for the oscillations. A sound wave is a characteristic hydrodynamic effect; it occurs for systems in which local thermodynamic equilibrium exists in consequence of the frequent collisions between the particles. When there is a local perturbation in the density, the particle collisions act as a restoring force to bring the density back to its average value. Sound waves thus require frequent collisions and exist only for frequencies ω, such that

$$\omega\tau << 1$$

where τ is the average time between particle collisions. For the collective modes, on the other hand, the averaged force field of a large number of other particles acts as a restoring force. Collisions between

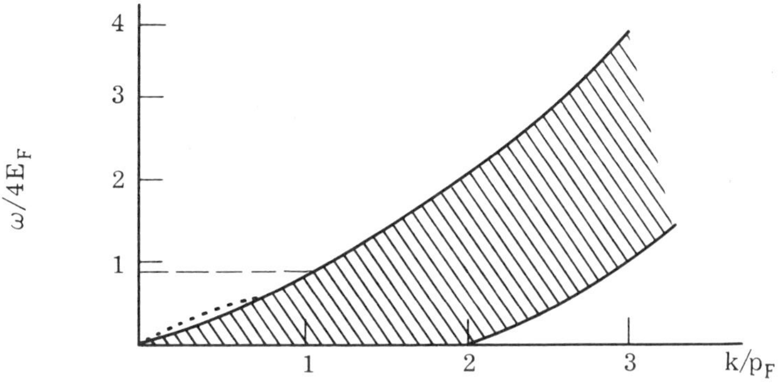

FIG. 1-2. Dashed lines, plasmons; dotted lines, zero sound.

particles tend to disrupt this averaged field and so act to damp the collective mode. As a result, collective modes exist only when their frequency satisfies the condition

$$\omega\tau >> 1$$

which may always be satisfied at sufficiently low temperatures. He^3 offers an example of a system in which it may be possible to observe the change from ordinary sound to zero sound by lowering the temperature of the system.

A further requirement for the collective modes to exist as an independent well-defined elementary excitation of the many-body system is that they possess a long lifetime against decay into quasi-particle pairs, a decay which is always energetically possible but which may be unlikely as a result of coherence effects on the relevant matrix elements. Thus, the collective-mode energies, like those of the quasi particles, will in general be complex.

In a superconductor there is a contribution to the density-fluctuation excitations from quasi-particle pairs; the energies of such excitations form a continuum, with a minimum excitation energy of 2Δ. There are, in addition, two kinds of collective modes; the plasmons, which possess energies only slightly different from their values in the normal state, and exciton-like states, which lie within the gap 2Δ which exists for quasi-particle pair excitation.

For a noninteracting boson gas, the density fluctuations possess a unique frequency, $k^2/2m$, corresponding to the possibility of taking a particle from state of momentum zero to the state k. Thus, the density-fluctuation excitation energy is identical with the free-particle energy. This feature persists to a considerable extent in interacting boson systems. For example, the elementary excitation-energy spectrum measured by inelastic scattering of slow neutrons agrees, within experimental error, with the quasi-particle spectrum proposed by Landau to explain the specific-heat measurements. This identity indicates that the dominant contribution to the density-fluctuation spectrum arises from the excitation of quasi particles out of the condensed state.

Collective modes exist in other systems as well. In a ferromagnet one has spin waves that correspond to oscillations of the spin-density fluctuations, $S_k^z = \Sigma_i \sigma_i^z \exp(-i\mathbf{k}\cdot\mathbf{r}_i)$. These are also a possible excitation of a Fermi liquid. In the electron-ion system there exists in addition to the high-frequency plasma mode, in which the electron and ion move out of phase with one another, a low-frequency mode, in which they move in phase. This is the longitudinal phonon mode of metals. By analogy with the excitation spectrum of ionic crystals, we may call the high-frequency mode an optical plasma mode; the low-frequency mode, an acoustic plasma mode. There exists the possibility of both optical and acoustic plasmons (the frequency of the latter being proportional to k at long wavelengths) for any system of coupled charged particles, e.g., electrons and holes in semiconductors.

PHYSICAL CONCEPTS

I should like to summarize briefly some of the principal physical concepts that one encounters in all facets of the many-body problem. One of the most important, which we have been discussing, is that of the correlations in particle positions brought about by their mutual interaction. Such correlations give rise to important coherence effects. For example, where a density-fluctuation collective mode exists as a well-defined elementary excitation, it is usually the dominant excitation mode of the system—this, despite the fact that there is but a single collective mode, compared to the continuum of quasi-particle excitations. The reason is that a large number of particles take part in the collective mode. The correlations responsible for the collective mode likewise give rise to coherence factors which act to reduce sharply the probability of exciting quasi-particle pairs, since the over-all energy transfer is governed by a simple sum rule.

Such correlations frequently take the form of screening—the alteration in the effective interaction of a pair of particles brought about by the remaining particles in the system. For example, a given electron in a dense electron gas acts to polarize its immediate surroundings; it pushes other electrons out of its way until its associated screening cloud possesses a charge nearly equal (and opposite) to its own. The quasi particles (electrons plus screening clouds) interact via an effective short-range interaction of the order of the interparticle spacing, in contrast to the original long-range Coulomb interaction. The plasma oscillation is a polarization wave associated with that screening action.

Other important correlations are those associated with the repeated scattering of a pair of particles against one another with no other particles in the medium playing a role. Such multiple-scattering effects are of special importance for low-density systems of either bosons or fermions.

In a superconductor there are strong correlations between particles of opposite spin and momentum in consequence of the effective attractive interaction between the electrons. These pairing correlations give rise to pronounced coherence effects, so that, for example, the absorption of electromagnetic radiation in a superconductor just below the transition temperature differs markedly from the absorption of sound waves.

Another concept of great importance is that of backflow, which was first introduced by Feynman and Cohen in their theory of excitations in

He^4. The longitudinal-current conservation law requires that for a stationary state of a many-body system the current associated with the motion of an excitation should be divergence-free. Hence, there must be associated with the motion of a particle-like excitation a return flow of other particles in such a way as to ensure longitudinal-current conservation. Backflow for a system of interacting particles corresponds to the cloud of virtual particle excitations which surround and move with a given particle in such a way that the resulting quasi particle represents a suitable elementary excitation of the system. Longitudinal-current conservation and backflow are of particular importance in the theory of superconductivity; it is only when these concepts are properly included in the theory that one obtains gauge-invariant results (Anderson,[32] Pines and Schrieffer, and Rickayzen[33]).

MODEL PROBLEMS AND PHYSICAL SYSTEMS

I should now like to discuss briefly several of the model calculations that have been carried out in recent years. In so doing, I shall attempt to summarize the cogent features of the model, including the diagrams that are summed in a field-theoretic formulation. I shall also try to indicate the relevance of the model calculation to the corresponding physical system.

High-Density Electron Gas

1. The ground-state energy per electron can be expressed in a series expansion in r_s, the interelectron spacing measured in units of the Bohr radius a_0,

$$r_s = (3/4\pi a_0^3 N)^{1/3}$$

The expansion is (Gell-Mann and Brueckner[6])

$$E_0\,(\text{ryd}) = 2.21/r_s^2 - 0.916/r_s + 0.062 \ln r_s + 0.096 + e r_s \ln r_s + f r_s + \cdots$$

where the coefficients e, f, etc., have not yet been determined. The first term corresponds to the average kinetic energy per electron, the second is the exchange energy that arises from the Pauli-principle correlations, and the remaining terms are called the correlation energy, to denote the reduction in energy brought about by charge-induced correlations in the electron motion.

2. The expansion parameter r_s represents the ratio of the lowest-order potential-energy term, the exchange energy, to the average kinetic energy of the electrons. The theory is thus a weak coupling theory. The expansion is valid for $r_s < 1$ (Nozières and Pines[12]).

3. The elementary excitations are

a. Quasi particles for which the energy versus momentum curve in the immediate vicinity of the Fermi surface is (Gell-Mann,[7] Quinn and Ferrell, DuBois)

$$\epsilon(p) = (p_F^2/2m)\{p^2/p_F^2 - 0.166r_s[p/p_F(\ln r_s + 0.203) - \ln r_s + 1.80] + i(0.252\, r_s^{1/2})[(p/p_F) - 1]^2\}$$

Corrections to this will be of higher order in r_s.

b. Plasmons for which the energy versus momentum curve at long wavelengths is given by (Bohm and Pines,[4] Nozières and Pines,[12] DuBois)

$$E(p) = \omega_p\{1 + (p^2/p_F^2)[(1.3/r_s) - (3/40) + i(4.4r_s^{1/2}) + \cdots]\}$$

Correction terms are again of higher order in r_s and involve only the terms of order p^2.

4. The principal physical effect is the polarization of the medium, that is, the screening of the effective interaction between a pair of particles as a result of the motion of the other particles in the medium. The screening can be simply expressed in terms of the longitudinal frequency and wave-vector-dependent dielectric constant, and corresponds to the polarization cloud that follows a given electron on its motion.

5. The principal approximation that describes the high-density electron gas is the random-phase approximation (RPA) (Bohm and Pines[4,5]), in which each momentum transfer in the interaction between the electrons is treated independently. For example, in the calculation of $S(k\omega)$ one would consider only the k-th component of the Coulomb interaction. The diagrams that are summed are the lowest-order polarization diagrams (Hubbard[11]). Thus the particles may be regarded as interacting via a screened Coulomb interaction, $v_k/\epsilon(k\omega)$, according to the diagrams of Fig. 1-3. Successive terms in the perturbation-series expansion for $1/\epsilon(k\omega)$ at low excitation frequencies are more and more divergent; it was this kind of divergence that led to serious difficulties with the use of second-order perturbation theory. However, it is easy to sum the series, corresponding to the summation of all chains of polarization bubbles. One finds

$$\epsilon(k\omega) = 1 + 4\pi\,\alpha_0(k\omega)$$

where $4\pi\alpha_0(k\omega)$ is the free electron polarizability, corresponding to the simple bubble diagram.

6. The electron gas serves as a model for the behavior of electrons in metals, a model in which the lattice of positive ions is replaced by a uniform background of positive charge. The density of metallic electrons is such that one encounters values of r_s given by

$$1.8 \lesssim r_s \lesssim 5.5$$

Thus, one cannot take over directly the results obtained in the RPA, or by any perturbation-theoretic calculation of the corrections thereto, since the high-density electron-gas-series expansion loses its validity for $r_s \gtrsim 1$. Various interpolation procedures (Hubbard,[11] Nozières and Pines[12]) have been proposed to deal with the ground-state energy of electron gas in the intermediate density region characteristic of metallic densities. It is estimated that these give results accurate to about 10 per cent. The experimental results on cohesion in the alkali metals are in good agreement with the values predicted with the aid of these interpolation formulas.

7. It is straightforward to develop a theory of plasmons in solids based on the RPA. The theoretical predictions are in good quantitative agreement with the results of experiments on the inelastic scattering of kilovolt electrons from thin solid films and confirm the existence of the plasmon as a well-defined elementary excitation in a large class of solids (Pines, Nozières and Pines).

Low-Density Electron Solid

1. At sufficiently low densities the Coulomb interaction exerts a dominating influence on the behavior of the electrons, and one has the strong

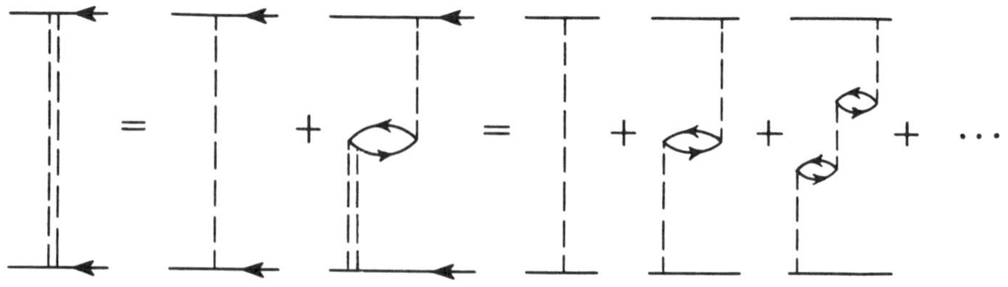

FIG. 1-3

coupling limit for the electron gas. The electrons may then be expected to form a stable lattice in a background of uniform positive charge, since the potential energy keeps the electrons apart and the kinetic energy is insufficient to prevent the electrons from becoming localized at fixed sites (Wigner).

2. The ground-state energy possesses a series expansion in $r_s^{1/2}$ which is (Wigner, Carr)

$$E_0 = (1.79/r_s) + (2.66/r_s^{3/2}) + (a/r_s^2) + (b/r_s^{5/2}) + \cdots$$

The first term is the potential energy of the electrons localized at fixed sites, the second the zero-point oscillations of the electron oscillations about these sites, the third and higher terms in the power series in $r_s^{-1/2}$ arise from anharmonicities in the lattice vibration. There are also terms proportional to $\exp(-cr_s^{1/2})$, which arise from exchange effects (Carr).

3. As the density of the electron gas is increased, the electron solid will melt. Estimates of the value of r_s for which melting would occur show $r_s \gtrsim 20$ (Nozières and Pines[12]). Thus the electron solid represents an extreme-low-density limiting case for the electron gas.

Dilute Fermion Gas with Arbitrary Repulsive Interactions of Finite Range

1. The ground-state energy particle can be expressed in a series expansion in f_0/r_0, where f_0 is the s-wave free-particle scattering length and r_0 is the average interparticle spacing. One has (Huang and Yang, de Dominicis and Martin, Galitskii[18])

$$E_0/N = (3/5)(p_F^2/2m)\,[1 + (10/9\pi)p_F a + (4/21\pi^2)(11 - 2\ln 2)\, p_F^2 a + \cdots]$$

2. The elementary excitations are of two kinds:

a. Quasi particles for which the energy versus momentum curve in the immediate vicinity of the Fermi surface is given by (Galitskii[18])

$$\epsilon(p) = (p_F^2/2m)\{1 + (4/3\pi)\, p_F a + (4/15\pi^2)\, p_F^2 a^2\,[7 + 26 \ln 2] + (p^2/p_F^2) - (16/15\pi^2)\, p_F^2 a^2 (p/p_F)(7 \ln 2 - 1) - i\,(p_F^2 a^2/\pi)\,[(p/p_F)^{-1}]^2 + \cdots\}$$

b. Zero-sound vibrations, whose energy is (Gottfried and Pičman)

$$E(p) = (pp_F/m)\{1 + 0.27\,[\exp - (\pi/p_F a)] \times [1 + 0.57(p^2/p_F^2)\exp(2\pi/p_F a)]\}$$

3. The expansion parameter a/r_0 is characteristic of a low-density system. It is a reflection of the fact that the principal effect of interest is the multiple scattering of a pair of isolated particles, with the other particles in the medium giving rise to small corrections in this scattering. Thus, once one excites a pair of particles, they prefer to scatter repeatedly against one another, instead of creating particle-hole pairs (screening), etc. By taking into account the multiple scattering, one can deal with arbitrarily strong repulsive interactions, including even hard spheres.

4. The principal approximation involves the summation of the ladder diagrams characteristic of multiple-scattering problems (Brueckner, Galitskii[18]). In lowest order one replaces the vertex for particle scattering by one in which the repeated scattering of a pair of particles is taken into account. Thus one has the diagram of Fig. 1-4. The resulting integral equation is simply solved in the low-density limit, and one finds in lowest order that $t = 4\pi a/m$. Thus multiple scattering is included if one replaces each vertex by the scattering length, an approach that is equivalent to the use of a pseudopotential (Lee, Huang, and Yang,[21] Lee and Yang).

5. The dilute fermion gas serves as a model for the behavior of nuclear matter and He^3. It is not a particularly good model, because neither of these systems is at all dilute; moreover, both involve attractive forces that act to alter the system behavior by bringing in the pairing correlations characteristic of superconductors. The principal calculations on nuclear matter and He^3 have been made by Brueckner and his collaborators, by methods equivalent to summing a selected set of higher-order diagrams in the perturbation theory. It is difficult to assess the validity of the Brueckner calculations.

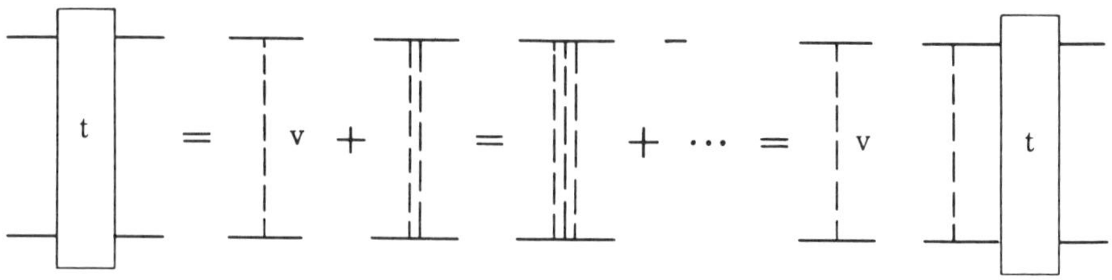

FIG. 1-4

The BCS Model for Superconductivity

1. Bardeen, Cooper, and Schrieffer[28] consider a system of fermions interacting via an effective interaction that is constant, and attractive, for particles located in the vicinity of the Fermi surface, and zero otherwise. They obtain an energy difference between the configurations of the normal and superconducting states which is given by

$$\Delta E = -2N(O)\omega^2 \exp[-2/N(O)V]$$

for the case of weak attractive interactions. Here ω is the width in energy by which the attractive region extends above the Fermi surface, V is the magnitude of the attractive force, and N(O) is the density of states in energy at the Fermi surface.

2. The elementary excitations are

a. Quasi particles for which one finds

$$E(p) = [\tilde{\epsilon}^2(p) + \Delta^2]^{1/2}$$

where $\tilde{\epsilon}(p)$ is the "normal-state" quasi-particle energy measured with respect to the energy of a particle on the Fermi surface, and Δ is the energy gap, given by

$$\Delta = 2\omega \exp[-1/N(O)V]$$

for weak interactions.

b. Collective excitations of two kinds (Anderson,[32] Rickayzen,[33] Bogoljubov et al.). The first are plasmons, for which the energy versus momentum curve is only very slightly altered by a change in the coefficient of p^2 in the superconducting state. The second are exciton states that correspond to a pair of quasi particles bound together moving with a net center of mass momentum p. The spectrum of the exciton states is strongly dependent on the effective two-body interaction (Bardasis and Schrieffer).

3. The principal physical effect to which the attractive force between fermions gives rise is the pairing of particles of opposite spin and momentum. Such particles are singled out because in suitable approximation two particles of opposite spin and momentum may form a bound state, no matter how weak the attraction between them (Cooper[27]). Thus, the Fermi sea is unstable against the formation of bound pairs. The BCS theory may be regarded as allowing in self-consistent fashion for the formation of virtual bound-pair states. The resultant ground state, which cannot be obtained by a perturbation-theoretic calculation based on "normal-state" wave functions, is stable against pair creation in virtue of the

and scattering length of the He atoms are such as to render the series expansion in $(Nf_0^3)^{1/2}$ meaningless. The elementary excitation curve observed in slow-neutron-scattering experiments is in good agreement with the variational calculation of Feynman and Cohen. An accurate microscopic calculation of the spectrum does not yet exist; the closest approach has been made by Brueckner and Sawada, who sum a selected class of higher-order terms in the perturbation series in a manner closely related to the Brueckner nuclear-matter calculations. Their calculation, which yields a qualitatively correct spectrum, is open to question because the depletion of the zero momentum state due to particle interaction has not been properly handled.

Nuclear Structure

The approximations developed for the many-body problem have recently found very successful application in the determination of the energy levels, magnetic moments, etc., of finite nuclei. The model used is that of a set of independent particles in shell-model configurations which interact weakly via some effective interaction. Both the pairing correlations of the BCS theory and the collective modes and screening characteristic of the random-phase approximation are found to have an important influence on the nuclear structure; when these are properly incorporated a surprisingly good account of the properties of the ground state and the low-lying excited states is obtained (Beliaev, Brown, Baranger, Mottelson).

PERTURBATION-THEORETIC EXPANSIONS AND STABILITY CONSIDERATIONS

I should like to conclude this survey with a few remarks about the character of the model problems we have discussed and of perturbation expansions in general.

First, I should like to call your attention to the fact that in the various series expansions considered, one always encounters a lack of analyticity in the dependence on the coupling constant (or expansion parameter). This is probably not an accident. It would seem that nature does not believe in power-series expansions in the many-body problem; one suspects the same may hold true in other areas of physics (quantum electrodynamics, for example).

The models we have considered, with the exception of superconductivity, are those for which perturbation theory has been applied. True, it has often been applied in sophisticated fashion, but that is no guarantee that it works. What I mean is that one cannot be sure, even when one is in a region in which a well-defined series expansion appears to be converging, that there does not exist a ground state of the system with quite different character (more order, for example) from that assumed in the perturbation-theoretic calculation. There are several well-known examples of this. One is the fact that one will not describe properly a nonspherical Fermi surface by starting with elements characteristic of spherical surfaces and applying perturbation theory (Kohn and Luttinger). Another is the failure of perturbation-theoretic methods in which one considers only "normal" kinds of intermediate states to yield the theory of superconductivity. Still another example would be the failure of perturbation theory to yield the aligned ferromagnetic ground state.

What one can hope for from an incorrect perturbation-theoretic approach is that it will contain built-in danger signals. These danger signals take the form of inconsistencies; a quantity that should, but does not, satisfy causality requirements, or an instability of a given excitation mode of the system. Thus, as we have already mentioned, Cooper found for a system of fermions with attractive interactions that the Fermi sea is unstable against the formation of pairs of particles of opposite spin and momentum. This instability both signals the inadequacy of perturbation theory and provides a clue to a better theory. It suggests the importance of assigning a special role to correlations between particles of opposite spin and momentum and taking them into account in nonperturbative fashion. Indeed, as first shown by Goldstone, one can determine the transition temperature for superconductivity by finding the point at which the instability first appears, in a finite-temperature version of the Cooper problem. Other examples include the possibility of antiferromagnetism in low-density fermion systems (Overhauser, Sawada and Fukuda) and the gas-liquid phase transition (Martin and Mermin). One expects that the generalization of these stability considerations might be equally promising for other problems, such as the general solid-liquid transition, the free electron gas-electron solid transition at low densities, and the liquid He^4-solid He^4 phase transition.

I have proposed caution in the acceptance of perturbation-theoretic results because we are at a rather primitive stage in our ability to recognize danger signals. Thus, the stability of rather simple kinds of collective modes has been investigated using rather simple approximations. The use of more-sophisticated modes or more-sophisticated approximations may well show up new kinds of instabilities. These considerations are of special concern to the solid-state physicist, who would like to know whether there exists a well-defined Fermi surface, characterized by a discontinuity in the single-particle distribution function at $p = p_F$. Luttinger has used perturbation theory to study this question and has shown that if perturbation theory applies, there will be such a discontinuity. Van Hove has studied a problem for fermions with repulsive interactions which is rather similar to that investigated by Cooper; he finds certain abnormalities in the scattering of two particles of opposite spin and momentum which may signal a subtle instability of the system and remove the discontinuity. The question is at present open.

2 MATHEMATICAL PRINCIPLES AND TECHNIQUES

A BRIEF HISTORY

A great variety of mathematical techniques have been applied to the many-body problem, both field-theoretic and non-field-theoretic in origin. The early calculations were almost entirely of a variational character. (Hartree and Hartree-Fock methods, Wigner's calculation of the correlation energy of a free electron gas). Variational calculations have continued to be extremely useful, as evidenced by the calculations of Jastrow and others for hard-sphere gases, and particularly by the Feynman-Cohen calculation of the excitation spectrum of liquid He^4, which is by far the best calculation we possess on that system. The calculations of Brueckner and his collaborators on nuclear matter may be thought of as falling into this category.

The next class of calculations to be developed were those based on the explicit introduction of collective coordinates to emphasize and describe properly the collective modes of the system (Tomonaga, Bohm and Pines,[4] Bogoljubov and Zubarev, Lipkin et al.). In the collective coordinate method, extra variables are introduced to describe the collective modes, and a series of canonical transformations are then carried out to decouple the collective modes from the individual particle motion. The collective coordinate method is useful both in the description of the collective modes and in providing an explicit demonstration of the coherence (screening) effects that accompany their appearance. It suffers from the disadvantage that it requires the introduction of extra variables, with a corresponding number of subsidiary conditions on the system wave function. The subsidiary conditions do not affect either the ground-state energy or the collective-mode excitation spectrum; they can act to change the individual particle spectrum somewhat. The method has found its principal application in the theory of the electron gas (and of electrons in solids), where an explicit demonstration of the equivalence of the ground-state energy so obtained to that obtained by other methods has been given (Nozières and Pines[12]).

The more-recent developments have tended to fall into several classes:

1. The equation-of-motion approach, in which one attempts to obtain the elementary excitations directly by approximately solving a set of coupled second-quantized operator equations. This method was proposed in the work of Bohm and Pines,[4] and used, in a slightly different form, by Lindhard to calculate the frequency- and wave-vector-dependent dielectric constant of an electron gas. It has been developed into a fine art by Sawada,[8] Brout,[9] Anderson,[32] Rickayzen,[33] Suhl, Boguljubov, and others. The equation-of-motion method is particularly useful in treating the electron gas and superconductivity. It represents what one might call a "poor man's version" of the elegant approach of Martin and Schwinger, in which one solves approximately a set of coupled equations for the various Green's functions of the system. A density-matrix version of the approach has been given by Ehrenreich and Cohen[14] and by Goldstone and Gottfried[19]; the latter authors indicate the relationship of the method to the Fermi-liquid theory of Landau. The method is also known as the time-dependent Hartree-Fock method (Bogoljubov).

2. The use of quantum-field-theoretic methods to derive expressions for the ground-state energy and excitation spectrum for a system of interacting fermions. Goldstone[1] used the Feynman formulation of time-dependent perturbation theory to prove the Brueckner linked-cluster expansion and obtain an expression for the ground-state energy. Hugenholtz derived the ground-state energy by use of the time-dependent perturbation-theory formulation of Van Hove, and Hugenholtz and Van Hove derived an expression for the single-particle excitation spectrum in this fashion. Hubbard[2] used the Feynman formulation to discuss collective motion from a field-theoretic viewpoint and to obtain an alternative expression for the ground-state energy. Galitskii and Migdal[3] made use of the Feynman formulation to introduce the one- and two-particle Green's functions to describe the single-particle and collective excitations, and showed how the ground-state energy could be obtained from a knowledge of the single-particle Green's function. More-recent developments along similar lines are due to Klein and Prange, Martin and Schwinger, and DuBois. Rules for calculating the single-particle Green's function in terms of the appropriate Feynman diagrams may be found in Galitskii,[18] Klein and Prange, and DuBois. Martin and Schwinger discuss ways to obtain this quantity by solving an infinite set of coupled integral equations. For the particular case of the free electron gas, Nozières and

Pines[13] showed that a knowledge of the frequency and wave-vector-dependent longitudinal dielectric constant sufficed to obtain the ground-state energy and density-fluctuation excitation spectrum, and discussed a method of calculation that does not depend explicitly on the use of perturbation theory.

3. The summation of a selected class of terms in the perturbation-series expansions for the ground-state energy and excitation spectrum. The first such calculations were those carried out using Rayleigh-Schrödinger perturbation theory by Brueckner and his collaborators on nuclear matter. Gell-Mann and Brueckner[6] used Feynman diagrams to calculate the ground-state energy of the free electron gas in the high-density limit. Subsequent calculations have included those of Hubbard[11] and DuBois on the ground-state energy and plasma oscillations of the free electron gas, and Galitskii[18] and Gottfried and Pičman on the low-density fermion gas.

4. Landau[15,16] has developed a semiphenomenological theory of normal fermion systems, which has as its basis the assumption that at sufficiently low temperatures there exists a well-defined set of quasi particles which obey Fermi-Dirac statistics. Landau and his collaborators have applied the theory to a wide range of physical problems, and Landau[17] has discussed certain aspects of the theory from a field-theoretic viewpoint. A review of the Landau theory, together with many of its applications, has been given by Abrikosov and Khalatnikov.

5. The special role played by the condensed state in interacting boson systems requires modification of the perturbation-series approach developed for fermions. Bogoljubov[20] in his classic paper showed that the creation and annihilation operators for the zero momentum state could be replaced by "c" numbers; this led to an effective Hamiltonian, a part of which can be diagonalized by a canonical transformation to new quasi-particle variables. The original method of Bogoljubov applied only to weak interactions; Lee, Huang, and Yang[21] used a pseudopotential approach to extend the theory to arbitrarily strong interactions in a low-density system; Brueckner and Sawada obtained the same result by summing the diagrams in the perturbation series. The modifications in a field-theoretical approach required by the special role of the zero momentum state have been developed and applied to the dilute boson gas by Beliaev[22,23] and Hugenholtz and Pines.[24]

6. The system of interacting electrons and ions in a metal represents an example of a problem in which one must sort out simultaneously the effect of particle-particle interactions and particle-field interactions. An understanding of this problem is of course basic for superconductivity. Bardeen and Pines[25] have used the collective coordinate method to obtain a solution within the random-phase approximation, and Migdal[26] has applied field-theoretic methods to sum a large class of diagrams in the

perturbation series for the restricted problem, first considered by Frölich, in which direct electron-electron interactions are neglected and one considers only electron-phonon interactions.

7. As we have mentioned, the usual perturbation-theoretic methods fail for the problem of superconductivity. Bardeen, Cooper, and Schrieffer[28] derived their theory originally as a variational calculation, in which a new kind of paired ground state was shown to possess a lower energy and an energy gap for quasi-particle excitations. Bogoljubov[29] and Valatin[30] showed that a canonical transformation to the quasi-particle variables led to substantial simplification of the mathematics of the BCS theory. Field-theoretic formulations of superconductivity have been developed by Gorkov,[31] Nambu, Beliaev, and Pines and Schrieffer. Bardeen and Rickayzen[34] have shown that for a reduced Hamiltonian the BCS solution is correct to order 1/N, where N is the number of particles in the system. Gauge invariance and the role of the collective excitations have been considered by Anderson,[32] Rickayzen,[33] Pines and Schrieffer, Nambu, Bogoljubov, Kadanoff and Martin, and others.

In the present lectures, there is obviously not enough time to describe all the different approaches to all the physical problems we wish to discuss. I shall begin, therefore, by summarizing the general mathematical description of the elementary excitations based on field-theoretic considerations. I shall then discuss briefly two approaches: the equation-of-motion method and the perturbation-theoretic approach based on the use of Feynman diagrams in a time-dependent formulation. In so doing I shall go into almost none of the mathematical details, but hope to convey the spirit of the methods, as well as offering a "do-it-yourself" kit for carrying out simple calculations and for reading the current literature.

QUASI-PARTICLE EXCITATIONS

Definition of Single-Particle Propagator

We have defined the single-particle Green's function, or propagator, as

$$G(p, \tau) = -i \langle \Psi_0 \mid T\{C_p(\tau)\, C_p^+(0)\} \mid \Psi_0 \rangle \tag{2-1a}$$

The operators and state vectors are given in the Heisenberg representation, in which the exact ground-state wave function, Ψ_0, is time-independent, and the time dependence of the creation and annihilation operators is defined according to

$$C_p(\tau) = e^{iH\tau}\, C_p(0)\, e^{-iH\tau} \tag{2-2}$$

$C_p(0)$ is time-independent, and H is the exact Hamiltonian of the many-body system. T is the Dyson chronological operator, which orders earlier times to the right. Thus, using (2-2), we have

$$G(p,\tau) = -i \langle \Psi_0 | C_p e^{-iH\tau} C_p^+ | \Psi_0 \rangle e^{iE_0\tau} \qquad \tau > 0 \qquad (2\text{-}3a)$$

$$G(p,\tau) = \pm i \langle \Psi_0 | C_p^+ e^{iH\tau} C_p | \Psi_0 \rangle e^{-iE_0\tau} \qquad \tau < 0 \qquad (2\text{-}3b)$$

where the (+) sign is for fermions, the (−) sign for bosons; the sign change in (2-3b) arises from the fermion commutation relations. (Throughout these lectures I shall suppress the spin index on fermion operators, unless its explicit appearance is required to prevent ambiguity.) $G(p,\tau)$ is the Fourier transform of the space-dependent single-particle Green's function

$$G(r,\tau) = -i \langle \Psi_0 | T\{\Psi(r,\tau)\Psi^+(0)\} | \Psi_0 \rangle \qquad (2\text{-}1b)$$

We are restricting our attention to time-independent, spatially homogeneous systems so that more-general single-particle propagators need not be considered. The Fourier transform of $G(p,\epsilon)$ of $G(p,\tau)$, defined by

$$G(p,\tau) = (1/2\pi) \int_{\infty}^{\infty} d\epsilon \; G(p,\epsilon) e^{-i\epsilon\tau} \qquad (2\text{-}4)$$

is of special importance, as we shall see that the analytic behavior of $G(p,\epsilon)$ near the real ϵ axis yields the quasi-particle energy and lifetime (Galitskii and Migdal[3]).

Some Simple Examples

Consider a free-particle system for which the exact ground-state wave function $|\Psi_0\rangle$ is the vacuum state $|0\rangle$. We then have, from (2-3a) and (2-3b),

$$G_f(p,\tau) = -ie^{-i\epsilon(p)\tau} \qquad \tau > 0$$

$$= 0 \qquad \tau < 0 \qquad (2\text{-}5)$$

where $\epsilon(p) = p^2/2m$ is the free-particle energy. The Fourier transform of $G_f(p,\tau)$ is

$$G_f(p,\epsilon) = 1/[\epsilon - \epsilon(p) + i\delta] \qquad (2\text{-}6)$$

where δ is an infinitesimal quantity, specifying the position of the pole in the complex ϵ plane. To see that (2-6) is correct, consider

$$G_f(p,\tau) = \int_{-\infty}^{\infty} (d\epsilon/2\pi)\{e^{-i\epsilon\tau}/[\epsilon - \epsilon(p) + i\delta]\}$$

We may do the ϵ integration as a contour integration, as shown in Fig. 2-1. For $\tau < 0$, we can close the contour above the real axis, since then the exponential factor guarantees convergence of the integral; we obtain zero. For $\tau > 0$ we can close the contour below the real axis. We find then a contribution from the pole at $\epsilon = \epsilon(p)$, with residue $-2\pi i e^{-i\epsilon(p)\tau}$, and so obtain (2-4).

Next consider the noninteracting fermion systems for which the ground state φ_0 is a filled Fermi sphere of momentum p_F. We have then

$$C_p\,\varphi_0 = 0 \qquad p > p_F$$

$$C_p^+\,\varphi_0 = 0 \qquad p < p_F$$

The Green's function $G_0(p,\tau)$ is therefore specified by

$$\begin{aligned} G_0(p,\tau) &= i n_p e^{-i\epsilon(p)\tau} && \tau < 0 \\ &= -i(1 - n_p) e^{-i\epsilon(p)\tau} && \tau > 0 \end{aligned} \qquad (2\text{-}7)$$

where

$$\begin{aligned} n_p &= 1 \qquad p < p_F \\ &= 0 \qquad p > p_F \end{aligned}$$

The Fourier transform, $G_0(p,\epsilon)$ is given by

$$G_0(p,\epsilon) = 1/[\epsilon - \epsilon(p) + i\delta_p] \qquad \begin{aligned} \delta_p &= +\delta,\ p > p_F \\ \delta_p &= -\delta,\ p < p_F \end{aligned} \qquad (2\text{-}8)$$

Again, one verifies that (2-8) is correct by carrying out the integration

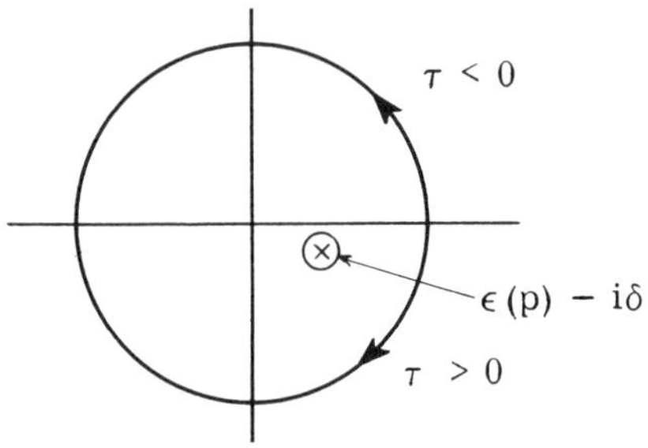

FIG. 2-1

over ϵ. The result (2-7) follows from closing the contour above the real axis for $\tau < 0$, below for $\tau > 0$; because δ_p changes sign as one goes through the Fermi surface, one finds (2-7).

We remark that in both (2-6) and (2-8) the singularities in $G(p, \epsilon)$ yield the quasi-particle energies. We further note that $G_0(p, \tau)$ describes hole propagation for $\tau < 0$. It vanishes for $p > p_F$; for $p < p_F$ it is the probability amplitude that if a hole is created at some negative time τ, it will be found in that same momentum state at time 0. For $\tau > 0$, $G_0(p, \tau)$ describes particle propagation just as did $G_f(p, \tau)$.

Distribution Function

We remark from the definition (2-1) that

$$G(p,0^-) = +i \langle \Psi_0 | C_p^+ C_p | \Psi_0 \rangle = iN_p$$

where by 0^- we mean letting $\tau \to 0$ from the side of minus values of τ. Here N_p is the expectation value of the number operator, $c_p^+ c_p$. We have directly from the definition (2-4) that

$$N_p = (1/2\pi i) \int_C d\epsilon \; G(p\epsilon) \tag{2-9a}$$

where the contour C must pass above the real axis (since we study the limit $\tau \to 0^-$). Hence from a knowledge of $G(p, \epsilon)$, we can determine the distribution function N_p for a system of interacting particles.

We further note that

$$G(p,0^+) = -i(1 - N_p) \tag{2-9b}$$

Thus $G(p, \tau)$ is discontinuous at $\tau = 0$; the magnitude of the discontinuity is $G(p,0^+) - G(p,0^-) = -i$.

Spectral Representation

I should now like to discuss some formal properties of the exact Green's function. The development I give is that found in Galitskii and Migdal.[3] I wish to reproduce it in some detail both because of the results one obtains and because the methods applied are equally useful in discussing the properties of other propagators, the dielectric constant of an electron gas, etc. I shall carry the development to the point at which a spectral representation is obtained for the single-particle Green's function and then refer to the article of Galitskii and Migdal for further details.

We begin by introducing an exact set of intermediate states into our definition (2-3) for $G(p, \tau)$; we find

$$G(p,\tau) = -i\sum_n \langle \Psi_0 | C_p | \Psi_n \rangle \langle \Psi_n | e^{-iH\tau} C_p^+ | \Psi_0 \rangle e^{iE_0\tau}$$

$$= -i\sum_n |(C_p^+)_{n0}|^2 e^{-i(E_n - E_0)\tau} \qquad \tau > 0 \tag{2-10}$$

by permitting the exact Hamiltonian to act on the exact intermediate states n, which possess energy E_n, momentum p, and correspond to a state of the N + 1 particle system. We now write

$$E_n(N+1) - E_0(N) = E_n(N+1) - E_0(N+1) + E_0(N+1) - E_0(N)$$

$$= \omega_{n0} + \mu \tag{2-11}$$

where ω_{n0} is the excitation energy in the N + 1 particle system and

$$\mu = E_0(N+1) - E_0(N) \tag{2-12}$$

is the change in ground-state energy on adding one particle, known as the chemical potential. One may do the same thing for $G(p,\tau)$ for $\tau < 0$, and so obtain

$$G(p,\tau) = +i\sum_n |(C_p)_{n0}|^2 e^{i(E_n - E_0)\tau} \qquad \tau < 0 \tag{2-13}$$

where the intermediate states now correspond to states of momentum $-p$ of the N − 1 particle system. Again one can write

$$E_n(N-1) - E_0(N) = \omega'_{n0} - \mu' \tag{2-14}$$

where ω'_{n0} is an excitation energy in the N − 1 particle system and μ' is the chemical potential for going from N − 1 to N particles. If we now assume that we are dealing with a large system (which is reflection-invariant) we can write

$$\omega_{n0} = \omega'_{n0} \qquad \mu' = \mu$$

to an accuracy of order 1/N. Hence we find

$$G(p,\tau) = -i\sum_n |(C_p^+)_{n0}|^2 e^{-i(\omega_{n0}+\mu)\tau} \qquad \tau > 0 \tag{2-15a}$$

$$G(p,\tau) = +i\sum_n |(C_p)_{n0}|^2 e^{i(\omega_{n0}-\mu)\tau} \qquad \tau < 0 \tag{2-15b}$$

where the excitation energies ω_{n0} are necessarily always positive.

We next remark that we deal with a large number of closely spaced states and therefore can convert the sums in (2-15) into integrals by introducing the spectral functions

$$A(p,\omega) = \sum_n |(C_p^+)_{n0}|^2 \delta(\omega - \omega_{n0})$$

$$B(p,\omega) = \sum_n |(C_p)_{n0}|^2 \delta(\omega - \omega_{n0}) \tag{2-16}$$

We then have

$$G(p,\tau) = -i \int_0^\infty d\omega \; A(p,\omega)\, e^{-i(\omega+\mu)\tau} \qquad \tau > 0 \tag{2-17a}$$

$$G(p,\tau) = +i \int_0^\infty d\omega \; B(p,\omega)\, e^{i(\omega-\mu)\tau} \qquad \tau < 0 \tag{2-17b}$$

We can furthermore obtain at once, by methods similar to those used to find $G_f(p,\epsilon)$ and $G_0(p,\epsilon)$, the result

$$G(p,\epsilon) = \int_0^\infty d\omega \left[\frac{A(p,\omega)}{\epsilon - (\omega + \mu) + i\delta} + \frac{B(p,\omega)}{\epsilon + \omega - \mu - i\delta}\right] \tag{2-18}$$

Equation (2-18) is exact. It is the representation of $G(p,\epsilon)$ in terms of the spectral functions $A(p\omega)$ and $B(p\omega)$ and is known as the Lehmann representation in quantum field theory (Galitskii and Migdal[3]). With the aid of (2-18), Galitskii and Migdal discuss the analytic properties of $G(p,\epsilon)$ and derive dispersion relations that connect its real and imaginary parts.

Definition of a Quasi Particle

I should now like to indicate how one goes about defining a quasi particle with the aid of (2-17) and (2-18). We first recall that $G(p,\tau)$ is the probability amplitude that if there is a particle with momentum p ($p > p_F$) at $t = 0$, then it will be found in that state at time τ; in other words $G(p,\tau)$ describes the propagation in time of a particle in that momentum state. Moreover, from (2-17) we see that this propagation is determined by the spectral function $A(p,\omega)$. For the case of a noninteracting fermion system, one finds directly, on comparing (2-8) and (2-18), that

$$A(p,\omega) = \delta\{\omega - [\epsilon(p) - \mu]\}(1 - n_p) \tag{2-19a}$$

The propagation is therefore undamped, and the frequency of oscillation of $G(p,\tau)$ is the free-particle energy $\epsilon(p)$.

In the presence of interaction, $A(p\omega)$ will no longer be a simple δ function, and in general its behavior, which is determined by the magnitude of the matrix elements and the spread in energies associated with the transitions $(C_p^+)_{n0}$ of (2-16), will be rather complicated. In fact the state $C_p^+ \Psi_0$ is not a proper eigenstate of the N + 1 particle system, but must consist of a superposition of a number of such states spread over a range of energies. As a result, the state must have a finite lifetime and its propagation will in general be damped. We are, therefore, led to ask the following questions: (1) Under what circumstances is the notion of a quasi particle fruitful? (2) Where the quasi-particle concept is justified, how does one pass from a knowledge of $A(p\omega)$ to the quasi-particle energy versus momentum relation?

We could answer the first question by saying that if $G(p, \tau)$ behaves like

$$G(p, \tau) = -iZ_p e^{-i\tilde{\epsilon}(p)\tau} e^{-\Gamma_p \tau} \qquad (2\text{-}19b)$$

for some range of times τ, then the notion of a quasi particle would appear justified, since (2-19b) describes propagation of a state of energy $\tilde{\epsilon}(p)$, of strength Z_p, which persists for a time $\tau \sim 1/\Gamma_p$. It is further clear from inspection of (2-17a) that in order to find (2-19b), $A(p\omega)$ must peak in the vicinity of $\omega = \tilde{\epsilon}(p) - \mu$. Suppose that peaking took the simple form

$$A(p\omega) = \frac{f(p\omega)}{(\omega - [\tilde{\epsilon}(p) - \mu])^2 + \Gamma_p^2}$$

$$= \frac{iZ_p/2\pi}{\omega - [\tilde{\epsilon}(p) - \mu] + i\Gamma_p} + \text{C.C.} \qquad (2\text{-}20)$$

If we then carry out the integration over ω in (2-17a), by closing the contour below the real axis, we obtain a contribution to $G(p, \tau)$ from the peak described by (2-20) which is just (2-19b).

Now, in general, we can consider evaluating (2-17a) by closing the integral below the real axis, as shown in Fig. 2-2. We have from Cauchy's theorem, that

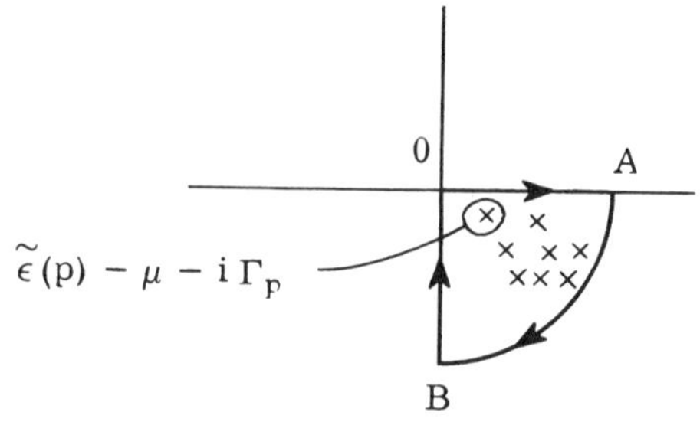

FIG. 2-2

$$\int_0^\infty d\omega\, A(p\omega)\, e^{-i(\omega+\mu)\tau} = 2\pi i \sum_n Z_n e^{-i\epsilon_n\tau} e^{-\Gamma_n\tau} + \int_{C'} d\omega\, A(p\omega)\, e^{-i(\omega+\mu)\tau}$$

where the first term arises from the sum of the residues of all the poles in the lower-right-hand portion of the complex ω plane, of which there may be many others than that at $\tilde{\epsilon}(p) - \mu - i\Gamma_p$. The second is associated with the contour C′ specified by OB. We see that in order to get the simple result (2-19b), we must be able to neglect the contribution from all the poles other than that at $\tilde{\epsilon}(p) - \mu - i\Gamma_p$, and we must be able to neglect the contribution from the contour C′. The first condition is simply satisfied by taking as the quasi-particle pole that which lies closest to the real ϵ axis and therefore damps most slowly; if one waits long enough, only it will persist. Detailed investigation shows that one can neglect the contribution from the integral over the contour C′ provided one chooses a time τ long enough that destructive interference between the different states contained in $C_p|\Psi_0\rangle$ acts to reduce this contribution, but not so long that the contribution from the pole at $\tilde{\epsilon}(p) - \mu - i\Gamma_p$ has itself become negligible.

Thus we conclude that the quasi-particle concept is justified for times τ which satisfy

$$\tau_C \lesssim \tau \lesssim 1/\Gamma_p$$

where τ_C would be of the order of $[\tilde{\epsilon}(p) - \mu]^{-1}$, this being the other time of interest in the problem. Moreover, we see that the pole in the analytic continuation of $A(p\omega)$ into the lower-right-hand quadrant of the complex ω plane which lies nearest the real ω axis yields the quasi-particle energy and lifetime. Finally, it is then easy to show that from the analytic properties of $G(p\epsilon)$ determined by (2-18) the pole in the analytic continuation of $G(p\epsilon)$ (into the lower half of the complex ϵ plane) which lies nearest the real ϵ axis likewise yields the energy and lifetime of the quasi particle of momentum p.

This result is reasonable; one simply passes from $G_0(p,\epsilon)$, which for $p > p_F$ has a pole at $\epsilon(p)$ that is only infinitesimally displaced into the lower half of the complex plane, to a $G(p,\epsilon)$, which can be written for $p \geq p_F$,

$$G(p,\epsilon) = Z_p/[\epsilon - \tilde{\epsilon}(p) + i\Gamma_p] + \text{correction terms} \qquad (2\text{-}21)$$

The quasi-particle properties derive from the first term in $G(p,\epsilon)$. We may also recall that Γ_p will be proportional to $(p - p_F)^2$ in the vicinity of the Fermi surface, as a result of Pauli-principle restrictions on quasi-particle scattering.

The same method may be applied to the quasi particles with $p < p_F$ (the "quasi holes"); one finds then that the pole that lies nearest the real axis in the analytic continuation of $G(p, \epsilon)$ *above* the real axis determines the quasi-particle energy and lifetime for $p < p_F$. We thus see that Γ_p must change sign as one goes through the Fermi surface. This fact was used by Migdal to define the Fermi surface by the discontinuity in the particle distribution as one goes through p_F. For p in the immediate vicinity of p_F, Migdal remarks that the correction terms in (2-21) can be written as $f(p, \epsilon)$, a function regular at $\epsilon_p - i\Gamma_p$. Hence one finds, on using the definition, (2-9a), of the particle momentum distribution N_p,

$$
\begin{aligned}
N_p &= Z_p + N_p' \qquad & p &= p_F - \delta \\
&= N_p' & p &= p_F + \delta
\end{aligned}
\tag{2-22}
$$

since the pole moves from above the ϵ axis to below it as p goes from below p_F to above p_F. Therefore from (2-21), Z_p, the strength of the quasi-particle pole at ϵ_p, is the discontinuity in the Fermi surface; that is,

$$
Z_p = N_{p_{F+}} - N_{p_{F-}}
$$

The perturbation-theoretic calculations on the dilute Fermi gas and the high-density electron gas yield results in agreement with (2-21), and the magnitude of Z_p has been calculated for these models (Galitskii,[18] Daniel and Vosko). As we have mentioned, the existence of the discontinuity is to be regarded as a property of "normal" fermion systems, in the sense that a perturbation-theoretic expansion which uses $G_0(p, \epsilon)$ as its primitive element is assumed to apply (Luttinger). We know that a superconductor does not possess this property, and, indeed, as we have already mentioned, whether any system does remains an open question.

DENSITY-FLUCTUATION EXCITATIONS AND COLLECTIVE MODES

Retarded Response Function

We turn now to a description of the response of the many-body system to external perturbations, and hence to a consideration of the elementary excitations one can hope to observe directly by suitable experiment. We shall confine our attention to longitudinal external probes that couple to the density fluctuations of the system; the methods used can easily be generalized to cover the coupling to any sort of external field. (For example, Galitskii and Migdal[3] compute the response to a weak electromagnetic field.) We consider a coupling of the form

$$H_{ext} = \sum_k V_k \rho_k r_k(t)$$

where $r_k(t)$ describes the time-dependent external probe coupled to the system by a matrix element V_k. We have already considered an example of such a probe—a single particle, for which

$$r_k(t) = \exp[i\mathbf{k} \cdot \mathbf{R}(t)]$$

where $\mathbf{R}(t)$ is the position of the particle. As another example we may imagine $r_k(t)$ to be a test charge interacting with the electron gas. In that case the response of the system can be expressed in terms of $\epsilon(k\omega)$, the longitudinal frequency and wave-vector-dependent dielectric constant (Nozières and Pines[13]).

We are interested in computing the time-dependent response of the system to the external probe. Thus we wish to calculate the expectation value of $\rho(rt)$ in the presence of the probe, $\langle \Psi_a | \rho(rt) | \Psi_a \rangle$, where Ψ_a refers to the system wave function in the presence of the probe. An external probe is principally useful when its coupling to the system is weak. It then does not appreciably perturb the energy levels of the many-body system, and the information it yields is characteristic of the many-body system alone, rather than of many-body system plus probe. We assume that condition to be well satisfied, in which case we are justified in using second-order perturbation theory in computing the response of the system to the probe. Under these circumstances, it is useful to Fourier-analyze the external perturbation $r_k(t)$ and the response of the system $\rho_k(t)$, because each frequency Fourier component may be treated independently.

Before carrying out the calculation, a few words about the boundary conditions are in order. In the present problem we are interested in determining the adiabatic response of the many-body system, that is, the response under circumstances in which the probe does not heat up the system. This we may do by turning the interaction with the probe on adiabatically at some time in the far past, and permitting it to build up gradually until the time $t = 0$, at which point we study the system response. With this boundary condition, and treating each frequency component independently, we are led to study the response $\langle \rho_{k\omega}(t) \rangle$ to an interaction

$$V_k r_k \rho_k e^{-i\omega t} e^{\delta t}$$

where δ is an infinitesimal quantity. The calculation is simply performed using time-dependent perturbation theory (see Nozières and Pines,[13] for example) and one finds for the expectation value of ρ_k in the presence of r_k the result

$$\langle \rho_{k\omega}(t) \rangle = V_k r_k e^{-i\omega t} \sum_n \left[\frac{|(\rho_k)_{n0}|^2}{\omega - \omega_{n0} + i\delta} - \frac{|(\rho_{-k})_{n0}|^2}{\omega + \omega_{n0} + i\delta} \right] \tag{2-23}$$

The ω_{n0} are the exact excitation frequencies of the states of the many-body system available to the density fluctuation ρ_k. If we now make use of the reflection invariance of the system (the matrix elements and excitation frequencies of states available to ρ_{-k} are identical to those for ρ_k), we have

$$\langle \rho_{k\omega}(t) \rangle = V_k r_k e^{-i\omega t} \sum_n |(\rho_k)_{n0}|^2 \left(\frac{1}{\omega - \omega_{n0} + i\delta} - \frac{1}{\omega + \omega_{n0} + i\delta} \right) \tag{2-24}$$

The properties of the many-body system are embodied in the function

$$F^r(k\omega) = \sum_n (\rho_k)^2_{n0} \left(\frac{1}{\omega - \omega_{n0} + i\delta} - \frac{1}{\omega + \omega_{n0} + i\delta} \right) \tag{2-25}$$

which may easily be seen to be the Fourier transform of the retarded commutator defined by

$$\begin{aligned} F^r_k(\tau) &= -i \langle \Psi_0 | [\rho_k(\tau), \rho_{-k}(0)] | \Psi_0 \rangle \qquad && \tau > 0 \\ &= 0 && \tau < 0 \end{aligned} \tag{2-26}$$

where the operators and exact state vectors are again specified in the Heisenberg representation. To go from (2-25) to (2-26) use the relation

$$\begin{aligned} F^r_k(\tau) &= \int_{-\infty}^{\infty} (d\omega/2\pi)\, F^r(k\omega)\, e^{-i\omega\tau} \\ &= -i \sum_n (\rho_k)^2_{n0} (e^{-i\omega_{n0}\tau} - e^{i\omega_{n0}\tau}) \qquad && \tau > 0 \\ &= 0 && \tau < 0 \end{aligned} \tag{2-27}$$

A Lehmann representation for $F^r(k\omega)$ is easily established by introducing the spectral function

$$S(k\omega) = \sum_n (\rho_k)^2_{n0}\,\delta(\omega - \omega_{n0}) \tag{2-28}$$

We have, directly from (2-25),

$$F^r(k\omega) = \int_{-\infty}^{\infty} d\omega'\, S(k\omega')\left(\frac{1}{\omega - \omega' + i\delta} - \frac{1}{\omega + \omega' + i\delta}\right) \tag{2-29}$$

We see from (2-29) that $F^r(k\omega)$ is analytic in the upper half-plane; this is scarcely surprising, as we have chosen causal boundary conditions, which guarantee that the response function $F^r_k(t)$ follows the onset of the driving function, the external probe $r_k(t)$. From this property, we have

$$\int_C d\omega'\, [F^r(k\omega')/(\omega - \omega')] = 0$$

where the contour C in the ω' plane is specified as shown in Fig. 2-3. The contribution from the semicircle BDA is zero. The contribution from AB may be written

$$\int_{-\infty}^{\infty} d\omega'\, F^r(k\omega')[P(1/\omega' - \omega) - i\pi\delta(\omega - \omega')] = 0$$

so that we have

$$\int_{-\infty}^{\infty} d\omega'\, F^r(k\omega')P(1/\omega' - \omega) = i\pi F^r(k\omega) \tag{2-30}$$

We get the Kramers-Kronig relations when the real and imaginary parts of (2-30) are taken.

The development we have given here is a slight generalization of that used by Nozières and Pines[13] for the introduction of the retarded dielectric constant of the free electron gas. To make the connection, let $r_k(t)$ correspond to a weak test charge and let V_k be the Fourier transform of the Coulomb interaction. The dielectric constant $\epsilon(k\omega)$ is defined by

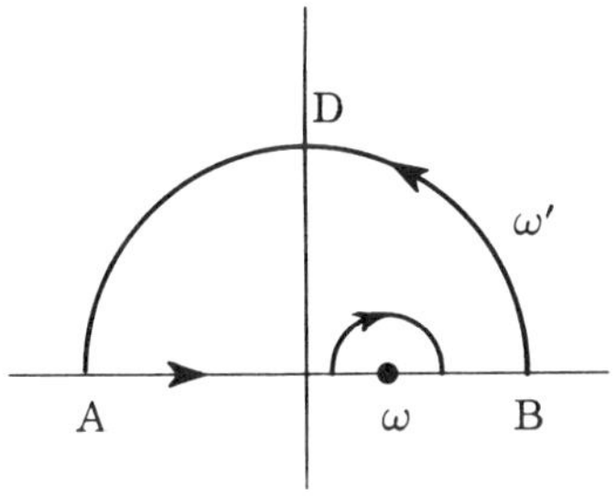

FIG. 2-3

taking the Fourier transform in space and time of the macroscopic Poisson equations.

$$\begin{aligned} \operatorname{div} D(k\omega) &= \epsilon(k\omega) \operatorname{div} E(k\omega) = 4\pi r(k\omega) \\ \operatorname{div} E(k\omega) &= 4\pi [r(k\omega) - e \langle \rho(k\omega) \rangle] \end{aligned} \tag{2-31}$$

One finds, in the present notation,

$$\frac{1}{\epsilon(k\omega)} - 1 = \frac{-e \langle \rho(k\omega) \rangle}{r(k\omega)} = \frac{4\pi e^2}{k^2} F^r(k\omega) \tag{2-32a}$$

If we take the imaginary part of both sides of (2-32a) and make use of (2-29), we have the useful relation

$$\operatorname{Im} \frac{1}{\epsilon(k\omega)} = \frac{-4\pi^2 e^2}{k^2} S(k\omega) \qquad (\omega > 0) \tag{2-32b}$$

We see that the properties of $1/\epsilon(k\omega)$ follow directly from those we have derived for $F^r(k\omega)$.

Properties and Physical Significance of S(kω)

We encountered the spectral function $S(k\omega)$ in our first lecture, where we saw that it is measured directly in an inelastic-scattering experiment for which the Born approximation applies. In particular one finds

$$d^2\sigma/d\Omega\, d\omega = (M^2/8\pi^3)(P_F/P_I)\, V_k^2 S(k\omega) \tag{2-33a}$$

for a particle of mass M, initial momentum $\mathbf{P}_I$ scattered to a final state $\mathbf{P}_F$, with energy transfer between ω and $\omega + d\omega$, and scattering angle specified by $d\Omega$. (The energy transfer ω is related to the momentum transfer and scattering angle by the relation $\omega = \mathbf{k} \cdot \mathbf{P}_I/M + k^2/2M$, where $\mathbf{P}_I$, $\mathbf{k}$, and $\mathbf{P}_F$ are given by

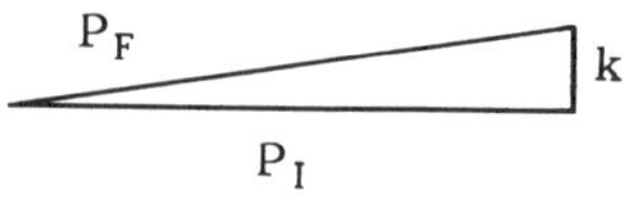

We summarize some properties of $S(k\omega)$ (Van Hove):

1. $2\pi S(k\omega)$ is the Fourier transform in space and time of the time-dependent pair-correlation function $S(r\tau)$, defined by

$$S(r\tau) = \langle \Psi_0 \,|\, \rho(r\tau)\, \rho(00) \,|\, \Psi_0 \rangle$$

where $\rho(r\tau)$ is the particle density in the Heisenberg representation.

2. The structure factor S(k) is related to S(kω) by

$$\int_{-\infty}^{\infty} d\omega \; S(k\omega) = S(k)N \tag{2-34}$$

$$S(k) = \langle \Psi_0 \mid \rho_k^+ \rho_k \mid \Psi_0 \rangle / N \tag{2-35}$$

and is the Fourier transform of the time-independent pair-distribution function

$$p(r) = \langle \Psi_0 \mid \rho(r)\,\rho(0) \mid \Psi_0 \rangle / N \tag{2-36}$$

We note that S(k) is directly measured in a scattering experiment in which one determines the differential cross section integrated over all energy transfers:

$$\frac{d\sigma}{d\theta} = \int_0^{\infty} d\omega \; \frac{d^2\sigma}{d\theta \; d\omega} = \frac{M^2}{8\pi^3} \; \frac{P_F}{P_I} \; V_k^2 S(k)N \tag{2-33b}$$

3. S(kω) obeys a sum rule

$$\int_0^{\infty} d\omega \; S(k\omega)\omega = Nk^2/2m \tag{2-37}$$

where m is the mass of the particles in the system. To derive the sum rule, (2-37), remark that from the definition of S(kω), (2-28), we have

$$\int_0^{\infty} d\omega \; \omega S(k\omega) = \sum_n |(\rho_k)_{n0}|^2 \, \omega_{n0}$$

Direct calculation of $\langle \Psi_0 \mid [\rho_k, [\rho_{-k}, H]] \mid \Psi_0 \rangle$ for a system that is reflection-invariant with interactions which are velocity-independent shows that

$$\sum_n (2m/k^2)\, \omega_{n0} \mid (\rho_k)_{n0} \mid^2 = \sum_n f_{0n} = N \tag{2-38}$$

from which (2-37) follows. The relation (2-38) is frequently called the f-sum rule; it is a longitudinal version of the Thomas-Reich-Kuhn sum rule, and is very useful in discussing the electron gas (Nozières and Pines[13]).

Collective Modes

As we have mentioned, collective modes may make their appearance in S(kω). It is intuitively obvious that they will occur as well-defined excitations if S(kω) is peaked about a particular energy ω(k) with width $\Gamma_k \ll \omega(k)$, because under these circumstances one will see a distinct peak in the inelastic-particle scattering curve. A measurement of the angular distribution of the inelastically scattered particles then furnishes a direct experimental measurement of ω(k).

We can make the analogy with the derivation of the preceding section complete by defining the density-fluctuation propagator

$$F_k(\tau) = -i \langle \Psi_0 \mid T\{\rho_k(\tau)\rho_k^+(0)\} \mid \Psi_0 \rangle \tag{2-39}$$

which possesses the Fourier transform

$$F(k\omega) = \sum_n \mid (\rho_k)_{n0} \mid^2 \left(\frac{1}{\omega - \omega_{n0} + i\delta} - \frac{1}{\omega + \omega_{n0} - i\delta} \right) \tag{2-40}$$

and the Lehmann representation

$$F(k\omega) = \int_0^\infty d\omega' \; S(k\omega') \left(\frac{1}{\omega - \omega' + i\delta} - \frac{1}{\omega + \omega' - i\delta} \right) \tag{2-41}$$

Suppose now that $S(k\omega') = S(k)\,\delta\,[\omega' - \omega(k)]$, i.e., that there is only a single density-fluctuation excitation at frequency $\omega(k)$. We find

$$F(k\omega) = S(k) \left(\frac{1}{\omega - \omega(k) + i\delta} - \frac{1}{\omega + \omega(k) - i\delta} \right) \tag{2-42}$$

Equation (2-42) is the expression for the propagation of a field quantum associated with $\rho(x, t)$, i.e., a density-fluctuation collective mode.

To see that, we pause for a moment to define the propagator for, say, the phonon field coordinate, $\varphi(x, t)$. The propagator

$$D(r, \tau) = -i \langle \Psi_0 \mid T\{\varphi^+(r, \tau)\varphi(0)\} \mid \Psi_0 \rangle \tag{2-43}$$

Its Fourier transform is

$$D_k(\tau) = -i \langle \Psi_0 \mid T\{\varphi_k^+(\tau)\,\varphi(0)\} \mid \Psi_0 \rangle \tag{2-44}$$

On comparing (2-44) with (2-39) we see that we can at once write for $D_k(\tau)$ a Fourier transform

$$D(k\omega) = \sum_n (\varphi_k)_{n0}^2 \left(\frac{1}{\omega - \omega_{n0} + i\delta} - \frac{1}{\omega + \omega_{n0} - i\delta} \right) \tag{2-45}$$

and a Lehmann representation,

$$D(k\omega) = \int_0^\infty d\omega' \; \varphi(k\omega') \left(\frac{1}{\omega - \omega' + i\delta} - \frac{1}{\omega + \omega' - i\delta} \right) \tag{2-46}$$

Finally, unlike ρ_k, we may always assume that φ_k may be expanded in terms of creation and annihilation operators according to

$$\varphi_k = \alpha_k (b_k + b_{-k}^+) \tag{2-47}$$

where α_k is a constant. On substituting (2-47) into (2-45) (or by direct calculation) we find for a noninteracting collection of phonons the propagator

$$D_0(k\omega) = \alpha_k^2 \left(\frac{1}{\omega - \omega_k^0 + i\delta} - \frac{1}{\omega + \omega_k^0 - i\delta} \right) \tag{2-48}$$

where ω_k^0 is the free-phonon frequency.

A comparison of (2-48) and (2-42) shows that if $S(k\omega)$ possesses a δ-function singularity, the ρ_k oscillate at a fixed frequency $\omega(k)$ and propagate without damping, just as if they were a collection of independent sound-wave coordinates. [The same conclusion could of course have been reached much more directly from the very definition of $S(k\omega)$.] Moreover, we can now apply to $F(k\omega)$ and $F(k, \tau)$ the same arguments that we developed for the quasi particles. We conclude that if $S(k\omega)$ possesses a well-developed peak in the vicinity of some $\omega(k)$, then it is both legitimate and useful to introduce the concept of a slightly damped collective mode, resembling a phonon. For example, for $\tau > 0$, the energy of the collective mode and its lifetime will be given by that pole in the analytic continuation of $F(k\omega)$ below the real axis which lies nearest the real ω axis.

$F_k(\tau)$ is simply related to the general two-particle Green's function, which we shall introduce presently. We note that there is a close connection between $F_k(\tau)$ and $F_k^r(\tau)$, and that from a knowledge of their analytic behavior it is easy to pass from one to the other. We see, on comparing (2-25) and (2-40), that the real parts of the two functions are identical. Moreover, since $\omega_{n0} \geq 0$, for $\omega > 0$, the imaginary parts are identical, while for $\omega < 0$ one is the negative of the other. Thus we have

$$\begin{aligned} &\text{Re } F(k\omega) = \text{Re } F^r(k\omega) \\ &\text{Im } F(k\omega) = \text{Im } F^r(k\omega) \qquad \omega > 0 \\ &\text{Im } F(k\omega) = -\text{Im } F^r(k\omega) \qquad \omega < 0 \end{aligned} \tag{2-49}$$

We further remark that in perturbation-theoretic calculations it is natural to define a propagating dielectric constant, according to

$$[1/\epsilon_p(k\omega)] - 1 = (4\pi e^2/k^2)\, F(k\omega) \tag{2-50}$$

The relationship between $\epsilon_p(k\omega)$ and $\epsilon(k\omega)$ is at once obtained from (2-49).

Finally, we comment on the connection between the collective mode we have defined and the definition that is natural for an electron gas,

$$\epsilon(k\omega) = 0 \tag{2-51}$$

The condition (2-51) corresponds to the possibility of a nonzero electric field and charge fluctuation in the absence of any external probe [see (2-31)] and thus describes the free resonant, or collective, modes of the system. But from (2-32) we see that the condition for a zero of $\epsilon(k\omega)$ is identical to that for a pole in $F^r(k\omega)$, or in $F(k\omega)$.

TWO-PARTICLE GREEN'S FUNCTIONS

Two-particle Green's functions can be defined in a fashion analogous to (2-1). Thus one can write

$$K(12;34) = -i \langle \Psi_0 | T\{\psi(1)\psi^+(2)\psi(3)\psi^+(4)\} | \Psi_0 \rangle$$

where the coordinates and times for each wave-field operator in the Heisenberg representation are denoted by 1, 2, 3, and 4. Galitskii and Migdal discuss the decomposition of K(12;34) and write down an integral equation that it must satisfy. Here I wish only to indicate the way in which K(12;34) reduces to the density-fluctuation propagator. If one takes the coordinate in 2 to be the same as that in 1, and the time in 2 to be infinitesimally later than in 1, a limiting procedure we may denote as $2 = 1^+$, and if we likewise take $4 = 3^+$, then we see that

$$K(11^+; 33^+) = -i \langle \Psi_0 | T\{\rho(1)\rho(3)\} | \Psi_0 \rangle$$

For a homogeneous system $K(11^+;33^+)$ depends only on the difference of coordinates and times 1-3, and possesses the Fourier transform $F(k\omega)$. Thus, the collective modes we have been considering may be regarded as arising from the poles of the two-particle Green's function in the limiting case in which it describes coupled hole-particle propagation.

GROUND-STATE-ENERGY THEOREMS

We consider the basic Hamiltonian:

$$H = \sum_p \epsilon(p) C_p^+ C_p + \sum_{pqk} (V_k/2) C_{p+k}^+ C_{q-k}^+ C_q C_p \tag{2-52}$$

for an assembly of interacting fermions. It can also be written in the form

$$H = \sum_p \epsilon(p) C_p^+ C_p + \sum_k (V_k/2)(\rho_k^+ \rho_k - N) \tag{2-53}$$

by expressing the potential energy in terms of the density fluctuations. We quote without detailed proof two alternative expressions for the ground-state energy E_0 of the system. The first, due to Galitskii and Migdal, is

$$E_0 = (1/2\pi i) \int_C d\epsilon \int [d^3p/(2\pi)^3](\epsilon + p^2/2m)\, G(p, \epsilon) \tag{2-54}$$

where C denotes a contour that is to be closed above the real ϵ axis. Equation (2-54) may be simply derived by calculating $\partial/\partial\tau[G(p, \tau)]$ for $\tau < 0$ and passing to the limit $\tau \to 0_-$. We have here included the factor of 2 required to sum over fermions of both kinds of spin.

The second form is based on first obtaining a simple expression for the interaction energy

$$E_{int} = \langle \Psi_0 | \sum_k [V_k(\rho_k^+ \rho_k - N)/2] | \Psi_0 \rangle \tag{2-55}$$

Using (2-35), (2-34), and the result

$$\begin{aligned} S(k\omega) &= -(1/\pi)\, \text{Im}\; F^r(k\omega) \\ &= -(1/\pi)\, \text{Im}\; F(k\omega) \qquad \omega > 0 \end{aligned} \tag{2-56}$$

we can write

$$\begin{aligned} E_{int} &= \sum_k{}' (V_k/2)\, N[S(k) - 1] \\ &= -\sum_k (V_k/2) \left[\int_0^\infty (d\omega/\pi)\, \text{Im}\; F^r(k\omega) + N \right] \\ &= -\sum_k (V_k/2) \left[\int_0^\infty (d\omega/\pi)\, \text{Im}\; F(k\omega) + N \right] \end{aligned} \tag{2-57}$$

For the electron gas, (2-57) takes an especially simple form (Nozières and Pines[13]),

$$E_{int} = -\sum_k \left\{ \int_0^\infty (d\omega/2\pi)\, \text{Im}\,[1/\epsilon(k\omega)] + (2\pi Ne^2/k^2) \right\} \tag{2-58}$$

The second step is to remark that one can pass directly from E_{int} to E_0 by regarding E_{int} as a function of some variable coupling constant α. This trick, which is apparently first due to Pauli, and has since been rediscovered by many people (Feynman, Sawada et al.), yields

$$E_0 = E(0) + \int_0^g (d\alpha/\alpha)\, E_{int}(\alpha) \tag{2-59}$$

where g is the actual coupling constant contained in V_k. We see that a knowledge of $S(k\omega)$ for all arbitrary coupling constants thus enables one to determine the ground-state energy. The expression (2-59) is useful because in some applications it is easier to calculate $S(k\omega)$ to a given order than to calculate $G(p\epsilon)$ to that same order; the reverse situation also sometimes applies.

THE EQUATION-OF-MOTION METHOD

In the equation-of-motion method one attempts to determine the excitation spectrum and ground state by an approximate self-consistent solution of coupled operator equations of motion (Sawada et al.,[8] Anderson,[32] Suhl and Werthamer, Sawada and Fukuda). The steps by which one proceeds may be outlined as follows:

1. Assume the ground-state wave function Ψ_0 to be known.

2. Find the creation and destruction operators for the elementary excitations of the system, O_p^+ and O_p, by requiring that they satisfy oscillatory equations of motion:

$$(H, O_p^+) = \omega_p O_p^+ \tag{2-60a}$$

$$(H, O_p) = -\omega_p O_p \tag{2-60b}$$

3. Applied to Ψ_0, O_p^+ creates an excitation of energy ω_p. Likewise, one must have

$$O_p \Psi_0 = 0 \tag{2-61}$$

if Ψ_0 is to be the ground-state wave function. The condition (2-61) is thus used to determine Ψ_0.

4. Once ω_p and Ψ_0 are determined, the other properties of the system (ground-state energy, etc.) can be determined more-or-less directly in a variety of ways.

The method is best appreciated by some simple examples. First, we investigate quasi-particle excitations by considering the possibility of taking for O_p the electron annihilation operator C_p. We take as our Hamiltonian

$$H' = H - \mu N = \sum_{p\sigma} [\epsilon(p) - \mu] C_{p\sigma}^+ C_{p\sigma} + \sum_{\substack{pqk \\ \sigma\sigma'}} (V_k/2)\, C_{p+k\,\sigma}^+ C_{q-k\sigma'}^+ C_{q\sigma'} C_{p\sigma} \tag{2-62}$$

We have added the term $-\mu N$, where μ is the chemical potential or, what is equivalent, the energy of a particle at the Fermi surface. The equation of motion satisfied by C_p^+ is

$$(H', C_p^+) = [\epsilon(p) - \mu] C_p^+ + \sum_{kq} V_k C_{p+k}^+ C_{q-k}^+ C_q \tag{2-63}$$

Thus the creation operator for a single excitation is coupled to a trilinear operator which in general acts to create a particle and a particle-hole pair. It is useful to rewrite the equation by separating out those parts of the trilinear term that are proportional to C_p^+, that is, the terms with $k = 0$ and $k = q - p$; one has

$$\begin{aligned}(H', C_p^+) = {} & [\epsilon(p) - \mu + NV_0 - \sum_q V_{q-p} N_q] C_p^+ \\ & - \sum_q V_{q-p} [C_q^+ C_q - N_q] C_p^+ \\ & - \sum_{\substack{k \neq 0 \\ q \neq p+k}} V_k C_{p+k}^+ C_{q-k}^+ C_q \end{aligned} \tag{2-64}$$

We can therefore linearize the equations of motion by neglecting the second fluctuation term and the triple term on the right-hand side of (2-64). We have then

$$(H', C_p^+) = [\tilde{\epsilon}(p) - \mu] C_p^+ \tag{2-65}$$

where $\tilde{\epsilon}(p)$, the quasi-particle excitation energy, is given by

$$\tilde{\epsilon}(p) = \epsilon(p) + NV_0 - \sum_q V_{q-p} N_q \tag{2-66}$$

We likewise have

$$(H', C_p) = -[\tilde{\epsilon}(p) - \mu] C_p \tag{2-67}$$

Equations (2-66) and (2-67) determine the ground-state wave function directly. The equations are only compatible with a Ψ_0 that possesses the property

$$\begin{aligned} \tilde{\epsilon}(p) < \mu \qquad & C_p^+ \Psi_0 = 0 \\ \tilde{\epsilon}(p) > \mu \qquad & C_p \Psi_0 = 0 \end{aligned} \tag{2-68}$$

Hence

$$\Psi_0 = \prod_{\substack{\tilde{\epsilon}(p) < \mu \\ p, \sigma}} C_n^+ \cdots C_{p_2 \sigma}^+ C_{p_1 \sigma}^+ | 0 \rangle \tag{2-69}$$

where $|0\rangle$ is the vacuum state. The number of particles is $N = \sum_p n(p)$, so that $p = p_F$, and hence

$$\mu = \epsilon(p_F) \tag{2-70}$$

Finally, the ground-state energy may be obtained from the formula

$$\mu = \partial E_0 / \partial N \tag{2-71}$$

One finds

$$E_0 / N = (3p_F^2/10m) + (1/2)NV_0 - (1/2)\sum_{pq} N_q V_{q-p} \tag{2-72}$$

We remark that N_q is taken to be the unperturbed particle-distribution function.

What we have found in this fashion is, of course, the Hartree-Fock approximation. The first term in the correction to the quasi-particle energy in (2-66) is the direct, or Hartree, term; the second is the exchange, or Hartree-Fock, term.

Suppose one wants to do better than the Hartree-Fock approximation. Suhl and Werthamer have shown that the corrections arising from the fluctuation term are of order $N^{-1/3}$ and are therefore negligible for a large system. What one does next, then, is to write down the equation of motion for the triple term; upon contraction it will involve terms proportional to C_p^+, triple terms, and quintuple terms. If one neglects the quintuple terms, for example, one has a set of linear equations involving only the linear and triple terms, which can then be solved. Suhl and Werthamer have carried through this procedure for the electron gas and show that it gives results equivalent to those obtained by summing the perturbation series for various properties of the electron gas. We refer to their paper for further details.

We may apply the same approach to obtain approximate density-fluctuation excitations (Bohm and Pines[4]). Consider the equation of motion of $C_{p+k}^+ C_p$, a coupled particle-hole pair operator. One finds

$$\begin{aligned}(H', C_{p+k}^+ C_p) = {} & [\epsilon(p+k) - \epsilon(p)]\, C_{p+k}^+ C_p \\ & - \sum_{k'} (V_{k'}/2)\,[(C_{p+k}^+ C_{p+k'} - C_{p+k-k'}^+ C_p)\rho_k^+ \\ & + \rho_k^+ (C_{p+k}^+ C_{p+k'} - C_{p+k-k'}^+ C_p)]\end{aligned} \tag{2-73}$$

If we neglect completely the interaction terms in (2-73) we see that the electron-hole pair oscillates at a frequency

$$\omega(kp) = \epsilon(p+k) - \epsilon(p)$$

This is the free-particle, or Hartree-Fock approximation, result, in which the density-fluctuation spectrum consists solely of electron-hole pairs, with energies in the band from zero (or $kv_F - k^2/2m$) up to $kv_F + k^2/2m$, which we have considered earlier in Fig. 1-1.

Again, one can take certain effects of the interaction into account by linearization. A particularly simple result is obtained if one makes the random-phase approximation: *In determining the motion of a pair of momentum k, keep only the interaction term associated with the k-th momentum transfer.* In this case (2-73) reduces to

$$(H', C^+_{p+k} C_p) = [\epsilon(p + k) - \epsilon(p)] C^+_{p+k} C_p - V_k (n_{p+k} - n_p) \sum_q C^+_{q+k} C_q \tag{2-74}$$

The resulting equation possesses oscillatory solutions at a frequency ω_k; the dispersion relation is found by setting the right-hand side equal to $\omega_k C^+_{p+k} C_p$ and requiring that the result be consistent. One finds

$$1 = V_k \sum_p \frac{n_p - n_{p+k}}{\omega_k - \epsilon(p + k) + \epsilon(p)} \tag{2-75}$$

There are two types of roots in (2-75). The first type is at

$$\omega_k = \epsilon(p + k) - \epsilon(p)$$

[the difficulty with the pole may be avoided by taking ω_k to have a small positive imaginary part corresponding to a retarded solution of (2-75)]. The particle-hole pairs continue to be a possible excitation in the RPA. The second type of root is displaced from the particle-hole continuum at long wavelengths—the amount of the displacement depending on the size of V_k. It is the collective mode; in the case of Coulomb interaction, $V_k = 4\pi e^2/k^2$, and (2-75) is the plasmon dispersion relation in the RPA. For a dilute fermion gas, one again finds the result (2-75) for a collective mode, with V_k replaced by m^{-1} times the scattering length f_0; (2-75) is then the dispersion relation for zero sound.

The examples we have considered demonstrate the utility of the equation-of-motion method. We shall later see that it provides a simple way to obtain the BCS solution for the case of attractive forces; it is also useful in determining the collective modes in superconductors, nuclear matter, and finite nuclei, and in studying simple stability problems. I believe, therefore, that it should be part of the "arsenal" of every physicist interested in the many-body problem. It is, however, not without its defects. In going beyond Hartree-Fock approximation for the particle motion, or the RPA for particle-hole motion, one must take considerable care to ensure that the higher-order terms are treated consistently, as may be seen from the discussion given in Suhl and Werthamer. It may well be that as physicists gain further familiarity with the method, a simple set of prescriptions for treating these terms will become available; in any event, it would seem likely that it will continue to supplement rather than supplant the perturbation-theoretic approach we now consider.

TIME-DEPENDENT PERTURBATION THEORY

The Use of Feynman Diagrams

Perturbation-theoretic calculations based on the use of Feynman diagrams have been developed for the ground-state energy (Goldstone,[1] Hubbard[2]), for the single-particle propagator (Galitskii,[18] DuBois), and for the propagating dielectric constant (Hubbard,[3] DuBois). In most of these papers the S-matrix expansion and the rules for calculating the appropriate diagrams are derived and discussed in some detail, so that I do not believe it useful to repeat them here. What I wish to do is to convey a sense of the physical significance of the diagrams. I shall therefore summarize the rules and demonstrate their application and utility by some simple examples.

In the calculation of $G(r\tau)$ we are concerned with the propagation of a single particle. In the absence of interaction we could represent that propagation by a straight line, drawn from right to left: $\tau \longleftarrow 0$. Because the unperturbed state is a filled Fermi sphere, we also have the possibility of propagation as a hole, which would then be a line drawn from left to right: $-\tau \longrightarrow 0$; this convention is consistent with the results we found for $G_0(r\tau)$ and the line can be regarded as equivalent to $G_0(r\tau)$. The effect of particle interaction is to modify $G_0(r\tau)$, corresponding to the scattering of the particle by the other particles in the system. The interactions derive from the interaction term in (2-52), which corresponds to particle-particle scattering. In lowest order of perturbation theory, two kinds of scattering processes contribute to $G(r\tau)$. The first looks as shown in Fig. 2-4. The particle propagates as a free particle until it undergoes forward scattering against the background particles at time τ_1; it then propagates freely from τ_1 to τ. The interaction is represented by a dashed line, the circle corresponds to a description of a background particle being knocked out of a given momentum state, then returning to it once more. The second process is as shown in Fig. 2-5 and corresponds to a particle that undergoes exchange scattering against the background particles.

In second order, a number of scattering processes can affect the particle propagation. It can forward-scatter twice, exchange-scatter twice,

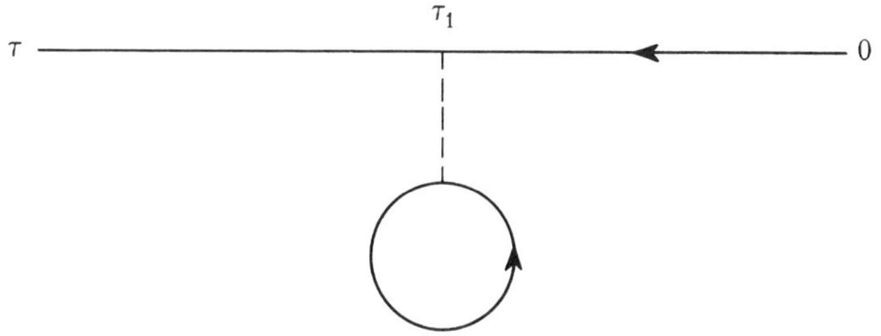

FIG. 2-4

or forward-scatter then exchange-scatter. It can also do as shown in Fig. 2-6. At time τ_1, the particle scatters against a background particle, knocking it into a different state. This corresponds to creating a particle-hole pair. At time τ_2 the particle scatters once more by annihilating the pair; that is, the background particle returns to its original state. The process corresponds to the polarization of the background particles.

One builds up the perturbation-series expansion for $G(r\tau)$ by considering the sum of all the scattering processes available. This is most simply done by drawing the set of diagrams in which a particle enters from the right, undergoes all topologically distinct interactions, then leaves at the left, and by giving the rules for the evaluation of the contribution from each diagram. It can be shown that there is a one-to-one correspondence between the terms in the perturbation-series expansion for $G(r\tau)$ and these diagrams. The rules for evaluating the diagrams can be given with respect to space-time propagation, which is what we have considered, or with respect to the Fourier transform of $G(r\tau)$, $G(p,\epsilon)$. The latter is more convenient for most applications; where it is used, the arrows no longer denote a time direction but simply serve to yield energy and momentum conservation at each scattering vertex.

The rules in momentum space are the following:

1. Assign a momentum and energy to each directed solid line; for each solid line of momentum p, energy ϵ, a factor $iG_0(p,\epsilon) = i[\epsilon - \epsilon(p) + i\delta_p]^{-1}$ enters.

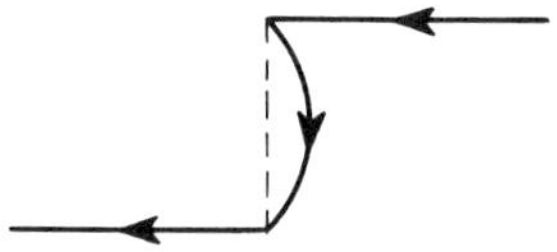

FIG. 2-5

2. For each interaction (represented by a dotted line), a factor $-iV_q$ if the interaction is local and instantaneous; otherwise a factor $-iV(q\omega)$.

3. Conserve energy and momentum at each vertex, with the convention that lines pointing toward the vertex enter with a (+) sign, those pointing away with a (−) sign.

4. For each free variable p, ϵ in the diagram, carry out an integration

$$[1/(2\pi)^4] \int d^3p \int d\epsilon$$

5. A factor of (− 1) for each internal closed loop.

6. Introduce a factor of 2 when spin sums are necessary.

7. A special rule for nonpropagating lines, such as enter in Figs. 2-4 and 2-5; a factor $-n_p$ for each such line of momentum p.

Some Simple Examples

We apply the rules to the diagrams we have thus far considered.

1. Forward scattering against the background (Fig. 2-4)

$$A = (i)^2 G_0(p,\epsilon) G_0(p,\epsilon) (-iV_0)\{[-2/(2\pi)^3] \int d^3p'(-n_{p'}) = N\}$$

so that the closed loop contributes a factor N.

2. Exchange scattering against the background (Fig. 2-5)

$$B = (i)^2 G_0(p,\epsilon) G_0(p,\epsilon)[1/(2\pi)^3] \int d^3q\,(-iV_q)(-n_{p+q})$$

3. Polarization of the background (Fig. 2-6)

$$C = [iG_0(p,\epsilon)]^2\,[1/(2\pi)^4] \int d^3q \int d\omega\,[iG_0(p-q,\epsilon-\omega)]$$

$$\times\,(-iV_q)^2\,[-i\pi_0(q\omega)]$$

where we have introduced the lowest-order polarization diagram, $-i\pi_0(q\omega)$, defined by

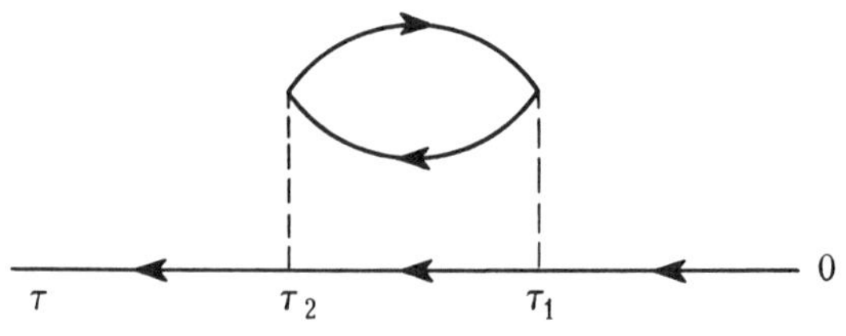

FIG. 2-6

$$-i\pi_0(q\omega) = -2 \int \frac{d^3p' \int d\epsilon'}{(2\pi)^4} (i)^2 \left[\frac{1}{\epsilon' + \omega - \epsilon(p'+q) + i\delta_{p'+q}} \times \frac{1}{\epsilon' - \epsilon(p') + i\delta_{p'}} \right]$$

It is instructive to evaluate $\pi_0(q, \omega)$. We consider separately the regions $p' > p_F$ and $p' < p_F$ and carry out the ϵ' integration first. For $p' > p_F$, we remark that we get a contribution to the integral only when $|\mathbf{p}' + \mathbf{q}| < p_F$; if we had both $p' > p_f$, $|\mathbf{p}' + \mathbf{q}| > p_F$, the singularities in the complex ϵ' plane would both lie below the real axis, so that closing the contour above yields zero. The contribution for $p' > p_F$, $|\mathbf{p}' + \mathbf{q}| < p_F$, is

$$\int \frac{d\epsilon'}{2\pi} \frac{1}{\epsilon' + \omega - \epsilon(p'+q) - i\delta} \frac{1}{\epsilon' - \epsilon(p') + i\delta} = \frac{-i}{\epsilon(p') - \epsilon(p'+q) + \omega - i\delta}$$

In the same way, on considering $p' < p_F$, we see that $|\mathbf{p}' + \mathbf{q}| > p_F$, in which case the ϵ' integration yields

$$\int \frac{d\epsilon'}{2\pi} \frac{1}{\epsilon' + \omega - \epsilon(p'+q) + i\delta} \frac{1}{\epsilon' - \epsilon(p') - i\delta} = \frac{i}{\epsilon(p') - \epsilon(p'+q) + \omega + i\delta}$$

We find, therefore,

$$\pi_0(q\omega) = \frac{-2}{(2\pi)^3} \left\{ \int\limits_{\substack{p' < p_F \\ |\mathbf{p}' + \mathbf{q}| > p_F}} d^3p' \frac{1}{\epsilon(p') - \epsilon(p'+q) + \omega + i\delta} - \int\limits_{\substack{p' > p_F \\ |\mathbf{p}' + \mathbf{q}| < p_F}} d^3p' \frac{1}{\epsilon(p') - \epsilon(p'+q) + \omega - i\delta} \right\}$$

This can be written more symmetrically by interchanging $\mathbf{p}'$ and $(\mathbf{p}' + \mathbf{q})$ in the second integral. We then find

$$\pi_0(q\omega) = -2 \int\limits_{\substack{p' < p_F \\ |\mathbf{p}' + \mathbf{q}| > p_F}} \frac{d^3p'}{(2\pi)^3} \left[\frac{1}{\omega + \epsilon(p') - \epsilon(p'+q) + i\delta} - \frac{1}{\omega + \epsilon(p'+q) - \epsilon(p') - i\delta} \right] \tag{2-76}$$

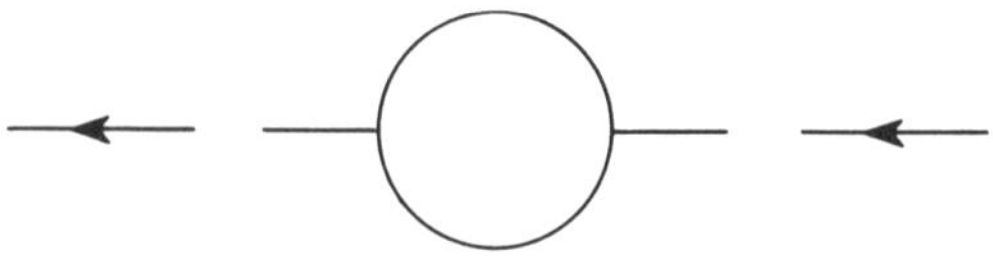

FIG. 2-7

The Self-Energy Operator and the Dyson Equation

The diagrams we have been considering all have the structure of self-energy diagrams; that is, they act to modify the energy $\epsilon(p)$ of the propagating particle. They fall into the general class of diagrams to which by attaching one line at the left, one at the right, we obtain a single-particle propagator (Fig. 2-7). A simple way to sum an infinite set of self-energy diagrams if one knows the contribution of a simple, primitive diagram is through the use of the integral equation of Fig. 2-8, where the double line denotes the exact Green's function, $iG(p, \epsilon)$, and we define the circle to represent the sum of all proper self-energy diagrams, $-i\Sigma(p, \epsilon)$.

One sees by iteration that in this fashion one sums the series for $G(p, \epsilon)$ given in Fig. 2-9. Hence, in order to prevent our counting terms in the perturbation series twice, it is necessary to define a proper self-energy diagram as one which cannot be split into two self-energy diagrams by cutting a line in two.

The Dyson equation,

$$G(p, \epsilon) = G_0(p, \epsilon) + G_0(p, \epsilon)\Sigma(p, \epsilon)G(p, \epsilon)$$

is simply solved, and yields

$$G(p, \epsilon) = 1/[\epsilon - \epsilon(p) - \Sigma(p, \epsilon) + i\delta_p]$$

We see that $\Sigma(p, \epsilon)$ describes in compact fashion the modification in particle propagation brought about by the fermion interaction. It will in general have both real and imaginary parts; the quasi-particle energy is obtained by solving the equation

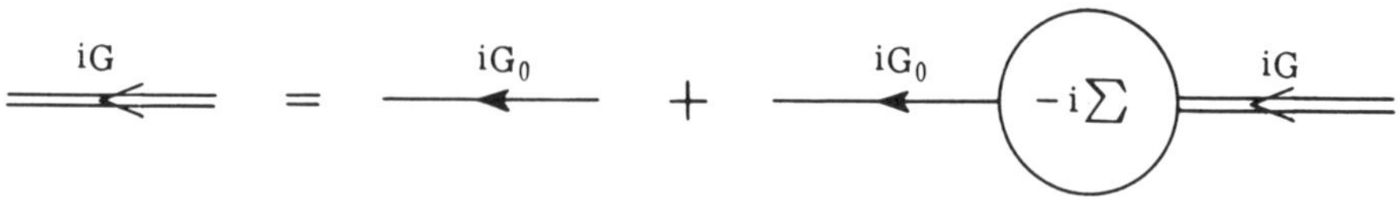

FIG. 2-8

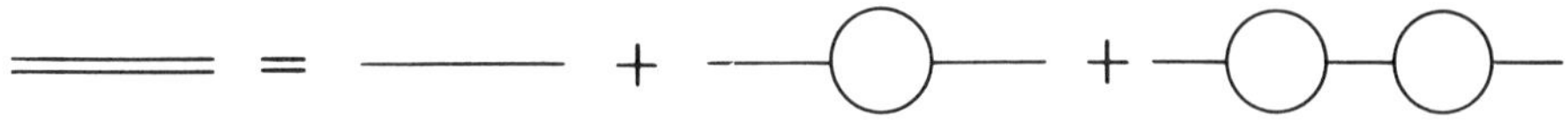

FIG. 2-9

$$\epsilon = \epsilon(p) + \Sigma(p, \epsilon)$$

which may be a very complicated affair indeed.

The self-energy parts of the diagrams we have considered are

$$\Sigma_A = NV_0$$

$$\Sigma_B = -[1/(2\pi)^3] \int d^3q\, V_q N_{p+q}$$

$$\Sigma_C = (-i)[1/(2\pi)^4] \int d^3q \int d\omega (V_q)^2 \pi_0(q, \omega) G_0(p - q, \epsilon - \omega)$$

We remark that Σ_A and Σ_B represent the Hartree and Hartree-Fock contributions, respectively, to the quasi-particle energy, as may be seen from (2-66).

Polarization Diagrams and $\epsilon_p(k\omega)$

The same kind of simple summation device is useful in considering the modification in particle interaction as a result of the polarization of the medium. We can define $1/\epsilon_p(k\omega)$ as the screened particle interaction when all polarization processes are taken into account. Thus, the effective particle interaction is taken to be as given in Fig. 2-10. An infinite class of terms which contribute to $\epsilon_p(k\omega)$ may then be summed in simple fashion if we write the equation (shown in Fig. 2-11)

$$-iv_k/\epsilon_p(k\omega) = (-)iv_k + \{(-iV_k)[-i\pi(k\omega)](-iV_k)\}/\epsilon_p(k\omega)$$

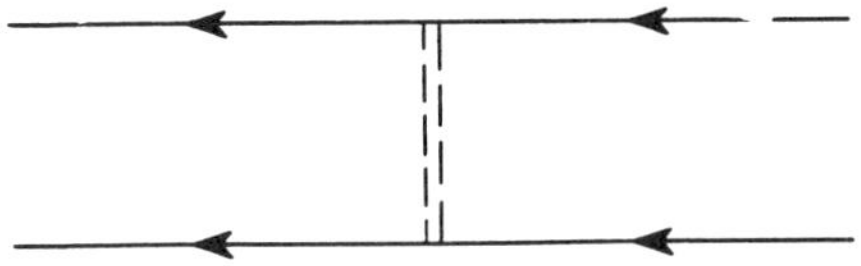

FIG. 2-10

where $-i\pi(k\omega)$ represents the sum of all proper polarization diagrams, that is, a polarization diagram that cannot be split into two

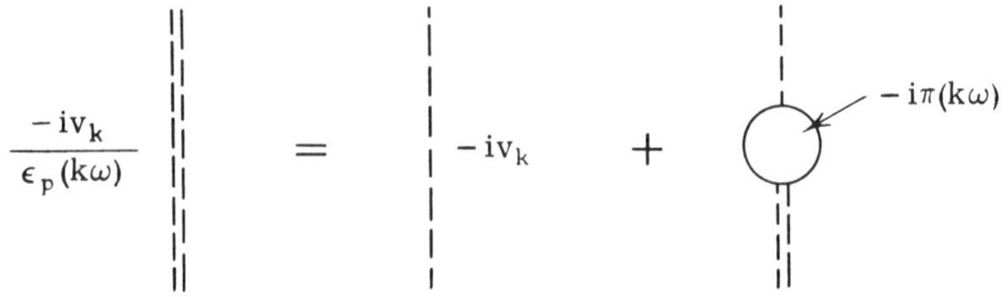

FIG. 2-11

polarization diagrams by cutting an interaction line. This equation possesses the simple solution

$$\epsilon_p(k\omega) = 1 + V_k\,\pi(k\omega)$$

We have already considered $\pi_0(k\omega)$, the lowest-order contribution to $\pi(k\omega)$.

3 SPECIFIC PROBLEMS

THE ELECTRON GAS

1. The basic Hamiltonian is that of Eq. (2-52), with $V_q = 4\pi e^2/q^2$, and the term with $q = 0$ omitted, because it cancels against the uniform background of positive charge. The Hartree approximation therefore yields the same results as the free-particle approximation; the ground-state energy in either is

$$E_0/N = (3/5)\,\epsilon(p_F) = 2.21/r_s^2 \qquad \text{ryd} \tag{3-1}$$

the average kinetic energy of the particles in the unperturbed Fermi sphere.

2. In the Hartree-Fock approximation, one takes into account the correlations in electron positions brought about by the Pauli principle. These keep electrons of parallel spin apart, and thereby reduce the ground-state energy of the system. It is illuminating to calculate this reduction, known as the exchange energy, from Eqs. (2-57) and (2-59), making use of the Hartree-Fock expression for the mean-square density fluctuation,

$$S_{HF}(k) = \sum_{p\sigma} \frac{n_{p\sigma}(1 - n_{p+k\sigma})}{N}$$

$$= \begin{cases} \dfrac{3}{4}\dfrac{k}{k_F} - \dfrac{3}{48}\dfrac{k^3}{k_F^3} & k < 2k_F \\[2ex] 1 & k > 2k_F \end{cases} \tag{3-2}$$

One has, per particle

$$E_{exch} = E_{int}/N = \sum_k (2\pi e^2/k^2)[S(k) - 1] = -0.916/r_s \quad \text{ryd} \qquad (3\text{-}3)$$

the first equality following because E_{int} is linear in e^2. We see from (3-1) and (3-2) that the reduction in S(k) for small k is responsible for the exchange energy. [Remark that S(k) = 1 in the Hartree approximation.] The Hartree-Fock pair-distribution function, g(r), which is related to s(k) and p(r) according to

$$g(r) = N^{-1} \int \frac{d^3k}{(2\pi)^3} [S(k) - 1] e^{+i\mathbf{k}\cdot\mathbf{r}} = N^{-1} [p(r) - \delta(r)]$$

is shown in Fig. 3-1. It is 1/2 at r = 0, corresponding to the fact that electrons of parallel spin cannot overlap (Fig. 3-1).

3. The quasi-particle energy in the Hartree-Fock approximation displays a logarithmic singularity in the vicinity of p_F:

$$\epsilon_{HF}(p) = \frac{p^2}{2m} - \frac{e^2 p_F}{2\pi}\left(2 + \frac{p_F^2 - p^2}{pp_F} \ln \left|\frac{p + p_F}{p - p_F}\right|\right) \qquad (3\text{-}4)$$

As a result the specific heat, which depends on $(\partial\epsilon/\partial p)_{p=p_F}$, is greatly altered, and varies as T/ln T, a result in contradiction with experiment. The origin of the difficulty with (3-4) is the long range of the Coulomb force. Other difficulties of similar origin appear when one tries to improve on the Hartree-Fock calculation by going to second-order perturbation theory; one finds a divergent result.

4. The difficulties are overcome in the random-phase approximation, in which the effects of the interaction that depend on a single momentum transfer k are summed in every order in the perturbation series; one obtains thereby a convergent result (Gell-Mann and Brueckner). For example, the series summation for the dielectric constant reads

$$\begin{aligned} 1/\epsilon_{RPA}(k\omega) &= 1 - V_k\pi_0(k\omega) + [V_k\pi_0(k\omega)]^2 - [V_k\pi_0(k\omega)]^3 + \cdots \\ &= 1/[1 + V_k\pi_0(k\omega)] \end{aligned} \qquad (3\text{-}5)$$

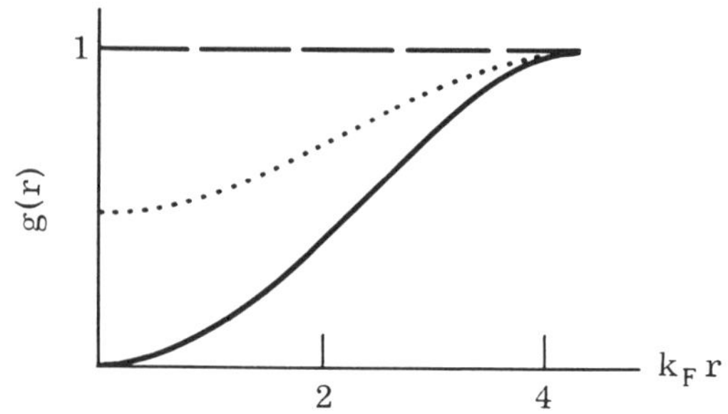

FIG. 3-1. Dashed line, Hartree approximation; dotted line, Hartree-Fock approximation; solid line, random-phase approximation; solid line, possible form for metallic densities.

The first term in the series is the Hartree-Fock result; the second, the contribution in second order from the polarization bubbles, leads to divergent results for the ground-state energy and quasi-particle excitation energies; the third order leads to still-more-divergent results, etc. The summation [which is performed automatically via the Dyson-Hubbard procedure (page 53)] yields an $\epsilon_p(k\omega)$ that is well behaved. The result (3-5) can also be obtained directly from a study of the equations of motion (if one adds a forcing term to the Hamiltonian) and by many other techniques as well.

5. The properties of the system are simply calculated from a knowledge of $\epsilon_{RPA}(k\omega)$ [Nozières and Pines[13]]. [For positive frequencies we need not distinguish between the retarded and the propagating $\epsilon(k\omega)$.] On making use of (2-76) and (3-5) we write $\epsilon(k\omega)$ in the form

$$\epsilon_{RPA}(k\omega) = \epsilon_1(k\omega) + i\epsilon_2(k\omega) \tag{3-6}$$

where

$$\epsilon_1(k\omega) = 1 - \frac{8\pi e^2}{k^2} \sum_p \frac{n(p)[\epsilon(p-k) - \epsilon(p)]}{\omega^2 - [\epsilon(p-k) - \epsilon(p)]^2} \tag{3-7}$$

$$\epsilon_2(k\omega) = (4\pi^2 e^2/k^2) \sum_p n(p)[1 - n(p-k)]\,\delta[\omega - \epsilon(p-k) + \epsilon(p)] \tag{3-8}$$

A plot of ϵ_1 and ϵ_2 at long wavelengths is given in Fig. 3-2.

We remark on the following properties of $\epsilon_{RPA}(k\omega)$ at long wavelengths:

a. At low frequencies and long wavelengths, $\epsilon_1(k\omega) \cong 1 + k_{FT}^2/k^2$, where k_{FT}^2 is the Fermi-Thomas screening wave vector,

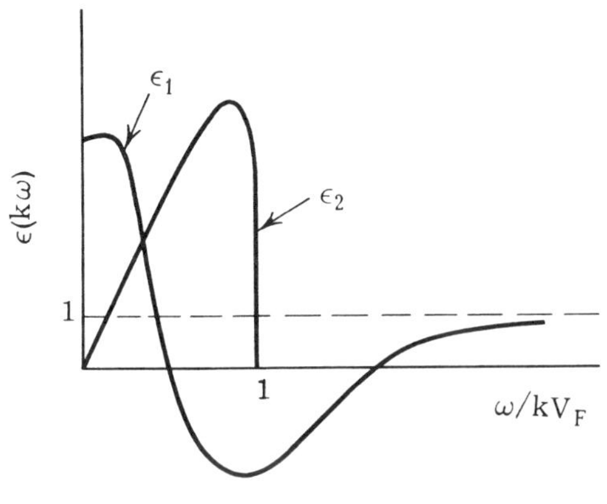

FIG. 3-2

$$k_{FT}/k_F = \sqrt{3}\ \omega_p/V_F k_F = 0.82 r_s^{1/2} \tag{3-9}$$

k_{FT}^{-1} is the characteristic screening wave vector for the electron gas; it is of the order of the interparticle spacing for electron densities of physical interest.

b.
$$\begin{aligned} \epsilon_2 &= (\pi/2)(k_{FT}^2/k^2)(\omega/kV_F) & \omega &\lesssim kV_F \\ &= 0 & \omega &\gtrsim kV_F \end{aligned} \tag{3-10}$$

Thus ϵ_2 cuts off at the maximum free particle-hole excitation frequency, as is obvious from (3-8).

c. $\epsilon_1(k\omega)$ possesses a zero at high frequencies for which $\epsilon_2 = 0$. At very high frequencies, one has the asymptotic result

$$\epsilon_1(k\omega) \to 1 - (\omega_p^2/\omega^2) \qquad \omega >> kV_F$$

6. The correlation properties of the electron gas in the RPA may be seen most directly from the expression

$$\begin{aligned} S(k) &= -\frac{k^2}{4\pi^2 Ne^2}\int_0^\infty d\omega\ \mathrm{Im}\ \frac{1}{\epsilon(k\omega)} \\ &= \frac{k^2}{4\pi^2 Ne^2}\int_0^\infty d\omega\ \frac{\epsilon_2(k\omega)}{\epsilon_1^2(k\omega) + \epsilon_2^2(k\omega)} \end{aligned} \tag{3-11}$$

After a bit of algebra, one can show that

$$S(k) = (1/N)\langle \varphi_0 | \rho_k^{s+} \rho_k^s | \varphi_0 \rangle \tag{3-12}$$

where ρ_k^s is a screened density-fluctuation operator, defined according to

$$(\rho_k^s)_{n0} = (\rho_k)_{n0}/\epsilon(k\omega_{n0}) \tag{3-13}$$

and the state labels in (3-12) and (3-13) refer to free-particle matrix elements and excitation frequencies. We thus see that the correlations are such that each electron is surrounded by a screening, or polarization cloud, which moves with the particle in such a way that the density fluctuation associated with the electron is screened by the RPA dielectric constant at the characteristic frequency of the excitation. An equivalent way of writing (3-13) is to replace the electron-hole-pair creation operator

$$\rho_{kp}^+ = C_{p+k}^+ C_p$$

by a screened-pair operator

$$\rho_{kp}^{s+} = C_{p+k}^{+} C_p \Big/ [\epsilon(k, k \cdot p/m + k^2/2m)]$$

The effect of this coherence factor is to reduce markedly the matrix elements for pair excitation, since both ϵ_1 and ϵ_2 are proportional to k_{FT}^2/k^2 at long wavelengths. We further remark that in the RPA the correlations between electrons of antiparallel spin play an important role. The pair distribution function, g(r), has been calculated by Glick and Ferrell in the RPA for a gas density $r_s = 2$; it is negative near the origin. This unphysical result shows clearly the inapplicability of the RPA at such a density. A possible realistic form for g(r) is shown in Fig. 3.1.

7. The plasmon energies are determined by the dispersion relation

$$\epsilon_1(k\omega_k) = 0$$

At long wavelengths, the plasmon energy is

$$\omega_k = \omega_p[1 + (3/10)(k^2 V_F^2/\omega_p^2) + \cdots] \tag{3-14}$$

The plasmons will be a distinct well-defined elementary excitation of the system until their energy overlaps with that of the pair continuum; this occurs at

$$\omega_p \cong k_c V_F \qquad \text{or} \qquad (k_c/k_F) \cong 0.47 r_s^{1/2} \tag{3-15}$$

Once decay of a plasmon into an electron-hole pair is energetically possible, the process will go quite fast, so that it is only reasonable to regard the plasmon as an independent long-lived excitation for $k \lesssim k_c$.

8. The role played by Coulomb correlations in giving rise to plasmons and screened-pair excitations is seen clearly by a study of S(kω). We have

$$S(k\omega) = -\frac{k^2}{4\pi^2 e^2} \,\mathrm{Im}\, \frac{1}{\epsilon(k\omega)} = \frac{k^2}{4\pi^2 e^2} \frac{\epsilon_2(k\omega)}{|\epsilon(k\omega)|^2} \tag{3-16}$$

For long wavelengths, we can separate S(kω) into two parts. One is the pair contribution, which comes from the frequency region in which $\epsilon_2(k\omega)$ is different from zero. This part may be written as

$$S_{\text{pair}}(k\omega) = \sum_{\mu} \frac{|(\rho_k)_{\mu 0}|^2 \delta(\omega - \omega_{\mu 0})}{|\epsilon(k, \omega_{\mu 0})|^2} \tag{3-17}$$

where the index μ refers to the plane-wave states of the noninteracting electron system. We see that S_{pair} possesses the same continuous energy spectrum as $S_{HF}(k\omega)$, the spectral function for the free electron gas, i.e., that calculated in the Hartree-Fock approximation; however, the strength of the pair excitations is reduced by the coherence factor, $1/|\epsilon(k\omega)|^2$. At long wavelengths the reduction is enormous; it is of order k^4/k_{FT}^4. Remark that the result (3-17) is just what we would have expected from the replacement (3-13); it is due to the screening clouds around each individual electron.

The second contribution to $S(k\omega)$ comes from the frequency region in which $\epsilon_2(k\omega) = 0$ and $\epsilon_1(k\omega) = 0$, that is, from the plasmons. Careful attention to limiting procedures shows that one can write, for this region,

$$\underset{\text{plasmon}}{S(k\omega)} \cong (Nk^2/2m\omega_k)\,\delta(\omega - \omega_k) \tag{3-18}$$

to an accuracy of order k^4/k_{FT}^4.

We see therefore that at long wavelengths the coherence effects are such that the plasmon mode is dominant, and the pair excitation modes are strongly screened. This conclusion applies equally well to $F(k\omega)$, the Fourier transform of the density-fluctuation propagator. One has, using (2-50),

$$F(k\omega) = \frac{Nk^2}{2m\omega_k}\left(\frac{1}{\omega - \omega_k + i\delta} - \frac{1}{\omega + \omega_k - i\delta}\right) + \sum_{\mu} \frac{(\rho_k)_{\mu 0}^2}{|\epsilon(k,\omega_{\mu 0})|^2}\left(\frac{1}{\omega - \omega_{\mu 0} + i\delta} - \frac{1}{\omega + \omega_{\mu 0} - i\delta}\right) \tag{3-19}$$

The first term in (3-19) is characteristic of a free boson propagator, the second that appropriate to propagation of screened density fluctuations. The separation we have given here (see also DuBois) is directly analogous to the separation made in the collective coordinate method of the density fluctuations into a plasmon part and an individual particle contribution, with collective coordinates introduced to describe the plasmons. We remark that the separation is only meaningful for wave vectors less than k_c for which the plasmons form an independent excitation mode. In higher-order approximations, even these long-wavelength plasmons are damped; nonetheless, the approximate separation continues to be useful for many applications.

9. We have seen earlier that if one measures the energy and angular distribution of inelastically scattered fast electrons, one thereby measures $S(k\omega)$ directly. The relevant expression for $(dP/dt)(k\omega)$, the probability per unit time that the fast electron transfer momentum $\mathbf{k}$ and energy ω to the electron gas, is

$$\frac{dP}{dt}(k\omega) = 2\pi V_k^2 S(k\omega) = \frac{8\pi e^2}{k^2} \frac{\epsilon_2(k\omega)}{\epsilon_1^2(k\omega) + \epsilon_2^2(k\omega)} \tag{3-20}$$

We thus expect from (3-17) and (3-18) that at long wavelengths (small scattering angles) the major energy and momentum transfer will be to the plasmons, with a rather negligible amount going to the screened pair excitations. Moreover, one verifies immediately that the plasmons exhaust the sum rule for $S(k\omega)$, (2-37), to order k^4/k_{FT}^4. For the electron gas, the sum rule is frequently written

$$\int_0^\infty d\omega\, \omega\, \mathrm{Im}\, [1/\epsilon(k\omega)] = -(\pi/2)\, \omega_p^2 \tag{3-21}$$

Upon comparison with (3-16) we see that it is a sum rule for the energy transfer per unit time at momentum k, which is

$$\frac{dW}{dt} = \int_0^\infty d\omega\, \omega\, \frac{dP}{dt}(k\omega) = \frac{4\pi^2 e^2}{k^2}\, \omega_p^2 \tag{3-22}$$

in which form it is the Bethe sum rule for the stopping power. We also remark that as one goes to shorter wavelengths, ($k \gtrsim k_c$) the plasmon ceases to be the dominant excitation mode. It fades away into the continuum, and one measures only the quasi-particle pair excitations (Glick and Ferrell).

10. The principal experimental evidence for the existence of plasmons as a well-defined excitation mode in electron systems comes from characteristic energy-loss experiments in solids. In such experiments, one measures the energy spectrum of kilovolt electrons either as they emerge from a thin solid film or after they are reflected from a solid surface. A typical transmission-experiment result is sketched in Fig. 3-3.

The losses frequently occur in multiples of a basic quantum which is close to the free-electron plasmon energy. The reason for this is simple: The energy of a plasmon in a solid is of the order of 10 to 25 ev; this energy is large compared to the energies characteristic of the periodic

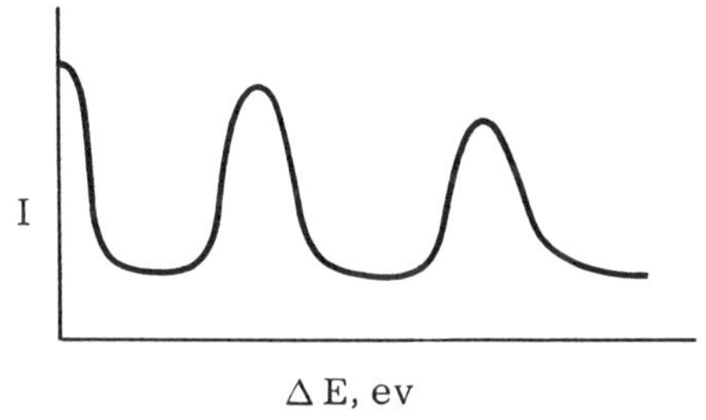

FIG. 3-3. I = intensity of the beam; ΔE = energy loss.

ionic potential, which are frequently only a few electron volts; hence, the electrons oscillating at ω_p do not "see" the periodicity characteristic of the solid and behave as if they were free. One can easily make an RPA theory for the plasmons in solids, and the results are in good qualitative and often quantitative agreement with experiment. For a further discussion of plasmons in solids, see Pines (1956), Nozières and Pines (1958), and Pines (1960).

11. The ground-state energy may be calculated from the expressions (2-58) and (3-6) (Hubbard[11]). The result one obtains agrees with that found by Gell-Mann and Brueckner.[6] The comparison between the two expressions is facilitated if one writes the Gell-Mann-Brueckner result as

$$\frac{E_0}{N} = \frac{2.21}{r_s^2} - \frac{3}{4\pi\alpha^2 r_s^2}\int_0^\infty dk\, k^3 \int_0^\infty du \ln\left[1 + \pi_0(k, iku)\right] - \sum_k \frac{2\pi e^2}{k^2} \tag{3-23}$$

A simple transformation of the dielectric-constant expression

$$\frac{E_0}{N} = \frac{2.21}{r_s^2} + \sum_k \frac{1}{2\pi}\int_0^{e^2} \frac{d\alpha}{\alpha}\int_0^\infty d\omega \,\mathrm{Im}\, \frac{1}{1 + \pi_0^\alpha(k,\omega)} - \sum_k \frac{2\pi e^2}{k^2} \tag{3-24}$$

combined with an appropriate contour integration (Sawada et al.[8]) shows the equivalence of the two expressions. The result is

$$E_0/N = (2.21/r_s^2) - (0.916/r_s) + 0.62 \ln r_s - 0.142 \quad \text{ryd} \tag{3-25}$$

It should also be noted that, as may be seen from (3-18) and (3-24), there is an explicit contribution to the ground-state energy from the zero-point energy of the well-defined plasmons, $\Sigma_k \omega_k/2$. Finally we note that the remarks by Sawada et al.[8] concerning the comparison of the Gell-Mann-Brueckner theory with the Bohm-Pines theory are not quite correct. The two theories possess identical expansions at long wavelengths, as has been shown by Nozières and Pines.[12] The contribution to the correlation energy from long-wavelength momentum transfers is given by

$$\begin{aligned} E_{corr}(\beta) &= \sum_{k < \beta k_F} E_{corr}(k) \\ &= -0.46\frac{\beta^2}{r_s} + 0.87\frac{\beta^3}{r_s^{3/2}} - 0.98\frac{\beta^4}{r_s^2} + 0.019\frac{\beta^4}{r_s} \\ &\quad + 0.71\frac{\beta^5}{r_s^{5/2}} + \cdots \quad \text{ryd} \end{aligned} \tag{3-26}$$

The terms which are even in β arise from the screened single-particle interactions; those which are odd are associated with the plasmon zero-point energy.

12. The physical picture of a quasi particle in the random-phase approximation is clear from the preceding discussion: It is an electron plus its associated screening cloud. The cloud of virtual excitations which move with the electron represents the backflow of the other particles about the electron as it moves along. (For a calculation of the contribution to the backflow from the virtual plasmon cloud, see Pines and Schrieffer.)

The energy of the quasi particle is simply calculated from the screened exchange self-energy diagram, Fig. 3-4. One has, from the rules for time-dependent perturbation theory,

$$\sum(p,\epsilon) = i\int \frac{d^3k}{(2\pi)^3}\int_{-\infty}^{\infty}\frac{d\omega}{(2\pi)}\,\frac{4\pi e^2/k^2}{\epsilon-\omega-\epsilon(p-k)+i\delta_{p-k}}\,\frac{1}{\epsilon_{\mathrm{RPA}}(k\omega)} \tag{3-27}$$

where one must use the RPA value of the propagating dielectric constant, and the integration is carried out by a contour which is closed in the lower half-plane. The calculation was first carried out by Quinn and Ferrell; they find for the energy of a quasi particle near the Fermi surface,

$$\tilde{\epsilon}(p) = \frac{p_F^2}{2m}\left\{\frac{p^2}{p_F^2} - 0.166r_s\left[\frac{p}{p_F}(\ln r_s + 0.203) + \ln r_s - 1.80\right]\right.$$
$$\left. + i(0.252r_s^{1/2})\left[\left(\frac{p}{p_F}\right) - 1\right]^2\right\} \tag{3-28}$$

This expression at once yields the Gell-Mann[7] result for the specific heat in the RPA,

$$C_{\mathrm{RPA}}/C_s = 1 + 0.083r_s(\ln r_s + 0.203) \tag{3-29}$$

where C_s is the Sommerfeld free-electron value.

13. Another problem of considerable physical interest is that of the screening of an impurity by the electron gas. For a fixed charge of strength Z (which is assumed to be weakly coupled to the electrons), one has directly from (2-32) that the induced polarization charge is

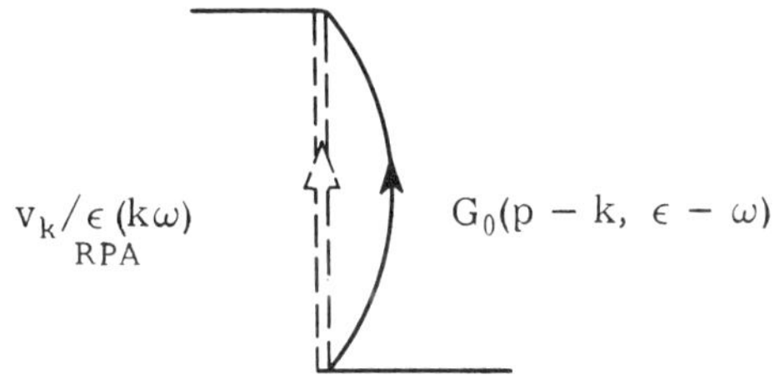

FIG. 3-4

$$-e\,\delta\rho(\mathbf{r}) = \sum_{k} Ze\{[1/\epsilon(k0)] - 1\}\, e^{i\mathbf{k}\cdot\mathbf{r}} \tag{3-30}$$

while the potential due to the charge is

$$\varphi(\mathbf{r}) = \sum_{k} (4\pi Ze/k^2)\,[1/\epsilon(k0)]e^{i\mathbf{k}\cdot\mathbf{r}} \tag{3-31}$$

The static dielectric constant in the RPA is given by

$$\epsilon(k0) = 1 + \frac{k_{FT}^2}{k^2}\left[\frac{1}{2} + \frac{k_F}{2k}\left(1 - \frac{k^2}{4k_F^2}\right)\ln\left|\frac{k + 2k_F}{k - 2k_F}\right|\right] \tag{3-32}$$

which for small k behaves like $1 + k_{FT}^2/k^2$. One would therefore expect from this small k behavior that at large distances the amount of screening charge would be negligible, and the potential due to the impurity would fall off as $e^{(-k_{FT}r)}/r$. Such is not the case because $\epsilon(k0)$ possesses a logarithmic singularity; $\partial\epsilon/\partial k$ is ∞ at $k = 2k_F$. The origin of this singularity is the sharpness of the Fermi surface (Kohn); as a result one finds that for large distances the induced charge density and the potential vary as $(\cos 2p_F r)/r^3$ (Langer and Vosko). The disturbance is therefore effective at rather large distances, a result borne out by nuclear magnetic resonance experiments on the influence of impurities on the Knight shift (Rowland).

14. The corrections to the RPA may be regarded as arising from the next-order irreducible polarization diagrams of importance for the dielectric constant. The diagrams are given in Fig. 3-5, corresponding to electron-hole scattering via a screened Coulomb interaction and an electron (or hole) self-energy correction. The influence of these terms on various properties of the electron gas has been considered by Nozières and Pines using the collective coordinate method and by DuBois using a field-theoretic approach. The ground-state energy, to lowest order in r_s, is altered by a constant term, which arises, in fact, from the second-order exchange term in the perturbation series. The value of this constant was determined by Gell-Mann and Brueckner to be -0.046, so that the ground-state energy for terms up to order r_s reads

$$E_0/N = (2.21/r_s^2) - (0.916/r_s) + 0.626 \ln r_s - 0.096 + \cdots \quad \text{ryd} \tag{3-33}$$

In this order one finds a shift in the energy of the plasmons (Nozières and

FIG. 3-5

Pines[12]) and plasmon damping, because there now exist matrix elements for the decay of a plasmon into two electron-hole pairs (DuBois). The energy of a long-wavelength plasmon becomes

$$\omega_k = \omega_p \{1 + (k^2/k_F^2)[(1.3/r_s) - (3/40) + i(4.4r_s^{1/2})] + \cdots\} \quad (3\text{-}34)$$

DuBois has calculated corrections for a number of other system properties.

15. Nozières and I came to the conclusion that it was not especially fruitful to pursue the systematic study of corrections to the RPA because it is intrinsically a high-density expansion, which would be limited to a region of validity $r_s \lesssim 1$ no matter how many correction terms one includes. I refer to our paper for the details of the argument (Nozières and Pines[12]). We concluded that for small momentum transfers the RPA is correct for metallic densities ($1.8 \lesssim r_s \lesssim 5.5$) and that for large momentum transfers the dominant effect is the interaction between electrons of antiparallel spin. Essentially the same conclusions were reached by Hubbard.[11] Hubbard argued that at large momentum transfers, the exchange contribution to a given diagram would tend to cancel one-half the direct contribution. He therefore suggested replacing the contribution from the basic polarization diagram, $\pi_0(k\omega)$, by

$$\frac{\pi_0(k\omega)}{1 + (1/2)[k^2/(k^2 + k_F^2)]\,\pi_0(k\omega)}$$

We see that for small k, $\pi_0(k\omega)$ is unaffected; for large k ($k \gg k_F$), $\pi_0(k\omega) \rightarrow \pi_0(k\omega)/2$.

Nozières and I obtained, by means of a simple interpolation method (between the high-momentum-transfer and low-momentum-transfer regions), the following approximate expression for the correlation energy:

$$E_{corr} \cong (-0.115 + 0.031 \ln r_s)\ \text{ryd} \quad (3\text{-}35)$$

a result that agrees well with the results obtained using Hubbard's interpolation procedure. Calculations presently underway at the University of Illinois indicate that a similar interpolation procedure yields results for the specific heat and spin susceptibility which are in good agreement with the experimental values (Silverstein).

FERMION SYSTEMS WITH SHORT-RANGE REPULSIVE INTERACTIONS

1. The solution for the ground-state energy and quasi-particle excitation spectrum of a low-density fermion gas is described very clearly in the paper by Galitskii.[18] I wish, therefore, to make only a few brief pedagogical and supplementary remarks.

2. As Galitskii shows, in order to treat the low-density limit it is necessary that one take into account the multiple scattering between a pair of particles. Indeed, the most important terms in lowest order are those in which a pair of particles repeatedly scatter against one another, without exciting or involving the background particles (the holes). These terms are simply summed with the aid of a t matrix defined according to the diagrammatic equation of Fig. 3-6. The equation, which sums all ladder diagrams, reads

$$t(pp';q\omega) = V_{p-p'} + i\int \frac{d^3p''}{(2\pi)^3}\int \frac{d\epsilon''}{(2\pi)} V_{p''-p}$$
$$\times \frac{1}{-\epsilon'' + \omega - \epsilon(q - p'') + i\delta_{q-p''}}$$
$$\times \frac{t(pp'';q\omega)}{\epsilon'' - \epsilon(p'') + i\delta_{p''}} \qquad (3\text{-}36)$$

Inspection of the poles in the ϵ'' plane shows that one gets a nonzero result only when $p'' > p_F$; $|p'' - q| > p_F$ and $p'' < p_F$; $|p'' - q| < p_F$, corresponding to intermediate-state propagation as either a particle-particle pair or a hole-hole pair. One finds on carrying out the ϵ'' integration,

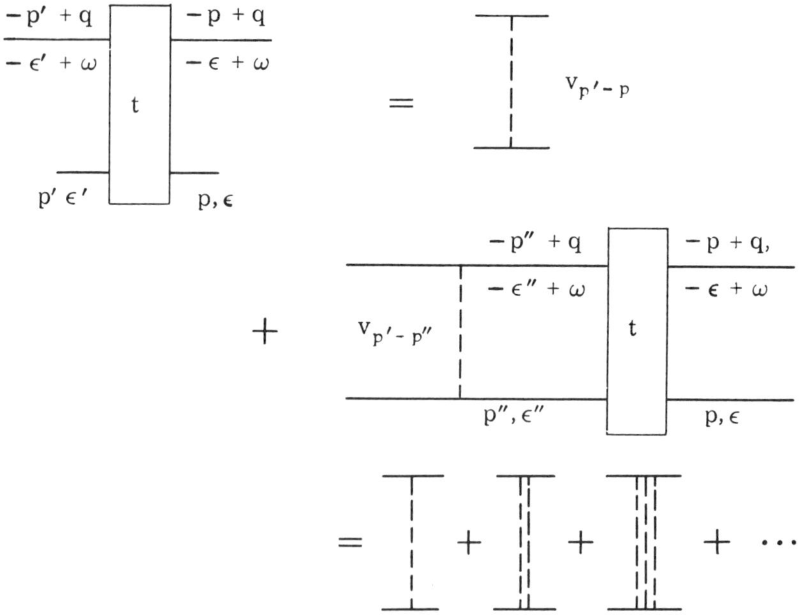

FIG. 3-6

$$t(pp';q\omega) = v_{p-p'}$$

$$- (2\pi)^{-3} \int_{\substack{p'' > p_F \\ |p''-q| > p_F}} dp'' \frac{V_{p''-p}\, t(pp'';q\omega)}{\omega - \epsilon(p'') - \epsilon(p''-q) - i\delta}$$

$$- (2\pi)^{-3} \int_{\substack{p'' < p_F \\ p''-q < p_F}} dp'' \frac{v_{p''-p}\, t(pp'';q\omega)}{\omega - \epsilon(p'') - \epsilon(p''-q) - i\delta} \tag{3-37}$$

3. Galitskii solves (3-37) by first expressing t in terms of t_{free}, the scattering matrix for a pair of particles (with no Fermi sea); the latter quantity may then be obtained by writing the equivalent two-body Schrödinger equation. Hence, one finds no difficulty in treating an arbitrarily strong repulsive interaction, including the case of hard-sphere interactions. One finds, to lowest order, that

$$t \cong 4\pi a/m + \cdots \tag{3-38}$$

where a is the s-wave free-particle scattering length. Corrections to (3-38) come from hole-hole scattering and off-the-energy-shell propagation.

4. The quasi-particle energy is calculated from the two lowest-order self-energy diagrams for the t matrix. Thus one has

$- i\,\Sigma(p,\epsilon) =$ 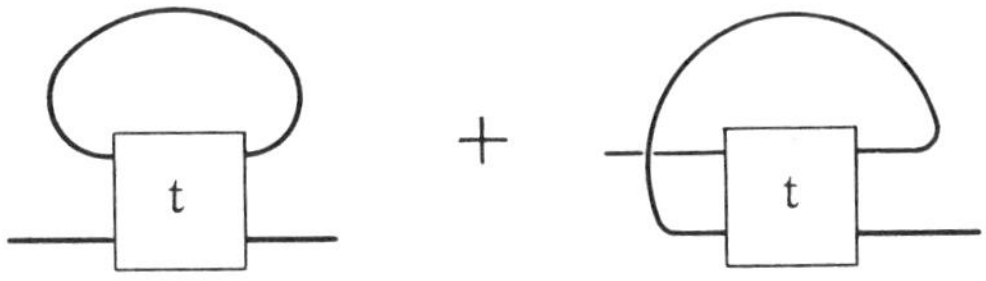

$$\Sigma(p,\epsilon) = -i\,2 \int \frac{d^3q}{(2\pi)^3} \int \frac{d\omega}{2\pi} \frac{t(pp;q\omega)}{\omega - \epsilon - \epsilon(q-p) + i\delta_{q-p}}$$

$$+ i \int \frac{d^3q}{(2\pi)^3} \int \frac{d\omega}{2\pi} \frac{t(pq-p;q\omega)}{\omega - \epsilon - \epsilon(q-p) + i\delta_{q-p}} \tag{3-39}$$

with the factor of 2 arising from the two spin states available for the direct contribution. For the lowest-order term, (3-38), one obtains

$$\Sigma_1(p,\epsilon) = 2\pi Na/m \tag{3-40}$$

a constant shift in the quasi-particle energy.

5. On taking into account the higher-order corrections to t, Galitskii obtains for the energy of a quasi particle near the Fermi surface,

$$\epsilon(p) = (p_F^2/2m)\{1 + (4/3\pi)\, p_F\, a + (4/15\pi^2)\, p_F^2 a^2 [7 + 26 \ln^2]$$

$$+ p^2/p_F^2 - (16/15\pi)^2 p_F^2 a^2 (p/p_F)(7 \ln 2 - 1)$$

$$- i(2p_F^2 a^2/\pi)[(p/p_F) - 1]^2\} + \cdots \qquad (3\text{-}41)$$

We remark that as was the case with the electron gas, and as one expects on quite general grounds, the imaginary part of $\epsilon(p)$ is proportional to $(p - p_F)^2$ in the vicinity of p_F.

6. The expansion parameter for the low-density fermion gas is $p_F a$ or a/r_0, the ratio of the scattering length to the interparticle spacing; one must have, therefore,

$$p_F a << 1 \qquad (3\text{-}42)$$

This condition is not well satisfied for either nuclear matter or He^3; in both systems one has $a \cong r_0$.

7. Galitskii obtains the ground-state energy by calculating $\mu = \epsilon(p_F)$ and making use of the relation, $\mu = \partial E_0/\partial N$. He finds

$$E_0/N = (3/5)(p_F^2/2m)[1 + (10/9\pi)\, p_F\, a$$

$$+ (4/21\pi^2)(11 - 2 \ln 2) p_F^2 a^2 + \cdots] \qquad (3\text{-}43)$$

This result has also been obtained by Huang and Yang and by de Dominicis and Martin with a pseudopotential method, and by Abrikosov and Khalatnikov from the Landau-Fermi liquid theory.

8. The collective modes of the system are the phonons corresponding to zero sound. The dispersion relation to lowest order may be determined by making the random-phase approximation with the effective potential taken to be $t \cong (4\pi a/m)$. [See, for example, (2-75).] The dispersion relation at long wavelengths (Abrikosov and Khalatnikov, Goldstone and Gottfried, Gottfried and Pičman) is given by

$$\omega(k) = k v_F \{1 + 0.270 e^{-1/\xi} [1 + 0.309 (k/k_c^0)^2 + \cdots]\} \qquad (3\text{-}44)$$

where

$$\xi = p_F\, a/\pi \qquad (3\text{-}45)$$

and

$$k_c^0 = 0.736 k_F e^{-1/\xi} \qquad (3\text{-}46)$$

is the wave vector for which the phonon merges with the particle-hole pair continuum. We note that in the low-density limit, the phonon is only very slightly displaced from the pair continuum.

9. Gottfried and Pičman have considered the particular higher-order terms in the low-density expansion which give rise to a shift in the frequency of zero sound and a damping of the phonons; we refer to their paper for the details on these phenomena.

10. Brueckner and his collaborators have attempted to go beyond the low-density limit by summing a selected set of higher-order diagrams in a way that is equivalent to the following procedure:

a. Let the intermediate-state propagators in the equation determining the t matrix be quasi-particle propagators, with self-energies given by (3-39).

b. The resulting equations are then highly nonlinear. One has, as the generalization of (3-37), an equation in which $\epsilon(p)$ is everywhere replaced by

$$\epsilon(p) = \tilde{\epsilon}(p) + \Sigma(p, \epsilon) \tag{3-47}$$

with $\Sigma(p, \epsilon)$ dependent on t through (3-39).

c. Brueckner solves the coupled set of equations by first making approximations which include:

(1) Neglect of hole-hole propagation (an approximation which is only obviously all right for sufficiently low densities).

(2) Neglect of any possible energy dependence of $\Sigma(p, \epsilon)$.

(3) Neglect of off-the-energy-shell propagation. What this approximation means may be seen in (3-37). ω, the energy of the scattering pair, is set equal to the center of mass energy, $q^2/4$. The energy denominator becomes, then, $(q^2/4) + p'' \cdot (p'' + q) + 2\Sigma(p'')$. If, moreover, one changes variable to the relative momentum of the intermediate state pair, $k'' = p'' + (q/2)$, the denominator reduces to $k''^2 + 2\Sigma(p'')$, which is much simpler to deal with.

I would summarize the situation by saying that the Brueckner idea of going beyond the low-density limit by means of a self-consistent coupled set of equations is an ingenious and fruitful one, but that at present we do not have enough experience to know to what extent either the particular selection of terms to be summed, or the approximations (a) to (c), correspond to physical reality for intermediate density systems.

BOSON SYSTEMS WITH REPULSIVE INTERACTIONS

1. The basic Hamiltonian is

$$H = \sum_p \epsilon(p) a_p^+ a_p + \sum_{kpq} (v_k/2)\, a_{p+k}^+ a_{q-k}^+ a_q a_p \tag{3-48}$$

where the boson creation and annihilation operators satisfy the commutation relations

$$[a_p, a_q^+] = \delta_{pq} \qquad [a_p, a_q] = [a_p^+, a_q^+] = 0 \tag{3-49}$$

In the ideal Bose gas, N_0, the number of particles in the zero-momentum condensed state, is a macroscopic number, and is equal to N. Suppose we assume, as was first done by Bogoljubov,[20] that in the presence of interactions, N_0 continues to be a finite fraction of N. We shall see that from this assumption there follows directly the prediction of superfluid behavior in Bose systems.

2. Consider a_0 and a_0^+ acting on the ground-state wave function Ψ_0. One has

$$a_0 \mid \Psi_0(N_0) \rangle = \sqrt{N_0} \mid \Psi_0 (N_0 - 1) \rangle$$

$$a_0^+ \mid \Psi_0(N_0) \rangle = (N_0 + 1)^{1/2} \mid \Psi_0 (N_0 + 1) \rangle \tag{3-50}$$

where $\Psi_0(N_0)$ is the ground-state wave function with N_0 particles in the zero-momentum state, $\Psi_0(N_0 - 1)$ is that same state with $N_0 - 1$ particles in the zero-momentum state, etc. We see that if $N_0 >> 1$, we may do two things:

a. Replace $(N_0 + 1)^{1/2}$ by $(N_0)^{1/2}$.

b. Assume that the difference between $\Psi_0(N_0 - 1)$ and $\Psi_0(N_0 + 1)$ is of order $1/N$; that is, removing or adding a particle to the zero-momentum state will not alter the physical properties of the system.

Under the circumstances, a_0 and a_0^+ will commute with each other; since they already commute with all a_p and a_p^+ for $p \neq 0$, they may simply be replaced by the "c" number $\sqrt{N_0}$. This is the famour Bogoljubov prescription; it amounts to neglecting the dynamic behavior of the condensed state.

3. With the Bogoljubov prescription, one no longer has a Hamiltonian which conserves particle number. For example, part of the interaction terms in (3-48) reduce to

$$\sum_p N_0 V_p a_p^+ a_{-p}^+ /2$$

corresponding to the process represented by

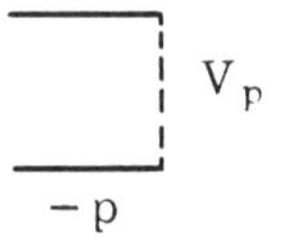

in which a pair of particles are created. Another way of saying this is that the number operator

$$N'_{op} = \sum_{p \neq 0} a_p^+ a_p$$

no longer commutes with the Hamiltonian, once one replaces a_0 and a_0^+ by $\sqrt{N_0}$. A consistent way out of the difficulty is to add a term, $-\mu N_{op}$, to the Hamiltonian, (3-48). This gives us, to borrow a phrase of Bogoljubov's, a "hunting license" to consider nonparticle conserving intermediate states, while offering a mechanism to conserve particle number on the average (Hugenholtz and Pines[24]). μ is, as before, the chemical potential. We find the following Hamiltonian:

$$H' = H_1 + H_2 + \tfrac{1}{2} N_0^2 V_0 \tag{3-51a}$$

where

$$H_1 = \sum_p{}' \tilde{\epsilon}(p)\, a_p^+ a_p + \sum_p{}' N_0 V_p (a_{-p}^+ a_p^+ + a_p a_{-p})/2 \tag{3-51b}$$

$$H_2 = \sum_p{}' \sqrt{N_0}\; V_k (a_{p+k}^+ a_k a_p + a_{p+k}^+ a_{-k}^+ a_p)$$
$$+ \sum_{kpq}{}' (V_k/2)\, a_{p+k}^+ a_{q-k}^+ a_q a_p \tag{3-51c}$$

and

$$\tilde{\epsilon}_p = \epsilon(p) + N_0 V_0 + N_0 V_p - \mu \tag{3-52a}$$

is the quasi-particle energy in the Hartree-Fock approximation. N_0 and μ are to be determined by the conditions

$$N_0 + \langle \Psi_0 | \sum_p{}' a_p^+ a_p | \Psi_0 \rangle = N \tag{3-53}$$

$$\mu = \partial E_0 / \partial N = \partial E'_0 / \partial N_0 \tag{3-54}$$

The primes on the summations indicate that the states of zero momentum are to be omitted. E_0 is the ground-state energy of H and E'_0 is that of H'_0. The condition (3-53) guarantees particle conservation on the average.

4. Bogoljubov next assumed that in suitable approximation (which turns out to be the weak coupling limit) the terms of H_1 in Eqs. (3-51) are large compared to those in H_2. H_1 can be diagonalized by the following canonical transformation:

$$a_k = u_k \alpha_k - v_k \alpha_{-k}^+$$
$$a_k^+ = u_k \alpha_k^+ - v_k \alpha_{-k}$$
$$u_k^2 - v_k^2 = 1 \qquad u_k^+ = u_k \qquad v_k^+ = v_k \tag{3-55}$$

The relation between u_k and v_k ensures that the transformation is canonical, i.e., that

$$[\alpha_k, \alpha_{k'}^+] = \delta_{kk'}$$

$$[\alpha_k, \alpha_{k'}] = [\alpha_k^+, \alpha_{k'}^+] = 0$$

One finds, by requiring that the resultant Hamiltonian be diagonal, that

$$H_1 = \sum_k{}' [\omega_k \alpha_k^+ \alpha_k + (\omega_k/2) - (\tilde{\epsilon}_k/2)] \tag{3-56}$$

with

$$u_k^2 = (1/2)[1 + (\tilde{\epsilon}_k/\omega_k)]$$

$$v_k^2 = (1/2)[-1 + (\tilde{\epsilon}_k/\omega_k)] \tag{3-57}$$

and

$$\omega_k^2 = \tilde{\epsilon}_k^2 - N_0^2 V_k^2 = \epsilon_k^2 + 2\epsilon_k N_0 V_k \tag{3-58}$$

upon making use of the relation

$$\mu = N_0 V_0$$

which leads to

$$\tilde{\epsilon}_k = \epsilon_k + N_0 V_k \tag{3-52b}$$

5. The Hamiltonian (3-56) describes a collection of noninteracting quasi particles with frequency

$$\omega_k = [(k^2 N_0 V_k/m) + (k^4/4m^2)]^{1/2} \tag{3-59}$$

At long wavelength or small k the quasi particles are phonons with velocity

$$s = (N_0 V_0/m)^{1/2}$$

At short wavelengths, the quasi particles behave essentially like free particles with an energy $k^2/2m$. The wave vector at which the transition from phonon to single-particle behavior occurs is

$$k_c = 2(m N_0 V_k)^{1/2} \cong 2ms$$

The inverse of this,

$$\lambda_c = 1/2ms$$

furnishes a natural measure of the distance over which coherence effects are important in the interaction between particles. One may call this a correlation length, provided one understands clearly that it refers to correlations between excitations in the system. This is quite different from the long-range correlations which lead to condensation in the $k = 0$ mode.

The essential point is that the single-particle excitations possess a phonon character, so that already the Bogoljubov approximation satisfies Landau's criterion for superfluidity; that no excitations can be created in a fluid moving with velocity $v < s$.

6. It is instructive to view the canonical transformation from the standpoint of many-body perturbation theory. One may show that it is equivalent to carrying out a summation of the most divergent terms in the perturbation-series expansion for the ground-state energy (Brueckner and Sawada). These have the following character:

Second order:

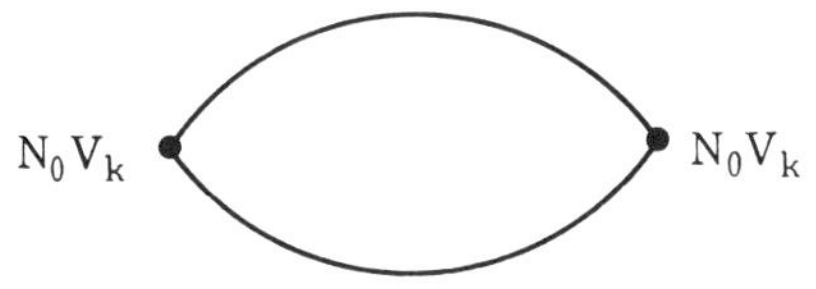

$$\Delta E = -\sum_k [N_0^2 V_k^2/(k^2/m)]$$

Third order:

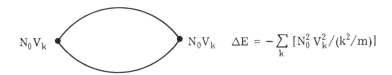

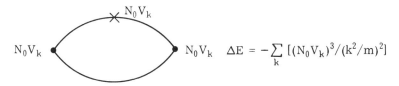

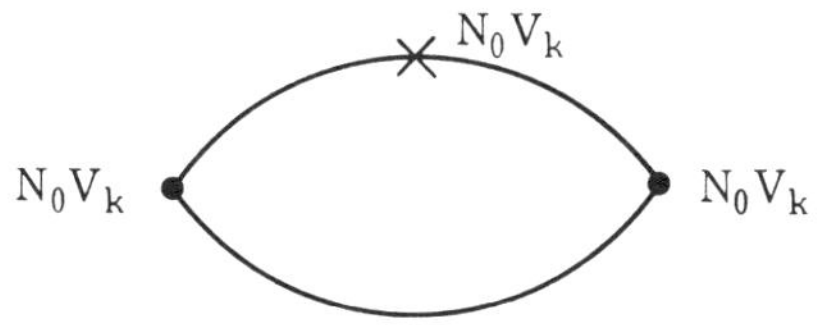

$$\Delta E = -\sum_k [(N_0 V_k)^3/(k^2/m)^2]$$

The cross, x, denotes a forward-scattering process proportional to $N_0 V_k$. As one goes to higher orders, the divergence for small k associated with iteration of the terms in the interaction involving the momentum transfer k becomes more and more pronounced. If one sums the series, however, one obtains a convergent result,

$$E_0 = \sum_k [(\omega_k/2) - (\tilde{\epsilon}_k/2)] + (1/2)N_0^2 V_0 \tag{3-60}$$

as may be obtained directly from (3-56), on adding the term $(1/2)N_0^2 V_0$ to pass from E_0' to E_0.

7. The summation of the series is equivalent to making the random-phase approximation, since the terms that are kept in each order are

those which involve a single momentum transfer k. Let us therefore consider the frequency and wave-vector-dependent screening constant, $\epsilon'(k\omega)$, which is the analogue of the dielectric constant for the boson gas (Nozières and Pines). One has

$$\begin{aligned}\epsilon'(k\omega) &= 1 + V_k\, \pi_0'(k\omega) \\ &= 1 - V_k \sum_\mu (\rho_k)_{\mu 0}^2 \left(\frac{1}{\omega - \omega_{\mu 0} + i\delta} - \frac{1}{\omega + \omega_{\mu 0} - i\delta} \right)\end{aligned} \tag{3-61}$$

where the states μ refer to the states of the noninteracting system. Hence, one has

$$\sum_\mu (\rho_k)_{\mu 0}^2 = N_0$$

$$\omega_{\mu 0} = k^2/2m$$

and

$$\epsilon'(k\omega) = 1 - [(N_0 k^2 V_k/m)/(\omega^2 - k^4/4m^2 + i\delta)] \tag{3-62}$$

The zeros of $\epsilon'(k\omega)$ yield a phonon spectrum in agreement with (3-59), so that in the Bogoljubov approximation the quasi-particle spectrum is identical to the collective-mode spectrum. If one writes the interaction energy according to (2-58) and carries out the coupling-constant integration, (2-59), one finds again (3-59)(Nozières and Pines).

8. Using either the Bogoljubov transformation or (3-62), one sees easily that

$$S(k\omega) = (N_0 k^2/2m\omega_k)\, \delta(\omega - \omega_k) \tag{3-63}$$

one further sees that the sum rule (2-37) is exhausted by the identical quasi-particle collective-mode excitation to the extent that $N_0 = N$.

9. To the extent that the Bogoljubov approximation is valid, one can take $N_0 = N$. The Bogoljubov approximation is a weak coupling theory, and

$$\begin{aligned}N - N_0 &= \langle 0 | \sum_p{}' a_p^+ a_p | 0 \rangle = \sum_p v_p^2 \\ &= \tfrac{1}{2} \sum_p [(\epsilon_p + N_0 V_p)/\omega_p] - 1 \cong 0(V_p)\end{aligned}$$

10. To summarize: The Bogoljubov approximation, although valid only in a weak coupling limit (as one sees by computing simple diagrams), offers the essential clue to a microscopic theory of superfluidity, in that the low-lying excitations are phonons which obey a linear dispersion relation at long wavelengths.

11. To obtain a satisfactory microscopic model, valid in a given range of densities, one must go beyond Bogoljubov, by taking into account the multiple-scattering diagrams as well as the RPA diagrams. This has been done in a number of ways—by a pseudopotential method (Lee, Huang, and Yang[21]), by summing ladder diagrams using the Rayleigh-Schrödinger perturbation theory (Brueckner and Sawada), and by field-theoretic methods (Beliaev,[23] Hugenholtz and Pines[24]). We shall consider the field-theoretic approach in more detail in a moment; let us here consider briefly the series-summation approach, in which we take multiple scattering into account by introducing a t matrix. We do this for the boson problem by means of the same integral equation described for fermions (Fig. 3-7), but we now take a bare line to represent free-particle propagation, according to

$$+iG_0(p,\epsilon) = i/[\epsilon - \epsilon(p) + i\delta] \tag{3-64}$$

The t-matrix equation is then [compare (3-36) and (3-37)]

$$t_{\text{free}}(pp';q\omega) = v_{p-p'} - (2\pi)^{-3}\int dp'' \times \frac{v_{p''-p}\,t_{\text{free}}(pp'';q\omega)}{\omega - \epsilon(p'') - \epsilon(p''+q) - i\delta} \tag{3-65}$$

t_{free} is the free-particle reaction matrix. The solution to (3-65), to lowest order in the density, is

$$t_{\text{free}} = f_0/m$$

where $f_0 = (4\pi a/m)$, a being the s-wave free-particle scattering length. We can proceed to substitute f_0/m for V_k everywhere in our Bogoljubov formulation; in so doing one sums simultaneously RPA and ladder diagrams. The procedure looks very much like the pseudopotential method; it is not, however, because the pseudopotential is taken to be

$$V_{\text{eff}}(r) = (f_0/m)(\partial/\partial r)[r\delta(r)] \tag{3-66}$$

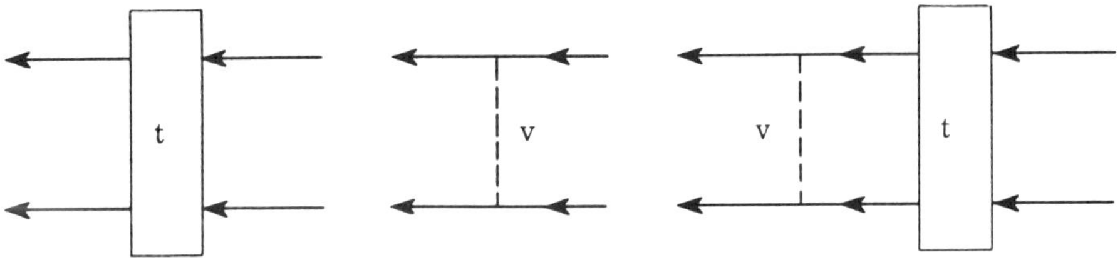

FIG. 3-7

rather than

$$V_{\text{eff}}(r) = (f_0/m)\delta(r) \tag{3-67}$$

the assumption we have used. The difference between (3-66) and (3-67) corresponds to allowing for the possibility of diagram duplication in the simultaneous summation of RPA and t-matrix diagrams (Nozières and Pines). Inspection shows that no excited-state diagrams are duplicated thereby, but that the ground-state-energy diagram , where $\bullet$ is now given by $-iV_k^{\text{eff}} = (-if_0/m)$, is counted twice: once as an RPA diagram, and once as a contribution from the Hartree approximation. The latter term is

$N_0^2 f_0/m = \bullet = \cdot +$  $+$  $+$  $+ \cdots$

$= \cdot +$ 

Hence the term $-\Sigma_k N_0 f_0^2/mk^2 = -$  must be subtracted from the ground-state-energy expression, (3-59), in which $V_k = f_0/m$.

12. The properties of the dilute boson gas are as follows:

Excitation spectrum: $\omega_k = [(N_0 k^2 f_0/m^2) + (k^4/4m^2)]^{1/2}$

Sound velocity: $s = (1/m)(N_0 f_0)^{1/2}$

Correlation length: $\lambda_c = 1/2ms = 1/2(N_0 f_0)^{1/2}$

Structure factor: $S(k) = k^2/2m\omega_k = k^2/2m\,[1/(N_0 k^2 f_0/m^2 + k^4/4m^2)]^{1/2}$

$= 1/(1 + k_c^2/k^2)^{1/2} \qquad \text{where } k_c = 2(N_0 f_0)^{1/2}$

Depletion of ground state: $N = N_0\{1 + [(Nf_0^3)^{1/2}/3\pi^2] + \cdots\}$

Ground-state energy: $E_0 = (N^2 f_0/2m)[1 + (16/15\pi^2)(Nf_0^3)^{1/2} + \cdots]$

13. The expansion parameter $(Nf_0^3)^{1/2}$ is the ratio of f_0 to the correlation length $(Nf_0)^{-1/2}$; it is this ratio that is important because one must *at the outset* include the RPA correlations in order to obtain a sensible theory for a boson system.

14. Calculations to the next highest order in $(Nf_0^3)^{1/2}$ have been made by Lee and Yang, Beliaev,[23] Sawada, Wu, and Hugenholtz and Pines.[24] The results are

$$E_0 = (N^2 f_0/2m)\{1 + (16/15\pi^2)(Nf_0^3)^{1/2}$$
$$+ (Nf_0^3/8\pi^2)[(4/3) - (3/\pi)] \ln (Nf_0^3) + \cdots\}$$
$$\omega_k = (k/m)(Nf_0)^{1/2}\{[1 + (1/\pi^2)(Nf_0^3)^{1/2}]$$
$$- i(3/640\pi)(Nf_0^3)^{1/2}[k/(Nf_0)^{1/2}]^4 + \cdots\}$$

for a low-momentum quasi particle $[p \ll (Nf_0)^{1/2}]$. The imaginary part in the sound-wave frequency results from the fact that in the low-density gas one phonon can decay into two, when higher-order terms are considered. It should also be noted that the term in $\ln (Nf_0^3)$ in the ground-state energy shows that there is not a simple power series expansion in Nf_0^3.

15. The sound velocity obtained from the single-particle excitation spectrum agrees with that calculated from the ground-state energy in accordance with the macroscopic relation

$$s = (\partial P/\partial \rho)^{1/2} \qquad \text{where } P = \rho^2 (d/d\rho)(E_0/\rho)$$

16. The perturbation-theoretic formulation (based on quantum field-theory methods) of the boson problem is greatly complicated, at first sight, by the macroscopic occupation of the zero-momentum state. One cannot simply apply the S-matrix development utilized for fermion systems, because the linked-cluster theorem fails for diagrams involving the zero-momentum particles (Beliaev,[23] Hugenholtz and Pines[24]). Several methods have been proposed for avoiding this difficulty (Beliaev,[23] Sawada) of which the simplest, I believe, is that of Hugenholtz and myself. We begin by adding the extra term $-\mu N_{op}$ to the Hamiltonian, and then follow the Bogoljubov prescription of replacing a_0 and a_0^+ by $\sqrt{N_0}$. The starting Hamiltonian is therefore (3-51a). There are no difficulties with the zero-momentum state because all dynamic effects associated with it are neglected. One can then proceed in the usual way with an S-matrix development, with the important difference that in place of a single scattering diagram, there are now six different kinds (Fig. 3-8). Here the interaction, represented by a dot, is taken to be one-half the sum of the direct and exchange terms, a useful simplification in dealing with short-range interactions. The details of the derivations of the basic formulas and the rules for carrying out perturbation-theoretic calculations may be found in Hugenholtz and Pines.[24]

17. Let us consider now the quite-different structure of the Dyson self-energy equation, and its implications for the boson problem. The different structure arises from the quite-different character of the interaction terms. The Dyson equation is replaced by the pair of Beliaev equations (Fig. 3-9), which read

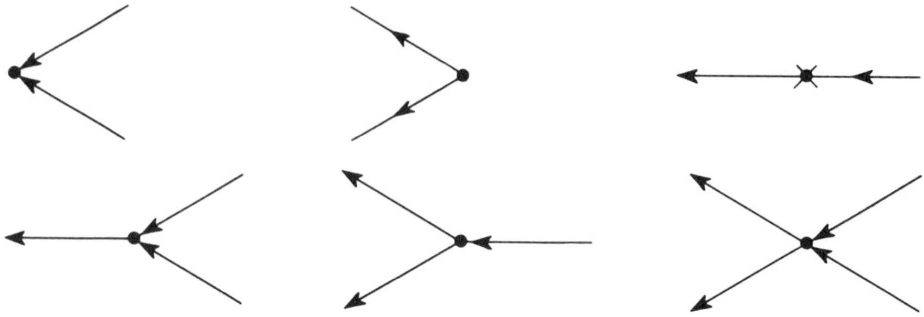

FIG. 3-8

$$G(p,\epsilon) = G_0(p\epsilon) + G(p\epsilon)\Sigma_{11}^{+}G_0(p\epsilon) + \widetilde{G}(p,\epsilon)\Sigma_{02}(p,\epsilon)\,G_0(p\epsilon)$$

$$\widetilde{G}(p\epsilon) = G_0(p\epsilon)\Sigma_{11}^{-}\widetilde{G}(p\epsilon) + G_0(p\epsilon)\Sigma_{20}G(p\epsilon) \qquad (3\text{-}68)$$

Σ_{11} is the irreducible self-energy part we have introduced earlier; $\Sigma_{02} = \Sigma_{20}$ is the new self-energy part required to take account of the new kind of interactions present in (3-51). $\widetilde{G}(p,\epsilon)$ is the Fourier transform of the propagator

$$\begin{aligned}\widetilde{G}(p,\tau) &= i\langle \Psi_0 \,|\, T[a_p^{+}(0)a_{-p}^{+}(\tau)] \,|\, \Psi_0 \rangle \\ &= i\langle \Psi_0 \,|\, T[a_{-p}(\tau)a_p(0)] \,|\, \Psi_0 \rangle \qquad (3\text{-}69)\end{aligned}$$

The pair of coupled equations, (3-68), are algebraic, and may be simply solved to yield

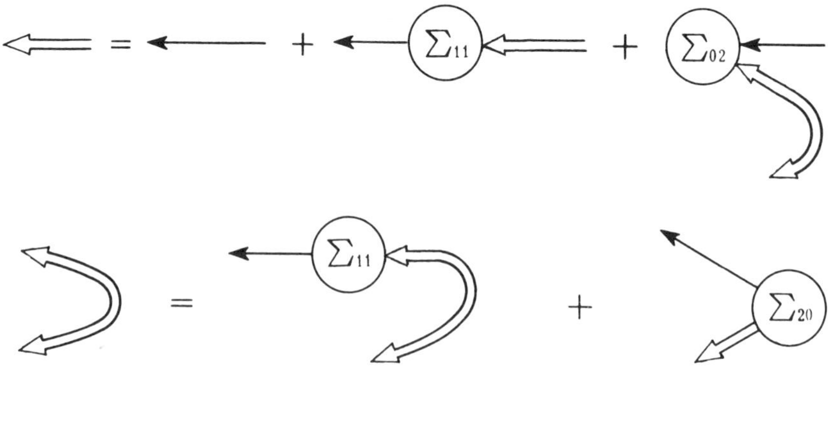

FIG. 3-9

$$G(p,\epsilon) = \frac{\epsilon + \epsilon(p) - \mu + \Sigma_{11}^{-}}{\left[\epsilon - \dfrac{(\Sigma_{11}^{+} - \Sigma_{11}^{-})}{2}\right]^2 - \left[\epsilon(p) - \mu + \dfrac{1}{2}(\Sigma_{11}^{+} + \Sigma_{11}^{-})\right]^2 + \Sigma_{02}^2} \tag{3-70}$$

where $\Sigma_{11}^{+} = \Sigma_{11}(p,\epsilon)$ and $\Sigma_{11}^{-} = \Sigma_{11}(-p-\epsilon)$.

18. The structure of $G(p,\epsilon)$ for the boson system is quite different from that for fermion systems. Hugenholtz and I were able to prove, from a term-by-term examination of the perturbation series, that

$$\Sigma_{11}(0,0) - \Sigma_{02}(0,0) = \mu \tag{3-71}$$

The content of the theorem is that the quasi-particle spectrum cannot exhibit a gap, and that, apart from possible pathological cases, one expects the low-momentum quasi-particle excitations will have a linear dispersion relation

$$\epsilon(p) = sp$$

Thus we see that the assumption of macroscopic occupation of the zero-momentum state leads directly to a linear phonon-dispersion relation, and hence, via Landau's famous argument, to the concept of superfluidity. The proof is dependent on the validity of the perturbation expansion.

19. By means of the summation procedures inherent in the field-theoretic formulation we have removed the divergences in the Rayleigh-Schrödinger perturbation series at low momentum transfer. For example, if in (3-70) we insert the weak coupling values,

$\Sigma_{11} =$  $= N_0(V_k + V_0)$

$\Sigma_{02} =$ 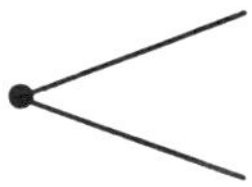 $= N_0 V_k$

$\mu = N_0 V_0$

we see that we at once obtain the Bogoljubov theory. We find

$$G(p,\epsilon) = [u_p^2/(\epsilon - \omega_p + i\delta)] - [v_p^2/(\epsilon + \omega_p - i\delta)] \tag{3-72}$$

where u_p, v_p, and ω_p are given by (3-57) and (3-58). Thus u_p measures the strength of the free-particle-like pole, and v_p measures the strength of the new pole introduced by the condensed state. The calculation of the low-density limit may be performed with equal ease (and with no difficulties about diagram duplication either). For that calculation, and for the calculation of the logarithmic term in the ground-state energy and the corrections to the phonon spectrum, I refer to the original papers (Beliaev,[23] Hugenholtz and Pines[24]).

20. There is an alternative way of deriving the basic equation, (3-70), which I should like to mention. Suppose instead of the Beliaev equations we consider the equation of Fig. 3-10, in which the double lines denote the true single-particle propagator, and the heavy lines denote a "normal-state" propagator

$$G_n(p,\epsilon) = 1/[\epsilon - \epsilon(p) + \mu - \Sigma_{11}(p\epsilon) + i\delta] \tag{3-73}$$

that is, one in which all self-energy effects associated with one line in and one line out have been included. The above equation then takes the form

$$G = G_n^+ + G_n^+ G_n^- \Sigma_{02}^2 G$$

$$= (G_n^-)^{-1} / [(G_n^- G_n^+)^{-1} - \Sigma_{02}^2] \tag{3-74}$$

which may easily be seen to reduce to our previous result (3-70). It is, in fact, not difficult to see from the diagrams, by comparing Beliaev's equation with the above, that the two must be identical. In some ways the above equation is more suggestive of the new features in the problem, in that it shows clearly that one need only account in consistent fashion for the scattering of pairs of particles in and out of the condensed state. We see, too, that the lack of particle conservation arises only in intermediate-state processes.

21. The Fourier transform $F(k\omega)$ of the density-fluctuation propagator may be calculated for the low-density boson system. In lowest approximation it yields just the results we have already described for $S(k\omega)$. A higher-order calculation has not yet been carried out. The calculation is of some interest because it is an open question whether "pure" collective modes exist in the boson system. Thus, one would expect, in general,

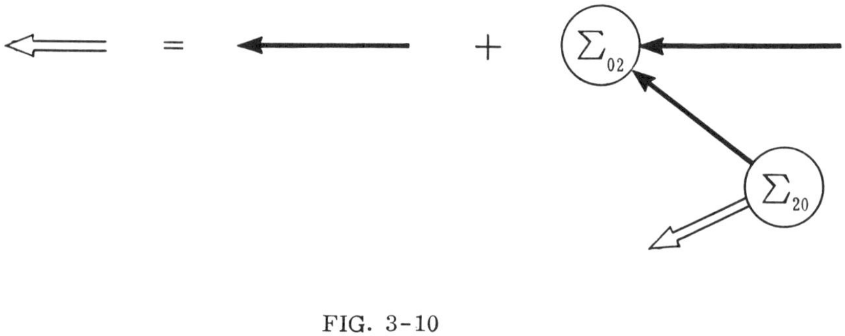

FIG. 3-10

poles in $F(k\omega)$ associated with the excitation of quasi particles from the condensed state. These are, in fact, the only poles thus far observed in the neutron-scattering experiments (Henshaw and Woods), since in the region of the roton minimum and in the phonon region, the energy of the rotons and the phonons agrees within experimental error with the quasi-particle energies determined from specific-heat data.

22. The dilute boson-gas calculation does not furnish a reasonable model for the behavior of actual liquid He^4, because the range of the repulsive interaction and the density are such that $Nf_0^3 >> 1$. Brueckner and Sawada have attempted an intermediate density calculation along the same lines of approximation as the Brueckner fermion calculations, by taking quasi-particle propagators as the intermediate-state propagators for the boson t matrix, which is then to be determined in self-consistent fashion. Their results are in fair qualitative agreement with the experimental excitation spectrum. However, the results are open to serious question, because Brueckner and Sawada did not allow for depletion of the ground state as a consequence of particle interaction. Recently Parry and Ter Haar have shown that the depletion of the zero-momentum state with the Brueckner-Sawada theory amounts to some 270 per cent! They then do a calculation which allows for the depletion, and find the resulting quasi-particle excitation curves no longer bend over, so that even the qualitative agreement with experiment is lost.

23. I should like to make some brief remarks on the extent to which microscopic calculations may draw clues and inspiration from the successful variational calculations of Feynman and Feynman and Cohen on the excitation spectrum of liquid He^4. The remarks are drawn, for the most part, from a paper by Miller, Nozières, and Pines.

a. The Feynman excitation spectrum follows directly if one assumes a single excitation exhausts the sum rule for $S(k\omega)$ and yields the experimental structure factor $S(k)$. Thus, if one takes $S(k\omega) = NS(k)\delta[\omega - \omega(k)]$, one finds at once from (2-34) that

$$\omega(k) = k^2/2mS(k) \tag{3-75}$$

which is the Feynman result for a trial wave function $\rho_k \mid \Psi_0 \rangle$.

b. The Feynman excitation spectrum agrees with experiment in the phonon region, but not in the roton region, as may be seen in Fig. 3-11. One can therefore conclude that at short wavelengths the spectral density $S(k\omega)$ cannot consist of a single discrete line, but must be broadened as a consequence of the interaction between the different excitations.

c. Feynman and Cohen gave arguments for improving the Feynman wave function by taking into account the backflow around a given particle as it moves through the system. With the improved wave function, considerably better agreement was found, as shown in Fig. 3-11.

d. One is therefore led to suspect that the backflow corresponds to taking into account the broadening of $S(k\omega)$ due to the interaction between the excitations. A given excitation can undergo virtual transitions to other configurations, and thus may be viewed as surrounded by a cloud of other excitations. The virtual cloud of excitations acts to increase the mass of the Feynman excitation (3-75), and so bring about the reduction in its energy toward the experimental value.

e. This picture is offered additional support by the fact that a calculation of the motion of an impurity atom through a boson system in the Bogoljubov approximation shows that the virtual phonon cloud around the atom gives rise to exactly the backflow proposed by Feynman and Cohen.

24. Finally, I should like to say a word about the role of the attractive forces in He^4 and about attractive forces in boson systems in general. If one considers V_q to be negative for small q, one finds that the quasi-particle energies, as well as the density fluctuations are then pure imaginary, signalling an instability of the system or an incorrect starting point for the calculation. What seems likely, since this is a very long wavelength instability, is that the system will reduce its volume to the point that the "effective" potential V_q^{eff} is positive for small momentum transfers. Thus, I am arguing that the condensed state only exists once the system has reached a size such that the long-wavelength phonons represent a stable excitation. It seems clear, on the other hand, that the attractive forces do play a considerable role in determining the short-wavelength excitation modes in liquid He^4, in that they are undoubtedly responsible for the observed structure factor and may also be responsible for much of the bending over of the excitation curve toward the roton minimum.

ELECTRON-PHONON INTERACTIONS

1. An understanding of the coupling between electrons and phonons in normal metals is essential to the development of a proper criterion

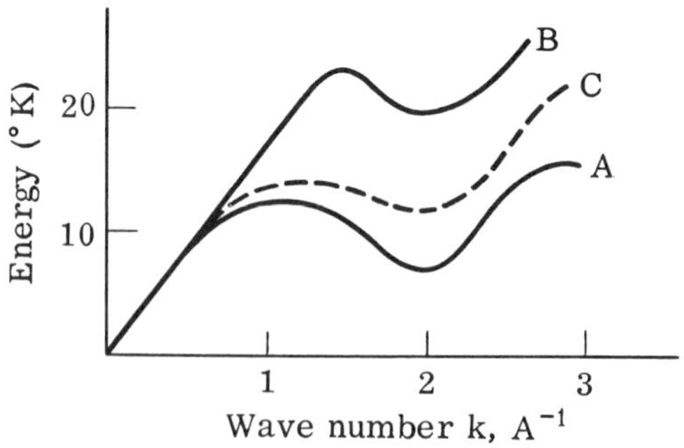

FIG. 3-11. Curve A, experiment (Henshaw and Woods); B, Feynman; C, Feynman and Cohen.

for the existence of superconductivity. Moreover, from an energetic point of view superconductivity is a small, indeed delicate effect, superimposed on the normal configurations of the electrons; it should, therefore, be formulated in terms of the best-possible normal-state solutions (Bardeen, Cooper, and Schrieffer[28]) for the system of coupled phonons and electrons. The electron-phonon coupling is, of course, interesting in and of itself, as a source of resistance and of ultrasonic attenuation in metals. In this lecture I wish to concentrate on discussion of the excitation modes, the effective electron-phonon interaction, and the effective electron interaction, both for a simple model and for real metals.

2. A simplified basic Hamiltonian to describe the electron-phonon interaction in metals has been derived by Bardeen and Pines.[25] It may be written

$$H_1 = H_{el} + H_{ph} + H_{int} + H_{Coul} \tag{3-76}$$

where

$$H_{el} = \sum_p \epsilon(p) C_p^+ C_p \tag{3-76a}$$

$$H_{ph} = (1/2) \sum_k [p_k^+ p_k + \Omega_k^2 q_k^+ q_k] \tag{3-76b}$$

$$H_{int} = \sum_k q_k v_k^i \rho_{-k} \tag{3-76c}$$

$$H_{Coul} = \sum_{kpq} (2\pi e^2/k^2) C_{p+k}^+ C_{q-k}^+ C_q C_p \tag{3-76d}$$

a. H_{el} describes a collection of electrons moving in the periodic field of the ion cores; the creation and annihilation operators are those for the Bloch waves appropriate to a single electron moving in a periodic potential field of energy $\epsilon(k)$.

b. H_{ph} describes the normal modes of the ions immersed in a uniform background of negative charge. If one assumes only Coulomb interactions, the frequencies Ω_k may be computed directly (Kohn, Clark, Coldwell-Horsfall and Maradudin). The sum of the squares of the longitudinal and transverse modes obeys a simple sum rule:

$$\sum_\mu \Omega_{k\mu}^2 = \Omega_p^2 = 4\pi N Z^2 e^2/M \tag{3-77}$$

A sketch of the dispersion relation for the frequencies is given in Fig. 3-12. The longitudinal mode must begin at Ω_p, the ionic plasma frequency (since the transverse waves have a linear dispersion relation near $k = 0$); the maximum wave vector for the phonons is k_D, where

$$k_D^3/6\pi^2 = N$$

The ions are taken to have charge Z and mass M. In (3-76) we have restricted our attention to the longitudinal phonon modes, since at long wavelengths only these are coupled to the electrons.

c. H_{int} describes the electron-phonon interaction. The matrix element, v_k^i, is given by

$$v_k^i = -(4\pi Ze^2 i/k)(N/M)^{1/2} \tag{3-78}$$

if one neglects the influence of periodicity and assumes a pure Coulomb interaction. Periodicity effects are, in fact, generally important. For example "Umklapprocesses," in which an electron goes from a state **p** to a state $\mathbf{p} - \mathbf{k} - \mathbf{K}_n$, are naturally included in (3-76c), since the wave vector **k** runs over all values of **k** for the density fluctuation and the matrix element v_k^i, while for the phonons q_k (where $k > k_D$) must be replaced by q_{k-K_n}, where K_n is the reciprocal lattice vector required to bring the phonon wave vectors back into their allowed region, specified by $k < k_D$. "U" processes play an important role in determining the conductivity and in determining the magnitude of the attractive electron interaction that gives rise to superconductivity.

d. H_{Coul} is the usual Coulomb interaction between the electrons.

3. The energies of the elementary excitations and the effective electron interaction for the system, (3-76), were determined in the random-phase approximation by Bardeen and Pines by introducing collective coordinates for the plasmons and carrying out a series of canonical transformations. Their results may be obtained in somewhat simpler fashion by using the dielectric formulation (Pines) or an equation of motion technique to calculate the longitudinal dielectric constant $\epsilon(k\omega)$ directly. One finds, by a simple extension of the methods we have described,

$$\epsilon(k\omega) = 1 + v_k \pi_0(k\omega) + v_k \pi_{ion}(k\omega) \tag{3-79}$$

where $\pi_0(k\omega)$ is the free-electron polarizability already introduced and

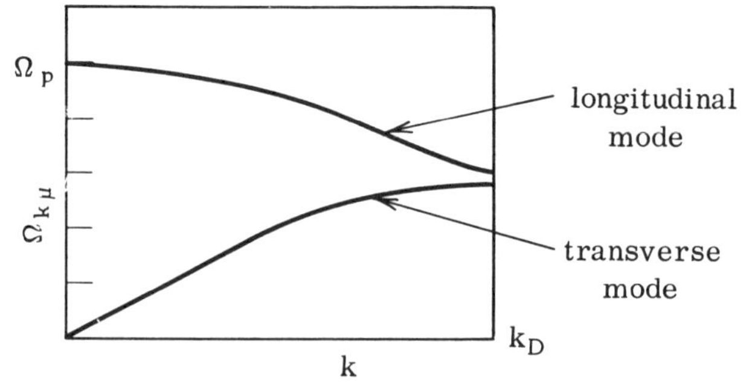

FIG. 3-12

$\pi_{ion}(k\omega)$ is the additional contribution from the ionic polarizability, which is given by

$$v_k \pi_{ion}(k\omega) = -\Omega_a^2/(\omega^2 - \Omega_0^2 + \Omega_a^2 + i\delta) \qquad (3\text{-}80)$$

with $v_k = 4\pi e^2/k^2$ and

$$\Omega_a^2 = (v_k^i)^2/v_k \qquad (3\text{-}81)$$

4. There are two branches to the excitation spectrum defined by $\epsilon(k\omega) = 0$. The first is the high-frequency branch, the plasmon solution we have already considered; for long wavelengths the ionic polarizability reduces to $-\Omega_p^2/\omega^2$, so that the ions act to shift the plasmon frequency from ω_p to $(\omega_p^2 + \Omega_p^2)^{1/2}$, a correction of order m/M. This mode is an "optical" mode (in analogy to the theory of sound-wave modes in polar crystals) in which the electrons and ions oscillate out of phase.

5. The second branch is a low-frequency acoustic mode with the dispersion relation

$$\omega_a^2(k) = \Omega_0^2 - \Omega_a^2 + \{\Omega_a^2/[1 + \pi_0(k\omega)v_k]\} \qquad (3\text{-}82)$$

in which the electrons follow the motion of the ions. To obtain a simple estimate of $\omega_a(k)$, we may neglect the effects of periodicity; one then has

$$\Omega_0^2 \cong \Omega_a^2 \cong \Omega_p^2$$

and

$$\omega_a^2 \cong \Omega_p^2/\epsilon_{RPA}(k\omega) \qquad (3\text{-}83)$$

If we now make use of the expression derived earlier for $\epsilon_{RPA}(k\omega)$, we find

$$\omega_a = \omega_1 + i\omega_2 \qquad (3\text{-}84a)$$

where

$$\omega_1 = \frac{\Omega_p^2}{1 + k_{FT}^2/k^2} = \frac{mZ}{3M} V_F^2 \frac{k^2}{1 + k^2/k_{FT}^2} = \frac{s^2 k^2}{1 + k^2/k_{FT}^2} \qquad (3\text{-}84b)$$

and

$$\omega_2 = -(\pi/4)(m/3M)\,\omega_1 \qquad (3\text{-}84c)$$

The phonon frequency is thus reduced, at long wavelengths, from Ω_p to sk as a result of the screening of the ions by the electron motion. The result (3-84b) agrees with a Fermi-Thomas calculation (Bardeen and Pines[25]) and with the equation-of-motion approach of Bohm and Staver. The sound velocities calculated from (3-84b) are in fair agreement with experiment. If one uses the more-accurate expression (3-82), which takes into account the effects of periodicity, one obtains good agreement with the elastic constants for the alkali metals (Bardeen and Pines[25]).

6. The imaginary part of the sound-wave frequency arises from the possibility of a phonon decaying into an electron-hole pair. One measures this quantity directly in an ultrasonic attenuation experiment, and the experimental results are again in good qualitative agreement with the simple theory.

7. The result, (3-79), follows directly from a field-theoretic calculation in which one allows for the influence of the phonons on the effective electron interaction. The Dyson-Hubbard equations are replaced by

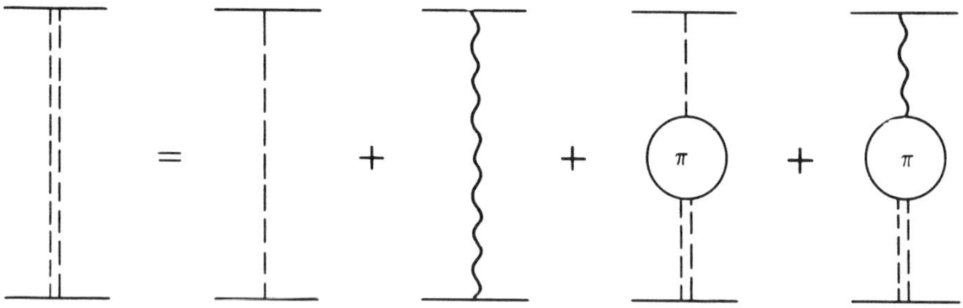

(3-85)

where the wavy line denotes the free-phonon propagator $iD_0(k\omega)$, which is coupled to the electrons by a matrix element $-ig_{\mathbf{k}}$; the box, $-i\pi(k\omega)$, is, as before, the sum of all the irreducible polarization diagrams for the electron gas, provided one neglects the influence of phonon vertex corrections on the electronic polarizability; that is, diagrams like .

By a generalization of an argument due to Migdal,[26] one can show these to be of order m/M, and hence negligible. If one takes for $\pi(k\omega)$ the RPA value, $\pi_0(k\omega)$, makes the identification,

$$g_k^2 = (v_k^i)^2/2\Omega_0(k) \tag{3-86}$$

and, takes for $D_0(k\omega)$, [compare (2-48)],

$$D_0(k\omega) = [1/(\omega - \Omega_0 + i\delta)] - [1/(\omega + \Omega_0 - i\delta)] \tag{3-87}$$

one recovers directly the result (3-79) for the dielectric constant and effective interaction between the electrons.

8. The effective electron interaction may be transformed by making use of the sound-wave dispersion relation. The results simplify considerably if one confines attention to low frequencies, makes the RPA, and further, approximates $\pi_0(k\omega)$ by $\pi_0(k0)$. One then finds

$$1/\epsilon(k\omega) = \{1/[1 + \pi_0(k,0)]\} \times \{1 + [(\omega_a^2 + \Omega_a^2 - \Omega_0^2)/(\omega^2 - \omega_a^2)]\} \tag{3-88}$$

The first term on the second line corresponds to the screened Coulomb interaction; the second term may be regarded as representing the effect of phonon exchange and is identical to that derived by Bardeen and Pines.[25]

9. We see that the effective interaction possesses the following properties:

a. If one neglects the effects of periodicity, i.e., takes $\Omega_a^2 = \Omega_0^2 = \Omega_p^2$, one finds an interaction which is zero at zero excitation frequency, attractive for excitation frequencies up to ω_a, repulsive beyond. [The infinity at $\omega = \omega_a$ is of course cancelled by taking into account the imaginary part of $\pi(k\omega)$.] This result, first obtained by Nozières, would say that all metals superconduct.

b. The effects of periodicity on Ω_a and Ω_0 can be included without too much difficulty for the alkali metals. One then finds that $\Omega_a < \Omega_0$; the net interaction possesses the following properties:

$0 < \omega_{exc}^2 < \Omega_0^2 - \Omega_a^2$	repulsive
$\Omega_0^2 - \Omega_a^2 < \omega_{exc}^2 < \omega_a^2$	attractive
$\omega_{exc}^2 > \omega_a^2$	repulsive

We return later to the effect which this might have on the superconducting properties of the alkali metals.

c. The attractive (in general) character of the part of the interaction associated with phonon exchange is characteristic of all systems studied just below a natural resonance frequency; it is an "antiscreening effect," which arises because the system overresponds to a disturbance (and, in fact, acts to change its sign). After one passes through the resonance, one finds the usual screening process, in which the responding system never quite cancels the influence of an external probe.

d. It seems likely that all metals will fall into category (a) or (b) above; that is, that one will never encounter metals for which the effective interaction is attractive at zero excitation frequency. As Nozières first emphasized, the latter situation would not be compatible with causality, and would therefore lead to an instability of the system for long-wavelength density fluctuations. Such an instability would in general imply that the system would change its spatial configuration in such a way that long-wavelength density fluctuations are stable. In other words, if we calculate $1/\epsilon(k,0)$ and find it negative, we may conclude that our calculation is not accurate, and attempt to take more careful account of effects of periodicity.

10. The electron-phonon interaction is screened by the electronic dielectric constant, ϵ_{RPA}, in the RPA; thus one has

$$v^{eff}(k\omega) = v_k^i/\epsilon_{RPA}(k\omega)$$

In higher orders there will be both screening effects associated with corrections to the RPA result, and vertex corrections arising from electron-electron interaction, of which the simplest takes the form

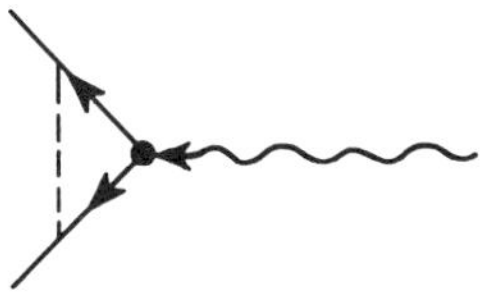

Migdal[26] has shown that the vertex corrections arising from phonon exchange, such as

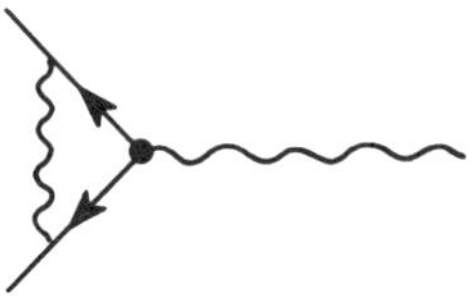

are of order m/M and may therefore be neglected.

11. The phonon frequencies may also be obtained directly from the following equation:

(3-89)

The double wavy line denotes $iD(k\omega)$, where $D(k\omega)$ is the Fourier transform of the exact-phonon Green's function, $D(k,\tau)$. The phonon polarization operator $-i\pi'(k\omega)$, which acts to shift the phonon energies and give the phonons a finite lifetime, may, to order m/M, be written

$$\pi'(k\omega) = \pi(k\omega)/[1 + V_k \pi(k\omega)] \tag{3-90}$$

where $\pi(k\omega)$ is the sum of all irreducible electron polarization diagrams. On solving the algebraic equation, (3-89), one finds the result

$$\omega_a^2(k) = \Omega_0^2 - \{V_k \pi(k\omega_a)\, \Omega_a^2/[1 + V_k \pi(k\omega_a)]\} \tag{3-91}$$

this result also follows directly from (3-85).

12. One interesting consequence of (3-91) is the Kohn effect. Kohn pointed out that if the Fermi surface is sharp (or nearly so), there will exist a logarithmic singularity in $\pi(k,\omega)$ at $k = 2k_F$, such that $\partial\pi/\partial k = -\infty$. (We have discussed one consequence of this singularity, the long-range effect of an impurity.) As a result, one may see "images" of the Fermi surface in the phonon spectrum, since there exist surfaces of $\omega_a(k)$ for which $|\text{grad}_k\, \omega(k)| = \infty$, the locus of the surface being determined by the shape of the Fermi surface. Recently Brockhouse et al. have reported an experimental finding of the Kohn effect in the scattering of neutrons by the lattice vibrations in Pb.

13. Migdal[26] has solved the problem of an electron interacting with a phonon field to order m/M, in the approximation that electron-electron interactions may be neglected. He showed that the phonon corrections to the vertex operator, $\Gamma(k\omega)$, were of order m/M. Thus one has

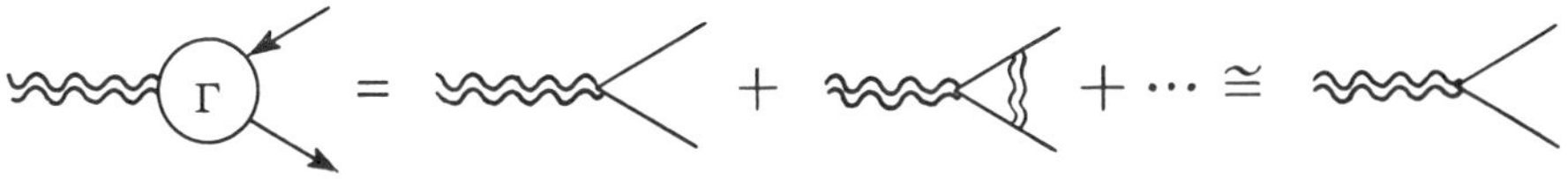

(3-92)

Moreover, for this problem the phonon polarization operator $\pi'(k\omega)$ is equal to $\pi_0(k\omega)$ to terms of order m/M. Migdal is therefore left with the following self-energy equation to solve:

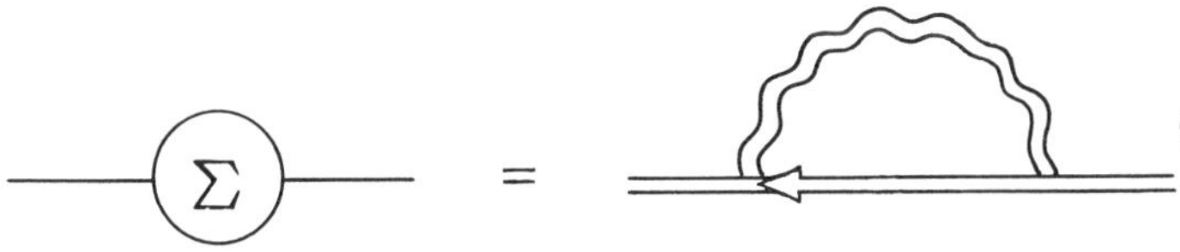

(3-93)

where the double line is the true-electron Green's function, defined according to the Dyson equation:

(3-94)

The nonlinear equation resulting from (3-93) and (3-94) is

$$\Sigma(p,\epsilon) = \int \frac{d^3k}{(2\pi)^3} \int \frac{d\omega}{2\pi} \frac{D(k\omega)g_k^2}{\epsilon - \omega - \epsilon(p-k) - \Sigma(p-k, \epsilon-\omega)} \tag{3-95}$$

This equation corresponds to the sum of all electron-phonon self-energy diagrams in which no lines cross; that is,

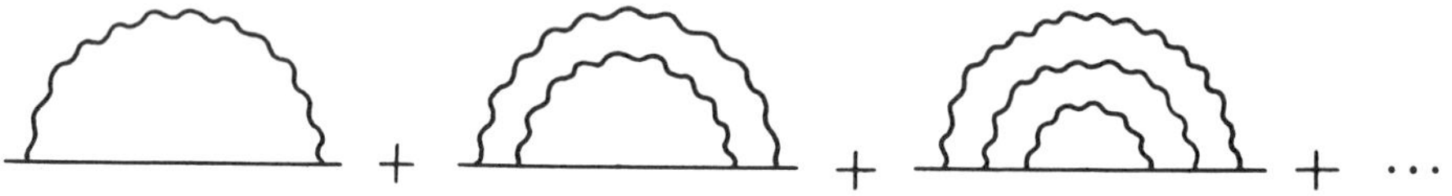

as may be seen easily by combining graphically (3-93) and (3-94). Migdal proceeds to solve (3-95) for $\Sigma(p,\epsilon)$. He finds a damping of the quasi particles due to phonon interaction which varies as η^3, where η is the excitation energy above the Fermi surface, and a correction to the effective mass, and hence specific heat, of the electron gas.

14. Migdal's results may be transposed, with certain qualifications, to the more-realistic case in which electron-electron interactions also obtain. To see this, let us consider the diagrams that we have summed in the RPA. For the phonon self-energy or polarization part, we have taken

$$\pi(k\omega) = \pi_0(k\omega)/[1 + V_k\pi_0(k\omega)]$$

We have replaced the bare interaction v_k by a screened interaction $v_k^i/\epsilon_{RPA}(k\omega)$, where $\epsilon_{RPA}(k\omega)$ is the electron dielectric constant. Thus we have

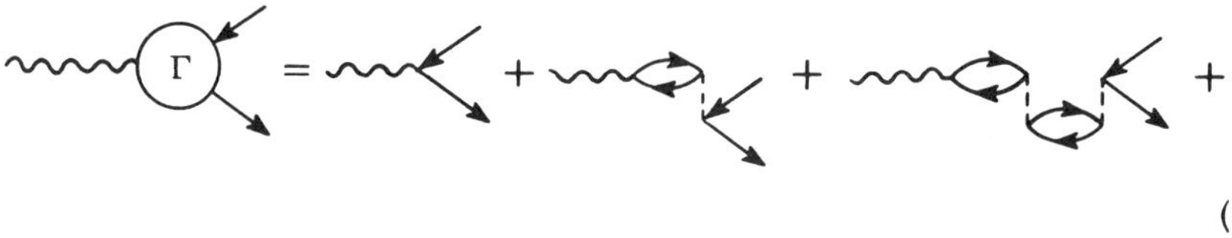

(3-96)

We could therefore incorporate these results directly into Midgal's formulation if we work directly with the equation

$$\Sigma(p,\epsilon) = \int \frac{d^3k}{(2\pi)^3} \int \frac{d\omega}{2\pi}$$

$$\times \frac{D_{RPA}(k\omega)\{[(v_k^i)^2/2\omega_0]/\epsilon(k,\omega_0)\}}{\epsilon - \omega - \epsilon(p-k) - \Sigma(p-k, \epsilon-\omega) + i\delta_{p-k}} \tag{3-97}$$

If one neglects periodicity, then one finds that in Migdal's terminology the bare coupling constant (for Coulomb interactions between electrons and ions) is $\lambda_0 = 1/2$, so that in this approximation Migdal's calculation is easily generalized.

15. Finally, it should be emphasized that all considerations of the present lectures are restricted to normal systems; for superconducting systems many of the results we have obtained will be modified.

SUPERCONDUCTIVITY

1. The success of the Bardeen, Cooper, and Schrieffer theory is by now so well known as not to require extensive recounting in these lectures. Bardeen and Schrieffer have written a detailed review of the present status of the theory and its comparison with experiment, and I refer to their original article with Cooper[28] and to their review article for a general survey of the field. In the present lectures I should like to consider the phenomenon of superconductivity within the general framework of the recent developments in the many-body problem. For the most part our discussion will apply to any system of fermions with attractive interactions; it will include the following points:

a. The instability of the normal state for fermion systems with attractive interactions.

b. A simple alternative derivation of the quasi-particle spectrum based on the equations of motion, which has been developed by Anderson and Schrieffer.

c. A scheme for extending the usual field-theoretic formulation of time-dependent perturbation theory, so that it will lead to the BCS theory and offer a systematic way of computing corrections to that theory.

d. A brief consideration of the collective modes.

e. The criterion for superconductivity.

Many of the new results I shall describe have been obtained in collaboration with J. R. Schrieffer and will appear in a paper by us.

2. The early theories of Fröhlich and Bardeen were based on an attempt to describe superconductivity as arising from the *self-energy* of the electrons interacting with the phonon field. Thus it was argued that in consequence of the phonon interaction it would be energetically favorable (from the point of view of the electronic kinetic energy plus the self-energy due to phonon interaction) to alter the character of the occupation of the electrons in the Fermi sphere. The theories were unsuccessful, in part because the interaction energies turned out to be far too large, but even more because they could not lead to a Meissner effect or (upon its experimental discovery) to an energy gap in the quasi-particle excitation spectrum.

3. The modern theory of superconductivity takes the *interaction* between the electrons (due to phonon exchange and screened Coulomb interaction) as its starting point. Where that interaction is attractive (and we have seen that it will be for a given frequency range in metals), Cooper[27] showed that the usual Fermi-sphere momentum distribution would be unstable against pair excitation. Let us recapitulate briefly Cooper's argument in our present notation. We consider a system of particles interacting via the following interaction:

$$V(p - p') = -\lambda w_p w_{p'} \tag{3-98}$$

where

$$\begin{aligned} w_p &= 1 \qquad \epsilon(p_F) - \omega_m \leq \epsilon(p) \leq \epsilon(p_F) + \omega_m \\ &= 0 \qquad \text{otherwise} \end{aligned}$$

Thus only pairs of particles lying within a region ω_m about the Fermi surface can interact with another. We consider the multiple scattering of a pair of particles lying within this region. For the model specified by (3-98), the t-matrix equation, (3-36), may be solved exactly. One finds

$$t(pp';q\omega) = -\lambda w_{p'} w_p / [1 + \lambda \varphi(q\omega)] \tag{3-99}$$

where $\varphi(q\omega)$ may be written in the form

$$\varphi(q\omega) = \sum_\mu{}' \frac{|(b_q^+)_{\mu 0}|^2}{\omega - \omega_{\mu 0} + i\delta} - \sum_\mu{}' \frac{|(b_q)_{\mu 0}|^2}{\omega + \omega_{\mu 0} - i\delta} \tag{3-100}$$

We have introduced the particle-pair creation and annihilation operators,

$$b_q^+ = \sum_p C^+_{\frac{q}{2}-p} C^+_{\frac{q}{2}+p} \qquad b_q = \sum_p C_{\frac{q}{2}+p} C_{\frac{q}{2}-p} \tag{3-101}$$

The excitation frequencies $\omega_{\mu 0}$ are $\pm[\epsilon(q + p) + \epsilon(q - p)]$ for pair creation and annihilation, respectively. (In this lecture we measure all energies with respect to the Fermi surface.) We have written (3-100) in this form to emphasize the similarity between the problem of particle-pair propagation and electron-hole propagation: Compare (3-100) with (2-40), with the latter calculated for plane waves. We remark that the result (3-99) may also be obtained directly from the equation-of-motion method, by studying the commutators (H,b_q) and (H,b_q^+) and keeping only those terms corresponding to repeated pair scattering (Sawada and Fukada).

We are interested in the solutions of the eigenvalue equation for the pair excitation frequency ω,

$$1 + \lambda \varphi(q\omega) = 0 \tag{3-102}$$

If we study this in the limit $q = 0$, and further, with Cooper, keep only the contribution to $\varphi(0\omega)$ from intermediate states involving particle propagation [i.e., the first term in (3-100)], we find (N − 1) continuum states with the positive energies $\omega_{pq} = \epsilon[(q/2) + p] + \epsilon[(q/2) - p]$ [like the particle-hole continuum states of the RPA dispersion relation, $1 + \pi_0(k\omega)V_k = 0$] and one state, which separates off below the continuum, and possesses an energy

$$\omega_r = -\omega_m e^{-1/2N(0)\lambda} \tag{3-103}$$

N(0) is the density of states per unit energy of particles of one kind of spin and we have assumed λ is small, so that $\omega_r \ll \omega_m$. This result is not consistent with the stability of the Fermi sphere, which requires that μ, the chemical potential, be greater than zero; in this example, one finds on adding two particles to the system, that $\mu = -\omega_r/2$. [One can see this more elegantly by going to the Lehmann representation for $B(q\omega)$, the Fourier transform of $B(q,\tau) = \langle \Psi_0 | T[b_q(\tau) b_q^+(0)] | \Psi_0 \rangle$, in direct analogy to our treatment of $F(q\omega)$ and $F(q,\tau)$.]

If we now go on to take hole-hole scattering into account [as was first done for (3-102) by Goldstone] we find that corresponding to the state which formerly separated off below the continuum there now exist two pure imaginary solutions of (3-102), at

$$\omega = \pm i\omega_r \tag{3-104}$$

This is equally disastrous to the stability of the normal-state Fermi distribution, for there now exists a growing wave solution for the pair-propagation function $B(0,\tau)$, that is, a solution that grows exponentially in time. This conclusion is not altered by going to pairs of particles of net momentum q; the growth rate is a maximum for $q = 0$ and falls off with increasing values of q.

4. It is interesting to study (3-102) at finite temperatures (Kadanoff and Martin). We can pass to the case of finite temperatures in (3-100) by writing the obvious finite-temperature analogues for the plane-wave states,

$$\begin{aligned} |(b_q^+)_{\mu 0}|^2 &\rightarrow f_{\frac{q}{2}+p} f_{\frac{q}{2}-p} \\ |(b_q)_{\mu 0}|^2 &\rightarrow (1 - f_{\frac{q}{2}+p})(1 - f_{\frac{q}{2}-p}) \end{aligned} \tag{3-105}$$

where the f_p are the Fermi-Dirac distributions functions, defined by

$$f_p = 1/[e^{\beta\epsilon(p)} + 1]$$

with $\beta = 1/KT$. Equation (3-102) is replaced by

$$1 + \lambda\varphi(0\omega) \tanh [\beta\epsilon(p)/2] = 0 \qquad (3\text{-}106)$$

in the limit of $q = 0$.

One then finds that as the temperature is increased from zero, the growth rate of the instability is decreased, until one reaches a temperature T_c such that one has "marginal" stability, i.e., $\omega = 0$; at T_c one has

$$1 + \lambda\varphi(\omega) \tanh [\beta_c \epsilon(p)/2] = 0 \qquad (3\text{-}107a)$$

or

$$1 = \lambda \sum_p [(w_p)^2/2\epsilon(p)] \tanh [\beta_c \epsilon(p)/2] \qquad (3\text{-}107b)$$

For $T > T_c$ there are no unstable solutions of (3-106). The result, (3-107b), is identical with the result obtained for the transition temperature by BCS, who found the temperature at which the energy gap goes to zero. We shall see the reason for this shortly.

5. We may say that the Cooper instability provides the essential clue for making a theory of superconductivity. It tells us there is something special about particles of opposite spin and momentum, and that these have a tendency to go into a bound state of some sort. It led Bardeen, Cooper, and Schrieffer to try a calculation of the ground-state energy and excitation spectrum with a wave function which emphasized the correlations between such pairs, viz.,

$$\Psi = \prod_{p\sigma} [u_p + v_p C^+_{p\sigma} C^+_{-p-\sigma}] \mid 0 \rangle \qquad (3\text{-}108)$$

where $\mid 0 \rangle$ is the vacuum-wave function, u_p^2 is the probability that a given pair state $p\sigma, -p-\sigma$ is empty, v_p^2 that it is filled, and $u_p^2 + v_p^2 = 1$. Note that the wave function, (3-108), does not correspond to a system with a fixed number of particles. This difficulty may be resolved by carrying through the calculations for a grand canonical ensemble (by adding $-\mu N$ to the Hamiltonian) and then guaranteeing particle conservation on the average. We shall here consider an alternative calculation, due to Anderson and Schrieffer, which utilizes the equation-of-motion method.

6. We consider the Hamiltonian (2-62), and once more write down the equation of motion for C_p^+. It is, from (2-64),

$$[H', C^+_{p\uparrow}] = \tilde{\epsilon}(p) C^+_{p\uparrow} + \sum_{\substack{k\neq 0 \\ p\neq k+q}} V_k C^+_{p+k\uparrow} C^+_{q-k\sigma'} C_{q\sigma'} \qquad (3\text{-}109)$$

where $\tilde{\epsilon}(p)$ is the quasi-particle energy in the Hartree-Fock approximation, measured with respect to the Fermi surface, μ_{HF}. We now keep an additional contribution from the triple term in (3-109); we *assume* that in the ground state, as a result of the pair correlations, there will be a nonzero amplitude for

$$\langle \Psi_0 | C_q^+ C_{-q}^+ | \Psi_0 \rangle = h_q^+$$

Such a nonvanishing amplitude could not exist for a "normal" fermion system; we here assume that it exists, and then pursue the consequences of that assumption. In assuming its existence we have abandoned *ab initio* the idea of particle conservation in intermediate states; particle conservation on the average is still guaranteed by putting $\langle N_{op} \rangle = N$. Our motivation is the Cooper instability, which suggests the possible existence of virtual-bound-pair states. With this assumption, we may write (3-109) as

$$[H', C_{p\uparrow}^+] = \tilde{\epsilon}(p) C_{p\uparrow}^+ + \Delta_p^+ C_{-p\downarrow} \qquad (3\text{-}110a)$$

where

$$\Delta_p^+ = \sum_q V_{q-p} h_q^+ \qquad (3\text{-}111a)$$

One likewise has

$$(H, C_{-p\downarrow}) = -\tilde{\epsilon}(p) C_{-p\downarrow} + \Delta_p C_{p\uparrow}^+ \qquad (3\text{-}110b)$$

$$\Delta_p = \sum_q V_{q-p} h_q = \sum_q V_{q-p} \langle \Psi_0 | C_{-q\downarrow} C_{q\uparrow} | \Psi_0 \rangle \qquad (3\text{-}111b)$$

We see that the equation of motion for $C_{p\uparrow}^+$ depends on $C_{-p\downarrow}$ and vice versa. The equations are consistent, and $C_{p\uparrow}^+$ and $C_{-p\downarrow}$ will oscillate at a frequency E_p provided that

$$E_p = \pm [\tilde{\epsilon}_p^2 + \Delta_p^2]^{1/2} \qquad (3\text{-}112)$$

There remains the problem of determining Δ_p.

This may be done by finding the proper quasi-particle creation and annihilation operators and then imposing the condition that the ground-state wave function be consistent with that identification based on the equations of motion; the procedure followed is thus a slight generalization of that discussed earlier. We study

$$\begin{aligned} \gamma_{p0}^+ &= u_p C_{p\uparrow}^+ - v_p C_{-p\downarrow} & \qquad \gamma_{p0} &= u_p C_{p\uparrow} - v_p C_{-p\downarrow}^+ \\ \gamma_{p1}^+ &= u_p C_{-p\downarrow}^+ + v_p C_{p\uparrow} & \qquad \gamma_{p1} &= u_p C_{-p\downarrow} + v_p C_{p\uparrow}^+ \end{aligned} \qquad (3\text{-}113)$$

where the γ's will satisfy the proper commutation rules if $u_p^2 + v_p^2 = 1$.

Note that Eq. (3-113) is just the Bogoljubov-Valatin transformation. Here we determine u_p, v_p, and Δ_p by writing down the equations of motion for the γ_p's using (3-110) and (3-111), and further imposing the condition

$$\gamma_{p0} \mid \Psi_0 \rangle = 0$$

$$\gamma_{p1} \mid \Psi_0 \rangle = 0 \qquad (3\text{-}114)$$

After a bit of algebra one finds

$$u_p^2 = (1 + \tilde{\epsilon}_p/E_p)/2 \qquad h_p^+ = u_p v_p = h_p$$
$$v_p^2 = (1 - \tilde{\epsilon}_p/E_p)/2 \qquad n_p = v_p^2 \qquad (3\text{-}115)$$

and the self-consistency condition,

$$\Delta_p = -\sum_q V_{q-p}(\Delta_q/2E_q) \qquad (3\text{-}116)$$

which is the BCS equation for the energy gap.

7. We summarize what we may learn about the BCS theory from the equation-of-motion method:

a. Begin by making a radical, nonperturbative assumption, that there exists an amplitude for bound-pair formation.

b. The quasi particles are then admixtures of particles and holes of opposite spin, the admixture being governed by the coefficients that appear in the Bogoljubov transformation, to which one is directly led from assumption (a).

c. The quasi-particle spectrum possesses an energy gap Δ_p, which is determined by a self-consistency condition. For the potential (3-98), the solution for weak coupling is

$$\Delta = 2\omega_m \exp - 1/N(0)\lambda \qquad (3\text{-}117)$$

d. The ground-state energy may be calculated directly in a variety of ways; one finds for weak coupling,

$$E_0 = -2N(0)\omega_m^2 \exp - 2/N(0)\lambda \qquad (3\text{-}118)$$

an obviously nonperturbative result.

e. There is no discontinuity at the Fermi surface; $n_p = v_p^2$ varies smoothly as one passes through $p = p_F$; further one may show that the ground-state wave function takes the form (3-108).

f. This equation-of-motion approach is directly related to the field-theoretic approach of Gorkov,[31] who solves the equations of motion of coupled propagators in similar fashion.

8. We remark that both because of the uncertainty in the precise character of the effective electron interaction in metals, and because of the existence of a law of corresponding states, BCS take the simple potential (3-98) in calculating properties of real superconductors. They then have a one-parameter theory, which, as we have said, leads to agreement with experiment for nearly all superconducting properties of interest.

9. The coherence effects in the theory arise because the quasi particle is an admixture of particles of opposite spin and momentum. As a result these electrons absorb coherently and, for a given phenomenon, their contributions may add or interfere. One finds destructive interference for a coupling of the system to a density fluctuation (e.g., a sound wave); constructive interference for coupling to an external spin density (as in a nuclear-spin-relation experiment). The predictions are in excellent agreement with experiment. The coherence length, which is also a measure of the size of the correlated pairs, is $\xi_0 \cong V_F/\Delta$, and is about 10^{-4} cm; for this reason the picture of the pairs as forming a Bose condensate is not very appropriate, since the size of a pair is enormous compared to the interparticle spacing.

10. One may ask at this point: Why have we taken pairs of particles of opposite spin? Would not triplet states with parallel spin do just as well? Bardeen, Cooper, and Schrieffer point out that exchange terms generally act to reduce the matrix elements, and hence energy gain, for parallel spin pairing. Another argument may be given along the following lines. In a sense, we may regard the potential $V(p - p')$ of (3-98) as realistic, provided we consider a low-density system, in which only s-wave scattering will be important for particles near the Fermi surface, for which the s-wave scattering amplitude $f_0 = -\lambda_m$. We could then argue that although metals are not low-density systems, there is still enough resemblance to a low-density system to make the s-wave, singlet, partial-wave amplitude more attractive than the triplet for particles near the Fermi surface; in this case the instability would be stronger for particles of opposite spin, and one would find the BCS pairing.

11. It should be emphasized that the interaction terms crucial for superconductivity are those which scatter particles of opposite spin and momentum. These terms have a negligible weight (of order 1/N) compared to all others, yet they yield the entire effect, because coherence effects make up what is lacking in a purely statistical argument. Bardeen and Rickayzen[34] and Bogoljubov have shown that the BCS solution for the "reduced" Hamiltonian,

$$H = \sum_p \epsilon(p)\, C_p^+ C_p + \sum_{pp'} [v(p - p')/2]\, C_{-p'\downarrow}^+ C_{+p'\uparrow}^+ C_{-p\downarrow} C_{p\uparrow} \tag{3-119}$$

is, in fact, exact to terms of order 1/N.

12. Let us now see how we might obtain superconductivity by a suitable extension of time-dependent perturbation theory. The character of the extension which is required is indicated by the Cooper instability; we need, in some sense, to permit particles of opposite spin and momentum to go into a bound state. We need, however, do this only in intermediate states, since there are no real bound states in the system. What we wish to do then is to construct a field-theoretic perturbation theory in which we allow for virtual bound states for particles of opposite spin and momentum. We may accomplish this aim by writing the following equation for the single-particle propagator:

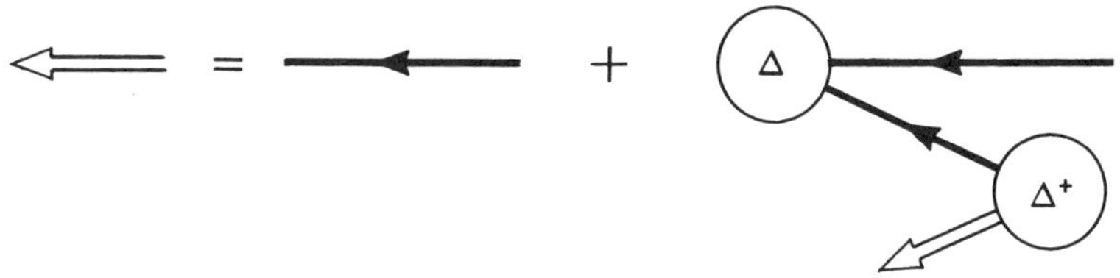

(3-120)

where the double line denotes the true propagator $iG(p, \epsilon)$ and the single heavy line is a "normal-state" propagator $iG_n(p, \epsilon)$, in which all self-energy processes $-i\Sigma_n(p, \epsilon)$ of the type 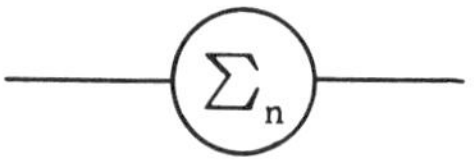 have been included:

$$G_n = \frac{1}{\epsilon - \epsilon(p) - \Sigma_n(p, \epsilon) + i\delta_p} \tag{3-121}$$

$-i\Delta(p, \epsilon)$ is the amplitude for a pair of particles of opposite spin and momentum to go into a bound intermediate state, and $i\Delta^*(p, \epsilon)$ is the amplitude for a pair of particles to come out of a bound intermediate state. The rules for working with (3-120) follow from an appropriate generalization of the S-matrix theory. Equation (3-120) possesses the solution

$$G = \frac{(G_n^-)^{-1}}{(G_n^+ G_n^-)^{-1} + \Delta^2}$$

$$= \frac{\epsilon + \epsilon(p) + \Sigma^- - \mu}{[\epsilon - \epsilon(p) - \Sigma^+ + \mu][\epsilon + \epsilon(p) + \Sigma^- - \mu] - \Delta^2 + i\delta} \tag{3-122}$$

How do we now calculate $\Delta(p, \epsilon)$? We cannot use perturbation theory, because to every order in perturbation theory for a normal fermion system, $\Delta(p, \epsilon) = 0$. What we must do is to determine it in self-consistent fashion, according to the equation

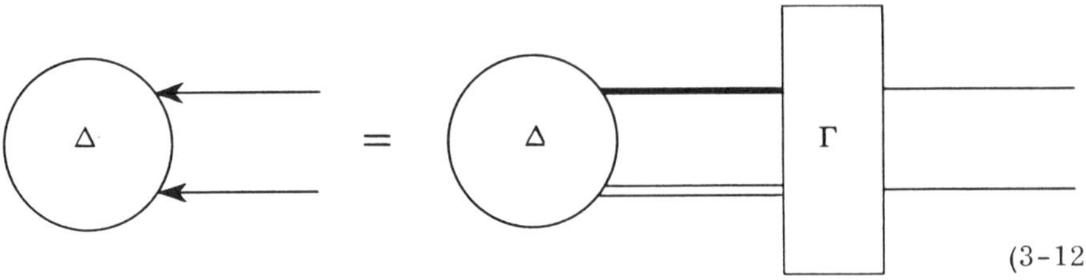

(3-123)

where $-i\Gamma$ is the sum of all interactions which scatter a pair of particles, with the proviso that duplication of diagrams must be avoided, since in (3-120) and (3-123) one has already summed automatically a certain class of diagrams. The somewhat-asymmetric form of (3-123), in which intermediate-state propagation is partly by G_n, partly by the true G, is a direct consequence of the prohibition against counting diagrams twice; the use of two G's would lead to diagram duplication.

13. Let us summarize some of the properties of the coupled equations (3-120) and (3-123):

a. The BCS solution is at once obtained if we take, for the simple scattering interaction, $v(p - p')$, and choose $\Sigma^+ = \Sigma^-$, with both quantities real. This approximation is equivalent to neglecting lifetime effects and wave-function renormalization for the normal-state quasi particles, and choosing a simple effective interaction between the quasi particles. Equation (3-123) yields the gap equation (3-116); the superconducting Green's function is

$$G_{BCS} = [u_p^2/(\epsilon - E_p + i\delta)] + [v_p^2/(\epsilon + E_p - i\delta)] \qquad (3\text{-}124)$$

where u_p, v_p, and E_p are given by (3-112) and (3-115).

b. Both the structure of the Green's function, (3-120) and (3-122), and the BCS result (3-124) bear a very close resemblance to the boson problem. The essential difference is that in the interacting boson gas we could calculate Σ_{02} by using perturbation theory; for the superconducting problem no such perturbation-theoretic solution exists. It is essential that Δ be obtained by a "bootstrap" method in which one seeks new, self-consistent, nonperturbative solutions.

c. The equations (3-120) and (3-122) are identical to those proposed by Nambu and Beliaev; Beliaev's equations, which Schrieffer and I also found, are identical in structure to (3-68) of our boson treatment; if one expands Nambu's two-component formulation, one again finds (3-122). I believe the form (3-120) is closest in spirit to the approach originally adopted by BCS: One tries to find the best-possible normal-state quasi particles (e.g., propagators), and then adds in the new, characteristically superconducting effect—the possibility of virtual-bound-pair states.

d. Equations (3-122) and (3-123) offer a simple systematic way to go beyond the BCS theory by:

(1) Taking more-realistic normal-state quasi-particle propagators; it is clear that the change in Σ_n on going from the normal to the superconducting state will not have much effect on the resulting gap equation (3-123), in which integrations are over all possible intermediate states.

(2) Considering more-complicated pair-scattering operators, Γ.

e. Study of (3-120) and (3-123) shows that the diagrams summed in the BCS theory are those in which one takes into account all multiple-scattering processes in and out of the virtual-bound-pair states. Thus, it would be incorrect to attempt to replace Γ in (3-123) by a t matrix; such diagrams are already included if one takes $\Gamma = v$.

f. Equation (3-123) permits us to see why the BCS calculation of the transition temperature agrees with that based on the Cooper instability. If we isolate the "bound-state" part of the t-matrix diagram, we write the equation

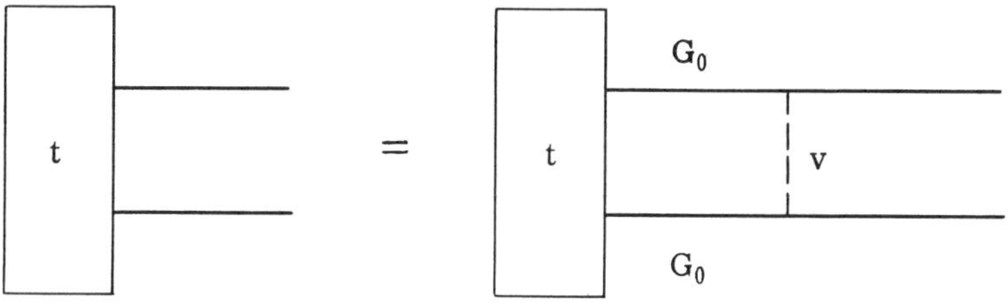

(3-125)

which may easily be seen to agree with (3-102). If we now define the transition temperature as that for which $\Delta = 0$, it follows at once that (3-123) and (3-125) must agree, since in this case $G = G_n$ in (3-123).

14. The collective modes in a superconductor are perhaps most simply calculated by the equation-of-motion method. As Anderson[32] first showed, the calculation may be carried out along the lines described in the equation-of-motion method (page 44) for the motion of a particle-hole pair; however, it is necessary that one allow for the existence of nonzero expectation values for $C_p^+ C_{-p}^+$ and $C_{-p} C_p$ in carrying out the resulting linearization. As a result the particle-hole pairs are coupled to particle-particle pairs $C_{p+k}^+ C_{-p}^+$ and hole-hole pairs of similar character. The solution of the resulting set of coupled equations yields the frequencies of the collective modes. The collective modes are of two kinds: plasmons, with energies little changed from their normal-state values, and excitons, which are bound-pair states with energies lying within the gap. The energy of the exciton states is strongly dependent on the angular dependence of the residual quasi-particle interaction; and we refer to the original literature (Anderson,[32] Rickayzen,[33] Bardasis and Schrieffer) for a further discussion.

It is necessary that the collective modes be included in order to obtain a gauge-invariant description of the Meissner effect. The reason is simple: A change in gauge generally corresponds to asking a "longitudinal" question of the electrons; collective modes are essential in obtaining the response to an external longitudinal field, as we have seen. The way in which the collective excitations make possible a gauge-invariant description follows closely along the lines first suggested by Bardeen in that the virtual plasmon response cancels out most of the effect of the gauge change. Papers that discuss the resolution of difficulties with gauge invariance include those of Anderson,[32] Bogoljubov, Nambu, and Pines and Schrieffer. Perhaps the most explicit and complete demonstration of the gauge invariance of the BCS theory when collective modes are properly included may be found in Rickayzen.[33]

15. The choice of a criterion for the occurrence of superconductivity is at present an open question. Because of the many complications with the role of lifetime effects in determining the criterion, and the additional complications introduced by periodic structure (especially the U processes), it would seem that perhaps the best one can do for metals is to explain why the alkali metals do not superconduct (assuming that they do not, in fact, so do). Here the situation does not seem completely hopeless. One has a fairly good knowledge of the influence of periodicity on the matrix elements v_k^i for these metals (Bardeen); as we have mentioned, the effective interaction then turns out to be repulsive for low excitation frequencies, then attractive, then repulsive. In these circumstances it would seem physically reasonable that if the quasi-particle lifetime is sufficiently short at the excitation frequencies for which the interaction is attractive, then superconductivity cannot occur. Detailed calculations are now under way to see whether this view is justified.

16. From a fundamental point of view, the most important remaining open question in the theory of superconductivity is the development of a model solution along the lines we already have for, say, superfluidity; that is, to show that the BCS theory is correct for a given region of densities, class of interactions, etc. We may expect that this will prove exceedingly difficult because, as we have mentioned, there is such a very small energy charge in going to the superconducting configuration. The most one can probably hope for is to demonstrate that under certain circumstances, all corrections to the quasi-particle excitation spectrum are small. To sum up, it seems very likely that the BCS solution for any such model problem will prove correct; it only remains to find the appropriate model, and prove it so.

PAPERS FOR SUPPLEMENTARY READING

GENERAL FORMULATION

[1] J. Goldstone, Derivation of Brueckner Many-Body Theory, Proc. Roy. Soc. (London), **A239**, 267-279 (1957).

[2] J. Hubbard, The Description of Collective Motions in Terms of Many-Body Perturbation Theory, Proc. Roy. Soc. (London), **A240**, 539-560 (1957).

[3] V. Galitskii and A. Migdal, Application of Quantum Field Theory Methods to the Many-Body Problem, Soviet Phys. JETP, **7**, 96-104 (1958).

ELECTRON GAS

[4] D. Bohm and D. Pines, A Collective Description of Electron Interactions: III. Coulomb Interactions in a Degenerate Electron Gas, Phys. Rev., **92**, 609-625 (1953).

[5] D. Pines, A Collective Description of Electron Interactions: IV. Electron Interaction in Metals, Phys. Rev., **92**, 626-636 (1953).

[6] M. Gell-Mann and K. A. Brueckner, Correlation Energy of an Electron Gas at High Density, Phys. Rev., **106**, 364-368 (1957).

[7] M. Gell-Mann, Specific Heat of a Degenerate Electron Gas at High Density, Phys. Rev., **106**, 369-372 (1957).

[8] K. Sawada, K. A. Brueckner, N. Fukada, and R. Brout, Correlation Energy of an Electron Gas at High Density: Plasma Oscillations, Phys. Rev., **108**, 507-514 (1957).

[9] R. Brout, Correlation Energy of a High-Density Gas: Plasma Coordinates, Phys. Rev., **108**, 515-517 (1957).

[10] G. Wentzel, Diamagnetism of a Dense Electron Gas, Phys. Rev., **108**, 1593-1596 (1957).

[11] J. Hubbard, The Description of Collective Motions in Terms of Many-Body Perturbation Theory: II. The Correlation Energy of a Free-Electron Gas, Proc. Roy. Soc. (London), **A243**, 336-352 (1957).

[12] P. Nozières and D. Pines, Correlation Energy of a Free Electron Gas, Phys. Rev., **111**, 442-454 (1958).

[13] P. Nozières and D. Pines, A Dielectric Formulation of the Many Body Problem: Application to the Free Electron Gas, Nuovo cimento, [X]**9**, 470-490 (1958).

[14] H. Ehrenreich and M. H. Cohen, Self-Consistent Field Approach to the Many-Electron Problem, Phys. Rev., **115**, 786-790 (1959).

INTERACTING FERMION SYSTEMS

[15] L. D. Landau, The Theory of a Fermi Liquid, Soviet Phys. JETP, **3**, 920-925 (1957).

[16] L. D. Landau, Oscillations in a Fermi Liquid, Soviet Phys. JETP, **5**, 101-108 (1957).

[17] L. D. Landau, On the Theory of the Fermi Liquid, Soviet Phys. JETP, **8**, 70-74 (1959).

[18] V. Galitskii, The Energy Spectrum of a Non-Ideal Fermi Gas, Soviet Phys. JETP, **7**, 104-112 (1958).

[19] J. Goldstone and K. Gottfried, Collective Excitations of Fermi Gases, Nuovo cimento, [X]**13**, 849-852 (1959).

INTERACTING BOSON SYSTEMS

[20] N. Bogoljubov, On the Theory of Superfluidity, J. Phys. (U.S.S.R.), **11**, 23-32 (1947).

[21] T. D. Lee, K. Huang, and C. N. Yang, Eigenvalues and Eigenfunctions of a Bose System of Hard Spheres and Its Low-Temperature Properties, Phys. Rev., **106**, 1135-1145 (1957).

[22] S. T. Beliaev, Application of the Methods of Quantum Field Theory to a System of Bosons, Soviet Phys. JETP, **7**, 289-299 (1958).

[23] S. T. Beliaev, Energy Spectrum of a Non-Ideal Bose Gas, Soviet Phys. JETP, **7**, 299-307 (1958).

[24] N. M. Hugenholtz and D. Pines, Ground-State Energy and Excitation Spectrum of a System of Interacting Bosons, Phys. Rev., **116**, 489-506 (1959).

ELECTRON-PHONON INTERACTION

[25] J. Bardeen and D. Pines, Electron-Phonon Interaction in Metals, Phys. Rev., **99**, 1140-1150 (1955).

[26] A. B. Migdal, Interaction between Electrons and Lattice Vibrations in a Normal Metal, Soviet Phys. JETP, **7**, 996-1001 (1958).

SUPERCONDUCTIVITY

[27] L. N. Cooper, Bound Electron Pairs in a Degenerate Fermi Gas, Phys. Rev., **104**, 1189-1190 (1956).

[28] J. Bardeen, L. N. Cooper, and J. R. Schrieffer, Theory of Superconductivity, Phys. Rev., **108**, 1175-1204 (1957).

[29] N. N. Bogoljubov, A New Method in the Theory of Superconductivity, Soviet Phys. JETP, **7**, 41-46 (1958).

[30] J. G. Valatin, Comments on the Theory of Superconductivity, Nuovo cimento, [X]**7**, 843-857 (1958).

[31] L. P. Gorkov, On the Energy Spectrum of Superconductors, Soviet Phys. JETP, **7**, 505-508 (1958).

[32] P. W. Anderson, Random-Phase Approximation in the Theory of Superconductivity, Phys. Rev., **112**, 1900-1916 (1958).

[33] G. Rickayzen, Collective Excitations in the Theory of Superconductivity, Phys. Rev., **115**, 795-808 (1959).

[34] J. Bardeen and G. Rickayzen, Ground State Energy and Green's Function for Reduced Hamiltonian for Superconductivity, Phys. Rev., **118**, 936-937 (1960).

REFERENCES

MATHEMATICAL PRINCIPLES AND TECHNIQUES

Bloch, C., Nuclear Phys., **7**, 451 (1958).
Bogoljubov, N. N., Soviet Phys. Uspekhi, **2**, 236 (1959).
Bogoljubov, N. N., and Zubarev, N., Soviet Phys. JETP, **1**, 83 (1955).
Brueckner, K. A., in "The Many-Body Problem," Dunod-Wiley, New York, 1959.
DuBois, D. F., Ann. Phys., **7**, 174 (1959).
Hugenholtz, N. M., Physica, **23**, 481 (1957).
Hugenholtz, N. M., in "The Many-Body Problem," Dunod-Wiley, New York, 1959.
Hugenholtz, N. M., and Van Hove, L., Physica, **24**, 363 (1958).
Jastrow, R., Phys. Rev., **98**, 1479 (1955).
Klein, A., Phys. Rev., **121**, 950 (1961).
Klein, A., and Prange, R., Phys. Rev., **112**, 994 (1958).
Kohn, W., and Luttinger, J., Phys. Rev., **118**, 41 (1960).
Lipkin, H., Ann. Phys., **9**, 272 (1960).
Luttinger, J. M., Phys. Rev., **121**, 942 (1961).
Luttinger, J., and Ward, J. C., Phys. Rev., **118**, 1417 (1960).
Martin, P. C., and Mermin, D. (to be published).
Martin, P. C., and Schwinger, J., Phys. Rev., **115**, 1342 (1959).
Sawada, K., and Fukuda, N., Progr. Theoret. Phys. (Kyoto), **25**, 653 (1961).
Suhl, H., and Werthamer, N. R., Phys. Rev., **122**, 359 (1961).
Tomonaga, S., Progr. Theoret. Phys. (Kyoto), **13**, 467 (1955).
Van Hove, L., Phys. Rev., **95**, 249 (1954).
Van Hove, L., Physica, **26**, 200 (1960).

INTERACTING ELECTRON SYSTEMS

Carr, W. J., Phys. Rev., 122, 1437 (1961).
Coldwell-Horsfall, R., and Maradudin, A. A., J. Math. Phys., **1**, 395 (1960).
Daniel, E., and Vosko, S., Phys. Rev., **120**, 2041 (1960).
DuBois, D. V., Ann. Phys., **8**, 24 (1959).
Glick, A. J., and Ferrell, R. A., Ann. Phys., **11**, 359 (1960).
Langer, J., and Vosko, S. J., Phys. and Chem. Solids, **12**, 196 (1960).
Lindhard, J., Kgl. Danske Videnskab. Selskab, Mat.-fys. Medd., **28**, 8 (1954).
Nozières, D., and Pines, D., Phys. Rev., **113**, 1254 (1959).
Pines, D., Revs. Mod. Phys., **28**, 184 (1956).
Pines, D., Physica, **26**, 103 (1960).
Quinn, J. J., and Ferrell, R. A., Phys. Rev., **112**, 812 (1958).
Sawada, K., Phys. Rev., **106**, 372 (1957).
Wigner, E. P., Phys. Rev., **46**, 1002 (1934).
Wigner, E. P., Trans. Faraday Soc., **34**, 678 (1938).

INTERACTING FERMION SYSTEMS

Abrikosov, A. A., and Khalatnikov, J. M., Repts. Progr. Phys., **22**, 329 (1959).
Brueckner, K. A., in "The Many-Body Problem," Dunod-Wiley, New York, 1959.
Brueckner, K. A., and Gammel, J. L., Phys. Rev., **109**, 1023 (1958); **109**, 1040 (1958).
De Dominicis, C., and Martin, P. C., Phys. Rev., **105**, 1417 (1957).
Gottfried, K., and Pičman, L., Kgl. Danske Videnskab. Selskab, Mat.-fys. Medd., **32**, 13 (1960).
Lee, T. D., and Yang, C. N., Phys. Rev., **117**, 12 (1960).
Overhauser, A. W., Phys. Rev. Letters, **4**, 341 (1960).

NUCLEAR STRUCTURE

Baranger, M., Phys. Rev., **120**, 957 (1960).
Beliaev, S. T., Kgl. Danske Videnskab. Selskab, Mat.-fys. Medd., **31**, 11 (1959).
Brown, G. E., and Thouless, D., Physica, **26**, 145 (1960).
Mottelson, B. R., in "The Many-Body Problem," Dunod-Wiley, New York, 1959.

INTERACTING BOSON SYSTEMS AND LIQUID HELIUM

Brueckner, K. A., and Sawada, K., Phys. Rev., **106**, 1117 (1959); **106**, 1128 (1959).
Cohen, M., and Feynman, R. P., Phys. Rev., **107**, 13 (1957).
Feynman, R. P., Phys. Rev., **94**, 262 (1954).
Feynman, R. P., and Cohen, M., Phys. Rev., **102**, 1189 (1956).
Henshaw, D. G., and Woods, A. D. B., Phys. Rev., **121**, 1266 (1961).
Landau, L. D., J. Phys. (U.S.S.R.), **5**, 71 (1941); **11**, 91 (1947).
Lee, T. D., and Yang, C. N., Phys. Rev., **112**, 1419 (1958).
Miller, A., Nozières, P., and Pines, D. (to be published).
Nozières, P., and Pines, D. (to be published).
Parry, W. E., and Ter Haar, D. (to be published).
Sawada, K., Phys. Rev., **116**, 1344 (1959).
Wu, T. T., Phys. Rev., **115**, 491 (1959).

ELECTRON-PHONON INTERACTIONS

Bardeen, J., Phys. Rev., **52**, 688 (1937).
Bohm, D., and Staver, T., Phys. Rev., **84**, 836 (1952).
Brockhouse, B. N., Rao, K. R., and Woods, A. D. B., Phys. Rev. Letters, **7**, 93 (1961).
Clark, C. B., Phys. Rev., **109**, 1133 (1958).
Fröhlich, H., Proc. Roy. Soc. (London), **A215**, 291 (1952).
Kohn, W., Phys. Rev. Letters, **2**, 393 (1959).
Pines, D., "The Many-Body Problem," Dunod-Wiley, New York, 1959.
Staver, T., Ph.D. thesis, Princeton University, 1952 (unpublished).

SUPERCONDUCTIVITY

Bardasis, A., and Schrieffer, J. R., Phys. Rev., **121**, 1050 (1961).
Bardeen, J., and Schrieffer, J. R., Progr. in Low Temp. Phys., **3**, 170-287 (1961).
Beliaev, S. T., Physica, **26**, 181 (1960).
Bogoljubov, N. N., Physica, **26**, 1 (1960).
Bogoljubov, N. N. Tolmachev, V. V., and Shirkov, D. V., "A New Method in the Theory of Superconductivity," Consultants Bureau, New York, 1959.
Eliashberg, G. M., Soviet Phys. JETP, **11**, 696 (1960).
Goldstone, J., Ph.D. thesis, Cambridge University, 1958 (unpublished).

Kadanoff, L., and Martin, P. C., Phys. Rev. (to be published).
Morel, P., Phys. and Chem. Solids, **10**, 277 (1959).
Nambu, Y., Phys. Rev., **117**, 648 (1960).
Pines, D., Phys. Rev., **109**, 280 (1958).
Pines, D., and Schrieffer, J. R., Nuovo cimento, **10**, 496 (1958).
Pines, D., and Schrieffer, J. R. (to be published).
Schrieffer, J. R., Physica, **26**, 124 (1960).

Derivation of the Brueckner many-body theory

By J. Goldstone
Trinity College, University of Cambridge‡

(*Communicated by N. F. Mott, F.R.S.—Received* 24 *August* 1956)

An exact formal solution is obtained to the problem of a system of fermions in interaction. This solution is expressed in a form which avoids the problem of unlinked clusters in many-body theory. The technique of Feynman graphs is used to derive the series and to define linked terms. The graphs are those appropriate to a system of many fermions and are used to give a new derivation of the Hartree–Fock and Brueckner methods for this problem.

1. Introduction

The Hartree–Fock approximation for the many-body problem uses a wave function which is a determinant of single-particle wave functions—that is, an independent-particle model. The single-particle states are eigenstates of a particle in a potential V, which is determined from the two-body interaction v by a self-consistent calculation. The Brueckner theory (Brueckner & Levinson 1955; Bethe 1956; Eden 1956) gives an improved method of defining V and shows why the residual effects of v not allowed for by V can be small. In particular, in the nuclear problem the corrections to the energy are small, even though the corrections to the wave function are large. The theory thus gives a reconciliation of the shell model, the strong two-nucleon interactions, and the observed two-body correlations in the nucleus. The smallness of the corrections is due to the operation of the exclusion principle. Bethe (1956) has shown that this same exclusion effect makes even the Hartree–Fock approximation good for quite strong interactions, such as an exponential potential fitted to low-energy nucleon-nucleon scattering.

The first problem on which calculations have been made is that of 'nuclear matter', that is, a very large nucleus with surface effects neglected (Brueckner 1955*a*; Wada & Brueckner 1956). In this problem the aim is to show that at a fixed density the energy is proportional to the number of particles, and that as the density is varied the energy per particle has a minimum at the observed density of large nuclei, and that this minimum value gives the observed volume energy of large nuclei. The single-particle wave functions are plane waves, and the potential V is diagonal in momentum space (in contrast to the ordinary Hartree potential which is diagonal in configuration space). The independent-particle model state is a 'Fermi gas' state with all the one-particle states filled up to the Fermi momentum k_F which depends only on the density.

Brueckner & Levinson's derivation, and that of Eden, is based on the multiple scattering formalism of Watson (Watson 1953). The proportionality of the energy of nuclear matter of a given density to the number of particles follows at once from the theory provided certain terms which represent several interactions occurring

‡ Author's present address: Institut for Teoretisk Fysik, Blegdamevej 17, Copenhagen.

independently are not present. There is no satisfactory proof of this in the usual presentation of the theory. It has been shown (Brueckner 1955*b*) that the usual perturbation theory for bound states can be recast so that these terms disappear from the first few orders. The present paper proves a new perturbation formula in which these terms are absent and so completely solves this problem of 'unlinked clusters'.

The method of Feynman graphs (Feynman 1949) is used to enumerate the terms of the perturbation series. To derive the 'linked cluster' result it is essential to describe states in a particular way explained later, which is equivalent to treating the independent-particle ground state as a 'vacuum' state. This description then emphasizes the important exclusion effects, and is used to give a derivation of the Hartree–Fock approximation which seems very natural in this context. The ideas of the Brueckner method for dealing with strong potentials are then introduced and are shown to fit naturally into the Feynman graph treatment.

2. Time-dependent perturbation theory and Feynman graph analysis

Consider A particles with the Hamiltonian

$$H = \sum_{i=1}^{A} T_i + \sum_{i<j} v_{ij}. \tag{2·1}$$

T_i is the kinetic energy of the ith particle and v_{ij} the interaction potential between particles i and j. Introduce the one-body potential V which is to be chosen later to give a reasonable independent-particle model of the system. Let V_i be this potential acting on particle i. Define

$$H_0 = \sum_i (T_i + V_i), \tag{2·2}$$

$$H_1 = \sum_{i<j} v_{ij} - \sum_i V_i, \tag{2·3}$$

so that

$$H = H_0 + H_1. \tag{2·4}$$

Expansions will be in powers of H_1, but the complete series obtained will finally be rearranged so that higher-order terms represent small effects when V is suitably defined. Let the solutions of the one-particle Schrödinger equation

$$(T+V)\psi = E\psi \tag{2·5}$$

be a series of one-particle eigenstates ψ_n with eigenvalues E_n. V must be a potential which gives a discrete series of bound eigenstates ψ_n. (From now on suffixes m, n, etc., will refer to these states, not to particles.)

The second-quantized formalism will be used. Let $\eta_n^\dagger$, η_n be creation and destruction operations for the state ψ_n with the usual anti-commutation relations. Define matrix elements of v and V by

$$\langle rs \mid v \mid mn \rangle = \int \psi_r^*(1)\,\psi_s^*(2)\,v_{12}\,\psi_m(1)\,\psi_n(2)\,d\tau_1\,d\tau_2, \tag{2·6}$$

$$\langle r \mid V \mid m \rangle = \int \psi_r^*(1)\,V_1\,\psi_m(1)\,d\tau_1. \tag{2·7}$$

The matrix element of v defined by (2·6) is not antisymmetrized and corresponds to an interaction in which one particle goes from state ψ_m to state ψ_r, while the other goes from state ψ_n to state ψ_s. With these definitions

$$H_0 = \sum_n E_n \eta_n^\dagger \eta_n, \tag{2·8}$$

$$\begin{aligned} H_1 &= \Sigma \langle rs \mid v \mid mn\rangle \eta_r^\dagger \eta_s^\dagger \eta_n \eta_m \\ &\quad - \Sigma \langle r \mid V \mid m\rangle \eta_r^\dagger \eta_m. \end{aligned} \tag{2·9}$$

The first sum in (2·9) is over all distinct matrix elements, a matrix element $\langle rs \mid v \mid mn\rangle$ being characterized by the pair of transitions (ψ_m to ψ_r) and (ψ_n to ψ_s). Thus $\langle sr \mid v \mid nm\rangle$ is not distinct from $\langle rs \mid v \mid mn\rangle$, but $\langle sr \mid v \mid mn\rangle$ is distinct. This way of introducing antisymmetry is the most suitable for graphical representation.

An eigenstate Φ of H_0 is a determinant formed from A of the ψ_n and can be described by enumerating these A one-particle states. A different description is necessary to obtain the results of this paper. It is supposed that H_0 has a non-degenerate ground state Φ_0 formed from the lowest A of the ψ_n. The proofs of this paper only apply to this case, that is, only to the ground state of a closed-shell nucleus or the ground state of 'nuclear matter'. The states ψ_n occupied in Φ_0 will be called unexcited states, and all the higher states ψ_n will be called excited states. Thus for 'nuclear matter' with a Fermi momentum k_F, an unexcited state means one with momentum $k < k_F$, an excited state one with $k > k_F$. An eigenstate Φ of H_0 can now be described by enumerating all the excited states which are occupied, and all the unexcited states which are not occupied. An unoccupied unexcited state is regarded as a 'hole', and the theory will deal with particles in excited states and holes in unexcited states. This treatment is analogous to the theory of positrons, with Φ_0 as the 'vacuum' state. An unexcited state is automatically regarded as occupied and so excluded for other particles, unless a hole in that state is introduced explicitly. Thus the chief effect of the exclusion principle is emphasized by this description. This is the essential difference from the theory of positrons, in which there is symmetry between particles and holes. In this theory the asymmetry between particles and holes is emphasized. To introduce this method formally equations (2·8) and (2·9) are retained, but the interpretation of $\eta_n^\dagger, \eta_n$ for unexcited ψ_n is altered. η_n will now be the operator creating a hole in state ψ_n, $\eta_n^\dagger$, the operator destroying a hole.

The following derivation of the perturbation formula uses time-dependent perturbation theory in the interaction representation. In this way certain of the results needed appear more naturally than in a completely time-independent presentation. Let Φ_0 be the ground state of H_0 as described above, assumed to be non-degenerate, and let Ψ_0 be the lowest eigenstate of H. Ψ_0 will be derived from Φ_0 by adiabatically switching on the interaction H_1 over the time interval $-\infty$ to 0. For this case of a discrete series of eigenstates with a unique ground state the adiabatic theorem can be proved in the following form (Gell-Mann & Low 1951).

Define
$$H_1(t) = e^{iH_0 t} H_1 e^{-iH_0 t} e^{\alpha t}, \tag{2·10}$$

and let
$$U_\alpha = \sum_{n=0}^{\infty} (-i)^n \int_{0>t_1>t_2\ldots>t_n} H_1(t_1) H_1(t_2) \ldots H_1(t_n)\, dt_1 \ldots dt_n. \tag{2·11}$$

As $\alpha \to 0$ the unitary operator U_α describes the adiabatic process.

Let
$$\Psi_0 = \lim_{\alpha \to 0} \frac{U_\alpha \Phi_0}{\langle \Phi_0 \mid U_\alpha \mid \Phi_0 \rangle}. \tag{2·12}$$

By using Feynman graphs this limit will be shown to exist and an explicit expression derived for it. Then the adiabatic theorem states that

$$H\Psi_0 = (E_0 + \Delta E)\Psi_0, \tag{2·13}$$

where
$$H_0 \Phi_0 = E_0 \Phi_0 \tag{2·14}$$

and
$$\Delta E = \langle \Phi_0 \mid H_1 \mid \Psi_0 \rangle = \lim_{\alpha \to 0} \frac{\langle \Phi_0 \mid H_1 U_\alpha \mid \Phi_0 \rangle}{\langle \Phi_0 \mid U_\alpha \mid \Phi_0 \rangle}. \tag{2·15}$$

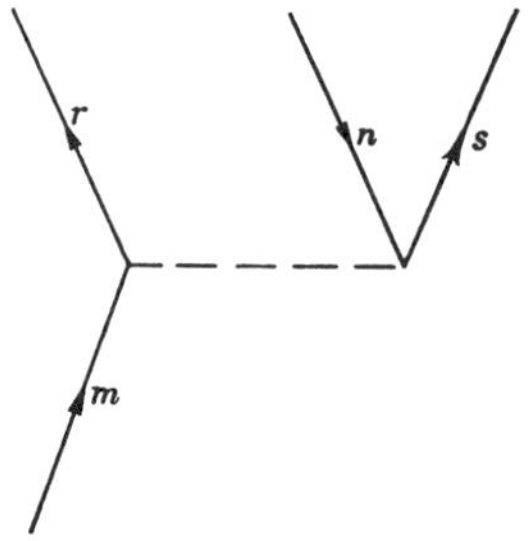

FIGURE 1. In all the graphs the direction of increasing time is upwards.

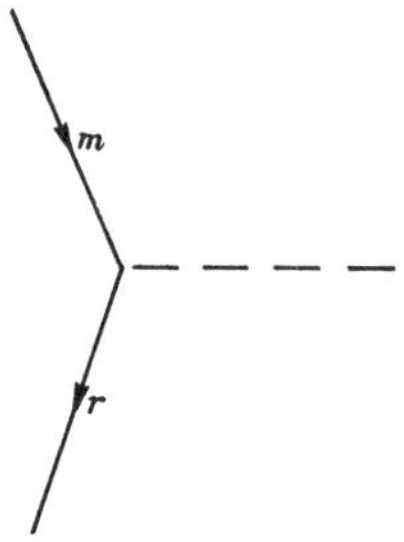

FIGURE 2

The required perturbation formulae for Ψ_0 and ΔE will be obtained on carrying out the time integrations in the expression for the limits in equations (2·12) and (2·15).

$H_1(t)$ is derived from equation (2·9) for H_1 by substituting $\eta_n(t)$ for η_n, where

$$\eta_n(t) = \eta_n \mathrm{e}^{-\mathrm{i}E_n t} \tag{2·16}$$

and then multiplying by $\mathrm{e}^{\alpha t}$. The expression (2·11) for U_α then becomes a sum of products of v and V matrix elements, $\mathrm{e}^{\mathrm{i}Et}$ and $\mathrm{e}^{\alpha t}$ factors and operators $\eta^\dagger$ and η. Analysis of the products of operators by the same algebra as is used in proving Wick's theorem (Wick 1950) leads to the following expression for $U_\alpha \Phi_0$ as a sum of terms represented by Feynman graphs. Each graph represents a series of $H_1(t)$ interactions. A particle in an excited state is represented by a line in the direction of increasing time. A hole in an unexcited state is represented by a line in the opposite direction. A matrix element $\langle rs \mid v \mid mn \rangle$ in $H_1(t)$ is represented as in figure 1. This is for the case in which ψ_m, ψ_r, ψ_s are excited states and ψ_n an unexcited state. It represents an interaction between two particles in which one is scattered from ψ_m to ψ_r while the other jumps from ψ_n into ψ_s leaving a hole in ψ_n. With this graph is associated a time factor $\mathrm{e}^{\mathrm{i}(E_r+E_s-E_m-E_n)t}\mathrm{e}^{\alpha t}$. The other combinations of excited and unexcited states $\psi_m \psi_n \psi_r \psi_s$ are represented similarly.

A matrix element $\langle r \mid V \mid m \rangle$ is represented as in figure 2. This shows a particle scattered from state ψ_m to ψ_r by V, both states unexcited. Initially there was a hole in state ψ_r, otherwise the interaction is excluded, and finally there is a hole in state ψ_m.

There are further possibilities which do not occur in positron theory. Here the unexcited states are occupied by real particles not explicitly represented in the graphs, but interacting with each other and with the particles represented in the graphs. These particles will be called passive unexcited particles. Their interactions are the most important ones present, and it is these interactions which must be

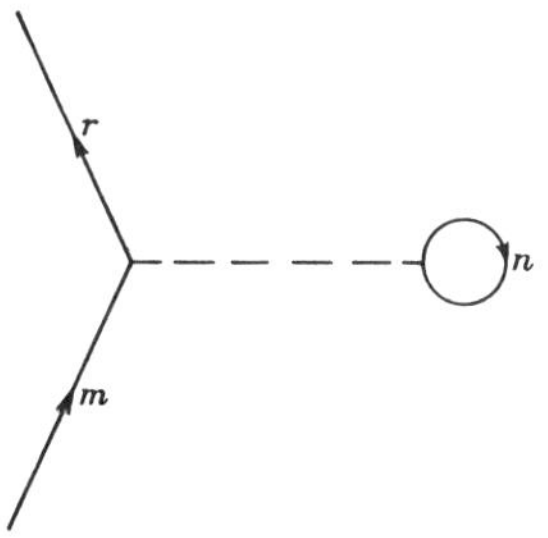

FIGURE 3

allowed for in the choice of V. They are represented as in figure 3. This shows a particle scattered from excited state ψ_m to excited state ψ_r by the particle in the unexcited state ψ_n which remains in the same state after the interaction. (In 'nuclear matter' this is 'forward' scattering.) Figure 3 corresponds to a factor

$$\langle rn \mid v \mid mn \rangle\, e^{i(E_r - E_m)t}.$$

The 'exchange' term corresponding to this contains the matrix element $\langle rn \mid v \mid nm \rangle$ and is represented as in figure 4. Finally, figure 5 shows the graphs representing interactions in which only passive unexcited particles take part. The matrix elements are for figure 5 (*a*), $\langle mn \mid v \mid mn \rangle$; for figure 5 (*b*) $\langle mn \mid v \mid nm \rangle$; for figure 5 (*c*) $\langle n \mid V \mid n \rangle$.

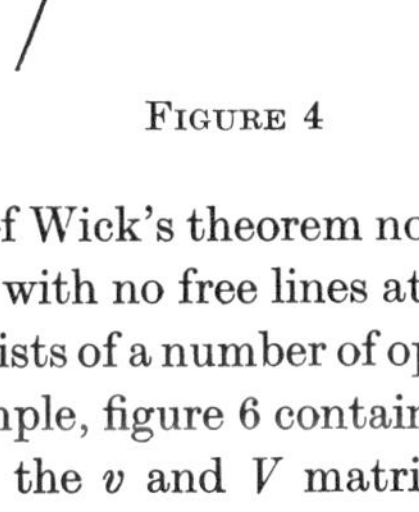

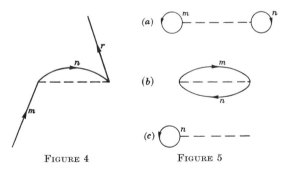

FIGURE 4

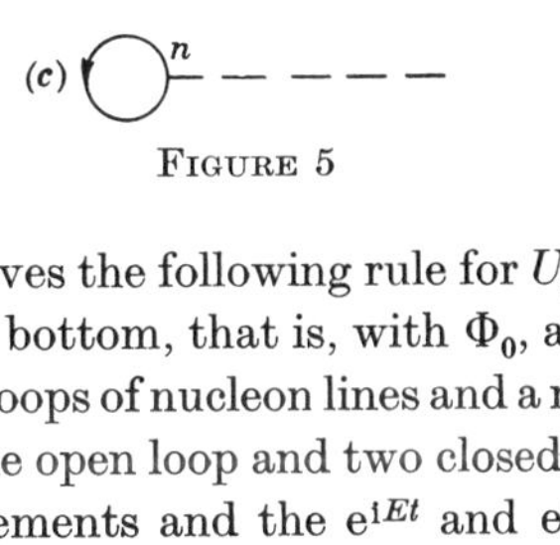

FIGURE 5

The algebra of Wick's theorem now gives the following rule for $U_\alpha \Phi_0$. All distinct graphs starting with no free lines at the bottom, that is, with Φ_0, are drawn. Each such graph consists of a number of open loops of nucleon lines and a number of closed loops. For example, figure 6 contains one open loop and two closed loops. For each graph multiply the v and V matrix elements and the e^{iEt} and $e^{\alpha t}$ factors and a factor $(-1)^{h+l}$, where h is the number of internal hole lines (four in figure 6; the line labelled m is an external line) and l the number of closed loops. A passive unexcited particle loop as in figure 5 (*c*) contributes a plus sign, counting as one hole line and one closed loop, while figure 5(*b*) has a minus sign having two hole lines and one closed loop. Each V matrix element has a minus sign attached since it occurs with a minus sign in H_1. Attach the pairs of creation operators corresponding to the external lines at the ends of each open loop with the hole operator to the right ($\eta_r^\dagger \eta_m$ for figure 6). Finally, carry out the time integrations. Then $U_\alpha \Phi_0$ is the sum of all these terms acting on Φ_0. It is important to note that

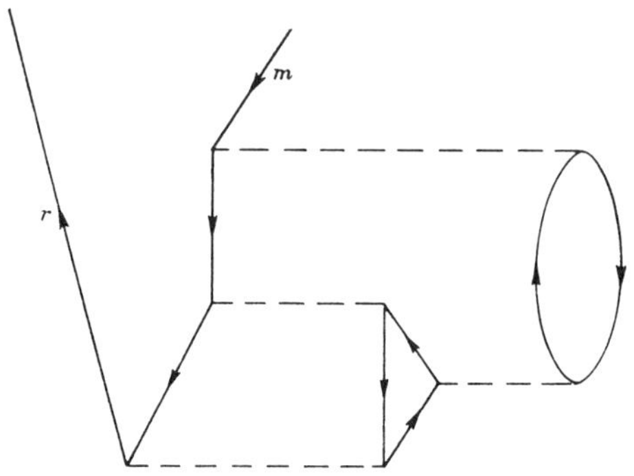

FIGURE 6

the exclusion principle is to be ignored in labelling the graphs. The major effects of exclusion are already taken into account by the 'hole' picture as described above. The rest must not be included if the results of § 3 are to be derived. The algebra of the η operators does give this result, which is merely a careful application to this case of the principle that intermediate states need not be anti-symmetrized. In fact all graphs which contradict the exclusion principle are exactly cancelled by the corresponding 'exchange' graphs. However, in § 3 certain graphs will be removed and then this cancellation will no longer occur and the graphs contradicting exclusion will represent important physical effects. This representation is essential for the derivation of the 'linked cluster' result.

3. The linked-cluster perturbation formula

Any part of a graph which is completely disconnected from the rest of the graph and which has no external lines attached will be called an unlinked part. In the expression for $U_\alpha \Phi_0$ before the time integrations are carried out the lines of a graph can be labelled independently of each other (this is where it is essential not to have to take exclusion into account), and the factors attached to the interaction lines are independent of each other. Now consider a graph containing unlinked parts and take together with it all the graphs which differ only by having the interactions in the unlinked parts in different positions relative to those in the rest of the graph. The order of the interactions in the two parts separately is kept fixed. Let the times of the interactions in the unlinked part be $t_1, t_2, \ldots, t_n$ and in the rest be $t'_1 t'_2 \ldots t'_m$, where the order of the two parts separately is given by $0 > t_1 > t_2 > \ldots > t_n$ and $0 > t'_1 > t'_2 > \ldots > t'_m$. The sum over all the different relative positions of the two parts is obtained by carrying out the time integrations with only these restrictions on the order of the times and so is the product of the expressions obtained from the two parts separately. A graph containing no unlinked parts will be called a linked graph. It follows that $U_\alpha \Phi_0$ is given by the rules of § 2 applied to the sum of linked graphs only, multiplied by a factor given by the sum of all graphs consisting only of unlinked parts. This factor is just what the rules of § 2 give for $\langle \Phi_0 | U_\alpha | \Phi_0 \rangle$. Thus Ψ_0 as defined by (2·12) is given by taking the limit $\alpha \to 0$ in the sum of linked graphs only.

The result of carrying out the time integrations in this sum may be written as

$$\Psi_0 = \lim_{\alpha \to 0} \sum_L \frac{1}{E_0 - H_0 + in\alpha} H_1 \cdots \frac{1}{E_0 - H_0 + 2i\alpha} H_1 \frac{1}{E_0 - H_0 + i\alpha} H_1 \Phi_0. \tag{3·1}$$

$\sum_L$ means that the terms are to be enumerated by the linked graphs described above. Φ_0 cannot occur as an intermediate state in a linked graph as the part of the graph below that intermediate state would be an unlinked part. Since all other

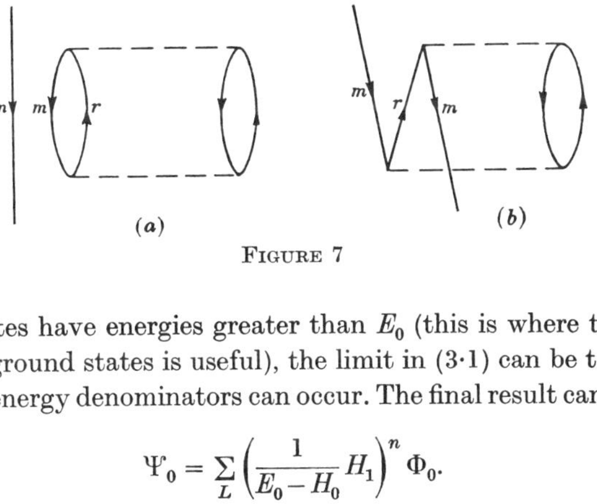

FIGURE 7

intermediate states have energies greater than E_0 (this is where the limitation to non-degenerate ground states is useful), the limit in (3·1) can be taken by putting $\alpha = 0$ as no zero energy denominators can occur. The final result can then be written

$$\Psi_0 = \sum_L \left(\frac{1}{E_0 - H_0} H_1 \right)^n \Phi_0. \tag{3·2}$$

The energy shift ΔE is given by (2·15), and using the same arguments as for $U_\alpha \Phi_0$,

$$\Delta E = \sum_L \left\langle \Phi_0 \middle| H_1 \left(\frac{1}{E_0 - H_0} H_1 \right)^n \middle| \Phi_0 \right\rangle, \tag{3·3}$$

where now $\sum_L$ means summed over all connected graphs leading from Φ_0 to Φ_0, that is, with no external lines. (3·2) and (3·3) are the linked-cluster perturbation formulae. They differ from the usual bound state perturbation formula by having E_0 in the denominator instead of the usual $E_0 + \Delta E$. This difference is compensated by the different enumeration of terms, that is, by summing only over linked graphs and by ignoring exclusion as described in § 2.

A typical graph contradicting the exclusion principle is figure 7 (*b*). Before the unlinked parts were removed this was cancelled by figure 7 (*a*) which has the same matrix elements and an extra minus sign. Figure 7 (*a*) represents an interaction of the passive particle in the unexcited state ψ_m. Many repetitions of figure 7 (*b*) combine to give the modification of the energy of state ψ_m due to this interaction.

These linked-cluster expansions can be derived without using time-dependent theory. Ψ_0 can be defined to be given by (3·2). It then follows that

$$(E_0 - H_0)\Psi_0 = \sum_L H_1 \left(\frac{1}{E_0 - H_0} H_1\right)^n \Phi_0 \qquad (3\cdot 4)$$

and

$$H_1 \Psi_0 = H_1 \sum_L \left(\frac{1}{E_0 - H_0} H_1\right)^n \Phi_0. \qquad (3\cdot 5)$$

The right-hand side of (3·5) is given by those graphs which are linked when the last H_1 is removed. Some care is needed to prove this, since Wick's theorem does not immediately apply to the time-integrated expression (3·4). Subtracting (3·4) from (3·5) gives

$$(H - E_0)\Psi_0 = \Sigma' H_1 \left(\frac{1}{E_0 - H_0} H_1\right)^n \Phi_0, \qquad (3\cdot 6)$$

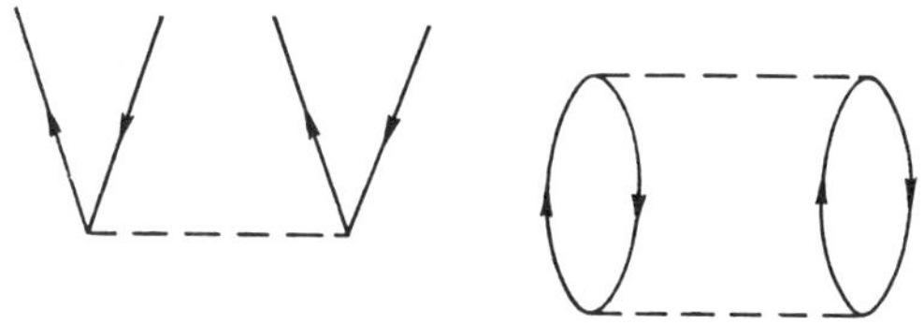

FIGURE 8

where Σ' means summed over all graphs containing an unlinked part but which are linked when the last H_1 line is removed. Such graphs must be of the type shown in figure 8. Now the last H_1 line in the unlinked part may be kept fixed and a sum taken over the different positions of the rest of the unlinked part relative to the rest of the graph. By using algebraic identities on the energy denominators which are equivalent to the separation of the time integrations in the time-dependent proof it can be shown that the right-hand side of (3·6) is equal to the product of Ψ_0 with the sum of all connected closed graphs, that is, with ΔE as defined by (3·3). Then (3·6) gives

$$H\Psi_0 = (E_0 + \Delta E)\Psi_0, \qquad (3\cdot 7)$$

the required result.

This method of proof has one advantage over the other in that it does not use time-dependent methods to prove a time-independent result. However, the time-dependent proof gives the easiest way of enumerating the terms correctly and of combining the contributions of different positions of unlinked parts. The adiabatic theorem used can be strictly proved under the conditions of this paper. The time-independent method has been used by the author to extend the results to excited and degenerate states.

4. Choice of V: the Hartree–Fock method

The simplest way to choose V is to make it allow for the first-order interactions with passive unexcited particles. This is done by making the graph parts in figure 9 cancel, that is, by defining

$$\langle r \mid V \mid m \rangle = \sum_n \{\langle rn \mid v \mid mn \rangle - \langle rn \mid v \mid nm \rangle\}. \qquad (4\cdot 1)$$

The sum is over all unexcited states ψ_n. The states ψ_n are determined by

$$(T+V)\,\psi_n = E_n \psi_n. \tag{4·2}$$

(4·1) and (4·2) are the Hartree–Fock self-consistent equations.

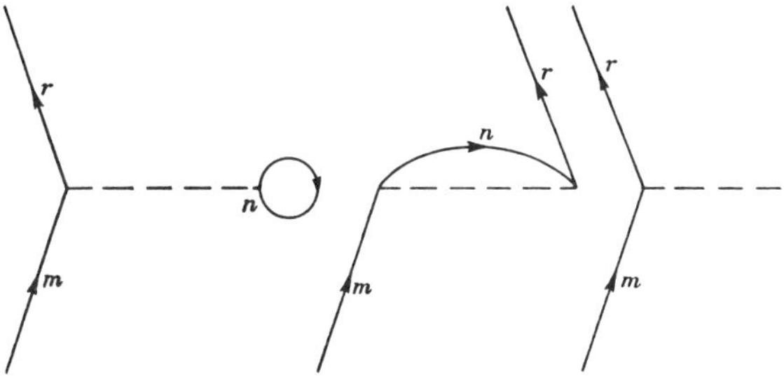

FIGURE 9

This definition ensures the complete disappearance of the V interaction and the interactions with passive unexcited states from all graphs except the connected closed parts in figure 5. These represent the first-order terms in ΔE. Figure 5 (c) contributes $-\sum_n \langle n \mid V \mid n\rangle$, while figures 5 ($a$) and ($b$) contribute

$$\tfrac{1}{2}\sum_{m,n}\{\langle mn \mid v \mid mn\rangle - \langle mn \mid v \mid nm\rangle\} = \tfrac{1}{2}\sum_n \langle n \mid V \mid n\rangle \tag{4·3}$$

when summed over all distinct possibilities. Also,

$$E_0 = \sum_n \{\langle n \mid T \mid n\rangle + \langle n \mid V \mid n\rangle\}. \tag{4·4}$$

Thus, to the first order in v,

$$E = E_0 + \Delta E = \sum_n \langle n \mid T \mid n\rangle + \tfrac{1}{2}\sum_n \langle n \mid V \mid n\rangle. \tag{4·5}$$

This factor of $\frac{1}{2}$ is familiar in the Hartree–Fock method.

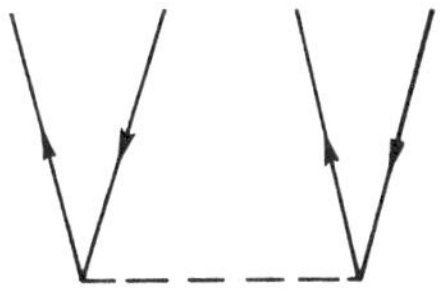

FIGURE 10

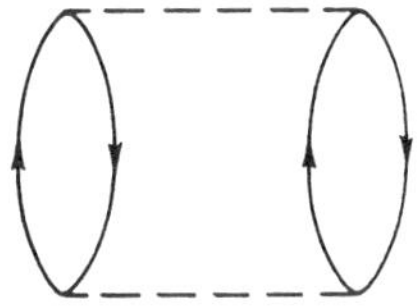

FIGURE 11

The higher-order corrections to E are given by the sum (3·3) over all connected graphs with no external lines and with no V interactions and no interactions with passive unexcited particles. The wave function Ψ_0 is given by the sum (3·2) again without the above interactions. The expression for Ψ_0 contains terms which are the product of many factors represented by graphs like figure 10. The result is that Φ_0 is only a very small component of Ψ_0. (Note that Ψ_0 is normalized to $\langle \Phi_0 \mid \Psi_0\rangle = 1$.) However, the corresponding correction to the energy can contain the factor represented by figure 11 once only. Bethe (1956) has shown that the exclusion principle which limits the particles in excited states to states with momentum $> k_F$ can make this correction fairly small even for strong potentials for the values of k_F of interest. Thus the Hartree–Fock method can give the energy quite well even for strong potentials. This is a quantitative version of the old argument that strong interactions would be inhibited by the exclusion principle. It applies to the energy but not to the wave function. There are certainly strong correlations between nucleons in a nucleus and the Brueckner theory can be used to explain them (Brueckner, Eden & Francis 1955).

5. The Brueckner theory

The nucleon-nucleon potential very probably has a steep repulsive core at small distances. (This will certainly ensure saturation but a proper theory is needed to obtain an energy minimum at the observed nuclear density.) For this v it is clearly impossible to choose V by the Hartree–Fock method, as the matrix elements of

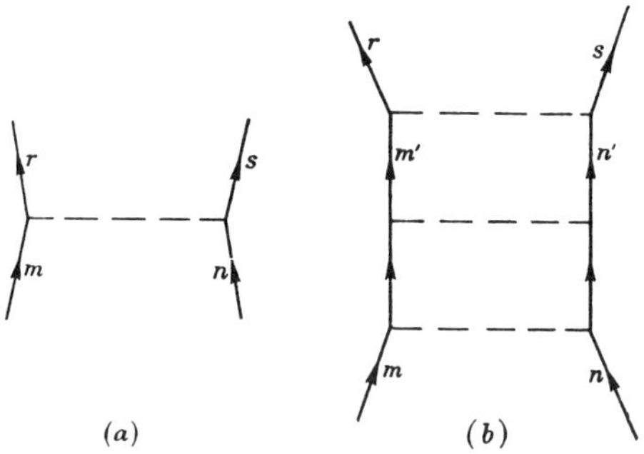

Figure 12

v will have a large contribution from the core. The Brueckner theory replaces v by a reaction matrix t calculated from a two-body equation of the type

$$t = v + v\frac{1}{E_0 - H_0}t. \tag{5·1}$$

The idea is to derive V from t instead of from v. Since H_0 contains V, V occurs in the energy denominator so that there is a further self-consistency requirement in addition to the Hartree–Fock condition on the wave functions. In fact, for 'nuclear matter' the Hartree–Fock self-consistency disappears since the wave functions must be plane waves. Brueckner (1955*a*) has shown that this new self-consistency is important.

The procedure in terms of graphs is as follows. Corresponding to any graph with a single v line in a certain position as in figure 12 (*a*), there are more complicated ones in which figure 12 (*a*) is replaced by the 'ladder' graph of figure 12 (*b*). In the intermediate states of figure 12 (*b*) both particles are in excited states. The sum of all such parts is given by an integral equation of the type (5·1). When figure 12 (*b*) occurs as part of a larger graph the energy denominator for the intermediate state containing $\psi_{m'}\psi_{n'}$ is

$$-E_{m'} - E_{n'} - E_R = E_m + E_n - E_{m'} - E_{n'} - \delta E, \tag{5·2}$$

where E_R is the excitation energy of the other particles present while the interaction represented in 12 (*b*) occurs and δE is the excitation energy of the complete intermediate state at the beginning of the interaction. The excitation energy of a state is the sum of the energies E_n of occupied excited states minus the sum of the energies of unexcited states in which there are holes. The integral equation for the sum of the terms represented by figure 12 is then

$$\langle rs \mid t \mid mn \rangle = \langle rs \mid v \mid mn \rangle + \sum_{m'n'} \frac{\langle rs \mid v \mid m'n' \rangle \langle m'n' \mid t \mid mn \rangle}{E_m + E_n - E_{m'} - E_{n'} - \delta E}, \tag{5·3}$$

where the sum is over $\psi_{m'}, \psi_{n'}$ excited states only. The solution is a matrix $t(\delta E)$ which can be used to replace v and which is finite even for a repulsive core potential.

A graph will be called irreducible if it contains no 'ladders' of the type of figure 12 (*b*). A sequence of v interactions as in figure 12 (*b*) only forms a 'ladder' if all the intermediate states are excited and if there are no other interactions in other

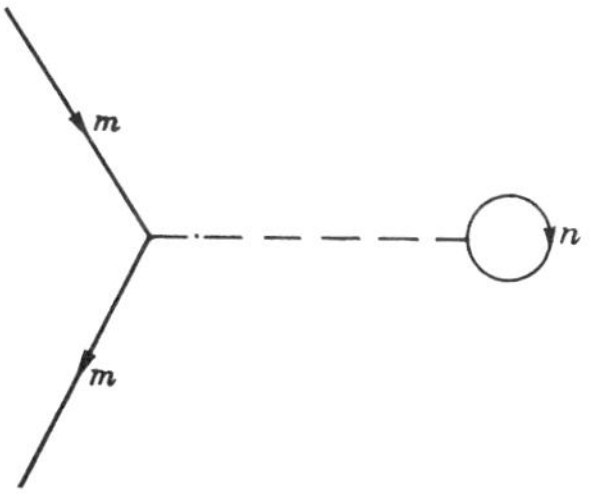

FIGURE 13

parts of the graph between the ends of the ladder. All graphs can be obtained by substituting independently 'ladders' for each v line in the irreducible graphs. The terms of the linked cluster expansion can thus be grouped together so that each v matrix element is replaced by a matrix element of $t(\delta E)$. δE is the excitation energy of the intermediate state to the right of the matrix element in the series (below it in the graph). The sums must now be taken over linked irreducible graphs only.

Figure 7 (*b*) is an important type of ladder graph and is absorbed into the t matrix element $\langle mn \mid t \mid mn \rangle$ represented in figure 13. Note that (5·3) is not antisymmetrized. The ladder graph in which the lines of figure 12 (*b*) cross over is counted in $\langle sr \mid t \mid mn \rangle$.

V can now be defined to cancel the t-interactions with passive unexcited states, that is, by (4·1) with v replaced by t. However, the cancellation cannot be complete because of the dependence of t on δE. (The procedure in this problem contrasts with that in field theory in which the time ordering and the dependence of one part of a graph on another are completely removed by introducing an extra energy variable for each particle. This does not seem appropriate here.) The best that can be done is to choose some average value of δE appropriate to the matrix element of t being evaluated.

Equation (4·5) is replaced by the following expression for the energy to first order in t:

$$E = \sum_n \langle n \mid T \mid n \rangle + \tfrac{1}{2} \sum_{m,n} \{\langle mn \mid t(0) \mid mn \rangle - \langle nm \mid t(0) \mid mn \rangle\}. \tag{5·4}$$

The $\langle n \mid V \mid n \rangle$ in E_0 is cancelled by the term represented by figure 5 (c) whatever the definition of V. The second term in (5·4) will equal $\frac{1}{2}\sum\limits_{n}\langle n \mid V \mid n\rangle$ only if $\langle n \mid V \mid n \rangle$ is derived from t with $\delta E = 0$. This is the most straightforward choice for the diagonal elements of V between unexcited states.

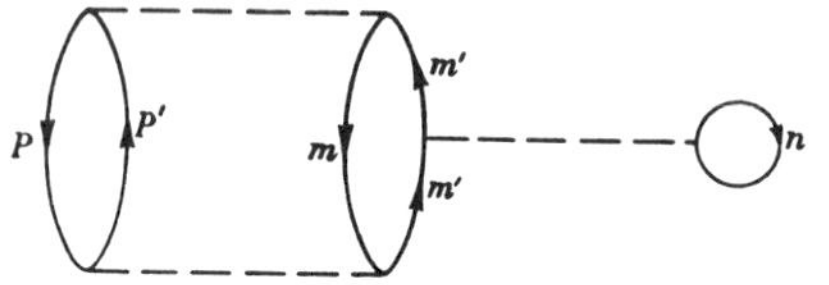

FIGURE 14

Figure 14 represents a term in the energy given by

$$\langle mp \mid t(\delta E) \mid m'p' \rangle \left(-\frac{1}{\delta E}\right) \langle m'n \mid t(\delta E) \mid m'n \rangle \left(-\frac{1}{\delta E}\right) \langle m'p' \mid t(0) \mid mp \rangle, \qquad (5\cdot5)$$

where
$$\delta E = E_{p'} - E_p + E_{m'} - E_m. \qquad (5\cdot6)$$

An average of this δE used in the definition of $\langle m' \mid V \mid m' \rangle$ will ensure as much cancellation of this term as possible. For 'nuclear matter' the conservation of momentum limits the possible values of $\psi_p \psi_{p'}$ and ψ_m given $\psi_{m'}$.

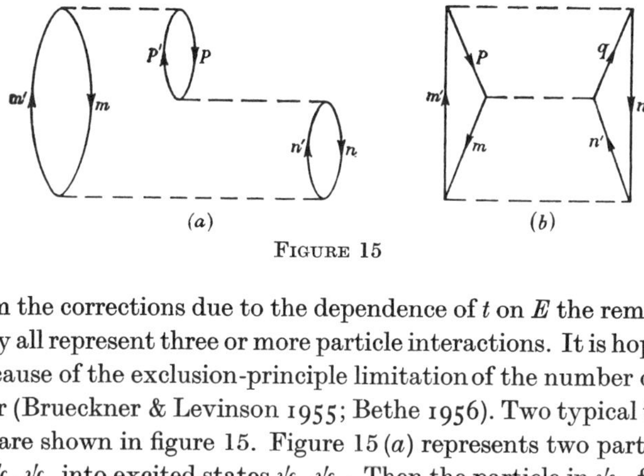

FIGURE 15

Apart from the corrections due to the dependence of t on E the remaining graphs for the energy all represent three or more particle interactions. It is hoped that these are small because of the exclusion-principle limitation of the number of states to be summed over (Brueckner & Levinson 1955; Bethe 1956). Two typical three-particle interactions are shown in figure 15. Figure 15 (a) represents two particles jumping from states $\psi_m \psi_n$ into excited states $\psi_{m'}\psi_{n'}$. Then the particle in $\psi_{n'}$ falls back into ψ_n while another particle jumps from ψ_p into $\psi_{p'}$. Finally the particles in $\psi_{p'}$ and $\psi_{m'}$ interact and fall back. The corresponding matrix elements are

$$\langle pm \mid t \mid p'm' \rangle \langle np' \mid t \mid n'p \rangle \langle m'n' \mid t \mid mn \rangle. \qquad (5\cdot7)$$

Figure 15 (b) represents two particles jumping from states $\psi_m \psi_n$ into states $\psi_{m'}\psi_{n'}$. The particle in $\psi_{n'}$ then interacts with the particle in the occupied state ψ_p. The particle in $\psi_{n'}$ is scattered into ψ_q while that in ψ_p jumps into the hole in ψ_m leaving a hole in ψ_p. Finally the particles in $\psi_{m'}$, ψ_q fall back into the holes in ψ_p, ψ_n. The corresponding matrix elements are

$$\langle np \mid t \mid qm' \rangle \langle qm \mid t \mid n'p \rangle \langle m'n' \mid t \mid mn \rangle. \qquad (5\cdot8)$$

I am very grateful to Professor H. A. Bethe and Dr R. J. Eden for much advice and discussion on this problem and for very helpful criticism of the first draft of this paper. I also thank the Department of Scientific and Industrial Research for a maintenance grant.

References

Bethe, H. A. 1956 *Phys. Rev.* **103**, 1353.
Brueckner, K. A. 1955*a* *Phys. Rev.* **97**, 1353.
Brueckner, K. A. 1955*b* *Phys. Rev.* **100**, 36.
Brueckner, K. A., Eden, R. J. & Francis, N. C. 1955 *Phys. Rev.* **98**, 1445.
Brueckner, K. A. & Levinson, C. A. 1955 *Phys. Rev.* **97**, 1344.
Eden, R. J. 1956 *Proc. Roy. Soc.* A, **235**, 408.
Feynman, R. P. 1949 *Phys. Rev.* **76**, 749.
Gell-Mann, M. & Low, F. 1951 *Phys. Rev.* **84**, 350 (Appendix).
Wada, W. W. & Brueckner, K. A. 1956 *Phys. Rev.* **103**, 1008.
Watson, K. M. 1953 *Phys. Rev.* **89**, 575.
Wick, G. C. 1950 *Phys. Rev.* **80**, 268.

The description of collective motions in terms of many-body perturbation theory

By J. Hubbard

Atomic Energy Research Establishment, Harwell, Didcot, Berkshire

(*Communicated by R. E. Peierls, F.R.S.—Received* 2 *February* 1957)

In this and a succeeding paper it is shown how a theory equivalent to the Bohm & Pines collective motion theory of the electron plasma can be derived directly from a perturbation series which gives in principle an exact solution of the many-body problem. This result is attained by making use of a diagrammatic method of analysis of the perturbation series. By a process analogous to the elimination of photon self-energy parts from the electrodynamic S matrix it is found possible to simplify the perturbation series, introducing a modified interaction between the particles. A useful integral equation for this modified interaction can be set up, and it is shown how the energy of the system can be expressed in terms of the modified interaction. The close connexion between this approach and the dielectric theory of plasma oscillations is indicated.

1. Introduction

Within recent years much attention has been given in the study of the quantum mechanical many-body problem to the collective modes of motion which may be present (Bohm & Pines 1953; Tomonaga 1955; Bohr & Mottelson 1953). Two main theories of collective motion have been developed, that of Tomonaga (1955), and the superfluous co-ordinate type of theory introduced by Bohm & Pines (1953). In the Tomonaga theory a transformation of variables is made in such a way that some of the new co-ordinates are directly related to the collective modes of motion, whilst the remaining new co-ordinates are associated with internal modes of motion. In the superfluous co-ordinate treatment certain auxiliary variables are introduced together with an equal number of subsidiary conditions to preserve the correct number of degrees of freedom, and a transformation is made in such a way that the new auxiliary variables are related to the collective motion, whilst the original co-ordinates when transformed are related to the internal motion. If the collective modes being studied have real physical significance, then it will be found in both these methods that the Hamiltonian is, to a good approximation, separable in the new co-ordinates, and a separation of the collective motion is thereby obtained.

Though these methods are quite successful, they have certain unsatisfactory features. In the Tomonaga method it is generally found that when the Hamiltonian has been separated the problem of finding the eigenvalues of the internal motion part is very difficult. In the superfluous co-ordinate treatment one does not meet with this difficulty but with an equivalent one; this is that it is difficult to find eigenfunctions satisfying the subsidiary conditions. In addition, both theories suffer from the difficulty of not being able to treat very easily the interaction between the collective and internal modes of motion, or the intimately related problem of the damping of the collective motion; where the damping is small this is not a very serious problem, but in certain cases, e.g. the application of the Bohm & Pines theory to metals, it may be quite serious. We shall see later that this problem of treating the damping is also intimately connected with the problem of the cut-off in the Bohm & Pines theory of the electron plasma (1953).

In the case of the electron plasma an alternative type of theory of a different character from those described above has been developed. In this 'dielectric' theory (Mott 1954; Fröhlich & Pelzer 1955; Hubbard 1955*a*, *b*) one argues along semi-classical lines, regarding the electron gas as a dielectric medium. From this point of view one thinks of the electrons as interacting with one another like particles in the dielectric medium represented by the remaining electrons; their interaction is therefore modified and screened. The plasma oscillations are thought of as being the polarization waves in the dielectric medium. This approach has advantages over the other two approaches in that it can easily treat the damping problem, and does not attempt so complete a separation of the collective and internal modes of motion, the one going smoothly over into the other; it suffers from the disadvantages of being a phenomenological theory and difficult to quantize satisfactorily.

It is the purpose of this and a succeeding paper (Hubbard 1957) to develop yet another approach to the collective motion problem which is applicable to the electron plasma, and to similar systems, and which we hope combines the advantages of the treatments described above and yet is free of their disadvantages. This theory is based upon an (infinite) perturbation series which provides in principle an exact solution to the many-body problem and therefore contains all the physical effects including the collective motion. The various contributions to this perturbation series can be conveniently analyzed making use of diagrams similar to Feynman diagrams (Goldstone 1957). It is now argued that we may be able to simplify our perturbation series by a process exactly analogous to the elimination of photon self-energy parts in the analysis of the S-matrix in quantum electrodynamics (see, for example, Dyson 1949). This is in fact so, and the analysis shows that we need only retain in our perturbation series terms corresponding to diagrams free from these parts, provided everywhere in the perturbation series we replace the ordinary interaction between the particles by a modified interaction.

To establish the connexion between this apparently formal device for simplifying the perturbation series and the collective motion problem let us now consider the physical interpretation of what we shall do. Just as in the electrodynamic case, we can regard the diagrams used in the analysis of the perturbation series as representing the actual physical process which gives rise to the corresponding contribution to the perturbation series. Interpreting the parts of diagrams analogous to photon self-energy parts in this way, we can see that they correspond to the modification of the ordinary interaction between particles by the polarization of the medium represented by the remaining particles; we shall refer to these as polarization parts. Thus, the elimination of these parts and the replacement of the ordinary interaction by the modified interaction in the perturbation series is exactly equivalent to going over to the viewpoint of the dielectric theory described above, so that we may expect our theory to be equivalent to the other theories of collective motion. This expectation is in fact borne out by detailed calculation.

The present paper develops the theory in a general form, and the detailed treatment of the electron plasma is reserved for a succeeding paper. Since the diagrammatic analysis is used in a different (although equivalent) form to that given by Goldstone (1957), it is developed afresh in §§ 2 to 4. In § 5 are discussed certain simplifications which arise when one considers the case of a uniform gas; in the remainder of the paper it is assumed that we are dealing with this case.

Section 6 proceeds with the main programme; the polarization parts are eliminated and the modified interaction introduced. In § 7 an integral equation is derived for the modified interaction which very greatly simplifies its calculation. Finally, in § 8 it is shown how the energy of the system can be expressed exactly in terms of the modified interaction. This result will enable us to calculate correlation energies directly from the modified interaction which in turn can be calculated easily using the integral equation.

2. The diagrammatic analysis of the perturbation series

We shall consider the problem of determining the energy spectrum and wave functions of a gas of Fermi–Dirac particles interacting with one another through an instantaneous two-body potential and moving so slowly that relativistic effects can be neglected; we include also the case in which the particles move in an external potential field. The Hamilton for such a system is

$$H = H_0 + H', \tag{1}$$

where H_0 includes the kinetic energy of the particles and their potential energy in the external field, and H' is the interaction energy of the particles. H' will be treated as a perturbation.

One way in which we may develop the perturbation series for H' is by making use of the adiabatic approximation. We consider the interaction H' to be slowly switched on between $t = -\infty$ and $t = 0$, and to be slowly switched off between $t = 0$ and $t = +\infty$; then a system which at $t = -\infty$ is in an eigenstate Ψ_0 of H_0 will between $t = -\infty$ and $t = 0$ slowly change into an eigenstate of H. This result has been proved by Gell-Mann & Low (1951) in the following form: if Ψ_0 is an eigenstate of H_0 belonging to the energy E_0, then

$$\Psi = \lim_{\alpha \to +0} S_\alpha(0, -\infty)\Psi_0/(\Psi_0 \,|\, S_\alpha(0, -\infty) \,|\, \Psi_0) \tag{2}$$

is an eigenstate of H belonging to the energy

$$E = E_0 + \Delta E = E_0 + \lim_{\alpha \to +0} (\Psi_0 \,|\, H' S_\alpha(0, -\infty) \,|\, \Psi_0)/(\Psi_0 \,|\, S_\alpha(0, -\infty) \,|\, \Psi_0), \tag{3}$$

where $S_\alpha(t, t')$ is the solution of the equation

$$\mathrm{i}\hbar \frac{\mathrm{d}}{\mathrm{d}t} S_\alpha(t, t') = H_\alpha(t)\, S_\alpha(t, t'), \tag{4}$$

satisfying the boundary condition $S_\alpha(t', t') = 1$, and

$$H_\alpha(t) = \mathrm{e}^{(\mathrm{i}t/\hbar)H_0} H' \mathrm{e}^{-(\mathrm{i}t/\hbar)H_0} \mathrm{e}^{-\alpha|t|}. \tag{5}$$

The parameter α is seen to govern the rate of switching on and off of the interaction; the limit $\alpha \to +0$ means that the potential is switched on infinitely slowly, the condition for the exactness of the adiabatic approximation.

We can now obtain a perturbation series for Ψ by solving (4) by iteration. To do this we replace (4) by the integral equation

$$S_\alpha(t) = 1 + \frac{1}{i\hbar}\int_{-\infty}^{t} H_\alpha(t')\, S_\alpha(t')\, dt', \tag{6}$$

incorporating the boundary condition at $t = -\infty$ (in future we shall for brevity write $S_\alpha(t, -\infty)$ as $S_\alpha(t)$). Iteration of (6) gives

$$\begin{aligned} S_\alpha(t) &= 1 + \sum_{n=1}^{\infty}\left(\frac{1}{i\hbar}\right)^n \int_{-\infty}^{t} dt_1 \int_{-\infty}^{t_1} dt_2 \ldots \int_{-\infty}^{t_{n-1}} dt_n\, H_\alpha(t_1)\, H_\alpha(t_2) \ldots H_\alpha(t_n) \\ &= 1 + \sum_{n=1}^{\infty}\frac{1}{n!}\left(\frac{1}{i\hbar}\right)^n \int_{-\infty}^{t} dt_1 \int_{-\infty}^{t} dt_2 \ldots \int_{-\infty}^{t} dt_n\, P[H_\alpha(t_1)\, H_\alpha(t_2) \ldots H_\alpha(t_n)], \end{aligned} \tag{7}$$

where P is the chronological ordering operator; this is the perturbation series we shall use.

Our interaction Hamilton can be written in the notation of field theory

$$H' = \frac{1}{2}\int \bar{\psi}(\mathbf{x}')\, \psi(\mathbf{x}')\, v(\mathbf{x}-\mathbf{x}')\, \bar{\psi}(\mathbf{x})\, \psi(\mathbf{x})\, d\mathbf{x}\, d\mathbf{x}' - \tfrac{1}{2}Nv(0), \tag{8}$$

where $v(\mathbf{x}-\mathbf{x}')$ is the mutual potential energy of two particles at $\mathbf{x}$ and $\mathbf{x}'$, N is the number of particles in the system, and $\bar{\psi}(\mathbf{x})$, $\psi(\mathbf{x})$ are the particle field operators. These can be written

$$\left.\begin{aligned} \psi(\mathbf{x}) &= \sum_i u_i(\mathbf{x})\, \eta_i, \\ \bar{\psi}(\mathbf{x}) &= \sum_i \bar{u}_i(\mathbf{x})\, \bar{\eta}_i, \end{aligned}\right\} \tag{9}$$

where the $u_i(\mathbf{x})$ are the eigenfunctions of the one-particle Hamiltonian,

$$\left[\frac{p^2}{2m} + U(\mathbf{x})\right] u_i(\mathbf{x}) = E_i\, u_i(\mathbf{x}), \tag{10}$$

in which $\mathbf{p}$ is the momentum and $U(\mathbf{x})$ the external potential. The operators $\bar{\eta}_i$ and η_i are creation and destruction operators for the particle in the state i; since our particles obey Fermi–Dirac statistics they satisfy the anticommutation relations

$$[\eta_i, \eta_j]_+ = [\bar{\eta}_i, \bar{\eta}_j]_+ = 0, \quad [\eta_i, \bar{\eta}_j]_+ = \delta_{ij}. \tag{11}$$

In (8) spinor indices are suppressed since their inclusion only requires a trivial generalization of the theory. The second term in (8) subtracts the self-energies of the particles due to the interaction v, since these are included together with the mutual interactions of the particles in the first term of (8). (If the potential is singular at the origin $v(0)$ is not defined. We can, however, easily introduce a suitable limiting procedure. If we Fourier transform v, $v(\mathbf{x}) = \int v(\mathbf{k})\, e^{i\mathbf{k}\cdot\mathbf{x}}\, d\mathbf{k}$, then we can work with $v_K(\mathbf{x}) = \int_{k<K} v(\mathbf{k})\, e^{i\mathbf{k}\cdot\mathbf{x}} d\mathbf{k}$, for which $v_K(0)$ is defined, and take the limit $K \to \infty$ at the end of the calculation.)

To obtain our perturbation series in a suitable form for computation we now have to substitute (8) into (5) and (5) into (7). Before doing this it is convenient to include the (constant) second term of (8) in H_0 and to symmetrize the first term; H' then becomes

$$H' = \frac{1}{4}\int v(\mathbf{x}-\mathbf{x}')\,[\overline{\psi}(\mathbf{x}')\,\psi(\mathbf{x}')\,\overline{\psi}(\mathbf{x})\,\psi(\mathbf{x})+\overline{\psi}(\mathbf{x})\,\psi(\mathbf{x})\,\overline{\psi}(\mathbf{x}')\,\psi(\mathbf{x}')]\,\mathrm{d}\mathbf{x}\,\mathrm{d}\mathbf{x}'. \quad (12)$$

Putting (12) into (5) we can easily obtain

$$H_\alpha(t) = \frac{1}{4}\int \mathrm{d}x\int \mathrm{d}x'\,v_\alpha(x-x')\,[\overline{\psi}(x')\,\psi(x')\,\overline{\psi}(x)\,\psi(x)+\overline{\psi}(x)\,\psi(x)\,\overline{\psi}(x')\,\psi(x')], \quad (13)$$

where x stands for $(\mathbf{x}, t)$ and x' for $(\mathbf{x}', t')$,

$$v_\alpha(x-x') = v(\mathbf{x}-\mathbf{x}')\,\delta(t-t')\,\mathrm{e}^{-\alpha|t|}, \quad (14)$$

and

$$\psi(x) = \psi(\mathbf{x},t) = \sum_i u_i(x)\,\eta_i = \sum_i u_i(\mathbf{x})\,\mathrm{e}^{-(\mathrm{i}t/\hbar)E_i}\,\eta_i,$$

$$\overline{\psi}(x) = \sum_i \overline{u}_i(x)\,\overline{\eta}_i. \quad (15)$$

Substituting (13) into (7) we obtain

$$S_\alpha(t) = 1+\sum_{n=1}^{\infty}\frac{1}{n!}\left(\frac{1}{4\mathrm{i}\hbar}\right)^n\int^t \mathrm{d}x_1\int^t \mathrm{d}x_2\ldots\int^t \mathrm{d}x_n\int^t \mathrm{d}x_1'\ldots\int^t \mathrm{d}x_n'$$
$$\times\, v_\alpha(x_1-x_1')\,v_\alpha(x_2-x_2')\ldots v_\alpha(x_n-x_n')$$
$$\times\, P[\overline{\psi}(x_1')\,\psi(x_1')\,\overline{\psi}(x_1)\,\psi(x_1)+\overline{\psi}(x_1)\,\psi(x_1)\,\overline{\psi}(x_1')\,\psi(x_1'),\ldots$$
$$\ldots,\overline{\psi}(x_n')\,\psi(x_n')\,\overline{\psi}(x_n)\,\psi(x_n)+\overline{\psi}(x_n)\,\psi(x_n)\,\overline{\psi}(x_n')\,\psi(x_n')], \quad (16)$$

where $\int^t \mathrm{d}x$ means the integral over all that part of space-time behind the surface t.

We can conveniently analyze the P product in (16) by making use of Wick's theorem (Wick 1950). This analysis can be simplified if we notice that we are interested only in the operation of S_α on the particular eigenstate Ψ_0 of H_0. We shall, henceforth, assume that Ψ_0 is a non-degenerate eigenstate of H_0 in which certain definite states i are occupied. Then we can conveniently take the state Ψ_0 as a 'redefined vacuum state' (Salam 1953) and resolve $\overline{\psi}$ and ψ according to

$$\psi = \psi^+ + \psi^-, \quad \overline{\psi} = \overline{\psi}^+ + \overline{\psi}^-, \quad (17)$$

where

$$\psi^-(x) = \sum_i^{\text{unocc.}} u_i(x)\,\eta_i \quad \text{(destroys particles)},$$

$$\psi^+(x) = \sum_i^{\text{occ.}} u_i(x)\,\eta_i \quad \text{(creates holes)},$$

$$\overline{\psi}^-(x) = \sum_i^{\text{occ.}} \overline{u}_i(x)\,\overline{\eta}_i \quad \text{(destroys holes)},$$

$$\overline{\psi}^+(x) = \sum_i^{\text{unocc.}} \overline{u}_i(x)\,\overline{\eta}_i \quad \text{(creates particles)}. \quad (18)$$

Here the terms occupied and unoccupied refer to the state of occupation of the state i in Ψ_0. We then have the result

$$\psi^-(x)\Psi_0 = \bar{\psi}^-(x)\Psi_0 = 0. \tag{19}$$

To take advantage of the result (19) we resolve the P product in (16) in such a way as to move all the operators ψ^+, $\bar{\psi}^+$ to the left and all the operators ψ^-, $\bar{\psi}^-$ to the right. This analysis can be performed quite straightforwardly using Wick's theorem. When this analysis has been completed we can reject all terms involving ψ^- and $\bar{\psi}^-$ by virtue of (19). Only one special point arises in this analysis. This is that, whereas in the case dealt with by Wick all operators with the same time argument either commuted or were already arranged as an S product, this is not so in the present case and one has to allow for contractions between operators with the same time argument.

When the analysis has been completed, it is found that the various terms contributing to $S_\alpha(t)\,\psi_0$ (the terms not involving ψ^- or $\bar{\psi}^-$) can be conveniently classified in terms of certain diagrams similar to Feynmann diagrams, there being a contribution to $S_\alpha(t)\,\psi_0$ corresponding to each diagram. We proceed at once to the prescription for drawing these diagrams and for calculating the corresponding contribution to $S_\alpha(t)$.

The diagrams will be of orders 1, 2, ..., corresponding to the contributions arising from different order terms of (16). The prescription for drawing an nth order diagram is as follows:

(i) Mark n points on the diagram and label these $x_1, x_2, \ldots, x_n$; mark a further n points and label with $x'_1, \ldots, x'_n$; join the pairs of points x_i, x'_i by 'interaction' lines (broken lines in the diagrams of this paper).

(ii) Draw directed 'particle' lines, one entering and one leaving each point; these lines may run between points or from a point to itself or from a point to the edge of the diagram or from the edge of the diagram to a point.

The different nth order diagrams are obtained by drawing in the particle lines in all possible ways. It will be noted that the particle lines form closed polygons and open polygonal arcs, so that to every particle line running inwards from the edge of the diagram there is one running out connected to it by a chain of particle lines.

The contributions to $S_\alpha(t)$ corresponding to a given diagram is a certain integral which can be written down from the following prescription:

(i) For every interaction line $x_i x'_i$ introduce a factor $v(x_i - x'_i)$ into the integrand.

(ii) For every particle line running from a point y(= some x_i or x'_i) to the edge of the diagram introduce a factor $\bar{\psi}^+(y)$ into the integrand, and for every particle line running from the edge of the diagram to a point y introduce a factor $\psi^+(y)$ into the integrand. The ψ^+ and $\bar{\psi}^+$ are to be arranged so that if $\psi^+(y)$ corresponds to the incoming line of one of the open polygonal arcs and $\bar{\psi}^+(z)$ to the corresponding outgoing line, then $\bar{\psi}^+(z)$ is adjacent to and on the left of $\psi^+(y)$.

(iii) For every particle line running from a point y to a point $z\,(\neq y)$ introduce a factor $S(z, y)$ into the integrand.

(iv) For every particle line running from a point y to itself introduce a factor $\rho(y)$ into the integrand.

(v) Integrate with respect to $dx_1 dx_2 \ldots dx_n dx'_1 \ldots dx'_n$ over the region of space-time behind the surface t.

(vi) Multiply the integral by $(2i\hbar)^{-n}(n!)^{-1}(-1)^p$, where p is the number of closed particle loops in the diagram.

The quantity $\rho(x)$ is just the charge density corresponding to Ψ_0, namely

$$\rho(x) = \sum^{\text{occ.}} \bar{u}_i(\mathbf{x})\, u_i(\mathbf{x}), \tag{20}$$

whilst the quantity $S(x', x)$ is a propagator given by

$$S(x', x) = \epsilon(t'-t) \sum_i^{\text{unocc.}} u_i(x')\,\bar{u}_i(x) - \epsilon(t-t') \sum_i^{\text{occ.}} u_i(x')\,\bar{u}_i(x), \tag{21}$$

where

$$\begin{aligned} \epsilon(t) &= 1 \quad \text{if} \quad t > 0 \\ &= 0 \text{ otherwise.} \end{aligned}$$

The function $S(x, x')$ is the solution of the equation

$$\left[i\hbar\frac{\partial}{\partial t} + \frac{\hbar^2}{2m}\nabla^2 - U(\mathbf{x})\right] S(x, x') = \delta(\mathbf{x}-\mathbf{x}')\,\delta(t-t'), \tag{22}$$

which reduces to $\frac{1}{2} - \rho(\mathbf{x}, \mathbf{x}')$ when $t = t'$, where $\rho(\mathbf{x}, \mathbf{x}')$ is the ordinary density matrix corresponding to the state Ψ_0:

$$\rho(\mathbf{x}, \mathbf{x}') = \sum_i^{\text{occ.}} u_i(\mathbf{x})\,\bar{u}_i(\mathbf{x}'). \tag{23}$$

As an example of the above prescriptions we give the contribution to $S_\alpha(t)$ corresponding to the diagram shown in figure 1 (a); it is

$$\begin{aligned} -\frac{1}{4!}\left(\frac{1}{2i\hbar}\right)^4 \int & dx_1\, dx_2\, dx_3\, dx_4\, dx'_1\, dx'_2\, dx'_3\, dx'_4 \; v_\alpha(x_1 - x'_1)\, v_\alpha(x_2 - x'_2)\, v_\alpha(x_3 - x'_3)\, v_\alpha(x_4 - x'_4) \\ & \times \bar{\psi}^+(x_4)\, \psi^+(x_1)\, \bar{\psi}^+(x'_3)\, \psi^+(x'_1)\, S(x_4, x'_4)\, S(x'_4, x_3)\, S(x_3, x_1)\, S(x'_3, x_2)\, S(x_2, x'_1)\, \rho(x'_2). \end{aligned} \tag{24}$$

3. The linked-cluster expansion

The linked-cluster expansion was first suggested by Brueckner (1955) and has been proved by Goldstone (1957) using the diagrammatic method of analysis of the perturbation series. The necessity for this result arises because the ordinary perturbation series for the energy, including that derived above, contain terms which diverge more strongly than N, the number of particles in the system, as $N \to \infty$. Such terms can have no physical significance and must cancel out against each other: we should, therefore, be able to eliminate them from the series, which is done in the linked-cluster expansion. This elimination will be carried out easily and naturally in this section using the diagrammatic analysis.

Before proceeding to develop the linked-cluster expansion it is first convenient to classify the diagrams in a certain way. We shall say that two diagrams belong to the same class if they have the same basic structure, i.e. if they have the same arrangement of vertices, interaction and particle lines and differ only in the labelling of their vertices; for example, the diagrams shown in figures 1 (a) and (b) belong to the same class since they have the same basic structure shown in figure 1 (*c*). It is seen that each class is associated with a certain basic structure.

It can be seen from the prescription of § 2 that if two diagrams G, G' belong to the same class their contributions to $S_\alpha(t)$ are integrals which differ only by a permutation of the variables of integration, and are therefore equal. Thus all the diagrams of the same class give an equal contribution, and the contribution of the whole class is the contribution of a typical member multiplied by the number of diagrams in the class.

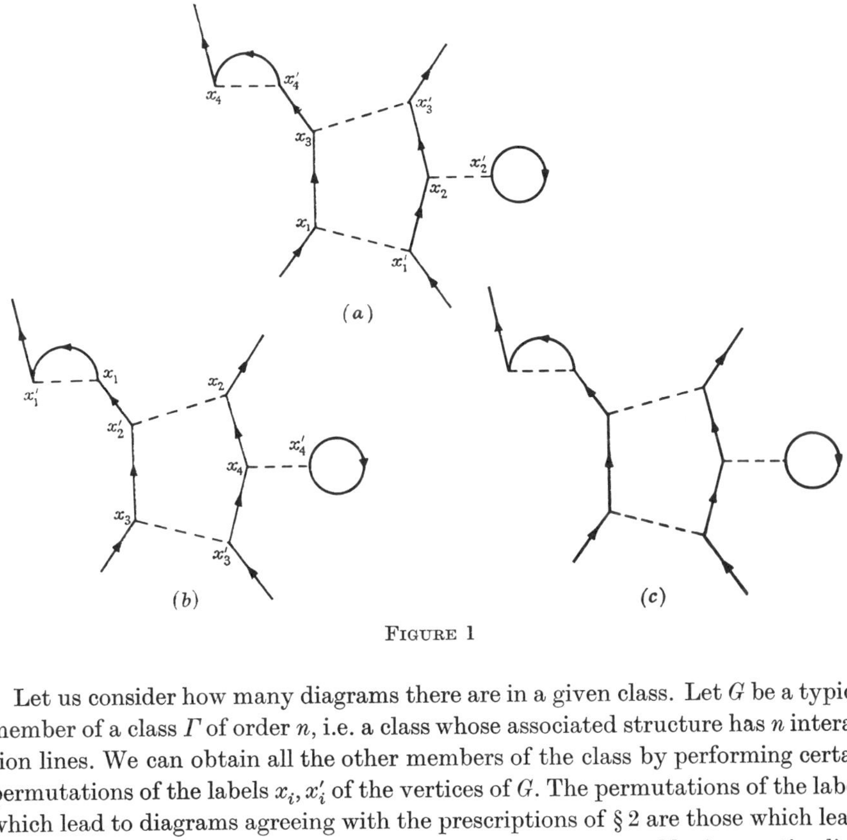

FIGURE 1

Let us consider how many diagrams there are in a given class. Let G be a typical member of a class Γ of order n, i.e. a class whose associated structure has n interaction lines. We can obtain all the other members of the class by performing certain permutations of the labels x_i, x_i' of the vertices of G. The permutations of the labels which lead to diagrams agreeing with the prescriptions of § 2 are those which leave the pairs x_i, x_i' connected by interaction lines in G still connected by interaction lines after the permutation. The only permutations which do this can be built up from the following types of permutation:

(i) simultaneous permutations of the x_i and x_i' in the same way;

(ii) the interchange of any pair x_i, x_i'.

The number of distinct permutations which can be built up from these is $2^n n!$. If the application of every one of these permutations to the labels of G led to a diagram distinct from G, then this would be the number of diagrams in the class. However, it may be that the application of some of these permutations to G leads to diagrams which are not topologically distinct from G; for example, the diagrams shown in figures 2(*a*) and (*b*) which are obtainable from one another by a permutation of labels are topologically equivalent and must not be counted separately. Suppose the number of permutations which take G into diagrams topologically equivalent to itself is $g(\Gamma)$ (this number is a function of the structure Γ rather than of the particular diagram); then we can easily show that the number of diagrams in the class is in fact $2^n n!/g(\Gamma)$.

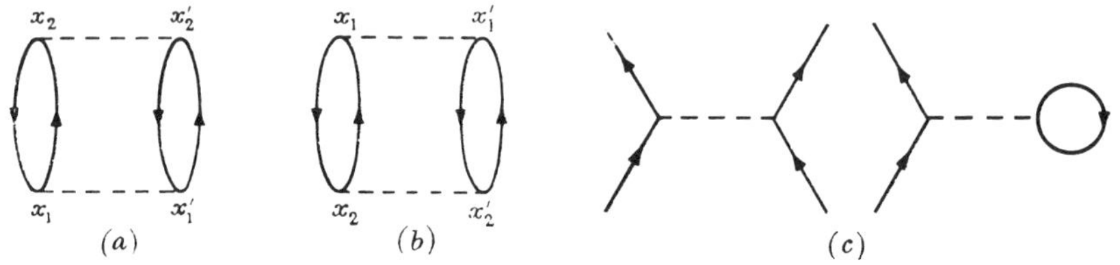

FIGURE 2

If we denote the contribution of a diagram G to $S_\alpha(t)$ by $S_\alpha(t, G)$ and the contribution of the class Γ to which G belongs by $S_\alpha(t, \Gamma)$, then we have

$$S_\alpha(t, \Gamma) = \frac{2^n n!}{g(\Gamma)} S_\alpha(t, G). \tag{25}$$

We can now develop the linked-cluster expansion. A structure Γ may or may not fall into two or more unconnected parts; in the former case we shall say that it is an unlinked, in the latter case linked. For example, the structure shown in figure 1 (*c*) is linked whilst that shown in figure 2(*c*) is unlinked. It should be noticed that this definition of linked and unlinked diagrams is not quite the same as that used by Goldstone (1957).

An unlinked structure Γ can be resolved into a set of linked structures; if the unlinked structure Γ is made up of p_1 linked structures Γ_1, p_2 linked structures Γ_2 etc., we shall write

$$\Gamma = p_1 \Gamma_1 + p_2 \Gamma_2 + \dots.$$

Let G be a typical diagram with the structure Γ. Using the prescription of § 2 we can easily prove that

$$S_\alpha(t, G) = \frac{1}{n!} [n_1!\, S_\alpha(t, G_1)]^{p_1} [n_2!\, S_\alpha(t, G_2)]^{p_2} \dots,$$

where $G_1, G_2 \dots$ are typical diagrams with the structures $\Gamma_1, \Gamma_2, \dots$, n is the order of Γ, and $n_1, n_2 \dots$ are the orders of Γ_1, Γ_2, etc. Using (25) we obtain

$$S_\alpha(t, \Gamma) = \frac{1}{g(\Gamma)} [g(\Gamma_1)\, S_\alpha(t, \Gamma_1)]^{p_1} [g(\Gamma_2)\, S_\alpha(t, \Gamma_2)]^{p_2} \dots. \tag{26}$$

Finally it can be seen that

$$g(\Gamma) = p_1!\, p_2! \dots [g(\Gamma_1)]^{p_1} [g(\Gamma_2)]^{p_2} \dots, \tag{27}$$

so that

$$S_\alpha(t, \Gamma) = \frac{1}{p_1!\, p_2! \dots} [S_\alpha(t, \Gamma_1)]^{p_1} [S_\alpha(t, \Gamma_2)]^{p_2} \dots. \tag{28}$$

Let $\Gamma_1, \Gamma_2, \ldots$ be the set of all linked structures; if in (26) we allow $p_1, p_2, \ldots$ to run over all the values $0, 1, 2, \ldots$, then we obtain all possible structures. Since $S_\alpha(t)$ is the sum of the contributions from all possible structures we have

$$\begin{aligned} S_\alpha(t) &= \sum_{p_1=0}^{\infty} \sum_{p_2=0}^{\infty} \cdots \frac{1}{p_1!\,p_2!\ldots}[S_\alpha(t,\Gamma_1)]^{p_1}[S_\alpha(t,\Gamma_2)]^{p_2}\ldots \\ &= \exp\{S_\alpha(t,\Gamma_1)\}\exp\{S_\alpha(t,\Gamma_2)\}\ldots \\ &= \exp\{S_{L\alpha}(t)\}, \end{aligned} \tag{29}$$

where

$$S_{L\alpha}(t) = \sum_{\Gamma}^{\text{linked}} S_\alpha(t,\Gamma), \tag{30}$$

and we have made use of the fact that the various $S_\alpha(t, G)$ commute with each other; this is because all the ψ^+ and $\bar{\psi}^+$ operators anticommute (see equation (18)), and each $S_\alpha(t, G)$ contains an even number of these. Thus we see that $S_\alpha(t)$ can be expressed in terms of $S_{L\alpha}$, the sum of the contributions from linked diagrams.

Let us now resolve $S_{L\alpha}(t)$ into two parts

$$S_{L\alpha}(t) = S_{L\alpha}^{(0)}(t) + S'_{L\alpha}(t), \tag{31}$$

$S_{L\alpha}^{(0)}(t)$ containing the contributions from all linked 'vacuum' diagrams, that is, diagrams with no particle lines running to or from the edge of the diagram, and $S'_{L\alpha}(t)$ the contributions from all linked diagrams with external lines; $S_{L\alpha}^{(0)}(t)$ contains no operators and is a c number.

Substituting (31) into (29) and (29) into (2) and dividing out by the c number $\exp\{S_{L\alpha}^{(0)}(0)\}$, we obtain

$$\Psi = \lim_{\alpha\to+0} \exp\{S'_{L\alpha}(0)\}\Psi_0/(\Psi_0\,|\exp\{S'_{L\alpha}(0)\}\,|\,\Psi_0). \tag{32}$$

It will be shown in the next section that $S'_{L\alpha}(t)$ is continuous as $\alpha \to +0$ whilst $S_{\alpha L}^{(0)}(t)$ diverges like $1/\alpha$; thus, dropping a normalization factor,

$$\Psi = \exp\{S'_L(0)\}\Psi, \tag{33}$$

where

$$S'_L(0) = \lim_{\alpha\to+0} S'_{L\alpha}(0).$$

We shall see in the next section that the energy shift ΔE can be derived from $S_{L\alpha}^{(0)}(t)$, i.e. from a series involving only the linked terms; this series does not contain terms diverging more strongly than N and so is free from the difficulty mentioned at the beginning of this section.

4. The evaluation of integrals

In this section we perform as an illustration the evaluation of the integral representing the contribution of a simple diagram to $S_\alpha(0)$. The result obtained is typical of the general case, and we can deduce from it certain properties of the $S_\alpha(0, G)$; furthermore, these results afford a link between the present formalism and that of Goldstone.

The diagram whose evaluation we shall perform is that shown in figure 3 (*a*). The corresponding contribution to $S_\alpha(0)$ is according to the prescription of § 2

$$\frac{1}{2}\left(\frac{1}{2\mathrm{i}\hbar}\right)^2\int_{-\infty}^{0}\mathrm{d}t_1\int_{-\infty}^{0}\mathrm{d}t_2\int\mathrm{d}\mathbf{x}_1\,\mathrm{d}\mathbf{x}_2\,\mathrm{d}\mathbf{x}_1'\,\mathrm{d}\mathbf{x}_2'\,v(\mathbf{x}_1-\mathbf{x}_1')\,v(\mathbf{x}_2-\mathbf{x}_2')\,\mathrm{e}^{\alpha(t_1+t_2)}$$
$$\times S(\mathbf{x}_2,t_2;\,\mathbf{x}_1,t_1)\,S(\mathbf{x}_2',t_2;\,\mathbf{x}_1',t_1)\,\overline{\psi}^+(\mathbf{x}_2,t_2)\,\psi^+(\mathbf{x}_1,t_1)\,\overline{\psi}^+(\mathbf{x}_2',t_2)\,\psi^+(\mathbf{x}_1',t_1), \quad (34)$$

where we have already performed the integrations over t_1' and t_2'. Because of the nature of the function S (equation (21)), it is convenient to divide the integration

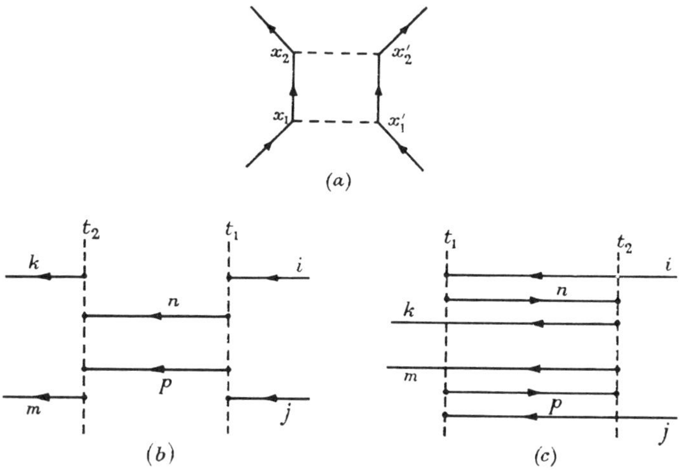

FIGURE 3

over t_1 and t_2 into two parts, one part arising from the region in which $t_2 > t_1$ and the other from the region in which $t_2 < t_1$. Substituting for S from (21) and for $\overline{\psi}^+$, ψ^+, from (18), we have for the first part of the integral

$$\frac{1}{2}\left(\frac{1}{2\mathrm{i}\hbar}\right)^2\sum_{i,j}^{\text{occ.}}\sum_{kmnp}^{\text{unocc.}}\int\overline{u}_k(\mathbf{x}_2)\,\overline{u}_m(\mathbf{x}_2')\,v(\mathbf{x}_2-\mathbf{x}_2')\,u_n(\mathbf{x}_2)\,u_p(\mathbf{x}_2')\,\mathrm{d}\mathbf{x}_2\,\mathrm{d}\mathbf{x}_2'$$
$$\times\int\overline{u}_n(\mathbf{x}_1)\,\overline{u}_p(\mathbf{x}_1')\,v(\mathbf{x}_1-\mathbf{x}_1')\,u_i(\mathbf{x}_1)\,u_j(\mathbf{x}_1')\,\mathrm{d}\mathbf{x}_1\,\mathrm{d}\mathbf{x}_1'\int_{-\infty}^{0}\mathrm{d}t_2\int_{-\infty}^{t_2}\mathrm{d}t_1$$
$$\times\exp\left\{\alpha(t_1+t_2)+\frac{\mathrm{i}}{\hbar}(E_k+E_m-E_n-E_p)\,t_2+\frac{\mathrm{i}}{\hbar}(E_n+E_p-E_i-E_j)\,t_1\right\}\overline{\eta}_k\eta_i\overline{\eta}_m\eta_j$$
$$=\frac{1}{8}\sum_{ij}^{\text{occ.}}\sum_{kmnp}^{\text{unocc.}}\frac{(km\,|\,v\,|\,np)\,(np\,|\,v\,|\,ij)}{(E_k+E_m-E_i-E_j+2\mathrm{i}\hbar\alpha)\,(E_n+E_p-E_i-E_j+\mathrm{i}\hbar\alpha)}\overline{\eta}_k\eta_i\overline{\eta}_m\eta_j, \quad (35)$$

where
$$(km\,|\,v\,|\,np)=\int\overline{u}_k(\mathbf{x})\,\overline{u}_m(\mathbf{x}')\,v(\mathbf{x}-\mathbf{x}')\,u_n(\mathbf{x})\,u_p(\mathbf{x}')\,\mathrm{d}\mathbf{x}\,\mathrm{d}\mathbf{x}'. \quad (36)$$

A similar evaluation of the second part of the integral gives

$$\frac{1}{8}\sum_{ijnp}^{\text{occ.}}\sum_{km}^{\text{unocc.}}\frac{(km\,|\,v\,|\,np)\,(np\,|\,v\,|\,ij)}{(E_k+E_m-E_i-E_j+2\mathrm{i}\hbar\alpha)\,(E_k+E_m-E_n-E_p+\mathrm{i}\hbar\alpha)}\overline{\eta}_k\eta_i\overline{\eta}_m\eta_j. \quad (37)$$

The similarity of these results to the terms of the ordinary perturbation series is at once recognizable.

The two terms (35) and (37) may be represented diagrammatically as in figures 3 (*b*), (*c*); these diagrams are obtained from (*a*) by arranging t_1 and t_2 in particular orders and labelling the lines rather than the vertices. Figure 3 (*b*) represents a process in which two particles in states i and j interact and scatter into states n and p, and re-interact and scatter into states m and k; thus the whole process is a scattering of particles out of the states i and j into the states k and m via the intermediate state n, p. Similarly, figure 3 (*c*) represents a process in which two particles scatter themselves into states k and m leaving holes in the states n and p into which the two particles in states i and j then scatter themselves; thus the final state is the same in both cases but the intermediate state is different. It will be seen that in each case the expressions (35) and (37) contain in their denominators the difference in energy of the final and initial state and the difference in energy of the initial and intermediate state.

These results are quite general. Any of the integrals representing contributions to $S_\alpha(0)$ can be evaluated by the above method, and leads to a series of terms of the form of (35) which can be interpreted as representing certain physical processes with the help of diagrams of the type of figures 3 (*b*), (*c*). In each case the denominators of the expression will contain the difference in energy of the final and initial state and the differences in energy between the initial and intermediate states.

We can now make certain deductions from these results. The first thing to notice is that if an expression of the form of (35) arises from a linked diagram, and in view of the results of the preceding section we need only consider such diagrams, then none of the intermediate states can coincide with the initial state. Since the latter has been assumed to be non-degenerate, it follows that the energy of none of the intermediate states can coincide with that of the initial state. If, further, the final state is different from the initial state, as it is in all the terms contributing to $S'_{L\alpha}(0)$, then the energy of the final state will be different from that of the initial state. Thus, in the case of terms contributing to $S'_{L\alpha}(0)$, none of the energy differences in the denominators is zero; it follows at once (see (35)) that each of these terms is continuous as $\alpha \to +0$, and that $S'_{L\alpha}(0)$ is continuous $\alpha \to +0$. Thus $\lim S'_{L\alpha}(0)$ as $\alpha \to +0$ exists, and may be evaluated putting $\alpha = 0$ at the beginning of the calculation.

The situation with $S^{(0)}_{L\alpha}(0)$ is different, however. In the case of terms contributing to $S^{(0)}_{L\alpha}(0)$ the initial and final states coincide (although the intermediate states are different from the initial state), and the corresponding energy difference vanishes giving rise to a factor α in the denominator. Thus $S^{(0)}_{L\alpha}(0)$ diverges like $1/\alpha$ as $\alpha \to +0$.

We should now like to consider the calculation of the energy shift ΔE. This is given by (3). It may, however, be more conveniently calculated using a formula given by Gell-Mann & Low; it is shown in an appendix that it can be derived from $S^{(0)}_{L\alpha}(\infty)$, and is given by the following prescription: *write down the sum of all integrals contributing to* $S^{(0)}_{L\alpha}(\infty)$ (*one from each vacuum diagram*); *introduce into the integrand of each of these integrals a factor* $i\hbar\delta(t_1)$ *and put* $\alpha = 0$; *the resulting sum gives* ΔE.

In the remainder of this paper we shall devote our attention to the calculation of ΔE.

5. The uniform gas case

The theory of the preceding sections was general in the sense that H_0 was supposed to include not only the kinetic energy of the particles but also their potential energy in an external field. In the remaining sections of this paper we shall for simplicity restrict ourselves to the case in which the external field is constant and the particles form a uniform gas; it is hoped to deal with the general case later.

In the case of a uniform gas it is convenient to resolve the H' of (8) into two parts. Suppose the Fourier transform of v is

$$v(\mathbf{x}) = \sum_{\mathbf{k}} u(\mathbf{k})\, e^{i\mathbf{k}.\mathbf{x}}. \tag{38}$$

Then we split v into two parts according to

$$v(\mathbf{x}) = \sum_{k\neq 0} u(\mathbf{k})\, e^{i\mathbf{k}.\mathbf{x}} + u(0) = v'(\mathbf{x}) + u(0). \tag{39}$$

We can correspondingly write H' in the form

$$H' = \frac{1}{2}\int d\mathbf{x}\, d\mathbf{x}'\, \bar{\psi}(\mathbf{x}')\, \psi(\mathbf{x}')\, v'(\mathbf{x}-\mathbf{x}')\, \bar{\psi}(\mathbf{x})\, \psi(\mathbf{x})\, d\mathbf{x}\, d\mathbf{x}' + \tfrac{1}{2}N^2 u(0) - \tfrac{1}{2}N v(0) \tag{40}$$

and include the last two (constant) terms in H_0. All the above theory then goes through as before except that v is replaced by v' everywhere. The potential v' has the useful property

$$\int v'(\mathbf{x}-\mathbf{x}')\, d\mathbf{x}' = 0. \tag{41}$$

We can now make an important deduction concerning the perturbation series in the case of a uniform gas. This is that *in the case of a uniform gas those diagrams which contain a part which is attached to the rest of the diagram by only a single interaction line give no contribution $S_\alpha(t)$ or any derived quantities.* To prove this let us consider the contribution of a diagram of the type shown in figure 4 (a), where the two parts Γ' and Γ'' are connected by only the single interaction line shown. If in the integral representing the contribution of this diagram we perform all the integrations except those over $\mathbf{x}$ and $\mathbf{x}'$, we must obtain an expression of the form

$$\int F(\mathbf{x})\, v'(\mathbf{x}-\mathbf{x}')\, G(\mathbf{x}')\, d\mathbf{x}\, d\mathbf{x}'. \tag{42}$$

However, since the gas is uniform and has no natural origin of co-ordinates, F and G must be independent of $\mathbf{x}$ and $\mathbf{x}'$, so that it reduces to an integral over $v'(\mathbf{x}-\mathbf{x}')$ which must vanish by virtue of (41). Thus we can omit all diagrams of this type.

In future we shall drop the prime on v', it being always understood that we are working with v'.

6. The elimination of polarization parts

As pointed out in the introduction, the main point of introducing the diagrammatic analysis in the present theory is that it enables us easily to recognize those parts of the perturbation series which represent polarization effects. We are now in a position to investigate this.

Let us consider the physical interpretation of the *subdiagram* shown in figure 4 (b). The net effect of this subdiagram is that the two incoming particles scatter off each other, not through the ordinary direct interaction but via a closed particle loop. More generally, the subdiagram shown in figure 4 (c), where the circle is meant to represent some set of closed particle loops and interaction lines connected to the rest of the diagram by only the two interaction lines shown, represents the scattering of two particles through some more complicated process which, however, returns all the particles involved to their original states. Our hypothesis, which will be borne out by further calculation, is that these subdiagrams represent the polarization effect referred to in the introduction, and that they can be eliminated from the perturbation series by replacing the ordinary interaction by a modified interaction.

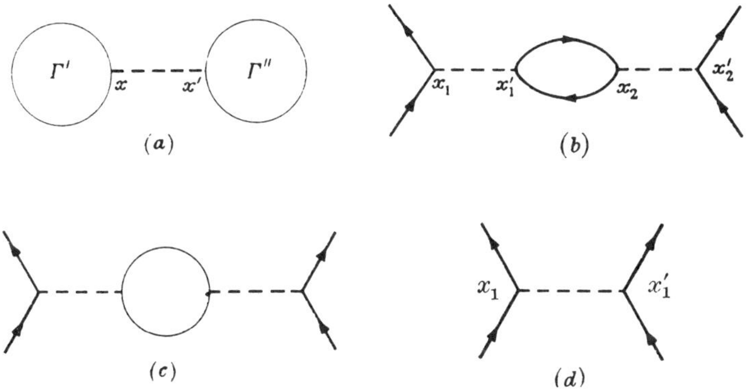

FIGURE 4

The way in which such subdiagrams can be eliminated from the perturbation series is already familiar in quantum electrodynamics, being exactly analogous to the elimination of photon self-energy parts (Dyson 1949). Let Γ be some structure. Then it may or may not be that Γ contains some polarization part, i.e. some connected part without external lines attached to the rest of the structure by only two interaction lines; in the former case we shall say that Γ is (polarization) reducible, in the latter case that Γ is irreducible. It is obvious that we can obtain all reducible structures by inserting polarization parts in place of interaction lines in irreducible structures.

Let us consider as an example the diagram shown in figure 4 (b), regarding this now as a complete diagram rather than as a subdiagram of some larger diagram. This is a reducible diagram contributing to $S'_L(t)$, obtainable from the irreducible diagram shown in figure 4 (d) by substituting the polarization part shown in figure 6 (a) for its interaction line. The contributions to $S'_L(t)$ of the diagrams of figures 4 (b), (d) are

$$\frac{1}{2!}\left(\frac{1}{2i\hbar}\right)^2\int^t dx_1\int^t dx_1'\int^t dx_2\int^t dx_2'\, v(x_1-x_1')\, v(x_2-x_2')$$
$$\times\,\overline{\psi}^+(x_1)\,\psi^+(x_1)\,\overline{\psi}^+(x_2')\,\psi^+(x_2')\,S(x_1',x_2)\,S(x_2,x_1') \quad (43)$$

and
$$\frac{1}{1!}\left(\frac{1}{2i\hbar}\right)\int^t dx_1\int^t dx_1'\, v(x_1-x_1')\,\overline{\psi}^+(x_1)\,\psi^+(x_1)\,\overline{\psi}^+(x_1')\,\psi^+(x_1'). \quad (44)$$

It will be seen that (43) is obtainable from (44) (apart from an unimportant change in the integration variable) by:

(i) multiplying by $1!/2!$, the ratio of the factorials of the orders of the diagrams;

(ii) by replacing $v(x-x')$ in (44) by the quantity

$$\frac{1}{2i\hbar}\int^t dx\int^t dx'\, v(x_1-x)\, v(x_1'-x')\, S(x,x')\, S(x',x) \tag{45}$$

which is just the contribution to (43) of the interaction and particle lines of the polarization part of figure 6 (a) with the integrations at the vertices performed. This result is typical; the contribution of a reducible diagram is the same as that of the corresponding irreducible diagram except that the v's are replaced by functions of the type (45) corresponding to the various polarization parts which have been inserted in place of interaction lines, and the whole integral has been multiplied by the ratio of the factorials of the orders.

We can enunciate this result more precisely as follows. If G is a reducible diagram obtained by substituting the polarization part Γ'_{s_1} for the interaction line x_1x_1', Γ'_{s_2} for x_2x_2', etc., of the irreducible diagram G' (introducing the convention that if the line x_ix_i' of G' is left unchanged we say it has been replaced by the polarization part Γ_0'), then

$$S(t,G) = \frac{n'!}{n!}S(t,G';\; W_t'(\Gamma'_{s_1}),\; W_t'(\Gamma'_{s_2}),\ldots), \tag{46}$$

where n and n' are the orders of G and G', $S(t,G';\; W_t'(\Gamma'_{s_1}),\; W_t'(\Gamma'_{s_2}),\ldots)$ means that integral for $S(t,G')$ with the $v(x_i-x_i')$ replaced by the functions $W_t'(x_i',x_i,\Gamma'_{si})$, and $W_t'(x',x,\Gamma')$ is the expression which arises from the polarization part Γ' in the same way that (45) arises from the polarization part of figure 6 (a), remembering that in the case of the special polarization part Γ_0 this is just $v(x-x')$.

Let Γ be a reducible structure contributing to $S_L'(t)$, i.e. with external lines. Then it can be seen that Γ arises from some unique irreducible structure Γ'. Consider the total contribution to $S_L'(t)$ of all the diagrams with structures which reduce to a given irreducible structure Γ'. A straightforward counting of diagrams shows now that this total contribution is given by the expression

$$\frac{n'!\,2^{n'}}{g(\Gamma')}\sum_{\Gamma'_{s_1}}\sum_{\Gamma'_{s_2}}\ldots S(t,G';W_t(\Gamma'_{s_1}),W_t(\Gamma'_{s_2}),\ldots), \tag{47}$$

where the sums run over all polarization parts, and

$$W_t(x',x,\Gamma') = \frac{2^m}{g(\Gamma')}\,W_t'(x',x,\Gamma'), \tag{48}$$

where $g(\Gamma')$ is the g factor of the polarization part Γ', defined to be the number of permutations of the *internal* interaction lines of Γ' which take it into itself (these g factors are exactly analogous to those of §3), and m is its order, defined to be the number of *internal* interaction lines of Γ' plus one.

Since $S(t,G';\; W_t(\Gamma'_{s_1}),\; W_t(\Gamma'_{s_2}),\ldots)$ depends linearly upon the $W_t(\Gamma'_s)$ we can write (47) as

$$S(t,\Gamma';\mathscr{V}_t,\mathscr{V}_t,\ldots), \tag{49}$$

using (25), where

$$\mathscr{V}_t(x',x) = \sum_{\Gamma'} W_t(x',x,\Gamma'_s). \tag{50}$$

Thus, we see that the contribution to $S_L(t)$ of all diagrams reducing to the structure Γ' can be expressed in terms of the contribution of the class Γ' by replacing v everywhere by $\mathscr{V}_t$. Since every diagram contributing to $S_L(t)$ is uniquely reducible, we see that $S_L(t)$ can be expressed as a sum over contributions from irreducible diagrams by replacing v by $\mathscr{V}_t$ in all the integrals. Thus $\mathscr{V}_t$ plays the part of a modified interaction. In order to calculate the perturbed wave function we shall be interested in $\mathscr{V}_0$.

We have shown above how the perturbation series for the perturbed wave function can be contracted by introducing a modified interaction. Let us now consider the perturbation series for the energy shift ΔE. This has been expressed as a sum over vacuum diagrams in §4. We could proceed as above and reduce this to a sum over irreducible vacuum diagrams by replacing v everywhere by $\mathscr{V}_\infty$ (since the time integrations run up to $t = \infty$ in the case of ΔE) except for one feature. Whereas every diagram contributing to $S'_L(t)$ is uniquely reducible, this is not true for vacuum diagrams in general. For example, the diagram shown in figure 5 (*a*) could be reduced to either of the diagrams shown in figures 5 (*b*) or (*c*) by regarding either the right hand or the left hand part as the polarization part.

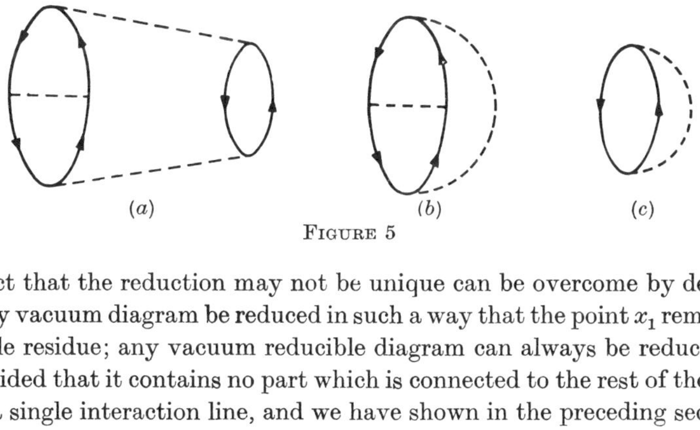

(*a*) (*b*) (*c*)

FIGURE 5

The fact that the reduction may not be unique can be overcome by demanding that every vacuum diagram be reduced in such a way that the point x_1 remain in the irreducible residue; any vacuum reducible diagram can always be reduced in this way provided that it contains no part which is connected to the rest of the diagram by only a single interaction line, and we have shown in the preceding section that such diagrams can be omitted. Furthermore, it is necessary to reduce the diagrams contributing to ΔE in this way, otherwise the factor $i\hbar\delta(t_1)$ would appear in some polarization part and spoil the theory.

Taking this point into account, we find by counting diagrams that the total contribution to ΔE of all diagrams which reduce to a given irreducible structure Γ' is

$$\frac{2^{n'}n'!}{g(\Gamma')}\sum_{\Gamma_{s_1}}\sum_{\Gamma_{s_2}}\cdots\frac{n'}{n'+n_{s_1}+n_{s_2}+\ldots}\Delta E(G'; W_\infty(\Gamma'_{s_1}), W_\infty(\Gamma'_{s_2}), \ldots), \tag{51}$$

where G' is a typical diagram with the structure Γ' and n_s is the order of the polarization part Γ_s. Because of the factor $(n'+n_{s_1}+n_{s_2}+\ldots)^{-1}$, we cannot perform the sum in the simple way we did in (47). However, we notice that if we regard the interaction v as being linearly proportional to some coupling constant λ, then $\Delta E(G'; W_\infty(\Gamma'_{s_1}), W_\infty(\Gamma'_{s_2}), \ldots)$ varies as $\lambda^{n'+n_{s_1}+n_{s_2}+\cdots}$. Thus we can write (51) as

$$\int_0^\lambda \frac{d\lambda}{\lambda}\frac{2^{n'}n'!}{g(\Gamma')}\sum_{\Gamma'_{s_1}}\sum_{\Gamma_{s_2}}\ldots n'\,\Delta E\,(G'; W_\infty(\Gamma'_{s_1}), W_\infty(\Gamma'_{s_2})\ldots). \tag{52}$$

We can now perform the sum as we did in (47) and obtain

$$\int_0^\lambda \frac{d\lambda}{\lambda} \frac{2^{n'} n'!}{g(\Gamma')} n' \Delta E(G'; \mathscr{V}_\infty, \mathscr{V}_\infty \ldots), \tag{53}$$

enabling us to express ΔE as a sum over irreducible diagrams. In this case we replace v by $\mathscr{V}_\infty$ (which will be written $\mathscr{V}$ for short in future); it is this quantity $\mathscr{V}_\infty$ that we regard as the real modified interaction rather than the more general $\mathscr{V}_t$, since in the calculation of all observable quantities, such as ΔE, $\mathscr{V}_\infty$ will turn up rather than $\mathscr{V}_t$.

7. An integral equation for $\mathscr{V}$

In the preceding section we have introduced the modified interaction and expressed it in terms of an infinite series. In this section we should like to consider some of its properties.

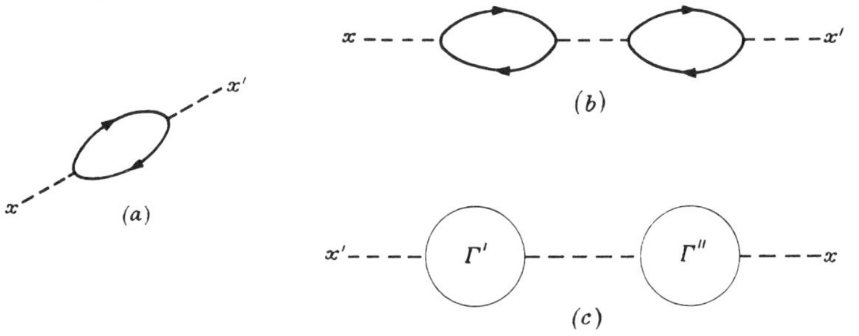

FIGURE 6

We first notice that in the case of a uniform gas $\mathscr{V}(x', x)$ is a function of $x - x'$ only. For so far as the space co-ordinates are concerned it can only be a function of $\mathbf{x} - \mathbf{x}'$ because of the spatial homogeneity of the system. Also each of the integrals in the series for $\mathscr{V}$ (equation (50) with $t = \infty$) is an integral whose time integrations run from $-\infty$ to $+\infty$ and whose integrands are products of factors each of which depends only upon time differences (see e.g. (45)); consequently $\mathscr{V}(x', x)$ can depend only upon $t' - t$, and is therefore a function of $x' - x$. It should further be noticed that there is no reason why $\mathscr{V}$ should vanish when $t \neq t'$, so that the modified interaction $\mathscr{V}$ will not in general be an instantaneous interaction like v but a sort of retarded interaction.

We shall now see how we can set up an integral equation for $\mathscr{V}$ which will enable us to express it in terms of a series more rapidly convergent than (50). Let Γ be any polarization structure. Then it may or may not be that Γ consists of two or more parts which are only connected by single interaction lines. In the first case we shall say that Γ is an improper polarization structure, in the latter case that it is a proper polarization structure. For example, the polarization structure shown in figure 6(*a*) is a proper structure whilst that shown in figure 6(*b*) is an improper one.

Let Γ be some improper polarization structure. Then Γ can be uniquely resolved as shown in figure 6(*c*) into a proper structure Γ' and some other polarization structure Γ'', proper or improper. Let us introduce as follows the quantity $\overline{W}(x', x, \Gamma)$

related to $W_\infty(\Gamma)$ (to be written $W(\Gamma)$ for brevity in future). It will be seen from the prescription of the previous section for $W(\Gamma)$ that it can be written in the form

$$W(x', x, \Gamma) = \int v(x' - x'')\, \overline{W}(x'', x''', \Gamma)\, v(x''' - x)\, dx''\, dx''', \tag{54}$$

the two v's being those corresponding to the two outgoing lines of the polarization structure, and $\overline{W}$ containing the contribution from all the internal interaction and particle lines with the appropriate integrations performed. For example, in the case of the polarization part shown in figure 6(*a*) (see (45)) $\overline{W}$ will be

$$\overline{W}(x', x) = \frac{1}{i\hbar} S(x', x)\, S(x, x'). \tag{55}$$

We can now prove that the contribution of the polarization part resolved as in figure 6(*c*) can be written

$$W(x', x, \Gamma) = \int v(x' - x_1)\, \overline{W}(x_1, x_2, \Gamma')\, v(x_2 - x_3)\, \overline{W}(x_3, x_4, \Gamma'')\, v(x_4 - x)\, dx_1\, dx_2\, dx_3\, dx_4. \tag{56}$$

If we introduce the quantity

$$W^*(x', x, \Gamma') = \int v(x' - x'')\, \overline{W}(x'', x, \Gamma')\, dx'', \tag{57}$$

we can write (56) as

$$W(x', x, \Gamma) = \int W^*(x', x'', \Gamma')\, W(x'', x, \Gamma'')\, dx''. \tag{58}$$

Let us now consider the contribution to $\mathscr{V}$ of all those improper polarization structures which give rise to a given proper structure Γ' when resolved as in figure 6(*c*). This is evidently obtained by summing over all structures Γ'' apart from Γ_0 on the right-hand side of (58), and is given by

$$\int W^*(x', x'', \Gamma') \sum_{\Gamma'' \neq \Gamma_0} W(x'', x, \Gamma'')\, dx'' \tag{59}$$

If we add to this the contribution of Γ' itself to $\mathscr{V}$, which from (54) and (57) is given by

$$\int W^*(x', x'', \Gamma')\, v(x'' - x)\, dx, \tag{60}$$

we obtain

$$\int W^*(x', x'', \Gamma')\, [v(x'' - x') + \sum_{\Gamma'' \neq \Gamma_0} W(x'', x, \Gamma'')]\, dx'' = \int W^*(x', x'', \Gamma')\, \mathscr{V}(x'', x)\, dx'' \tag{61}$$

for the total contribution to $\mathscr{V}$ of the proper diagram Γ' and all the improper ones which give Γ' on resolution. To obtain $\mathscr{V}$ we now have to sum this expression over all proper polarization parts Γ' and add the contribution of the polarization part Γ_0 which is just $v(x - x')$. Defining

$$\mathscr{V}^*(x', x) = \sum_{\Gamma' \neq \Gamma_0}^{\text{proper}} W(x', x, \Gamma'), \tag{62}$$

we have

$$\mathscr{V}(x', x) = v(x' - x) + \int \mathscr{V}^*(x', x'')\, \mathscr{V}(x'', x)\, dx''. \tag{63}$$

Thus we have derived an integral equation which gives $\mathscr{V}$ in terms of $\mathscr{V}^*$. This is desirable because the series (62) is very much more rapidly convergent than (50). The solution of (63) presents no particular difficulty in the uniform gas because all the quantities involved depend upon the differences in their arguments and the equation may be solved by Fourier transformation.

We should like to complete this section with a comment on the physical interpretation of the result (62). The modified interaction between two particles can be regarded as a superposition of their direct interaction and the interaction of each with the polarization field of the other. These two interactions are represented by the first and second terms on the right-hand side of (62). The polarization produced at any point by one of the particles depends, however, not upon the direct field of the particle at that point, but upon the modified field; this is represented in (63) by the dependence of the second term upon $\mathscr{V}$, $\mathscr{V}^*(x', x'')$ representing the field at x' due to the polarization at the point x''.

8. An expression for ΔE in terms of V

We have seen in §6 how ΔE can be expressed as a sum of terms corresponding to irreducible vacuum diagrams. Our result can be written

$$\Delta E = \sum_{G}^{\substack{\text{irreducible}\\ \text{vacuum}}} \int_0^\lambda \frac{d\lambda}{\lambda} n(G)\, \Delta E_v(G), \tag{64}$$

where $n(G)$ is the order of G and $\Delta E_v(G)$ means the contribution of the diagram G to ΔE with v replaced everywhere by $\mathscr{V}$.

Suppose now that G is some irreducible vacuum diagram. If we break the interaction line $x_1 x_1'$, the diagram may or may not fall into two separate parts. If it does, then it is of the form shown in Figure 4 (a), and we have seen in §5 that such diagrams give no contribution. If it does not, then breaking this interaction line leads to some proper irreducible polarization part Γ (since G was assumed to be irreducible).

It can now be proved that if on breaking the line $x_1 x_1'$ in the diagram G we obtain the polarization part Γ, then

$$\Delta E_v(G) = \int dx_1 \int dx_1'\, i\hbar\delta(t_1)\, \mathscr{V}(x_1 - x_1')\, \overline{W}_v(x_1, x_1', \Gamma) \frac{g(\Gamma)}{n(G)!\, 2^{n(G)}}, \tag{65}$$

where $\overline{W}_v(x', x, \Gamma)$ is the same as $\overline{W}(x', x, \Gamma)$ except that v has been replaced everywhere by $\mathscr{V}$.

If we substitute (65) into (64), our sum over diagrams reduces to a sum over irreducible polarization structures, each structure being counted a certain number of times. Counting of diagrams shows that this sum reduces to

$$\Delta E = \frac{1}{2}\int_0^\lambda \frac{d\lambda}{\lambda} \int dx_1 \int dx_1'\, i\hbar\delta(t_1)\, \mathscr{V}(x_1 - x_1') \sum_{\Gamma \neq \Gamma_0}^{\substack{\text{proper}\\ \text{irreducible}}} \overline{W}_v(x_1', x_1, \Gamma). \tag{66}$$

Defining the quantity

$$\overline{\mathscr{V}}(x', x) = \sum_{\Gamma \neq \Gamma_0}^{\substack{\text{proper}\\ \text{irreducible}}} \overline{W}_v(x', x, \Gamma) = \sum_{\Gamma \neq \Gamma_0}^{\text{proper}} \overline{W}(x', x\Gamma), \tag{67}$$

we see that (66) can be written

$$\Delta E = \frac{1}{2}\int_0^\lambda \frac{\mathrm{d}\lambda}{\lambda}\int \mathrm{d}x_1 \int \mathrm{d}x_1' \, \mathrm{i}\hbar\delta(t_1)\, \overline{\mathscr{V}}(x_1, x_1')\, \mathscr{V}(x_1', x_1). \tag{68}$$

The quantity $\overline{\mathscr{V}}$ is closely related to $\mathscr{V}$. We have in fact (57), (62) and (67)

$$\mathscr{V}^*(x', x_1) = \int \overline{\mathscr{V}}(x', x'')\, v(x''-x)\, \mathrm{d}x, \tag{69}$$

and $\mathscr{V}$ is related to $\mathscr{V}^*$ by (62); thus (68) is in effect an expression for ΔE in terms of $\mathscr{V}$. Using the result (68), we obtain the expression

$$E = \sum_i^{\text{occ.}} E_i - \tfrac{1}{2}Nv(0) + \tfrac{1}{2}N^2 u(0) + \frac{\mathrm{i}\hbar}{2}\int_0^\lambda \frac{\mathrm{d}\lambda}{\lambda}\int \mathrm{d}x_1 \int \mathrm{d}x_1' \, \delta(t_1)\, \mathscr{V}(x_1, x_1')\, \overline{\mathscr{V}}(x_1', x_1) \tag{70}$$

for the energy of a uniform interacting gas, where the E_i are those of equation (10). In the case of a uniform gas the first term of (70) is just the kinetic energy of the particles in the occupied states together with their potential energy in any constant background potential which is present.

9. Summary of results

In this paper the following results have been obtained.

(i) It has been shown how the perturbation series for the many-body problem may be expressed as a sum of terms each of which is associated with a certain diagram.

(ii) This has been reduced to a series of terms (linked-cluster expansion) corresponding to linked diagrams.

(iii) A series has been obtained for the energy shift ΔE due to the interaction, each term of which is associated with a certain 'vacuum' diagram.

(iv) In the case of a uniform gas it has been shown that some of the terms of this series give no contribution and may be omitted.

(v) This series has been reduced to one over irreducible vacuum diagrams, i.e. over diagrams containing no polarization parts, by introducing a modified interaction $\mathscr{V}$ in place of the ordinary interaction v. An infinite series has been given for $\mathscr{V}$.

(vi) An integral equation has been obtained for $\mathscr{V}$ in terms of a quantity $\mathscr{V}^*$ which is given by an infinite series much more rapidly convergent than that for $\mathscr{V}$.

(vii) Finally, the energy shift and the energy of the whole system have been expressed in terms of the modified interaction $\mathscr{V}$; the latter is calculated from the integral equation the rapidly convergent series for $\mathscr{V}^*$ replaces the original perturbation series.

In a later paper these results will be applied to calculate the correlation energy of a free-electron gas, and it will be shown that essentially the same results as those of Bohm & Pines (1953) can be obtained by taking only the first term of the series for $\mathscr{V}^*$ and making allowance for certain exchange terms.

The author wishes to express his thanks to Dr J. S. Bell for much helpful advice and criticism.

References

Bohm, D. & Pines, D. 1953 *Phys. Rev.* **92**, 609.
Bohr, A. & Mottelson, B. R. 1953 *Dan. Mat. Fys. Medd.* **27**, no. 16.
Brueckner, K. A. 1955 *Phys. Rev.* **100**, 36.
Dyson, F. J. 1949 *Phys. Rev.* **75**, 1736.
Fröhlich, H. & Pelzer, H. 1955 *Proc. Phys. Soc.* A, **68**, 525.
Gell-Mann, M. & Low, F. 1951 *Phys. Rev.* **84**, 350 (appendix).
Goldstone, J. 1957 *Proc. Roy. Soc.* A **239**, 267.
Hubbard, J. 1955*a* *Proc. Phys. Soc.* A, **68**, 441.
Hubbard, J. 1955*b* *Proc. Phys. Soc.* A **68**, 976.
Hubbard, J. 1957 (In preparation).
Mott, N. F. 1954 *Proc. 10th Solvay Congr.*, Bruxelles.
Salam, A. 1953 *Progr. Theor. Phys., Osaka*, **9**, 550.
Tomonaga, S. 1955 *Progr. Theor. Phys., Osaka*, **13**, 467.
Wick, G. C. 1950 *Phys. Rev.* **80**, 268.

Appendix. The calculation of ΔE

To calculate ΔE we start from a formula given by Gell-Mann & Low (1951), namely

$$\Delta E = \lim_{\alpha\to+0} i\hbar\alpha\lambda\frac{\partial}{\partial\lambda}\ln(\Psi_0\,|\,S_\alpha(0,-\infty)\,|\,\Psi_0), \tag{A 1}$$

where λ is the coupling constant of the interaction. Using the same method as Gell-Mann & Low we can also derive the formula

$$\Delta E = -\lim_{\alpha\to+0} i\hbar\alpha\lambda\frac{\partial}{\partial\lambda}\ln(\Psi_0\,|\,S_\alpha^{-1}(\infty,0)\,|\,\Psi_0). \tag{A 2}$$

Furthermore, we have

$$S_\alpha(\infty,-\infty) = S_\alpha(\infty,0)\,S_\alpha(0,-\infty), \tag{A 3}$$

from which

$$S_\alpha^{-1}(\infty,0) = S_\alpha(0,-\infty)\,S_\alpha^{-1}(\infty,-\infty) \tag{A 4}$$

It was shown in § 3 that

$$S_\alpha(\infty,-\infty)\Psi_0 = \exp\{S_{L\alpha}^{(0)}(\infty)+S'_{L\alpha}(\infty)\}\Psi_0. \tag{A 5}$$

Since Ψ_0 is non-degenerate and $S_\alpha(\infty,-\infty)$ connects only states with the same energy in the limit as $\alpha\to+0$, we see that $S'_{L\alpha}(\infty,-\infty)\to 0$ as $\alpha\to+0$. From (A 1), (A 2), (A 4) and (A 5) we have

$$\begin{aligned}\Delta E &= \lim_{\alpha\to+0} -i\hbar\alpha\lambda\frac{\partial}{\partial\lambda}\ln(\Psi_0\,|\,S_\alpha(0)\,[\exp\{S'_{L\alpha}(\infty)\}]^{-1}\exp\{-S_{L\alpha}^{(0)}(\infty)\}\,|\,\Psi_0)\\ &= \lim_{\alpha\to+0} -i\hbar\alpha\lambda\frac{\partial}{\partial\lambda}[\ln(\Psi_0\,|\,S_\alpha(0)\,|\,\Psi_0)-S_{L\alpha}^{(0)}(\infty)]\\ &= -\Delta E+\lim_{\alpha\to+0} i\hbar\alpha\lambda\frac{\partial}{\partial\lambda}S_{L\alpha}^{(0)}(\infty).\end{aligned} \tag{A 6}$$

Thus

$$\Delta E = \tfrac{1}{2}\lim_{\alpha\to+0} i\hbar\alpha\lambda\frac{\partial}{\partial\lambda}S_{L\alpha}^{(0)}(\infty). \tag{A 7}$$

Now $S^{(0)}_{L\alpha}(\infty)$ is a sum of terms corresponding to vacuum diagrams. When all the integrations in an nth order diagram except those over $t_1, t_2, \ldots, t_n$ have been performed, we shall have an expression of the form

$$\int \mathrm{d}t_1 \ldots \int \mathrm{d}t_n \, U_\alpha(t_1, t_2, \ldots, t_n). \tag{A 8}$$

We can split this up into $n!$ terms by making different time orderings. In each of these terms we can perform the time integrations in any particular order; let us agree to always perform the integration over t_1 last. When we have performed all the integrations apart from that over t_1 in any term we end up with an expression of the form

$$\int_{-\infty}^{\infty} \mathrm{d}t_1 \, \mathrm{e}^{-\alpha|t_1| + \mathrm{i}Et_1} M \, \mathrm{e}^{-(n-1)\alpha|t_1| - \mathrm{i}Et_1} + O(1), \tag{A 9}$$

where the factor $\exp\{-\alpha|t_1| + \mathrm{i}Et_1\}$ arises from the corresponding factor in U_α and the term $\exp\{-(n-1)a|t_1| - \mathrm{i}Et_1\}$ arises from the performance of the remaining integrations. Performing the integration in (A 9) gives $2M/n\alpha + O(1)$. The operation $\lambda(\mathrm{d}/\mathrm{d}\lambda)$ on the contribution to $S^{(0)}_{L\alpha}(\infty)$ of an nth order term just multiplies it by n. Thus the contribution of the term (A 9) to ΔE is just

$$\lim_{\alpha \to +0} \tfrac{1}{2}\mathrm{i}\hbar\alpha n \left\{\frac{2M}{n\alpha} + O(1)\right\} = \mathrm{i}\hbar M. \tag{A 10}$$

Let us now compare this result with

$$\int_{-\infty}^{\infty} \mathrm{d}t_1 \ldots \int_{-\infty}^{\infty} \mathrm{d}t_n \mathrm{i}\hbar\delta(t_1) \, U_0(t_1, t_2, \ldots, t_n), \tag{A 11}$$

which arises from (A 8) by inserting a factor $\mathrm{i}\hbar\delta(t_1)$ into the integrand and putting $\alpha = 0$. If we evaluate this by splitting it into terms with different time orderings and perform the integration over t_1 last, we find that the term which corresponds to (A 9) is when all the integrations except that over t_1 have been performed

$$\int_{-\infty}^{\infty} \mathrm{d}t_1 \mathrm{i}\hbar\delta(t_1) \, \mathrm{M} \, \mathrm{e}^{\mathrm{i}Et_1 - \mathrm{i}Et_1} = \mathrm{i}\hbar M, \tag{A 12}$$

which agrees with (A 11), the terms written $O(1)$ in (A 9) disappearing when $\alpha = 0$. Thus the insertion of a factor $\mathrm{i}\hbar\delta(t_1)$ into the integrand of (A 8) and putting $\alpha = 0$ has the same effect as operating with $\mathrm{i}\hbar\lambda(\mathrm{d}/\mathrm{d}\lambda)\alpha$ and taking the limit as $\alpha \to +0$. Thus we obtain the prescription of § 4.

SOVIET PHYSICS JETP VOLUME 34(7), NUMBER 1 JULY, 1958

APPLICATION OF QUANTUM FIELD THEORY METHODS TO THE MANY BODY PROBLEM

V. M. GALITSKII and A. B. MIGDAL

Moscow Engineering-Physics Institute

Submitted to JETP editor July 12, 1957; resubmitted October 24, 1957.

J. Exptl. Theoret. Phys. (U.S.S.R.) **34**, 139-150 (January, 1958).

It is shown that the energy and damping of quasiparticles are determined by the poles of a single particle propagation function. The relation between the two-particle Green's function and the kinetic equation is established.

INTRODUCTION

In many cases, weakly-excited states of a system of interacting particles can be described approximately as an aggregate of elementary excitations — quasiparticles. In such a treatment the excited state of the system is described by fewer parameters than are needed for an exact description. Thus the elementary excitation is not a stationary state, but is rather a packet of stationary states with a narrow energy spread. The washing out of the packet leads to damping of the excitation. A description of the states of a system in terms of elementary excitations is possible if the energy spread of the packet, which determines its damping, is small compared to the excitation energy.

We shall consider the case of a homogeneous unbounded system. In such a system, the momentum operator commutes with the Hamiltonian, so that the excited states are characterised by the value of the momentum of the system, in addition to the other parameters.

Apparently, in all Fermi systems there are excitations analogous to the excitations in an ideal Fermi gas. The energy of such an excitation is $E(p_1, p_2) = \epsilon(p_1) - \epsilon(p_2)$, where p_1, $\epsilon(p_1)$, and p_2, $\epsilon(p_2)$ are the momenta and energies of the particle and hole which constitute the excitation. Here $p_1 > p_0 > p_2$, and p_0 is the limiting Fermi momentum for the quasiparticles.

A quasiparticle with momentum p, near to p_0, can reduce its energy, transferring another quasiparticle from the Fermi sphere to a state with $p' > p_0$. From the limitations imposed by the Pauli principle and the laws of conservation of energy and momentum, it follows that the probability for such a process, which determines the damping of the quasiparticles, is proportional to $(p - p_0)^2$. Thus the description of excited states of a Fermi system by means of quasiparticles is the more exact the closer the momentum of the quasiparticles to p_0.

The properties of the excitations are conveniently studied by the methods of quantum field theory, by introducing the Green's function of the system. Then the single particle Green's function determines the energy and the damping of the quasiparticles. However there may be excitations in the system whose energy is not describable as a sum of energies of quasiparticles. The energy spectrum of such excitations can be found from the two-particle Green's function. The two-particle Green's function, as we shall show later, also enables us to determine the behavior of the system in a weak external field.

In addition to the Green's function of the particles, we can also introduce the propagation function for the interaction between the particles. For example, for the problem of electrons in a metal interacting with the lattice, this propagation function is the Green's function of the phonon. The phonon Green's function determines the energy and damping of excitations of the lattice.

SINGLE PARTICLE GREEN'S FUNCTION AND ENERGY SPECTRUM

1. The single particle Green's function is defined, as usual, by

$$G(\mathbf{r}_1 t_1, \mathbf{r}_2 t_2) = i\langle T\{e^{iHt_1}\psi(\mathbf{r}_1)e^{-iH(t_1-t_2)}\psi^+(\mathbf{r}_2)e^{-iHt_2}\}\rangle, \quad (1)$$

where $\psi(\mathbf{r}) = \sum_{\mathbf{p}} a_{\mathbf{p}} e^{i\mathbf{pr}}$; the average is taken over the ground state function of the Hamiltonian system H.

The Green's function of the field φ which provides the interaction between the particles is defined similarly:

$$D(\mathbf{r}_1 t_1, \mathbf{r}_2 t_2) = i\langle P\{e^{iHt_1}\varphi(\mathbf{r}_1)e^{iH(t_1-t_2)}\varphi(\mathbf{r}_2)e^{-iHt_2}\}\rangle. \quad (2)$$

In the case of the phonon field,

$$\varphi(\mathbf{r}) = \sum_{\mathbf{q}} \alpha_{\mathbf{q}} (b_{\mathbf{q}} + b^{+}_{-\mathbf{q}}) e^{i\mathbf{q}\mathbf{r}},$$

where $b_{\mathbf{q}}$ and $b^{+}_{\mathbf{q}}$ are the phonon annihilation and creation operators.

In the absence of external fields, the functions G and D depend only on $r = |\mathbf{r}_1 - \mathbf{r}_2|$ and $\tau = t_1 - t_2$. Expanding the functions $G(r, \tau)$ and $D(r, \tau)$ in Fourier integrals, we get

$$\begin{aligned} G(r,\tau) &= \int \frac{dp d\varepsilon}{(2\pi)^4} G(p, \varepsilon) e^{i(\mathbf{p}\mathbf{r} - \varepsilon\tau)}, \\ D(r,\tau) &= \int \frac{dq d\omega}{(2\pi)^4} D(q, \omega) e^{i(\mathbf{q}\mathbf{r} - \omega\tau)}, \end{aligned} \tag{3}$$

where $G(p, \epsilon)$ and $D(q, \omega)$ are the Green's functions in momentum representation.

In the absence of interactions between the particles, we easily find from (1), for Fermi systems,*

$$G_0(r,\tau) = i\int \frac{dp}{(2\pi)^3} e^{i(\mathbf{p}\mathbf{r} - \varepsilon^0_p \tau)} \begin{cases} 1 - n_p, & \tau > 0 \\ -n_p, & \tau < 0, \end{cases}$$

where $n_p = a^{+}_p a_p$. Going over to momentum representation by means of formula (3), we get

$$\begin{aligned} G_0(p, \varepsilon) &= 1/(\varepsilon^0_p - \varepsilon - i\Delta), \\ \Delta &\to \begin{cases} +0 & p > p_0 \\ -0 & p < p_0. \end{cases} \end{aligned} \tag{4}$$

Similarly, we find from (3),

$$D_0(q, \omega) = \alpha^2_q \left\{ \frac{1}{\omega^0_q - \omega - i\delta} + \frac{1}{\omega^0_q + \omega - i\delta} \right\}, \quad \delta \to +0. \tag{5}$$

2. We now go on to the Fourier transforms with respect to $\mathbf{r} = \mathbf{r}_1 - \mathbf{r}_2$. From (1), we find for

$$G(p, \tau) = \frac{1}{2\pi} \int d\varepsilon G(d, \varepsilon) e^{-i\varepsilon\tau}$$

the expression

$$G(p,\tau) = i \begin{cases} \langle a_p e^{-iH\tau} a^{+}_p \rangle e^{iE_0\tau} & \tau > 0 \\ -\langle a^{+}_p e^{iH\tau} a_p \rangle e^{-iE_0\tau} & \tau < 0. \end{cases} \tag{6}$$

If we express the operators which appear in $G(p, \tau)$ in the energy representation, we have

$$G(p,\tau) = \begin{cases} i \sum_s |(a^{+}_p)_{s0}|^2 \exp\{-i(E_s - E_0)\tau\} & \tau > 0, \\ -i \sum_s |(a_p)_{s0}|^2 \exp\{i(E_s - E_0)\tau\} & \tau < 0. \end{cases} \tag{7}$$

Since the operator a^{+}_p increases the momentum of the system by the amount $\mathbf{p}$, and the number of particles in the system by unity, the summation for $\tau > 0$ extends over all states with momentum $\mathbf{p}$ and particle number $N + 1$, if the number of particles in the ground state was N and the momentum was equal to zero. Similarly, the summation for $\tau < 0$ is taken over states with particle number $N - 1$ and momentum $-\mathbf{p}$. We use the notation

$$\begin{aligned} &E_s(N+1) - E_0(N) \\ &= \varepsilon_s(N+1) + E_0(N+1) - E_0(N) = \varepsilon_s + \mu, \end{aligned}$$

where $\mu = E_S(N+1) - E_0(N)$ is the chemical potential. The excitation energy $\epsilon_S = E_S(N+1) - E_0(N+1)$ is, by definition, positive. Similarly

$$\begin{aligned} &E_s(N-1) - E_0(N) \\ &= \varepsilon_s(N-1) - E_0(N) + E_0(N-1) = \varepsilon'_s - \mu'. \end{aligned}$$

The quantities ϵ'_S and μ' are identical with ϵ_S and μ to terms of order $1/N$. We introduce the functions

$$\begin{aligned} A(p, E)\,dE &= \sum_s |(a^{+}_p)_{s0}|^2, \quad E < \varepsilon_s < E + dE, \\ B(p, E)\,dE &= \sum_s |(a_p)_{s0}|^2, \quad E < \varepsilon_s < E + dE \end{aligned} \tag{8}$$

and carry out a Fourier transformation with respect to τ in (7). We get

$$G(p, \varepsilon) = \int_0^\infty dE \left\{ \frac{A(p,E)}{E - \varepsilon + \mu - i\delta} - \frac{B(p,E)}{E + \varepsilon - \mu - i\delta} \right\}. \tag{9}$$

Formula (9) is Lehmann's[1] expansion for the single particle Green's function of a system consisting of a finite number of fermions. Using it, we can obtain some relations between the real and imaginary parts of the function $G(p, \epsilon)$. In fact, from the equality

$$\frac{1}{E - \varepsilon + \mu - i\delta} = \mathrm{P}\frac{1}{E - \varepsilon + \mu} + i\pi\delta(E - \varepsilon + \mu)$$

if follows that

$$\operatorname{Im} G(p, \varepsilon) = \pi \begin{cases} A(p, \varepsilon - \mu) & \varepsilon > \mu \\ -B(p, \mu - \varepsilon) & \varepsilon < \mu, \end{cases} \tag{10}$$

i.e., the imaginary part of the Green's function changes sign at the point $\epsilon = \mu$. Using (9) and (10), it is easy to obtain the formula* which gives the relation between the real and imaginary parts of G:

$$\operatorname{Re} G(p, \varepsilon) = \frac{1}{\pi} \int_{-\infty}^{\infty} \mathrm{P}\frac{\operatorname{Im} G(p, \varepsilon')}{\varepsilon' - \varepsilon} d\varepsilon' \tag{11}$$

*We shall give the formulas for Fermi systems. As is easily seen, most of the results also are valid for the case of Bose particles.

*This formula was obtained by L. D. Landau.

We remark that in the case of a system of bosons the expression for $G(p, \epsilon)$ differs from (9) only in a change in sign of $B(p, \epsilon)$. Thus for bosons, unlike (10), the imaginary part of $G(p, \epsilon)$ is positive for all p and ϵ. Formula (11) remains valid for bosons.

Similar formulas can be gotten for $D(q, \omega)$:

$$D(q, \tau) = \begin{cases} i \sum_s |(\varphi_{-q})_{s0}|^2 \exp\{-i(E_s - E_0)\tau\} & \tau > 0 \\ i \sum_s |(\varphi_q)_{s0}|^2 \exp\{i(E_s - E_0)\tau\} & \tau < 0. \end{cases} \tag{12}$$

Here we have made use of the reality of the field φ : $\varphi_q^+ = \varphi_q$. Unlike (7), the number of particles, N, in the states of the sum (12) is equal to the number of particles in the ground state. We use the notation

$$\Phi(q, \omega')\, d\omega' = \sum_s |(\varphi_{-q})_{s0}|^2 = \sum_s |(\varphi_q)_{s0}|^2, \quad \omega' < E_s - E_0 < \omega' + d\omega', \tag{13}$$

Then

$$D(q, \tau) = \int_0^\infty d\omega' \Phi(q, \omega')\, e^{-i\omega'|\tau|}. \tag{14}$$

Taking the Fourier transform of (14) with respect to τ, we get

$$D(q, \omega) = \int_0^\infty d\omega' \Phi(q, \omega') \left\{ \frac{1}{\omega' - \omega - i\delta} + \frac{1}{\omega' + \omega - i\delta} \right\}. \tag{15}$$

From (15) we have

$$\operatorname{Im} D(q, \omega) = \pi \Phi(q, |\omega|) > 0; \tag{16}$$

$$\operatorname{Re} D(q, \omega) = \frac{1}{\pi} \int_0^\infty d\omega' \operatorname{Im} D(q, \omega')\, P\left\{ \frac{1}{\omega' - \omega} + \frac{1}{\omega' + \omega} \right\}. \tag{17}$$

3. Let us examine the properties of the Green's function in the complex ϵ plane. Replacing E by $-E$ in the second term of (9), we get

$$G(p, \varepsilon) = \int_C \frac{F(p, E)}{E - \varepsilon + \mu}\, dE. \tag{18}$$

The integration contour C is shown in Fig. 1. The expression on the right side of (18) is an integral of the Cauchy type. Functions defined by such integrals are known[2] to be analytic throughout the plane except for the points on the contour of integration. In our case the integration contour C divides the plane of the complex variable $\epsilon - \mu$ into two regions, and the integral (18) defines two different functions: $f_I(\epsilon - \mu)$ which is analytic in region I, and $f_{II}(\epsilon - \mu)$ which is analytic in region II. The Green's function $G(p, \epsilon)$, which is defined by the values of the integral (12) on the real axis, coincides with f_{II} for $\epsilon < \mu$, and with f_I for $\epsilon > \mu$. Thus $G(p, \epsilon)$ is not an analytic function of ϵ, but has a singularity at $\epsilon = \mu$.

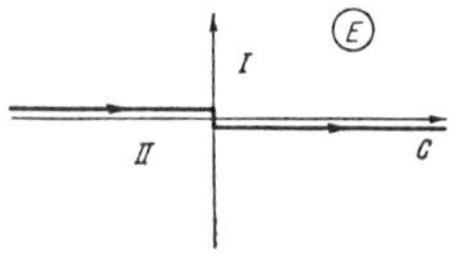

FIG. 1

Using the reality of the function $F(p, E)$, it is not difficult to show that the values of the integral (18) for points lying infinitely close to one another on opposite sides of the contour C are complex conjugates. Thus the function f_I, for negative values of $\epsilon - \mu$ lying above the contour C, is the complex conjugate of $G(p, \epsilon)$:

$$f_I(\varepsilon - \mu + 2i\delta) = G^*(p, \varepsilon), \quad \varepsilon < \mu. \tag{19}$$

Thus $G(p, \epsilon)$ for $\epsilon > \mu$, when continued analytically into the upper half-plane, coincides for $\epsilon < \mu$ with $G^*(p, \epsilon)$, or in other words, $G(p, \epsilon)$ for $\epsilon > \mu$ and $G^*(p, \epsilon)$ for $\epsilon < \mu$ comprise the analytic function f_I. Similarly, $G(p, \epsilon)$ for $\epsilon < \mu$, when continued analytically into the lower half-plane, coincides for $\epsilon > \mu$ with $G^*(p, \epsilon)$.

4. We shall now establish the connection of the single-particle Green's function with the spectrum of excitations. The Green's function $G(p, \tau)$ has a simple physical meaning. Suppose that initially the system is in the state $\Psi(0) = a_p^+ \Phi_0$, where Φ_0 is the ground state of the system of N particles (the physical "vacuum"). At the time $\tau > 0$, the wave function of the system is

$$\Psi(\tau) = e^{iH\tau} a_p^+ \Phi_0.$$

The function $G(p, \tau)$ is the probability for finding the system in state $\Psi(0)$ at time τ.

In fact,

$$(\Psi(0), \Psi(\tau)) = (\Phi_0 a_p e^{-iH\tau} a_p^+ \Phi_0) = -iG(p, \tau). \tag{20}$$

A similar relation holds for $\tau < 0$. According to (7) and (8), for $\tau > 0$

$$(\Psi(0), \Psi(\tau)) = e^{-i\mu\tau} \int_0^\infty A(p, E)\, e^{-iE\tau} dE. \tag{21}$$

In the presence of interaction, for values of p greater than p_0,

$$A(p, E) = \delta(E + \mu - \varepsilon_p^0) \text{ and } (\Psi(0), \Psi(\tau)) = e^{-i\varepsilon_p^0 \tau}.$$

When we switch on the interaction between particles, the δ function in $A(p, E)$ is replaced by a function having a sharp maximum near $E = \epsilon_p - \mu$, where ϵ_p is the energy of the quasiparticles.

Let us look at the behavior of the Green's function for large positive times. Suppose that the singularity of the analytic continuation of A(p, E) into the lower half-plane, which is closest to the real axis is a simple pole at $E = \epsilon_p - \mu - i\Gamma$. Then, by shifting the contour of integration in (21) into the lower half-plane, we get

$$G(p, \tau) = ie^{-i\mu\tau}\int_C Ae^{-iE\tau}dE. \qquad (21')$$

The integration contour C is shown in Fig. 2. The non-exponential term in the function $G(p, \tau)$,

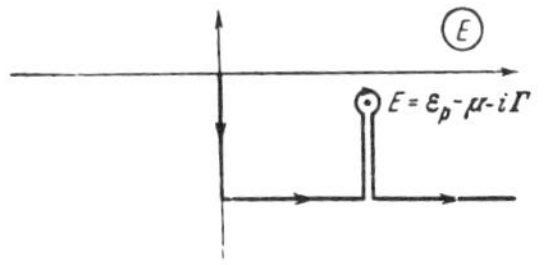

FIG. 2

which arises from the integration along the imaginary axis near E = 0, is of order $(\Gamma/\epsilon_p)^2$ for $\tau \gg 1/\Gamma$. Thus

$$G(p, \tau) = c_p e^{-i\varepsilon_p\tau - \Gamma\tau} + O[(\Gamma/\varepsilon_p)^2]. \qquad (22)$$

This result can be interpreted in the following way: the state $\Psi(0)$ contains, with amplitude c_p, a packet describing a quasiparticle with energy ϵ_p and damping Γ. The values of ϵ_p and Γ are determined by the position of the pole of A(p, E), i.e., by the imaginary part of $G(p, \epsilon)$ in the lower half-plane. The poles of Im G coincide with the poles of G or G*, but the analytic continuation of the latter is $f_{II}(\epsilon - \mu)$, which is analytic in the lower half-plane. Thus the energy and damping of the excitations are determined by the real and imaginary parts of the poles of the analytic continuation of $G(p, \epsilon)$ for $\epsilon > \mu$ in the lower half-plane. Similarly, the energy and damping of holes in the Fermi distribution are given by the poles of the analytic continuation of $G(p, \epsilon)$ for $\epsilon < \mu$ into the upper half-plane.

Let us introduce the irreducible part of the proper energy of the particles, $\Sigma(p, \epsilon)$:

$$G^{-1}(p, \varepsilon) = \varepsilon_p^0 - \varepsilon - \Sigma(p, \varepsilon)$$
$$= \varepsilon_p^0 - \varepsilon - \Sigma_0(p, \varepsilon) - i\,\Sigma_1(p, \varepsilon), \qquad (23)$$

where Σ_0 and Σ_1 are the real and imaginary parts of Σ. The energy and damping of the quasiparticle are determined from the equation

$$\varepsilon_p^0 - (\varepsilon_p - i\Gamma) - \widetilde{\Sigma}(p, \varepsilon_p - i\Gamma) = 0, \qquad (24)$$

where $\widetilde{\Sigma}(p, \epsilon)$ is the analytic continuation of $\Sigma(p, \epsilon)$ for $\epsilon > \mu$. For $\Gamma/\epsilon_p \ll 1$, we have approximately:

$$\varepsilon_p^0 - \varepsilon_p - \Sigma_0(p, \varepsilon_p) = 0,$$
$$\Gamma_p = \Sigma_1(p, \varepsilon_p)\Big/\left[1 + \left(\frac{\partial\Sigma_0}{\partial\varepsilon}\right)_{\varepsilon=\varepsilon_p}\right]. \qquad (24')$$

Analogous results hold for $D(q, \omega)$. Let $\Pi(q, \omega)$ be the irreducible part of the proper energy of the phonon

$$D(q, \omega) = \alpha_q^2/[\omega_q^{02} - \omega^2 - 2\omega_q^0\alpha_q^2\Pi(q, \omega)]$$
$$= \alpha_q^2/[\omega_q^{02} - \omega^2 - 2\omega_q^0\alpha_q^2\Pi_0(q, \omega) - 2i\omega_q^0\alpha_q^2\Pi_1(q, \omega)].$$

As above, the energy and damping of the phonon excitation are given by the poles of the analytic continuation of $D(q, \omega)$ for $\omega > 0$:

$$\omega_q^{02} - (\omega_q - i\gamma)^2 - 2\omega_q^0\alpha_q^2\widetilde{\Pi}(q, \omega_q - i\gamma) = 0, \qquad (25)$$

where $\widetilde{\Pi}(q, \omega)$ is the analytic continuation of $\Pi(q, \omega)$ for $\omega > 0$. Approximately, for $\omega_q \gg \gamma$:

$$\omega_q^2 = \omega_q^{02} - 2\omega_q^0\alpha_q^2\Pi_0(q, \omega_q),$$
$$\gamma_q = \alpha_q^2\frac{\omega_q^0}{\omega_q}\Pi_1(q, \omega_q)\Big/\left[1 + 2\omega_q^0\alpha_q^2\left(\frac{\partial\Pi_0}{\partial\omega}\right)_{\omega=\omega_q}\right]. \qquad (25')$$

We determine the momentum p_0 from the condition $\mu = \epsilon_p$, where μ is the chemical potential of the system. Then, assuming that Im G is continuous at $\epsilon = \mu$, we find from (10) and (24′) that $\Gamma(p_0) = 0$. Thus the damping of the excitations goes to zero at the point p_0, which is determined by the equation $\epsilon_{p_0} = \mu$.

5. The single-particle Green's function also enables us to find other characteristics of the system. Thus the momentum distribution of the particles is related to the Green's function by[3]

$$n_p = i\int_C G(p, \varepsilon)\frac{d\varepsilon}{(2\pi)}, \qquad (26)$$

where the contour C consists of the real axis and a semicircle of infinite radius in the upper half-plane.

As was shown in Ref. 3, when we pass through the point $p = p_0$, the pole of $G(p, \epsilon)$ which lies nearest to the real axis moves into the lower half of the ϵ plane, and is thus outside of the contour C of formula (26). Thus the jump in n_p at $p = p_0$ remains even in the presence of an arbitrary interaction.

Let us express the energy of the ground state of the system in terms of the function G. Differentiating $G(p, \tau)$ with respect to the time t, it is not difficult to obtain the formula

$$\left(i\frac{\partial}{\partial t} - \varepsilon_p^0\right)G(p, \tau) = -\delta(\tau) - i\langle T\{[H'(t), a_p(t)]\,a_p^+(t')\}\rangle, \qquad (27)$$

where $H'(t)$ is the interaction Hamiltonian between particles, and $[H', a_p]$ is the commutator of the operators H' and a_p. Comparing (27) in the (p, ϵ) representation with the expression relating G and G_0, we find the general form of the product $\Sigma(p, \epsilon) G(p, \epsilon)$:

$$\Sigma(p, \varepsilon) G(p, \varepsilon) = i \int d\tau e^{i\varepsilon\tau} \langle T\{[H'(t), a_p(t)] a_p^+(t')\}\rangle. \tag{28}$$

By integrating (28) with the factor $e^{i\epsilon\Delta}$ and going to the limit $\Delta \to +0$, we get the formula

$$\lim_{\Delta\to+0} \int \frac{d\varepsilon}{2\pi} e^{i\varepsilon\Delta} \Sigma(p, \varepsilon) G(p, \varepsilon)$$

$$\equiv \int_C \frac{d\varepsilon}{2\pi} \Sigma(p, \varepsilon) G(p, \varepsilon) = -i \langle a_p^+ [H', a_p]\rangle,$$

or,

$$i \int_C \frac{d^4p}{(2\pi)^4} \Sigma(p, \varepsilon) G(p, \varepsilon) = \int \langle a_p^+ [H', a_p]\rangle \frac{dp}{(2\pi)^3}, \tag{29}$$

where the contour C coincides with the integration contour in (26).

In the case of pair interaction between particles, the right side of (29) reduces to the average value of the interaction Hamiltonian

$$\langle H'\rangle = -\frac{i}{2} \int_C \frac{d^4p}{(2\pi)^4} \Sigma(p, \varepsilon) G(p, \varepsilon). \tag{30}$$

Adding the average value of the Hamiltonian of the non-interacting particles to (30), we get finally

$$E_0 = \frac{i}{(2\pi)^4} \int_C \left\{\varepsilon_p^0 - \frac{1}{2}\Sigma(p, \varepsilon)\right\} G(p, \varepsilon) d^4p. \tag{31}$$

Differentiation of E_0 with respect to the number of particles N gives the chemical potential, μ, of the system.

TWO-PARTICLE GREEN'S FUNCTION. KINETIC EQUATION

To study the energy spectrum and the behavior of the system in weak external fields, we must consider the two-particle Green's function. The two-particle Green's function $K(1, 2; 3, 4)$ is defined by

$$K(1,2;\ 3,4) = i \langle T\{\psi(1)\psi^+(2)\psi(3)\psi^+(4)\}\rangle, \tag{32}$$

where 1, 2; 3, 4 stand for the sets of coordinates of the space-time points. If $t_1, t_2 > t_3, t_4$, K can be written in the form

$$K(1, 2;\ 3,4) = i \sum_s \chi_s(1,2) \tilde{\chi}_s(3,4), \tag{33}$$

where

$$\chi_s(1,2) = (T\{\psi(1)\psi^+(2)\})_{0s},$$
$$\tilde{\chi}_s(3,4) = (\tilde{T}\{\psi(4)\psi^+(3)\})_{0s}^{*}; \tag{34}$$

$\tilde{T}$ orders the operators in the reverse order from T. The functions $\chi_s(1, 2)$ for simultaneous times $t_1 = t_2$ have the physical meaning of wave functions describing the behavior of a particle and a hole in the state s. In the absence of external fields, the dependence on the coordinates of the "center of gravity" $X = (x_1 + x_2)/2$ can be separated off from χ:

$$\chi_s(x_1, x_2) = e^{ikX} f_{\mathbf{k},\omega}(x),\quad kX = \mathbf{k}\mathbf{R} - \omega T,$$
$$x = x_1 - x_2,\quad \mathbf{k} = \mathbf{p}_s - \mathbf{p}_0,\quad \omega = E_s - E_0, \tag{35}$$

where $\mathbf{k}$ and ω are the momentum and energy of the excitation.

As will be shown later, the function $f_{\mathbf{k},\omega}(x)$ for $t_1 = t_2$ in momentum representation, i.e. $f_{\mathbf{k},\omega}(p)$, is the Fourier component of the distribution function $f(\mathbf{r}, \mathbf{p}, t)$. As a matter of fact, the density matrix, normalized to the total number of particles:

$$P(\mathbf{r}, \mathbf{r}', t)$$
$$= \sum_{i=1}^{N} \int \varphi^*(\mathbf{r}_1, \ldots, \mathbf{r}', \ldots, \mathbf{r}_N; t)\, \varphi(\mathbf{r}_1, \ldots, \mathbf{r}, \ldots, \mathbf{r}_N; t) \prod_{h\neq i} dV_h \tag{36}$$

(where $\varphi(\mathbf{r}_1, \ldots \mathbf{r}_N; t)$ is the wave function of the system in configuration space), can be written as the average value of an integral operator with the kernel

$$(\hat{P}(\mathbf{r}, \mathbf{r}', t))_{\mathbf{r}_i', \mathbf{r}_i} = \sum_i \delta(\mathbf{r}' - \mathbf{r}_i')\, \delta(\mathbf{r} - \mathbf{r}_i), \tag{37}$$

which we may call the density matrix operator. In the occupation number representation, this operator has the form

$$\hat{P}(\mathbf{r}, \mathbf{r}', t)$$
$$= \int \psi^+(\mathbf{r}_1')\, \delta(\mathbf{r}' - \mathbf{r}_1')\, \delta(\mathbf{r} - \mathbf{r}_1)\, \psi(\mathbf{r}_1)\, dV_1 dV_1' = \psi^+(\mathbf{r}')\, \psi(\mathbf{r}). \tag{37'}$$

$\chi_s(1, 2)$ for $t_1 = t_2$ is, to within a factor which is independent of points 1 and 2, the part of the density matrix which oscillates with frequency $\omega = E_s - E_0$. The Fourier component of the density matrix with respect to the coordinate $\mathbf{x} = \mathbf{r}_1 - \mathbf{r}_2$ is related, as we know, to the distribution function (the density matrix in mixed representation). Therefore the function f, to within a normalization factor, coincides with the Fourier component of the distribution function

$$f_{\mathbf{k},\omega}(\mathbf{p}) = c\int f(\mathbf{r}, \mathbf{p}, t)\, e^{-i(\mathbf{kr}-\omega t)}\, dv dt. \tag{38}$$

From (34) we see that the two-particle Green's function K is suitable for studying excited states of a system of N particles, in which there are particles and holes, while the single particle Green's function G enables us to investigate states of a system of N + 1 particles which differ from the ground state of the N particle system by the presence (or absence) of one quasiparticle. The essential feature of the states described by the two-particle Green's function K is the interaction between particle and hole. If this interaction leads only to scattering of the particle by the hole, then the energy of the excitation is equal to the energy of the particle and hole at infinity, $E = \epsilon(\mathbf{p}_1) - \epsilon(\mathbf{p}_2)$, $\mathbf{p} = \mathbf{p}_1 - \mathbf{p}_2$. In this case, the two-particle Green's function gives no new information concerning the energy spectrum of the system, beyond that from the single-particle function. In some cases the interaction can lead to the presence of excited states which can be interpreted as bound states of a particle and a hole. Such excited states were studied in the papers of Klimontovich and Silin[4,5] and Landau,[6] and were called zeroth sound.

As was shown in the papers of Schwinger,[7] and Gell-Mann and Low,[8] the equation for the function K has the form

$$K(x_1x_2;\ x_3x_4) = iG(x_1 - x_4)\, G(x_3 - x_2) - iG(x_1 - x_2)\, G(x_3 - x_4)$$
$$+\, i\int G(x_1 - x_5)\, G(x_6 - x_2)\, \Gamma(x_5x_6,\ x_7x_8)\, K(x_7x_8;\ x_3x_4)\, d^4x_5 d^4x_6 d^4x_7 d^4x_8, \tag{39}$$

Γ is a compact four-pole diagram, i.e., a set of graphs which start and end with a pair of solid lines, while the graphs cannot be split into parts which are joined only by a pair of solid lines. The free term of this equation describes the propagation of non-interacting particle and hole, and does not contain the frequencies corresponding to bound states. Therefore, extraction of the function χ_s, which describes bound states, leads, as was shown in Ref. 8, to the following homogeneous equation for χ:

$$\chi_s(x_1x_2) = i\int G(x_1 - x_5)\, G(x_6 - x_2)\, \Gamma(x_5x_6,\ x_7x_8)\, \chi_s(x_7x_8)\, d^4x_5 d^4x_6 d^4x_7 d^4x_8. \tag{40}$$

Substituting χ_s in the form (35) into this equation, we get an eigenvalue problem whose solution gives us the frequency of zeroth sound and the functions χ_s. In the following, we shall limit ourselves to the case of a system of particles interacting with one another via a weak non-retarded potential V. To first order in the strength of the interaction, the zeroth order Green's functions should be used as the Green's function, while the compact four-pole Γ is given by the pair of graphs shown in Fig. 3. (The dotted lines on these graphs refer to the propagation function of the interaction, $-iV_q$. For our further work, we must include the spin of the particles. In the most usual case, of spin $\frac{1}{2}$, we can construct four functions χ_s:

$$\chi_s(1,2;\sigma_1,\sigma_2) = (T\{\psi_{\sigma_1}(1),\psi^+_{\sigma_2}(2)\})_{0s}. \tag{41}$$

It is not hard to see that the functions $\chi_s(1,2;\frac{1}{2},-\frac{1}{2})$ and $\chi_s(1,2;-\frac{1}{2},\frac{1}{2})$ correspond to excitations with spin 1 and projections +1 and −1, respectively. As the compact four-pole in the equation for these χ_s, we can use the Γ which is described by only the first of the graphs in Fig. 3; because the potential is independent of the spin

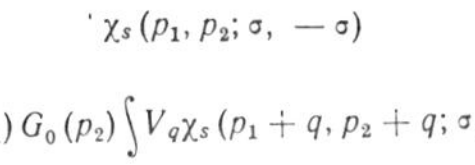

FIG. 3

variables, only a particle and hole with total spin equal to zero can participate in the second of the interactions. Thus the equation for these functions has the following form in momentum representation:

$$\chi_s(p_1, p_2;\sigma, -\sigma)$$
$$= iG_0(p_1)\, G_0(p_2)\int V_q \chi_s(p_1 + q, p_2 + q;\sigma, -\sigma)\, \frac{dq}{(2\pi)^4}. \tag{42}$$

In contrast to this case, the compact four-pole in the equation for the functions $\chi_s(1,2;\frac{1}{2},\frac{1}{2})$ and $\chi_s(1,2;-\frac{1}{2},-\frac{1}{2})$ is described by both of the graphs in Fig. 3. The equation for these functions is

$$\chi_s(p_1, p_2; \sigma, \sigma) = iG_0(p_1)G_0(p_2)\left\{\int V_q \chi_s(p_1+q, p_2+q; \sigma, \sigma)\frac{dq}{(2\pi)^4} - V_{p_1-p_2}\sum_{\sigma'}\int \chi_s\left(p+\frac{p_1-p_2}{2}, p-\frac{p_1-p_2}{2}; \sigma', \sigma'\right)\frac{dp}{(2\pi)^4}\right\}. \quad (43)$$

We introduce functions $\chi_s^+(p_1, p_2)$ and $\chi_s^-(p_1, p_2)$:

$$\chi_s^+(p_1, p_2) = \frac{1}{\sqrt{2}}\left\{\chi_s\left(p_1, p_2; \frac{1}{2}, \frac{1}{2}\right) + \chi_s\left(p_1, p_2; -\frac{1}{2}, -\frac{1}{2}\right)\right\}, \quad (44)$$

$$\chi_s^-(p_1, p_2) = \frac{1}{\sqrt{2}}\left\{\chi_s\left(p_1, p_2; \frac{1}{2}, \frac{1}{2}\right) - \chi_s\left(p_1, p_2; -\frac{1}{2}, -\frac{1}{2}\right)\right\}. \quad (44')$$

The function $\chi_s^+(p_1, p_2)$ corresponds to excitation with spin zero, the function $\chi_s^-(p, p)$ to excitation with spin 1 and projection 0. The equations for these functions can be gotten by adding and subtracting Eq. (43) with spin values $\sigma = \frac{1}{2}$ and $\sigma = -\frac{1}{2}$. The equation for χ_s^- is the same as Eq. (42) for the function with spin 1 and projections ± 1. The equation for χ_s^+ is:

$$\chi_s^+(p_1, p_2) = iG_0(p_1)G_0(p_2)\left\{\int V_q\chi_s^+(p_1+q, p_2+q)\frac{dq}{(2\pi)^4} - 2V_{p_1-p_2}\int\chi_s^+\left(p+\frac{p_1-p_2}{2}, p-\frac{p_1-p_2}{2}\right)\frac{dp}{(2\pi)^4}\right\}. \quad (45)$$

Transforming to the "relative" momentum $k = p_1 = p_2$, which is equal to the momentum of the excited state, and the "total" momentum $p = (p_1 + p_2)/2$, we get the following equations for excitations with total spin zero and one, respectively:

$$\chi^0_{k\omega}(p) = iG_0\left(p+\frac{k}{2}\right)G_0\left(p-\frac{k}{2}\right) \times \left\{\int V_q\chi^0_{k\omega}(p+q)\frac{dq}{(2\pi)^4} - 2V_k\int\chi^0_{k\omega}(p)\frac{dp}{(2\pi)^4}\right\}, \quad (46)$$

$$\chi^1_{k\omega}(p) = iG_0\left(p+\frac{k}{2}\right)G_0\left(p-\frac{k}{2}\right)\int V_q\chi^1_{k\omega}(p+q)\frac{dq}{(2\pi)^4}. \quad (47)$$

The function $\chi^0_{k\omega}$ corresponds to excitation with total spin 0, the function $\chi^1_{k\omega}$ to excitation with total spin 1.

In the absence of retardation, the Fourier component of the potential does not depend on the fourth component of $q: V_q \equiv V(\mathbf{q})$, so that Eqs. (46) and (47) can be integrated with respect to ϵ (the fourth component of p). Integration of the function χ with respect to ϵ corresponds in coordinate representation to equating the times t_1 and t_2, so that as a result of the integration we get an equation for $f_{k\omega}(\mathbf{p})$, the Fourier components of the distribution function:

$$f^0_{\mathbf{k}\omega}(\mathbf{p}) = \frac{n_0(\mathbf{p}+\mathbf{k}/2) - n_0(\mathbf{p}-\mathbf{k}/2)}{\omega - \mathbf{kp} - i\delta[n_0(\mathbf{p}+\mathbf{k}/2) - n_0(\mathbf{p}-\mathbf{k}/2)]}\left\{\int V(\mathbf{q})f^0_{\mathbf{k}\omega}(\mathbf{p}+\mathbf{q})\frac{d\mathbf{q}}{(2\pi)^3} - 2V(\mathbf{k})\int f^0_{\mathbf{k}\omega}(\mathbf{p})\frac{d\mathbf{p}}{(2\pi)^3}\right\}, \quad (46')$$

$$f^1_{\mathbf{k}\omega}(\mathbf{p}) = \frac{n_0(\mathbf{p}+\mathbf{k}/2) - n_0(\mathbf{p}-\mathbf{k}/2)}{\omega - \mathbf{kp} - i\delta[n_0(\mathbf{p}+\mathbf{k}/2) - n_0(\mathbf{p}-\mathbf{k}/2)]}\int V(\mathbf{q})f^1_{\mathbf{k}\omega}(\mathbf{p}+\mathbf{q})\frac{d\mathbf{q}}{(2\pi)^3}, \quad (47')$$

where $n_0(\mathbf{p})$ are the occupation numbers for non-interacting particles.

Let us consider the case of short range forces ($ap_0 \ll 1$, where a is the range of the potential). For excitations with low momentum k, the function $f_{k\omega}(\mathbf{p})$ differs from zero over a narrow range of momenta near p_0. Because of this, the potential $V(\mathbf{q})$ can be taken out from under the integral sign and, like $V(\mathbf{k})$, replaced by $V(0)$. Writing the difference $n_0(\mathbf{p}+\mathbf{k}/2) - n_0(\mathbf{p}-\mathbf{k}/2)$ in the form $\frac{1}{2}\mathbf{k}\partial f_0/\partial\mathbf{p}$ ($f_0 = 2n_0(\mathbf{p})$ is the distribution function of the non-interacting particles in the ground state), we find

$$f^0_{\mathbf{k}\omega}(\mathbf{p}) = -\frac{1}{2}\frac{\mathbf{k}\,\partial f_0/\partial\mathbf{p}}{\omega - \mathbf{kp} - i\delta[n_0(\mathbf{p}+\mathbf{k}/2) - n_0(\mathbf{p}-\mathbf{k}/2)]}V(0)\int f^0_{\mathbf{k}\omega}(\mathbf{p})\,d\mathbf{p}/(2\pi)^3, \quad (46'')$$

$$f^1_{\mathbf{k}\omega}(\mathbf{p}) = \frac{1}{2}\frac{\mathbf{k}\,\partial f_0/\partial\mathbf{p}}{\omega - \mathbf{kp} - i\delta[n_0(\mathbf{p}+\mathbf{k}/2) - n_0(\mathbf{p}-\mathbf{k}/2)]}V(0)\int f^1_{\mathbf{k}\omega}(\mathbf{p})\,d\mathbf{p}/(2\pi)^3 \quad (47'')$$

Equation (46″) coincides with the kinetic equation in the self-consistent field approximation, but with the number of particles reduced by a factor of two. This equation has a solution only for $V(0) > 0$, i.e., in the case of repulsion between the particles.

The formal solution of equation (47″) for the case of attraction is not justified, because of the readjustment of the Fermi sphere caused by the formation of correlated pairs. Thus for a repulsive short-range potential, propagation of spinless zeroth sound is possible.

In the case of long-range repulsive forces, V(k) has a pole for $k \to 0$, so that the second term in (46′) is much greater than the first, in which the integration makes the pole in V unimportant (we note that this result remains true in all approximations). Neglecting the first term, we write (46′) as

$$f^0_{\mathbf{k}\omega} = \frac{-\mathbf{k}\,\partial f_0/\partial \mathbf{p}}{\omega - \mathbf{kp} - i\delta\,[n_0(\mathbf{p}+\mathbf{k}/2) - n_0(\mathbf{p}-\mathbf{k}/2)]} \times V(\mathbf{k}) \int f^0_{\mathbf{k}\omega}(\mathbf{p}) \frac{d\mathbf{p}}{(2\pi)^3}. \tag{48}$$

This equation coincides with the kinetic equation for the **k**, ω Fourier components of the distribution function in the self-consistent field approximation.

Let us treat the behavior of the system in a weak electromagnetic field A(r, t). The Hamiltonian for the interaction of the system with the field is

$$H' = \frac{1}{c}\int j^\alpha(\mathbf{r},t)\,A^\alpha(\mathbf{r},t)\,dv. \tag{49}$$

The summation over α extends from 1 to 4.

After the field is switched on at time t_0, the wave function of the system varies in time according to the law

$$\Phi(t) = T\left\{\exp\left(-i\int_{t_0}^{t} H'dt'\right)\right\}\Phi_0 = T\left\{\exp\left[-\frac{i}{c}\iint_{t_0}^{t} j^\alpha(\mathbf{r},t')A^\alpha(\mathbf{r},t')\,dv\,dt'\right]\right\}\Phi_0, \tag{50}$$

or, in first approximation in powers of the external field,

$$\Phi(t) = \left\{1 - \frac{i}{c}\int_{t_0}^{t}\int j^\alpha(\mathbf{r},t')\,A^\alpha(\mathbf{r},t')\,dv\,dt'\right\}\Phi_0. \tag{50'}$$

In (50) and (50′) the current operator is taken in the Heisenberg representation for the unperturbed Hamiltonian of the system. The current of the system at time t is determined by the average value of the operator j(r, t) over the function Φ(t). Making use of the fact that the current of the system is equal to zero in the unperturbed state Φ_0, we easily find

$$j^\beta(\mathbf{r},t) = -\frac{i}{c}\iint_{t_0}^{t} dt'dv' \langle [j^\beta(\mathbf{r},t),\, j^\alpha(\mathbf{r}',t')] \rangle A^\alpha(\mathbf{r}',t'), \tag{51}$$

where ⟨ ⟩ denotes an average over the ground state of the system. For the **k**-Fourier components, the relation (51) takes the form:

$$j^\beta_{\mathbf{k}}(t) = -\frac{i}{c}\int_{t_0}^{t} dt' \langle [j^\beta_{\mathbf{k}}(t),\, j^\alpha_{-\mathbf{k}}(t')]\rangle A^\alpha_{\mathbf{k}}(t')$$

$$= -\frac{i}{c}\int_{t_0}^{t} dt' \iint d\mathbf{p}\,d\mathbf{p}'\, j^\beta(\mathbf{p})\, j^\alpha(\mathbf{p}') \times \langle [a^+_{\mathbf{p}-\mathbf{k}/2}(t)\,a_{\mathbf{p}+\mathbf{k}/2}(t),\; a^+_{\mathbf{p}'+\mathbf{k}/2}(t')\,a_{\mathbf{p}'-\mathbf{k}/2}(t')]\rangle A^\alpha_{\mathbf{k}}(t'), \tag{51'}$$

where $j^\alpha(\mathbf{p}) = p^\alpha$ for α = 1, 2, 3, and $j^4(\mathbf{p}) = 1$. The average value of the commutator under the integral sign in (51) can be expressed in terms of the functions $f_{\mathbf{k}\omega}(\mathbf{p})$

$$\langle [a^+_{\mathbf{p}-\mathbf{k}/2}(t)\,a_{\mathbf{p}+\mathbf{k}/2}(t),\; a^+_{\mathbf{p}'+\mathbf{k}/2}(t')\,a_{\mathbf{p}'-\mathbf{k}/2}(t')]\rangle = \sum_s \{e^{-i\omega_s(t-t')} f_{\mathbf{k},\omega_s}(\mathbf{p})\, f^*_{\mathbf{k},\omega_s}(\mathbf{p}') - e^{i\omega_s(t-t')} f_{-\mathbf{k},\omega_s}(\mathbf{p}')\, f^*_{-\mathbf{k},\omega_s}(\mathbf{p})\}. \tag{52}$$

Thus the knowledge of this system of functions is sufficient for determining the current of the system. On the other hand, this commutator can be expressed directly in terms of the two-particle Green's function K. Denoting by $\widetilde{K}$ the two-particle Green's function in momentum representation for $t_1 = t_2 = t$ and $t_3 = t_4 = t'$ ($t - t' = \tau$):

$$K\left(\mathbf{p}+\frac{\mathbf{k}}{2},\, t,\, \mathbf{p}-\frac{\mathbf{k}}{2},\, t;\; \mathbf{p}'-\frac{\mathbf{k}}{2},\, t',\, \mathbf{p}'+\frac{\mathbf{k}}{2},\, t'\right) \equiv \widetilde{K}(\mathbf{p},\mathbf{p}',\mathbf{k};\tau), \tag{53}$$

we easily find

$$\langle [a^+_{\mathbf{p}-\mathbf{k}/2}(t)\,a_{\mathbf{p}+\mathbf{k}/2}(t),\; a^+_{\mathbf{p}'+\mathbf{k}/2}(t')\,a_{\mathbf{p}'-\mathbf{k}/2}(t')]\rangle = -i\widetilde{K}(\mathbf{p},\mathbf{p}',\mathbf{k};\tau) + i\widetilde{K}^*(\mathbf{p},\mathbf{p}',-\mathbf{k};\tau). \tag{54}$$

Substituting (54) in (51′), we have

$$j^\alpha_{\mathbf{k}}(t) = -\frac{1}{c}\int^{t} dt' \iint d\mathbf{p}\,d\mathbf{p}'\, j^\alpha(\mathbf{p})\, j^\beta(\mathbf{p}')\,\{\widetilde{K}(\mathbf{p},\mathbf{p}',\mathbf{k};\tau) - \widetilde{K}^*(\mathbf{p},\mathbf{p}',-\mathbf{k};\tau)\}\, A^\beta_{\mathbf{k}}(t'). \tag{55}$$

Going to the limit of $t_0 \to -\infty$, we get the relation between the time Fourier components of j(t) and A(t):

$$j^\alpha_{\mathbf{k},\omega} = \varkappa_{\alpha,\beta}(\mathbf{k},\omega)\, A^\beta_{\mathbf{k},\omega},$$

$$\varkappa_{\alpha,\beta}(\mathbf{k},\omega) = \frac{i}{2\pi c}\int d\mathbf{p}\,d\mathbf{p}' \frac{d\omega'}{\omega-\omega'+i\delta}\{\widetilde{K}(\mathbf{p},\mathbf{p}';\mathbf{k},\omega') - \widetilde{K}^*(\mathbf{p},\mathbf{p}';-\mathbf{k},-\omega')\}, \tag{56}$$

where $\widetilde{K}(\mathbf{p},\mathbf{p}';\mathbf{k},\omega)$ is the time Fourier component of the function $\widetilde{K}(\mathbf{p},\mathbf{p}';\mathbf{k},\tau)$.

In conclusion, the authors express their thanks to L. D. Landau and S. T. Beliaev for interesting discussions.

[1] H. Lehmann, Nuovo cimento **11**, 342 (1954).

[2] M. Ia. Lavrent' ev and B. V. Shabat, Методы теории функчий комплексного переменного

(Methods of the Theory of Functions of a Complex Variable), GITTL, 1951, p. 257.

[3] A. B. Migdal, J. Exptl. Theoret. Phys. (U.S.S.R.) **32**, 399 (1957), Soviet Phys. JETP **5**, 333 (1957).

[4] Iu. L. Klimontovich and V. P. Silin, J. Exptl. Theoret. Phys. (U.S.S.R.) **23**, 151 (1952).

[5] V. P. Silin, J. Exptl. Theoret. Phys. (U.S.S.R.) **23**, 641 (1952).

[6] L. D. Landau, J. Exptl. Theoret. Phys. (U.S.S.R.) **32**, 59 (1957), Soviet Phys. JETP **5**, 101 (1957).

[7] J. Schwinger, Proc. Nat. Acad. Sci. **37**, 452 (1951).

[8] M. Gell-Mann and F. Low, Phys. Rev. **84**, 350 (1951).

Translated by M. Hamermesh
22

A Collective Description of Electron Interactions: III. Coulomb Interactions in a Degenerate Electron Gas

DAVID BOHM, *Faculdade de Filosofia, Ciencias e Letras, Universidade de Sao Paulo, Sao Paulo, Brazil*

AND

DAVID PINES, *Department of Physics, University of Illinois, Urbana, Illinois*

(Received May 21, 1953)

The behavior of the electrons in a dense electron gas is analyzed quantum-mechanically by a series of canonical transformations. The usual Hamiltonian corresponding to a system of individual electrons with Coulomb interactions is first re-expressed in such a way that the long-range part of the Coulomb interactions between the electrons is described in terms of collective fields, representing organized "plasma" oscillation of the system as a whole. The Hamiltonian then describes these collective fields plus a set of individual electrons which interact with the collective fields and with one another via short-range screened Coulomb interactions. There is, in addition, a set of subsidiary conditions on the system wave function which relate the field and particle variables. The field-particle interaction is eliminated to a high degree of approximation by a further canonical transformation to a new representation in which the Hamiltonian describes independent collective fields, with n' degrees of freedom, plus the system of electrons interacting via screened Coulomb forces with a range of the order of the inter electronic distance. The new subsidiary conditions act only on the electronic wave functions; they strongly inhibit long wavelength electronic density fluctuations and act to reduce the number of individual electronic degrees of freedom by n'. The general properties of this system are discussed, and the methods and results obtained are related to the classical density fluctuation approach and Tomonaga's one-dimensional treatment of the degenerate Fermi gas.

I.

IN this paper we wish to develop a collective description of the behavior of the electrons in a dense electron gas which will be appropriate when a quantum-mechanical treatment of the electronic motion is required, as is the case for the electrons in a metal. Our collective description is based on the organized behavior of the electrons brought about by their long-range Coulomb interactions, which act to couple together the motion of many electrons. In the first paper of this series[1] hereafter referred to as I, we developed a collective description of the organized behavior in an electron gas due to the transverse electromagnetic interactions between the electrons. This was done by means of a canonical transformation to a set of transverse collective coordinates which were appropriate for a description of this organized behavior. Here we shall develop an analogous canonical transformation to a set of longitudinal collective coordinates which are appropriate for a description of the organization brought about by the Coulomb interactions.

In the preceding paper[2] hereafter referred to as II, we developed a detailed physical picture of the electronic behavior (due to the Coulomb interactions). Although the electron gas was treated classically, we shall see that most of the conclusions reached there are also appropriate (with certain modifications) in the quantum domain. Let us review briefly the physical picture we developed in II, since we shall have occasion to make frequent use of it in this paper.

We found that, in general, the electron gas displays both collective and individual particle aspects. The primary manifestations of the collective behavior are organized oscillation of the system as a whole, the so-called "plasma" oscillation, and the screening of the field of any individual electron within a Debye length by the remainder of the electron gas. In a collective oscillation, each individual electron suffers a small periodic perturbation of its velocity and position due to the combined potential of all the other particles. The cumulative potential of all the electrons may be quite large since the long range of the Coulomb interaction permits a very large number of electrons to contribute to the potential at a given point. The screening of the electronic fields may be viewed as arising from the Coulomb repulsion, which causes the electrons to stay apart, and so leads to a deficiency of negative charge in the immediate neighborhood of a given electron. The collective behavior of the electron gas is decisive for phenomena involving distances greater than the Debye length, while for smaller distances the electron gas is best considered as a collection of individual particles which interact weakly by means of a screened Coulomb force.

These conclusions were reached by analyzing the behavior of the electrons in terms of their density fluctuations. It was found that these density fluctuations could be split into two approximately independent components, associated with collective and individual particle aspects of the electronic motion. The collective component is present only for wavelengths greater than the Debye length and represents the "plasma" oscillation. It may be regarded as including the effects of the long range of the Coulomb force which leads to the simultaneous interaction of many particles. The individual particles component is associated with the random thermal motion of the electrons and shows no collective behavior; it represents a collection of individual electrons surrounded by co-moving clouds of

[1] D. Bohm and D. Pines, Phys. Rev. 82, 625 (1951).
[2] D. Pines and D. Bohm, Phys. Rev. 85, 338 (1952).

charge which act to screen their fields as described above. The individual particles component thus includes the effects of the residual short-range screened Coulomb force, which leads only to two-body collisions.

A quantum-mechanical generalization of the density fluctuation method is quite straightforward and is sketched briefly in Appendix I. However, we do not choose to adopt this point of view, because although it is quite useful in establishing the existence of collective oscillations and describing certain related phenomena, it does not enable one to obtain a satisfactory over-all description of the electron gas. Quantum-mechanical calculations aimed at solving for the wave functions and the energy levels of the system are much more conveniently done in terms of a Hamiltonian formalism through the use of appropriate canonical transformations.

Our general approach in this series of papers has been to analyze the collective oscillatory motion first, since this is associated with the long-range aspects of the interaction which, in a sense, are responsible for the major complications in the many-electron problem. Once the collective motion is accounted for, we then investigate the aspects of the electronic behavior which are independent of the collective behavior, and which, if our method is successful, should turn out to be simple. Thus we are led to seek a canonical transformation to a representation in which the existence of the collective oscillations is explicitly recognized, and in which these oscillations are independent of the individual electronic behavior. In this representation, which we shall call the collective representation, we do not expect that the electron gas can be described entirely in terms of the collective coordinates which describe the organized oscillations, since we know that the gas also displays individual particle behavior. We shall see that in the collective representation, the individual electronic coordinates correspond to the electrons plus their associated screening fields, and that as might be anticipated from II, these screened electrons interact rather weakly via a screened Coulomb force.

In this paper we shall be primarily concerned with obtaining the canonical transformation to the collective representation. We shall discuss the approximations involved and, in a general way, the resultant wave functions of our electron system in the collective representation. Our development of a quantum-mechanical description of the electron assembly makes possible a treatment of the effects of electron interaction in metallic phenomena which utilizes at the outset the simplicity brought about by the organized oscillatory behavior. The detailed application of the collective description to the electrons in a metal is given in the following paper,[3] hereafter referred to as IV.

Historically, the first utilization of the 'plasma' aspects of the electron gas in a metal is due to Kronig and Korringa,[4] who treated the effect of electron-electron interaction on the stopping power of a metal for fast charged particles. However, their treatment is open to objection, in that they describe the electron gas as a classical fluid, with an artificially introduced coefficient of internal friction. A more satisfactory treatment of electron-electron interaction in the stopping power problem is due to Kramers[4] and Bohr.[4] The quantum treatment of this problem from the viewpoint of the collective description is given in Paper IV.

Tomonaga[5] has independently investigated the extent to which a degenerate Fermi gas can be described in terms of longitudinal oscillations. Tomonaga's treatment is, however, confined to a one-dimensional system, and as we shall see, there are certain essential difficulties associated with its generalization to a three-dimensional system which make the direct extension of this approach to three dimensions impossible. The relationship between our approach and that of Tomonaga is discussed in Appendix II.

II.

We consider an aggregate of electrons embedded in a background of uniform positive charge, whose density is equal to that of the electrons. The Hamiltonian for our system may be written

$$\sum_i \frac{p_i^2}{2m} + 2\pi e^2 \sum_{kij}{}' \frac{e^{i\mathbf{k}\cdot(\mathbf{x}_i-\mathbf{x}_j)}}{k^2} - 2\pi n e^2 \sum_k{}' \frac{1}{k^2}, \tag{1}$$

where the first term corresponds to the kinetic energy of the electrons, the second to their Coulomb interaction and the third to a subtraction of their self energy. The prime in the summations over k denotes a sum in which $k=0$ is excluded, and this takes into account the uniform background of positive charge, and hence the over-all charge neutrality of our system.[6] In obtaining (1) we have used the fact that the Coulomb interaction between the ith and jth electrons may be expanded as a Fourier series in a box of unit volume, and is $(e^2/|\mathbf{x}_i-\mathbf{x}_j|)=4\pi e^2\sum_k(1/k^2)e^{i\mathbf{k}\cdot(\mathbf{x}_i-\mathbf{x}_j)}$. n is the total number of electrons and is numerically equal to the mean density (since we are working in a box of unit volume).

Instead of working directly with the Hamiltonian of Eq. (1), we shall find it convenient to introduce an equivalent Hamiltonian which is expressed in terms of the longitudinal vector potential of the electromagnetic field, $\mathbf{A}(\mathbf{x})$, where $\mathbf{A}(\mathbf{x})$ may be Fourier-analyzed as

$$\mathbf{A}(\mathbf{x}) = (4\pi c^2)^{\frac{1}{2}} \sum_k q_k \boldsymbol{\epsilon}_k e^{i\mathbf{k}\cdot\mathbf{x}}, \tag{2}$$

[3] D. Pines, following paper [Phys. Rev. 92, 626 (1953)].

[4] R. Kronig and J. Korringa, Physica 10, 406 (1943). See also H. A. Kramers, Physica 13, 401 (1947); A. Bohr, Kgl. Danske Videnskab. Selskab, Mat.-fys. Medd. 24, No. 19 (1948); and R. Kronig, Physica 14, 667 (1949).

[5] S. Tomonaga, Prog. Theor. Phys. 5, 544 (1950).

[6] We shall drop this prime in the remainder of this paper since we have no further occasion to make explicit use of the fact that the term with $k=0$ is excluded.

and ε_k denotes a unit vector in the $\mathbf{k}$ direction. The electric field intensity, $\mathbf{E}(\mathbf{x})$ is

$$\begin{aligned}\mathbf{E}(\mathbf{x}) &= -(4\pi)^{\frac{1}{2}} \sum_k \dot{q}_k \varepsilon_k e^{+i\mathbf{k}\cdot\mathbf{x}} \\ &= (4\pi)^{\frac{1}{2}} \sum_k p_{-k} \varepsilon_k e^{i\mathbf{k}\cdot\mathbf{x}}.\end{aligned} \tag{3}$$

To ensure that $\mathbf{A}(\mathbf{x})$ and $\mathbf{E}(\mathbf{x})$ are real, we take

$$q_k = -q_{-k}^*, \quad p_k = -p_{-k}^*. \tag{4}$$

Our equivalent Hamiltonian is then given by

$$H = \sum_i \left[\mathbf{p}_i + \frac{e}{c}\mathbf{A}(\mathbf{x}_i)\right]^2 \Big/ 2m + \int [E^2(\mathbf{x})/8\pi] d\mathbf{x} - 2\pi n e^2 \sum_k \frac{1}{k^2}, \tag{5}$$

which using (2) and (3) may be shown to become

$$\begin{aligned} H = \sum_i \frac{p_i^2}{2m} &+ \frac{e}{m}(4\pi)^{\frac{1}{2}} \sum_{ik} \varepsilon_k \cdot (\mathbf{p}_i - \hbar\mathbf{k}/2) q_k e^{i\mathbf{k}\cdot\mathbf{x}_i} \\ &+ (2\pi e^2/m) \sum_{ikl} \varepsilon_k \cdot \varepsilon_l q_k q_l e^{i(\mathbf{k}+\mathbf{l})\cdot\mathbf{x}_i} - \sum_k \tfrac{1}{2} p_k p_{-k} \\ &- 2\pi n e^2 \sum_k 1/k^2. \end{aligned} \tag{6}$$

This Hamiltonian, when used in conjunction with a set of subsidiary conditions acting on the wave function of our system,

$$\Omega_k \Phi = 0 \quad (\text{for all } k), \tag{7}$$

where

$$\Omega_k = p_{-k} - i\left(\frac{4\pi e^2}{k^2}\right)^{\frac{1}{2}} \sum_i e^{-i\mathbf{k}\cdot\mathbf{x}_i} \tag{8}$$

will lead to the correct electron equations of motion. Ω_k is proportional to the kth fourier component of $\operatorname{div}\mathbf{E}(\mathbf{x}) - 4\pi\rho(\mathbf{x})$, and hence these subsidiary conditions guarantee that Maxwell's equations are satisfied. It may easily be verified that the subsidiary condition operator Ω_k commutes with the Hamiltonian (6), so that if the subsidiary condition (7) is satisfied at some initial time, it will be true at all subsequent times.[7]

The equivalence of our Hamiltonian (6) with the Hamiltonian expressed by (1) may be seen by applying the unitary transformation[8] $\Phi = S\psi$, where,

$$S = \exp\left[-(1/\hbar) \sum_{ki} (4\pi e^2/k^2)^{\frac{1}{2}} q_k e^{i\mathbf{k}\cdot\mathbf{x}_i}\right]. \tag{9}$$

With this transformation, we find

$$\begin{aligned} p_i &\to S^{-1} p_i S = p_i - (4\pi e^2)^{\frac{1}{2}} \sum_k q_k \varepsilon_k e^{i\mathbf{k}\cdot\mathbf{x}_i} \\ p_k &\to S^{-1} p_k S = p_k + i(4\pi e^2/k^2)^{\frac{1}{2}} \sum_i e^{i\mathbf{k}\cdot\mathbf{x}_i}, \end{aligned}$$

and

$$\begin{aligned} H \to \mathcal{H} = S^{-1}HS &= \sum_i \frac{p_i^2}{2m} + 2\pi e^2 \sum_{ij} \frac{e^{i\mathbf{k}\cdot(\mathbf{x}_i - \mathbf{x}_j)}}{k^2} \\ &- \sum_k \frac{p_k p_{-k}}{2} - i \sum_{ki} \left(\frac{4\pi e^2}{k^2}\right)^{\frac{1}{2}} p_k e^{-i\mathbf{k}\cdot\mathbf{x}_i} - 2\pi n e^2 \sum_k \left(\frac{1}{k^2}\right). \end{aligned}$$

The subsidiary condition (8) becomes

$$p_{-k}\psi = 0 \quad (\text{for all } k).$$

If we choose a ψ which is independent of q_k, we may satisfy the new subsidiary condition identically, the terms involving p_k in the Hamiltonian will drop out, and $\mathcal{H}$ is seen to be equivalent to (1). We note that the term $-2\pi n e^2 \sum_k (1/k^2)$ was included in (6) so that this Hamiltonian might be numerically equivalent to (1), as well as leading to equivalent equations of motion, since this term is just what is needed to cancel the terms with $i=j$ in the Coulomb energy.

The introduction of the longitudinal decrees of freedom, q_k, and the subsidiary conditions (7) provides a convenient means of introducing the concept of independent collective oscillation within the framework of the Hamiltonian formalism. The utility of this representation lies in the fact that (7) introduces in a simple way a relationship between the fourier components of the electronic density, $\rho_k = \sum_i e^{-i\mathbf{k}\cdot\mathbf{x}_i}$, and a set of field variables p_k. We shall see that there is, in consequence, a very close parallel between the behavior of the ρ_k, as analyzed in II, and the behavior of our field coordinates. In this representation we find that the field variables (just as did the ρ_k) oscillate with a frequency equal to the plasma frequency, provided we neglect a small coupling between the collective motion and the individual electronic behavior (characterized by their random thermal motion). Furthermore, just as we found it possible in II to find a purely oscillatory component of the density fluctuations, which is approximately independent of the individual electronic behavior, so we shall here be able to carry out a canonical transformation to a new set of field variables, which describe pure collective behavior and do not interact with the individual electrons to a good degree of approximation. In this section we shall analyze the approximate oscillatory behavior of the (q_k, p_k), while in the next section we carry out the canonical transformation to the pure collective coordinates.

Before beginning our analysis, we find it desirable to modify somewhat our Hamiltonian (6). We found in Paper II that in the classical theory there is a minimum wavelength λ_c (which classically is the Debye length), and hence a maximum wave vector k_c, beyond which organized oscillation is not possible. We may anticipate that in the quantum theory a similar (but not identical) limit arises, so that there is a corresponding limit on the extent to which we can introduce collective coordinates to describe the electron gas.

[7] This may be contrasted with the customary gauge [corresponding to $\operatorname{div}\mathbf{A} = (1/c)\partial\varphi/\partial t$], in which the commutator of the subsidiary condition with H is proportional to the subsidiary condition itself, and is therefore zero only when the subsidiary condition is satisfied.

[8] See G. Wentzel, *Quantum Theory of Wave Fields* (Interscience Publishers, New York, 1949), p. 131.

Since this is the case, rather than introduce the full spectrum of longitudinal field coordinates (and associated subsidiary conditions) as we do in (6), we might as well confine our attention to only as many p_k and q_k as we expect to display collective behavior, i.e., (p_k, q_k) for $k<k_c$. The number of collective coordinates, n', will then correspond to the number of k values lying between $k=0$ and $k=k_c$, and so will be given by

$$n'=\frac{4\pi}{3}\frac{k_c^3}{(2\pi)^3}=\frac{k_c^3}{6\pi^2}. \tag{10}$$

One might expect that there is a natural upper limit to n', *viz.*, the total number of longitudinal degrees of freedom n (for a system of n electrons), since at most n independent longitudinal degrees of freedom may be introduced. In practice we find that n' is considerably less than this theoretical maximum.

The modification of (6) to include only terms involving (p_k, q_k) with $k<k_c$ may be conveniently carried out by applying a unitary transformation similar to (9), but involving only q_k for which $k>k_c$. Thus we take $\Phi=S\psi$ where,

$$S=\exp[-(1/\hbar)\sum_{i,k>k_c}(4\pi e^2/k^2)^{\frac{1}{2}}q_k e^{i\mathbf{k}\cdot\mathbf{x}_i}], \tag{9a}$$

and where ψ is chosen to be independent of all q_k with wave numbers greater than k_c. We then obtain for our Hamiltonian

$$H=\sum_i\frac{p_i^2}{2m}+(4\pi)^{\frac{1}{2}}\frac{e}{m}\sum_{ik<k_c}\boldsymbol{\varepsilon}_k\cdot(\mathbf{p}_i-\hbar\mathbf{k}/2)q_k e^{i\mathbf{k}\cdot\mathbf{x}_i}$$

$$+(2\pi e^2/m)\sum_{\substack{ik<k_c\\ l<k_c}}\boldsymbol{\varepsilon}_k\cdot\boldsymbol{\varepsilon}_l q_k q_l e^{i(\mathbf{k}+\mathbf{l})\cdot\mathbf{x}_i}-\sum_{k<k_c}\frac{p_k p_{-k}}{2}$$

$$+2\pi e^2\sum_{\substack{k>k_c\\ i\neq j}}\frac{e^{i\mathbf{k}\cdot(\mathbf{x}_i-\mathbf{x}_j)}}{k^2}-2\pi n e^2\sum_{k<k_c}\frac{1}{k^2}, \tag{11}$$

with the associated set of subsidiary conditions:

$$\Omega_k\psi=0 \quad (k<k_c). \tag{12}$$

We shall find it convenient, in dealing with this Hamiltonian, to split up the third term into two parts. That part for which $\mathbf{k}+\mathbf{l}=0$ is independent of the electron coordinates and is given by

$$\frac{2\pi n e^2}{m}\sum_{k<k_c}q_k q_{-k}=\frac{\omega_p^2}{2}\sum_{k<k_c}q_k q_{-k}, \tag{13}$$

where we have introduced ω_p, the so-called plasma frequency, defined by

$$\omega_p=(4\pi n e^2/m)^{\frac{1}{2}}. \tag{14}$$

The remaining part, for which $\mathbf{k}+\mathbf{l}\neq 0$, we shall denote by U where,

$$U=\frac{2\pi e^2}{m}\sum_{\substack{ik<k_c\\ l<k_c\\ l\neq -k}}\boldsymbol{\varepsilon}_k\cdot\boldsymbol{\varepsilon}_l q_k q_l e^{i(\mathbf{k}+\mathbf{l})\cdot\mathbf{x}_i}. \tag{15}$$

U is much smaller than (13), for it always depends on the electron coordinates, and since these are distributed over a wide variety of positions, there is a strong tendency for the various terms entering into U to cancel. Let us for the time being neglect U, a procedure which we have called the random phase approximation in our earlier papers, and which we shall presently justify.

With this approximation we see that the third and fourth terms in our Hamiltonian (11) reduce to

$$H_{\text{osc}}=-\tfrac{1}{2}\sum_{k<k_c}(p_k p_{-k}+\omega_p^2 q_k q_{-k}) \tag{16}$$

the Hamiltonian appropriate to a set of harmonic oscillators, representing collective fields, with a frequency ω_p. The first term in (11) represents the kinetic energy of the electrons, while the second term,

$$H_I=(4\pi)^{\frac{1}{2}}\frac{e}{m}\sum_{ik<k_c}\boldsymbol{\varepsilon}_k\cdot\left(\mathbf{p}_i-\frac{\hbar\mathbf{k}}{2}\right)q_k e^{i\mathbf{k}\cdot\mathbf{x}_i}, \tag{17}$$

represents a simple interaction between the electrons and the collective fields, which is linear in the field variables. The fifth term,

$$H_{\text{s.r.}}=2\pi e^2\sum_{\substack{k>k_c\\ i\neq j}}\frac{e^{i\mathbf{k}\cdot(\mathbf{x}_i-\mathbf{x}_j)}}{k^2}, \tag{18}$$

represents the short-range part of the Coulomb interaction between the electrons. If we carry out the indicated summation, we find

$$H_{\text{s.r.}}=\tfrac{1}{2}\sum_{i\neq j}\frac{e^2}{|\mathbf{x}_i-\mathbf{x}_j|}\left\{1-\frac{2}{\pi}\operatorname{Si}(k_c|\mathbf{x}_i-\mathbf{x}_j|)\right\}, \tag{19}$$

where

$$\operatorname{Si}(y)=\int_0^y dx\frac{\sin x}{x}.$$

$\operatorname{Si}(y)=\pi/2$ for $y=2$ and oscillates near $\pi/2$ for larger values of y, so that $H_{\text{s.r.}}$ describes screened electron interaction with a range $\sim k_c$. A plot of $H_{\text{s.r.}}$ is given in Fig. 1.

Thus we see that in using (11) we have redescribed the long-range part of the Coulomb interactions between the electrons in terms of the collective oscillations (16), which interact with the electrons via H_I, (17). Our problem has now been reduced to one quite analogous to that encountered in I, *viz.*, a set of particles interacting with collective fields; the only new complications are the short-range interaction $H_{\text{s.r.}}$, and the subsidiary conditions on the system wave function. We shall see that as was the case in I with the transverse collective oscillations, the coupling between the fields and particles described by H_I is not very strong, so that it is possible to obtain a good qualitative understanding of the behavior of the system by neglecting this term. In this section, we shall make this approximation, and then investigate to what extent it applies, while in Sec. III we will give a more accurate treatment which includes the effects of the electron-field interaction.

If we neglect H_I, we may write the stationary state wave function as

$$\psi = \psi_{\text{osc}}\chi(\mathbf{x}_i \cdots \mathbf{x}_n). \quad (20)$$

ψ_{osc} represents the wave functions of the collective fields, and may be written as a product of harmonic oscillator wave functions like

$$h_n(p_k)\exp[-(|p_k|^2/2\hbar\omega_p)],$$

where h_n is the nth Hermite polynomial, and we are using the *momentum* representation of the oscillator wave functions. $\chi(x_i \cdots x_n)$ represents the eigenfunction for a set of particles interacting through $H_{\text{s.r.}}$. For the lowest state, we then get

$$\psi_0 = [\exp - \{\sum_{k<k_c} |p_k|^2/2\hbar\omega_p\}]\chi_0(\mathbf{x}_1 \cdots \mathbf{x}_n), \quad (21)$$

where $\chi_0(\mathbf{x}_1 \cdots \mathbf{x}_n)$ is the lowest state electron wave function.

In general χ_0 will be quite complex. However, just because the long-range part of the Coulomb potential is included in the oscillator energy, the remaining part $H_{\text{s.r.}}$ is considerably reduced in effectiveness. In fact it will often be of so short a range that for many purposes the free particle wave functions will constitute an adequate approximation. In this case, the lowest state wave function is

$$\psi_0 = \{\exp[-\sum_{k<k_c} |p_k|^2/2\hbar\omega_p]\}D_0(\mathbf{x}_1 \cdots \mathbf{x}_n), \quad (22)$$

where D_0 is the usual Slater determinantal wave function composed of the free electron wave functions appropriate to the ground state of the individual electrons. Our wave function ψ_0 then satisfies the exclusion principle.

Let us now consider the effects of the subsidiary conditions (12). In the representation in which p_k and $\mathbf{x}_i$ are diagonal these reduce to n' algebraic relations. We can view these relations in either of the following ways:

(a) They permit us to eliminate the p_k in terms of the $\mathbf{x}_i$.

(b) They permit us to eliminate n' of the $\mathbf{x}_i$ in terms of the p_k.

Let us begin with the first way. Our wave function (22) then becomes

$$\psi_0 = \exp[-\tfrac{1}{2}(\sum_{ij} F(\mathbf{x}_i - \mathbf{x}_j)/\hbar\omega_p)D_0(\mathbf{x}_1 \cdots \mathbf{x}_n), \quad (23)$$

where

$$F(\mathbf{x}_i - \mathbf{x}_j) = 2\pi e^2 \sum_{k<k_c} e^{i\mathbf{k}\cdot(\mathbf{x}_i - \mathbf{x}_j)}/k^2 \quad (24)$$

represents the long-range part of the Coulomb potential. $F(\mathbf{x}_i - \mathbf{x}_j) = e^2/|\mathbf{x}_i - \mathbf{x}_j|$ for $|\mathbf{x}_i - \mathbf{x}_j| \gg 1/k_c$ but approaches a constant $4\pi e^2 \sum_{k<k_c}(1/k^2)$, when $|\mathbf{x}_i - \mathbf{x}_j| \ll 1/k_c$. Thus in (23) we have the usual free electron wave function D_0 modified by a factor which describes long-range electron correlation, such that the probability that two electrons are found a given distance apart is less than that calculated by neglecting the Coulomb interactions or by including the short-range interaction $H_{\text{s.r.}}$. In fact in consequence of this correlation term, each electron tends to keep apart from the others, in a manner quite similar to that obtained in the classical treatment of II. A similar result has been

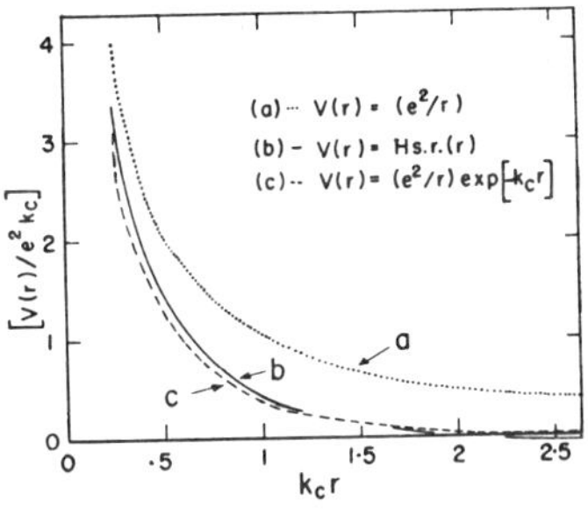

FIG. 1. $H_{\text{s.r.}}(r)$ compared with (e^2/r) and $(e^2/r)\exp(-k_c r)$.

obtained by Tomonaga in his one-dimensional treatment.

Let us now consider method (b), in which we seek to eliminate n' of the particle variables in terms of the field variables p_k. As is clear from the form of (12), this is a much more formidable task, one which we are not able to carry out explicitly. However, as we shall see throughout this paper, we can still draw a number of useful conclusions concerning the effect of such an elimination without actually solving for the $\mathbf{x}_i$ in terms of the p_k. In particular, we shall see in Sec. III how one may use a canonical transformation to replace (to lowest order in the field-particle coupling constant) n' of the individual particle degrees of freedom by as many collective degrees of freedom.

We now wish to justify our neglect of U and to investigate to what extent corrections arising from the inclusion of H_I will be of importance. In the remainder of this section we confine our attention to the lowest state of the system. We first show that the exact lowest state eigenfunction ψ_0 of our Hamiltonian (11) automatically satisfies the subsidiary conditions (12). For as we have noted the subsidiary condition operators Ω_k commute with the Hamiltonian H, so that the wave function ψ_0 can, in general, be expressed in terms of a series of simultaneous eigenfunctions of H and the Ω_k. The lowest state of the system is nondegenerate and hence corresponds to a single eigenvalue of the operator Ω_k, which we may call α_k. To determine the value of α_k, we consider a space displacement of the entire system (field plus electrons) through a distance $\Delta\mathbf{x}$, so that

$$\mathbf{x}\rightarrow\mathbf{x}'+\Delta\mathbf{x},\quad \mathbf{p}_i\rightarrow\mathbf{p}_i',$$
$$\mathbf{x}_i\rightarrow\mathbf{x}_i'+\Delta\mathbf{x}.$$

From Eqs. (2) and (3) we see that the effect of this displacement on the field coordinates is given by

$$p_k\rightarrow p_k'e^{-i\mathbf{k}\cdot\Delta\mathbf{x}},$$
$$q_k\rightarrow q_k'e^{i\mathbf{k}\cdot\Delta\mathbf{x}}.$$

The Hamiltonian is thus invariant under this displacement, while the subsidiary condition applied to our lowest state wave function becomes

$$e^{-i\mathbf{k}\cdot\Delta\mathbf{x}}\Omega_k'\psi_0=\alpha_k\psi_0.$$

However, since the lowest state is nondegenerate, it is not changed by this displacement, and we must have $\Omega_k'\psi_0=\alpha_k\psi_0$. Thus we find $\alpha_k=\alpha_k e^{i\mathbf{k}\cdot\Delta\mathbf{x}}$, which can only be satisfied if $\alpha_k=0$.

Thus, if we could obtain an exact solution for the lowest state eigenfunction ψ_0, we would automatically satisfy the subsidiary condition $\Omega_k\psi_0=0$. We may, in general, expect that if we obtain an approximate solution for ψ_0, we will not be able to satisfy the subsidiary condition, but that any error we make in determining the energy of the lowest state will not be increased by our failure to satisfy this subsidiary condition, since an exact solution satisfies the subsidiary condition and leads to the lowest possible energy state. The situation with regard to the excited states of the system will be somewhat different, and we will return to this question later.

Let us take as our approximate ψ_0, the wave function (22). In this approximation the energy of the lowest state is given by

$$\begin{aligned}E&=\tfrac{3}{5}E_0+\sum_{k<k_c}\frac{\hbar\omega_p}{2}-\frac{2\pi ne^2}{k^2}+\langle H_{\mathrm{s.r.}}\rangle_{\mathrm{Av}}\\&=\tfrac{3}{5}E_0+n'\frac{\hbar\omega_p}{2}-\frac{ne^2}{\pi}k_c+\langle H_{\mathrm{s.r.}}\rangle_{\mathrm{Av}},\end{aligned}\qquad(25)$$

where E_0 is the energy of an electron at the top of the Fermi distribution, and $\langle H_{\mathrm{s.r.}}\rangle_{\mathrm{Av}}$ is the exchange energy arising from the screened Coulomb interaction term, $H_{\mathrm{s.r.}}$, Eq. (18). We will not be concerned with evaluating $\langle H_{\mathrm{s.r.}}\rangle_{\mathrm{Av}}$ at present (reserving this for Paper IV), as we are here primarily interested in evaluating the corrections arising from U and H_{I}. We estimate these terms using perturbation theory. With the wave function (22) the average values of U and H_{I} vanish. U, in fact, has non-vanishing matrix elements only between the lowest state (with zero quanta) and a two-quantum state, while H_{I} connects the lowest state and a one-quantum state. From second-order perturbation theory, we have

$$\Delta U=-\sum_n\frac{|U_{0n}|^2}{E_n-E_0},\qquad(26)$$

where

$$U_{0n}=\frac{2\pi e^2}{m}\frac{\hbar}{2\omega_p}\boldsymbol{\varepsilon}_k\cdot\boldsymbol{\varepsilon}_l,\qquad(27)$$

if the state n has two quanta of momentum $\mathbf{k}$ and $\mathbf{l}$, respectively; and

$$E_n-E_0=2\hbar\omega_p+\frac{\hbar(\mathbf{k}+\mathbf{l})^2}{2m}-\frac{\hbar(\mathbf{k}+\mathbf{l})}{m}\cdot\mathbf{p}_i,\qquad(28)$$

if the electron in the initial state has momentum $\mathbf{p}_i$. In (28) we may, for the purpose of this rough estimate, approximate $E_n-E_0=2\hbar\omega_p$ since, as we shall see in Paper IV, $\hbar(\mathbf{k}+\mathbf{l})^2/2m-[\hbar(\mathbf{k}+\mathbf{l})/m]\cdot\mathbf{p}_i$ is always appreciably less than $2\hbar\omega_p$ as long as k, $l<k_c$. We then find

$$\begin{aligned}\Delta U&=-\left(\frac{\pi e^2\hbar}{m\omega_p}\right)^2 n\sum_{\substack{k<k_c\\l<k_c}}\frac{(\boldsymbol{\varepsilon}_k\cdot\boldsymbol{\varepsilon}_l)^2}{2\hbar\omega_p}\\&=-\left(\frac{\hbar\omega_p}{4}\right)^2\frac{1}{n}\frac{(n')^2}{6\hbar\omega_p}=-\frac{1}{48}\left(\frac{n'}{n}\right)\frac{n'\hbar\omega_p}{2}.\end{aligned}\qquad(29)$$

Thus ΔU introduces a fractional change in the zero point energy, per oscillator, of $(1/48)(n'/n)$, and since n' is never greater than n (and is, in fact, often quite a bit smaller), this change is negligible. Thus, we are justified in neglecting completely the term U.

We may estimate the corrections arising from H_{I} in similar fashion. We have

$$\Delta H_{\mathrm{I}}=-\sum_n\frac{|H_{\mathrm{I}0n}|^2}{E_n-E_0},\qquad(30)$$

where

$$(H_{\mathrm{I}})_{n0}=(2\pi\hbar/\omega_p)^{\frac{1}{2}}(e/m)\boldsymbol{\varepsilon}_k\cdot(\mathbf{p}_i-\hbar\mathbf{k}/2),\qquad(31)$$

if the state n has one quantum of momentum $\mathbf{k}$ present, and

$$E_n-E_0=\hbar\omega_p+\frac{\hbar k^2}{2m}-\frac{\hbar\mathbf{k}\cdot\mathbf{p}_i}{m}\cong\hbar\omega_p.\qquad(32)$$

We then find

$$\begin{aligned}\Delta H_{\mathrm{I}}&=-\frac{2\pi\hbar}{\omega_p}\frac{e^2}{m^2}\sum_{i,k<k_c}\frac{[\boldsymbol{\varepsilon}_k\cdot(\mathbf{p}_i-\hbar\mathbf{k}/2)]^2}{\hbar\omega_p}\\&=-\frac{n'}{3n}\left\{\sum_i\frac{p_i^2}{2m}-\frac{9n}{40}\frac{\hbar^2k_c^2}{m}\right\}.\end{aligned}\qquad(33)$$

Since, as we shall see, $k_c \lesssim k_0$, the wave vector of an electron at the top of the Fermi distribution, we see that the second term in the parenthesis in (33) is generally somewhat smaller than the first, and the first term corresponds to a fractional correction in the kinetic energy per electron (and thus in its effective mass) of $\sim n'/3n$. This may be appreciable if $n' \sim n$ but otherwise is small. This term implies a similar order of magnitude correction for the frequency of the collective oscillations, since $\sum_i p_i^2/2m$ and $n'\hbar\omega_p/2$ are roughly of the same order of magnitude. Thus we find that we are justified in neglecting $H_{\rm I}$ in order to obtain a qualitative and rough quantitative understanding of the behavior of our system, but that the effects arising from $H_{\rm I}$ should definitely be taken into account in a careful quantitative treatment. This we shall give in Sec. III.

Thus far we have not specified the value of k_c, and hence the number of collective degrees of freedom we find it desirable to introduce in our treatment. We may obtain a rough qualitative estimate of n' by minimizing our approximate expression for the lowest state energy (25) with respect to k_c (or n'). For the purpose of this rough estimate, let us neglect the dependence of $\langle H_{\rm s.r.}\rangle_{\rm Av}$ on k_c. We then note that the second term in (25) will be negative for those k for which $(2\pi ne^2/k^2) > \hbar\omega_p/2$. Hence we obtain the minimum value for (25) if we include in this summation, only those k for which this inequality is satisfied. This criterion yields

$$k_c^2 = \frac{4\pi ne^2}{\hbar\omega_p} = \frac{k_0^2}{2.14}\left(\frac{r_0}{a_0}\right)^{\frac{1}{2}}, \tag{34}$$

where r_0 is the interelectronic spacing, defined by

$$n = (4\pi r_0^3/3)^{-1}, \tag{35}$$

and a_0 is the Bohr radius. For a typical metal like Na, we have $(r_s/a_0) \sim 4$ and hence $k_c \sim k_0$. From (10) we see that in this case $n' \sim n/2$. In Paper IV where we give a more detailed treatment of that choice of k_c which minimizes the energy, including the effects of $H_{\rm I}$, and $\langle H_{\rm s.r.}\rangle_{\rm Av}$, we find for Na, $k_c \sim 0.68k_0$, and $n' \sim n/8$ in fair agreement with this rough estimate.

Finally we may remark that with the choice of k_c(34), the energy of the lowest state is

$$\begin{aligned} E &= \tfrac{3}{5}E_0 + \frac{n'\hbar\omega_p}{2} - \frac{ne^2k_c}{\pi} + \langle H_{\rm s.r.}\rangle_{\rm Av} \\ &= \tfrac{3}{5}E_0 - \frac{2}{3}\frac{ne^2}{\pi}k_c + \langle H_{\rm s.r.}\rangle_{\rm Av}. \end{aligned} \tag{36}$$

The energy $-\frac{2}{3}(ne^2/\pi)k_c$ represents a long-range correlation energy, i.e., that energy associated with the long-range correlations in electronic positions described by the wave function (23). In contrast to the exchange energy, this term represents Coulomb correlations between electrons of both kinds of spin. For Na it is, per electron,

$$-\frac{2}{3\pi}e^2k_c \cong -\frac{2}{3\pi}e^2k_0 \cong -0.4\frac{e^2}{r_0}, \tag{37}$$

a not inconsiderable energy. In Paper IV we return to a more careful estimate of the long-range correlation energy.

III.

In this section we wish to consider the effect of the field particle interaction term $H_{\rm I}$ on the motion of the electrons and the collective oscillations. We do this with the aid of a canonical transformation which is chosen to eliminate $H_{\rm I}$ in first approximation. Thus we seek a canonical transformation to a new representation in which the coupling between the fields and the electrons is described by a term $H_{\rm II}$, which is appreciably smaller than $H_{\rm I}$, and may consequently be neglected to a good degree of approximation (comparable, say, with our neglect of U). We shall then see that the effects of the coupling between the electrons and the collective field variables, as described by $H_{\rm I}$, are threefold: there is an increase in the electronic effective mass, the frequency of the collective oscillations is increased and becomes k dependent, and the effective electron-electron interaction is modified. As we anticipated on the basis of our perturbation-theoretic estimate of $H_{\rm I}$ in the preceding section, none of these effects is so large as to destroy the qualitative conclusions we reached there, although the quantitative estimates of the energy and wave functions of our system are somewhat altered.

The measure of the smallness of $H_{\rm II}$, and hence the extent to which we are successful in carrying out our canonical transformation, is the expansion parameter

$$\alpha = \left\langle\left(\frac{\mathbf{k}\cdot\mathbf{p}_i}{m\omega}\right)^2\right\rangle_{\rm Av}, \tag{38}$$

where we average over the particle momenta and the collective field wave vectors, and ω is the frequency of the collective oscillations. We find

$$\alpha \cong \frac{9}{25} \times \frac{1}{3}\frac{k_c^2p_0^2}{m^2\omega_p^2} \cong \tfrac{1}{2}\beta^2\frac{a_0}{r_s}, \tag{39}$$

where we have replaced ω by its approximate value ω_p and

$$\beta = k_c/k_0. \tag{40}$$

It is clear that by choosing β or k_c small enough, our expansion parameter α may be made as small as we like. We shall assume throughout the remainder of this paper that such a choice has been made, i.e., that $\alpha \ll 1$. In Paper IV we show that this criterion is satisfied in that $\alpha \sim 1/16$ for the electronic densities encountered in metals, if we take for β that value which minimizes the total energy. Another parameter of whose smallness we shall have occasion to make use is the ratio of the number of collective degrees of freedom, n', to the total number of degrees of freedom, $3n$. For most metals, with the above choice of β, we find $(n'/3n)\sim 1/25$.

We shall make the further approximation of neglecting the effects of our canonical transformation on $H_{s.r.}$, the short-range Coulomb interaction between the electrons. From Eq. (11), we see that if we neglect H_I, the collective oscillations are not affected at all by $H_{s.r.}$. Thus $H_{s.r.}$ can influence the q_k only indirectly through H_I. But, as we shall see, the *direct* effects of H_I on the collective oscillations are small. Thus, it may be expected that the *indirect* effects of $H_{s.r.}$ on the q_k through H_I are an order of magnitude smaller and may be neglected in our treatment which is aimed at approximating the effects of H_I. We will justify this procedure in greater detail in the following section.

With regard to the subsidiary conditions (11), we shall find that to order α, the subsidiary conditions in our new representation involve only the new particle coordinates $(\mathbf{X}_i, \mathbf{P}_i)$. Thus we may write our new wave function in terms of products like

$$\Phi_{\text{field}}\chi(\mathbf{X}_1\cdots\mathbf{X}_n),$$

and the subsidiary conditions will only act on the $\chi(X_i)$. The n' subsidiary conditions may thus be viewed as consisting of n' relationships among the particle variables, which effectively reduce the number of individual electronic degrees of freedom from $3n$ to $3n-n'$. This reduction is necessary, since in this new representation the n' collective degrees of freedom must be regarded as independent. For the field coordinates no longer appear in the subsidiary conditions, and hence describe real collective oscillation, which is independent of the electronic motion in this new representation.

There is a close resemblance between our Hamiltonian (11), which describes a collection of electrons interacting via longitudinal fields, and the Hamiltonian we considered in I, which described a collection of electrons interacting via the transverse electromagnetic fields. In fact, we shall see that our desired canonical transformation is just the longitudinal analog of that used in Paper I to treat the organized aspects of the transverse magnetic interactions in an electron gas. In order to point up this similarity and to simplify the commutator calculus, we introduce the creation and destruction operators for our longitudinal photon field, a_k and a_k^*, which are defined by[9]

$$q_k=(\hbar/2\omega)^{\frac{1}{2}}(a_k-a_{-k}^*),\qquad p_k=i(\hbar\omega/2)^{\frac{1}{2}}(a_k^*+a_{-k}),\tag{41}$$

and which possess the commutation properties

$$[a_k, a_{k'}]=[a_k^*, a_{k'}^*]=0,\qquad [a_k, a_{k'}^*]=\delta_{kk'},\tag{42}$$

[9] ω is here unspecified, but will later be chosen to be the frequency of the collective oscillations.

in virtue of (41). In terms of these variables, we then write our Hamiltonian and supplementary conditions schematically as

$$H=H_{\text{part}}+H_I+H_{\text{field}}+H_{s.r.},\qquad \Omega_k\Phi=0\quad(k<k_c),\tag{43}$$

where, using (11), (12), and (41) and neglecting U (Eq. 15),

$$H_{\text{part}}=\sum_i\frac{p_i^2}{2m}-\sum_{k<k_c}\frac{2\pi e^2n}{k^2},\tag{44a}$$

$$H_I=(e/m)\sum_{i,k<k_c}(2\pi\hbar/\omega)^{\frac{1}{2}}\{\boldsymbol{\varepsilon}_k\cdot(\mathbf{p}_i-\hbar\mathbf{k}/2)a_ke^{i\mathbf{k}\cdot\mathbf{x}_i}+e^{-i\mathbf{k}\cdot\mathbf{x}_i}a_k^*\boldsymbol{\varepsilon}_k\cdot(\mathbf{p}_i-\hbar\mathbf{k}/2)\},\tag{44b}$$

$$H_{\text{field}}=\sum_{k<k_c}\frac{\hbar\omega}{2}(a_k^*a_k+a_ka_k^*)+\frac{\hbar\omega}{4}(\omega_p^2-\omega^2)\times(a_k^*a_k+a_ka_k^*-a_ka_{-k}-a_{-k}^*a_k^*),\tag{44c}$$

$$H_{s.r.}=2\pi e^2\sum_{\substack{k>k_c\\ i\neq j}}\frac{e^{i\mathbf{k}\cdot(\mathbf{x}_i-\mathbf{x}_j)}}{k^2},\tag{44d}$$

$$\Omega_k=a_k^*+a_{-k}-\left(\frac{8\pi e^2}{k^2\hbar}\right)^{\frac{1}{2}}\sum_i e^{+i\mathbf{k}\cdot\mathbf{x}_i}.\tag{44e}$$

We note that H_{field} takes the form (44c), because we have expanded in terms of creation operators of frequency ω rather than ω_p.

We now consider a transformation from our operators $(\mathbf{x}_i, \mathbf{p}_i, a_k, a_k^*)$ to a new set of operators $(\mathbf{X}_i, \mathbf{P}_i, A_k, A_k^*)$, which possess the same eigenvalues and satisfy the same commutation rules as our original set.[10] The relation between these two sets may be written as

$$\mathbf{x}_i=e^{-iS/\hbar}\mathbf{X}_ie^{+iS/\hbar}\tag{45}$$

(with similar equations for $\mathbf{p}_i$, a_k, and a_k^*); (45) may be viewed as an operator equation, and we may take S the generating function of our canonical transformation to be a function of the new operators $(\mathbf{X}_i, \mathbf{P}_i, A_k, A_k^*)$ only. The operator relationship between the old and new Hamiltonians is

$$H=e^{-iS/\hbar}\mathcal{H}e^{iS/\hbar}=H_{\text{new}},\tag{46}$$

where $\mathcal{H}$ represents that Hamiltonian which is the same function of the new coordinates as H is of the old, and H_{new} denotes the Hamiltonian expressed in terms of the new coordinates.

[10] Quantum mechanical transformation theory is developed in, for instance, P. A. M. Dirac, *Principles of Quantum Mechanics* (Oxford University Press, London, 1935), second edition.

The problem of finding the proper form of S to realize our program was solved by a systematic study of the equations of motion. We do not have space to go into the details of this study here but confine ourselves to giving the correct transformation below. We shall then demonstrate that it leads to the desired results. Our canonical transformation is generated by

$$S=-(ei/m)\sum_{ik<k_c}(2\pi\hbar/\omega)^{\frac{1}{2}}\left\{\frac{\boldsymbol{\varepsilon}_k\cdot(\mathbf{P}_i-\hbar\mathbf{k}/2)A_k}{\omega-\mathbf{k}\cdot\mathbf{P}_i/m+\hbar k^2/2m}\times\exp(i\mathbf{k}\cdot\mathbf{X}_i)-\exp(-i\mathbf{k}\cdot\mathbf{X}_i)A_k^*\times\frac{\boldsymbol{\varepsilon}_k\cdot(\mathbf{P}_i-\hbar\mathbf{k}/2)}{\omega-\mathbf{k}\cdot\mathbf{P}_i/m+\hbar k^2/2m}\right\}. \quad (47)$$

On comparison with Eq. (45) of I, this generating function may be seen to be just the longitudinal analog of the "transverse" generating function given there. [The additional term in $\hbar k/2$ arises because $\mathbf{k}\cdot\mathbf{P}_i$ does not commute with $\exp(i\mathbf{k}\cdot\mathbf{X}_i)$.] Since H_{inter} and H_{field} are also analogous to the transverse terms encountered in I, we may expect that many of the results obtained there may be directly transposed to this longitudinal case. The differences in the treatments will arise from a consideration of $H_{\text{short-range}}$ and the subsidiary conditions.

We find it convenient to write the relationship between any old operator, O_{old} and the corresponding new operator O_{new} as

$$\begin{aligned}O_{\text{old}}&=\exp(-iS/\hbar)O_{\text{new}}\exp(iS/\hbar)\\&=O_{\text{new}}+(i/\hbar)[O_{\text{new}},S]\\&\quad-(1/2\hbar^2)[[O_{\text{new}},S],S]+\cdots,\end{aligned} \quad (48)$$

and we will classify terms in this series according to the power of S they contain; i.e., $[O,S]$ is the first-order commutator of O and S. We then find, keeping only first-order commutators, that

$$\mathbf{p}_i=\mathbf{P}_i+(e/m)\sum_{k<k_c}\left(\frac{2\pi\hbar}{\omega}\right)^{\frac{1}{2}}\mathbf{k}\left\{\frac{\boldsymbol{\varepsilon}_k\cdot(\mathbf{P}_i-\hbar\mathbf{k}/2)}{\omega-\mathbf{k}\cdot\mathbf{P}_i/m+\hbar k^2/2m}\times A_k\exp(i\mathbf{k}\cdot\mathbf{X}_i)+\exp(-i\mathbf{k}\cdot\mathbf{X}_i)A_k^*\times\frac{\boldsymbol{\varepsilon}_k\cdot(\mathbf{P}_i-\hbar\mathbf{k}/2)}{\omega-\mathbf{k}\cdot\mathbf{P}_i/m+\hbar k^2/2m}\right\}+\cdots, \quad (49)$$

$$a_k=A_k-(e/m)\sum_i\left(\frac{2\pi}{\hbar\omega}\right)^{\frac{1}{2}}\times\exp(-i\mathbf{k}\cdot\mathbf{X}_i)\frac{\boldsymbol{\varepsilon}_k\cdot(\mathbf{P}_i-(\hbar\mathbf{k})/2)}{\omega-\mathbf{k}\cdot\mathbf{P}_i/m+\hbar k^2/2m}+\cdots, \quad (50)$$

$$\begin{aligned}e^{i\mathbf{k}\cdot\mathbf{x}_i}=\exp(i\mathbf{k}\cdot\mathbf{X}_i)&+\sum_{l<l_c}\left(\frac{2\pi e^2\omega}{\hbar l^2}\right)^{\frac{1}{2}}\\&\times\left\{\frac{1}{\omega-\mathbf{l}\cdot\mathbf{P}_i/m+\hbar l^2/2m+\hbar\mathbf{l}\cdot\mathbf{k}/m}-\frac{1}{\omega-\mathbf{l}\cdot\mathbf{P}_i/m+\hbar l^2/2m}\right\}A_l\exp[i(\mathbf{l}+\mathbf{k})\cdot\mathbf{X}_i]\\&-\sum_{l<l_c}\left(\frac{2\pi e^2\omega}{\hbar l^2}\right)^{\frac{1}{2}}\exp(-i\mathbf{l}\cdot\mathbf{X}_i)A_l^*\\&\times\left\{\frac{1}{\omega-\mathbf{l}\cdot\mathbf{P}_i/m+\hbar l^2/2m+\hbar\mathbf{l}\cdot\mathbf{k}/m}-\frac{1}{\omega-\mathbf{l}\cdot\mathbf{P}_i/m+\hbar l^2/2m}\right\}\exp(i\mathbf{k}\cdot\mathbf{X}_i)+\cdots,\end{aligned} \quad (51)$$

and we shall use these relationships in determining H_{new}.

We now proceed in a manner directly analogous to that of Paper I. We classify terms in H_{new} by considering the corresponding schematic terms in H [Eq. (44a)–(44d)] from which they may be considered to arise. Every term, $\mathcal{T}$, in H, leads to a zero-order (commutator) term, $\mathcal{T}$, which is the same function of the new variables as it was of the old variables, and in addition, a first order commutator, $+(i/\hbar)[\mathcal{T},S]$, a second-order term, $-(1/2\hbar^2)[\mathcal{T}.[\mathcal{T},S]]$, etc. A convenient grouping of the terms in H exists which considerably simplifies the calculation of H_{new}. To demonstrate this grouping, we consider

$$H_a=\sum_iP_i^2/2m+\sum_{k<k_c}(\hbar\omega/2)(a_k^*a_k+a_ka_k^*).$$

The first-order commutator arising from H_a is

$$\begin{aligned}+(i/\hbar)[\mathcal{H}_a,S]=-(e/m)&\sum_{i,k<k_c}(2\pi\hbar/\omega)^{\frac{1}{2}}\\&\times\{\boldsymbol{\varepsilon}_k\cdot(\mathbf{P}_i-\hbar\mathbf{k}/2)A_k\exp(i\mathbf{k}\cdot\mathbf{X}_i)\\&+\exp(-i\mathbf{k}\cdot\mathbf{X}_i)A_k^*\boldsymbol{\varepsilon}_k\cdot(\mathbf{P}_i-\hbar\mathbf{k}/2)\}.\end{aligned} \quad (52)$$

By Eq. (44c) we see that the above term is just the negative of H_{I}, expressed in terms of the new variables. Thus, the first-order commutator of $\mathcal{H}_a$ with S cancels the term arising from the zero-order commutator of $\mathcal{H}_{\text{I}}$. $\mathcal{H}_{\text{I}}$ and $\mathcal{H}_a$ are thus "connected" in that a simple relationship exists between the various order commutators arising from these terms; in fact, the nth order commutator of $\mathcal{H}_a$ with S is equal to the negative of the $(n-1)$th-order commutator of $\mathcal{H}_{\text{I}}$ with S. The terms in H_{new} arising from the connected terms H_a+H_{I}, may consequently be written in the following series:

$$H'=\mathcal{H}_a+\sum_{n=1}^{\infty}[\mathcal{H}_{\text{I}},S]_n\left\{\frac{1}{n!}-\frac{1}{(n+1)!}\right\}(i/\hbar)^n, \quad (53)$$

where $[\mathcal{H}_I, S]_n$ is the nth-order commutator of $\mathcal{H}_I$ with S.

We shall see that the effects of the field-particle interaction (up to order α) are contained in the first correction term to $\mathcal{H}_a$, $[(i/2\hbar)S, \mathcal{H}_I]$. The higher-order commutators will be shown to lead to effects of order α^2 or $\alpha(n'/3n)$ and may hence be neglected. The evaluation of our lowest-order term, $(i/2h)[S, \mathcal{H}_I]$ is lengthy, but straightforward. We find, after some rearrangement of terms, that

$$(i/2h)[\mathcal{H}_I, S] = \frac{4\pi e^2}{m}\sum_{ki}\left(\frac{\hbar}{4\omega}\right) \times\left\{\frac{2\omega(\mathbf{k}\cdot\mathbf{P}_i/m)-(\mathbf{k}\cdot\mathbf{P}_i/m)^2+(\hbar^2k^4/4m^2)}{(\omega-\mathbf{k}\cdot\mathbf{P}_i/m)^2-(\hbar^2k^4/4m^2)}\right\}$$
$$\times(A_kA_k^*+A_k^*A_k)+\frac{4\pi e^2}{m}\sum_{ki}\left(\frac{\hbar}{4\omega}\right) \times\left\{\frac{2\omega(\mathbf{k}\cdot\mathbf{P}_i/m)+(\mathbf{k}\cdot\mathbf{P}_i/m)^2-(\hbar^2k^4/4m^2)}{(\omega+\mathbf{k}\cdot\mathbf{P}_i/m)^2-(\hbar^2k^4/4m^2)}\right\}$$
$$\times(A_kA_{-k}+A_k^*A_{-k}^*)-(\pi e^2/m^2) \times\sum_{\substack{k<k_c\\ i,j;i\neq j}}\frac{[\boldsymbol{\varepsilon}_k\cdot(\mathbf{P}_i-\hbar\mathbf{k}/2)][\boldsymbol{\varepsilon}_k\cdot(\mathbf{P}_j+\hbar\mathbf{k}/2)]}{\omega[\omega-\mathbf{k}\cdot\mathbf{P}_j/m-\hbar k^2/2m]}$$
$$\times\exp[i\mathbf{k}\cdot(\mathbf{X}_i-\mathbf{X}_j)]+\exp[-i\mathbf{k}\cdot(\mathbf{X}_i-\mathbf{X}_j)] \times\frac{\boldsymbol{\varepsilon}_k\cdot(\mathbf{P}_i-\hbar\mathbf{k}/2)\boldsymbol{\varepsilon}_k\cdot(\mathbf{P}_j+\hbar\mathbf{k}/2)}{\omega(\omega-\mathbf{k}\cdot\mathbf{P}_j/m-\hbar k^2/2m)}$$
$$-\frac{2\pi e^2}{m^2}\sum_{ik<k_c}\frac{(\boldsymbol{\varepsilon}_k\cdot\mathbf{P}_i)^2}{(\omega-\hbar k^2/2m)^2-(\mathbf{k}\cdot\mathbf{P}_i/m)^2}. \quad (54)$$

In obtaining (54) we have neglected a number of terms which are quadratic in the field variables and are multiplied by a phase factor with a nonvanishing argument, $\exp[i(\mathbf{k}+\mathbf{l})\cdot\mathbf{X}_i]$. These are terms like

$$\frac{4\pi e^2}{m^2}\sum_{\substack{k<k_c,i\\ l<k_c\\ l\neq -k}}\left(\frac{\hbar}{4\omega}\right)\frac{\boldsymbol{\varepsilon}_l\cdot(\mathbf{P}_i-\hbar\mathbf{l}/2)A_lA_k}{\omega-\mathbf{l}\cdot\mathbf{P}_i/m+\hbar l^2/2m} \times\exp[i(\mathbf{k}+\mathbf{l})\cdot\mathbf{X}_i]. \quad (55)$$

Such terms are of exactly the same character as those we considered earlier in U [Eq. (15)], except that they are smaller by a factor of $\sim(\mathbf{l}\cdot\mathbf{P}_i/m\omega)$. Exactly the same arguments that we applied in showing that U could be neglected may be applied to terms like (55), with the result that we find that these terms are also completely negligible, leading in fact to an energy correction which is smaller than that arising from U by a factor of α [Eq. (39)].

The remaining lowest-order term in H_{new} is just the zero-order term from $H_{\text{field}}-\sum_{k<k_c}(\hbar\omega/2)(a_k^*a_k+a_ka_k^*)$,

$$\mathcal{H}_{\text{field}}-\tfrac{1}{2}\sum_{k<k_c}(\hbar\omega/2)(A_k^*A_k+A_kA_k^*)=\sum_{k<k_c}(\hbar/4\omega) \times(\omega_p^2-\omega^2)[A_k^*A_k+A_kA_k^*-A_kA_{-k}-A_{-k}^*A_k^*]. \quad (56)$$

We will now show that if we define ω by the dispersion relation,

$$1=\frac{4\pi e^2}{m}\sum_i\frac{1}{(\omega-(\mathbf{k}\cdot\mathbf{P}_i/m))^2-\hbar^2k^4/4m^2}, \quad (57)$$

then the sum of (56) and the first two terms of (54) vanishes. To see this we note that multiplying (57) by $\omega^2-\omega_p^2$ on both sides, and rearranging terms on the right-hand side, yields

$$\omega^2-\omega_p^2=\frac{4\pi e^2}{m}\sum_i\frac{\omega^2-[\omega-(\mathbf{k}\cdot\mathbf{P}_i/m)]^2+\hbar^2k^4/4m^2}{[\omega-(\mathbf{k}\cdot\mathbf{P}_i/m)]^2-\hbar^2k^4/4m^2}$$
$$=\frac{4\pi e^2}{m}\sum_i\frac{2\omega(\mathbf{k}\cdot\mathbf{P}_i/m)+(\hbar^2k^4/4m^2)-(\mathbf{k}\cdot\mathbf{P}_i/m)^2}{[\omega-(\mathbf{k}\cdot\mathbf{P}_i/m)]^2-\hbar^2k^4/4m^2}, \quad (58)$$

from which the above statement follows for the $(A_k^*A_k+A_kA_k^*)$ terms in (54). The $(A_kA_{-k}$ and $A_k^*A_{-k}^*)$ terms likewise go out when we replace $\mathbf{k}$ by $-\mathbf{k}$ in (57) and (58) and use the resulting relations to compare (56) and (54).

The results in lowest order of our canonical transformation on the Hamiltonian may thus be expressed schematically as follows:

$$H_{\text{new}}^{(0)}=H_{\text{electron}}+H_{\text{coll.}}+H_{\text{res part}}, \quad (59)$$

where

$$H_{\text{electron}}=\sum_i\frac{P_i^2}{2m}-\frac{2\pi e^2}{m^2}\sum_{ik<k_c}\frac{(\boldsymbol{\varepsilon}_k\cdot\mathbf{P}_i)^2}{(\omega-\hbar k^2/2m)^2-(\mathbf{k}\cdot\mathbf{P}_i/m)^2}$$
$$+2\pi e^2\sum_{\substack{i,jk>k_c\\ i\neq j}}\frac{\exp[i\mathbf{k}\cdot(\mathbf{X}_i-\mathbf{X}_j)]}{k^2}-2\pi ne^2\sum_{k<k_c}\frac{1}{k^2}, \quad (60)$$

$$H_{\text{coll.}}=\tfrac{1}{2}\sum_{k<k_c}(\hbar\omega)(A_k^*A_k+A_kA_k^*), \quad (61)$$

and

$$H_{\text{res part}}=-\frac{\pi e^2}{m^2}\sum_{\substack{k<k_c\\ ij:i\neq j}}\frac{[\boldsymbol{\varepsilon}_k\cdot(\mathbf{P}_i-\hbar\mathbf{k}/2)][\boldsymbol{\varepsilon}_k\cdot(\mathbf{P}_j+\hbar\mathbf{k}/2)]}{\omega[\omega-\mathbf{k}\cdot\mathbf{P}_j/m-\hbar k^2/2m]}$$
$$\times\exp[i\mathbf{k}\cdot(\mathbf{X}_i-\mathbf{X}_j)]+\exp[-i\mathbf{k}\cdot(\mathbf{X}_i-\mathbf{X}_j)] \times\frac{[\boldsymbol{\varepsilon}_k\cdot(\mathbf{P}_i-\hbar\mathbf{k}/2)][\boldsymbol{\varepsilon}_k\cdot(\mathbf{P}_j+\hbar\mathbf{k}/2)]}{\omega[\omega-(\mathbf{k}\cdot\mathbf{P}_j/m)-(\hbar k^2/2m)]}. \quad (62)$$

The effect of our transformation on the subsidiary conditions may be obtained in similar fashion. Our new subsidiary conditions are given by

$$(\Omega_k)_{\rm new}\psi=\exp(-iS/\hbar)\Omega_k'\exp(iS/\hbar)\psi$$
$$=\left\{\Omega_k'-\frac{i}{\hbar}[S,\Omega_k']-\frac{1}{2\hbar^2}[S,[S,\Omega_k']]+\cdots\right\}$$
$$\times\psi=0,\quad (k<k_c),\qquad (63)$$

where ψ is our new system wave function, and Ω_k' is the same function of the new variables that Ω_k was of the old variables. We find

$$(\Omega_k)_{\rm new}=A_k^*\left(1-\frac{4\pi e^2}{m}\sum_i\frac{1}{[\omega-(\mathbf{k}\cdot\mathbf{P}_i/m)]^2-\hbar^2k^4/4m^2}\right)$$
$$+A_{-k}\left(1-\frac{4\pi e^2}{m}\sum_i\frac{1}{[\omega+(\mathbf{k}\cdot\mathbf{P}_i/m)]^2-\hbar^2k^4/4m^2}\right)$$
$$-\left(\frac{8\pi e^2}{k^2\hbar\omega}\right)^{\frac{1}{2}}\sum_i\frac{\omega^2}{\omega^2-(\mathbf{k}\cdot\mathbf{P}_i/m-\hbar k^2/2m)^2}$$
$$\times\exp(ik\cdot\mathbf{X}_i)--\left(\frac{8\pi e^2}{k^2\hbar\omega}\right)^{\frac{1}{2}}\sum_{\substack{l<l_c,i\\ l\neq k}}\left(\frac{2\pi e^2\omega}{\hbar l^2m^2}\right)^{\frac{1}{2}}$$
$$\times\left\{\left[\frac{1}{\omega+(\mathbf{l}\cdot\mathbf{P}_i/m)+\hbar l^2/2m-\hbar\mathbf{l}\cdot\mathbf{k}/m}\right.\right.$$
$$\left.-\frac{1}{\omega+\mathbf{l}\cdot\mathbf{P}_i/m+\hbar l^2/2m}\right]A_{-l}\exp[+i(\mathbf{k}-\mathbf{l})\cdot\mathbf{X}_i]$$
$$-\exp(-i\mathbf{l}\cdot\mathbf{X}_i)A_l^*\left[\frac{1}{\omega-\mathbf{l}\cdot\mathbf{P}_i/m+\hbar l^2/2m+\hbar\mathbf{l}\cdot\mathbf{k}/m}\right.$$
$$\left.\left.-\frac{1}{\omega-\mathbf{l}\cdot\mathbf{P}_i+\hbar l^2/2m}\right]\exp(i\mathbf{k}\cdot\mathbf{X}_i)\right\}$$
$$-\frac{1}{2\hbar^2}[S,[S,\Omega_k']]+\cdots.\qquad (64)$$

$(\Omega_k)_{\rm new}$ is considerably simplified when we note that the first two terms vanish when we apply the dispersion relation [Eq. (57)] for both plus and minus k. The fourth term consists of a linear term in the field coordinates multiplied by a nonvanishing phase factor, and the effect of such a term in the subsidiary condition is the same as that of a term like (55) in the Hamiltonian. Since there is no point in obtaining the subsidiary condition to a higher order of accuracy than is maintained in our Hamiltonian, we may neglect this term. With this approximation, our subsidiary condition reduces to one which does not involve (in lowest orders) the field variables, and is given by

$$(\Omega_k)_{\rm new}\psi=\sum_i\frac{\omega^2}{\omega^2-[(\mathbf{k}\cdot\mathbf{P}_i/m)-\hbar k^2/2m]^2}$$
$$\times\exp(i\mathbf{k}\cdot\mathbf{X}_i)\psi=0,\quad (k<k_c).\qquad (65)$$

IV.

The physical consequences of our canonical transformation follow from the lowest-order Hamiltonian, $H_{\rm new}^{(0)}$ [Eq. (59)] and the associated set of subsidiary conditions on our system wave function [Eq. (65)]. We discuss these briefly and then show that the higher-order terms in $H_{\rm new}$ and $(\Omega_k)_{\rm new}$ are actually negligible. We first note that our field coordinates occur only in $H_{\rm coll.}$, and thus describe a set of uncoupled fields which carry out real independent longitudinal oscillations, since the subsidiary conditions no longer relate field and particle variables, and since there are no field-particle interaction terms in $H_{\rm new}$. The frequency of these collective oscillations is given by the dispersion relation [Eq. (57)], which is the appropriate quantum-mechanical generalization of the classical dispersion relation derived in II, as well as being the longitudinal analog of the quantum-dispersion relation for organized transverse oscillation, which we obtained in I.

This dispersion relation plays a key role in our collective description, since it is only for $\omega(k)$ which satisfy it that we can eliminate the unwanted terms in the Hamiltonian [Eq. (54)] and the unwanted field terms in the subsidiary condition. For sufficiently small k, we may expand (57) in powers of $(\mathbf{k}\cdot\mathbf{P}_i/m\omega)$ and $(\hbar k^2/m\omega)$ and so always obtain a solution for $\omega(k)$. If we do this, and assume an isotropic distribution of $\mathbf{P}_i$, we find

$$\omega^2=\omega_p^2+\frac{k^2}{nm^2}\sum_i P_i^2+\frac{\hbar^2k^4}{4m^2},\qquad (66)$$

and hence

$$\omega=\omega_p\left(1+\frac{k^2}{2nm^2}\sum_i\frac{P_i^2}{\omega_p^2}+\frac{\hbar^2k^4}{8m^2\omega_p^2}\right).\qquad (67)$$

These appropriate dispersion relations are, in fact, quite sufficient for our purpose, since the expansions involved in obtaining them are the same that we have used in obtaining $H_{\rm new}$.

We have treated ω as a pure number thus far, although we see from (57) or (67) that ω is, in fact, an operator, since it contains $\mathbf{P}_i$. We have ignored this fact, for instance, in working out our commutation relations and obtaining $H_{\rm new}$. This treatment of ω as a pure number is only strictly justified if our system wave function is an eigenfunction of $\mathbf{P}_i$, which is not the case. Thus ω contains and, in turn, can contribute to the Hamiltonian, off-diagonal terms which cause transitions between states of different energy. These terms could then, in principle, be climinated from the

Hamiltonian by a further canonical transformation. However because the dependence of ω on P_i is already or order α, this elimination would produce terms of order α^2 which are truly negligible. We are justified in neglecting the off-diagonal elements of the operator ω.

According to (67), in consequence of the electron-field interaction the frequency of the collective oscillations has become k dependent. We may obtain an order-of-magnitude estimate of the fractional change in this frequency by averaging the dispersion relation (67) over all $k<k_c$ and carrying out the indicated sum over particle momenta. In obtaining this mean value of $\sum_i P_i^2$, we should use the appropriate eigenfunctions of our new Hamiltonian (59). However, as we shall see later, the correct particle eigenfunctions can be replaced for many applications by plane waves, so that $\sum_i P_i^2$ may be approximately evaluated by assuming a Fermi distribution of electrons at absolute zero. We then find

$$\langle\omega\rangle_{\text{Av}}=\omega_p\left(1+\frac{3}{10}\frac{\langle k^2\rangle_{\text{Av}}P_0^2}{m^2\omega_p^2}+\frac{\hbar^2\langle k^4\rangle_{\text{Av}}}{8m^2\omega_p^2}\right)$$

$$=\omega_p(1+3\alpha[1+(3/10)\beta^2]), \quad (68)$$

where α is given by (39) and β by (40). Since $\beta\lesssim 1$, we see that the effect of the k^4 term is small compared to the k^2 term. This result holds true quite generally, in that where an expansion in powers of $\alpha=\langle(\mathbf{k}\cdot\mathbf{P}_i/m\omega)^2\rangle_{\text{Av}}$ is justified, the terms of order $(\hbar^2k^4/4m^2\omega^2)$ are negligible. The average fractional increase in the frequency is thus of order 3α. As we have remarked, for the electronic densities encountered in metals, α turns out to be $\sim 1/16$, so that this constitutes at most a 20 percent correction in the collective oscillation frequency.

The effect on the electrons of the elimination (in lowest order) of the electron-field interaction may be seen by considering the second term in H_{electron} and $H_{\text{res part}}$. We first note that in the approximation of small α, the second term in H_{electron} becomes

$$E_{\text{red}}=-\frac{k_c^3}{18\pi^2 n}\sum_i\frac{P_i^2}{2m}=-\frac{n'}{3n}\sum_i\frac{P_i^2}{2m}.$$

If we combine this with the first term, $\sum_i P_i^2/2m$, we obtain

$$E_{\text{red}}+\sum_i\frac{P_i^2}{2m}\left(\frac{3n-n'}{3n}\right)=\sum_i\frac{P_i^2}{2m^*},$$

where

$$m^*=m\times 3n/(3n-n'). \quad (69)$$

Thus the "new" electrons behave as if they had an effective mass m^*, which is given by (69), and which is slightly greater than the "bare" electron mass m. This increase in the effective electronic mass has a simple physical interpretation. For we note that according to Eqs. (47) and (48),

$$\mathbf{X}_i=\mathbf{x}_i-\frac{ei}{m}\sum_{k<k_c}\left(\frac{2\pi\hbar}{\omega}\right)^{\frac{1}{2}}\left\{\frac{\boldsymbol{\varepsilon}_k\omega A_k}{\omega-\mathbf{k}\cdot\mathbf{P}_i/m+\hbar k^2/2m}\times\exp(i\mathbf{k}\cdot\mathbf{x}_i)-\exp(-i\mathbf{k}\cdot\mathbf{x}_i)\times\frac{\boldsymbol{\varepsilon}_k\omega A_k^*}{\omega-\mathbf{k}\cdot\mathbf{P}_i/m+\hbar k^2/2m}\right\}. \quad (70)$$

The $\mathbf{X}_i$ thus represents the "bare" electron plus an associated cloud of collective oscillation; the increased effective mass may be regarded as an inertial effect resulting from the fact that these electrons carry such a cloud along with them.

$H_{\text{res part}}$, in the approximation of small α, may be written as

$$H_{\text{r.p.}}=\frac{-2\pi e^2}{m^2}\sum_{\substack{k<k_c\\ i,j;i\neq j}}\frac{(\boldsymbol{\varepsilon}_k\cdot P_i)(\boldsymbol{\varepsilon}_k\cdot P_j)}{\omega_p^2+k^2\langle V^2\rangle_{\text{Av}}}\exp[ik\cdot(\mathbf{X}_i-\mathbf{X}_j)]$$

$$=-2\pi e^2\sum_{\substack{k<k_c\\ i,j;i\neq j}}\frac{(\boldsymbol{\varepsilon}_k\cdot\mathbf{P}_i)(\boldsymbol{\varepsilon}_k\cdot\mathbf{P}_j)/(m^2\langle V^2\rangle_{\text{Av}})}{k^2+K^2}\times\exp[i\mathbf{k}\cdot(\mathbf{X}_i-\mathbf{X}_j)], \quad (71)$$

where $\langle V^2\rangle_{\text{Av}}=\sum_i P_i^2/m^2 n$ and $K^2=\omega_p^2/\langle V^2\rangle_{\text{Av}}$. If we assume that the electrons form a completely degenerate gas, then for most metals,

$$K^2=(5/3)(\omega_p^2/v_0^2)\cong k_0^2. \quad (72)$$

Thus

$$H_{\text{r.p.}}\cong-2\pi e^2\sum_{\substack{i\neq j\\ k<k_c}}\frac{(\boldsymbol{\varepsilon}_k\cdot\mathbf{P}_i)(\boldsymbol{\varepsilon}_k\cdot\mathbf{P}_j)/(m^2\langle V^2\rangle_{\text{Av}})}{k^2+k_0^2}\times\exp[i\mathbf{k}\cdot(\mathbf{X}_i-\mathbf{X}_j)]. \quad (73)$$

$H_{\text{res part}}$ thus describes an extremely weak attractive velocity dependent electron-electron interaction. For if the summation in (73) were over all k, it would correspond to a screened interaction of range $\sim(1/k_0)$; however, the summation is only for $k<k_c$, where $k_c<k_0$, so that we are describing here that part of a screened interaction beyond the screening length. A more detailed analysis confirms that this qualitative estimate, and justifies our neglecting $H_{\text{r.p.}}$ in comparison with $H_{\text{s.r.}}$ in considering the effects of electron-electron interaction.

Let us now consider the effect of the higher-order terms, such as $[S,[S,\mathcal{H}_{\text{I}}]]$. {The higher-order commutators arising from $\mathcal{H}_{\text{field}}-\sum_{k<k_c}(\hbar\omega/2)(A_k^*A_k+A_kA_k^*)$ will be of this same type, since the zero-order commutator from this term cancelled part of $[S,\mathcal{H}_{\text{I}}]$.} The calculation of $[S,[S,\mathcal{H}_{\text{I}}]]$ is quite straightforward, but scarcely worth going into here, since by comparison of Eqs. (49), (50), (51), and (44), it may easily be seen that the lowest order non-negligible terms terms will resemble H_{I} but will be at least of order $(\mathbf{k}\cdot\mathbf{P}_i/m\omega)$ smaller. These terms which we earlier de-

noted by H_{II} could be eliminated by a further transformation. However, since as we have seen, the elimination of H_I led to effects of order α (or $n'/3n$), the effects so obtained would then be of order α^2, and we may neglect them entirely in our approximation of small α. Exactly the same conclusions apply with respect to the higher-order commutators of the subsidiary condition operator, $(\Omega_k)_{\text{new}}$, since it is not fruitful for us to evaluate $(\Omega_k)_{\text{new}}$ to any greater accuracy than that obtaining for H_{new}.

It is interesting to note that included in these higher-order terms is the influence of our effective mass correction, Eq. (69), on the frequency of the collective oscillations. Thus, on evaluating these terms, one finds

$$\omega^2=\omega_p^2+\frac{k^2}{n(m^*)^2}\sum_i P_i^2+\frac{\hbar^2k^4}{4(m^*)^2},$$

instead of the dispersion relation (66). This is, of course, just what might be expected, since the successive elimination of the field-particle interaction terms leads to a mass renormalization, familiar from quantum electrodynamics, in that everywhere m appears, it should properly be replaced by m^*.[11] This correction is here quite negligible, usually leading to a fractional change in the collective oscillation frequency of less than 1 percent. For this change is $\sim(\alpha n'/n)$, and for the electronic densities encountered in metals,

$$\alpha(n'/n)\cong(1/16)(3/25)=(3/400).$$

Our only other approximation has been to neglect the effect of the canonical transformation on $H_{\text{s.r.}}$, which will lead, indirectly, to the effect of $H_{\text{s.r.}}$ on the collective oscillations. Suppose we consider a typical first-order term arising from $[S, H_{\text{s.r.}}]$. This will be like

$$2\pi e^2 \sum_{\substack{k>k_c\\ l<k_c\\ i,j}} \left(\frac{2\pi e^2}{\hbar l^2}\right)^{\frac{1}{2}} \frac{\exp(-i\mathbf{l}\cdot\mathbf{X}_j)}{k^2} \frac{(\hbar\mathbf{l}\cdot\mathbf{k}/m)A_l^*\exp(i\mathbf{k}\cdot\mathbf{X}_i)\exp(-i\mathbf{k}\cdot\mathbf{X}_j)}{(\omega-\mathbf{l}\cdot\mathbf{P}_i/m+\hbar l^2/2m)(\omega-\mathbf{l}\cdot\mathbf{P}_i/m+\hbar l^2/2m+\hbar\mathbf{l}\cdot\mathbf{k}/m)}. \tag{74}$$

These terms thus consist of a nonvanishing phase factor multiplying a field variable and a short-wavelength density fluctuation. The structure of (74) is quite similar to that of U [Eq. (15)], the difference being that the short-wavelength density fluctuation $\sum_j \exp(-i\mathbf{k}\cdot\mathbf{X}_j)$ here plays the same role as the collective field variable (which is essentially a long-wavelength density fluctuation) did in U. If we had a term for which $\mathbf{k}=\mathbf{l}$, (74) would reduce to a term like H_I, just as the third term in (11) reduced to $\omega_p^2\sum_k \times(q_kq_{-k}/2)$. Thus we might expect that (74) bears about the same relationship to H_I, as U does to $(\omega_p^2/2) \times\sum_k q_kq_{-k}$. However, it is quite a bit more difficult to establish the smallness of (74) mathematically than it was for U, since a perturbation theoretic estimate involves the consideration of intermediate states in which two electrons are excited. We note that the main effect of $H_{\text{s.r.}}$ is to produce short-range correlations in particle positions, analogous to the long-range correlations produced by the long-range part of the Coulomb potential, in the sense that the particles tend to keep apart and thus tend to reduce the effectiveness of $H_{\text{s.r.}}$. Because of the analytical difficulties involved in a justification along these lines we prefer to justify our neglect of (74) in a more qualitative and physical fashion.

We see that (74) describes the effect of the collective oscillations on the short range collisions between the electrons, and conversely, the effect of the short-range collisions on the collective oscillations. We may expect that these effects will be quite small, since $H_{\text{s.r.}}$ is itself a comparatively weak interaction. The short-range electron-electron collisions arising from $H_{\text{s.r.}}$ will act to damp the collective oscillations, a phenomenon which has been treated in some detail classically by Bohm and Gross.[12] A test for the validity of our approximation in neglecting terms like (74) is that the damping time from the collisions be small compared with the period of a collective oscillation. In this connection we may make the following remarks:

(1) Electron-electron collisions are comparatively ineffective in damping the oscillations, since momentum is conserved in such collisions, so that to a first approximation such collisions produce no damping. [Such collisions produce damping only in powers of $(\mathbf{k}\cdot\mathbf{P}_i/m\omega)$ higher than the first.]

(2) The exclusion principle will further reduce the cross section for electron-electron collision.

(3) If H_I is neglected, collisions have no effect on the collective oscillations. This means that the major part of the collective energy is unaffected by these short-range collisions, since only that part coming from H_I, (which is of order α relative to $\hbar\omega_p$) can possibly be influenced. Thus at most 20 percent of the collective energy can be damped in a collision process.

All of these factors combine to reduce the rate of damping, so that we believe this rate is not more than 1 percent per period of an oscillation and probably is quite a bit less. A correspondingly small broadening of the levels of collective oscillation is to be expected. It is for these reasons that we feel justified in neglecting the effects of our canonical transformation on $H_{\text{s.r.}}$.

[11] Note, however, that m is not replaced by m^* in our expression for ω_p, Eq (14), since the collective oscillations are not affected by the field-electron interaction in this order.

[12] D. Bohm and E. P. Gross, Phys. Rev. 75, 1864 (1949).

V.

The motion of the electrons in our new representation is considerably more complicated than that of the collective fields. The major reason for this complication is our set of subsidiary conditions (65), which essentially act to reduce the number of individual electron degrees of freedom from $3n$ to $3n-n'$, where n' is the number of collective degrees of freedom and is given by

$$n'=\frac{k_c^3}{6\pi^2}=\frac{\beta^3k_0^3}{6\pi^2}=\frac{\beta^3n}{2}. \quad (75)$$

We may obtain a better understanding of the role of these subsidiary conditions by making use of the density fluctuation concept which we developed in Paper II. There we saw that classically the collective component of the density fluctuation ρ_k was proportional to

$$R_k{}^c=\sum_i\frac{1}{\omega^2-(\mathbf{k}\cdot\mathbf{P}_i/m)^2}e^{i\mathbf{k}\cdot\mathbf{x}_i}.$$

In a quantum-theoretical treatment of the density fluctuations, the collective component is found to be proportional to

$$R_k{}^q=\sum_i\frac{1}{\omega^2-[(\mathbf{k}\cdot\mathbf{P}_i/m)-\hbar k^2/2m]^2}e^{i\mathbf{k}\cdot\mathbf{x}_i}. \quad (76)$$

This result may be seen to follow directly from the quantum generalization of the methods of II given in Appendix I. In the preceding expressions, $\mathbf{x}_i$ and $\mathbf{p}_i$ of course refer to the "original" position and momentum of the electron, i.e., the Hamiltonian in terms of these variables is given by Eq. (1). On the other hand, our "new" electron variables (X_i, P_i) describe electron motion in the absence of any collective oscillation, since there are no terms in our Hamiltonian (59) which couple the electrons and the collective oscillation. Consequently we should expect that the collective component of the density fluctuation when expressed in terms of these "new" variables should vanish, since these variables are chosen to describe "pure" individual electron motion and are incapable of describing, or taking part in, collective oscillation. But this is just what our subsidiary conditions assert, as may be seen by comparing (76) and (65). Thus, if we carry out a transformation to "individual" electron variables, we must expect a set of subsidiary conditions given by (65), since these guarantee that we have developed a consistent description of the state of the electron gas in the absence of collective oscillation.

The physical content of the subsidiary condition also follows from the density fluctuation concept. For we may rewrite the subsidiary condition, Eq. (65) as

$$\sum_i\exp(+i\mathbf{k}\cdot\mathbf{X}_i)\psi=\sum_i\frac{(\mathbf{k}\cdot\mathbf{P}_i/m-\hbar k^2/2m)^2\exp(i\mathbf{k}\cdot\mathbf{X}_i)}{\omega^2-[(\mathbf{k}\cdot\mathbf{P}_i/m)-\hbar k^2/2m]^2}. \quad (77)$$

Since we are dealing with $k<k_c$, for which $(\mathbf{k}\cdot\mathbf{P}_i/m\omega)^2\ll1$, we see that the subsidiary condition asserts that in terms of our new coordinates and momenta, the density fluctuations of long wavelength are greatly reduced. This reduction is due to the fact that the major portion of the long-wavelength density fluctuations is associated with the collective oscillations, and described in terms of these in our collective description.

In our new representation, the subsidiary conditions (65) continue to commute with the Hamiltonian (59). This follows since the commutation relations are unchanged by a canonical transformation; it may easily be directly verified from (65) and (59) that these commute within the approximations we have made. Consequently, just as was the case with (11) and (12), if we correctly solved for the exact lowest state eigenfunction of our Hamiltonian H_{new}, we would automatically satisfy the subsidiary conditions (65), since the ground state of our system is nondegenerate. For this reason, the energy of the lowest state of our system is relatively insensitive to whether we satisfy the subsidiary conditions or not For since the lowest state wave function does satisty the subsidiary condition, moderate changes in this wave function, involving corresponding failures to satisfy the subsidiary conditions, will provide quite small changes in the energy. Conversely, because of this insensitivity of the ground state energy to the degree of satisfaction of the subsidiary condition, it will take a quite good approximation to the lowest eigenfunction of H_{new} to satisfy the subsidiary conditions to a fair degree of approximation.

It should be noted that the lowest state wave function satisfies the subsidiary condition because of the effects of the term $H_{\text{r.p.}}$ in the Hamiltonian. For as we have seen, the subsidiary condition describes a long range correlation in the particle positions, which is independent of the amplitude of collective oscillation. In the approximation that we are using, this correlation has to be due to the residual interaction between the particles, since the subsidiary conditions will automatically be satisfied if we solve for the lowest state wave function. At first sight, it might be thought that the short-range potential $H_{\text{s.r.}}$ might also play an important role in establishing these correlations, since it corresponds to a fairly strong interaction potential when the particles are close to each other. However, from the definition of $H_{\text{s.r.}}$ in Eq. (18), we see that it has no Fourier components corresponding to $k<k_c$. As a simple perturbation theoretical calculation shows, the only effect of $H_{\text{s.r.}}$ in the first approximation is to turn a plane wave function.

$$\psi_0=\exp(i\sum_n\mathbf{k}_n\cdot\mathbf{x}_n),$$

into the function

$$\psi=\psi_0(1+\sum_{\substack{m,n\\k>k_c}}C_{mn}\exp[i\mathbf{k}\cdot(\mathbf{x}_m-\mathbf{x}_n)]),$$

where C_{mn} is a suitable expansion coefficient, which can be obtained by a detailed calculation.[13] But since the sum is restricted to $k>k_c$, $H_{s.r.}$ introduces only *short-range* correlations, which have nothing to do with the subsidiary conditions. On the other hand, $H_{r.p.}$ which has only *long-range* fourier components (i.e., $k<k_c$) introduces only corresponding long-range correlations. Thus, in the present approximation, it is $H_{r.p.}$ that is responsible for the long-range correlations implied by the subsidiary condition.

On the basis of the above conclusions, we may deduce the following physical picture. The long range Coulomb forces produce a tendency for electrons to keep apart, as a result of which the Coulomb force itself tends to be neutralized. But this neutralization could not be perfect; for if it were, then there would be no force left to produce the necessary correlations in particle positions. Our calculations show that $H_{r.p.}$ is the small residual part of the Coulomb force which must remain unneutralized in order to produce the long-range correlations needed for agreeing. Because this force is so small, it will produce only correspondingly small changes in the particle momenta, so that in most applications a set of plane waves will provide a good approximation to the particle wave function (in the new representation, of course).

All of the above applies rigorously only in the ground state. In the excited states, similar conclusions apply; but the application of the subsidiary conditions is more difficult, because the wave functions of the excited states are no longer now degenerate. Here, we could in general expand an arbitrary eigenfunction of $H_{\text{new}}^{(0)}$ as a series of eigenfunctions of $(\Omega_k)_{\text{new}}$. To satisfy the subsidiary conditions, we then retain only those terms in this series for which $(\Omega_k)_{\text{new}}=0$. This reduction in the number of possible eigenfunctions corresponds to the reduction in the number of individual electron degrees of freedom implied by (65). The exact treatment of the problem of the excited states is quite complex and will be reserved for a later paper by one of us. However, we may expect that if the reduction in the number of individual electron degrees of freedom is comparatively small [i.e., $(n'/n)\ll 1$], then their effect on the energy spectrum of the electron gas will be correspondingly reduced.

We conclude this section by summing up the results of our canonical transformation to the collective description. We have obtained a Hamiltonian describing collective oscillation plus a system of individual electrons interacting via a screened Coulomb force, with a screening radius of the order of the inner-electronic distance. Although the individual electron wave functions are restricted by a set of n' subsidiary conditions, which act to reduce the number of individual electron degrees of freedom and to inhibit the long-range density fluctuations associated with the individual electron motion, for many purposes the effect of these subsidiary conditions may be neglected. In Paper IV we examine the physical conclusions we are led to by the use of the collective description for the motion of electrons in metals. We shall see that these are in good agreement with experiment and enable us to resolve a number of hitherto puzzling features of the usual one-electron theory.

One of us (D.P.) would like to acknowledge the support of the Office of Ordnance Research, U. S. Army, during the writing of this paper.

[13] The additional terms describe correlations in particle positions.

APPENDIX I

In this appendix we treat the collective fluctuations in charge density by finding the equations of motion of the associated operators, thus developing a direct quantum-mechanical extension of the methods used in Paper II. We use the electron field second-quantization formalism, in order to facilitate comparison with the work of Tomonaga and to take into account explicitly the fact that the electrons obey Fermi statistics.

Following the usual treatments,[14] we describe the electrons by the field quantities $\psi_\sigma(\mathbf{x})$ which satisfy the anti-commutation relations $[\psi_\sigma(\mathbf{x}),\psi_\sigma(\mathbf{x}')]_+=[\psi_\sigma^*(\mathbf{x}),\psi_{\sigma'}^*(\mathbf{x}')]_+=0$ and $[\psi_\sigma(\mathbf{x}),\psi_{\sigma'}^*(\mathbf{x}')]_+=\delta(\mathbf{x}-\mathbf{x}')\delta_{\sigma\sigma'}$. σ refers to the electron spin and takes on two values corresponding to the two orientations of the electron spin. We work in the Heisenberg representation. The Hamiltonian which determines the equation of motion of the ψ's is

$$H=-\int\psi^*(\mathbf{x})\frac{\hbar^2}{2m}\Delta\psi(\mathbf{x})d\mathbf{x}+\frac{e^2}{2}\int\int\frac{\rho(\mathbf{x})\rho(\mathbf{x}')}{|\mathbf{x}-\mathbf{x}'|}d\mathbf{x}d\mathbf{x}', \quad \text{(A1)}$$

where

$$\rho(\mathbf{x})=\textstyle\sum_\sigma\psi_\sigma^*(\mathbf{x})\psi_\sigma(\mathbf{x}). \quad \text{(A2)}$$

It is convenient to Fourier-analyze $\psi_\sigma(\mathbf{x})$ and $\psi_\sigma^*(\mathbf{x})$ by

$$\psi_\sigma(\mathbf{x})=\sum_k c_{k\sigma}e^{i\mathbf{k}\cdot\mathbf{x}},\qquad \psi_\sigma^*(\mathbf{x})=\sum_k c_{k\sigma}^*e^{-i\mathbf{k}\cdot\mathbf{x}}, \quad \text{(A3)}$$

where the $c_{k\sigma}$ and $c_{k\sigma}^*$ obey the anticommutation relations

$$[c_{k\sigma},c_{k'\sigma'}]_+=[c_{k\sigma}^*,c_{k'\sigma'}^*]_+=0,\qquad [c_{k\sigma},c_{k'\sigma'}^*]_+=\delta_{kk'}\delta_{\sigma\sigma'}, \quad \text{(A4)}$$

in virtue of the anticommutation relations satisfied by the $\psi_\sigma(x)$. We also find

$$\rho(x)=\textstyle\sum_k\rho_k e^{i\mathbf{k}\cdot\mathbf{x}}, \quad \text{(A5)}$$

where, using (A2) and (A3),

$$\rho_K=\textstyle\sum_k c_{k\sigma}^*c_{k+K\sigma}. \quad \text{(A6)}$$

In terms of $c_{k\sigma}$ and ρ_K, our Hamiltonian (A1) becomes

$$H=\sum_k c_{k\sigma}^*c_{k\sigma}\frac{\hbar^2k^2}{2m}+2\pi e^2\sum_k\frac{\rho_k\rho_{-k}}{k^2}. \quad \text{(A7)}$$

[14] See, for instance, G. Wentzel, *Quantum Theory of Wave Fields* (Interscience Publishers, New York, 1949).

The second quantization formalism we are using here is of course equivalent to the use of an antisymmetrized many-electron wave function in the usual configuration-space representation (which we use elsewhere in this paper). For instance, the density fluctuation operator ρ_k is equivalent to the configuration space operator $\sum_i \exp(-i\mathbf{k}\cdot\mathbf{X}_i)$ we introduce earlier. Thus the results obtained in this appendix may be directly compared to those obtained in the previous sections of this paper, and in Paper II.

In Paper II, we saw that classically ρ_k could be split into an oscillatory part q_k, and an additional part which represented the charge density of a set of screened electrons moving at random. We shall now show that a similar q_k can be introduced quantum mechanically, and is proportional to

$$q_K=\sum_{k\sigma}\frac{c_{k\sigma}{}^*c_{k+K\sigma}}{\omega^2-(\hbar\mathbf{k}\cdot\mathbf{K}-\hbar^2K^2/2m)^2}. \tag{A8}$$

In the usual coordinate representation, this operator is

$$q_K=\sum_i\frac{1}{\omega^2-(\mathbf{K}\cdot\mathbf{P}_i/m-\hbar K^2/2m)^2}\exp(-i\mathbf{K}\cdot\mathbf{X}_i). \tag{A8a}$$

In the limit of $\hbar\to 0$, this reduces to the q_k of Paper II (Eq. 16).

As in Paper II, Eq. (17) we find it convenient to introduce the quantities $\xi_{K,\omega}$, which are, quantum mechanically

$$\xi_{K,\omega}=\sum_k\frac{c_{k\sigma}{}^*c_{k+K\sigma}}{\omega-(\hbar\mathbf{k}\cdot\mathbf{K}-\hbar K^2/2m)}, \tag{A9}$$

and are related to q_K by

$$q_K=(1/2)[(\xi_{K,\omega}-\xi_{K,-\omega})/\omega]. \tag{A10}$$

If the $\xi_{K,\omega}$ satisfy

$$\dot{\xi}_{K,\omega}+i\omega\xi_{K,\omega}=0, \tag{A11}$$

then it immediately follows on differentiation of (A10) that

$$\ddot{q}_K+\omega^2q_K=0.$$

We have $\dot{\xi}_{K,\omega}=(1/i\hbar)[\xi_{K,\omega},H]$. If we use the commutation properties [Eq. (A4)], we find that

$$\begin{aligned}\dot{\xi}_{K,\omega}+i\omega\xi_{K,\omega}&=\sum_{k\sigma}c_{k\sigma}{}^*c_{k+K\sigma}\\&+2\pi e^2\sum_{\alpha k\sigma}c_{k\sigma}{}^*\frac{c_{k+K-\alpha,\sigma}\rho_{-\alpha}}{\alpha^2}\left\{\frac{1}{\omega-\hbar\mathbf{k}\cdot\mathbf{K}+\hbar K^2/2m}-\frac{1}{\omega-\hbar(\mathbf{k}-\boldsymbol{\alpha})\cdot\mathbf{K}+\hbar K^2/2m}\right\}+2\pi e^2\sum_{\sigma k\sigma}\\&\times\frac{\rho_\alpha c_{k\sigma}{}^*c_{k+K+\alpha,\sigma}}{\alpha^2}\frac{1}{\omega-\hbar\mathbf{k}\cdot\mathbf{K}+\hbar K^2/2m}-\frac{1}{\omega-\hbar(\mathbf{k}+\boldsymbol{\alpha})\cdot\mathbf{K}+\hbar K^2/2m}\end{aligned} \tag{A12}$$

We now split the sums over $\boldsymbol{\alpha}$ and $\mathbf{K}$ into two parts. In the second term on the right hand side of (A12), we see that those terms for which $\boldsymbol{\alpha}=\mathbf{K}$ give us a factor of n, the total number of particles, while the remaining terms, with $\alpha\neq K$ lead to nonlinear contributions, since there appear here effectively two factors, each of order ρ_K. It can be shown that the neglect of those terms for which $\alpha\neq K$ is equivalent to the "random phase approximation," as applied for instance in the neglect of U Eq. (15). Similarly, in the third term on the right-hand side of (A12) we find the terms for which $\boldsymbol{\alpha}=-\mathbf{K}$ give us a factor of n, while those with $\boldsymbol{\alpha}\neq-\mathbf{K}$ may be neglected in the random phase approximation.

With these approximations, we then obtain

$$\dot{\xi}_{K,\omega}+i\omega\xi_{K,\omega}=\rho_K\left\{1-\sum_{k\sigma}\frac{4\pi e^2}{m}\frac{c_{k\sigma}{}^*c_{k\sigma}}{(\omega-\mathbf{k}\cdot\mathbf{K}\hbar)^2-\hbar^2K^4/4m}\right\}. \tag{A13}$$

Thus we see that $\xi_{K,\omega}$ and hence q_K, oscillates harmonically provided ω satisfies the dispersion relation

$$1=\frac{4\pi e^2}{m}\sum_k{}'\frac{1}{(\omega-\hbar\mathbf{k}\cdot\mathbf{K})^2-\hbar^2K^4/4m^2}, \tag{A14}$$

where $\sum_k{}'$ here denotes the sum over all occupied electronic states. This dispersion relation is, however, identical with that we found in Sec. II Eq. (57). Thus we see that the same results can be obtained by solving for the operator equations of motion as can be obtained by the canonical transformation method.

However, a word of caution should be injected at this point. For if one naively diagonalizes the terms on the right-hand side on (A12), assuming the electrons occupy a Fermi distribution at $T=0$, one obtains additional "exchange" terms which apparently contribute to order k^2 in the dispersion relation (A14). This in turn introduces an apparent contradiction between the results herein obtained and the dispersion relation (57). The resolution of this contradiction lies in the fact that the electrons in consequence of the Coulomb interactions do not behave like a gas of free particles (as is tacitly assumed in diagonalizing A12), but rather exhibit long-range correlations in positions which act to reduce the long-wavelength density fluctuations. This reduction in the long-wavelength density fluctuations has the result that no "exchange" contributions to the dispersion relation appear up to order k^4. Physically this result follows from the fact that the long-range correlations act to keep the particles far apart, so that they have less chance to feel the effects of the exclusion principle. This result follows quite simply in our treatment in the body of this paper where we take into account the exclusion principle by antisymmetrizing the individual electronic wave functions. However, it is rather difficult to establish the equivalent result in the above second-quantization formalism, so we do not enter on this question farther here.

APPENDIX II

Tomonaga[5] has developed a very interesting one-dimensional treatment of the degenerate gas of Fermi particles in which the excitations are described in terms of a Bose field, and in which he obtains plasma oscillations for the degenerate electron gas. His method, however, appears to be intrinsically restricted to this one-dimensional case. It also involves the approximation that the wave function of the electron gas is not very different from that of a collection of free electrons with a Fermi distribution at absolute zero. In this appendix we shall exhibit the relationship between Tomonaga's methods and ours.

To do this let us first find the equation of motion of the operator ρ_K. We find it convenient to make the simple transformation[15]

$$\rho_K=\sum_k c_k^* c_{k+K}=\sum_k c_{k-K/2}^* c_{k+K/2}. \tag{A15}$$

The equations of motion of ρ_K may be obtained by commuting it with the Hamiltonian (A7). We find

$$\dot{\rho}_K=-i\sum_k (\hbar k\cdot K/m)c_{k-K/2}^* c_{k+K/2}, \tag{A16}$$

$$d^2\rho_K/dt^2=-\sum_k (\hbar k\cdot K/m)^2 c_{k-K/2}^* c_{k+K/2}-\omega_p^2\rho_K - \sum_{K'\neq K}\frac{(\mathbf{K}\cdot\mathbf{K}')}{(K')^2}\rho_{K'}\rho_{K-K'}. \tag{A17}$$

If we neglect the nonlinear terms on the right-hand side of the above equation, we see that ρ_K still does not quite oscillate harmonically. This is because of the term $-\sum_k(h\mathbf{k}\cdot\mathbf{K}/m)^2 c_{k-K/2}^* C_{k+K/2}$, which is the quantum-analog of the term $\sum_i(\mathbf{k}\cdot\mathbf{V}_i)^2 e^{-i\mathbf{k}\cdot\mathbf{x}_i}$ appearing in Paper II, Eq. (9). As in the classical treatment, this term arises from the fact that we have a collection of different electrons, each moving with a different velocity and each therefore contributing differently to ρ_K. Hence, for the same reasons given in Paper II, it is necessary to seek the function q_K [given in (A8)] which oscillates harmonically in spite of the random motions of the individual electrons.

However in the *one-dimensional* case a considerable simplification is possible when the wave function is approximately that associated with a Fermi distribution of electrons at absolute zero. For in this case, either the operator $c_{k-K/2}^*$ or the operator $c_{k+K/2}$ will be zero except in a small region of width K at the top of the distribution. If K is small, then the term $(K\cdot k)^2 = K^2k^2$ can be approximated as $K^2k_0^2$, where k_0 is the wave vector of an electron at the top of the distribution. We then get

$$d^2\rho_K/dt^2=-(h^2K^2k_0^2/m^2+\omega_p^2)\rho_K, \tag{A18}$$

and we see that ρ_K oscillates harmonically, which is the result of Tomonaga.

In the three-dimensional case, such a simplification is not possible. For the Fermi distribution is now spherical, and the factor $\mathbf{k}\cdot\mathbf{K}\cong k_0K\cos\vartheta$, where ϑ is the angle between $\mathbf{k}$ and $\mathbf{K}$ for the electrons at the top of the Fermi distribution. Thus the various terms $c_{k-K/2}^* \times c_{k+K/2}$ can no longer be given a common factor, and the simple result (A18) can no longer be obtained. The reason for this change may be given a simple physical interpretation. In a one-dimensional problem, the electrons at the top of the Fermi distribution have only one velocity, and therefore all electrons contribute approximately in unison to ρ_K. In the three-dimensional case, each electron contributes differently, so that the function ρ_K is altered in time, and a new function is introduced which cannot be expressed as a simple function of ρ_K.

It should also be noted that our criterion for the validity of the collective approximation is different from that of Tomonaga. For we require the smallness of $\alpha=\langle(\mathbf{k}\cdot\mathbf{P}_i/m\omega)^2\rangle_{\text{Av}}$, while Tomonaga requires the smallness of $(\Delta W/\hbar\omega)$ where ΔW is the mean excitation energy of the electron distribution over the ground state Fermi energy.

Finally, we shall demonstrate explicitly the relationship between Tomonaga's variables and ours. From (A8), (A9), and (A10) we may write for our collective variables in the one-dimensional case:

$$q_K=\sum_k \frac{c_{k-K/2}^* c_{k+K/2}\omega_p^2}{\omega^2-(\hbar kK/m)^2}, \tag{A19}$$

$$p_K=\frac{i}{\omega(K)}\sum_k \frac{c_{k-K/2}^* c_{k+K/2}}{\omega^2-(\hbar kK/m)^2}(\hbar kK/2m\omega)\omega_p^2. \tag{A20}$$

Now Tomonaga breaks his sums over k into two parts, corresponding to positive and negative values of k. We shall do the same, noting that the only nonzero contributions are in a small region of width K near the top of the distribution. We get

$$q_K=q_K^+ + q_K^- = \left\{\frac{\omega_p^2}{2[\omega^2-(\hbar k_0K/m)^2]}\right\} \times\sum_{+k} c_{k-K/2}^* c_{k+K/2}+\sum_{-k} c_{k-K/2}^* c_{k+K/2},$$

$$p_K=p_K^+ + p_K^- = \left\{\frac{\omega_p^2 i}{2[\omega^2-(\hbar k_0K/m)^2]}\right\} \times\sum_{+k} c_{k-K/2}^* c_{k+K/2}+\sum_{-k} c_{k-K/2}^* c_{k+K/2}.$$

If we note from (A18) that $\omega^2-(hk_0K/m)^2=\omega_p^2$, we then obtain for the operators

$$a_k=q_K-i\omega P_K=\sum_{+k} c_{k-K/2}^* c_{k+K/2},$$

$$a_k^*=q_K+i\omega P_K=\sum_{-k} c_{k-K/2}^* c_{k+K/2}.$$

These are just the Eqs. (2.5) of Tomonaga.

Thus, in the one-dimensional case, with Tomonaga's assumption of an approximate Fermi distribution of free electron momenta, we obtain the same results as Tomonaga.

[15] We here suppress the spin index, since this will play no role in what follows.

A Collective Description of Electron Interactions: IV. Electron Interaction in Metals

David Pines
Department of Physics, University of Illinois, Urbana, Illinois
(Received May 21, 1953)

The effects of the Coulomb interaction between free electrons in an electron gas are considered for a variety of phenomena. The analysis is based on the collective description, which describes the long-range correlations in electronic positions (due to the Coulomb force) in terms of the collective oscillations of the system as a whole. It is shown that an independent electron model should provide a good description of the electrons in a metal in many cases of interest. The ground state energy of the free electron gas is determined, and an estimate of the correlation energy is obtained, with results in good agreement with those of Wigner. The exchange energy is shown to be greatly reduced by the long-range correlations, so that its effect on the level density and the specific heat is comparatively slight, leading to an electronic specific heat for Na which is approximately 80 percent of the free-electron value. The possible ferromagnetism of a free-electron gas is investigated, and it is found that the long-range Coulomb correlations are such that a free-electron gas will never become ferromagnetic (no matter how low the density). The excitation of the collective oscillations by a fast charged particle is studied, and the semiclassical results obtained by Bohm and Pines are verified by a quantum-mechanical calculation. The results are applied to the experiments of Ruthermann and Lang on the scattering of electrons by thin metallic films and to experiments on the stopping power of light metals for fast charged particles, with resulting good agreement between theory and experiment.

I.

In the preceding paper,[1] a quantum-mechanical collective description of the electrons in a dense-electron gas has been developed. In this paper we wish to apply this collective description to the motion of the conduction electrons in metals. In so doing, we shall assume that the effect of the positive ions in the metal may be represented by a smeared out uniform background of positive charge. This assumption should be quite a good one for the alkali metals (in which the electronic wave functions are almost plane waves), and we may expect it to apply generally for any metallic phenomenon in which the periodicity of the lattice plays no important role. In assuming a uniform positive charge, we are also neglecting the ionic charge density fluctuations and so cannot consider the interaction of the electrons with the lattice vibrations. Actually, a collective description of the ionic motion is also possible and offers a promising approach to the treatment of problems in which the electron-lattice interaction plays an important role.[2,3]

We also assume that the only interactions of importance for the conduction electrons in a metal are those with the other conduction electrons. If this is not true, as might be the case if, for instance, the exchange interaction with the core electrons is large, then the collective description may well become inapplicable. For the validity of the collective description requires that the mean collision time for electron collisions which tend to disrupt the collective motion should be large compared to the period of a collective oscillation. This follows from the fact that the effect of these disruptive collisions is to cause damping of the collective oscillation, and the criterion,

$$\tau_{\text{coll}} \gg 2\pi/\omega_p, \tag{1}$$

is just the criterion that such damping be small.[4] If (1) is not valid then the damping is large, and the whole concept of collective oscillation loses its significance in a description of electron interaction.

The criterion (1) will be satisfied for a free-electron gas, since as was shown in Paper III, the only collisions which act to disrupt organized motion are via the short-range screened Coulomb force, and for an almost degenerate Fermi gas these lead to a collision time considerably larger than $2\pi/\omega_p$. This will also be the case for the collisions between the electrons and the lattice vibrations in metals, since the mean free time between such collisions is $\gtrsim 10^{-12}$ sec, which is long compared to the period of a collective oscillation. Whether the criterion (1) is satisfied for other disruptive effects requires detailed investigation for the metal in consideration, and we shall not enter on such questions here.

We first apply the collective description to a consideration of the widespread success of the independent electron model for the motion of electrons in metals. In this model, the motion of a given electron is assumed, in first approximation, to be independent of the motion of all the other electrons. The effect of the other electrons on this electron is then represented by a smeared-out potential, which can be determined by using the self-consistent field methods of Hartree and Fock. In this one-electron approximation, the correlations in the position and energy of the electrons due to their Coulomb interactions are treated as small perturbations, and often entirely neglected. It is rather puzzling that such an independent electron model should have been so successful qualitatively, and in many cases, quantitatively, since the Coulomb inter-

[1] D. Bohm and D. Pines, preceding paper [Phys. Rev. **92**, 609 (1953)]. This paper will hereafter be referred to as Paper III. The earlier papers in this series, hereafter referred to as I and II, respectively, are D. Bohm and D. Pines, Phys. Rev. **82**, 625 (1951) and D. Pines and D. Bohm, Phys. Rev. **85**, 338 (1952).

[2] D. Bohm and T. Staver, Phys. Rev. **84**, 836 (1951).

[3] T. Staver, Ph.D. thesis, Princeton University, 1952 (unpublished).

[4] D. Bohm and E. P. Gross, Phys. Rev. **75**, 1864 (1949).

action is a long-range interaction and might be expected to affect profoundly the electron motion in metals. In fact, as we have seen, it does bring about long-range correlations in the electron positions and so leads to organized oscillation of the electron system as a whole, a phenomenon which cannot be described in terms of an independent electron model.

The introduction of the collective description enables us to investigate in some detail just what physical phenomena are associated with the long-range aspects of the Coulomb force. We may sum up the mathematical results obtained in Paper III by writing down our Hamiltonian in the collective description. If we use Eqs. (59), (60), and (61) of III, we find

$$H=H_{\text{part}}+H_{\text{coll}}+H_{\text{s.r.}}, \tag{2}$$

where

$$H_{\text{part}}=\sum_i \frac{P_i^2}{2m}\left(1-\frac{n'}{3n}\right)-2\pi ne^2\sum_{k<k_c}\frac{1}{k^2}, \tag{3}$$

$$H_{\text{coll}}=\sum_{k<k_c}(\hbar\omega/2)(A_k{}^*A_k+A_kA_k{}^*), \tag{4}$$

$$H_{\text{s.r.}}=2\pi e^2\sum_{\substack{i\neq j\\ k>k_c}}\frac{\exp[i\mathbf{k}\cdot(\mathbf{X}_i-\mathbf{X}_j)]}{k^2}. \tag{5}$$

We also have a set of subsidiary conditions on our electronic wave functions which are [Eq. (65), III]

$$\sum_i \frac{\omega^2}{\omega^2-(\mathbf{k}\cdot\mathbf{P}_i/m-\hbar k^2/2m)^2}\exp(i\mathbf{k}\cdot\mathbf{X}_i)\Psi=0. \tag{6}$$

The above Hamiltonian and subsidiary conditions are accurate to order

$$\alpha=(\mathbf{k}\cdot\mathbf{P}_i/m\omega)^2{}_{\text{Av}} \tag{7}$$

or $(n'/3n)$ with respect to the electronic kinetic energy and the zero-point energy of the collective oscillations. As we shall see, this constitutes quite an accurate approximation. We have also neglected $H_{\text{r.p.}}$ [Eq. (73), III] since as pointed out in III, it will produce negligible effects when compared with $H_{\text{s.r.}}$, and this latter term itself is small.

We see from (2)–(5) that the long-range part of the Coulomb interactions has been effectively redescribed in terms of the collective oscillations of the system as a whole. The frequency of these oscillations is, from [Eq. (67), III],

$$\omega=\omega_p\left(1+\frac{k^2}{2nm^2}\sum_i\frac{P_i^2}{\omega^2{}_p}+\frac{\hbar^2k^4}{8m^2\omega_p{}^2}\right). \tag{8}$$

It may easily be seen that the energy of a quantum of collective oscillation is so high (being greater than the energy of an electron at the top of the Fermi distribution) that these will not be excited in metals at ordinary temperatures, and hence may not be expected to play an important role in our description of a metal under ordinary conditions. (We discuss the excitation of these oscillations by an external fast charged particle in Sec. V.)

The remainder of our Hamiltonian corresponds to a collection of individual electrons interacting via a comparatively weak short-range force $H_{\text{s.r.}}$. These electrons differ from the usual "free" electrons in that they possess a slightly larger effective mass,

$$m^*=(m/[1-(n'/3n)]), \tag{9}$$

and their wave functions are subject to a set of n' restrictions, as given by the subsidiary conditions (6). However, in the limit of small $n'/3n$, we may expect that both of these changes are unimportant qualitatively (and in some cases quantitatively). Furthermore, since the effective electron-electron interaction is so greatly reduced in our collective description, we should expect that it is quite a good approximation to neglect it for many applications. Thus we are led directly to the independent electron model for a metal.

The use of the collective description not only enables us to understand qualitatively the general success of the independent electron model, but it also enables us to clear up a number of quantitative difficulties arising from the application of this model to problems in which the electron-electron interaction is taken into account. We consider these questions in Sec. III.

II.

In this section we calculate, on the basis of the collective description, the ground-state energy for our free-electron gas. In so doing, we shall determine the maximum collective oscillation wave vector k_c (and hence the number of collective degrees of freedom n'), by minimizing the resultant energy with respect to this hitherto arbitrary parameter. We then apply our results to a consideration of the correlation energy correction to the calculation of the cohesive energies of the alkali metals.

Our wave equation for the ground state is

$$H\psi_0=\epsilon_0\psi_0, \tag{10}$$

where ψ_0 and ϵ_0 are the ground-state eigenfunction and energy, respectively. We will also have a set of subsidiary conditions on our wave function,

$$\sum_i \frac{\omega^2\exp(i\mathbf{k}\cdot\mathbf{X}_i)}{\omega^2-(\mathbf{k}\cdot\mathbf{P}_i-\hbar k^2/2m)^2}\psi_0=0\quad(k<k_c) \tag{11}$$

but, as was emphasized in III, for the ground state the exact eigenfunction ψ_0 which satisfies (10) will automatically satisfy (11). Thus, we may concentrate on obtaining the best possible solution of our eigenvalue equation (10). In doing so, we may expect that an approximate ψ_0 will not satisfy the subsidiary conditions, but that any errors we make in determining the energy of the lowest state will not be increased by our failure to satisfy this subsidiary condition, since an

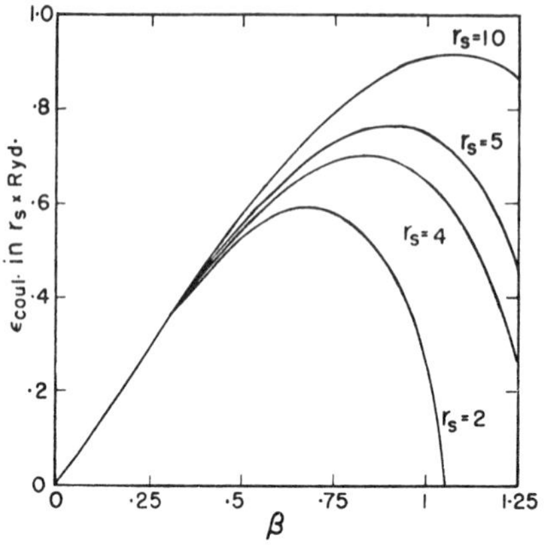

FIG. 1. ϵ_{Coul} *vs* β for several values of r_s.

exact solution satisfies it and leads to the lowest possible energy state.

We may obtain an approximate solution for ψ_0 by treating the comparatively weak short-range electron-electron interaction term $H_{\text{s.r.}}$ as a small perturbation in our expression for H, Eq. (2). In this case, we have as a first approximation,

$$\psi_0=\psi_0^{\text{osc}}\chi_0, \tag{12}$$

where ψ_0^{osc} represents a product of simple harmonic oscillator lowest-state wave functions, one for each collective oscillation wave-vector up to $k=k_c$, and χ_0 is the usual Slater determinant made up of the free-electron wave functions appropriate to the ground state of our system. The foregoing treatment may then be improved by taking into account the short-range correlations in electronic positions brought about by $H_{\text{s.r.}}$. We return to this question later.

In using the wave function (12), we are describing to a high degree of approximation the long-range correlations in electronic positions brought about by the Coulomb interactions. The great virtue of the collective description is that these quite complex correlations may be simply described in terms of the collective oscillations of the electron gas, and for the ground state of our system, are for the most part contained in ψ_0^{osc}. We note that if we attempted to express (12) in terms of our original "bare" electron coordinates $(\mathbf{x}_i\mathbf{p}_i)$, in a manner similar to that used in obtaining Eq. (23, III), we find that ψ_0 would be a complicated many-electron wave function, which would bear no simple resemblance to a determinant composed of single particle wave-functions and would not easily lend itself to a computation of the system energy.

Before evaluating the lowest-state energy ϵ_0, we find it convenient to re-express our Hamiltonian (2) by introducing the dimensionless parameter β, where

$$\beta=k_c/k_0. \tag{13}$$

We obtain

$$H=\sum_i P_i^2/2m(1-\beta^3/6)-(ne^2/\pi)\beta k_0 +\tfrac{1}{2}\sum_{k<\beta k_0}\hbar\omega(A_k^*A_k+A_kA_k^*) +2\pi e^2\sum_{\substack{i\neq j\\ k>\beta k_0}}\frac{\exp[i\mathbf{k}\cdot(\mathbf{X}_i-\mathbf{X}_j)]}{k^2}. \tag{14}$$

If we now evaluate ϵ_0 using (14) and ψ_0 as given in (12), we find that the ground-state energy per electron is given by

$$\epsilon_0'=(\epsilon_0/n)=\epsilon_F+\epsilon_{\text{Coul}}+\epsilon_{\text{exch}}, \tag{15}$$

where

$$\epsilon_F=E_F(1-\beta^3/6)=\tfrac{3}{10}(\hbar^2k_0^2/m)(1-\beta^3/6), \tag{16}$$

$$\epsilon_{\text{Coul}}=-(e^2/\pi)\beta k_0+(n'/n)\left\langle\frac{\hbar\omega}{2}\right\rangle_{\text{Av}}, \tag{17}$$

$$\epsilon_{\text{exch}}=-2\pi e^2\sum_{\substack{k,k'<k_0\\ |\mathbf{k}-\mathbf{k}'|>\beta k_0}}\frac{1}{|\mathbf{k}-\mathbf{k}'|^2}. \tag{18}$$

In Eq. (15), ϵ_F represents the average Fermi energy, differing from the usual expression in that our electrons have the effective mass m^*. ϵ_{Coul} represents the difference in energy (per electron) between the zero-point energy of the collective oscillations $(n'/n)(\langle\hbar\omega\rangle_{\text{Av}}/2)$ and the usual self-energy of the charge distribution the oscillations here describe $-e^2/\pi\beta k_0$. It may be regarded as arising from the reduction in the long-range density fluctuations of the electron gas, as described, for instance, by Eq. (77, III). $\langle\omega\rangle_{\text{Av}}$ is the average frequency of collective oscillation, obtained by averaging our dispersion relation over all $k<k_c$. One finds, from (68, III) that

$$\langle\omega\rangle_{\text{Av}}=\omega_p(1+3\alpha[1+(3/10)\beta^2]). \tag{19}$$

In Fig. 1 we give a plot of ϵ_{Coul} (in units of $r_s\times$rydbergs) *vs* β for several values of

$$r_s=r_0/a_0=(3n/4\pi)^{1/3}(me^2/\hbar^2), \tag{20}$$

the interelectron spacing measured in units of the Bohr radius.

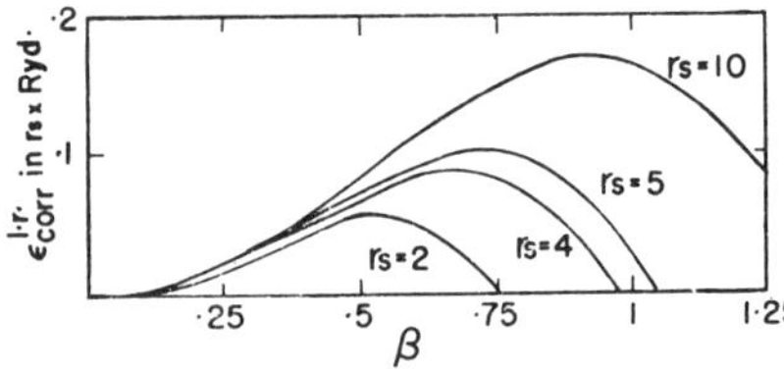

FIG. 2. $\epsilon_{\text{corr}}^{\text{l.r.}}(=-\epsilon_0'+E_F+E_{\text{exch}})$ *vs* β for several values of r_s.

ϵ_{exch} is the exchange energy arising from $H_{\text{s.r.}}$, the short-range electron interaction energy. The sums over $\mathbf{k}$ and $\mathbf{k}'$ in (18) are to be carried out for all electrons of parallel spin in the Fermi distribution (i.e., $k<k_0$, $k'<k_0$) such that $|\mathbf{k}'-\mathbf{k}|>\beta k_0$. This latter restriction arises from the short-range character of the interaction, as expressed by the restriction of k values in (5) to $k>\beta k_0$. We evaluate ϵ_{exch} in Appendix I, and there show that

$$\epsilon_{\text{exch}}=-(0.916/r_s)(1-(4/3)\beta+\beta^2/2-\beta^4/48) \text{ ry}. \quad (21)$$

The results of our calculation of the ground-state energy are given in Figs. 2 and 3. In Fig. 2 we plot ϵ_0' (actually $\epsilon_{\text{corr}}^{\text{l.r.}}=\epsilon_0'-E_{\text{exch}}-E_F$ [see Eq. (24)] as a function of β for various values of r_s. From this we may easily obtain $\beta_{\min}$, that value of β for which the ground-state energy is a minimum. In Fig. 3 we plot $\beta_{\min}$ as a function of r_s.

On the basis of these calculations we see that for the electronic densities encountered in metals (r_s roughly between 2 and 5), $\beta_{\min}$ runs between 0.5 and 0.75. We may now verify the validity of our perturbation theory expansions in powers of α and $(n'/3n)$. In terms of β these are given by

$$\alpha \cong \beta^2/2r_s, \quad (22)$$

and

$$n'/3n=\beta^3/6. \quad (23)$$

We find from (22), and Fig. 4, that $\alpha\sim 1/16$ for $r_s=2$ and $\alpha\sim 1/20$ for $r_s=5$, so that expansions in powers of α should be quite accurate. Similarly, we find $(n'/3n)$ is (1/48) for $r_s=2$ and $\sim(1/17)$ for $r_s=5$, so that it constitutes an equally valid expansion parameter.

The preceding results are conveniently analyzed by the introduction of the concept of the correlation energy of the free-electron gas. This energy may be defined as the difference between the energy calculated by means of suitable many-electron wave functions and the energy calculated in the Hartree-Fock one-electron approximation. The latter energy is[5]

$$E=E_F+E_{\text{exch}}, \quad (24)$$

TABLE I. Correlation energy in the free-electron gas model for the alkali metals. Units: Rows (c), (d), and (e), $r_s\times$ry; Rows (f) and (g), kcal/mole.

Metal	Li	Na	K	Rb	Cs
(a) r_s	3.22	3.96	4.87	5.18	5.57
(b) $\beta_{\min}$	0.63	0.68	0.73	0.75	0.77
(c) $\epsilon_{\text{corr}}^{\text{l.r.}}$	−0.076	−0.086	−0.100	−0.106	−0.112
(d) $\epsilon_{\text{corr}}^{\text{s.r.}}$	−0.164	−0.183	−0.206	−0.213	−0.221
(e) ϵ_{corr}	−0.240	−0.269	−0.308	−0.319	−0.333
(f) $\epsilon_{\text{corr}}\frac{\text{kcal}}{\text{mole}}$	23.3	21.1	19.5	19.1	18.6
(g) ϵ_{corr}(Wigner)	21.7	19.7	17.9	17.4	16.8

[5] F. Seitz, *The Modern Theory of Solids* (McGraw-Hill Book Company, Inc., New York, 1940), p. 341.

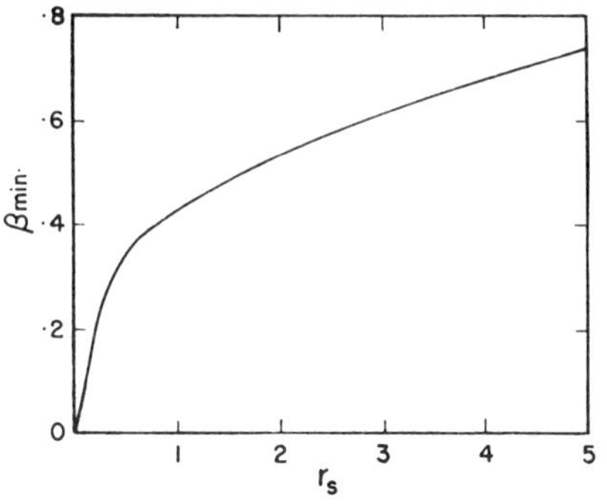

FIG. 3. $\beta_{\min}$ *vs* r_s.

where

$$E_F=\tfrac{3}{5}(\hbar^2k_0^2/2m)=(2.21/r_s^2) \text{ ry}. \quad (25)$$

and

$$E_{\text{exch}}=-(0.916/r_s) \text{ ry}. \quad (26)$$

The correlation energy for the free-electron gas was first calculated by Wigner,[6] who used a perturbation theory method in which the wave function of the electron of a given spin was assumed to depend on the positions of all the electrons of opposite spin. Wigner extended the results of his calculation, which was only valid for very high electronic densities ($r_s \lesssim 1$), to lower densities in such a way as to approach the correct value of the correlation energy for very low densities ($r_s \gg 1$). He obtained the result,

$$E_{\text{corr}}=-0.576/(r_s+5.1) \text{ ry}, \quad (27)$$

which he estimated to be accurate to within 20 percent.

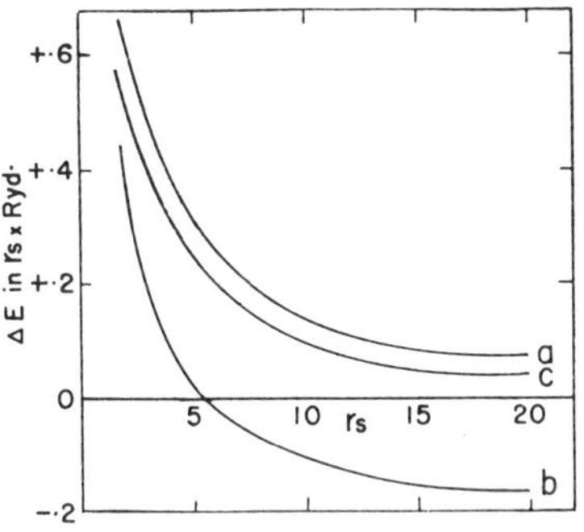

FIG. 4. Energy difference between ferromagnetic and nonferromagnetic states of free electron gas on three different models: (a) Simple theory ($E=E_F$); (b) Hartree-Fock theory ($E=E_F+E_{\text{exch}}$); (c) Hartree-Fock theory, with inclusion of long-range correlation effects.

[6] E. P. Wigner Phys. Rev. 46. 1002 (1934).

We shall regard the correlation energy, when calculated in the collective description, as split into two parts, representing long-range and short-range effects. The long-range correlation energy is associated with the long-range correlations in the electronic positions which are essentially described by our introduction of the collective degrees of freedom. This is therefore given by the difference between the ground-state energy we calculated above and the expression (24). The short-range correlation energy then arises from the modification in electronic wave functions brought about by $H_{s.r.}$, an effect we consider below.

The long-range correlation energy is thus from (24) and (15),

$$\epsilon_{corr}^{l.r.} = -(\beta^3/6)E_F + \epsilon_{Coul} + [\epsilon_{exch} - E_{exch}]. \quad (28)$$

The first two terms in (28) are both negative and represent the energy gain arising from the long-range Coulombic correlations we have considered above. The third term is positive, and represents the difference between the effect of the exclusion principle in a gas of *interacting* electrons and in a gas of *free* electrons. In Table I we give the calculated values of $\epsilon_{corr}^{l.r.}$ for electron gases of the same density as those found in alkali metals.

We may understand the origin of the terms contributing to $\epsilon_{corr}^{l.r.}$ in the following way. In the collective description we consider *ab initio* the correlations brought about by the Coulomb interactions. We find that the electrons tend to stay out of one another's way, and that in fact these long-range Coulomb correlations are such that the effective electron interaction is given by $H_{s.r.}$. The energy gain from such Coulombic correlations is $\epsilon_{Coul} - (\beta^3/6)E_F$. We then put in the exclusion principle in the usual way and find an additional energy gain coming from the "accidental" correlations due to the exclusion principle in our gas of *interacting electrons*, which is ϵ_{exch}. On the other hand, E_{exch} represents the exchange energy for a gas of *free electrons*, in which Coulombic effects are otherwise neglected. The use of this term represents an overestimate of the role of the exclusion principle, since the long-range Coulomb forces tend to keep the electrons apart, and would do so even if the electrons had integral spin. In consequence the exclusion principle should properly play no role in such long-range correlations, and we see that this is properly accounted for by our correction to the exchange energy $(\epsilon_{exch} - E_{exch})$. Other effects associated with the different role of the exclusion principle in a gas of interacting electrons are considered in the following section.

The short-range electron-electron interaction term, $H_{s.r.}$ may be consistently treated as a perturbation in the Hamiltonian (2), since it is of such a short-range and weak strength that it does not affect the individual electronic motion appreciably. The short-range correlation energy may then be calculated according to perturbation-theoretic techniques. When this is done one finds that the short-range correlation energy is associated almost entirely with correlations between electrons of antiparallel spin, and it is this latter energy which we list in Table I. The details of this calculation, together with a more extensive investigation of correlation phenomena, will be published in the near future.

We have summarized the results of our correlation-energy calculations in Table I, where we give those values corresponding to the electron densities encountered in the alkali metals. It may be seen that our results are in quite close agreement with those of Wigner, an agreement which is well within the accuracy of the two methods. The major source of error in our calculation lies in our perturbation theoretic estimate of the effects of $H_{s.r.}$, and a rough estimate indicates that the use of perturbation theory, together with certain other approximations, may be responsible for errors of as much as 20 percent in $\epsilon_{corr}^{s.r.}$. Our expansion parameters α and $(n'/3n)$ are both quite small, so that the errors introduced by our neglect of higher-order terms in these parameters will be less than 10 percent of $\epsilon_{corr}^{l.r.}$. An order of magnitude estimate of the exchange energy associated with $H_{r.p.}$ (a term we neglected in comparison to $H_{s.r.}$) shows that this is less than $(0.01/r_s)$ ry for Na, so that the neglect of this term is justified within the accuracy of the present treatment. If we combine the above estimates, we are led to estimate the over-all accuracy of our calculated correlation energy as approximately 15 percent.

III.

In this section we consider briefly the result of including electron-electron interactions in a description of certain metallic phenomena. In general, the most simple form of the theory of metals, in which the electron-electron interactions are entirely neglected, has led to qualitative, and in many cases, quantitative agreement with experiment. If the electron-electron interactions are then included in the one-electron picture, the corrections to the simple theory arise from the exchange energy contributions to the phenomenon under consideration. These corrections, far from improving the agreement with experiment, tend to worsen it, both qualitatively and quantitatively. Wigner[7] suggested that correlations brought about by the Coulomb interactions may counteract the difficulties arising from these exchange contributions. We here wish to show that the long-range correlations we have considered in the collective description have just this property.

In the one-electron approximation, the exchange energy contribution to the Coulomb interaction energy of two electrons of wave vectors k_1 and k_2 is

$$W_{1,2} = -4\pi e^2/|\mathbf{k}_1 - \mathbf{k}_2|^2. \quad (29)$$

For a given electron, the total energy of interaction

[7] E. P. Wigner, Trans. Faraday Soc. **34**, 678 (1938).

with all the other electrons is then

$$W_i=\sum_{j\neq i} W_{i,j}=-\frac{e^2k_0}{2\pi}\left(2+\frac{k_0^2-k^2}{k_0k}\ln\frac{k+k_0}{k-k_0}\right), \tag{30}$$

if we assume the electrons form a completely degenerate gas. As a consequence of (30) the *E versus k* curve for a given electron becomes very steep as k approaches k_0, so that the density-in-energy of electron levels at low temperatures will be greatly reduced. Bardeen[8] and Wohlfarth[9] calculated the modification in the specific heat of a free-electron gas due to the inclusion of the exchange energy (30) and found that at low temperatures the specific heat should vary as $T/\ln T$. This result is in contradiction with the experimental results, which show a linear dependence on T—a result that follows from theory if exchange effects are entirely ignored. This large reduction in the density of electron energy levels might also be expected to affect profoundly other metallic phenomena which are sensitive to its magnitude, such as paramagnetism, conductivity, and the optical properties of metals; such an effect on these phenomena has also not been found experimentally.

The origin of these difficulties lies in the long-range of the Coulomb interaction which leads to a very large exchange energy for electrons of nearly equal momentum; this, in turn, is responsible for the undesirable behavior of (30). If we have, instead, an effective short-range, screened Coulomb interaction between the electrons, these large exchange energy contributions are greatly reduced, and the energy term corresponding to (30) takes a more satisfactory form.

Actually, a screened Coulomb interaction between electrons was proposed empirically by Landsberg,[10] who found it necessary to introduce such an effective interaction in order to obtain agreement between theory and experiment for the width of the tail of the soft x-ray emission spectrum of sodium. Wohlfarth[9] then showed that the effect of the exchange energy on the specific heat of the electron gas is greatly reduced, provided the electron-electron interaction potential is empirically taken as $(e^2/r_{ij})\exp[-(r_{ij}/\lambda)]$, with a screening radius λ of the order of 10^{-8} cm, the value introduced by Landsberg.

In the collective description, the exchange energy contribution to the Coulomb interaction energy of two electrons of wave vectors k_1 and k_2 arises from $H_{s.r.}$, and is given by

$$\begin{aligned} W_{1,2}&=-4\pi e^2/|\mathbf{k}_1-\mathbf{k}_2|^2, \quad (|\mathbf{k}_1-\mathbf{k}_2|>k_c);\\ W_{1,2}&=0, \quad (|\mathbf{k}_1-\mathbf{k}_2|<k_c). \end{aligned} \tag{31}$$

In Appendix I, we show that for a given electron of wave vector $\mathbf{k}$, the total energy of interaction with all the other electrons is

$$W_i=-\frac{e^2k_0}{2\pi}\left\{1+\frac{k_0^2-k^2}{kk_0}\ln\left(\frac{k+k_0}{\beta k_0}\right)+\frac{3k^2-k_0^2}{2kk_0}-2\beta+\frac{\beta^2k_0}{2k}\right\}, \tag{32a}$$

when $(k_0-k_c)<k<k_0$, and

$$W_i=-\frac{e^2k_0}{2\pi}\left\{2+\frac{k_0^2-k^2}{kk_0}\ln\left(\frac{k+k_0}{k_0-k}\right)-4\beta\right\}, \tag{32b}$$

when $k<(k_0-k_c)$. It may easily be verified that the *E versus k* curve given by (32) no longer displays singular behavior as k approaches k_0.

With the aid of (32), we may now obtain an estimate of the exchange-energy contributions to the electronic specific heat in metals. We are limited in the accuracy of our estimate, because in considering the specific heat we should properly take into account the effect of our subsidiary conditions (6) on the excited states of the electron gas. As was pointed out in Paper III, this effect is in the direction of reducing the effective number of degrees of freedom of the electron gas from $3n$ to $3n-n'$. Since, as we have seen, $(n'/3n)\ll 1$ for electrons in metals, we might expect that the neglect of the subsidiary conditions should be a reasonably good approximation, and it is this approximation we adopt in what follows.

We may obtain the influence of the exchange energy on the specific heat of the electron gas by considering its effect on the density of levels at the top of the Fermi distribution. Lidiard[11] has shown by the use of a variational Fermi-Dirac distribution function, that provided the free energy may be sensibly expanded in powers of (KT/E_0) near $T=0$ according to the method of Sommerfeld and Bethe,[12] the specific heat per electron may be written as

$$C_e=\frac{\pi^2K^2T}{3n}\left(\frac{dn}{d\epsilon}\right)_{\epsilon=E_0}, \tag{33}$$

where $(dn/d\epsilon)_{\epsilon=E_0}$ is the density of electronic levels at the top of the Fermi distribution, and K is Boltzmann's constant. We may write

$$\frac{dn}{n}=\frac{k^2}{\pi^2}\frac{3\pi^2}{k_0^3}dk=\frac{3k^2}{k_0^3}\frac{d\epsilon}{(d\epsilon/dk)}, \tag{34}$$

so that

$$C_e=\frac{\pi^2K^2T}{k_0}\frac{1}{(\partial\epsilon/\partial k)_{k=k_0}}. \tag{35}$$

Now, from (32) we find that near the top of the Fermi

[8] J. Bardeen, Phys. Rev. **50**, 1098 (1936).

[9] E. P. Wohlfarth, Phil. Mag. **41**, 534 (1950).

[10] P. T. Landsberg, Proc. Phys. Soc. (London) **A162**, 49 (1949).

[11] A. B. Lidiard, Phil. Mag. **42**, 1325 (1951) and private communication. We should like to thank Dr. Lidiard for communicating his results to us.

[12] A. Sommerfeld and H. Bethe, *Handbuch der Physik* (J. Springer, Berlin, 1934), Vol. 24, p. 12.

distribution ($k>k_c$),

$$\epsilon=\frac{\hbar^2k_0^2}{2m}-\frac{e^2k_0}{2\pi}\left\{1+\frac{k_0^2-k^2}{kk_0}\ln\left(\frac{k+k_0}{\beta k_0}\right)+\frac{3(k^2-k_0^2)}{2kk_0}-2\beta+\frac{\beta^2k_0}{2k}\right\}, \quad (36)$$

and hence

$$C_e=\frac{\pi^2K^2T}{2E_0}\frac{1}{1+\frac{me^2}{2\pi\hbar^2k_0}\left[2\ln\left(\frac{2}{\beta}\right)+\frac{\beta^2}{2}-2\right]}. \quad (37)$$

Thus the ratio of the above electronic specific heat to the usual electronic specific heat $C_0=(\pi^2K^2T/2E_0)$ is

$$\frac{C_e}{C_0}=\left[1+\frac{r_s}{12}\left\{2\ln\frac{2}{\beta}+\frac{\beta^2}{2}-2\right\}\right]^{-1}. \quad (38)$$

For Na, for which $r_s=4$ and $\beta=0.65$, we find

$$C_e/C_0=1/1.22=0.82. \quad (39)$$

Thus the effect of taking the exchange energy into account is to reduce the electronic specific heat to about 80 percent of its free electron value for Na. We note from (38) that our calculated reduction in the density of energy levels at the top of the Fermi distribution is now comparatively small, so that the influence of the exchange effects on other metallic phenomena may be expected to be correspondingly small. The experimental accuracy in the determination of electronic specific heats does not at present appear to be sufficiently great to check the validity of the above formulas.

There is probably a slight further reduction in the electronic specific heat arising from the effect of the subsidiary conditions in reducing the number of individual electronic degrees of freedom. In addition, we may expect the short-range correlations produced by $H_{s.r.}$ to affect the density of energy levels and the electronic specific heats. This effect will be considered in a later paper.

We may also use our results on the long-range correlation energy to investigate the possible ferromagnetism of a free electron gas. If the energy of an electron in the gas in the nonmagnetic state is given by the expression (24)

$$E=+E_F+E_{\text{exch}}=2.21/r_s^2-0.916/r_s,$$

then it is clear that for sufficiently large r_s, the electron gas should become ferromagnetic. For the cost in kinetic energy which results from lining the spins up (an increase in k_0 to $k_0\sqrt[3]{2}$) will eventually be more than compensated by the gain in the exchange energy. Thus the energy for the magnetic state is given by

$$E=3.52/r_s^2-1.156/r_s,$$

and one finds that electron gases for which $r_s>5.47$ should be ferromagnetic. This is not the case (e.g., Cs), and the reason that it is not lies in the Coulomb correlation energy, as Wigner[7] has pointed out. We will now show that the long-range correlation energy is actually sufficient to prevent the free-electron gas from becoming ferromagnetic.

Qualitatively it is easy to understand why this is so. We have seen that the Coulomb interaction keeps the electrons sufficiently far apart so that the exchange-energy attraction which acts to line up the electron spins is greatly reduced, even for $r_s\sim 4$. Then as we go to higher r_s and lower electronic density, the screening cloud around each electron due to long-range correlations becomes even more efficient (corresponding to a higher value of β), so that the exchange energy is, in fact, further reduced, rather than having its relative strength increase. In Fig. 4 we give r_s times the energy difference between the ferromagnetic and nonferromagnetic states using our energy expression (15). This result has been obtained by calculating ϵ_0' as a function of β for both the nonmagnetic and ferromagnetic states, and choosing an optimum value of β (for which ϵ_0' is a minimum) in each case. For comparison we have plotted this energy difference as calculated using the simple theory, $E=E_F$, and using the Hartree-Fock approximation, $E=E_F+E_{\text{exch}}$. A result similar to this has been obtained by Wigner[7] on the basis of a somewhat different model for the free-electron gas.

The above results are subject to corrections arising from the effect of the subsidiary conditions (6) on our lowest-state wave function, since the lowest-state wave function is no longer nondegenerate and hence will no longer satisfy the subsidiary conditions automatically. However this should not alter our results appreciably, since the relative number of subsidiary conditions is small even for very low densities ($n'/3n\sim 15$ percent for $r_s\sim 10$), and since the subsidiary conditions involve only long-wavelength density fluctuations ($k<k_c$) while the exchange energy depends primarily on short-wavelength density fluctuations ($k>k_c$). Further corrections, which are in the direction of making ferromagnetism even less likely, will come from the short-range correlation energy. This follows from the fact that our energy $\epsilon_{\text{corr}}^{s.r.}$ will be absent in the magnetized state. We have not included these corrections here, because in the region of possible ferromagnetism ($r_s>5.47$), our perturbation theoretic estimates are beginning to become unreliable.

IV.

In the preceding sections we have considered the low-lying states of electrons in metals in the collective description. We have seen that at ordinary temperatures we should not expect the collective oscillations to be excited, since $\hbar\omega_p$ lies several electron volts higher than E_F for all metals, so that no electron in the metal will have sufficient energy to excite a collective oscillation. (Temperature excitation is clearly out of the question.) Thus, the only way that collective oscillations can be excited in a metal is by bombardment of the metal with charged particles of sufficient energy to excite the oscillations. In this section, we will apply our canonical transformation to the investigation of the interaction of a charged particle with the electron gas.

This problem was treated classically in II by the density fluctuation method. There it was shown that a fast charged particle would excite collective oscillation, and the results for this excitation, together with some semiclassical arguments, were applied to the experiments of Ruthemann[13] and Lang[14] on the scattering of kilovolt electrons by thin metallic films. By so doing, excellent agreement was obtained between theory and experiment. The density fluctuation method has also been applied to a determination of the contribution of the conduction electrons in a metal to its stopping power for a fast charged particle.[15] In this section, we shall obtain the appropriate quantum-mechanical results and verify the validity of the results obtained by the semiclassical application of the density fluctuation method.[16]

Let us consider a charged particle of mass M, charge Ze, position and momentum $(\mathbf{r}_0, \mathbf{P}_0)$. We can describe its motion and interaction with our electron gas by adding the following terms to the original Hamiltonian of Paper III, Eq. (43, III):

$$\frac{P_0^2}{2M}+-4\pi Ze^2\sum_{ik}\frac{e^{i\mathbf{k}\cdot(\mathbf{x}_i-\mathbf{r}_0)}}{k^2}. \quad (40)$$

The effect of the canonical transformation to the collective description on these terms may then be obtained by using Eq. (51, III) and applying the random phase approximation and the dispersion relation [Eq. (57, III)] to the resulting terms. We find that the above terms then become

$$H_{\text{add}}=-4\pi Ze^2\sum_{ik>k_c}\frac{\exp[i\mathbf{k}\cdot(\mathbf{X}_i-\mathbf{r}_0)]}{k^2}$$
$$-Ze\sum_{k<k_c}\left(\frac{2\pi\hbar\omega}{k^2}\right)^{\frac{1}{2}}(A_{-k}+A_k^*)e^{-i\mathbf{k}\cdot\mathbf{r}_0}+\frac{P_0^2}{2m}$$
$$-4\pi Ze^2\sum_{ik<k_c}\frac{\exp[i\mathbf{k}\cdot(\mathbf{X}_i-\mathbf{r}_0)]}{k^2}. \quad (41)$$

The first term in H_{add} describes a short-range screened Coulomb interaction between the charged particle and the individual electrons in our electron gas. This interaction is of the same form as that found for the interaction between the individual electrons $H_{\text{s.r.}}$ [Eq. (5)]. The second term describes the interaction between the charged particle and the collective oscillations of the system and may lead to the excitation of collective oscillations by the particle. The last term in H_{add} will be the only term affected by the subsidiary conditions on our system wave function. We see that when (6) is applied, this term reduces to

$$-4\pi Ze^2\sum_{ik<k_c}\frac{(\mathbf{k}\cdot\mathbf{P}_i/m)-(\hbar k^2/2m)^2}{\omega^2-(\mathbf{k}\cdot\mathbf{P}_i/m-\hbar k^2/2m)^2}\times\exp[i\mathbf{k}\cdot(\mathbf{X}_i-\mathbf{r}_0)], \quad (42)$$

which may be neglected in the approximation of small $(\mathbf{k}\cdot\mathbf{P}_i/m\omega)$.

Thus, we see that the use of our canonical transformation to the collective description provides us with a simple, natural splitup of the interaction between a charged particle and the electron gas into two parts: a short-range interaction with the individual electrons, and the interaction with the collective oscillations of the system as a whole (which has its origin in the long-range electron-electron and electron-particle interactions). This splitup is analogous to the splitup of the density fluctuations into individual particle and collective components, which was carried out in II.

Let us now consider the interaction between the charged particle and the collective oscillations. For this purpose, it is convenient to introduce the canonical collective variables, P_k and Q_k, defined by

$$Q_k=(\hbar/2\omega)^{\frac{1}{2}}(A_k-A_k^*),$$
$$P_k=i(\hbar\omega/2)^{\frac{1}{2}}(A_k^*+A_{-k}). \quad (43)$$

In terms of these variables, the collective interaction term in (4) becomes

$$iZe\sum_{k<k_c}(4\pi/k^2)^{\frac{1}{2}}P_ke^{-i\mathbf{k}\cdot\mathbf{r}_0}, \quad (44)$$

and

$$H_{\text{coll.}}=\tfrac{1}{2}\sum_{k<k_c}(P_k^2+\omega^2Q_k^2).$$

The equations of motion of our charged particle and the collective field are then given by

$$M\ddot{\mathbf{r}}_0=\mathbf{P}_0=-Ze\sum_{k<k_c}(4\pi)^{\frac{1}{2}}\boldsymbol{\varepsilon}_kP_ke^{-i\mathbf{k}\cdot\mathbf{r}_0} \quad (45)$$

and

$$\ddot{P}_k+\omega^2P_k=+Zei\omega^2(4\pi/k^2)^{\frac{1}{2}}e^{+i\mathbf{k}\cdot\mathbf{r}_0}. \quad (46)$$

Equation (46) describes forced harmonic oscillation of the collective fields and is directly analogous to Eq. (47) of II. Because of the similarity between these equations, and because the latter equation was analyzed in some detail in II, we will merely quote the results of our solutions for (46) here.

We can obtain a straightforward solution of (46) provided the velocity of the charged particle $\mathbf{V}_0=\mathbf{P}_0/M$ may be taken as constant. (This will be a good approxi-

[13] G. Ruthemann, Ann. Phys. 2, 113 (1948).
[14] W. Lang, Optik 3, 233 (1948).
[15] D. Pines, Phys. Rev. 85, 931 (1952). For earlier work on this subject, see references 18–21.
[16] D. Gabor [Phil. Mag. (to be published)], has obtained results in substantial agreement with ours by the use of a somewhat different method. We should like to thank Dr. Gabor for communicating his results to us prior to publication.

mation as long as the change in velocity of the charged particle during the period of oscillation is small compared to V_0, which will be true for all applications of interest to us here.) Then if $\mathbf{k}\cdot\mathbf{V}_0$ is not equal to ω, Eq. (46) has the steady-state solution

$$P_k=\left(\frac{4\pi}{k^2}\right)^{\frac{1}{2}}\frac{Zei\omega^2}{\omega^2-(\mathbf{k}\cdot\mathbf{V}_0)^2}e^{i\mathbf{k}\cdot\mathbf{r}_0}. \tag{47}$$

This solution corresponds to the particle moving through the electron gas accompanied by a co-moving cloud of collective oscillation. This co-moving cloud leads only to a somewhat larger effective mass for the charged particle, which may be calculated by substituting (47) into (44).

The more physically interesting case occurs when $\mathbf{k}\cdot\mathbf{V}_0=\omega$. In this case the steady-state solution (47) is no longer appropriate, and the correct solution corresponds to resonant excitation of the collective oscillation. This oscillation, which was discussed in some detail in II, will take the form of a wake of collective oscillation trailing behind the particle, a phenomenon which resembles closely the Čerenkov radiation produced by fast electrons in passing through dielectric materials. We calculate the energy loss per unit length to the collective oscillations by obtaining the force due to the wake at the position of the particle, under the boundary condition that the energy loss occurs behind the charged particle. We find

$$\left(\frac{dT}{dX}\right)_{\text{coll}}=\frac{2\pi nZ^2e^4}{MV_0^2}\left\{\ln\left[\frac{\beta^2k_0^2(V_0^2-\langle V^2\rangle_{\text{Av}})}{\omega_p^2}\right]+\frac{\langle V^2\rangle_{\text{Av}}}{V_0^2}\beta^2k_0^2(V_0^2-\langle V^2\rangle_{\text{Av}})\right\}, \tag{48}$$

where $\langle V^2\rangle_{\text{Av}}$ is the mean-square velocity of the electrons in the metal, and we have used $k_c=\beta k_0$. This result differs slightly in two respects from that obtained in II [Eq. (59a, II)]. The logarithmic term is slightly altered because in obtaining (48) we have chosen as the maximum value of the collective oscillation wave vector perpendicular to V_0 a more accurate value than that used in II, *viz.* $[\beta^2k_0^2(1-\langle V^2\rangle_{\text{Av}}/V_0^2)-\omega_p^2/V_0^2]^{\frac{1}{2}}$. The appearance of the second term in the brackets (multiplying $2\pi nZ^2e^4/MV_0^2$) is due to the fact that in (48) we have ω^2 as a factor on the right-hand side of the equation, as compared to the factor ω_p^2 appearing in the similar position in the analogous equation of II. Both of these differences become negligible in the limit of $V_0^2\gg\langle V^2\rangle_{\text{Av}}$.

As pointed out in II, our picture of collective oscillations of the electrons in a metal and the excitation of collective oscillations by a fast charged particle finds experimental confirmation in the experiments of Ruthemann[12] and Lang[13] on the scattering of kilovolt electrons by thin metallic films. They found that for Be and Al the electrons lose energy in integral multiples of a well-defined basic quantum. Theoretically we should expect this quantum to be very nearly $\hbar\omega_p$, since as we saw in II, the long-wavelength quanta play a major role in the stopping power if V_0 is considerably greater the mean velocity of the metallic electrons, as is the case in these experiments. The experimental values of this quantum of energy loss are 14.7 ev for Al and 19.0 ev for Be, and these agree very well with our calculated $\hbar\omega_p$ (under the assumption that all the valence electrons are free) of 15.9 ev for Al and 18.8 ev for Be.

We may also calculate the mean free path for the emission of a quantum of collective oscillation. In the limit of $V_0^2\gg\langle V^2\rangle_{\text{Av}}$, this is

$$\lambda=\frac{\hbar\omega_p}{(dT/dx)_{\text{coll}}}=\frac{\hbar\omega_pMV_0^2}{4\pi nZ^2e^4\ln(\beta k_0V_0/\omega_p)}. \tag{49}$$

From the data given by Lang on the thickness of his Al films, one may obtain an experimental estimate for λ. This turns out to be somewhat less than 185A. The theoretical value of λ is, from (49), ~160A (for the 7.6-kev electrons used by Lang) and is in good agreement with the above experimental estimate.

Lang and Ruthemann did not find a similar set of discrete energy losses in Ag, Cu, and Ni. This is probably due to the fact that the valence electrons in these metals are not sufficiently free (in the sense of Sec. IV) to take part in undamped collective motion. In the cases of Cu and Ag there is some evidence for a large exchange interaction with the core electrons, which is probably responsible for the damping of the collective oscillation.[17] Experiments have not yet been performed on the alkali metals, where we should expect to find collective oscillation and the appearance of discrete energy losses.

The interaction between the charged particle and the individual electrons in our electron gas, which is described by the term

$$-4\pi Ze^2\sum_{ik>k_c}(1/k^2)\exp[i\mathbf{k}\cdot(\mathbf{X}_i-\mathbf{r}_0)],$$

in our Hamiltonian (41), provides an alternate mechanism for the energy loss of the particle in traversing the gas. This term is not appreciably affected by our subsidiary condition (6), since the latter involves only long-wavelength density fluctuations ($k<k_c$). Thus the energy loss per unit length due to these individual electron collisions may be obtained by the usual methods of collision theory and is, in the nonrelativistic limit,

$$\left(\frac{dT}{dx}\right)_{\text{i.p.}}=\frac{4\pi nZ^2e^4}{MV_0^2}\ln\left(\frac{1}{b\beta k_0}\right) \tag{50}$$

[17] See reference 5. Recently P. Wolff [Bull. Am. Phys. Soc. 28, No. 2, 35 (1953)] has used the Hartree approximation to investigate the effect of the binding of the electrons in the lattice on collective oscillation. He finds that this may account for the single line of rather considerable width found in these metals.

for collisions involving impact parameters greater than b. The appropriate choice of a minimum impact parameter depends on the details of the collision considered (through Z, M, and V_0), but in general the energy loss to the individual electrons is roughly comparable to that given up to the collective oscillations. The total energy loss per unit path length of the charged particle is the sum of (48) and (50) and is

$$\frac{dT}{dx}=\frac{4\pi nZ^2e^4}{MV_0^2}\left\{\left(\ln\frac{V_0}{\omega_p b}\left[1-\frac{\langle V^2\rangle_{\text{Av}}}{V_0^2}\right]^{\frac{1}{2}}\right)+\frac{\beta^2k_0^2\langle V^2\rangle_{\text{Av}}}{2\omega_p^2}\left(1-\frac{\langle V^2\rangle_{\text{Av}}}{V_0^2}\right)\right\}. \tag{51}$$

We see that, in the limit of $V_0^2 \gg \langle V^2\rangle_{\text{Av}}$, this expression essentially is independent of our choice of screening parameter β.

It is of some interest to compare our description of the energy loss of a charged particle traversing a free-electron gas with those due to Kramers[18] and Bohr.[19] Both Kramers and Bohr take into account the effects of electron-electron interaction in determining this energy loss, and although their methods are rather different from ours, as we shall see their results are essentially equivalent to (51). Kramers[18] used a macroscopic description in which the electrons were treated as a continuum characterized by an effective dielectric constant. His method of treating the polarization effects associated with electron interaction is closely related to that used by Fermi in treating the analogous polarization effects for very fast particles interacting with bound electrons (the "density effect"). Bohr[19] has given a very interesting microscopic description of the collisions between the charged particle and the individual electrons (both free and bound) in which the influence of the electron-electron interactions is taken into account explicitly. Bohr has shown that the energy loss of a fast charged particle to a system of bound electrons may be considered to take place in two different modes. One mode corresponds to the Čerenkov radiation, which may be regarded as corresponding to the organized behavior of the system brought about by the electron-electron interaction. The other essentially corresponds to the interaction of the particle with the individual electrons, and displays no collective aspect. This microscopic separation of the mechanisms of energy loss of the charged particle to the bound electrons is directly analogous to that we have given above for the free-electron gas. We might add that, as is the case with the Čerenkov radiation for bound electrons, the energy given up by the particle to the collective oscillations does not constitute an additional source of energy loss from a microscopic viewpoint; for a collective oscillation is due to the cumulative contributions arising from the displacement of the individual electrons by the charged particle and represents the organized motion associated with these displacements.

A somewhat different method of treating this problem, which is in some respects similar to that used in the collective description, is due to Kronig and Korringa.[20] They have given a treatment by the methods of classical hydrodynamics in which the effects of electron-electron interactions are described in part in terms of an artificially introduced internal friction of the conduction electron fluid. Kronig[21] has suggested that this theory bears a relationship to that of Kramers which is similar to that obtaining between hydrodynamics and a kinetic theory of fluids.

For the free electron gas, both Kramers and Bohr find, in the nonrelativistic limit,

$$\frac{dT}{dx}=\frac{4\pi nZ^2e^4}{MV_0^2}\ln\left(\frac{V_0}{\omega_p b}\right). \tag{42}$$

Our expression (41) differs from (42) due to the fact that we have taken into account the dependence of the frequency of organized oscillation on the electron kinetic energy. This correction is rather small and may be neglected in the limit of high particle velocity $V_0^2 \gg \langle V^2\rangle_{\text{Av}}$.

The total stopping power of a metal for a fast charged particle is the sum of that due to the conduction electrons and to the core electrons. Thus, in order to obtain experimental verification of our expression (41) as applied to the conduction electrons, we must consider metals in which the number of core and valence electrons is roughly comparable, e.g., Li and Be. Bohr has used the expression (42) together with the appropriate theoretical expression for the core electrons to obtain a theoretical average excitation potential of 45 ev for Li and 60 ev for Be. (The corresponding excitation potentials using (41) are 44 ev and 57 ev, respectively.) These values are in good agreement with the experimental values of Bakker and Segrè, who found excitation potentials of 34 ev for Li and 60 ev for Be, when one considers the fact that an experimental uncertainty of $\sim$10 percent in the stopping power for Li corresponds to an uncertainty in the Li excitation potential of $\sim$50 percent.[19]

The author wishes to thank Professor D. Bohm, Professor J. Bardeen, Professor J. Blatt, Professor F. Low, and Dr. A. B. Lidiard for stimulating discussions of subjects related to this paper. He would like to acknowledge the partial support of the Office of Ordnance Research, U. S. Army, during this work.

[18] H. A. Kramers, Physica **13**, 401 (1947).
[19] A. Bohr, Kgl. Danske Videnskab. Selskab, Mat.-fys. Medd. **24**, No. 19 (1948).
[20] R. Kronig and J. Korringa, Physica **10**, 406 (1943).
[21] R. Kronig, Physica **15**, 667 (1949).

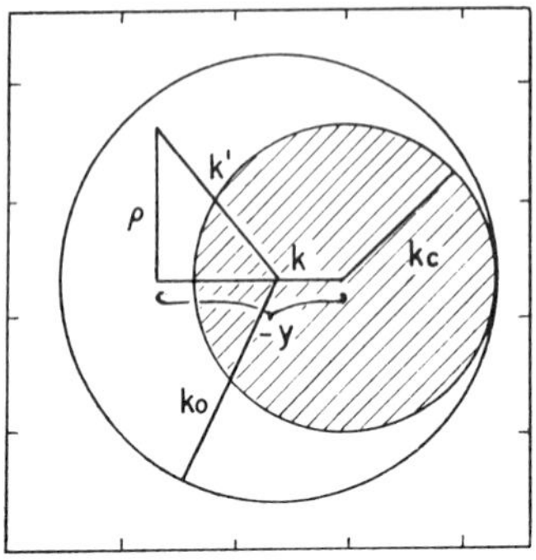

FIG. 5. Allowed regions for $\mathbf{k}'$ integration.

APPENDIX I

In this appendix we wish to evaluate ϵ_{exch} which may be written, according to (18), as

$$\epsilon_{\text{exch}} = -2\pi e^2 \sum_{\substack{k,\,k'<k_0 \\ |\mathbf{k}'-\mathbf{k}|>k_c}} \frac{1}{|\mathbf{k}'-\mathbf{k}|^2}, \tag{A1}$$

where the sum is to be carried out for all electrons of parallel spin in the Fermi distribution, such that $|\mathbf{k}'-\mathbf{k}|>k_c$, this latter restriction arising from the short-range character of the interaction $H_{\text{s.r.}}$. Let us first obtain the exchange energy of interaction with all the other electrons for a given electron of wave vector k. This is, on changing our sum over $\mathbf{k}'$ to an integral,

$$W_i = -\frac{4\pi e^2}{(2\pi)^3}\int d\mathbf{k}' \frac{1}{|\mathbf{k}'-\mathbf{k}|^2}, \tag{A2}$$

where the integration must be carried out in such a way as to exclude the shaded portion of the k_0 sphere (corresponding to $|\mathbf{k}'-\mathbf{k}|<k_c$) as illustrated in Fig. 5.

It is convenient to introduce for our $\mathbf{k}'$ coordinate system, cylindrical coordinates (ρ, y, φ) centered at k, as shown in Fig. 5, for we then find that

$$(\mathbf{k}'-\mathbf{k})^2 = \rho^2 + y^2. \tag{A3}$$

The requirement that $|\mathbf{k}'-\mathbf{k}|$ be greater than k_c then leads to the following regions of integration for our $\mathbf{k}'$ integration:

$k > k_0 - k_c$

$$\int d\mathbf{k}' = \int_{-(k+k_0)}^{-k_c} dy \int_0^{[k_0^2-(y+k)^2]^{\frac{1}{2}}} \rho d\rho \int_0^{2\pi} d\varphi + \int_{-k_c}^{[k_0^2-k_c^2-k^2]/2k} dy \int_{[k_c^2-y^2]^{\frac{1}{2}}}^{[k_0^2-(y+k)^2]^{\frac{1}{2}}} \rho d\rho \int_0^{2\pi} d\varphi,$$

$k < k_0 - k_c$

$$\int d\mathbf{k}' = \int_{-(k+k_0)}^{-k_c} dy \int_0^{[k_0^2-(y+k)^2]^{\frac{1}{2}}} \rho d\rho \int_0^{2\pi} d\varphi + \int_{-k_c}^{k_c} dy \int_{[k_c^2-y^2]^{\frac{1}{2}}}^{[k_0^2-(y+k)^2]^{\frac{1}{2}}} \rho d\rho \int_0^{2\pi} d\varphi + \int_{k_c}^{k_0-k} dy \int_0^{[k_0^2-(y+k)^2]^{\frac{1}{2}}} \rho d\rho \int_0^{2\pi} d\varphi.$$

The integrations for W_i are quite straightforward and yield for $k_0-k_c<k<k_0$

$$W_i = -\frac{e^2k_0}{2\pi}\left\{1+\frac{k_0^2-k^2}{kk_0}\ln\left(\frac{k+k_0}{\beta k_0}\right) + \frac{3k^2-k_0^2}{2kk_0} - 2\beta + \frac{\beta^2 k_0}{2k}\right\},$$

$k<(k_0-k_c)$

$$W_i = -\frac{e^2k_0}{2\pi}\left\{2+\frac{k_0^2-k^2}{kk_0}\ln\left(\frac{k+k_0}{k_0-k}\right) - 4\beta\right\}.$$

We may then evaluate ϵ_{exch} by summing over all $\mathbf{k}$ within the Fermi distribution (and dividing by two so that no interactions are counted twice). We find

$$\epsilon_{\text{exch}} = \tfrac{1}{2}\sum_k W_i = \frac{1}{(2\pi)^3}\int d\mathbf{k} W_i = -\frac{3e^2k_0 n}{4\pi}\left\{1-\frac{4}{3}\beta+\frac{\beta^2}{2}+\frac{\beta^4}{48}\right\},$$

where $\beta = k_c/k_0$. The exchange energy per electron may then be written as

$$\epsilon_{\text{exch}}' = -\frac{0.916}{r_s}\left\{1-\frac{4}{3}\beta+\frac{\beta^2}{2}-\frac{\beta^4}{48}\right\}\text{ry}.$$

PHYSICAL REVIEW VOLUME 106, NUMBER 2 APRIL 15, 1957

Correlation Energy of an Electron Gas at High Density*

MURRAY GELL-MANN, *Department of Physics, California Institute of Technology, Pasadena, California*
AND
KEITH A. BRUECKNER, *Department of Physics, University of Pennsylvania, Philadelphia, Pennsylvania*

(Received December 14, 1956)

The quantity ϵ_c is defined as the correlation energy per particle of an electron gas expressed in rydbergs. It is a function of the conventional dimensionless parameter r_s, where r_s^{-3} is proportional to the electron density. Here ϵ_c is computed for small values of r_s (high density) and found to be given by $\epsilon_c = A \ln r_s + C + O(r_s)$. The value of A is found to be 0.0622, a result that could be deduced from previous work of Wigner, Macke, and Pines. An exact formula for the constant C is given here for the first time; earlier workers had made only approximate calculations of C. Further, it is shown how the next correction in r_s can be computed. The method is based on summing the most highly divergent terms of the perturbation series under the integral sign to give a convergent result. The summation is performed by a technique similar to Feynman's methods in field theory.

WE consider the idealized problem of the ground-state energy of a gas of electrons in the presence of a uniform background of positive charge that makes the system neutral. For most practical problems, of course, the uniform positive charge must be replaced by a lattice of positive ions, but we shall not treat this more realistic case.

We have, then, a fully degenerate Fermi-Dirac system with Coulomb interactions. Let us employ the conventional notation for the problem. The inverse density or volume per electron is set equal to $\frac{4}{3}\pi r_0^3$. The dimensionless parameter r_s is defined as r_0 divided by the Bohr radius. The ground state energy per particle in rydbergs is called ϵ and is a function of r_s only, since there are no other dimensionless quantities involved.

We shall compute ϵ in the case of high density or small r_s. Since r_s is proportional to e^2, an expansion in powers of r_s is essentially an expansion in powers of e^2, that is, the perturbation expansion. Unfortunately, a straightforward perturbation expansion leads to divergences, but let us ignore that difficulty for a moment.

The leading term in the perturbation series is evidently the Fermi energy, the kinetic energy of the degenerate free electron gas. The maximum electron momentum P, the radius of the Fermi sphere in momentum space, is given by $P = (9\pi/4)^{\frac{1}{3}}\hbar r_0^{-1}$. The Fermi energy per particle in rydbergs is thus

$$\epsilon_F = \frac{3}{5}\left(\frac{P^2}{2m}\right)\Big/\left(\frac{e^4 m}{2\hbar^2}\right) = \frac{3}{5}\left(\frac{9\pi}{4}\right)^{\frac{2}{3}}\frac{1}{r_s^2} \approx \frac{2.21}{r_s^2}. \quad (1)$$

The next term in the series is the exchange energy, the expectation value of the potential energy in the ground state of a free electron gas. It is one higher order in e^2 or r_s than the Fermi energy and so is proportional to $1/r_s$. It is easily evaluated to be

$$\epsilon_x = -\frac{3}{4}\left(\frac{e^2}{\pi}\right)P\Big/\left(\frac{e^4 m}{2\hbar^2}\right) = -\frac{3}{2\pi}\left(\frac{9\pi}{4}\right)^{\frac{1}{3}}\frac{1}{r_s} \approx -\frac{0.916}{r_s}. \quad (2)$$

If we now calculate the effect of the potential in second-order perturbation theory, we should expect a term of one higher order in e^2 or r_s than (2), that is, a constant independent of r_s. However, the second-order perturbation formula diverges logarithmically at small momentum transfers on account of the long-range character of the Coulomb force. Thus some refinement of perturbation theory is necessary in order to carry the computation further.

The terms in the energy beyond (1) and (2) are called collectively the "correlation energy,"

$$\epsilon_c \equiv \epsilon - \epsilon_F - \epsilon_x. \quad (3)$$

This name was introduced by Wigner,[1] who called attention to the importance of the correlation energy in solid-state problems. Following Wigner's lead, calculations were made by Macke[2] and by Pines[3] that led essentially to expressions of the form

$$\begin{aligned}\epsilon_c &= (2/\pi^2)(1-\ln 2)\ln r_s + C \\ &\qquad + \text{terms that vanish as } r_s \to 0, \\ &\approx 0.0622 \ln r_s + C + \cdots,\end{aligned} \quad (4)$$

where in each case the constant C was calculated approximately.[4]

We give here an *exact* evaluation of the constant C by a method that should permit also the calculation of higher corrections in r_s. The basic idea of the method is to examine the increasingly divergent terms of the perturbation series and to notice that they fall into

* This study was performed by the authors as consultants to the RAND Corporation, Santa Monica, California, and was sponsored entirely by the U. S. Atomic Energy Commission.

[1] E. P. Wigner, Phys. Rev. **46**, 1002 (1934).

[2] W. Macke, Z. Naturforsch. **5a**, 192 (1950).

[3] D. Pines, Phys. Rev. **92**, 626 (1953); D. Pines, in *Solid State Physics*, edited by F. Seitz and D. Turnbull (Academic Press, Inc., New York, 1955), Vol. 1, p. 367.

[4] Pines' result is actually not of the form of Eq. (4). However, he has neglected a term which he calls the exchange correlation energy and which adds to his result for ϵ the quantity $0.0311 \ln r_s - 0.0905 + \epsilon_b^{(2)} + O(r_s)$, where $\epsilon_b^{(2)}$ is defined in our Eq. (9). When we supply this term, his final answer takes on the form of Eq. (4). When we quote Pines' result later, we mention this modification.

Macke's result does agree with Eq. (4). However, he seems to have made certain unnecessary approximations in his calculation of C, and we have therefore recomputed C, using his method, in Appendix I.

subseries that can be summed under the integral sign to give convergent results. The logarithmic divergence in second order is then automatically replaced by a logarithmic dependence on the expansion parameter r_s, as in (4). In this respect, our method is similar to Macke's.[2] However, Macke fails to sum *all* of the processes that contribute to the constant C. In our work, we are able to exhibit all of them and then to sum them by a procedure similar to Feynman's methods in field theory.

Let us discuss, then, the behavior of the formal perturbation series for ϵ. The coefficient of each power of r_s can be written as an integral over various dimensionless vectors $\mathbf{q}_i$, which are virtual momentum transfers divided by P, the Fermi momentum. The term independent of r_s then diverges logarithmically, as we have said. The next term, formally linear in r_s, diverges quadratically, the succeeding one quartically, etc. Now, since the correlation energy is finite, these integrals, when summed, must cut themselves off at some characteristic value of the dimensionless momentum transfers q. Moreover, the nature of the cutoff is clear from the results of work on the plasma vibrations of an electron gas, especially that of Bohm and Pines.[5] It has been shown that collective electron motions effectively screen the Coulomb field at a distance of the order of

$$r_{\max}\sim(\text{const.})\, r_0^{\frac{1}{2}}a^{\frac{1}{2}}+\text{higher terms in } r_0, \tag{5}$$

where a is the Bohr radius. The effective cutoff for q is then

$$q_{\min}\sim(\text{const.})\, r_s^{\frac{1}{2}}+\text{higher terms in } r_s. \tag{6}$$

We use this estimate in conjunction with our estimate of the correlation energy

$$\epsilon_c\sim(\text{log divergence})+r_s\,(\text{quadratic div.}) +r_s^2\,(\text{quartic div.})+\cdots, \tag{7}$$

and we deduce the following results:

(i) $\epsilon_c=A\ln r_s+C+$terms that vanish as $r_s\to 0$.

(ii) The coefficient A can be found merely from the strength of the logarithmic divergence in second-order perturbation theory. This leads to the value of 0.0622 quoted above.

(iii) The only virtual processes contributing to C beyond the second order are those which contribute the highest divergence in each order of perturbation theory. Those processes leading to lower divergences will give higher powers of r_s in the final expression for ϵ.

Now the processes that lead to the highest divergences are easily identified. The divergences are caused by the piling-up of factors $1/q^2$ coming from Coulomb interactions in momentum space. Evidently the greatest piling-up occurs when only a single momentum transfer is involved and this single q is handed from electron to electron, contributing a factor $1/q^2$ each time. Moreover, we may distinguish momentum transfers with and without exchange; when exchange occurs the factor is no longer $1/q^2$ but $1/(\mathbf{p}_1-\mathbf{p}_2+\mathbf{q})^2$, where $\mathbf{p}_1$ and $\mathbf{p}_2$ are the initial electron momenta. In the case of exchange, then, the singularity at $q=0$ is not enhanced. We may therefore ignore exchange entirely beyond the second order in computing C.

We may now list the processes that contribute to C in the first few orders of perturbation theory. In second order we must include everything; we have two terms, the logarithmically divergent one we have mentioned and a finite one coming from exchange. They may be written as follows:

$$\epsilon_a^{(2)}=-\frac{3}{8\pi^5}\int\frac{d^3q}{q^4}\int_{\substack{p_1<1\\|\mathbf{p}_1+\mathbf{q}|>1}}d^3p_1\int_{\substack{p_2<1\\|\mathbf{p}_2+\mathbf{q}|>1}}d^3p_2 \times\frac{1}{q^2+\mathbf{q}\cdot(\mathbf{p}_1+\mathbf{p}_2)}, \tag{8}$$

and

$$\epsilon_b^{(2)}=\frac{3}{16\pi^5}\int\frac{d^3q}{q^2}\int_{\substack{p_1<1\\|\mathbf{p}_1+\mathbf{q}|>1}}d^3p_1\int_{\substack{p_2<1\\|\mathbf{p}_2+\mathbf{q}|>1}}d^3p_2 \times\frac{1}{(\mathbf{q}+\mathbf{p}_1+\mathbf{p}_2)^2}\,\frac{1}{q^2+\mathbf{q}\cdot(\mathbf{p}_1+\mathbf{p}_2)}. \tag{9}$$

In these processes two electrons in the Fermi sea with initial momenta $\mathbf{p}_1P$ and $-\mathbf{p}_2P$ have undergone a collision with momentum transfer $\mathbf{q}P$, emerging into unoccupied states with momenta $(\mathbf{p}_1+\mathbf{q})P$ and $(-\mathbf{p}_2-\mathbf{q})P$ and then returning to their original states. The factor $1/[q^2+\mathbf{q}\cdot(\mathbf{p}_1+\mathbf{p}_2)]$ comes from the energy denominator. In (8) there is a factor $1/q^2$ for each collision, while in the exchange correlation energy (9) one factor is $1/q^2$ and the other $1/(\mathbf{q}+\mathbf{p}_1+\mathbf{p}_2)^2$.

In writing higher order integrals, we shall not indicate explicitly the conditions $|\mathbf{p}+\mathbf{q}|>1$ and $p<1$, but these must still be obeyed by all vectors $\mathbf{p}$ to insure that initial states are occupied and others unoccupied. In third order the processes involving a single momentum transfer are these: Two electrons with momenta $\mathbf{p}_1P$ and $-\mathbf{p}_2P$ emerge from the sea into states with momenta $(\mathbf{p}_1+\mathbf{q})P$ and $(-\mathbf{p}_2-\mathbf{q})P$, as before. One of them now returns to its original state and transfers its excess momentum $\mathbf{q}P$ or $-\mathbf{q}P$ to a third electron, which emerges from a state with momentum $\mathbf{p}_3P$. This third one and the one still outstanding now return to their original states. Both third-order processes contribute equal amounts to the energy, since the first and second electrons are quite equivalent and it does not matter which one first interacts with the third electron.

The processes involved may be represented diagrammatically as in Fig. 1. In second order, two electrons called 1 and 2 are excited and then de-excited. In third order, electrons 1 and 2 are excited, one of them is de-

[5] D. Bohm and D. Pines, Phys. Rev. 92, 609 (1953).

excited while exciting a third electron 3, and then the outstanding electrons are de-excited. And so forth.

The third-order contribution is given by

$$\epsilon^{(3)}=2\left(\frac{\alpha r_s}{\pi^2}\right)\left(\frac{3}{8\pi^5}\right)\int\frac{d^3q}{q^6}\int d^3p_1\int d^3p_2\int d^3p_3 \times\frac{1}{q^2+\mathbf{q}\cdot(\mathbf{p}_1+\mathbf{p}_2)}\,\frac{1}{q^2+\mathbf{q}\cdot(\mathbf{p}_1+\mathbf{p}_3)}, \quad (10)$$

where α is $(4/9\pi)^{\frac{1}{3}}$.

In fourth order, rather complicated processes begin to appear, as one may see from Fig. 1. In the first diagram in fourth order, electron 1 is twice replaced before de-exciting with electron 2. In the next, electron 1 is replaced, then electron 2 is replaced, and then the de-excitation takes place. The following two diagrams are similar with electrons 1 and 2 exchanging roles. In the last two diagrams the excitation of 1 and 2 is followed by the excitation of another pair 3 and 4; then electrons 1 and 3 de-excite together and electrons 2 and 4 de-excite together, in either order.

The fourth-order contributions are given by

$$\epsilon^{(4)}=-2\left(\frac{\alpha r_s}{\pi^2}\right)^2\left(\frac{3}{8\pi^5}\right)\int\frac{d^3q}{q^8}\int d^3p_1\int d^3p_2\int d^3p_3\int d^3p_4$$
$$\times\left\{\frac{1}{q^2+\mathbf{q}\cdot(\mathbf{p}_1+\mathbf{p}_2)}\,\frac{1}{q^2+\mathbf{q}\cdot(\mathbf{p}_1+\mathbf{p}_3)}\,\frac{1}{q^2+\mathbf{q}\cdot(\mathbf{p}_1+\mathbf{p}_4)}\right.$$
$$+\frac{1}{q^2+\mathbf{q}\cdot(\mathbf{p}_1+\mathbf{p}_2)}\,\frac{1}{q^2+\mathbf{q}\cdot(\mathbf{p}_1+\mathbf{p}_3)}\,\frac{1}{q^2+\mathbf{q}\cdot(\mathbf{p}_3+\mathbf{p}_4)}$$
$$+\frac{1}{q^2+\mathbf{q}\cdot(\mathbf{p}_1+\mathbf{p}_2)}\,\frac{1}{2q^2+\mathbf{q}\cdot(\mathbf{p}_1+\mathbf{p}_2+\mathbf{p}_3+\mathbf{p}_4)}$$
$$\left.\times\frac{1}{q^2+\mathbf{q}\cdot(\mathbf{p}_1+\mathbf{p}_3)}\right\}. \quad (11)$$

It is easy now to write down the contributions from any order. The problem then is to sum all these contributions before performing the integral over q. In order to see how to sum them, we note the similarity to diagrams in field theory. There, of course, pairs of *electrons* and *holes* (positrons) are under consideration, interacting with an external field. Here we consider pairs of *electrons* in interaction with the Fermi sea and ignore the holes. Nevertheless the similarity is sufficient for the application of Feynman's artifice of considering a pair as composed of one electron traveling forward in time and one backward in time. The creation or annihilation of a pair is interpreted as the turning-around of a particle in time. If we look back at Fig. 1 with Feynman's point of view in mind, we see that in each order all the diagrams are merely versions of one single diagram in which a single electron starts out, is replaced over and over again, and finally returns to its starting point. The various forms of the diagram come from the choice that faces the electron each time it is replaced, a choice of being replaced by an electron going forward in time or one going backward in time.

If we introduce "time variables," then, and let the electrons propagate either forward or backward in time with suitable propagators, we should be able to represent the sum of all diagrams in each order by a single integral. The integrals over the time-variables should give us the various energy denominators we need. We try as the propagator the function

$$F_q(t)=\int d^3p\,\exp[-|t|(\tfrac{1}{2}q^2+\mathbf{q}\cdot\mathbf{p})], \quad (12)$$

which is arranged so that integration over positive or negative time will introduce into the energy denominator plus or minus $(\frac{1}{2}q^2+\mathbf{q}\cdot\mathbf{p})$, respectively. Now we integrate around a loop using this propagator. In second order we look at

$$A_2\equiv\frac{1}{2}\int_{-\infty}^{\infty}dt_1\int_{-\infty}^{\infty}dt_2F_q(t_1)F_q(t_2)\delta(t_1+t_2), \quad (13)$$

where the δ function insures that the electron comes

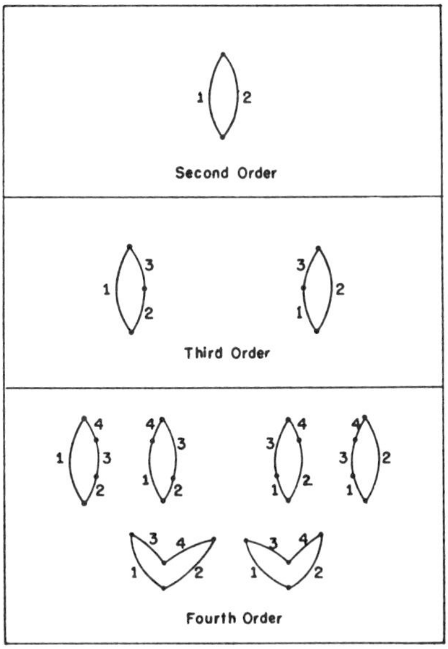

FIG. 1. The relevant second-, third-, and fourth-order processes represented diagrammatically. In second order the process may take place with or without exchange. In higher orders, exchange is neglected.

back to its starting point. Then we have

$$A_2=\int d^3p_1\int d^3p_2\int_0^\infty dt \times\exp\{-(\tfrac{1}{2}q^2+\mathbf{q}\cdot\mathbf{p}_1)t+(\tfrac{1}{2}q^2+\mathbf{q}\cdot\mathbf{p}_2)(-t)\} \tag{14}$$
$$=\int d^3p_1\int d^3p_2\frac{1}{q^2+\mathbf{q}\cdot(\mathbf{p}_1+\mathbf{p}_2)},$$

just what we need for Eq. (8).

In third order, we look at

$$A_3\equiv\frac{1}{3}\int_{-\infty}^{\infty}dt_1\int_{-\infty}^{\infty}dt_2\int_{-\infty}^{\infty}dt_3 \times F_q(t_1)F_q(t_2)F_q(t_3)\delta(t_1+t_2+t_3). \tag{15}$$

We find

$$A_3=2\int d^3p_1\int d^3p_2\int d^3p_3\int_0^\infty dt_1\int_{-t_1}^0 dt_2 \times\exp\{-(\tfrac{1}{2}q^2+\mathbf{q}\cdot\mathbf{p}_1)t_1+(\tfrac{1}{2}q^2+\mathbf{q}\cdot\mathbf{p}_2)t_2 +(\tfrac{1}{2}q^2+\mathbf{q}\cdot\mathbf{p}_3)(-t_1-t_2)\}=2\int d^3p_1\int d^3p_2\int d^3p_3 \times\frac{1}{q^2+\mathbf{q}\cdot(\mathbf{p}_1+\mathbf{p}_2)}\frac{1}{q^2+\mathbf{q}\cdot(\mathbf{p}_1+\mathbf{p}_3)}, \tag{16}$$

which is what we need for Eq. (10).

The agreement evidently extends to all orders. It should be noted that there is a direct correspondence between the various forms of the diagram in a given order and the various time-orderings in the corresponding integral A_n.

Now let us perform the Fourier transform of the δ function in the expression for A_n. We obtain

$$A_n=\frac{q}{2\pi n}\int_{-\infty}^{\infty}du[Q_q(u)]^n, \tag{17}$$

where

$$Q_q(u)=\int d^3p\int_{-\infty}^{\infty}e^{ituq}\exp\{-|t|[\tfrac{1}{2}q^2+\mathbf{q}\cdot\mathbf{p}]\}dt. \tag{18}$$

The terms in the correlation energy contributing to the constant C may then be summed, leaving aside the exchange term $\epsilon_b^{(2)}$. We have

$$\epsilon'\equiv\epsilon_a^{(2)}+\epsilon^{(3)}+\epsilon^{(4)}+\cdots=-\frac{3}{8\pi^5}\int\frac{d^3q}{q^3}\frac{1}{2\pi} \times\int_{-\infty}^{\infty}du\sum_{n=2}^{\infty}\frac{(-1)^n}{n}[Q_q(u)]^n\left(\frac{\alpha r_s}{\pi^2q^2}\right)^{n-2}. \tag{19}$$

This expression may now be enormously simplified when we realize that beyond the second order we are interested only in the most highly divergent part of the q integral, that is, the leading term near $q=0$. We may thus take, beyond the second order, just this leading term and put the upper limit of the q integral equal to some arbitrary number, which we take to be 1.

At small values of q, we may approximate $Q_q(u)$ as follows: We apply the restrictions on $\mathbf{p}$ that $p<1$ and $|\mathbf{p}+\mathbf{q}|>1$. If x is the direction cosine between $\mathbf{p}$ and $\mathbf{q}$ we see that at small q the variable x is restricted to the range $0\leq x\leq 1$ and p to the range $1-qx\leq p\leq 1$. Thus we have, for $q\ll 1$,

$$Q_q(u)\approx 2\pi q\int_0^1 xdx\int_{-\infty}^{\infty}dte^{ituq}e^{-|t|qx} =2\pi\int_0^1 xdx\int_{-\infty}^{\infty}d^qe^{isu}e^{-|s|x}=4\pi R(u), \tag{20}$$

independent of q. We have put

$$R(u)=1-u\operatorname{arc\,tan}u^{-1}. \tag{21}$$

We shall thus approximate $Q_q(u)$ by a function which is equal to $4\pi R(u)$ when $0\leq q\leq 1$ and which vanishes for $q>1$. In second order, we must supply a correction term that restores $\epsilon_a^{(2)}$ to its exact value, but in higher orders the approximation is sufficient. We have, then,

$$\epsilon'\approx-\frac{12}{\pi^3}\int_{-\infty}^{\infty}du\int_0^1\frac{dq}{q}\sum_{n=2}^{\infty}\frac{(-1)^n}{n} \times[R(u)]^n\left(\frac{4\alpha r_s}{\pi q^2}\right)^{n-2}+\delta, \tag{22}$$

where we have put

$$\delta\equiv\epsilon_a^{(2)}-\left[-\frac{12}{\pi^3}\int_{-\infty}^{\infty}du\int_0^1\frac{dq}{q}(\tfrac{1}{2}R^2)\right]$$
$$=\lim_{\beta\to 0}\left\{-\frac{3}{8\pi^5}\int_\beta^\infty\frac{d^3q}{q^4}\int_{\substack{p_1<1\\|\mathbf{p}_1+\mathbf{q}|>1}}d^3p_1\int_{\substack{p_2<1\\|\mathbf{p}_2+\mathbf{q}|>1}}d^3p_2 \times\frac{1}{q^2+\mathbf{q}\cdot(\mathbf{p}_1+\mathbf{p}_2)}+\frac{6}{\pi^2}\int_\beta^1\frac{dq}{q}\int_0^1 xdx\int_0^1 ydy\frac{1}{x+y}\right\}, \tag{23}$$

which is a finite number, the logarithmic divergences cancelling.

We comment here on a difficulty which occurs in carrying out the summation over n. For large values of q, the series converges and the summation is straightforward. For small q, however, the series diverges, with large contributions arising from large n. We shall, however, *assume* that the result valid for large q may be continued into the region of divergence. This procedure cannot be justified without a detailed investigation of the behavior of the series for large n; we argue, however, that any corrections must either vanish with large N or contribute to higher powers of r_s.

We may now perform the sum over n and the integral over q in Eq. (22), remarking that the integral really does cut itself off at a value of q proportional to $r_s^{\frac{1}{2}}$, as predicted in Eq. (6). Dropping terms that vanish as $r_s \to 0$, we have

$$\begin{aligned}\epsilon' &= \frac{3}{\pi^3}\int_{-\infty}^{\infty} du[R(u)]^2\left[\ln\left(\frac{4\alpha r_s}{\pi}\right)+\ln R(u)-\tfrac{1}{2}\right]+\delta \\ &= \frac{2}{\pi^2}(1-\ln 2)\left[\ln\left(\frac{4\alpha r_s}{\pi}\right)+\langle \ln R\rangle_{Av}-\tfrac{1}{2}\right]+\delta\end{aligned} \quad (24)$$

where

$$\langle \ln R\rangle_{Av} = \int_{-\infty}^{\infty} duR^2 \ln R \bigg/ \int_{-\infty}^{\infty} duR^2. \quad (25)$$

We see that Eq. (24) confirms our value of $(2/\pi^2)\times(1-\ln 2)$ for the constant A. For the constant C, we have

$$C = \frac{2}{\pi^2}(1-\ln 2)\left\{\ln\left[\frac{4}{\pi}\left(\frac{4}{9\pi}\right)^{\frac{1}{3}}\right]-\tfrac{1}{2}+\langle \ln R\rangle_{Av}\right\}+\delta. \quad (26)$$

Now Pines[3] has found the value -0.0508 for δ, and numerical integration yields the value -0.551 for $\langle \ln R\rangle_{Av}$. The multiple integral (9) for $\epsilon_b^{(2)}$ has been evaluated by the Monte Carlo method, with the result $\epsilon_b^{(2)} = 0.046 \pm 0.002$. Substituting these numbers into (26), we find

$$C = -0.096 \pm 0.002, \quad (27)$$

$$\epsilon_c = 0.0622 \ln r_s - 0.096 + O(r_s). \quad (28)$$

The expression given by Pines[3] is $0.0311 \ln r_s - 0.114 + O(r_s)$, although if the correction mentioned in footnote 4 is taken into account one gets by Pines' method the result $0.0622 \ln r_s - 0.158 + O(r_s)$.

In the Appendix we evaluate ϵ_c by Macke's method and obtain $0.0622 \ln r_s - 0.128 + O(r_s)$.

We see that the approximations of Macke and Pines tend to overestimate the magnitude of the constant term C in the correlation energy.

In conclusion, let us discuss the calculation of the next correlation to ϵ. In order to include all terms that are genuinely of order r_s or $r_s \ln r_s$, we must improve the present calculation in three ways;

(i) The contribution to ϵ' from Eq. (19) must be treated more carefully than in Eq. (22), so that terms of order r_s are retained.

(ii) We must calculate the contribution from the diagrams in Fig. 1 beyond the second order when *one* exchange is permitted in each process. Beyond the third order we may employ the crudest approximation that preserves the leading divergence.

(iii) The remaining third-order processes are the following:

$$\left.\begin{aligned}&1+2\to 1'+2'\\ &1'+2'\to 1''+2''\\ &1''+2''\to 1+2\end{aligned}\right\}\text{rescattering.}$$

$$\left.\begin{aligned}&1+2\to 1'+2'\\ &1'+3\to 1'+3\\ &1'+2'\to 1+2\end{aligned}\right\}\text{direct and exchange scattering with unexcited particles.}$$

Those are both logarithmically divergent and must be combined with a sequence of terms similar to those summed to remove the second-order divergence. The methods presented above are easily generalized to this case.

The authors would like to thank Dr. Richard Latter for many valuable discussions and Mr. J. I. Marcum and Mr. H. Kahn for the Monte Carlo computation of $\epsilon_b^{(2)}$.

APPENDIX. APPROXIMATION OF MACKE

The method of Macke[2] is suggested by the earlier work of Wigner.[1] It consists of summing, instead of the complete set of diagrams indicated in Fig. 1, just the first diagram in each order. Under the integral sign these form a simple geometric series. We obtain, in place of Eq. (19), the following:

$$\epsilon' \approx \frac{3}{8\pi^5}\int\frac{d^3q}{q^4}\int_{\substack{p_1<1\\|\mathbf{p}_1+\mathbf{q}|>1}} d^3p_1 \sum_{n=1}^{\infty}(-1)^n \times\left(\int_{\substack{p_2<1\\|\mathbf{p}_2+\mathbf{q}|>1}} d^3p_2\frac{1}{q^2+\mathbf{q}\cdot(\mathbf{p}_1+\mathbf{p}_2)}\right)^n\left(\frac{\alpha r_s}{\pi^2 q^2}\right)^{n-1}. \quad (A.1)$$

Making an approximation analogous to (20) in the orders beyond the second, we find instead of (22) the expression

$$\epsilon' \approx \delta + \frac{6}{\pi^2}\int_0^1\frac{dq}{q}\int_0^1 xdx\sum_{n=1}^{\infty}(-1)^n \times\left(\int_0^1\frac{ydy}{x+y}\right)^n\left(\frac{2\alpha r_s}{\pi q^2}\right)^{n-1}. \quad (A.2)$$

We may now, as before, perform the sum over n and the integral over q, dropping terms that vanish as $r_s \to 0$. Putting $I = \int_0^1 ydy/(x+y)$, we find

$$\epsilon' \approx \delta + \frac{3}{\pi^2}\int_0^1 xdxI\ln\left(\frac{2\alpha r_s I}{\pi}\right). \quad (A.3)$$

The approximate value of C is then

$$\begin{aligned}C &\approx \epsilon_b^{(2)} + \delta + \frac{3}{\pi^2}\int_0^1 xdxI\ln\left(\frac{2\alpha I}{\pi}\right) \\ &\approx \epsilon_b^{(2)} - 0.174.\end{aligned} \quad (A.4)$$

PHYSICAL REVIEW VOLUME 106, NUMBER 2 APRIL 15, 1957

Specific Heat of a Degenerate Electron Gas at High Density

MURRAY GELL-MANN*
Department of Physics, California Institute of Technology, Pasadena, California
(Received January 14, 1957)

The methods of the preceding paper of Gell-Mann and Brueckner are generalized so that not only the ground state but also the low excited states of an electron gas can be discussed. The energy levels relative to the ground state are the same as those of a gas of independent particles where the energy of each particle (in rydbergs) is a certain function $W(p)$ of its momentum (expressed in units of the Fermi momentum). The specific heat of the gas at low temperature is proportional to the density of single particle levels at the surface of the Fermi sea, or inversely proportional to $(dW/dp)_{p=1}$. This last quantity is calculated for high density (small r_s, where density is proportional to r_s^{-3}) and compared to the corresponding quantity for a free electron gas. The ratio is found to be $1+0.083r_s[-\ln r_s-0.203]+$higher terms in r_s. The expansion is exact and may be compared with the approximate result of Pines, who finds $1+0.083r_s[-\ln r_s+1.47]+\cdots$.

IN the preceding paper,[1] to be referred to as I, the energy of the ground state of an electron gas is computed at high density. Here we shall treat the energies of low excited states by the same method and thus calculate the low-temperature specific heat at high density

We may label the states by referring them to corresponding states of a free-electron gas. The familiar methods of perturbation theory (suitably modified to fit the problem) will convert a given energy eigenstate Φ_n of the system of free electrons into an energy eigenstate Ψ_n of the system with Coulomb interactions among the electrons. We shall be concerned, of course, with the perturbed energy E_n associated with Ψ_n, and not with the unperturbed (purely kinetic) energy associated with Φ_n. Nevertheless, we shall refer to the nth state by describing the *unperturbed* state Φ_n.

Thus we shall speak of the ground state as that in which all one-particle momentum states inside the Fermi sphere are occupied and all those outside are unoccupied. This is, of course, a description of the unperturbed ground state Φ_0. But when we speak of the energy E_0 of the ground state, we shall mean the perturbed energy associated with the exact eigenstate Ψ_0 of the interacting system.

Similarly, we shall specify an excited state by listing the momenta and spins $\{\mathbf{p}_j,s_j\}$ (with $j=1,\cdots\nu$) of the vacated one-particle states inside the Fermi sphere and the momenta $\{\mathbf{k}_j\}$ of the occupied one-particle states outside the sphere. (The momenta are expressed in units of the Fermi momentum P.) Again we are describing an unperturbed state; and again, in speaking of the energy, we shall mean the energy E of the perturbed state.

As in I, we consider a macroscopic sample of gas; the number N of electrons is very large, say $\sim 10^{23}$. We now restrict ourselves to treating those states of the gas in which the number ν of excited particles is small compared to N. At low temperatures, only such states will be important. For these states of the electron gas, the energy E (in units of the rydberg) may be written in the form

$$E=E_0+\sum_{j=1}^{\nu}\{W(k_j)-W(p_j)\}+O(\nu/N). \quad (1)$$

This can be done because the interaction energy of the excited particles among themselves is of order ν/N compared to their interaction energy with the rest of the gas. Under our assumptions the term of order ν/N can be neglected, and the specific heat at low temperatures depends only on the form of the function $W(p)$.

Thus, for the computation of the specific heat at low temperatures, we are justified in treating the interacting electron gas as a system of independent particles obeying Fermi-Dirac statistics and with the energy (in rydbergs) of each given as a function of momentum (in units of P) by $W(p)$. It is well known[2] that under these conditions the specific heat C per electron at constant volume is proportional to the density of "one-particle energy levels" at the surface of the Fermi sphere, that is, proportional to $[(dW/dp)_{p=1}]^{-1}$.

For a free-electron gas, we have

$$W_F(p)=\frac{p^2P^2}{2m}\left[\frac{e^4m}{2\hbar^2}\right]^{-1}=p^2/\alpha^2r_s^2, \quad (2)$$

and

$$[(dW_F/dp)_{p=1}]^{-1}=\alpha^2r_s^2/2, \quad (3)$$

where, as in I, $(4/3)\pi r_s^3$ is the volume per electron (in units of the Bohr radius cubed) and α is $(4/9\pi)^{\frac{1}{3}}$. The specific heat of a free-electron gas at low temperature is given by a familiar formula,[3] which we write as follows:

$$C_F(T,r_s)=m^{-1}\hbar^2e^{-4}K^2T\cdot\alpha^2r_s^2, \quad (4)$$

where T is temperature and K is Boltzmann's constant.

Evidently the formula for the low-temperature spe-

* This study was performed by the author as consultant to the RAND Corporation, Santa Monica, California, and was sponsored entirely by the U. S. Atomic Energy Commission.

[1] M. Gell-Mann and K. A. Brueckner, Phys. Rev. 106, 364 (1957), preceding paper.

[2] D. Pines, in *Solid State Physics* (Academic Press, Inc., New York, 1955), Vol. 1, p. 367.

[3] F. Seitz, *Modern Theory of Solids* (McGraw-Hill Book Company, Inc., New York 1940), p. 150.

cific heat of an electron gas with interactions is

$$C(T,r_s)=m^{-1}\hbar^2e^{-4}K^2T\cdot 2[(dW/dp)_{p=1}]^{-1}. \quad (5)$$

and in (4) we have the special case of free electrons, for which $W=W_F$. Our remaining task is to calculate $W(p)$ (or at least dW/dp at $p=1$) at high density for an electron gas with Coulomb interactions.

We may think of $W(p)$ in the following way: We consider the ground state of the gas, in which the filled one-particle states are those inside the Fermi sphere. We now examine the change in the energy (exact or perturbed energy) of the system when one electron, with momentum $\mathbf{p}$, is annihilated. (Clearly $p\leq 1$ here.) Then $W(p)$ is the energy lost in the annihilation, to within any additive constant. Similarly, if we start with the gas in its ground state and create an electron of momentum $\mathbf{k}$ ($k\geq 1$), then $W(k)$ is the energy gained in the creation. This interpretation of the quantity W is evidently consistent with Eq. (1).

Now in I we have treated the perturbation series for the ground state energy per particle, ϵ (in rydbergs). The fact that the terms of the series diverge is of no importance, since we have shown how to cure the divergence. In each term of the series for the ground state energy, there are sums over occupied one-particle states with momenta $\mathbf{p}_i$ such that $p_i\leq 1$ and over unoccupied states with momenta $\mathbf{k}_i$ (equal, say, to $\mathbf{p}_i+\mathbf{q}$) such that $k_i>1$. For example, see Eqs. (9), (10), and (11) of reference 1. (We shall refer to these as I-9, I-10, etc.)

If a particle with momentum $\mathbf{p}$ and spin s is annihilated, the only change in the perturbation series for the ground state energy consists in amending the condition "$p_i\leq 1$" for an occupied state by inserting the exception "$\mathbf{p}_i\neq\mathbf{p}$ if $s_i=s$" and amending the condition "$k_i>1$" for an occupied state by allowing $\mathbf{k}_i$ to equal $\mathbf{p}$ if $s_i=s$. The number of occupied states is decreased by one and that of unoccupied ones correspondingly increased. The resulting energy change can best be discussed by means of an example.

One term in the series for the ground state energy per particle is given in Eq. (I-8):

$$\epsilon_a^{(2)}=-\frac{3}{8\pi^5}\int\frac{d^3q}{q^4}\int_{\substack{p_1<1\\|\mathbf{p}_1+\mathbf{q}|>1}}d^3p_1\int_{\substack{p_2<1\\|\mathbf{p}_2+\mathbf{q}|>1}}d^3p_2\times\frac{1}{q^2+\mathbf{q}\cdot(\mathbf{p}_1+\mathbf{p}_2)}. \quad \text{(I-8)}$$

If we multiply $\epsilon_a^{(2)}$ by N, the number of electrons, we obtain the corresponding term in the total energy of the ground state, at least to order N. We may write N as $2\Omega(2\pi)^{-3}4/3\pi$, where Ω is the volume of the gas in units of $\hbar^3P^{-3}$ and the factor of 2 comes from the two spin states. Furthermore, we may replace the integral over $\mathbf{p}_1$, say, by $\Omega^{-1}(2\pi)^3$ times a sum over $\mathbf{p}_1$. We have, then, for the part of the total ground state energy corresponding to $\epsilon_a^{(2)}$:

$$E_a^{(2)}=-2\cdot\tfrac{4}{3}\pi\left(\frac{3}{8\pi^5}\right)\int\frac{d^3q}{q^4}\sum_{\substack{p_1<1\\|\mathbf{p}_1+\mathbf{q}|>1}}\int_{\substack{p_2<1\\|\mathbf{p}_2+\mathbf{q}|>1}}d^3p_2\times\frac{1}{q^2+\mathbf{q}\cdot(\mathbf{p}_1+\mathbf{p}_2)}. \quad (6)$$

Now let us remove from the gas a particle with spin up and momentum $\mathbf{p}$. There are four contributions to the loss of energy:

(a) For *one* of the spin states, the term $\mathbf{p}_1=\mathbf{p}$ is dropped from the sum. (The one-particle state with momentum $\mathbf{p}$ and spin up is no longer occupied.)

(b) For *one* of the spin states, a term $\mathbf{p}_1+\mathbf{q}=\mathbf{p}$ is added to the sum. (The one-particle state with momentum $\mathbf{p}$ and spin up is now among the unoccupied ones.)

(c) and (d) Corresponding contributions from the sum or integral over $\mathbf{p}_2$ rather than $\mathbf{p}_1$. These are evidently equal to (a) and (b).

We have, then, for the contribution $W_a^{(2)}$ of this process to the total energy removed, the expression

$$W_a^{(2)}=-2\cdot\tfrac{4}{3}\pi\frac{3}{8\pi^5}\left\{\int_{|\mathbf{p}+\mathbf{q}|>1}\frac{d^3q}{q^4}\int_{\substack{p_2<1\\|\mathbf{p}_2+\mathbf{q}|>1}}d^3p_2\times\frac{1}{q^2+\mathbf{q}\cdot(\mathbf{p}+\mathbf{p}_2)}-\int_{|\mathbf{p}-\mathbf{q}|<1}\frac{d^3q}{q^4}\int_{\substack{p_2<1\\|\mathbf{p}_2+\mathbf{q}|>1}}d^3p_2\times\frac{1}{\mathbf{q}\cdot(\mathbf{p}+\mathbf{p}_2)}\right\}, \quad (7)$$

where the factor 2 in front is really the product of three factors: 2 from the two spin states as in (6); $\frac{1}{2}$ from the selection of spin up only; and 2 from the existence of (c) and (d) in addition to (a) and (b).

By an obvious generalization of this method, we can compute the contribution to $W(p)$ corresponding to each term in ϵ that we investigated in I. The contribution $W_F(p)$ corresponding to the Fermi energy ϵ_F is already given by (2). We must take up next the term $W_x(p)$ corresponding to ϵ_x, the exchange energy. The expression (I-2) for ϵ_x comes from the more explicit formula

$$\epsilon_x=-\frac{3}{8\pi^3}\frac{1}{\alpha r_s}\int_{p_1<1}d^3p_1\int_{p_2<1}d^3p_2\frac{1}{(\mathbf{p}_1+\mathbf{p}_2)^2}. \quad (8)$$

The method outlined above then gives at once

$$W_x(p)=-2\cdot\tfrac{4}{3}\pi\cdot\frac{3}{8\pi^3}\frac{1}{\alpha r_s}\int_{p_2<1}d^3p_2\frac{1}{(\mathbf{p}+\mathbf{p}_2)^2} \quad (9)$$

Now let us attempt to compute $(dW/dp)_{p=1}$ as a series of ascending powers of r_s. The leading term is of course $(dW_F/dp)_{p=1}=2/(\alpha^2 r_s^2)$ and the next one should be $(dW_x/dp)_{p=1}$, presumably of order $1/r_s$. However, this comes out logarithmically divergent.[4] In fact, the situation encountered in I is repeated here, a series of increasing divergences but each occurring one order earlier than in I. Corresponding to Eq. (I-7), we have

$$\left(\frac{dW}{dp}\right)_{p=1}=\frac{2}{\alpha^2 r_s^2}+\frac{1}{r_s}\text{ (log divergence)}+\frac{1}{r_s^2}\text{ (quadratic div)}+\cdots. \quad (10)$$

Again, as in I, we expect that the divergent integrals, when summed, will cut themselves off at $q_{\min}\sim r_s^{\frac{1}{2}}$, so that in place of (10) we shall find

$$\left(\frac{dW}{dp}\right)_{p=1}=\frac{2}{\alpha^2 r_s^2}+\frac{1}{r_s}(B\ln r_s+D)+\text{higher terms in } r_s. \quad (11)$$

In order to obtain this form, we must, as in I, sum the leading divergence in each order. Moreover, the leading divergence in each order beyond the first is supplied by precisely the same process as in I, since the cause of divergence is still the piling-up of factors $1/q^2$ coming from successive Coulomb interactions with the same momentum transfer. So from just the processes considered in I, we can obtain *exact* values of B and D. Moreover, in all orders beyond the first, we may employ the crudest approximation that preserves the leading divergence. In particular, in second order, we may ignore $W_b^{(2)}$ compared to $W_a^{(2)}$, which alone has a quadratic divergence in $(dW/dp)_{p=1}$.

The series that we must examine, then, is $W_x+W_a^{(2)}+W^{(3)}+W^{(4)}+\cdots$, where these terms correspond precisely to ϵ_x, $\epsilon_a^{(2)}$, $\epsilon^{(3)}$, $\epsilon^{(4)}$, etc., of I. We have already calculated W_x and $W_a^{(2)}$, and we may now study the derivatives of these terms at $p=1$. For the first-order term W_x we have, from (9),

$$W_x=-\frac{1}{\pi^2\alpha r_s}\int\frac{d^3q}{q^2}\eta(1-|\mathbf{p}+\mathbf{q}|), \quad (12)$$

where $\eta(x)$ is unity for positive x and zero for negative x. The derivative is then

$$\left(\frac{dW_x}{dp}\right)_{p=1}=\frac{1}{\pi^2\alpha r_s}\int\frac{d^3q}{q^2}\delta(1-|\mathbf{n}+\mathbf{q}|)\mathbf{n}\cdot(\mathbf{n}+\mathbf{q}), \quad (13)$$

where $\mathbf{n}$ is a unit vector in the direction of $\mathbf{p}$. We may put $\mathbf{n}+\mathbf{q}$ equal to a unit vector $\mathbf{n}'$ pointed into the element of solid angle $d\Omega'$. Then we have

$$\left(\frac{dW_x}{dp}\right)_{p=1}=\frac{1}{\pi^2\alpha r_s}\int\frac{d\Omega'}{(\mathbf{n}'-\mathbf{n})^2}\mathbf{n}\cdot\mathbf{n}'=\frac{2}{\pi\alpha r_s}\int_{-1}^{1}\frac{xdx}{2(1-x)}. \quad (14)$$

We have put $x=\mathbf{n}\cdot\mathbf{n}'$. Note that the result is indeed logarithmically divergent.

We have treated $(dW_x/dp)_{p=1}$ without approximation. In the higher terms, however, we keep only the leading divergence. Now in differentiating formula (7) for $W_a^{(2)}$, we keep the leading divergence if we differentiate with respect to p only the limits of integration on $\mathbf{q}$: "$|\mathbf{p}+\mathbf{q}|>1$" and "$|\mathbf{p}-\mathbf{q}|<1$." We find, in fact,

$$\left(\frac{dW_a^{(2)}}{dp}\right)_{p=1}\approx-\frac{2}{\pi^4}\int_{p_2<1}d^3p_2\int_{|\mathbf{p}_2+\mathbf{q}|>1}d^3q\frac{1}{q^4}\times\frac{1}{q^2+\mathbf{q}\cdot(\mathbf{n}+\mathbf{p}_2)}\delta(1-|\mathbf{n}+\mathbf{q}|)\mathbf{n}\cdot(\mathbf{n}+\mathbf{q}), \quad (15)$$

where the factor of two comes from the existence of two terms in (7), which make equal contributions to (15). As before, we put $\mathbf{n}+\mathbf{q}=\mathbf{n}'$ and obtain

$$\left(\frac{dW_a^{(2)}}{dp}\right)_{p=1}\approx-\frac{2}{\pi^4}\int\frac{d\Omega'}{[(\mathbf{n}'-\mathbf{n})^2]^2}\times\mathbf{n}\cdot\mathbf{n}'\int_{\substack{p_2<1\\|\mathbf{n}'-\mathbf{n}+\mathbf{p}_2|>1}}d^3p_2\frac{1}{(\mathbf{n}'-\mathbf{n})\cdot(\mathbf{n}'+\mathbf{p}_2)}. \quad (16)$$

We are still interested only in the leading divergence at $\mathbf{n}'\approx\mathbf{n}$ and so we may take the limit $\mathbf{n}'-\mathbf{n}\rightarrow 0$ in the integral over $\mathbf{p}_2$. The limit is 2π. Thus we have

$$\left(\frac{dW_a^{(2)}}{dp}\right)_{p=1}\approx-\frac{8}{\pi^2}\int_{-1}^{1}\frac{xdx}{[2(1-x)]^2}. \quad (17)$$

Adding together the terms calculated so far in (3), (14), and (17), we have

$$\left(\frac{dW}{dp}\right)_{p=1}=\frac{2}{\alpha^2 r_s^2}+\frac{2}{\pi\alpha r_s}\int_{-1}^{1}\frac{xdx}{2(1-x)}\times\left\{1-\frac{4\alpha r_s}{\pi}\frac{1}{2(1-x)}+\cdots\right\}. \quad (18)$$

To complete this series to the desired accuracy, we must look at formula (I-19) for $\epsilon^{(n)}$ and use it to calculate $W^{(n)}$ for $n>2$. If we do that and compute the leading divergence in $(dW^{(n)}/dp)_{p=1}$, we find that the series in (18) becomes simply a geometric series. We have, thus, with sufficient accuracy to obtain exact values of B

[4] J. Bardeen, Phys. Rev. **50**, 1098 (1936). That the divergences would disappear in a correct calculation was pointed out by E. P. Wigner, Trans. Faraday Soc. **34**, 678 (1938).

and D in Eq. (11), the result:

$$\left(\frac{dW}{dp}\right)_{p=1} \approx \frac{2}{\alpha^2 r_s^2} + \frac{2}{\pi\alpha r_s}\int_{-1}^{1}\frac{x\,dx}{2(1-x)} \times\left[1+\frac{4\alpha r_s}{\pi}\frac{1}{2(1-x)}\right]^{-1}. \quad (19)$$

or

$$\left(\frac{dW}{dp}\right)_{p=1} = \frac{2}{\alpha^2 r_s^2} + \frac{1}{\pi\alpha r_s}\{\ln(\pi/\alpha r_s) - 2\} + \cdots. \quad (20)$$

At low temperature, then, the specific heat of a free electron gas is modified through Coulomb interactions by the factor

$$C/C_F = \left[1 + \frac{\alpha r_s}{2\pi}[-\ln r_s + \ln(\pi/\alpha) - 2] + \cdots\right]^{-1}, \quad (21)$$

where the expansion is valid at high densities. Inserting numerical values, we have

$$C/C_F = [1 + 0.083 r_s(-\ln r_s - 0.203) + \cdots]^{-1}. \quad (22)$$

This is to be compared with the approximate result of Pines,[2] who finds

$$C/C_F = [1 + 0.083 r_s(-\ln r_s + 1.47) + \cdots]^{-1}. \quad (23)$$

The method given here permits the computation of higher terms in the series, and the next correction is now being calculated. Applications to the specific heats of metals are also being studied.

ACKNOWLEDGMENTS

The author would like to thank Dr. Richard Latter and Dr. W. J. Karzas of the RAND Corporation and Dr. Keith A. Brueckner of the University of Pennsylvania for many interesting discussions.

Correlation Energy of an Electron Gas at High Density: Plasma Oscillations

K. SAWADA,* K. A. BRUECKNER, AND N. FUKUDA,* *University of Pennsylvania, Philadelphia, Pennsylvania*

AND

R. BROUT, *Cornell University, Ithaca, New York*

(Received May 29, 1957)

The contribution from zero-point plasma oscillations to the correlation energy of an electron gas at high density is considered, using the exact high-density theory of Gell-Mann and Brueckner and of Sawada. The plasmon energy is determined as a function of q by an eigenvalue equation identical with the dispersion relation of Bohm and Pines. The plasma solutions are stable only below the energy-momentum values at which they merge with the continuum spectrum arising from particle excitation, thus introducing a natural cutoff into the theory. At high density, however, it is shown that this cutoff can be allowed to become infinite without affecting the correlation energy.

The contribution from the plasma energy is exactly re-expressed in terms of the contribution from the scattering states by making use of the analytic properties of the scattering amplitudes. This transformation also establishes the connection between the Gell-Mann-Brueckner and Sawada results.

Some remarks are finally made on the relation between these results and those of Bohm and Pines.

I. INTRODUCTION

IN two previous papers[1,2] the exact correlation energy of an electron gas has been determined at high density. This was done first by Gell-Mann and Brueckner[2] who showed by examination of the structure of the perturbation series that the infrared divergence appearing in this series could be removed by formal summation of the most divergent terms of the series, the summed series then giving correctly the screening of the long range Coulomb interaction. In its original form this theory did not exhibit explicitly the well-known features of the collective or plasma degrees of freedom of the electron gas. This led to some questions concerning the contribution from the excited bound states (plasma oscillations) to the correlation energy since this might be overlooked in the perturbation theoretic approach.

Following this work, one of us (K. Sawada)[1] showed that the selective series summation of G-B was equivalent to the solution for the eigenvalues of a reduced form of the Coulomb Hamiltonian. It was further noted that the identity in structure of this reduced Hamiltonian to that of scalar-pair meson theory made it possible to diagonalize the Hamiltonian directly following closely the methods used by Wentzel[3] in his solution of the pair theory. The significance of the plasma solutions in Sawada's result was later pointed out by one of us (R. Brout); the discussion of the plasma properties forms the principal content of this paper.

In Sec. II, the eigenvalues and eigenfunctions of the reduced Hamiltonian corresponding to plasma oscillations are obtained and their contributions to the correlation energy is discussed. The high-momentum cutoff of the plasma degrees of freedom is naturally derived from the theory. In Sec. III the correspondence between the present results and those obtained in G-B is demonstrated directly by making use of the Wentzel transformation. It is shown at the same time that the reason that the perturbation theoretic approach of G-B includes the contribution from the excited bound states (plasma oscillations) is due to the analytic behavior of the scattering amplitudes. In Sec. IV some comments are made on the Bohm-Pines theory[4] of plasma oscillation.

* On leave of absence from the Tokyo University of Education.

[1] K. Sawada, Phys. Rev. **106**, 372 (1957), hereafter referred to as (I).

[2] M. Gell-Mann and K. A. Brueckner, Phys. Rev. **106**, 364 (1957), hereafter referred to as G-B.

[3] G. Wentzel, Helv. Phys. Acta **15**, 111 (1942).

[4] D. Bohm and D. Pines, Phys. Rev. **92**, 609 (1953); D. Pines, in *Solid State Physics* (Academic Press, Inc., New York, 1955), Vol. 1, p. 367, hereafter referred to as B-P.

II. PLASMA SOLUTIONS AND PLASMA ZERO POINT ENERGY

The reduced Hamiltonian considered by Sawada is

$$H=H_0+H_c, \tag{1}$$

where

$$H_0=\sum_k(p_k^2/2m)(a_k^*a_k-b_k^*b_k), \tag{2}$$

$$H_c=\sum_q \frac{2\pi\hbar^2e^2}{\Omega q^2}\sum_p(a_{p+q}^*b_p^*+b_{p+q}a_p) \times\sum_{p'}(a_{p'-q}^*b_{p'}^*+b_{p'-q}a_{p'}). \tag{3}$$

The approximations involved in obtaining this form are discussed in detail in I. A similar approximation is involved in the commutation rules, which are

$$[a_{p+q}^*b_p^*,H_c]_-=-\frac{4\pi\hbar^2e^2}{\Omega q^2}\sum_{p'}(a_{p'+q}^*b_{p'}^*+b_{p'+q}a_{p'}), \tag{4a}$$

$$[b_{p+q}a_p,H_c]_-=\frac{4\pi\hbar^2e^2}{\Omega q^2}\sum_{p'}(a_{p'+q}^*b_{p'}^*+b_{p'+q}a_{p'}). \tag{4b}$$

The terms discarded in reducing the original Hamiltonian and commutation rules to these forms, except certain obviously negligible terms, are those where the interactions corresponding to different momentum transfers are involved. These terms give a contribution which has been shown to vanish in the high-density limit (except the second order exchange energy defined as $\epsilon_{(b)}^{(2)}$ in G-B and I) compared to the leading terms retained. It is interesting to note that the so-called "random phase approximation" of Bohm and Pines is very similar to the approximations made in obtaining Eqs. (3) and (4) and is exact in the high-density region.

To proceed, we next consider the eigenvalue equation for the excitations. This is directly obtained by an application of the commutation rules of Eq. (4), as derived in (I). The result is

$$1=\frac{4\pi\hbar^2e^2}{\Omega q^2}\left\{\sum_{p\;|\mathbf{p}|<p_F,\,|\mathbf{p}+\mathbf{q}|>p_F}-\sum_{p\;|\mathbf{p}|>p_F,\,|\mathbf{p}+\mathbf{q}|<p_F}\right\} \times\frac{1}{E-E_0+E_{\mathbf{p}}^{(0)}-E_{\mathbf{p}+\mathbf{q}}^{(0)}}. \tag{5}$$

Writing $E-E_0=\hbar\omega$ and making the transformation $\mathbf{p}+\mathbf{q}\to-\mathbf{p}$ in the second term, we have[5]

$$1=\frac{8\pi\hbar^2e^2}{\Omega q^2}\sum_{|\mathbf{p}|<p_F,\,|\mathbf{p}+\mathbf{q}|>p_F}\frac{(\mathbf{p}\cdot\mathbf{q}/m)+(q^2/2m)}{(\hbar\omega)^2-(\mathbf{p}\cdot\mathbf{q}/m+q^2/2m)^2}. \tag{6}$$

This eigenvalue equation has two types of solutions; the simplest are those for which the eigenvalues lie in the continuum of solutions for which

$$\hbar\omega=(\mathbf{p}\cdot\mathbf{q}/m)+(q^2/2m);\quad |\mathbf{p}|<p_F,\quad |\mathbf{p}+\mathbf{q}|>p_F. \tag{7}$$

These correspond to the energies of free pair excitation and may be called scattering solutions. These scattering solutions are given explicitly in I. They are of a standard form, i.e., an incoming wave together with a scattered wave. The continuum of solutions resulting from pair excitations terminates at the maximum value of energy possible, which for a given value of q is $\hbar\omega_{\max}=(p_Fq/m)+(q^2/2m)$. Another type of solution lies above this continuum; this solution is of the plasma type, as we show in more detail below.

Integration of Eq. (6) gives the relationship between the plasma frequency $\omega_{\mathrm{pl}}(q)$ and r_s as follows:

$$1=\frac{\alpha r_s}{4\pi}\frac{p_F(2m)^2}{q^5}\left[\left\{\left(\frac{q^2}{2m}\right)^2+(\hbar\omega)^2-\left(\frac{qp_F}{m}\right)^2\right\} \times\ln\left\{\frac{[(q^2/2m)+(qp_F/m)]^2-(\hbar\omega)^2}{[(q^2/2m)-(qp_F/m)]^2-(\hbar\omega)^2}\right\} +2\frac{q^2}{2m}\hbar\omega\ln\left\{\frac{(q^2/2m)^2-(\hbar\omega+qp_F/m)^2}{(q^2/2m)^2-(\hbar\omega-qp_F/m)^2}\right\} -4\frac{q^2}{2m}\frac{qp_F}{m}\right], \tag{8}$$

where

$$\alpha=(4/9\pi)^{\frac{1}{3}}.$$

It is easy to obtain from this equation the values of $q_{\max}$ and $\omega_{\mathrm{pl}}(q_{\max})$ at which the plasma solutions cross over into the continuum; this occurs at $\hbar\omega_{\mathrm{pl}}(q_{\max})=(q_{\max}^2/2m)+(q_{\max}p_F/m)$, which gives[6]

$$\frac{q_{\max}^2}{p_F^2}=\frac{\alpha r_s}{\pi}\left[\left(2+\frac{q_{\max}}{p_F}\right)\ln\left(1+2\frac{p_F}{q_{\max}}\right)-2\right]. \tag{9}$$

For momenta above this value, the plasma solutions are unstable and quickly transfer their energy into particle excitation. Such strongly-damped solutions do not contribute to the high density energy. Consequently, a characteristic high-momentum cutoff for the plasma oscillations appears in the theory.

We next consider the properties of the plasma solutions and their contributions to the energy. The first result of interest is the explicit form of the wave function for a plasma oscillation or "plasmon" of momentum q and energy $\hbar\omega_{\mathrm{pl}}(q)$; this is

$$\Psi_{\mathrm{pl}}(q)=N_q\sum_p\left[\frac{a_{p+q}^*b_p^*}{\hbar\omega_{\mathrm{pl}}(q)-E_{p+q}^{(0)}+E_p^{(0)}} +\frac{b_{p+q}a_p}{\hbar\omega_{\mathrm{pl}}(q)-E_{p+q}^{(0)}+E_p^{(0)}}\right]\Psi_0, \tag{10}$$

[5] By a simple transformation this can also be rewritten as

$$1=\frac{4\pi e^2\hbar^2}{m}\sum_{\mathbf{p},\,|\mathbf{p}|<p_F}\frac{1}{(\hbar\omega-\mathbf{q}\cdot\mathbf{p}/m)^2-(q^2/2m)^2}$$

similar to the B-P dispersion formula, except that p's are c numbers here instead of operators.

[6] This equation has been independently derived by R. Ferrell, Bull. Am. Phys. Soc. Ser. II, 2, 146 (1957).

where N_q is the normalization constant determined in the Appendix. This form of the plasmon wave function exhibits the simplicity of the particle excitations which give rise to the plasma oscillations, these being simply pair excitations with a fixed momentum transfer, summed with proper phase relations.

The contribution of the plasma zero-point oscillations to the correlation energy is now easily obtained using a method similar to that used in I in obtaining the scattering-states contribution.

The details are given in the appendix. The result is

$$E_{\rm pl}=\sum_q \frac{1}{2}\int_0^{e^2}\frac{\partial\hbar\omega_{\rm pl}(q)}{\partial e'^2}d(e'^2)$$

$$=\tfrac{1}{2}\sum_q[\hbar\omega_{\rm pl}(q)-\hbar\omega_{\rm pl}(q)_{(e^2=0)}]. \quad (11)$$

Thus the correlation energy arising from the plasma oscillation is given as the difference between the zero-point energy of this oscillation and the value this energy approaches as the coupling is switched off.[7] This value is simply

$$\hbar\omega_{\rm pl}(q)_{e^2=0}=(q^2/2m)+(qp_F/m), \quad (12)$$

which is the upper limit of the continuum of pair excitation energy (at given q). One consequence of the appearance of this difference of energies is that in the high density limit the plasma cutoff $q_{\max}$ can be allowed to become infinite without affecting the contribution from the plasma energy, since the plasma energy $\hbar\omega_{\rm pl}(q)$ lies very close to the continuum limit for large q and small r_s. This is shown in Fig. 1.

We make use of this result to obtain explicitly the plasma energy in the high density limit. In this limit the dispersion relation [Eq. (8)] becomes, neglecting everywhere q/p_F compared to unity,

$$1=\frac{2\alpha r_s}{\pi}\frac{p_F^2}{q^2}f(x), \quad (13)$$

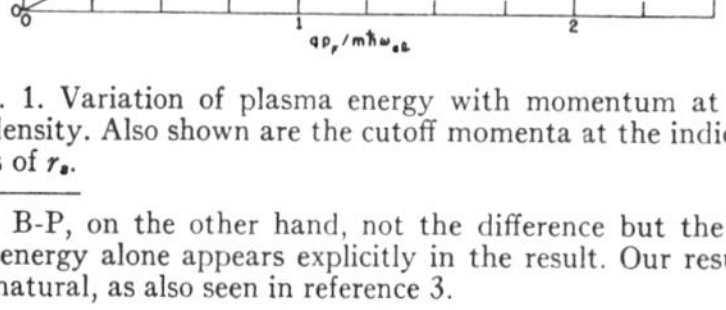

FIG. 1. Variation of plasma energy with momentum at very high density. Also shown are the cutoff momenta at the indicated values of r_s.

[7] In B-P, on the other hand, not the difference but the zero point energy alone appears explicitly in the result. Our result is more natural, as also seen in reference 3.

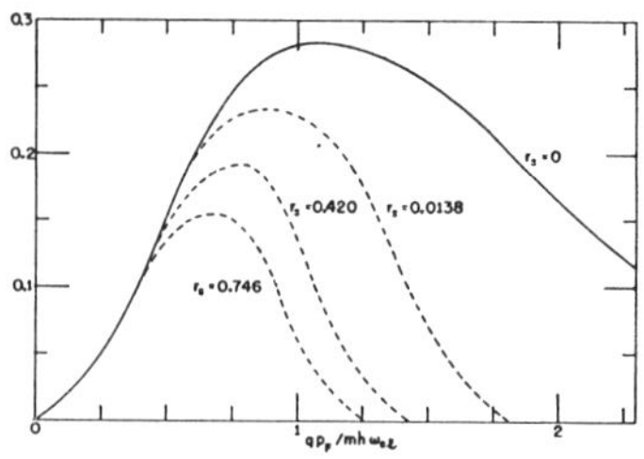

FIG. 2. Variation of the integrand of Eq. (15) determining the plasma energy. The function shown as a solid curve is the high-density limit of $\{[\hbar\omega_{\rm pl}(q)-qp_F/m]/\hbar\omega_{\rm cl}\}(qp_F/m\hbar\omega_{\rm cl})^2$. Also shown are approximate curves giving the above function at the indicated values of r_s.

where $x=qp_F/m\hbar\omega_{\rm cl}$ and

$$f(x)=\frac{1}{x}\ln\frac{1+x}{1-x}-2. \quad (14)$$

The plasma energy then is, summing over all momentum transfer q up to $q_{\max}$,

$$E_{\rm pl}=\frac{1}{2}\frac{4\pi\Omega}{(2\pi\hbar)^3}\int_0^{q_{\max}}q^2dq[\hbar\omega_{\rm pl}(q)-(qp_F/m)]. \quad (15)$$

To show precisely from what values of q the plasma energy arises, we give the integrand of Eq. (15) in Fig. 2, measuring the plasma energy in units of the classical plasma frequency

$$\omega_{\rm cl}=\lim_{q\to 0}\omega_{\rm pl}(q)=\frac{1}{\hbar}\left(\frac{4\alpha r_s}{3\pi}\frac{p_F^4}{m^2}\right)^{\frac{1}{2}}=(4\pi\rho e^2/m)^{\frac{1}{2}}, \quad (16)$$

and the momentum in units of $m\hbar\omega_{\rm cl}/p_F$. Figure 2 shows that the contribution comes largely from $(qp_F/m\hbar\omega_{\rm cl})$ of the order of or larger than unity, which corresponds to

$$(q/p_F)\gtrsim[(4/3\pi)\alpha r_s]^{\frac{1}{2}}=0.470r_s^{\frac{1}{2}}. \quad (17)$$

It is interesting to notice that the correlation energy at high density arising from the plasma oscillations comes mainly from value of (q/p_F) much larger than the limit obtained by Bohm and Pines,[4] which is

$$(q_{\max}/p_F)=0.353r_s^{\frac{1}{2}}. \quad (18)$$

We return to a discussion of this discrepancy in more detail in a later section.

Now returning to Eq. (15), we use the relation

$$(2qdq/p_F^2)=(2\alpha r_s/\pi)f'(x)dx, \quad (19)$$

and get

$$E_{\rm pl}=\frac{1}{2}\frac{4\pi\Omega}{(2\pi\hbar)^3}\int_0^1\frac{\alpha r_s}{\pi}f'(x)\left(\frac{1}{x}-1\right)\frac{p_F^5}{m}\frac{2\alpha r_s}{\pi}f(x)dx. \quad (20)$$

In the conventional notation, we measure the energy in Rydbergs and express it per particle; the result is

$$\epsilon_{\rm pl}=E_{\rm pl}/(me^2N/2\hbar^2)$$

$$=\frac{3}{\pi^2}\int_0^1 f(x)f'(x)\left[\frac{1}{x}-1\right]dx=\frac{3}{2\pi^2}\int_0^1\left[\frac{f(x)}{x}\right]^2 dx. \quad (21)$$

This integral cannot be evaluated analytically but numerical evaluation yields

$$\epsilon_{\rm pl}=0.133, \quad (22)$$

which, when combined with the scattering contribution

$$\epsilon_{\rm sc}=-0.229+0.0622\ \ln r_s, \quad (23)$$

gives as the total correlation energy

$$\epsilon_{\rm corr}=-0.096+0.0622\ \ln r_s. \quad (24)$$

This agrees with the result given in G-B.†

III. USE OF THE WENTZEL TRANSFORMATION

The correspondence between the results of the last section and those obtained by G-B can be demonstrated directly by using a transformation introduced by Wentzel in his treatment of the scalar pair theory. The total correlation energy (neglecting second order exchange) corresponding to a momentum transfer q is now given by (see appendix)

$$\frac{\partial E_{\rm corr}(q)}{\partial e^2}=\frac{1}{2}\frac{\partial\hbar\omega_{\rm pl}(q)}{\partial e^2}+\frac{2\pi\hbar^2}{\Omega q^2}\sum_{\mathbf{p},\ |\mathbf{p}|<p_F,\ |\mathbf{p}+\mathbf{q}|>p_F}\times\left(\frac{1}{|1+f(\omega_p+i\epsilon)|^2}-1\right), \quad (25)$$

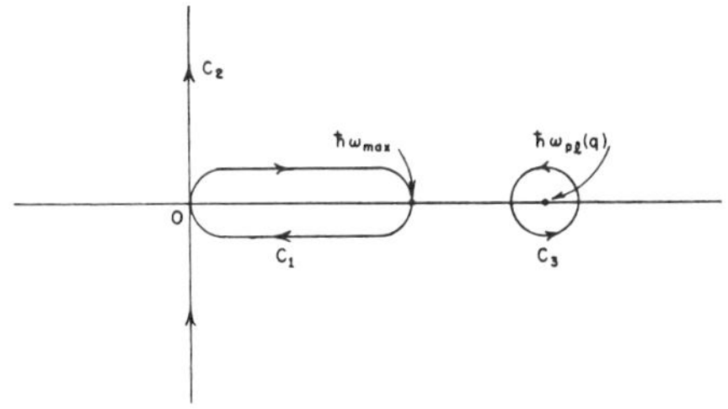

FIG. 3. Contours for analytic continuation. C_1 and C_3 circle around the continuum and the plasma energy, respectively. C_2 runs along the imaginary axis from $-i\infty$ to $+i\infty$.

† *Note added in proof.*—The integral in Eq. (21) has been evaluated by Professor Lars Onsager, who obtains the result for the plasma energy (in Rydbergs)

$$\epsilon_{\rm pl}=\frac{1}{3}-\frac{2}{\pi^2}=0.130691.$$

Re-evaluation of the scattering contribution, which must be done by numerical integration, gives $\epsilon_{\rm sc}=-0.227+0.0622\ \ln r_s$ so that the correlation energy is unchanged. We are indebted to Professor Onsager for informing us of his result.

where

$$f(\omega_p+i\epsilon)=\frac{4\pi\hbar^2e^2}{\Omega q^2}\sum_{p'}\left(\frac{1}{\omega_{p'}-\omega_p-i\epsilon}+\frac{1}{\omega_{p'}+\omega_p+i\epsilon}\right), \quad (26)$$

and

$$\omega_p=E_{p+q}^{(0)}-E_p^{(0)};\quad |\mathbf{p}|<p_F,\quad |\mathbf{p}+\mathbf{q}|>p_F. \quad (27)$$

We have added for the sake of convenience $+i\epsilon$ in the second denominator of f.

Making use of the identity[8]

$$\mathrm{Im}f(\omega_p+i\epsilon)=\pi\frac{4\pi\hbar^2e^2}{\Omega q^2}\sum_{p'}\delta(\omega_p-\omega_{p'}), \quad (28)$$

the 2nd term of Eq. (25) which is denoted by J is transformed into

$$J=\frac{1}{2\pi e^2}\int d\omega_p\ \mathrm{Im}f(\omega_p+i\epsilon)\left(\left|\frac{1}{1+f(\omega_p+i\epsilon)}\right|^2-1\right)$$

$$=\frac{1}{2\pi e^2}\ \mathrm{Im}\int d\omega_p\left(\frac{f(\omega_p+i\epsilon)}{1+f(\omega_p+i\epsilon)}-f(\omega_p+i\epsilon)\right). \quad (29)$$

The integral over ω_p runs from zero up to the upper bound of the continuous spectrum given by Eq. (7). We can write the above expression as the contour integral along C_1 shown in Fig. 3:

$$J=\frac{1}{4\pi e^2 i}\int_{C1}dz\left(\frac{f(z)}{1+f(z)}-f(z)\right). \quad (30)$$

Now, using the analytic property of $f(z)$, it is easily shown, that

$$\int_{C1}=\int_{C2}+\int_{C3}=\int_{C2}+\text{Residue at }[z=\hbar\omega_{\rm pl}(q)], \quad (31)$$

where the contour C_2 runs along the imaginary z axis from $-i\infty$ to $+i\infty$. The residue at $\hbar\omega_{\rm pl}(q)$ is

$$\mathrm{Res}=2\pi i\lim_{z\to\hbar\omega_{\rm pl}(q)}\left[\frac{f(z)}{1+f(z)}-f(z)\right][z-\hbar\omega_{\rm pl}(q)]$$

$$=-2\pi i\frac{1}{(\partial/\partial z)f(z)}\bigg|_{z=\hbar\omega_{\rm pl}(q)}=-2\pi i e^2\frac{\partial\hbar\omega_{\rm pl}(q)}{\partial e^2}, \quad (32)$$

where use is made of Eq. (A-18) in the appendix and

[8] The scattering amplitude, or the T matrix introduced in I, [Eq. (A-3) in I], is given by

$$T_{p'+q,p';p+q,p}=\frac{4\pi\hbar^2e^2}{\Omega q^2}[1+f(\omega_{p'}+i\epsilon)]^{-1}.$$

Then the unitarity condition for the S matrix given in terms of this T matrix

$$\mathrm{Im}T_{p+q,p;p+q,p}=\pi\sum_{p'}|T_{p'+q,p';p+q,p}|^2\delta(\omega_p-\omega_{p'}),$$

is essentially identical with the identity, Eq. (28).

the eigenvalue equation

$$f(\omega_{p1})+1=0. \tag{33}$$

Then, we get

$$\frac{\partial E_{\rm corr}(q)}{\partial e^2}=\frac{1}{4\pi e^2 i}\int_{C_2} dz\left(\frac{f(z)}{1+f(z)}-f(z)\right).$$

Putting $z=iv$ and integrating over e^2, we find

$$E_{\rm corr}=\sum_q \frac{1}{4\pi}\int_{-\infty}^{\infty} dv\left[\ln\left(1+\frac{8\pi\hbar^2e^2}{\Omega q^2}\sum_p \frac{\omega_p}{\omega_p{}^2+v^2}\right)-\frac{8\pi\hbar^2e^2}{\Omega q^2}\sum_p\frac{\omega_p}{\omega_p{}^2+v^2}\right]. \tag{34}$$

Now, to compare this result with that given by G-B, we make the change in variable

$$v=2u\frac{q}{p_F}\frac{p_F{}^2}{2m}, \quad \left(\frac{p_F{}^2}{2m}=\frac{1}{\alpha^2 r_s{}^2}\text{ Rydbergs}\right), \tag{35}$$

and then use the function $Q_q(u)$ defined in G-B (Eq. 18). We find

$$\frac{8\pi\hbar^2e^2}{\Omega q^2}\sum_p\frac{\omega_p}{\omega_p{}^2+v^2}=\frac{\alpha r_s}{\pi^2 q^2}Q_q(u), \tag{36}$$

in dimensionless unit. In this unit $\sum_q$ is equal to $(3/8\pi)N\int d\mathbf{q}$, N being the number of electrons. Hence the correlation energy per electron becomes

$$\epsilon_{\rm corr}=\frac{E_{\rm corr}}{N}=\frac{3}{4\pi}\int_0^\infty q^3dq\int_{-\infty}^{\infty}\left[\ln\left(1+\frac{\alpha r_s}{\pi^2q^2}Q_q(u)\right)-\frac{\alpha r_s}{\pi^2q^2}Q_q(u)\right]du\frac{1}{\alpha^2 r_s{}^2}\text{ Rydbergs}, \tag{37}$$

which gives just the same expression as obtained by G-B as a series [Eq. (19) in G-B]. This shows that the result obtained by G-B includes automatically the effect of bound states or plasma oscillations.

IV. COMPARISON WITH THE BOHM-PINES THEORY

The theory of Bohm and Pines has in the past been the only theory of the electron gas which has attempted to determine both the correlation energy and the collective properties of the system. As for the first point the B-P theory was quite successful since the correlation energy obtained agreed approximately with that obtained earlier by Wigner[9] and also with experiment. The plasma properties predicted also were highly reasonable since they were closely connected with the classical behavior of an ionized medium and experimentally confirmed. In spite of these successes which showed the soundness of the underlying physical concepts of the theory, it seems to us that the accurate quantitative aspects of the theory have been still somewhat in question. The uncertainties of the B-P results arose from certain approximations essential to their procedure, namely

(*a*) the random phase approximation,
(*b*) the perturbation theoretic treatment of the electron-plasma coupling,
(*c*) the determination of the cutoff momentum for the plasma oscillations,
(*d*) the neglect of the subsidiary conditions on the wave function.

We shall discuss these approximations and attempt to cast some light on their validity, making use of the results of the exact high-density theory given by the techniques of this and earlier papers.

The first approximation essential to the B-P theory is the random phase approximation which neglects the coupling between the excitations corresponding to different momentum transfers q and q'. As we have seen in our theory, this approximation is *in fact exact* in the high-density limit and is the *only approximation* required to obtain an *exact* high density result.

The validity of the perturbation treatment of the electron-plasma coupling is most readily examined by actually considering the structure of the B-P results in comparison with ours. Our result (taken in the G-B form),[10] disregarding $\epsilon_{(b)}{}^{(2)}$ and δ which are independent of r_s, is

$$\epsilon_{\rm corr}=-\frac{3}{4\pi}\int_0^1 q^3dq\int_{-\infty}^{\infty}\left[\frac{4\alpha r_s}{\pi q^2}R(u)-\ln\left(1+\frac{4\alpha r_s}{\pi q^2}R(u)\right)\right]du\frac{1}{\alpha^2r_s{}^2}\text{ Rydbergs}. \tag{38}$$

The B-P result is separated into two parts, that arising from long wavelengths $(q/p_F\leqslant\beta)$:

$$\begin{aligned}\epsilon_{\rm L.R.}&=\frac{0.866}{r_s{}^{\frac{1}{2}}}\beta^3-\frac{0.458}{r_s}\beta^2+\frac{0.019}{r_s}\beta^4\text{ ry},\\ \Delta\epsilon_{\rm L.R.}&=\frac{0.708}{r_s{}^{5/2}}\beta^5\left(1+\frac{3}{10}\beta^2\right)-\frac{0.517}{r_s{}^2}\beta^4-\frac{0.058}{r_s{}^2}\beta^6\text{ ry}.\end{aligned} \tag{39}$$

and that given in second order perturbation theory applied to the screened Coulomb interaction[11] $(q/p_F\geqslant\beta)$,

$$\epsilon_{\rm S.R.}=-(0.0254-0.0626\ln\beta+0.00637\beta^2)\text{ ry}. \tag{40}$$

In the exact theory the sum $\epsilon_{\rm L.R.}+\Delta\epsilon_{\rm L.R.}+\epsilon_{\rm S.R.}$ should be independent of β. This is not only manifestly not so

[9] E. P. Wigner, Phys. Rev. 46, 1002 (1934).

[10] Here $4\pi R(u)$ is the value of $Q_q(u)$ at $q=0$ and is given by
$$R(u)=1-u\tan^{-1}(1/u).$$

[11] The parallel spin correlation energies are omitted. See G-B, footnote 4.

for the B-P result, but also the polynomial expansion of $\epsilon_{L.R.}$ in powers of β is incompatible with the appearance of the term $\ln\beta$ in $\epsilon_{S.R.}$. In fact, the appearance of this term shows that a polynomial expansion for the long range part of the G-B-S energy, cutting off the integral at $q=\beta$ is not possible as is obvious from Eq. (38) since a direct expansion in powers of β^2 diverges at the term β^4.

The problems encountered above in the power series expansion in terms of β also cast some doubt on the B-P determination of the cut-off momentum $q_{max}=\beta p_F$. This is evaluated by them by minimizing the energy given by part of their transformed Hamiltonian [$\epsilon_{L.R.}$ of Eq. (39)], assuming all the remaining terms dependent on β to be neglected. This procedure seems to us for several reasons to be only a semiquantitative procedure. First, since the actual correlation energy is independent of β, it is not possible to decouple a *small part* of the Hamiltonian giving a β-dependent energy and to minimize it with respect to β, neglecting the variation of the remaining larger terms. This criticism is equivalent to the statement that it is not possible to decouple the various terms in the Hamiltonian in such a way as to treat the β variation of $\epsilon_{L.R.}$ separately from that of much larger terms in $\Delta\epsilon_{L.R.}$ and $\epsilon_{S.R.}$. Even if such a decoupling could be qualitatively justified by physical rather than mathematical argument, the value of β so determined can give at best only a rough approximation to the actual magnitude of $\Delta\epsilon_{L.R.}$ and $\epsilon_{S.R.}$.

The actual value of the cutoff obtained by B-P, $\beta=0.353r_s^{\frac{1}{2}}$, is considerably below the point at which the plasma solutions start to merge with the pair excitation continuum, which occurs for values of q/p_F larger than $\beta=0.470r_s^{\frac{1}{2}}$, particularly at high density. Consequently an important part of the plasma oscillations (important since the contribution to the energy varies roughly as β^3) is omitted if the B-P cutoff is used. It is to be emphasized that the plasma solutions lying above the low B-P cutoff have a perfectly real physical meaning since the plasma can in fact oscillate stably for these frequencies, particularly at high density.

We finally wish to comment briefly on the question of the B-P neglect of the subsidiary condition. We believe, although we have not been able to prove this in detail, that the B-P subsidiary condition is equivalent to our definition of the ground state of the system. As shown in the Appendix, the ground state can be defined as that state in which no plasmons are present, i.e., it satisfies the condition

$$A\Psi_0=0, \tag{41}$$

where A is the annihilation operator for plasmon defined by Eq. (A-10) and Ψ_0 is the exact ground-state wave function. Written explicitly in terms of the particle creation and annihilation operators, this condition is

$$\sum_p\left[\frac{b_pa_{p+q}}{\hbar\omega_{\rm pl}(q)-E_{p+q}^{(0)}+E_p^{(0)}}+\frac{a_p{}^*b_{p+q}{}^*}{\hbar\omega_{\rm pl}(q)-E_{p+q}^{(0)}+E_p^{(0)}}\right]\Psi_0=0, \tag{42}$$

which is reminiscent of the B-P subsidiary condition

$$\sum_i\frac{\omega^2}{\omega^2-[(\mathbf{k}\cdot\mathbf{p}_i/m)-(\hbar k^2/2m)]^2}\Psi=0,\qquad k<kc\quad(k=q/\hbar). \tag{43}$$

We wish here to emphasize that our condition stated in Eq. (41) or Eq. (42) is exactly satisfied at high density where our solution is exact; the B-P subsidiary condition will therefore also be exactly satisfied at high density by our ground-state wave function. The actual techniques used by B-P in obtaining their solution lead to some violation of the subsidiary condition, but almost certainly, as B-P emphasized, no serious error arises from this aspect of their approximations.

In conclusion we would like to say that the differences between our results and those of B-P are only in techniques and mathematical detail and that the latter theory, although it seems to be only semiquantitative in nature, provides an excellent physical insight into the properties of the electron gas.

ACKNOWLEDGMENTS

The authors wish to express their thanks to Dr. P. Nozières, Professor R. Ferrell, and particularly to Professor D. Pines for many stimulating discussions on this work.

APPENDIX A: CORRELATION ENERGY ARISING FROM PLASMA OSCILLATION

We first determine the plasma wave function. Since this is a one-pair state as discussed in Sec. II of I, it must have the form

$$\Psi_{\rm pl}(q)=A_q{}^*\Psi_0, \tag{A-1}$$

where

$$A_q{}^*=\sum_p(\alpha_pa_{p+q}{}^*b_p{}^*+\beta_pb_{p+q}a_p). \tag{A-2}$$

The constants α_p and β_p must be chosen so that $\Psi_{\rm pl}(q)$ is a properly normalized eigenfunction of the Hamiltonian, i.e.,

$$(H_0+H_c)\Psi_{\rm pl}(q)=[E_0+\hbar\omega_{\rm pl}(q)]\Psi_{\rm pl}(q), \tag{A-3}$$

where

$$(H_0+H_c)\Psi_0=E_0\Psi_0. \tag{A-4}$$

This can also be written as an operator equation

$$[(H_0+H_c),A^*]_-=\hbar\omega_{\rm pl}A^*. \tag{A-5}$$

Once this is satisfied it then follows that

$$[(H_0+H_c),A]_-=-\hbar\omega_{\rm pl}A, \tag{A-6}$$

so that

$$(H_0+H_c)A\Psi_0=(E_0-\hbar\omega_{\rm pl})A\Psi_0. \quad \text{(A-7)}$$

Since Ψ_0 is the state of lowest energy, this equation can be satisfied only if

$$A\Psi_0=0. \quad \text{(A-8)}$$

This condition allows us to fix the normalization since

$$(\Psi_{\rm pl},\Psi_{\rm pl})=(\Psi_0,AA^*\Psi_0)=(\Psi_0,[A,A^*]_-\Psi_0)=\sum_p(|\alpha_p|^2-|\beta_p|^2)=1. \quad \text{(A-9)}$$

The coefficients α_p and β_p may now be determined by making use of the commutation rules given in Eq. (4); the result is that

$$\alpha_p=\frac{1}{\hbar\omega_{\rm pl}-E_{p+q}^{(0)}+E_p^{(0)}}\times\text{const} \quad \text{for } |\mathbf{p}|<p_F,\quad |\mathbf{p}+\mathbf{q}|>p_F;$$

$$\beta_p=\frac{1}{\hbar\omega_{\rm pl}-E_{p+q}^{(0)}+E_p^{(0)}}\times\text{const.} \quad \text{for } |\mathbf{p}|>n_F,\quad |\mathbf{p}+\mathbf{q}|<p_F. \quad \text{(A-10)}$$

Combining this result with Eq. (A-9), we find

$$\Psi_{\rm pl}(q)=N_q\sum_p\left[\frac{a_{p+q}{}^*b_p{}^*}{\hbar\omega_{\rm pl}-E_{p+q}^{(0)}+E_p^{(0)}}+\frac{b_{p+q}a_p}{\hbar\omega_{\rm pl}-E_{p+q}^{(0)}+E_p^{(0)}}\right]\Psi_0, \quad \text{(A-11)}$$

where

$$\{|N_q|^2\}^{-1}=\Bigg(\sum_{\substack{|\mathbf{p}|<p_F\\|\mathbf{p}+\mathbf{q}|>p_F}}{}_p-\sum_{\substack{|\mathbf{p}|>p_F\\|\mathbf{p}+\mathbf{q}|<p_F}}{}_p\Bigg)\times\frac{1}{(\hbar\omega_{\rm pl}(q)-E_{p+q}^{(0)}+E_p^{(0)})^2}. \quad \text{(A-12)}$$

We now determine the plasma energy. Since the eigenvalue equation has a bound-state plasma solution, Eq. (9) in I should be modified to become

$$\sum_p(a_{p+q}{}^*b_p{}^*+b_{p+q}a_p)=\text{scattering solutions}+C_{\rm pl}(q)\Psi_{\rm pl}(q), \quad \text{(A-13)}$$

where $\Psi_{\rm pl}(q)$ is the normalized plasma solution given above. The plasma contribution to the energy then is

$$e^2\frac{\partial E_{\rm pl}}{\partial e^2}=\frac{2\pi\hbar^2e^2}{\Omega}\sum_q\frac{1}{q^2}|C_{\rm pl}(q)|^2. \quad \text{(A-14)}$$

To obtain the expansion coefficient $C_{\rm pl}(q)$, we take the scalar product of Eq. (A-13) with $\Psi_{\rm pl}(q)$, obtaining

$$C_{\rm pl}(q)=(\Psi_{\rm pl}(q),\sum_p(a_{p+q}{}^*b_p{}^*+b_{p+q}a_p)\Psi_0). \quad \text{(A-15)}$$

Using Eqs. (A-1), (A-3), and (A-8), this becomes

$$C_{\rm pl}(q)=(\Psi_0,[A,\sum_p(a_{p+q}{}^*b_p{}^*+b_{p+q}a_p)]_-\Psi_0)=\sum_p(\alpha_p-\beta_p).$$

Using Eqs. (A-9) and (A-10) for α_p and β_p, this is

$$C_{\rm pl}(q)=N_q\Bigg(\sum_{\substack{|\mathbf{p}|<p_F\\|\mathbf{p}+\mathbf{q}|>p_F}}{}_p-\sum_{\substack{|\mathbf{p}|>p_F\\|\mathbf{p}+\mathbf{q}|<p_F}}{}_p\Bigg)\times\frac{1}{\hbar\omega_{\rm pl}(q)-E_{p+q}^{(0)}+E_p^{(0)}}=\left(\frac{4\pi\hbar^2e^2}{\Omega q^2}\right)^{-1}N_q, \quad \text{(A-16)}$$

where we have used the dispersion formula to eliminate the sum over p. To bring this to final form, we differentiate the dispersion relation with respect to e^2, which gives

$$0=1-\frac{4\pi\hbar^2e^2}{\Omega q^2}\Bigg(\sum_{\substack{|\mathbf{p}|<p_F\\|\mathbf{p}+\mathbf{q}|>p_F}}{}_p-\sum_{\substack{|\mathbf{p}|>p_F\\|\mathbf{p}+\mathbf{q}|<p_F}}{}_p\Bigg)\times\frac{1}{(\hbar\omega_{\rm pl}-E_{p+q}^{(0)}+E_p^{(0)})^2}e^2\frac{\partial\hbar\omega_{\rm pl}}{\partial e^2}, \quad \text{(A-17)}$$

i.e., from (A-12),

$$|N_q|^2=\left(\frac{4\pi\hbar^2e^2}{\Omega q^2}\right)e^2\frac{\partial\hbar\omega_{\rm pl}(q)}{\partial e^2}. \quad \text{(A-18)}$$

Thus we find

$$C_{\rm pl}(q)=\left(\frac{4\pi\hbar^2e^2}{\Omega q^2}\right)^{-\frac{1}{2}}\left(e^2\frac{\partial\hbar\omega_{\rm pl}(q)}{\partial e^2}\right)^{\frac{1}{2}}, \quad \text{(A-19)}$$

so that Eq. (A-14) can be written as the simple result

$$\frac{\partial E_{\rm pl}}{\partial e^2}=\frac{1}{2}\sum_q\frac{\partial\hbar\omega_{\rm pl}(q)}{\partial e^2}. \quad \text{(A-20)}$$

which is the desired answer.

Finally it is interesting to note that the plasma energy is expressed in terms of the contributions from the scattering states alone. To this end, we make use of the Chew-Low-Wick[12] equation for the T matrix introduced in Eq. (A-3) of I, from which a term arising from the bound state (plasma state) is separated. It turns out that this equation becomes a differential equation [by means of (A-19)] to determine the plasma energy in terms of the scattering amplitudes, which has the

[12] G. F. Chew and F. E. Low, Phys. Rev. **101**, 1570 (1956); G. C. Wick, Revs. Modern Phys. **27**, 339 (1955).

solution

$$\hbar\omega_{\rm pl}(q)=\left\{\left(\frac{qp_F}{m}+\frac{q^2}{2m}\right)^2+4\int_0^{e^2}\frac{2\pi\hbar^2}{\Omega q^2}\sum_{\substack{|\mathbf{p}|<p_F\\|\mathbf{p}+\mathbf{q}|>p_F}}\!\!{}_{p}\times\left(1-\frac{1}{|1+f(\omega_p+ie)|^2}\right)\omega_p de^2\right\}^{\frac{1}{2}},$$

which f defined in (26).

APPENDIX B

In order to exhibit the formal correspondence of our calculation with that of B-P, we may write our result as

$$E=E_{\rm ex}+E_{\rm pl}+E_{\rm sc}, \tag{B-1}$$

where $E_{\rm ex}$ is the exchange energy given by

$$E_{\rm ex}=-\sum_{\substack{p<p_F\\|\mathbf{p}+\mathbf{q}|<p_F}}\sum_q\frac{2\pi\hbar^2e^2}{\Omega q^2}+E_b^{(2)}. \tag{B-2}$$

$E_{\rm sc}$ as given in I, Eq. (19), is of the form

$$E_{\rm sc}=\sum_{\substack{p<p_F\\|\mathbf{p}+\mathbf{q}|>p_F}}\sum_q\frac{2\pi\hbar^2e^2}{\Omega q^2}[F_q(p)-1], \tag{B-3}$$

and $E_{\rm pl}$ is the plasma energy. In the perturbation limit, $E_{\rm sc}$ is simply the (divergent) second order interaction energy. For small q the nonperturbation function $F_q(p)$ varies as q^2 and hence the low-momentum transfers in the scattering are screened out. Thus, in some sense, Eq. (B-3) represents a screened Coulomb interaction term.

To make the correspondence with B-P more apparent, we arrange the terms in (B-1) in a different fashion. First, we note that the first term in $E_{\rm ex}$ and the second term of $E_{\rm sc}$ combine. We next make the replacement in the sum of the first term of Eq. (B-3):

$$\sum_{\substack{p<p_F\\|\mathbf{p}+\mathbf{q}|>p_F}}\sum_q\to\sum_{p<p_F}\sum_q-\sum_{\substack{p<p_F\\|\mathbf{p}+\mathbf{q}|<p_F}}\sum_q.$$

We thus can rearrange Eq. (B-1) into the form

$$E=E_{\rm pl}-\sum_{p<p_F}\sum_q\frac{2\pi\hbar^2e^2}{\Omega q^2}+\sum_{p<p_F}\sum_q\frac{2\pi\hbar^2e^2}{\Omega q^2}F_q(p)-\sum_{\substack{p<p_F\\|\mathbf{p}+\mathbf{q}|>p_F}}\sum_q\frac{2\pi\hbar^2e^2}{\Omega q^2}F_q(p)+E_b^{(2)}. \tag{B-4}$$

As it stands, the second and third terms in Eq. (B-4) each diverge at large q. However, the sum converges since $F_q(p)\to1$ for large q. On the other hand, the two terms behave differently for small q, the $1/q^2$ behavior in the third term being cut off at small q. This suggests that the high-q part of the second term be separated and combined with the third. It also is convenient to choose the point of separation, which we denote as $q_{\rm max}$, to approximate as well as possible to the natural cutoff in $F_q(p)$. We thus are led to the final ordering:

$$E=\left[E_{\rm pl}-\sum_{p<p_F}\sum_{q<q_{\rm max}}\frac{2\pi\hbar^2e^2}{\Omega q^2}\right]-\left[\sum_{\substack{p<p_F\\|\mathbf{p}+\mathbf{q}|<p_F}}\sum_q\frac{2\pi\hbar^2e^2}{\Omega q^2}F_q(p)\right]+\left[\sum_{p<p_F}\sum_q\frac{2\pi\hbar^2e^2}{\Omega q^2}F_q(p)-\sum_{p<p_F}\sum_{q>q_{\rm max}}\frac{2\pi\hbar^2e^2}{\Omega q^2}\right]+E_b^{(2)}. \tag{B-5}$$

These terms are now in complete correspondence with the structure of the B-P result. The first bracketed terms correspond to the plasma energy of B-P, the second to the screened Coulomb exchange energy, and the third to Pines' screened short-range correlation energy.

PHYSICAL REVIEW VOLUME 108, NUMBER 3 NOVEMBER 1, 1957

Correlation Energy of a High-Density Gas: Plasma Coordinates

R. Brout
Department of Physics, Cornell University, Ithaca, New York
(Received June 24, 1957)

The model Hamiltonian of Sawada which describes electron correlation at high density is examined. It is shown that the set of scattering modes for momentum transfers below a certain q_{max} is not complete. It is completed by the plasma mode. $(q_{max})^{-1}$ is the natural Debye length of the theory.

I. INTRODUCTION

It is the purpose of this article to supplement the mathematical methods of the preceding paper[1] by a somewhat more detailed analysis. The problem is treated here in the language of continuous spectra. It is then shown that for $q < q_{max}$, where q_{max} is given by Eq. (9) of reference 1, the set of scattering states does not form a complete set. The set is completed by the plasma mode. For $q > q_{max}$, the set of scattering states is complete. This is true only in the infinite limit, for only then will the plasma mode for $q > q_{max}$ completely dissipate itself into the scattering modes. A similar situation arises in the theory of particle decay as carried out on a simplified model by Glaser and Kallën.[2]

This paper represents work carried out by the author after his remark on the existence of plasma modes in the Sawada theory.[3] The previous paper is, in the main, the work of Sawada, Brueckner, and Fukuda. Though the two independent investigations led to identical results, it was thought to be instructive to the reader to present the two lines of argument concurrently.

2. FORMULATION OF THE THEORY

We shall adopt Sawada's Hamiltonian[3] with some modifications in notation.

We define the creation operator of a pair (excited particle $\mathbf{p}+\mathbf{q}$, hole $\mathbf{p}$) as $d_\mathbf{q}^*(\mathbf{p})$, i.e.,

$$d_\mathbf{q}^*(\mathbf{p}) = a_{\mathbf{p}+\mathbf{q}}^* b_\mathbf{p}^*, \tag{2.1}$$

in Sawada's notation. Then defining the operator (p_F = Fermi momentum)

$$\sigma_\mathbf{q}^* = \frac{1}{(2\pi)^{\frac{3}{2}}} \int_{\substack{|\mathbf{p}|<p_F \\ |\mathbf{p}+\mathbf{q}|>p_F}} d\mathbf{p} [d_\mathbf{q}^*(\mathbf{p}) + d_{-\mathbf{q}}(-\mathbf{p})]. \tag{2.2}$$

The Hamiltonian taken by Sawada is ($\hbar=1$, sums on p include spins)

$$H = H_0 + H_c,$$

$$H_0 = \frac{1}{(2\pi)^3} \int_{p>p_F} d\mathbf{p}\, \epsilon_\mathbf{p} a_\mathbf{p}^* a_\mathbf{p} - \frac{1}{(2\pi)^3} \int_{p<p_F} d\mathbf{p}\, \epsilon_\mathbf{p} b_\mathbf{p}^* b_\mathbf{p} + \frac{1}{(2\pi)^3} \int_{p<p_F} d\mathbf{p}\, \epsilon_\mathbf{p}, \tag{2.3}$$

$$H_c = \frac{1}{(2\pi)^3} \int d\mathbf{q} \frac{2\pi e^2}{q^2} \left[\sigma_\mathbf{q}^* \sigma_\mathbf{q} - \frac{1}{(2\pi)^3} \int_{\substack{p<p_F \\ |p+q|>p_F}} d\mathbf{p} \right].$$

(Note that H_c is the usual Coulomb Hamiltonian but with scatterings of excited states to excited states and holes to holes omitted.) The usual operator,

$$\rho_\mathbf{q} = \frac{1}{(2\pi)^{\frac{3}{2}}} \int_{p<p_F} a_{\mathbf{p}+\mathbf{q}}^* a_\mathbf{p},$$

is now replaced by $\sigma_\mathbf{q}$.

In evaluating the commutator of $d_\mathbf{q}(\mathbf{p})$ with the Hamiltonian, one makes the further approximation of neglecting the commutator of $d_q(p)$ with all $d_{\mathbf{q}'}^*(\mathbf{p}')$ but for $\mathbf{q}=\mathbf{q}'$. This eliminates the exchange scattering diagrams in the Gell-Mann and Brueckner scheme[4] and is the direct analog of the random phase approximation in the Bohm and Pines theory.[5] With this approximation, the excitations decouple for different $\mathbf{q}$ and behave like bosons, i.e., one may take

$$[d_\mathbf{q}^*(\mathbf{p}), d_{\mathbf{q}'}(\mathbf{p})] = (2\pi)^3 \delta(\mathbf{q}-\mathbf{q}') \delta(\mathbf{p}-\mathbf{p}'). \tag{2.4}$$

Equations (2.3) and (2.4) define the problem.

The solution runs as follows. One finds the commutators

$$[H, d_{-\mathbf{q}}(-\mathbf{p})] = \omega_\mathbf{q}(\mathbf{p}) d_{-\mathbf{q}}(-\mathbf{p}) + \frac{1}{(2\pi)^{\frac{3}{2}}} \left(\frac{4\pi e^2}{q^2}\right) \sigma_\mathbf{q},$$
$$[H, d_\mathbf{q}^*(\mathbf{p})] = -\omega_\mathbf{q}(\mathbf{p}) d_\mathbf{q}^*(\mathbf{p}) - \frac{1}{(2\pi)^{\frac{3}{2}}} \left(\frac{4\pi e^2}{q^2}\right) \sigma_\mathbf{q}. \tag{2.4'}$$

Equations (2.4′) have the property that although H_0 is not a function of the pair operators, the commutator $[H_0, d_\mathbf{q}(\mathbf{p})]$ is nevertheless a function of $d_\mathbf{q}(\mathbf{p})$. This

[1] Sawada, Brueckner, Fukuda, and Brout [Phys. Rev. **108**, 507 (1957)], preceding paper.
[2] W. Glaser and G. Kallën, Nuclear Phys. **2**, 706 (1957).
[3] K. Sawada, Phys. Rev. **106**, 372 (1957).
[4] M. Gell-Mann and K. A. Brueckner, Phys. Rev. **106**, 364 (1957).
[5] D. Bohm and D. Pines, Phys. Rev. **92**, 609 (1953).

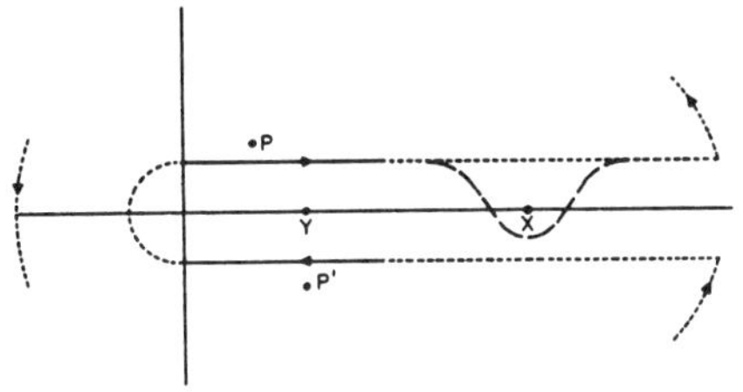

FIG. 1. Contour for the integration of Eq. (2.14).

makes commutators (2.4′) linear functions of $d_{\mathbf{q}}(\mathbf{p})$. One may then introduce the concept of normal modes (i.e., linear combinations of $d_{\mathbf{q}}(\mathbf{p})$, say η, which have the property that $[H,\eta]=\Omega\eta$. We shall work always in the continuous limit though the problem may be equally formulated for the discrete case. In that case consider the set of operators which create the real scattering states corresponding to the excitation $d_q^*(p)$. These are

$$\eta_{\mathbf{q}}^*(\mathbf{p})=d_{\mathbf{q}}^*(\mathbf{p})+\frac{4\pi e^2}{q^2}\left(\frac{1}{(2\pi)^3}\right)\int_{\substack{p'<p_F\\|\mathbf{p}'+\mathbf{q}|>p_F}} d\mathbf{p}' \times\left[\frac{1}{\varphi_+(\omega_{\mathbf{q}}(\mathbf{p}))}\frac{d_{\mathbf{q}}^*(\mathbf{p}')}{[\omega_{\mathbf{q}}(\mathbf{p})-\omega_{\mathbf{q}}(\mathbf{p}')+i\epsilon]}+\frac{1}{\varphi_+(\omega_q(p))}\frac{d_{-\mathbf{q}}(-\mathbf{p}')}{[\omega_{\mathbf{q}}(\mathbf{p})+\omega_{\mathbf{q}}(\mathbf{p}')]}\right], \tag{2.5}$$

and its complex conjugate. Here

$$\varphi_\pm=1-\frac{4\pi e^2}{q^2}\left(\frac{1}{(2\pi)^3}\right)\int_{\substack{p'<p_F\\|\mathbf{p}'+\mathbf{q}|>p_F}} d\mathbf{p}' \times\left[\frac{1}{\omega_{\mathbf{q}}(\mathbf{p})-\omega_{\mathbf{q}}(\mathbf{p}')+i\epsilon}-\frac{1}{\omega_{\mathbf{q}}(\mathbf{p})+\omega_{\mathbf{q}}(\mathbf{p}')}\right]. \tag{2.6}$$

With (2.6), it follows from (2.4′) that

$$[H,\eta_{\mathbf{q}}^*(\mathbf{p})]=-\omega_{\mathbf{q}}(\mathbf{p})\eta_{\mathbf{q}}^*(\mathbf{p}). \tag{2.7}$$

Equation (2.7) says that the real pairs have zero self-energy.

The next task is to investigate whether the states (2.5) constitute a complete orthonormal set. That the $\eta_q(p)$ are orthonormal follows immediately from the evaluation of the commutator $[\eta_q^*(p),\eta_q(p')]$. Using Eqs. (2.4) and (2.5), one finds that

$$[\eta_q^*(p),\eta_q(p')]=\delta(p-p')(2\pi)^3. \tag{2.8}$$

The only remaining question, completeness, would be established if the transformation (2.5) is unitary. Let us symbolize (2.5) by

$$\eta=Ud, \tag{2.9}$$

where an annihilation operator in the transformation carries a negative energy in its coefficient in accord with (2.5). The establishment of (2.8) is equivalent to $UU^+=1$. The completeness part of the theorem concerns U^+U and it is here that the plasma mode will appear. For this reason we shall enter into some detail upon the calculation (see also Klein and McCormick[6]).

The off-diagonal element of U^+U is, by direct evaluation,

$$\frac{1}{\varphi_+(\omega_{\mathbf{q}}(\mathbf{p}))}\frac{1}{[\omega_{\mathbf{q}}(\mathbf{p})-\omega_{\mathbf{q}}(\mathbf{p}')+i\epsilon]}+\frac{1}{[\omega_{\mathbf{q}}(\mathbf{p}')-\omega_{\mathbf{q}}(\mathbf{p})-i\epsilon]}\frac{1}{\varphi_-(\omega_{\mathbf{q}}(\mathbf{p}'))}$$
$$+\frac{1}{(2\pi)^3}\left(\frac{4\pi e^2}{q^2}\right)\int_{\substack{p''<p_F\\|\mathbf{p}''+\mathbf{q}|>p_F}} d\mathbf{p}''\left\{\frac{1}{[\omega_{\mathbf{q}}(\mathbf{p}'')-\omega_{\mathbf{q}}(\mathbf{p})-i\epsilon]}\frac{1}{\varphi_-(\omega_{\mathbf{q}}(\mathbf{p}''))\varphi_+(\omega_{\mathbf{q}}(\mathbf{p}''))}\frac{1}{[\omega_{\mathbf{q}}(\mathbf{p}'')-\omega_{\mathbf{q}}(\mathbf{p}')+i\epsilon]}\right.$$
$$\left.-\frac{1}{[\omega_q(p'')+\omega_q(p)]}\frac{1}{\varphi_-(\omega_q(p''))\varphi_+(\omega_q(p''))}\frac{1}{[\omega_q(p'')+\omega_q(p')]}\right\}. \tag{2.10}$$

The integral in the second half of (2.10) is transformed by using the algebraic relations

$$\frac{1}{[\omega_{\mathbf{q}}(\mathbf{p}'')-\omega_{\mathbf{q}}(\mathbf{p})-i\epsilon]}\frac{1}{[\omega_{\mathbf{q}}(\mathbf{p}'')-\omega_{\mathbf{q}}(\mathbf{p}')+i\epsilon]}=\left\{\frac{1}{\omega_{\mathbf{q}}(\mathbf{p}'')-\omega_{\mathbf{q}}(\mathbf{p})+i\epsilon}-\frac{1}{\omega_{\mathbf{q}}(\mathbf{p}'')-\omega_{\mathbf{q}}(\mathbf{p}')+i\epsilon}\right\}\frac{1}{[\omega_{\mathbf{q}}(\mathbf{p})-\omega_{\mathbf{q}}(\mathbf{p}')+i\epsilon]}, \tag{2.11}$$

$$\frac{1}{\varphi_+\varphi_-}=\left(\frac{1}{\varphi_+}-\frac{1}{\varphi_-}\right)\left(\frac{1}{\varphi_--\varphi_+}\right),$$

$$\varphi_-(\omega_{\mathbf{q}}(\mathbf{p}''))-\varphi_+(\omega_{\mathbf{q}}(\mathbf{p}''))=-\frac{2\pi i}{(2\pi)^3}\left(\frac{4\pi e^2}{q^2}\right)\int\delta(\omega_{\mathbf{q}}(\mathbf{p}''))-\omega_{\mathbf{q}}(\mathbf{p}'''))d\mathbf{p}'''$$
$$\equiv-2\pi i\left(\frac{4\pi e^2}{q^2}\right)E(\omega_{\mathbf{q}}(\mathbf{p}'')). \tag{2.12}$$

[6] A. Klein and B. McCormick, Phys. Rev. 98, 1428 (1955).

Writing

$$\frac{1}{(2\pi)^3}\int d\mathbf{p}''=\int E(\omega_{\mathbf{q}}(\mathbf{p}''))d\omega_{\mathbf{q}}(\mathbf{p}''),$$

where the $\omega_{\mathbf{q}}(\mathbf{p}'')$ are arranged in monotonic sequence, we have for the integral in (2.10) upon using (2.11) and (2.12),

$$-\frac{1}{2\pi i}\left(\frac{1}{\omega_q(p)-\omega_q(p')+i\epsilon}\right)\int d\omega_q(p'')\left[\frac{1}{\varphi_+(\omega_q(p''))}-\frac{1}{\varphi_-(\omega_q(p''))}\right]\left\{\frac{1}{\omega_q(p'')-\omega_q(p)-i\epsilon}+\frac{1}{\omega_q(p'')+\omega_q(p)}-\frac{1}{\omega_q(p'')-\omega_q(p')+i\epsilon}-\frac{1}{\omega_q(p'')+\omega_q(p')}\right\}. \quad (2.13)$$

Changing variable to $\zeta=\omega_q^2(p'')$, (2.13) becomes

$$-\frac{1}{\omega_{\mathbf{q}}(\mathbf{p})-\omega_{\mathbf{q}}(\mathbf{p}')+i\epsilon}\left(\frac{1}{2\pi i}\right)\int_C\frac{d\zeta}{\varphi(\sqrt{\zeta})}\left[\frac{1}{\zeta-\omega_{\mathbf{q}}^2(\mathbf{p})-i\epsilon}-\frac{1}{\zeta-\omega_{\mathbf{q}}^2(\mathbf{p}')+i\epsilon}\right]. \quad (2.14)$$

The specified contour is shown by the solid lines in Fig. 1.

We must now consider the singularities of φ. Define $\bar{\varphi}$ as the value of φ obtained from taking the principal-value parts of $\varphi_\pm$, i.e., $\bar{\varphi}=\mathrm{Re}\,\varphi_\pm$, and let $\bar{\omega}_{\mathbf{q}}$ be the root defined by

$$\bar{\varphi}(\bar{\omega}_{\mathbf{q}})=0. \quad (2.15)$$

Equation (2.15) is the dispersion relation of Bohm and Pines for the plasma frequency.[7] For $q\to 0$, the solution of (2.15) is $\bar{\omega}_{\mathbf{q}}=\omega_{\mathrm{pl}}=(4\pi ne^2/m)^{\frac{1}{2}}$.

Now two cases arise:

(a) $\bar{\omega}_{\mathbf{q}}$ is in the continuum, i.e., $\bar{\omega}_q<(1/2m)(2p_Fq+q^2)$,
(b) $\bar{\omega}_{\mathbf{q}}$ is out of the continuum, i.e., $\bar{\omega}_q>(1/2m)(2p_Fq+q^2)$.

If one has case (a), then $\lim_{\epsilon\to 0}\varphi(\bar{\omega}_{\mathbf{q}}\pm i\epsilon)\neq 0$ and the point $\zeta=\bar{\omega}_{\mathbf{q}}$ will not have pole-like behavior (the point Y in Fig. 1). The deformation indicated by the dotted lines in Fig. 1 is then permitted and one picks up the two poles P and P' as indicated. The result is that (2.14) exactly cancels the first part of (2.10).

If one has case (b), then $\lim_{\epsilon\to 0}\bar{\varphi}(\bar{\omega}_q\pm i\epsilon)=0$ and the point $\zeta=\bar{\omega}_q$ is a pole. In this case a convenient deformation is given by the dashed lines in Fig. 1 around pole X. The remaining poles cancel the first part of (2.10), leaving a residue from X. The final result is that

Case (a):

$$UU^+=U^+U=1,\quad \bar{\omega}_q<\frac{1}{2m}(2p_Fq+q^2).$$

Case (b):

$$UU^+=1,$$

$$(U^+U)_{\mathbf{pp}'}=\frac{4\pi e^2}{q^2}\left(\frac{2\bar{\omega}_{\mathbf{q}}}{\varphi'(\bar{\omega}_{\mathbf{q}})}\right)\left[\frac{1}{\bar{\omega}_{\mathbf{q}}^2-\omega_{\mathbf{q}}^2(\mathbf{p})}-\frac{1}{\bar{\omega}_{\mathbf{q}}^2-\omega_{\mathbf{q}}^2(\mathbf{p}')}\right]+\delta(p-p'),\quad \bar{\omega}_q>\frac{1}{2m}(2p_Fq+q^2). \quad (2.16)$$

The only remaining question is to find the coordinate which, when added to the set $\eta_q(p)$, will complete it. This is the plasma mode for $q<q_{\max}$, where $q_{\max}$ is defined by

$$\bar{\omega}_{q_{\max}}=\frac{1}{2m}[2p_Fq_{\max}+q_{\max}^2]. \quad (2.17)$$

These modes are given by

$$v_{\mathbf{q}}=\left[\frac{4\pi e^2/q^2}{\varphi'(\bar{\omega}_{\mathbf{q}})}\right]^{\frac{1}{2}}\left(\frac{1}{(2\pi)^{\frac{3}{2}}}\right)\int_{\substack{p<p_F\\|\mathbf{p}+\mathbf{q}|>p_F}}d\mathbf{p}\left[\frac{d_{\mathbf{q}}^*(\mathbf{p})}{\bar{\omega}_{\mathbf{q}}-\omega_{\mathbf{q}}(\mathbf{p})}+\frac{d_{-\mathbf{q}}(-\mathbf{p})}{\bar{\omega}_{\mathbf{q}}+\omega_{\mathbf{q}}(\mathbf{p})}\right]. \quad (2.18)$$

(Notice that as $q\to 0$, $v_{\mathbf{q}}\sim\sigma_{\mathbf{q}}$.) With the addition of $v_{\mathbf{q}}$, it is readily verified that the $\eta_{\mathbf{q}}(\mathbf{p})$ set is completed. This completes the proof. The calculation of the energy is given in reference 1.

[7] The fact that the inequality $|\mathbf{p}+\mathbf{q}|>p_F$ that arises in (2.15) [see (2.6)] drops by symmetry was pointed out by P. Nozières. This insures that (2.15) is the dispersion relation of Bohm and Pines.

PHYSICAL REVIEW VOLUME 108, NUMBER 6 DECEMBER 15, 1957

Diamagnetism of a Dense Electron Gas

GREGOR WENTZEL
Enrico Fermi Institute for Nuclear Studies, University of Chicago, Chicago, Illinois
(Received August 16, 1957)

The theory of Coulomb interactions in a dense electron gas at zero temperature is formally simplified by the introduction of an equivalent Hamiltonian which gives the correct high-density value for the correlation energy. In the corresponding approximation, the diamagnetism of a dense electron gas is found to be the same as that of noninteracting electrons. Coupling with longitudinal sound waves, or a periodic potential, fails to produce a Meissner effect.

1. INTRODUCTION

MUCH progress has been made lately in the study of the Coulomb interactions in a dense electron gas.[1–3] Particularly noteworthy is a paper by Sawada[2] which succeeds in formulating the high-density problem in such a fashion that the pertinent solutions can be constructed in closed form, avoiding the perturbation expansion of Gell-Mann and Brueckner[1] and reproducing their result for the correlation energy. The suppression of long-range Coulomb interactions manifests itself in a simple damping factor (depending on the momentum transfer q), in formal analogy with the so-called damping effect in "scalar pair theory"[4] (equivalent to a renormalization of the coupling constant). Indeed, Sawada invokes this formal analogy as a guide for the construction of his rigorous solutions. Bloch's phonon-like pairs[5] (excited electron plus hole) are the analogs to the "mesons" in pair theory.

It is the purpose of this paper to re-examine some properties of a dense electron gas, taking account of Coulomb interactions in Sawada's approximation. The emphasis will be on the diamagnetic properties, for instance the question: can a dense electron gas behave like a superconductor in a magnetic field (Meissner effect)? Schafroth[6] arrived at a negative answer by treating the Coulomb interaction as a perturbation, in arbitrary order. But this procedure is unreliable since the nonmagnetic energy diverges in every approximation but the first. We shall replace the dubious perturbation expansion by Sawada's high-density approximation and derive the same negative result (for zero temperature). It will also be shown that, in this respect, nothing is changed by coupling the electrons with longitudinal sound waves,[7] or by admitting a periodic lattice potential (Bloch wave functions).

In order to simplify the calculations, we shall first reformulate Sawada's theory in terms of an equivalent Hamiltonian, allowing us to make fuller use of the analogy with meson pair theory. This procedure will prove particularly helpful in the study of diamagnetism.

2. EQUIVALENT HAMILTONIAN

Regarding H_C, the Coulomb interaction potential, we follow Sawada[2] entirely and adopt his approximations as he justified them by comparison with Gell-Mann's and Brueckner's work. In the first place, we discard from H_C all matrix elements which do not contribute to the ground state energy in the limit of infinite electron density.[8] Then, we can rewrite H_C in the following abbreviated form:

$$H_C=\tfrac{1}{2}\sum_q \lambda_q C_q^* C_q+\text{const},$$
$$\lambda_q=4\pi e^2 q^{-2}\Omega^{-1},\quad (\hbar=1), \tag{1}$$
$$C_q=\sum_p (c_{p,q}+c_{-p,-q}^*),\quad c_{p,q}=a_p^* a_{p+q},$$

where a, a^* are the fermion absorption and emission operators, and it is essential that p always stands for a momentum vector *inside* the Fermi sphere, and $p+q$ always for one *outside* (the spin will be ignored for the

[1] M. Gell-Mann and K. A. Brueckner, Phys. Rev. **106**, 364 (1957).
[2] K. Sawada, Phys. Rev. **106**, 372 (1957).
[3] Sawada, Brueckner, Fukuda, and Brout, Phys. Rev. **108**, 507 (1957).
[4] G. Wentzel, Helv. Phys. Acta **15**, 111 (1942).
[5] F. Bloch, Helv. Phys. Acta **7**, 385 (1934).
[6] M. R. Schafroth, Nuovo cimento **9**, 291 (1952).
[7] See Schafroth's criticism of H. Fröhlich's theory of superconductivity: M. R. Schafroth, Helv. Phys. Acta **24**, 645 (1951).
[8] Example for an omitted matrix element: two excited electrons make transitions both remaining outside the Fermi sphere.

sake of simplicity). In other words, *whenever the symbol* p, *or* $\sum_p$, *occurs in the following,* p *is restricted to that part of the inside Fermi sphere* ($|p|<p_F$) *where* $|p+q|>p_F$, for a given pair momentum q. Note that automatically $|-p|<p_F$, $|-p-q|>p_F$, and

$$C_{-q}=\sum_p(c_{-p,-q}+c_{p,q}^*)=C_q^*.$$

A further simplification, justified in the high-density limit, is the omission, in the commutators $[H_C,c_{p,q}^{(*)}]$, of all terms $q'\neq q$. Again following Sawada, we adopt the approximate commutation rules

$$[H_C,c_{p,q}^*]=\lambda_q C_q^*,\quad [H_C,c_{p,q}]=-\lambda_q C_q. \qquad (2)$$

Note that this is equivalent to treating c^*, c as boson emission and absorption operators:

$$[c_{p,q},c_{p',q'}^*]=\delta_{pp'}\delta_{qq'},\quad [c_{p,q},c_{p',q'}]=0. \qquad (3)$$

The kinetic energy, H_K, enters into Sawada's calculation only through the commutators

$$[H_K,c_{p,q}^*]=\omega_{p,q}c_{p,q}^*,\quad [H_K,c_{p,q}]=-\omega_{p,q}c_{p,q},$$
$$\omega_{p,q}=E_{p+q}-E_p,\quad E_p=p^2/2m. \qquad (4)$$

These equations follow from the rigorous expression for H_K and the rigorous commutators $[a^*,c^*]$, etc. However, since Sawada's argument is based entirely on Eqs. (1), (2), and (4), one can safely replace H_K by an expression which yields precisely the same commutators (4) in conjunction with the boson commutation rules (3) [which are also compatible with (2)]. Such a substitute expression for H_K is

$$H_K=\sum_q\sum_p\omega_{p,q}c_{p,q}^*c_{p,q}, \qquad (5)$$

representing just the sum of the individual pair energies. With this choice, the mathematics of the problem becomes completely equivalent to that of a "pair theory,"[9] and Sawada's result can be rederived more simply, allowing a comparison at each stage.

To substantiate this, let us introduce canonical field variables

$$(2\omega_{p,q})^{-\frac{1}{2}}(c_{p,q}+c_{-p,-q}^*)=\varphi_{p,q}\equiv\varphi_{-p,-q}^*,$$
$$i(\omega_{p,q}/2)^{\frac{1}{2}}(c_{p,q}^*-c_{-p,-q})=\pi_{p,q}\equiv\pi_{-p,-q}^*. \qquad (6)$$

As a consequence of (3), the φ and π obey canonical commutation rules. Equations (5) and (1) may be rewritten as

$$H_K=\tfrac{1}{2}\sum_q\sum_p(\pi_{p,q}^*\pi_{p,q}+\omega_{p,q}^2\varphi_{p,q}^*\varphi_{p,q}-\omega_{p,q}), \qquad (5a)$$

$$H_C=\sum_q\lambda_q(\sum_p\omega_{p,q}^{\frac{1}{2}}\varphi_{p,q}^*)(\sum_{p'}\omega_{p',q}^{\frac{1}{2}}\varphi_{p',q}) +\text{const.} \qquad (1a)$$

The Hamiltonian H_K+H_C is that of a system of linearly coupled oscillators, and analysis in terms of normal modes is straightforward. The secular equation for the proper frequencies ν reads

$$\phi_q(\nu^2)\equiv1+2\lambda_q\sum_p\omega_{p,q}(\omega_{p,q}^2-\nu^2)^{-1}=0 \qquad (7)$$

(for any q). The roots ν [$=E-E_0$ in Sawada's notation, see his Eq. (7) in reference 2] are the excitation energies of the one-pair states (they form a continuous spectrum as $\Omega\to\infty$). The corresponding eigenvectors describe the scattering (one-pair to one-pair states) in accordance with Sawada. The "correlation energy" (including the contribution of the plasma vibrations[3]) is nothing but the lowest eigenvalue of H_K+H_C,[10] or the "zero-point energy" of the normal vibrations:

$$E_0=\tfrac{1}{2}\sum_q\sum_p(\nu_{p,q}-\omega_{p,q})+\text{const.}$$

Indeed, the function ϕ_q defined in (7) can be written,[4] in terms of its zeros and poles ($\xi=\nu^2$):

$$\phi_q(\xi)=\prod_p(\xi-\nu_{p,q}^2)[\prod_p(\xi-\omega_{p,q}^2)]^{-1}.$$

Hence:

$$E_0=(4\pi i)^{-1}\int\xi^{\frac{1}{2}}d\xi\sum_q\frac{\partial\ln\phi_q(\xi)}{\partial\xi}+\text{const},$$

where the closed contour in the complex ξ plane is taken to encircle all zeros ($\nu_{p,q}^2$) and poles ($\omega_{p,q}^2$). Integrating by parts, and writing $\xi=\nu^2$ again:

$$E_0=-(4\pi i)^{-1}\int d\nu\sum_q\ln\phi_q(\nu^2)+\text{const.}$$

Now, one can go to the continuum limit ($\Omega\to\infty$) and replace $\sum_p$ in (7) by the appropriate integral ($|p|<p_F$, $|p+q|>p_F$) whereby the function $\phi_q(\nu^2)$ (viz., its imaginary part) becomes discontinuous along part of the real ν axis. Carrying the contour along both sides of this cut, $\frac{1}{2}[\ln\phi_q(\nu^2+i\delta)-\ln\phi_q(\nu^2-i\delta)]$ becomes an arc tangent (the scattering phase shift), and one obtains precisely Sawada's result [Eq. (19) in reference 2], in a more condensed form. Of course, the contribution of the poles corresponding to the plasma vibrations [$\nu_{p,q}>\max(\omega_{p,q})$] must be added [see Eq. (11) in reference 3].

All this confirms that the use of the "equivalent Hamiltonian," with (5), is legitimate if the commutation relations (2) and (4) define the problem adequately. Needless to say, only in a very restricted class of problems will it be possible to treat the electron-hole pairs as bosons. If too many (real) pairs are present, as in thermal excitation, our method is, of course, inapplicable.

3. COUPLING WITH LONGITUDINAL SOUND

Before turning to the magnetic effects, we want to examine very briefly the interaction of the electrons with longitudinal sound waves carried by some con-

[9] The "meson rest mass" is here zero. λ_q plays the role of the coupling constant, and the "source function" for each q is essentially a step function.

[10] Strictly speaking, the second-order exchange energy, "$\epsilon_b^{(2)}$," should be added here.

tinuous elastic medium. The results will not be new, but for the later discussion it will be useful to have the problem restated in our notation.

For the free phonon energy, we write

$$H_S=\tfrac{1}{2}\sum_q(\pi_q^*\pi_q+\nu_q^2\varphi_q^*\varphi_q).$$

Then, a phonon-electron interaction of the Bloch type may be represented as

$$H_I=\sum_q\mu_q\varphi_q^*\sum_p\omega_{p,q}^{\frac{1}{2}}\varphi_{p,q}+\text{c.c.} \tag{8}$$

The problem defined by $H=H_K+H_C+H_S+H_I$ can again be solved by a normal modes analysis. Instead of (7), one has the secular equation

$$\phi_q(\nu^2)\equiv 1+\left(2\lambda_q-\frac{|\mu_q|^2}{\nu_q^2-\nu^2}\right)\sum_p\frac{\omega_{p,q}}{\omega_{p,q}^2-\nu^2}=0, \tag{7a}$$

and the eigenvectors are now mixtures of one-pair and one-phonon states. The term $|\mu_q|^2(\nu_q^2-\nu^2)^{-1}$ in (7a) indicates how the spectrum of (7) is altered by "switching on" the Bloch interaction H_I. Consider in particular that root ν of (7a) which in the limit $\mu_q\to 0$ becomes ν_q. In the continuum limit $(\Omega\to\infty)$, this root is complex†; the real part represents the sound frequency as modified by H_I, and the imaginary part determines the decay rate of the phonon q into one-pair states. To first order in $|\mu_q|^2$:

$$\nu-\nu_q=-(|\mu_q|^2/2\nu_q)f_q(1+2\lambda_qf_q)^{-1},$$

$$f_q=\lim_{\Omega\to\infty}\sum_p\omega_{p,q}(\omega_{p,q}^2-\nu_q^2\pm i\delta)^{-1}.$$

The "damping factor" $(1+2\lambda_qf_q)^{-1}$ (the same as appears in the correlation energy and scattering amplitudes) indicates the effect of the Coulomb interaction H_C: it amounts to a strong quenching of the phonon-electron interaction for small values of q (because $\lambda_q\propto q^{-2}$).[11]

It should be remarked that, although H_K was tacitly taken to have the substitute form (5), the same results may be derived by using, instead of (5), the rigorous commutation rules (4).

4. DIAMAGNETISM

For our purpose it suffices to take the vector potential of the external magnetic field as

$$A_x=A\cos kz,\quad A_y=A_z=0.$$

The additional terms in the Hamiltonian, linear and quadratic in A, will be called H' and H''. H' involves the electronic current whose "main part," namely that involving the operators of pair creation and annihilation, is expressible in terms of the operators π of (6):

$$H'=A(e/2mc)\sum_p p_x(\pi_{p,k}-\pi_{p,k}^*)i(\omega_{p,k}/2)^{-\frac{1}{2}}. \tag{9}$$

(Here, of course, the vector k has the components $0, 0, k$; and $|p|<p_F$, $|p+k|>p_F$ is again implied.) As before, we ignore the electronic spin and accordingly omit the spin-field interaction in H'. In H'', only the diagonal part matters:

$$H''=A^2(e^2/4mc^2)N, \tag{10}$$

where $N=p_F^3\Omega/6\pi^2=$ total number of electrons.

We now make explicit use of the simplification afforded by our "equivalent Hamiltonian," viz., the expression (5) or (5a) for H_K. It allows us to eliminate H' by the simple device of a translation in π space. This will be achieved by the following unitary transformation:

$$U=\exp[A(e/2mc)\sum_p p_x(\varphi_{p,k}-\varphi_{p,k}^*)(\omega_{p,k}/2)^{-\frac{1}{2}}].$$

Then

$$U^*\pi_{p,k}U=\pi_{p,k}-A(e/2mc)p_xi(\omega_{p,k}/2)^{-\frac{1}{2}}.$$

In H_K (5a), the only terms affected by the U transformation are $q=k$ and $q=-k$ (for $q=-k$, replace p by $-p$ for convenience of notation), with the result

$$U^*H_KU=H_K+H'+H_2,\qquad H_2=A^2(e^2/2m^2c^2)\sum_p p_x^2\omega_{p,k}^{-1}. \tag{11}$$

All other terms in the field-free Hamiltonian H^0, viz., H_C, H_S, H_I (if they are assumed present), are independent of the $\pi_{p,q}$ and commute with U. Hence

$$U(H^0+H'+H'')U^*=H^0-H_2+H''. \tag{12}$$

As the diamagnetic energy, $(-H_2+H'')$, is entirely unaffected by either Coulomb interaction or longitudinal sound, we cannot expect anything new. Indeed, evaluation of the integral in (11) gives

$$\sum_p\frac{p_x^2}{\omega_{p,k}}=\frac{\Omega}{32\pi^2}\frac{m}{k}\left[(p_F^2-\tfrac{1}{4}k^2)^2\ln\left(\frac{2p_F+k}{2p_F-k}\right)+\frac{5}{3}p_F^3k-\tfrac{1}{4}p_Fk^3\right],$$

and expanding this into powers of k^2/p_F^2, one confirms that the leading term $(\propto k^0)$ in $(-H_2)$ [Eq. (11)] just cancels H'' [Eq. (10)] (no Meissner effect),[12] whereas the next term in $(-H_2)$ gives precisely the Landau diamagnetism. (No de Haas-van Alphen fluctuations are found, presumably because our Hamiltonian is incomplete.) The conclusion is that the diamagnetism of

† *Note added in proof.*—For a full discussion of a similar situation (Lee model), see Glaser and Kallen, Nuclear Phys. 2, 706 (1956/57).

[11] Our result is in substantial agreement with that of Hayakawa, Kitano, Nakajima, and Nakano [*Proceedings of the International Conference of Theoretical Physics, Kyoto and Tokyo, 1953* (Science Council of Japan, Tokyo, 1954), p. 916] who comment on this problem in connection with Fröhlich's theory of superconductivity.

[12] The corresponding cancellation occurs, of course, in the expectation value of the current density. Note also that if p_x, in (9) and (11), is replaced by $p_x+\frac{1}{2}k$, the cancellation of $(-H_2)$ and H'' is complete, for all values of k. This proves that a longitudinal vector potential is spurious, and that our "truncated" Hamiltonian is gauge-invariant.

a very dense electron gas, treated in the approximation of Gell-Mann and Brueckner,[1] or Sawada,[2] is the same as for a gas of noninteracting electrons.[13]

5. PERIODIC POTENTIAL

Is it possible that this negative result is altered by the introduction of a periodic lattice potential and the use of Bloch wave functions? To carry out a simple generalization, let us replace $E_p=p^2/2m$ by some more general even function $E(p)$ of the "quasi-momentum" p. In a magnetic field, according to Luttinger's theorem,[14] the single-particle states are approximately determined by the Hamiltonian $E(-i\nabla-(e/c)\mathbf{A})+e\varphi$; according to Adams,[15] this is a "one-band" approximation, certainly good enough for our purposes.

Few changes are then required in our previous equations. The restriction imposed on p, for a given q, is now obviously

$$E(p)<E_F, \quad E(p+q)>E_F.$$

In (4), (5), etc., $\omega_{p,q}$ must now be interpreted as $E(p+q)-E(p)$. Expanding the Hamiltonian into powers of A, the linear and quadratic terms generalizing (9) and (10) are, to the lowest order in k:

$$H'=A(e/2c)\sum_p[\partial E(p)/\partial p_x](\pi_{p,k}-\pi_{p,k}^*)i(\omega_{p,k}/2)^{-\frac{1}{2}}+\cdots, \quad (9a)$$

$$H''=A^2(e^2/4c^2)\sum_{p \atop E(p)<E_F}\partial^2E(p)/\partial p_x^2+\cdots. \quad (10a)$$

With an obvious change in the transformation U, and assuming that all nonmagnetic interactions commute with U, Eq. (12) follows again, but with

$$H_2=A^2(e^2/2c^2)\sum_p[\partial E(p)/\partial p_x]^2\omega_{p,k}^{-1}+\cdots. \quad (11a)$$

Since we assume $k\ll p_F$, the p integral in (11a) extends over a thin layer bordering on the Fermi surface; the vertical distance between the bounding surfaces equals $\omega_{p,k}|\nabla_pE|^{-1}$. In the limit $k\rightarrow 0$, the volume integral reduces to a surface integral over half the Fermi surface, namely that half where $\partial E/\partial p_x>0$ ($d\sigma$=element of area in p space):

$$\lim_{k=0}H_2=A^2\left(\frac{e^2}{2c^2}\right)\left(\frac{\Omega}{8\pi^3}\right)\int_{F/2}d\sigma\left[\frac{\partial E}{\partial p_x}\right]^2|\nabla_pE|^{-1}.$$

The integral may be written as $\frac{1}{2}$ the integral over the entire Fermi surface and can then be transformed, by means of Gauss' theorem, into a volume integral over the inside:

$$\int_{F/2}d\sigma\left[\frac{\partial E}{\partial p_x}\right]^2|\nabla_pE|^{-1}=\frac{1}{2}\int_{E(p)<E_F}d^3p\frac{\partial^2E}{\partial p_x^2}.$$

Comparing with (10a), we find again

$$\lim_{k=0}H_2=H''.$$

For nonvanishing but small k, the leading term in $(-H_2+H'')$ varies as k^2.[16] Thus, again, there is no Meissner effect, but just ordinary diamagnetism.

One has to conclude that the phenomenon of superconductivity must result from interactions which we have disregarded. Since superconductivity preferably occurs in metals with high electron density, Sawada's method, suitably extended, may still be helpful in further analysis. In particular, one should try to include those matrix elements of H_C and H_I (omitted in (1) and (8), see reference 8) which Bardeen *et al.*[17] claim to be responsible for creating an "energy gap." This, however, will mean giving up, or amending, the boson commutation rules (3), and then our equivalent Hamiltonian will probably lose its usefulness.

---

[13] H. Kanazawa [Progr. Theoret. Phys. Japan **15**, 273 (1956); **17**, 1 (1957)] uses the Bohm-Pines method to compute the effect of long-range Coulomb interactions on the diamagnetic susceptibility, and he obtains a small correction to the Landau value. Such a discrepancy is not surprising since even the correlation energy is not quite correctly calculated by the Bohm-Pines theory (see reference 1, p. 368).

[14] J. M. Luttinger, Phys. Rev. **84**, 814 (1951).

[15] E. N. Adams, II, Phys. Rev. **85**, 41 (1952).

[16] It gives the Peierls susceptibility χ_3, as we have confirmed for the case that E depends quadratically on p.

[17] Bardeen, Cooper, and Schrieffer, Phys. Rev. **106**, 162 (1957).

The description of collective motions in terms of many-body perturbation theory

II. The correlation energy of a free-electron gas

By J. Hubbard

Atomic Energy Research Establishment, Harwell, Didcot, Berkshire

(*Communicated by R. E. Peierls, F.R.S.—Received* 16 *July* 1957)

The theory developed in a previous paper is applied to calculate the correlation energy of a free-electron gas. The theory involves no cut-off and gives a uniform description of collective motion effects in the long-range limit and of particle motion effects in the short-range limit. It is shown that in the lowest order the theory agrees with Bohm & Pines's plasma oscillation theory in the long-range limit, but is inadequate in the short-range limit. The theory is approximately evaluated to the next order, which is correct in the short-range limit, and is applied to calculate the correlation energy at several gas densities; the results are in good agreement with those of Bohm & Pines.

1. Introduction

In a previous paper (Hubbard 1957), which will be referred to as I, a method of analyzing the perturbation series for the many-body problem was described, which, it was stated, would lead to a theory equivalent to Bohm & Pines's (1953) plasma oscillation theory when applied to a free electron gas. In this paper the general theory of I is applied to a free-electron gas and this claim in part substantiated.

Bohm & Pines's theory has had three principal applications. These are

(i) to the calculation of the correlation energy of a free-electron gas;

(ii) to the interpretation of the energy loss spectra of fast electrons passing through thin metal films;

(iii) to give a theoretical basis to the 'independent particle' picture normally used in the discussion of metallic properties.

The present paper is concerned purely with the first application, i.e. the application of the theory developed in I to calculate the correlation energy of a free-electron gas. It is hoped in a later paper to deal with the items (ii) and (iii) above in terms of the theory developed in I.

In §2 of this paper the expression given in I for the total energy of the system is transformed into a form more suitable for calculation and the integral equation giving $\mathscr{V}$ in terms of $\mathscr{V}^*$ is solved. In §3 $\mathscr{V}^*$ is calculated in the lowest order for the ground state of an electron gas, and in §4 the nature of the results obtained is investigated; it is found that the theory is adequate in the long-range limit but inadequate in the short-range limit. In §5 it is shown that to make the theory correct in the short-range limit one has to carry the calculation of $\mathscr{V}^*$ to the next lowest order, and this calculation is performed approximately. Finally, in §6 the results of numerical computations are quoted and compared with Bohm & Pines's theory.

337 *Collective motions in terms of many-body perturbation theory. II*

2. The calculation of the correlation energy

The system considered is the same as that discussed by Bohm & Pines (1953), namely an infinite electron gas moving in a uniform background of positive charge of such a density as to make the system as a whole electrically neutral. The Hamiltonian of this system can be written

$$H = \sum_i \frac{1}{2m} p_i^2 + \sum_{i \neq j} \sum_{k \neq 0} \frac{2\pi e^2}{k^2} e^{i\mathbf{k}.(\mathbf{x}_i - \mathbf{x}_j)}, \tag{1}$$

in which the first term represents the kinetic energy of the electrons and the second term their mutual Coulomb interactions. The term with $\mathbf{k} = 0$ in the Fourier expansion of the Coulomb interaction has been removed because it is exactly cancelled out by the potential energy of the electrons in the background of positive charge.

In I it was shown that each eigenstate of the perturbed system (regarding the interaction between the electrons as the 'perturbation') could be associated with an eigenstate of the unperturbed system from which it evolves as the interaction is slowly switched on. The Hamiltonian of the unperturbed system consists of the first term of (1), the eigenfunctions of which are determinants of plane waves. Thus, each eigenstate of the unperturbed system is completely specified by saying which plane waves are present in the determinant, i.e. which plane-wave states are occupied. If the quantities $N_\mathbf{k}$ are defined by

$$\left.\begin{aligned} N_\mathbf{k} &= 1 \quad \text{if state } \mathbf{k} \text{ is occupied,} \\ N_\mathbf{k} &= 0 \quad \text{if state } \mathbf{k} \text{ is unoccupied,} \end{aligned}\right\} \tag{2}$$

then any eigenstate of the unperturbed system is specified by giving its occupation numbers $N_\mathbf{k}$. Further, each state of the perturbed system can be specified by giving the occupation numbers in the unperturbed state from which it arises when the interaction is slowly switched on.

In the present paper the correlation energy of that state of the perturbed system which arises from the ground state of the unperturbed system is calculated. The latter state is that in which all the plane-wave states below the Fermi surface are occupied and all those above are empty, i.e. the ordinary Fermi ground-state distribution. Whilst it is not quite obvious that the state which arises adiabatically from this one is the ground state of the perturbed system, this is taken to be so for the moment and we assume that we are calculating the correlation energy in the ground state of the perturbed system. Although it is our main purpose in this paper to calculate this ground-state correlation energy, much of the work of this section and the next section up to equation (24) is general and applicable to any distribution $N_\mathbf{k}$.

In I the expression (equation (70))

$$E = \sum_\mathbf{k} N_\mathbf{k} E_\mathbf{k} - \tfrac{1}{2} N v(0) + \tfrac{1}{2} N^2 u(0) + \frac{i\hbar}{2} \int_0^\lambda \frac{d\lambda}{\lambda} \int dx \int dx' \, \delta(t) \mathscr{V}(x', x) \bar{\mathscr{V}}(x, x') \tag{3}$$

for the energy of the state arising from a given unperturbed state was given. In this expression $N = \sum_\mathbf{k} N_\mathbf{k}$ is the total number of particles in the system (the system is now assumed to be enclosed in a very large box and surface effects are neglected), the

$E_{\mathbf{k}}$ are the energies of the individual particle states of the unperturbed system, $v(0)$ is the interaction energy at zero separation and $u(0)$ is the component with $\mathbf{k} = 0$ in the Fourier expansion of the interaction. The quantity λ is the interaction coupling constant, and $\mathscr{V}$ and $\bar{\mathscr{V}}$ are the modified interaction and a related quantity discussed extensively in I; these latter quantities depend implicitly on the occupation numbers $N_{\mathbf{k}}$ (see § 3).

In the present application the individual particle states are the plane waves $e^{i\mathbf{k}.\mathbf{x}}$, the corresponding energies being given by $E_{\mathbf{k}} = \hbar^2k^2/2m$. Thus for a ground-state Fermi distribution the first term of (2) is $(3/5)NE_F$, where E_F is the Fermi energy. Further, in the present case $u(0) = 0$ since the Fourier component with $\mathbf{k} = 0$ in the interaction term of (1) has been removed, whilst

$$v(0) = \int \frac{2\pi e^2}{k^2} \frac{d\mathbf{k}}{(2\pi)^3}.$$

This integral diverges, but the divergence is cancelled by a corresponding divergence in the last term of (3); as was pointed out in I these divergences lead to no difficulty since a suitable limiting procedure can easily be found. Combining these results one can write (3) as

$$E = \tfrac{3}{5}NE_F - \int \frac{2\pi Ne^2}{k^2} \frac{d\mathbf{k}}{(2\pi)^3} + \frac{i\hbar}{2}\int_0^\lambda \frac{d\lambda}{\lambda}\int dx \int dx'\delta(t)\,\mathscr{V}(x,x')\,\bar{\mathscr{V}}(x',x), \tag{4}$$

which can be rewritten

$$E = \tfrac{3}{5}NE_F + N\epsilon_{\text{exch.}} + \left\{\frac{i\hbar}{2}\int_0^\lambda \frac{d\lambda}{\lambda}\int dx \int dx'\delta(t)\,\mathscr{V}(x,x')\,\bar{\mathscr{V}}(x',x) - \int \frac{2\pi Ne^2}{k^2}\frac{d\mathbf{k}}{(2\pi)^3} - N\epsilon_{\text{exch.}}\right\}, \tag{5}$$

where $\epsilon_{\text{exch.}}$ is the ordinary exchange energy per particle given by

$$\epsilon_{\text{exch.}} = -\frac{1}{N}\sum_{\mathbf{k}',\mathbf{k}''}^{\text{occ.}} \frac{2\pi e^2}{|\mathbf{k}'-\mathbf{k}''|^2} = \frac{-0{\cdot}916}{r_s}\,\text{Ry}$$

and r_s is the radius of a sphere whose volume equals the volume per electron. The terms in the bracket in (5) constitute what is normally defined to be the correlation energy; to calculate this quantity it is necessary to calculate $\mathscr{V}$ and $\bar{\mathscr{V}}$.

The quantities $\mathscr{V}$, $\mathscr{V}^*$ and $\bar{\mathscr{V}}$ are related (see I) by the equations

$$\mathscr{V}(x',x) = v(x'-x) + \int \mathscr{V}^*(x',x'')\,\mathscr{V}(x'',x)\,dx'', \tag{6}$$

$$\mathscr{V}^*(x',x) = \int \bar{\mathscr{V}}(x',x'')\,v(x''-x)\,dx'', \tag{7}$$

where $v(x-x') = v(\mathbf{x}-\mathbf{x}')\,\delta(t-t')$ and $v(\mathbf{x}-\mathbf{x}')$ is the ordinary interaction, the Coulomb interaction in the case of an electron gas. The quantity $\mathscr{V}^*$ can be expressed as an infinite series of terms, each term being associated with a proper polarization part (see I); $\mathscr{V}$ can also be expressed as an infinite series of terms corresponding to polarization parts (proper and improper), but since this series is much less rapidly convergent than that giving $\mathscr{V}^*$, $\mathscr{V}$ is most conveniently calculated from (6). Since in the case of a uniform gas the quantities $\mathscr{V}(x', x)$, $\mathscr{V}^*(x', x)$ and $\overline{\mathscr{V}}(x', x)$ are functions of the differences of their arguments only, one can solve (6) at once by introducing their Fourier transforms:

$$\mathscr{V}(x', x) = \frac{1}{2\pi}\int d\mathbf{k}\int d\omega\, \mathscr{V}(\mathbf{k}, \omega)\, e^{i\mathbf{k}.(\mathbf{x}'-\mathbf{x})+i\omega(t'-t)} \tag{8}$$

with similar definitions for $\mathscr{V}^*(\mathbf{k}, \omega)$ and $\overline{\mathscr{V}}(\mathbf{k}, \omega)$. When these transforms are introduced into equations (6) and (7) one obtains

$$\mathscr{V}(\mathbf{k}, \omega) = v(\mathbf{k}) + \mathscr{V}^*(\mathbf{k}, \omega)\,\mathscr{V}(\mathbf{k}, \omega) \tag{9}$$

and

$$\mathscr{V}^*(\mathbf{k}, \omega) = v(\mathbf{k})\,\overline{\mathscr{V}}(\mathbf{k}, \omega), \tag{10}$$

where $v(\mathbf{k})$ is the Fourier component of the ordinary potential $v(\mathbf{x}-\mathbf{x}')$, given by $v(\mathbf{k}) = 4\pi e^2/k^2$ in the case of a Coulomb potential. Equation (9) can be solved immediately to give

$$\mathscr{V}(\mathbf{k}, \omega) = \frac{v(\mathbf{k})}{1-\mathscr{V}^*(\mathbf{k}, \omega)}. \tag{11}$$

Introducing the Fourier transforms of $\mathscr{V}$ and $\overline{\mathscr{V}}$ into (5), carrying out the integrations and dividing by N, one obtains making use of (10) and (11) the expression

$$\epsilon_{\text{corr.}} = \frac{i\hbar}{4\pi n}\int_0^\lambda \frac{d\lambda}{\lambda}\int \frac{d\mathbf{k}}{(2\pi)^3}\int d\omega\, \frac{\mathscr{V}^*(\mathbf{k}, \omega)}{1-\mathscr{V}^*(\mathbf{k}, \omega)} - \int \frac{2\pi e^2}{k^2}\frac{d\mathbf{k}}{(2\pi)^3} - \epsilon_{\text{exch.}} \tag{12}$$

for the correlation energy per electron, where n is the number of electrons per unit volume.

If one now writes
$$\mathscr{V}^*(\mathbf{k}, \omega) = A(\mathbf{k}, \omega) + i\Sigma(\mathbf{k}, \omega), \tag{13}$$

where $A(\mathbf{k}, \omega)$ and $\Sigma(\mathbf{k}, \omega)$ are the real and imaginary parts of $\mathscr{V}^*(\mathbf{k}, \omega)$, and notices that the expression (12) must be real, one obtains

$$\epsilon_{\text{corr}} = \frac{\hbar}{4\pi n}\int_0^\lambda \frac{d\lambda}{\lambda}\int \frac{d\mathbf{k}}{(2\pi)^3}\int d\omega\, \frac{-\Sigma(\mathbf{k}, \omega)}{[1-A(\mathbf{k}, \omega)]^2 + \Sigma^2(\mathbf{k}, \omega)} - \int \frac{2\pi e^2}{k^2}\frac{d\mathbf{k}}{(2\pi)^3} - \epsilon_{\text{exch.}}. \tag{14}$$

This is a general expression for the correlation energy. It remains now only to calculate A and Σ; in the next section this calculation is performed approximately.

3. The modified interaction

In I the quantity $\overline{\mathscr{V}}$ was defined (equation (67)) to be the sum over all proper polarization diagrams apart from Γ_0 of certain quantities $\overline{W}$

$$\overline{\mathscr{V}}(x', x) = \sum_{\Gamma \neq \Gamma_0}^{\text{proper}} \overline{W}_\Gamma(x', x).$$

Taking the Fourier transform of this relation, one has

$$\overline{\mathscr{V}}(\mathbf{k}, \omega) = \sum_{\Gamma \neq \Gamma_0}^{\text{proper}} \overline{W}_\Gamma(\mathbf{k}, \omega), \tag{15}$$

where $\overline{W}_\Gamma(\mathbf{k}, \omega)$ is the Fourier transform of $\overline{W}_\Gamma(x', x)$, the latter in a uniform gas being a function of the difference of its arguments only. It is hoped that this latter series is rapidly convergent; it has therefore been approximated to by its leading term, that corresponding to the simplest of all polarization diagrams, shown in figure 1 (*a*). It is actually found that this approximation is sufficient to give the effects associated with the plasma oscillations to the same accuracy as Bohm & Pines calculated them, but that the retention of only this single term leads to significant inaccuracies in the calculation of the particle aspects due to the omission of certain exchange terms. This latter point and the way in which it can be approximately allowed for is dealt with in § 5 of this paper.

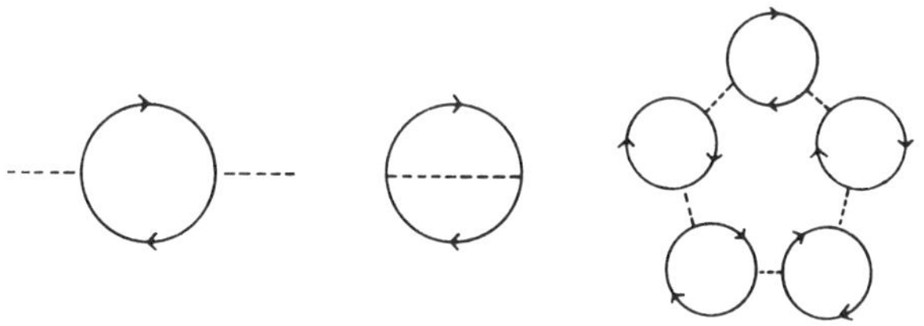

(*a*) (*b*) (*c*)

FIGURE 1

In I the expression (equation (55)) for the $\overline{W}$ corresponding to the polarization part shown in figure 1 (*a*) was given; it is

$$\overline{W}(x', x) = -\frac{1}{\mathrm{i}\hbar} S(x', x)\, S(x, x'), \tag{16}$$

where $S(x', x)$ is the particle propagator given in I, equation (21). In the case of a free electron gas this becomes

$$S(x', x) = \epsilon(t'-t) \sum_{\mathbf{k}}^{\text{unocc.}} \exp\left\{\mathrm{i}\mathbf{k}\,.\,(\mathbf{x}'-\mathbf{x}) - \frac{\mathrm{i}}{\hbar} E_{\mathbf{k}}(t'-t)\right\}$$
$$- \epsilon(t-t') \sum_{\mathbf{k}}^{\text{occ.}} \exp\left\{\mathrm{i}\mathbf{k}\,.\,(\mathbf{x}'-\mathbf{x}) - \frac{\mathrm{i}}{\hbar} E_{\mathbf{k}}(t'-t)\right\}, \tag{17}$$

where

$$E_{\mathbf{k}} = \frac{\hbar^2 k^2}{2m}. \tag{18}$$

Putting (17) into (16), one obtains

$$\overline{W}(x'-x) = \frac{1}{\mathrm{i}\hbar}\Bigg[\epsilon(t'-t) \sum_{\mathbf{k}'}^{\text{unocc.}} \sum_{\mathbf{k}''}^{\text{occ.}} \exp\left\{\mathrm{i}(\mathbf{k}'-\mathbf{k}'')\,.\,(\mathbf{x}'-\mathbf{x}) - \frac{\mathrm{i}}{\hbar}(E_{\mathbf{k}'}-E_{\mathbf{k}''})\,(t'-t)\right\}$$
$$+ \epsilon(t-t') \sum_{\mathbf{k}'}^{\text{occ.}} \sum_{\mathbf{k}''}^{\text{unocc.}} \exp\left\{\mathrm{i}(\mathbf{k}'-\mathbf{k}'')\,.\,(\mathbf{x}'-\mathbf{x}) - \frac{\mathrm{i}}{\hbar}(E_{\mathbf{k}'}-E_{\mathbf{k}''})\,(t'-t)\right\}\Bigg], \tag{19}$$

from which, on Fourier transformation, one has

$$\overline{W}(\mathbf{k}, \omega) = \frac{2\pi}{\mathrm{i}\hbar}\Bigg[\sum_{\mathbf{k}'}^{\text{unocc.}} \sum_{\mathbf{k}''}^{\text{occ.}} \delta(\mathbf{k}'-\mathbf{k}''-\mathbf{k})\, \delta_{-}\left\{\frac{E_{\mathbf{k}'}-E_{\mathbf{k}''}}{\hbar} + \omega\right\}$$
$$+ \sum_{\mathbf{k}'}^{\text{occ.}} \sum_{\mathbf{k}''}^{\text{unocc.}} \delta(\mathbf{k}'-\mathbf{k}''-\mathbf{k})\, \delta_{+}\left\{\frac{E_{\mathbf{k}'}-E_{\mathbf{k}''}}{\hbar} + \omega\right\}\Bigg], \tag{20}$$

341 *Collective motions in terms of many-body perturbation theory. II*

where δ_+ and δ_- are the positive and negative frequency parts of the δ function given by

$$\delta_+(x) = -\frac{1}{2\pi i}\frac{P}{x} + \tfrac{1}{2}\delta(x), \tag{21}$$

$$\delta_-(x) = \frac{1}{2\pi i}\frac{P}{x} + \tfrac{1}{2}\delta(x), \tag{22}$$

and P means principal value.

It is convenient now to introduce the quantities $N_{\mathbf{k}}$ defined in (2) and re-arrange (20) into the form

$$\overline{W}(\mathbf{k},\omega) = \frac{2\pi}{i\hbar}\left[\sum_{\mathbf{k}'} N_{\mathbf{k}'}(1-N_{\mathbf{k}'+\mathbf{k}})\,\delta_-\left\{\frac{E_{\mathbf{k}'+\mathbf{k}}-E_{\mathbf{k}'}}{\hbar}+\omega\right\} + \sum_{\mathbf{k}'} N_{\mathbf{k}'}(1-N_{\mathbf{k}'-\mathbf{k}})\,\delta_+\left\{\frac{E_{\mathbf{k}'}-E_{\mathbf{k}'-\mathbf{k}}}{\hbar}+\omega\right\}\right]. \tag{23}$$

The expression (23) gives the lowest-order approximation to $\mathscr{V}(\mathbf{k},\omega)$. According to (10), one obtains the lowest-order approximation to $\mathscr{V}^*$ by multiplying this by $v(\mathbf{k}) = 4\pi e^2/k^2$; if one does this, substitutes for the δ functions in (23) from (21) and (22), and re-arranges, one obtains the expression

$$\mathscr{V}^*(\mathbf{k},\omega) = \frac{4\pi e^2}{m}\sum_{\mathbf{k}'} N_{\mathbf{k}'}\frac{P}{\left(\omega-\dfrac{\hbar}{m}\mathbf{k}.\mathbf{k}'\right)^2-\dfrac{\hbar^2k^4}{4m^2}} - i\,\frac{4\pi^2e^2}{\hbar k^2}\sum_{\mathbf{k}'} N_{\mathbf{k}'}(1-N_{\mathbf{k}'+\mathbf{k}})\left[\delta\left\{\frac{E_{\mathbf{k}'+\mathbf{k}}-E_{\mathbf{k}'}}{\hbar}-\omega\right\}+\delta\left\{\frac{E_{\mathbf{k}'+\mathbf{k}}-E_{\mathbf{k}'}}{\hbar}+\omega\right\}\right] \tag{24}$$

for $\mathscr{V}^*(\mathbf{k},\omega)$ in the lowest-order approximation. The real part of this expression gives A and the imaginary part Σ to the present degree of approximation. It will be noticed that the real part coincides with the polarizability of a free electron gas calculated in the lowest order (Hubbard 1955) and that Σ differs from the corresponding conductivity in that the sum rather than the difference of the two δ functions appears.

In the case when the electrons occupy a ground-state Fermi distribution the sums in (24) can be evaluated quite straightforwardly. The results can be conveniently expressed in terms of the reduced variables

$$\left.\begin{aligned} x &= k/k_F, \\ y &= \hbar\omega/E_F. \end{aligned}\right\} \tag{25}$$

It is found that $\Sigma(x,y)$ (Σ and A are functions of $k = |\mathbf{k}|$ only since the Fermi distribution is isotropic) is given for $y > 0$ by

$$\begin{aligned} \Sigma(x,y) &= 0 \quad \text{when} \quad y > x(x+2), \\ &= 0 \quad \text{when} \quad x > 2 \quad \text{and} \quad y < x(x-2), \\ &= -\frac{\pi}{2}\frac{\xi}{x^3}\left[1-\frac{1}{4}\left(\frac{y}{x}-x\right)^2\right] \\ &\qquad \text{when} \quad x>2 \quad \text{and} \quad x(x-2)<y<x(x+2), \\ &\qquad \text{or when} \quad x<2 \quad \text{and} \quad x(2-x)<y<x(x+2); \\ &= -\frac{\pi}{2}\frac{\xi y}{x^3} \quad \text{when} \quad x<2 \quad \text{and} \quad 0<y<x(2-x) \end{aligned} \tag{26}$$

whilst for $y < 0$ Σ is given by $\Sigma(x, -y) = \Sigma(x, y)$. A is given for all x and y by

$$A(x,y) = -\frac{\xi}{x^3}\left[x + \frac{1}{2}\left\{1 - \frac{1}{4}\left(\frac{y}{x} - x\right)^2\right\}\ln\left|\frac{y - x(x+2)}{y - x(x-2)}\right| + \frac{1}{2}\left\{1 - \frac{1}{4}\left(\frac{y}{x} + x\right)^2\right\}\ln\left|\frac{y + x(x+2)}{y + x(x-2)}\right|\right], \tag{27}$$

where ξ is the quantity
$$\xi = \frac{e^2 k_F}{\pi}\Big/ E_F = 0{\cdot}33\, r_s \tag{28}$$

and r_s is the radius in Bohr units (B.U. $= \hbar^2/me^2$) of a sphere whose volume is equal to the volume per electron. It will be noticed that to the present degree of approximation the gas density only enters Σ and A as a multiplicative constant.

For small x and $y \gg x(x+2)$, the expression (27) reduces approximately to

$$A(x,y) = -y_P^2/y^2, \tag{29}$$

where y_P is the reduced plasma frequency given by

$$y_P = \frac{\hbar\omega_P}{E_F} = \frac{\hbar}{E_F}\left(\frac{4\pi n e^2}{m}\right)^{\frac{1}{2}}. \tag{30}$$

4. The general nature of the results

To evaluate the correlation energy to the present degree of approximation one must now substitute (26) and (27) into (14) and evaluate the integral. If the reduced variables x and y are introduced into (14) and the angular part of the $\mathbf{k}$ integration performed (Σ and A depend only upon the length of $\mathbf{k}$), then one obtains

$$\epsilon_{\text{corr.}} = \frac{3}{4\pi}E_F\int_0^\lambda \frac{d\lambda}{\lambda}\int_0^\infty x^2\,dx\int_0^\infty dy\,\frac{-\Sigma(x,y)}{[1 - A(x,y)]^2 + \Sigma^2(x,y)} - \int \frac{2\pi e^2}{k^2}\frac{d\mathbf{k}}{(2\pi)^3} - \epsilon_{\text{exch.}}. \tag{31}$$

It is useful to investigate the contributions to the integral in (31) arising from different parts of the xy plane; this can be conveniently done with the aid of the diagram of the plane given in figure 2. On this diagram the lines *oc* and *bd* whose equations are $y = x(x+2)$ and $y = x(x-2)$ have been drawn together with that part of the line $A(x,y) = 1$ lying outside the region *cobd* for a typical value of ξ, namely $\xi = 1{\cdot}33$ (*ef*). According to (26) Σ vanishes except in the region between the lines *oc* and *obd*. It might be thought therefore that it is only this region of the plane which gives any contribution to the integral in (31). However, one must be cautious because along the line $A(x,y) = 1$ the integrand becomes indeterminate.

If the calculation had been carried to a higher degree of approximation, i.e. in the calculation of Σ one had included terms corresponding to more complicated polarization parts than that shown in figure 1(*a*), then it would have been found that Σ was different from zero everywhere in the plane. Nevertheless, if it is supposed that the calculation retaining only the first term is a good approximation, then it follows that although Σ does not vanish outside of the region *cobd*, it is small

343 *Collective motions in terms of many-body perturbation theory. II*

outside this region. One can then sensibly approximate by taking the limit of the integrand of (14) as $\Sigma \to 0$ in the region outside *cobd*. Taking this limit one obtains

$$\lim_{\Sigma \to -0} \frac{-\Sigma}{[1-A]^2 + \Sigma^2} = \pi\delta[1-A]. \tag{32}$$

Thus the integrand has a δ function singularity along the part of the line $A(x, y) = 1$ lying in the region outside *cobd*.

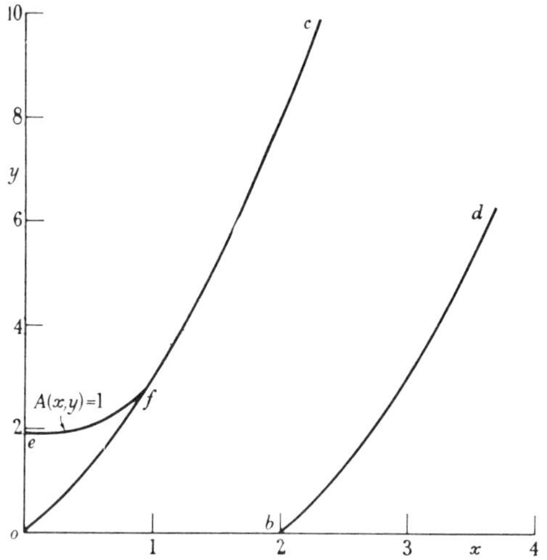

FIGURE 2. Diagram of the xy plane; the lines $y = x(x+2)$ and $y = x(x-2)$ (*oc* and *bd*) are shown, together with that part of the line $A(x, y) = 1$ lying outside *cobd* for $r_s = 4$ (*ef*).

For small x one can evaluate the contribution of this singularity to the integral by using the approximation (29) for A; one obtains

$$\begin{aligned}
&\frac{\hbar}{4\pi n}\int_0^\lambda \frac{d\lambda}{\lambda}\int \frac{d\mathbf{k}}{(2\pi)^3}\int_{-\infty}^{\infty} \frac{E_F}{\hbar}\, dy\, \pi\delta\left\{1 - \frac{y_P^2}{y^2}\right\} \\
&\quad = \frac{1}{4n}\int_0^\lambda \frac{d\lambda}{\lambda}\int \frac{d\mathbf{k}}{(2\pi)^3}\int_{-\infty}^{\infty} dy\, E_F \frac{y_P}{2}[\delta(y-y_P) + \delta(y+y_P)] \\
&\quad = \frac{1}{4n}\int_0^\lambda \frac{d\lambda}{\lambda}\int \frac{d\mathbf{k}}{(2\pi)^3}\hbar\omega_P.
\end{aligned} \tag{33}$$

Now $\omega_P \propto e \propto \lambda^{\frac{1}{2}}$ since $\lambda = e^2$ is the coupling constant, so that the performance of the λ integration yields a factor 2, giving

$$\frac{1}{n}\int \frac{d\mathbf{k}}{(2\pi)^3}\frac{\hbar\omega_P}{2}. \tag{34}$$

Thus, for small x the contribution of the singularity to the correlation energy is just what one would ordinarily call the zero-point energy of the plasma oscillations. Since $1-A$ is always the real part of the dielectric constant, whose vanishing gives the plasma frequency, one may interpret the total contribution of the singularity as representing the contribution to the energy of the collective oscillations of the gas.

Having pointed out the existence of a contribution to the integral from the region of the plane outside *obcd* and its connexion with the collective motion, we now consider a simpler method of evaluating the integral (31). According to the calculation of §3, Σ and A are linearly proportional to the coupling constant λ. Using this fact one can immediately perform the λ integration in (31) to obtain

$$\epsilon_{\text{corr.}} = \frac{3}{4\pi} E_F \int_0^\infty x^2\, \mathrm{d}x \int_0^\infty \mathrm{d}y \tan^{-1} \frac{-\Sigma(x,y)}{1-A(x,y)} - \int \frac{2\pi e^2}{k^2} \frac{\mathrm{d}\mathbf{k}}{(2\pi)^3} - \epsilon_{\text{exch.}}, \tag{35}$$

where one must always take the branch lying between 0 and π of the many-valued inverse tangent. In those regions where $\Sigma = 0$ in the first approximation we take the limit as $\Sigma \to -0$ of the integrand; this limit will be 0 in the region in which $A < 1$ and will be π in the region in which $A > 1$. Thus, in figure 2 the integrand in (35) will be between 0 and π in the region *cobd* since Σ is non-zero in this region, will be π in the region *efo* since $\Sigma = 0$ and $A > 1$ in this region, and will be zero in all other parts of the plane. There are good reasons for thinking of the contribution from the region *efo* as being associated with the collective motion and of the contribution from the region *cobd* as being associated with the particle motion; it will be noticed that the contribution from the region *efo* for small x is again of the form (34), and that for small x this is almost the whole of the contribution.

From (24) one can obtain

$$\begin{aligned} -\frac{\hbar}{4\pi n} \int_0^\lambda \frac{\mathrm{d}\lambda}{\lambda} \int \frac{\mathrm{d}\mathbf{k}}{(2\pi)^3} \int \mathrm{d}\omega \,.\, \Sigma(\mathbf{k},\omega) &= \frac{1}{n} \int_0^\lambda \frac{\mathrm{d}\lambda}{\lambda} \int \frac{\mathrm{d}\mathbf{k}}{(2\pi)^3} \int \mathrm{d}\omega \frac{\pi e^2}{k^2} \\ &\quad \times \sum_{\mathbf{k}'} N_{\mathbf{k}'}(1-N_{\mathbf{k}'+\mathbf{k}}) \left[\delta\left(\frac{E_{\mathbf{k}'+\mathbf{k}} - E_{\mathbf{k}'}}{\hbar} - \omega \right) + \delta\left(\frac{E_{\mathbf{k}'+\mathbf{k}} - E_{\mathbf{k}'}}{\hbar} + \omega \right) \right] \\ &= \frac{1}{n} \int \frac{\mathrm{d}\mathbf{k}}{(2\pi)^3} \frac{2\pi e^2}{k^2} \sum_{\mathbf{k}'} N_{\mathbf{k}'}(1-N_{\mathbf{k}'+\mathbf{k}}) \\ &= \int \frac{2\pi e^2}{k^2} \frac{\mathrm{d}\mathbf{k}}{(2\pi)^3} - \epsilon_{\text{exch.}}, \end{aligned} \tag{36}$$

since

$$\sum_{\mathbf{k}'} N_{\mathbf{k}'} = n$$

and

$$\frac{1}{n} \int \frac{\mathrm{d}\mathbf{k}}{(2\pi)^3} \frac{2\pi e^2}{k^2} \sum_{\mathbf{k}'} N_{\mathbf{k}'} N_{\mathbf{k}'+\mathbf{k}} \tag{37}$$

is the exchange energy per particle. Thus one can substitute the left-hand side of (36) for the last two terms of (35). If the left-hand side is written in terms of the reduced variables x and y one obtains

$$\epsilon_{\text{corr.}} = -\frac{3}{4\pi} E_F \int_0^\infty x^2\, \mathrm{d}x \int_0^\infty \mathrm{d}y \left\{ \tan^{-1} \frac{\Sigma(x,y)}{1-A(x,y)} - \Sigma(x,y) \right\}. \tag{38}$$

As $x \to \infty$, Σ and $A \to 0$ like $1/x^3$, so that the integrand of (38) tends to 0 like $1/x^4$ and the integral is convergent for large x; thus the divergences in the first and second terms on the right-hand side of (31) cancel.

We should like to complete this section by commenting on how the theory behaves in the limits of small and large x. Going back to equation (35), it has already been pointed out that the contribution of the first term is given for small x by equation (34) and represents the zero point of the plasma oscillations. The second term represents a subtraction of the self-energies of the electrons, whilst the third term cancels out the long-range part of the exchange energy. Thus, in the limit of small x, the results agree exactly with those Bohm & Pines obtain in the long-range limit (Pines 1953).

For large x, Σ and A become small and one can expand the denominator in the first term of the integrand of (38) in powers of Σ and A. If one carries out this expansion and performs the various integrations, one obtains an expansion for the correlation energy which consists of a subset of the terms of the original perturbation series, namely, all those terms which correspond to diagrams of the type shown in figure 1 (c), consisting of rings of the polarization parts shown in figure 1 (a). One has, of course, only obtained a subset of the terms of the original perturbation series because only the leading term corresponding to the polarization part of figure 1 (a) was retained in calculating $\mathscr{V}^*$.

If one performs the expansion of the first term of the integrand of (38) described in the preceding paragraph, then the two leading terms correspond to the diagrams of figures 1 (b) and 3 (a). The leading term, corresponding to figure 1(b), is

$$-\frac{3}{4\pi} E_F \int_0^\infty x^2 \, dx \int_0^\infty dy \, \Sigma(x, y), \tag{39}$$

which cancels out the second term in the integrand of (38). Thus, in the limit of large x the leading term in the correlation energy is that corresponding to the diagram of figure 3 (a). If one evaluates the contribution of this diagram to the correlation energy directly, one obtains the expression

$$64\pi^2 \sum_{\mathbf{k}} \frac{1}{k^4} \sum_{\substack{k_1 < k_F \\ |\mathbf{k}_1 + \mathbf{k}| > k_F}} \sum_{\substack{k_2 < k_F \\ |\mathbf{k}_2 - \mathbf{k}| > k_F}} \frac{1}{\mathbf{k} . (\mathbf{k}_1 - \mathbf{k}_2 + \mathbf{k})} \text{ Ry/electron}, \tag{40}$$

which is just the direct part (as opposed to the exchange part) of the second-order perturbation term (Macke 1950).

5. Allowance for the exchange correction

At the end of the previous section it was shown that in the short-range limit (38) reduces to the direct part of the second-order perturbation term. This is unsatisfactory because it is known that in this limit half of the direct term is cancelled by the exchange part of the second-order perturbation, since only correlations between electrons with opposite spins are important in this limit. Thus the simple theory based upon taking into account only the polarization part shown in figure 1 (a) must be corrected to obtain a theory which is correct in the short-range limit; this correction will be discussed in the present section.

The exchange part of the second-order perturbation term corresponds to the diagram shown in figure 3 (b). One would automatically include the contribution of this diagram if one included in the calculation of $\mathscr{V}^*$ not only the contribution of the polarization part of figure 1(a), but also the contribution of the next simplest polarization part, that shown in figure 3 (c). Unfortunately, however, the contribution of this polarization part is very difficult to compute, and one can therefore only allow for the exchange correction in an approximate way.

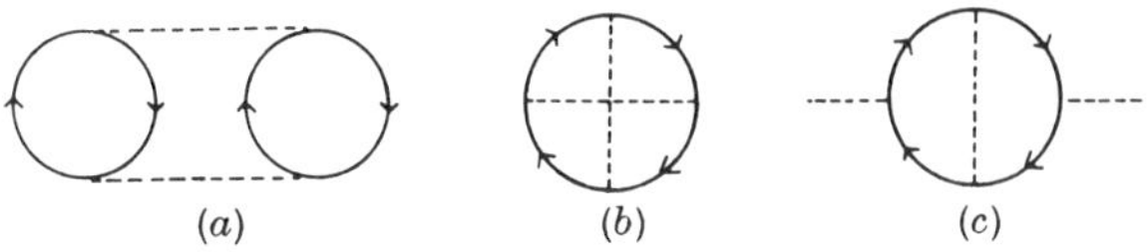

FIGURE 3

The simplest way in which one could allow for the exchange term would be to calculate the contribution to the correlation energy from the exchange part of the second-order perturbation term, and add this energy to the correlation energy calculated from (38). The exchange part of the second-order perturbation term is given by the expression

$$\epsilon^{(2)}_{\text{exch.}} = 32\pi^2 \sum_{\mathbf{k}} \sum_{\substack{\mathbf{k}_1 \\ k_1<k_F \\ |\mathbf{k}_1+\mathbf{k}|>k_F}} \sum_{\substack{\mathbf{k}_2 \\ k_2<k_F \\ |\mathbf{k}_2-\mathbf{k}|>k_F}} \frac{1}{k^2} \frac{1}{(\mathbf{k}+\mathbf{k}_1-\mathbf{k}_2)^2} \frac{1}{\mathbf{k}\,.\,(\mathbf{k}_1-\mathbf{k}_2+\mathbf{k})} \text{Ry} \tag{41}$$

(Macke 1950); the magnitude of this energy, which is independent of the gas density, has been calculated by Brueckner & Gell-Mann (1957) using a Monte Carlo method and is found to be $0{\cdot}046 \pm 0{\cdot}002$ Ry per electron. Although this method would be simple, it would also be rather inaccurate. This is because, although the expression (41) has no divergence for small k, nevertheless, a large part of the contribution to the k sum comes from small k; however, for small k the screening of the interaction which is not taken into account in (41) is important. For this reason, simply to add the energy $\epsilon^{(2)}_{\text{exch.}}$ would considerably overestimate the importance of the exchange correction and lead to too low a value of the correlation energy.

The method used to correct for the exchange effects was to make an approximate estimate of the contribution of $\mathscr{V}^*$ of the polarization part shown in figure 3 (c) and of certain higher-order terms of the same type. This method has the advantage that since it is $\mathscr{V}^*$ itself which is corrected the effect of the screening is automatically taken into account. The results obtained using this method are, however, only accurate to the extent to which the method of estimating the contribution of the diagram 3 (c) is accurate. It is shown in § 6 that owing to this cause one overestimates the correlation energy by about $0{\cdot}004$ Ry per electron.

From a study of the terms involved (see Appendix for further discussion) it was concluded that the contribution to $\mathscr{V}^*$ of the diagram of figure 3 (c) together with certain higher-order terms could be approximately represented by replacing the $\mathscr{V}^*(\mathbf{k}, \omega)$ calculated in § 3 by

$$\mathscr{V}^*(\mathbf{k}, \omega) = \frac{\mathscr{V}_0^*(\mathbf{k}, \omega)}{1+f(\mathbf{k})\mathscr{V}_0^*(\mathbf{k}, \omega)}, \tag{42}$$

where $f(\mathbf{k})$ is the function
$$f(\mathbf{k}) = \frac{1}{2}\frac{k^2}{k^2+k_F^2}, \tag{43}$$

and $\mathscr{V}_0^*(\mathbf{k}, \omega)$ is the $\mathscr{V}^*(\mathbf{k}, \omega)$ calculated in § 3. If the expression (42) for $\mathscr{V}^*$ is substituted into (12), then it is found that in (14) one must replace the Σ and A in the *denominator* of the integrand of the first term by Σ' and A', where Σ' and A' are the same as Σ and A but are multiplied by a factor

$$1-\frac{1}{2}\frac{k^2}{k^2+k_F^2}. \tag{44}$$

Although the method of estimating the additional contribution to $\mathscr{V}^*$ from the diagram 3 (*c*) is rather crude, it has a certain physical reasonableness. For the effect of multiplying the Σ and A in the denominator of (14) by the factor (44) is equivalent to replacing the factor $v(k) = 4\pi e^2/k^2$ which they contain by a factor

$$\frac{4\pi e^2}{k^2}-\frac{1}{2}\frac{4\pi e^2}{k^2+k_F^2}. \tag{45}$$

However, the effect of the exchange terms is to reduce the correlation effects between electrons with parallel spins because they are not permitted to approach one another more closely than a distance of the order r_s by the exclusion principle. This is roughly equivalent to saying that the correlation effects due to the interaction of electrons with parallel spins at short distances are much reduced. But the introduction of (45) in place of $v(k) = 4\pi e^2/k^2$ is equivalent to this since half the interactions of an electron are with electrons of opposite spin, and these are cut off in (45) on the short-range side at about the right distance.

Further, if one replaces Σ and A by Σ' and A' in (12) and expands in powers of Σ' and A' for large x, one now obtains in the short-range limit instead of (40) the short-range part of

$$64\pi^2 \sum_{\mathbf{k}} \frac{1}{k^2}\left\{\frac{1}{k^2}-\frac{1}{2}\frac{1}{k^2+k_F^2}\right\} \sum_{\substack{\mathbf{k}_1 \\ k_1<k_F \\ |\mathbf{k}_1+\mathbf{k}|>k_F}} \sum_{\substack{\mathbf{k}_2 \\ k_2<k_F \\ |\mathbf{k}_2-\mathbf{k}|>k_F}} \frac{1}{\mathbf{k}.(\mathbf{k}_1-\mathbf{k}_2+\mathbf{k})} \text{ Ry/electron}, \tag{46}$$

which in the limit of large k is one-half of (40), as it should be. Furthermore, comparing (46) with (40), it is seen that the present method estimates the exchange part of the second-order perturbation term to be

$$32\,\pi^2 \sum_{\mathbf{k}} \frac{1}{k^2}\frac{1}{k^2+k_F^2} \sum_{\substack{\mathbf{k}_1 \\ k_1<k_F \\ |\mathbf{k}_1+\mathbf{k}|>k_F}} \sum_{\substack{\mathbf{k}_2 \\ k_2<k_F \\ |\mathbf{k}_2-\mathbf{k}|>k_F}} \frac{1}{\mathbf{k}.(\mathbf{k}_1-\mathbf{k}_2+\mathbf{k})} \text{ Ry/electron}, \tag{47}$$

which is to be compared with the exact expression (41). If one calculates the value of the expression (47) and compares it with the 0·046 Ry per electron of the expression (41) some idea of how good the approximation (42) is can be obtained. In fact, calculation shows that (47) has the value 0·036 Ry per electron, which is rather smaller than Brueckner and Gell-Mann's estimate of (41). It can be concluded that the approximation (42) underestimates the effect of the exchange term by about 20 %.

If one replaces Σ and A by Σ' and A' in (14) then it is found that (38) has to be replaced by

$$-\frac{3}{4\pi}E_F\int_0^\infty x^2\,dx\int_0^\infty dy\left\{\frac{\Sigma}{\Sigma'}\tan^{-1}\frac{\Sigma'}{1-A'}-\Sigma\right\}, \tag{48}$$

where Σ' and A' are given by (26) and (27) except that they are to be multiplied by a factor

$$1-\frac{1}{2}\frac{x^2}{1+x^2}. \tag{49}$$

6. Numerical results

The integrals in equations (38) and (48) have been evaluated numerically for values of ξ corresponding to gas densities $r_s = 2, 3, 4$ and 5 (in B.U.), making use of the Ferranti Mark I* Computer at the Atomic Weapons Research Establishment, Aldermaston. The results are quoted in the first and third lines of table 1. In the last line of this table the differences between the correlation energies calculated from (38) and (48) are quoted. These differences represent the contributions to the correlation energy of the exchange effects when screening is taken into account. Comparison of these contributions with the value 0·046 Ry per electron of the expression (41) shows the importance of taking the screening into account when correcting for the exchange effects. Since in the preceding section it was shown that the approximation (42) underestimated the exchange effects by about 20 %, it is to be expected that the estimates of the correlation energy are too big by about 20 % of the energies in the last line of table 1, i.e. are overestimates by about 0·004 Ry per electron.

Table 1. Correlation energies (Ry per electron)

gas density, r	2·0	3·0	4·0	5·0
$\epsilon_{\text{corr.}}$ calculated from (48)	−0·099	−0·086	−0·074	−0·067
$\epsilon_{\text{corr.}}$ calculated by Bohm & Pines	−0·093	−0·081	−0·072	−0·066
$\epsilon_{\text{corr.}}$ calculated from (38)	−0·121	−0·104	−0·090	−0·081
exchange correction	0·022	0·018	0·016	0·014

It will be seen that the correlation energies calculated from (48) are in good agreement with those calculated by Bohm & Pines (Pines 1955), the latter being quoted in the second line of table 1.

Not only is it possible to compare the total correlation energies obtained here with those obtained by Bohm & Pines, but one can also compare the contributions arising from the long- and short-range parts of the interaction and from the intermediate region between these. This is possible because one can in the present method, as in Bohm & Pines's method, determine the contribution to the correlation energy arising from a single Fourier component of the Coulomb interaction.

Consider a general interaction whose Fourier expansion is

$$v(\mathbf{x}-\mathbf{x}') = \sum_{\mathbf{k}} v(\mathbf{k})\, e^{i\mathbf{k}\cdot(\mathbf{x}-\mathbf{x}')}. \tag{50}$$

Then the total correlation energy will be given by (12). However, $\mathscr{V}^*(\mathbf{k},\omega)$ contains a factor $v(\mathbf{k})$; if, therefore, one were to omit a given $v(\mathbf{k})$ from the expansion (50), then the corresponding contribution to the $\mathbf{k}$ integration in the first term of (12) would drop out. Thus the contribution of this Fourier component of the interaction to the correlation energy through the first term of (12) is

$$\frac{i\hbar}{4\pi n}\int_0^\lambda \frac{d\lambda}{\lambda}\int d\omega\, \frac{\mathscr{V}^*(\mathbf{k},\omega)}{1-\mathscr{V}^*(\mathbf{k},\omega)}. \tag{51}$$

Similarly, one can find the contribution of this Fourier component of the interaction to the second and third terms of (12). In this way it can be shown that the

349 *Collective motions in terms of many-body perturbation theory. II*

contribution of those Fourier components of the interaction whose length is between xk_F and $(x+\mathrm{d}x)\,k_F$ to the correlation energy is given by

$$\frac{3}{4\pi}E_F I(x)\,\mathrm{d}x = -\frac{3}{4\pi}E_F\int_0^\infty \mathrm{d}y\left\{\frac{\Sigma}{\Sigma'}\tan^{-1}\frac{\Sigma'}{1-A'}-\Sigma\right\}x^2\,\mathrm{d}x. \tag{52}$$

In figure 4 plotted $I(x)$ has been plotted together with the corresponding quantity in Bohm & Pines's theory, $I_{BP}(x)$, for a gas density $r_s = 4$. $I_{BP}(x)$ was calculated from the expressions

$$\frac{3}{4\pi}E_F I_{BP}(x)\,\mathrm{d}x = \left[-\frac{1}{3n}\sum_i\frac{p_i^2}{2m}-\frac{2\pi e^2}{k^2}+\frac{\hbar\omega_p}{2}\left\{1+\frac{k^2}{2nm^2\omega_p^2}\sum_i p_i^2+\frac{\hbar^2k^4}{8m^2\omega_p^2}\right\}\right]\frac{4\pi k^2\,\mathrm{d}k}{(2\pi)^3}$$

$$+\frac{0{\cdot}916}{r_s}\left[\frac{4}{3}+\frac{k}{k_F}-\frac{1}{16}\left(\frac{k}{k_F}\right)^3\right]\mathrm{d}k \qquad \text{for} \quad x<\beta;$$

$$=\frac{4\pi k^2\,\mathrm{d}k}{8\pi^2}\frac{32\pi^2}{k^4}\sum_{\substack{\mathbf{k}_1\\k_1<k_F\\|\mathbf{k}_1+\mathbf{k}|>k_F}}\sum_{\substack{\mathbf{k}_2\\k_2<k_F\\|\mathbf{k}_2-\mathbf{k}|>k_F}}\frac{1}{\mathbf{k}.(\mathbf{k}_1-\mathbf{k}_2+\mathbf{k})} \quad \text{for} \quad x>\beta. \tag{53}$$

which give the contributions to the correlation energy of Fourier components of the interaction with lengths between xk_F and $(x+\mathrm{d}x)k_F$ in Bohm & Pines's theory; the notation used on the right-hand sides of these expressions is that of Bohm & Pines (1953).

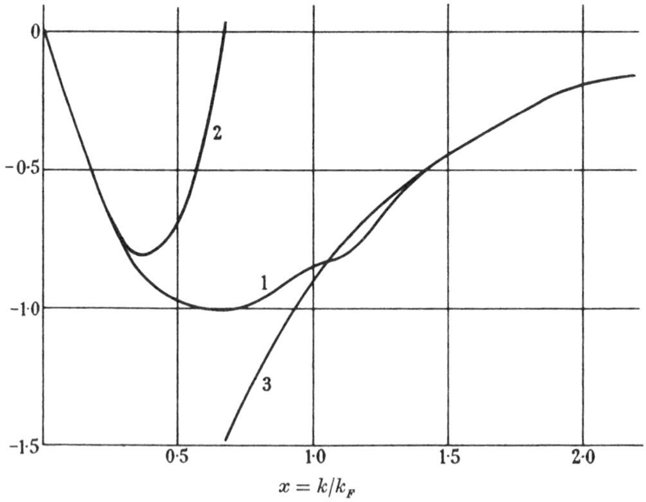

FIGURE 4. Graph showing the contributions of various Fourier components of the interaction to the correlation energy in the present and in Bohm & Pines's theory. 1, $I(x)$: 2, $I_{BP}(x)$, long-range part: 3, $I_{BP}(x)$, short-range part.

It will be seen that at $k = \beta k_F$ there is a discontinuity in $I_{BP}(x)$ due to the fact that the two methods of calculation used by Bohm & Pines for the long- and short-range parts of the interaction do not agree at the cut-off. On the other hand, the curve $I(x)$ shows no such discontinuity because the treatment given here applies throughout the whole range of k, giving a description of collective motion at one end of the range and of particle motion at the other end. It will be noticed that the curve $I(x)$ shows a point of inflexion at $k \approx k_F$; we have not yet been able to give any simple explanation of this effect.

7. Summary and conclusions

In the present paper the following results have been obtained:

(i) The correlation energy of a free-electron gas has been expressed in terms of $\mathscr{V}(\mathbf{k},\omega)$, the Fourier transform of the modified interaction discussed in I, and the integral equation giving $\mathscr{V}(\mathbf{k},\omega)$ in terms of $\mathscr{V}^*(\mathbf{k},\omega)$ has been solved.

(ii) The quantity $\mathscr{V}^*(\mathbf{k},\omega)$ has been calculated in the lowest order for the ground state of an electron gas.

(iii) It has been shown that in the long-range limit the results obtained in this lowest-order calculation agree with Bohm & Pines's plasma oscillation theory, and the singularity which represents the plasma oscillations in the present theory has been identified.

(iv) It has been shown that in the short-range limit these lowest-order calculations are inadequate because they fail to give the exchange part of the second-order perturbation term; failure to allow for this term can lead to significant errors.

(v) An approximate method of correcting the theory to allow for these exchange terms has been suggested and applied to calculate the correlation energy at several gas densities; the error due to the approximations in this calculation is estimated to be about 0·004 Ry per electron. The results have been compared to Bohm & Pines and are to be in good agreement with them as regards total correlation energies.

I should like to thank Dr J. Corner and the staff of the Theoretical Physics Division at the Atomic Weapons Research Establishment, Aldermaston for permission to use the Ferranti Mark I* Computer and for their general assistance in this part of the work.

References

Bohm, D. & Pines, D. 1953 *Phys. Rev.* **92**, 609.
Brueckner, K. A. & Gell-Mann, M. 1957 *Phys. Rev.* **106**, 364.
Hubbard, J. 1955 *Proc. Phys. Soc.* A, **68**, 976.
Hubbard, J. 1957 *Proc. Roy. Soc.* A, **240**, 539.
Macke, W. 1950 *Z. Naturf.* **5***a*, 192.
Pines, D. 1953 *Phys. Rev.* **92**, 625.
Pines, D. 1955 *Solid state physics*, vol. 1, p. 368. New York: Academic Press Inc.

Appendix

In this appendix the rough intuitive arguments which lie behind the adoption of the approximation (42) for $\mathscr{V}^*$ are briefly outlined. It is first convenient to introduce the idea of exchange conjugate diagrams. Let us compare the diagrams shown in figures 5 (a) and 5 (b); in each of these diagrams one interaction line and its vertices have been shown explicitly; the parts of the diagram not shown explicitly are supposed to be identical. Then the only difference between these diagrams is that the outcoming lines from the interaction shown have been crossed over in one diagram relative to the other. This corresponds physically to the fact that the two diagrams represent essentially the same physical process except that in one of them the scattering shown is an exchange scattering process. Such diagrams will therefore be called exchange conjugates of one another. In general one can obtain exchange conjugates of any particular diagram by crossing over the outcoming particle lines at any number of interactions.

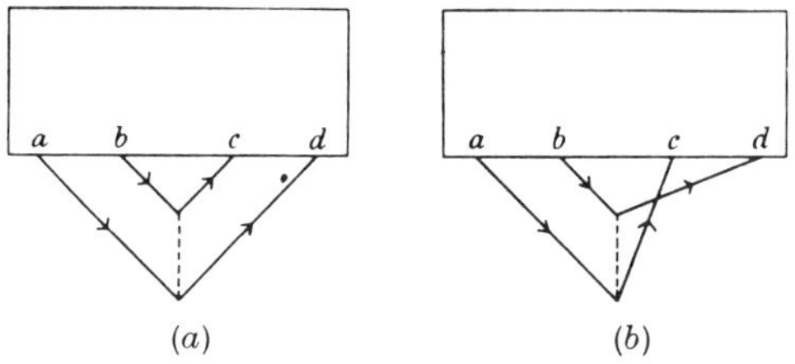

FIGURE 5

If one writes down the contribution of a diagram G to the S matrix of energy shift in the form of an integral over the particle momenta attached to the particle lines of the diagram, then one can obtain the corresponding integrals giving the contributions of its exchange conjugate diagrams by simple substitutions. For if in the integral giving the contribution of G a certain interaction line gives a factor $v(\mathbf{k})$, then the integral representing the contribution of the diagram obtained from G by exchange conjugation at this interaction line is the same as that giving the contribution of G except that $v(\mathbf{k})$ must be replaced by $v(\mathbf{k}+\mathbf{k}_1-\mathbf{k}_2)$, where $\mathbf{k}_1$ and $\mathbf{k}_2$ are the momenta of the incoming particle lines at the interaction line, and one must multiply by a factor (-2) or $(-\frac{1}{2})$ to allow for the change in the number of closed particle loops which results from exchange conjugation (the factor 2 or $\frac{1}{2}$ arises from spin summations since on carrying out the spin summation each particle loop yields a factor 2).

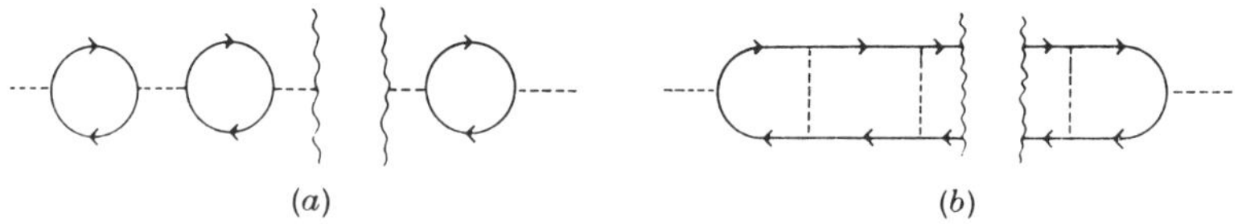

FIGURE 6

Consider now the polarization part shown in figure 6 (*a*), consisting of n polarization parts of the type shown in figure 1 (*a*) strung together. If one exchanges conjugates at all the interaction lines one obtains the polarization part shown in figure 6 (*b*) and $(n-1)$ interaction lines crossing the loop; this is a proper polarization part contributing to $\mathscr{V}^*$. Now according to the preceding paragraph, the integral representing the contribution of this polarization part will be the same as that representing the contribution of diagram 6 (*a*) except that we must replace the $v(\mathbf{k})$ corresponding to the interaction lines of the latter by $v(\mathbf{k}+\mathbf{k}_1-\mathbf{k}_2)$ and must multiply by $(-\frac{1}{2})^{n-1}$. If one now makes the essential approximation of replacing the $v(\mathbf{k}+\mathbf{k}_1-\mathbf{k}_2)$ by $4\pi e^2/(k^2+k_F^2)$ everywhere, arguing that $|\mathbf{k}_1-\mathbf{k}_2| \approx k_F$ and therefore

$$v(\mathbf{k}+\mathbf{k}_1-\mathbf{k}_2) = \frac{4\pi e^2}{|\mathbf{k}+\mathbf{k}_1-\mathbf{k}_2|^2} \approx \frac{4\pi e^2}{k^2+k_F^2} \equiv v'(\mathbf{k});$$

on the average, then since the contribution to $\mathscr{V}(\mathbf{k},\omega)$ of the diagram 6 (*a*) is

$$[\mathscr{V}_0^*(\mathbf{k},\omega)]^n,$$

where $\mathscr{V}_0^*(\mathbf{k},\omega)$ is that corresponding to the diagram 1 (*a*) which was calculated in § 3, the contribution of diagram 6 (*b*) to $\mathscr{V}^*(\mathbf{k},\omega)$ will be

$$\mathscr{V}_0^*(\mathbf{k},\omega)\left[-\frac{1}{2}\frac{v'(\mathbf{k})}{v(\mathbf{k})}\mathscr{V}_0^*(\mathbf{k},\omega)\right]^{n-1} = \mathscr{V}_0^*(\mathbf{k},\omega)\left[-f(\mathbf{k})\mathscr{V}_0^*\right]^{n-1}$$

where
$$\frac{1}{2}\frac{v'(\mathbf{k})}{v(\mathbf{k})} = \frac{1}{2}\frac{k^2}{k^2+k_F^2} \equiv f(\mathbf{k}).$$

Summing over $n = 0, 1, 2, \ldots$, one finds that the total contribution of diagrams of the type 6 (*b*) to $\mathscr{V}^*(\mathbf{k},\omega)$ is

$$\mathscr{V}_0^*(\mathbf{k},\omega)\sum_{n=0}^{\infty}[-f(\mathbf{k})\mathscr{V}_0^*(\mathbf{k},\omega)]^n = \frac{\mathscr{V}_0^*(\mathbf{k},\omega)}{1+f(\mathbf{k})\mathscr{V}_0^*(\mathbf{k},\omega)},$$

which is the result (42). It will be noticed that the simplest diagram of the type 6 (*b*) is that shown in figure 3 (*c*), which has therefore been included.

Correlation Energy of a Free Electron Gas

P. Nozières and D. Pines*
Laboratoire de Physique, École Normale Supérieure, rue Lhomond, Paris, France

(Received February 24, 1958)

The limits of validity of the correlation-energy calculations in the regions of high density, low density, and actual metallic electron densities are discussed. Simple physical arguments are given which show that the high-density calculation of Gell-Mann and Brueckner is valid for $r_s \lesssim 1$ while the low-density calculation of Wigner is valid for $r_s \gtrsim 20$. For actual metallic densities it is shown that the contribution to the correlation energy from long-wavelength momentum transfers ($k < \beta k_0 < 0.47 r_s^{1/2} k_0$) may be accurately calculated in the random phase approximation. This contribution is calculated using the Bohm-Pines extended Hamiltonian, and is shown to be

$$E(\beta)=\left(-0.458\frac{\beta^2}{r_s}+0.866\frac{\beta^3}{r_s^{1/2}}-0.98\frac{\beta^4}{r_s^2}+0.019\frac{\beta^4}{r_s}+0.706\frac{\beta^5}{r_s^{5/2}}+\cdots\right)\text{ ry}.$$

An identical result is obtained by a suitable expansion of the result of Gell-Mann and Brueckner; the validity of the Bohm-Pines neglect of subsidiary conditions in the calculation of the ground-state energy is thereby explicitly established. The contribution to the correlation energy from sufficiently high momentum transfers ($k \gtrsim k_0$) will arise only from the interaction between electrons of antiparallel spin, and may be estimated using second-order perturbation theory. The contribution arising from intermediate momentum transfers ($0.47 r_s^{1/2} k_0 \lesssim k \lesssim k_0$) cannot be calculated analytically; the interpolation procedures for this domain proposed by Pines and Hubbard are shown to be nearly identical, and their accuracy is estimated as ~15%. The result for the over-all correlation energy using the interpolation procedure of Pines is

$$E_c \cong (-0.115+0.031 \ln r_s)\text{ ry}.$$

I. INTRODUCTION

THE ground-state energy of a free electron gas has now been calculated accurately in both the high- and the low-density limit. It depends only on the inter-electron spacing r_s, which is defined by $(4\pi r_s^3 a_0^3/3) = n^{-1}$, where n is the electron density.[1] The results of such calculations may be conveniently expressed in terms of the extent to which they represent an improvement over the Hartree-Fock calculation of the system energy. Thus we may write

$$E_0=\left(\frac{2.21}{r_s^2}-\frac{0.916}{r_s}+E_c\right)\text{ ry},$$

where E_0 is the ground-state energy per electron, $2.21/r_s^2 + 0.916/r_s$ is that quantity calculated in the Hartree-Fock approximation, and E_c is the correlation energy.

The behavior of the electron gas is simple in the limit of high densities ($r_s \ll 1$) because here the Coulomb interaction is a relatively small perturbation on the motion of the electrons. Gell-Mann and Brueckner[2] have shown that in this case the correlation energy may be expressed as a series of the following type:

$$E_c=(A \ln r_s + C + D r_s \ln r_s + E r_s + \cdots)\text{ ry}, \qquad (1)$$

and have calculated the coefficients A and C. Results which are equivalent to those of GB in the high-density limit have recently been obtained by a number of investigators.[3–5]

The behavior of the electron gas in the low-density limit is again simple because here the Coulomb interaction exerts a dominating influence on the electrons. As Wigner[6] first remarked, at sufficiently low densities the electrons may be expected to form a stable lattice in the sea of uniform positive charge. The potential energy keeps the electrons apart, and the kinetic energy for large r_s ($r_s \gg 10$) is insufficent to prevent the electrons becoming localized at fixed sites. The correlation energy may then be expanded as a power series in $(1/r_s)^{1/2}$,

$$E_c=\left(\frac{U}{r_s}+\frac{V}{r_s^{3/2}}+\frac{W}{r_s^2}+\cdots\right)\text{ ry}. \qquad (2)$$

The coefficients U and V have been estimated by Wigner.[6]

The region of actual metallic densities ($1.8 \lesssim r_s \lesssim 5.6$) is essentially an intermediate density regime. The kinetic energy and the potential energy play roughly comparable roles in determining the electron behavior. *There exists no simple rigorous series expression for the correlation energy.* In some ways the behavior of the system resembles that of the high-density regime, in some ways that of the low-density regime. A detailed physical discussion of the electron behavior at metallic densities may be found in the work of Bohm and Pines.[7] Pines[8,9] has given a calculation of the correlation energy in this regime, together with calculations of the influence of electron interaction on the various one-electron

* National Science Foundation Senior Post-Doctoral Fellow, on leave of absence from Princeton University, 1957–1958.

[1] We adopt the notation of reference 7, unless we explicitly indicate otherwise.

[2] M. Gell-Mann and K. A. Brueckner, Phys. Rev. **106**, 364 (1957), hereafter referred to as GB.

[3] K. Sawada, Phys. Rev. **106**, 372 (1957); Sawada, Brueckner, Fukuda, and Brout, Phys. Rev. **108**, 507 (1957), hereafter referred to as SB; R. Brout, Phys. Rev. **108**, 515 (1957).

[4] J. Hubbard, Proc. Roy. Soc. (London) **A240**, 539 (1957); Proc. Roy. Soc. (London) **A243**, 336 (1958).

[5] P. Nozières and D. Pines, Phys. Rev. **109**, 1009 (1958).

[6] E. P. Wigner, Trans. Faraday Soc. **34**, 678 (1938).

[7] D. Bohm and D. Pines, Phys. Rev. **85**, 332 (1952), hereafter referred to as BP II; D. Bohm and D. Pines, Phys. Rev. **92**, 609 (1953), hereafter referred to as BP III. Where no specific reference is required, we refer to the approach developed in these papers, and in reference 8, as the BP approach.

[8] D. Pines, Phys. Rev. **92**, 625 (1953).

[9] D. Pines, in *Solid State Physics*, edited by F. Seitz and D. Turnbull (Academic Press, Inc., New York, 1955), Vol. 1, p. 373, hereafter referred to as SSP.

properties (specific heat, spin susceptibility, transport properties, etc.). The correlation-energy calculation is based on an approximate interpolation between the contribution to the system energy arising from the long wavelength part and that arising from the short wavelength part of the Coulomb interaction, both limits being accurately calculated.

The BP calculation of the correlation energy gives results which are in good agreement with the interpolation formula proposed by Wigner[6,10]:

$$E_c = -0.88/(r_s+7.8)\text{ ry}. \tag{3}$$

Recently Hubbard has proposed a somewhat different interpolation procedure which yields results in close agreement with those of Pines. The results of Pines, Hubbard, or the Wigner interpolation formula, when combined with the calculation of the remaining contributions to the cohesive energy of the alkali metals, are in satisfactory agreement with experiment for these simple metals.[9,11]

In a sense, then, the question of the correlation energy at actual metallic densities might be regarded as satisfactorily solved. However, because the procedures adopted by Pines and Hubbard are interpolation procedures, which do not yield exact results in either the high-density or the low-density regimes, it seems desirable to explore further the relationship between the various approaches. In particular, one would like to answer the following questions:

(1) What is the region of validity of the high-density result of Gell-Mann and Brueckner? Of the low-density result of Wigner?

(2) May one hope to obtain an accurate result for the correlation energy at metallic densities by calculating the next terms in either the high-density or low-density series expressions?

(3) What is the relationship between the interpolation procedures of Pines and Hubbard? How accurate are they for metallic densities?

It is the aim of the present paper to discuss the foregoing questions with a minimum of mathematical detail. We use simple physical arguments to estimate the region of validity of the present calculations of the correlation energy in the high- and low-density limits. We conclude that for $r_s \lesssim 1$, the GB result should be accurate, while for $r_s \gtrsim 20$, the Wigner result should apply. We establish the close relationship of the interpolation procedures of Pines and Hubbard and estimate the accuracy of either method as no worse than 15% for actual metallic densities. We do not give a definitive answer to question (2), but we give plausibility arguments which lead us to conclude that such a systematic extension does not appear profitable.

In Sec. II, we review the different methods for the high-density limit calculation, all of which are based on the random phase approximation (RPA) introduced by Bohm and Pines. We identify the region of electron density for which the breakdown of the RPA may be expected to alter appreciably the calculation of the correlation energy, and discuss the role played by the exchange diagrams which lie outside the RPA. The RPA works best for the contribution to the correlation energy coming from the long-range part of the Coulomb interaction. Therefore, in order to make a quantitative study of the breakdown of the RPA it is necessary to study this long-range part of the correlation energy in some detail. We carry out such an investigation in Sec. IV, and summarize there the detailed results of our studies, which are based on the BP collective description.

We find that the contribution to the correlation energy from a given momentum transfer k of the Coulomb interaction may be expressed as a power series in k in the form obtained by BP. We first calculate the coefficients of the leading terms of this expansion within the RPA. In an appendix we show that, contrary to the recent opinion expressed by Sawada *et al.*, the GB result may likewise be expressed as a power series in k. The resulting expression is identical with that obtained using the BP method. This result establishes the validity of the BP scheme as a method for calculating the individual particle contribution to the correlation energy, and represents an explicit justification of their use of an extended Hamiltonian for this problem. We then proceed to estimate the alteration of this power series expansion arising from the breakdown of the RPA by calculating two leading correction terms which arise outside the RPA. We find these terms are small, so that it is not unlikely that the RPA calculation of the long-range part of the correlation energy is valid even in the region of metallic densities.

In Sec. III we discuss the region of validity of the low-density correlation energy calculation of Wigner. In Sec. V we consider the relationship between the interpolation schemes of Pines and Hubbard, and the accuracy these possess for metallic densities.

We do not here consider the applicability of any of the above approaches to the problem of electron interaction in actual solids. We refer the interested reader to a series of recent papers which deal with the generalization of the BP collective approach to electrons in solids.[12] A similar generalization may be carried out in the high-density regime, as indicated in reference 5.

II. HIGH-DENSITY ELECTRON GAS

The Hamiltonian for the free electron gas may be written as

$$H=\sum_i \frac{p_i^2}{2m}+\sum_k{}' \frac{2\pi e^2}{k^2}(\rho_k{}^*\rho_k - n), \tag{4}$$

where ρ_k is the density fluctuation of momentum k

[10] E. P. Wigner, Phys. Rev. **46**, 1002 (1934).

[11] H. Brooks, Phys. Rev. **91**, 1027 (1953).

[12] P. Nozières and D. Pines, Phys. Rev. **109**, 741 (1958); **109**, 762 (1958); **109**, 1062 (1958), hereafter referred to as NP I, NP II, NP III.

defined by

$$\rho_k = \int d\mathbf{x}\, \rho(\mathbf{x}) e^{-i\mathbf{k}\cdot\mathbf{x}} = \sum_i e^{-i\mathbf{k}\cdot\mathbf{x}_i}. \qquad (5)$$

The prime in the summation in (4) indicates that we leave out the Fourier component with momentum zero; this part of the electron interaction is cancelled by the uniform background of positive charge. The prime will be understood in all similar sums over k which follow.

The total system interaction energy in the ground state is defined by

$$E_{\text{int}}(e^2) = \sum_k \frac{2\pi e^2}{k^2} \{ \langle \Psi_0{}^*(e) | \rho_k{}^* \rho_k | \Psi_0(e) \rangle - n \}, \qquad (6)$$

where $\Psi_0(e)$ is the ground-state wave function for the interacting electron system. One may obtain the ground-state energy, E_G, from (6) by regarding the charge as a variable parameter and making use of the relation, valid for any normalized $\Psi_0(e)$,[13]

$$E_G = \tfrac{3}{5} n E_F + \int_0^{(e')^2} \frac{d(e')^2}{(e')^2} E_{\text{int}}[(e')^2], \qquad (7)$$

where E_F is the energy of an electron at the top of the Fermi distribution. The correlation energy is then given by

$$E_C = \left(\frac{E_G}{n} - \frac{2.21}{r_s{}^2} + \frac{0.916}{r_s} \right) \text{ry}, \qquad (8)$$

provided E_G is expressed in rydbergs. The form of Eqs. (6) and (7) shows that, in principle, one can compute the contribution to the system energy from each interaction momentum, k, separately.

The correlation energy in the high-density limit may now be obtained by a number of different methods. We review these briefly here. The first accurate calculation was that of Gell-Mann and Brueckner (GB)[2] who used a method deriving from an earlier calculation of Macke.[14] From an examination of the perturbation theory expansion they show that the energy may be expanded as a power series of the form (1) in the high-density limit. The basic GB procedure involves the summation, under the integral sign, of the most divergent terms of the perturbation theory expansion of the energy. The summation was carried out with the aid of techniques similar to those used by Feynman in quantum electrodynamics. In lowest order, this procedure is equivalent to making the random phase approximation (RPA) of Bohm and Pines for all interaction momenta. In the RPA the contribution to the system energy from each interaction momentum k depends only on the effect of the kth component of the interaction on $\Psi_0(e)$, and may be computed independently of all the other interaction momenta.

The result obtained by GB, on neglecting contributions to the coefficients, D, E, etc. in (1) which arise within the RPA, is

$$A = 0.0622, \quad C = -0.142 \qquad (9)$$

for the leading coefficients of the correlation-energy expansion, (1). The result for A is exact. There is a further contribution to C from the exchange part of the second-order interaction energy for electrons of parallel spin

$$E_b{}^{(2)} = +\frac{3}{16\pi^5} \frac{\int d\mathbf{k} \int d\mathbf{p} \int d\mathbf{q}}{k^2(\mathbf{k}+\mathbf{p}+\mathbf{q})^2 \mathbf{k}\cdot(\mathbf{p}+\mathbf{q}+\mathbf{k})} \text{ ry}. \qquad (10)$$

In (10), all momenta are measured in units of k_0, the wave-vector of an electron at the top of the Fermi distribution. The limits of integration are $p < k_0$, $q < k_0$, and $|\mathbf{p}+\mathbf{k}| > k_0$, $|\mathbf{q}+\mathbf{k}| > k_0$, GB report a numerical integration of $E_b{}^{(2)}$ by the Monte-Carlo technique, which yields the result

$$E_b{}^{(2)} = (0.046 \pm 0.02) \text{ ry}. \qquad (11)$$

Their final result for C is then

$$C = -0.096 \text{ ry}. \qquad (12)$$

The actual procedure which Gell-Mann and Brueckner followed was somewhat open to question, in that the series which they sum is convergent only for large momentum transfers [actually $(k/k_0) \gtrsim 0.814 r_s{}^{\frac{1}{2}}$]. They assumed that the result thereby obtained could be analytically continued into the region of low momentum transfer. This procedure is difficult to justify directly. However, Sawada *et al.*[3] (SB) have obtained the result (9) by a different method. They make use of the fact that the RPA, as emphasized by BP, is equivalent to linearizing the equations of motion of the density fluctuations. They then find the normal modes of the electron gas (the q_k of BP II and BP III) and use field-theoretic techniques to calculate the system energy. Their calculation, like that of BP, includes an explicit plasmon contribution to the energy. The SB form of the correlation energy [from which (9) may be calculated] differs from that of GB; the equivalence of the two expressions is established by SB.

Hubbard has developed a method for the electron gas which involves the regrouping of the perturbation series expansion into a summation over certain polarization diagrams. With the aid of a procedure analogous to that used by Dyson for summing the vacuum polarization terms in quantum electrodynamics, Hubbard obtains a set of coupled integral equations from which the energy of the system may be calculated. Hubbard's first approximation, that of keeping only the simplest set of polarization diagrams, is equivalent to making the RPA, and yields the correct high-density result.[5] This same high-density result may also be obtained by a slight modification of the basic BP technique, in which the stopping power is calculated using the minority carrier technique of NP II, and use is made of a simple

[13] See, for instance, K. Sawada, Phys. Rev. **106**, 372 (1957).
[14] W. Mäcke, Z. Naturforsch. **5a**, 192 (1950).

relationship between the stopping power and the ground-state energy of the system.[5]

A derivation of the high-density result which does not involve the use of field-theoretic techniques may be given along the following lines. We have recently shown (details will be given elsewhere)[15] that the *exact* interaction energy in the ground state is given by

$$E_{\text{int}}=\frac{\hbar}{2\pi}\,\text{Im}\int_0^\infty d\Omega\frac{1}{\epsilon(k,\Omega)},\tag{13}$$

where $\epsilon(k,\Omega)$ is the *exact* dielectric constant at wave-vector k and frequency Ω for the free electron gas. The exact ground-state energy may then be obtained from (7). The correct high-density result for the correlation energy then follows if we substitute for $\epsilon(k,\Omega)$ in (13) its value calculated in the random phase approximation,

$$\epsilon(k,\Omega)=1+\frac{8\pi e^2}{\hbar k^2}\sum_{\substack{p<k_0\\|\mathbf{k}+\mathbf{p}|>k_0}}\left\{\frac{[(\mathbf{k}\cdot\mathbf{p}/m)+(k^2\hbar/m)]}{[(\mathbf{k}\cdot\mathbf{p}/m)+(\hbar k^2/2m)]^2-\Omega^2}+\frac{i\pi}{2}\delta\left[\Omega-\left(\frac{\mathbf{k}\cdot\mathbf{p}}{m}+\frac{\hbar k^2}{2m}\right)\right]\right\}.\tag{14}$$

The use of the RPA in obtaining the dielectric constant is discussed in some detail in NP II, where it is shown that it corresponds to neglecting local field corrections. With the substitution (14), the result (13) is in the form given by Hubbard.

How far may the high-density result (9) and (12) be extrapolated toward the region of actual metallic densities? The validity of the result depends on the applicability of the RPA. Physically, as is discussed in some detail in BP, it is clear that the RPA can succeed only when the wavelengths of interest are long compared to the inter-particle spacing. In the high-density limit, the dominant contributions to the energy come from wave-vectors less than or of the order of magnitude of

$$k_c=(0.814r_s^{\frac{1}{2}})k_0,\tag{15}$$

where k_0 is the wave-vector of an electron at the top of the Fermi distribution. This result follows by inspection of either the expression (14) or the direct perturbation series expansion. It corresponds to the fact that the correlations induced by the Coulomb interaction act over a characteristic screening length, λ_c, here k_c^{-1}. Each electron may thus be regarded as surrounded by a correlation hole, which corresponds to the effective range over which electron interaction takes place, that is, the screening length. The gain in energy from this screening process is the correlation energy. The screening length corresponding to (15) is simply the Fermi-Thomas screening length, a not unsurprising result in view of the fact that we here deal with long wavelengths and high kinetic energies.

The breakdown of the RPA arises from the exchange diagrams in the perturbation series expansion of the system energies. In order for this approximation to be satisfactory, these exchange diagrams should not be important for $k\sim k_c$. Now exchange effects occur only for electrons of parallel spin, over a range λ_{exch}, equal to the diameter of the exchange hole. Practically, λ_{exch} is of the order of the interparticle spacing[16]

$$\lambda_{\text{exch}}\simeq(0.5k_0)^{-1}\sim r_s.\tag{16}$$

In the high-density limit, with the estimates (15) and (16), we find that $(\lambda_{\text{exch}}/\lambda_c)\propto r_s^{\frac{1}{2}}$, so that this limit is characterized by an exchange length which is small compared to the screening length. Physically this result means that the most effective interaction present is that brought about by the Pauli principle between electrons of parallel spin, a result which is equivalent to saying that the exchange energy is large compared to the correlation energy in the high-density limit. It is this smallness of the exchange length which also makes possible the neglect of exchange diagrams, since the dominant contribution to the higher order terms in the perturbation series expansion of the energy come from momentum transfers for which k is small compared to $\lambda_{\text{exch}}^{-1}$.

We are led to the following picture of the high-density correlation energy results. For $r_s\ll 1$, the RPA is certainly valid; the dominant contribution to the correlation energy is the $A\ln r_s$ term which is calculated exactly within the RPA. As r_s increases, exchange diagrams become of importance, being certainly important for values of r_s for which there are contributions to the energy from momentum transfers comparable to λ_{exch}. This will be the case for $r_s\lesssim 1$. In this region, however, there is still the possibility of carrying out a tolerably accurate calculation of the correlation energy by including the simplest set of exchange terms, those which arise in second-order perturbation series, and yield the $E_b^{(2)}$ of Eq. (10). These exchange terms act to reduce the random phase value of C to some $\frac{2}{3}$ of its value, and therefore represent a not inappreciable correction to the correlation energy.

As we further increase r_s, the higher order exchange terms become important. One might try to calculate the exact contribution made by processes involving one more exchange, which GB have shown gives rise to the r_s and $r_s\ln r_s$ terms in (1), two more exchange terms, etc. In fact such a procedure appears hopeless for actual metallic densities, for the GB series shows little sign of convergence there. It may easily be seen that for momentum transfers which are large compared to $\lambda_{\text{exch}}^{-1}$, that is $k>0.5k_0$, exchange terms of all orders in a perturbation series expansion will play an important role. Fortunately, in this limit that role tends to be simple: the exchange diagrams roughly cancel one-half the contribution to the energy from the direct diagrams.

[15] P. Nozières and D. Pines (to be published). Equation (13) has been independently derived by J. Hubbard (private communication).

[16] See F. Seitz, *The Modern Theory of Solids* (McGraw-Hill Book Company, Inc., New York, 1940), pp. 241–242.

Hence, in a given order, there remains approximately only half the direct interaction, which corresponds then to an interaction between electrons of antiparallel spin only.

Physically this result is a consequence of the fact that the Pauli principle renders it unlikely that electrons of parallel spin approach closely to one another; therefore they cannot further interact through a short-range interaction which gives rise to high momentum transfers. The electrons of antiparallel spin are under no such inhibition of course, and thus the major influence of the high momentum transfer part of the interaction is on electrons of antiparallel spin. We return to this question in a later section.

III. LOW-DENSITY ELECTRON SOLID

The electron gas in the low-density limit represents an extreme example of the breakdown of the RPA. The electrons will be found in a periodic array in the sea of positive charge. Hence a typical term which in the high-density limit may be assumed small compared to n because of the RPA,

$$\rho_k=\sum_i e^{-i\mathbf{k}\cdot\mathbf{x}_i}, \quad (k\neq 0)$$

will assume the value n whenever $\mathbf{k}$ is equal to the reciprocal lattice vector $\mathbf{K}$. As another indication of the quite different physical behavior in the low-density limit, the series expansion for the correlation energy assumes the form, (2),

$$E_c=\left(\frac{U}{r_s}+\frac{V}{r_s^{\frac{3}{2}}}+\frac{W}{r_s^2}+\cdots\right)\text{ ry.}$$

The first term represents the difference between the potential energy of the electrons on fixed lattice sites and the exchange energy. The second arises from the zero-point oscillations of the electrons about their equilibrium positions. The third comes from higher order terms in the expansion of the electronic vibrations about their equilibrium positions.

The constant U may be calculated by carrying out an Ewald sum for the assumed lattice structure. If one is not concerned with the actual lattice structure, one may calculate U by making the Wigner-Seitz approximation of replacing the actual unit cell by a sphere. A simple electrostatic calculation then yields[9]

$$U=-0.088\text{ ry.} \tag{17}$$

A precise determination of the constant V requires consideration of the spectrum of the oscillations of the electrons, a spectrum which depends upon the particular form of the lattice one assumes.[17] We may, however, place upper and lower bounds on V in the following way. The sum of the squares of the phonon frequencies satisfies the relation[17,18]:

$$\sum_\mu \Omega_k^2(\mu)=\omega_p^2, \tag{18}$$

where ω_p is the plasma frequency,

$$\omega_p=(4\pi ne^2/m)^{\frac{1}{2}}. \tag{19}$$

If one takes an Einstein model for the oscillations, as did Wigner, one finds one longitudinal and two transverse oscillations at $\omega_p/\sqrt{3}$ for each value of k. The constant V would then be 3. The alternative extreme, also consistent with the sum rule (18), would be to assume only longitudinal phonons (plasmons) with a constant frequency ω_p. In the latter case one finds $V=1.73$. We therefore have

$$1.73\text{ ry}\lesssim V\lesssim 3\text{ ry.} \tag{20}$$

The constant W has not yet been evaluated.

For what values of r_s may we expect the energy to be well represented by a power series expansion of the form (2)? As we start from the low-density side, and reduce r_s, we may expect to reach a value of r_s for which the electronic solid will "melt," in that the electrons will no longer be bound in their equilibrium positions. This melting as a result of increasing pressure could take place at absolute zero; it is therefore not clear whether the phase transition would in fact be a sharp one. We may estimate the density at which it occurs in the following way.

In ordinary solids melting may be regarded as arising from the increase with temperature of the vibrational amplitudes of the atoms oscillating about their equilibrium positions. This is the underlying physical basis of the notably successful Lindemann melting-point formula.[19] The Lindemann formula may be interpreted to state that any solid will melt when the mean vibrational amplitude $\langle\delta R^2\rangle_{\text{Av}}^{\frac{1}{2}}$ reaches a certain critical fraction of the interatomic spacing R_0,[20]

$$(\langle\delta R^2\rangle_{\text{Av}}^{\frac{1}{2}}/R_0)=\delta. \tag{21}$$

The constant δ varies somewhat from one solid to another but is of the order of $\frac{1}{4}$ for most simple lattice types. For our electronic solid at $T=0$, $\langle\delta R^2\rangle_{\text{Av}}$ is determined solely by the zero-point vibrations of the electrons. We underestimate their efficacy if we assume only longitudinal phonons at a frequency ω_p. We then have

$$\frac{\langle\delta R^2\rangle_{\text{Av}}^{\frac{1}{2}}}{R_0}=\left(\frac{\hbar}{2m\omega_p}\right)^{\frac{1}{2}}\frac{1}{r_s}=\left(\frac{1}{12r_s}\right)^{\frac{1}{2}},$$

and we expect that below $r_s\sim 20$ the electronic solid will not be stable.

Of course, even after the electronic solid has been transformed to an electronic liquid, there may still be a

[17] An investigation of the phonon modes for simple lattice structures has been carried out by W. Kohn and T. Kjeldaas (private communication).

[18] D. Pines, lecture notes in Solid State Physics, Princeton University, 1957 (unpublished).

[19] N. F. Mott and H. Jones, *The Theory of Metals and Alloys* (Clarendon Press, Oxford, 1936).

[20] See reference 18 for a further discussion of this interpretation, together with a derivation of (21).

considerable range of r_s for which the expansion (2) is approximately valid for the correlation energy. Some indication that this may be the case may be found in an approximate calculation by one of us,[9] in which the BP approach was applied to the low-density electron gas. In general, however, we tend to believe that the low-density limit is not a particularly illuminating guide to the behavior of electrons at metallic densities.

IV. CONTRIBUTIONS TO THE CORRELATION ENERGY FROM DIFFERENT REGIONS OF MOMENTUM TRANSFER

Thus far we have concentrated on the r_s behavior of the correlation energy in various limiting cases. We have seen on the basis of our qualitative discussion in Sec. II that the concept of the contribution to the correlation energy arising from a given set of momentum transfers is particularly useful in assessing the accuracy of a given method for computing the correlation energy. We now study this aspect of the problem in some detail. As we have remarked, it is in principle possible to evaluate separately the contribution, $E_c(k)$, to the correlation energy from each interaction momentum k of the Coulomb interaction energy. We shall find it convenient to express our results in terms of the long-range correlation energy

$$E_c^{\mathrm{lr}}(\beta)=\sum_{k<\beta k_0} E(k), \tag{22a}$$

and the short-range correlation energy

$$E_c^{\mathrm{sr}}(\beta)=\sum_{k>\beta k_0} E(k). \tag{22b}$$

We begin our study with a consideration of the long-range correlation energy. Since the RPA breaks down as one goes to larger values of k, it should be possible to study the onset of the breakdown by considering the way in which exchange diagrams affect the long-range correlation energy. Within the RPA, as one of us has shown,[9] the long-range correlation energy may be expressed as a simple series expansion in β. The following terms were explicitly calculated:

$$E_c^{\mathrm{lr}}(\beta)=a\frac{\beta^2}{r_s}+c\frac{\beta^3}{r_s^{\frac{1}{2}}}+d\frac{\beta^4}{r_s^2}+d'\frac{\beta^4}{r_s} +e\frac{\beta^5}{r_s^{\frac{5}{2}}}+f'\frac{\beta^6}{r_s^2}+g'\frac{\beta^7}{r_s^{\frac{5}{2}}}\ \mathrm{ry}. \tag{23}$$

The effective limiting value of β is proportional to $r_s^{\frac{1}{2}}$, so that the unprimed terms contribute to the constant C in the GB expansion, (1), while the primed terms contribute to E. In going beyond the RPA we shall consider two of the lowest order (in β) contributions of the exchange diagrams: the alteration of d' and a new term

$$e'\beta^5/r_s^{\frac{3}{2}} \tag{24}$$

so introduced into $E_c^{\mathrm{lr}}(\beta)$. A comparison of d and d', e and e', then enables us to estimate the values of r_s for which $E_c^{\mathrm{lr}}(\beta)$ may be reliably calculated within the RPA.

We shall make use of the collective description of Bohm and Pines, which is well-suited to the present investigation. Bohm and Pines did not work directly with the Hamiltonian, (4). Instead they developed a method designed to take advantage of the fact that at long wavelengths the electron interactions give rise to organized oscillations of the electron system as a whole, the plasma oscillations. The plasmons (the quantized modes of plasma oscillation) are the dominant low momentum elementary excitations of the electron gas; they possess a minimum frequency which is the classical frequency of plasma oscillation, ω_p. BP describe the plasmons explicitly in terms of a suitable set of field coordinates. This is done by introducing a Hamiltonian which is equivalent to (4) and which corresponds to a collection of plasmons of wave-vector less than some βk_0 interacting with the electron system. In this Hamiltonian the density fluctuations with $k<\beta k_0$ no longer appear; their effect resides entirely in the plasmon coordinates. The contribution to the system energy coming from the plasmons and their interaction with the electrons is then evaluated within certain well-defined approximations described below.

The equivalent Hamiltonian is given by

$$H=\sum_i \frac{p_i^2}{2m}+\sum_{k<\beta k_0}\left\{\frac{\pi_k^*\pi_k+\omega_p^2 Q_k^* Q_k}{2}-\frac{2\pi n e^2}{k^2}\right\} +H_{\mathrm{int}}+U+H_{\mathrm{sr}}, \tag{25}$$

where π_k and Q_k are the momentum and coordinate of a plasmon of wave-vector $\mathbf{k}$.[21] In (25)

$$H_{\mathrm{int}}=-i\sum_{i,\,k<\beta k_0}\left(\frac{4\pi e^2}{k^2}\right)^{\frac{1}{2}}\left(\frac{\mathbf{k}\cdot\mathbf{p}_i}{m}+\frac{\hbar k^2}{2m}\right)Q_k e^{-i\mathbf{k}\cdot\mathbf{x}_i}, \tag{26a}$$

and describes a linear coupling between the plasmons and the electrons;

$$U=\sum_{\substack{i,\,k<\beta k_0\\ l<\beta k_0\\ k\neq l}}\frac{2\pi e^2(\mathbf{k}\cdot\mathbf{l})^2}{k^2l^2m}Q_kQ_l\exp[i(\mathbf{k}-\mathbf{l})\cdot\mathbf{x}_i], \tag{26b}$$

and describes a nonlinear coupling between plasmons and electrons;

$$H_{\mathrm{sr}}=\sum_{\substack{k>\beta k_0\\ i\neq j}}\frac{2\pi e^2}{k^2}\exp[i\mathbf{k}\cdot(\mathbf{x}_i-\mathbf{x}_j)], \tag{26c}$$

and describes the short-range part of the Coulomb interaction.

Within the RPA, the long-range correlation energy is obtained by solving the reduced Hamiltonian given by the first three terms of (25). The effect of U and H_{sr} on the low momentum transfer part of the Hamiltonian

[21] For a detailed discussion of this equivalence, and of the role played by the BP subsidiary conditions (which we here omit) we refer the interested reader to Bohm, Huang, and Pines, Phys. Rev. **107**, 71 (1957).

gives rise to corrections to the RPA. There exists a simple series expansion (23) for $E_c^{\rm lr}(\beta)$ because the reduced Hamiltonian has a simple structure for long wavelengths. The coupling between free electrons and free plasmons induced by $H_{\rm int}$ is weak, and may be treated by means of a systematic perturbation series expansion. The coupling constant which measures the strength of the interaction between a plasmon of momentum $\hbar\mathbf{k}$ and an electron of momentum $\mathbf{p}$ is[22]

$$g_k^2(p)=\frac{(\mathbf{k}\cdot\mathbf{p}+\frac{1}{2}k^2\hbar)^2}{m^2\omega_p^2}=\frac{3\pi}{4\alpha r_s k_0^4}[(\mathbf{k}\cdot\mathbf{p})^2+\tfrac{1}{4}\hbar k^4], \quad (27)$$

where

$$\alpha=(4/9\pi)^{\frac{1}{3}}.$$

The expansion (23) is a simple expansion in powers of $g_k^2(p)$, suitably averaged over electron momentum $\mathbf{p}$.

The leading terms, which are of zeroth order in g_k^2, are obtained by neglecting $H_{\rm int}$ entirely. The ground-state energy is then given by the difference between the zero-point vibrational energy of the plasmons and the self-energy of the charge distribution they have replaced,

$$\sum_{k<\beta k_0}\left(\frac{\hbar\omega_p}{2}-\frac{2\pi Ne^2}{k^2}\right). \quad (28)$$

To get the correlation energy, subtract from (28) the exchange energy associated with the interaction momenta $k<\beta k_0$; (28) then becomes

$$\sum_{k<k_c}\left(\frac{\hbar\omega_p}{2}-\frac{2\pi e^2}{k^2}\langle\rho_k{}^*\rho_k\rangle_{\rm Av}\right), \quad (29)$$

where the second term is now the sum of the exchange and self-energies of the charge distribution which has been described by the plasmons. From (29), we find

$$E_c^{\rm lr}=\left[-\frac{3\beta^2}{2\pi\alpha r_s}+\frac{\sqrt{3}}{2}\frac{\beta^3}{r_s^{\frac{3}{2}}}+\left(\frac{3}{96\pi\alpha}\right)\frac{\beta^4}{r_s^2}\right]\text{ry},$$

or

$$E_c^{\rm lr}\cong\left(-0.458\frac{\beta^2}{r_s}+0.866\frac{\beta^3}{r_s^{\frac{3}{2}}}+0.019\frac{\beta^4}{r_s}\right)\text{ ry}. \quad (30)$$

The next terms in (23) are of order g_k^2. They may be obtained directly by transforming from (25) to a Hamiltonian in which there is no longer any coupling between the electrons and the plasmons to this order. The canonical transformation required is given in BP III; the new Hamiltonian is

$$H=\sum_i\frac{p_i^2}{2m}+\sum_{k<\beta k_0}\left\{\frac{P_k{}^*P_k+\omega^2Q_k{}^*Q_k}{2}-\frac{2\pi ne^2}{k^2}\right\}+H_{\rm rp}+H_{\rm sr}. \quad (31)$$

The plasmon frequency has been modified to ω, and is defined through the dispersion relation

$$1=\frac{4\pi e^2}{m}\sum_{p<k_0}\left\{\frac{1}{\left[\left(\omega-\frac{\mathbf{k}\cdot\mathbf{p}}{m}\right)^2-\frac{\hbar^2k^4}{4m^2}\right]}\right\}, \quad (32)$$

which reads, to order g_k^2,

$$\omega^2=\omega_p^2+\sum_{p<k_0}\left[3\left(\frac{\mathbf{k}\cdot\mathbf{p}}{m\omega_p}\right)^2+\frac{\hbar^2k^4}{4m^2\omega_p^2}\right]. \quad (33)$$

The electrons interact through a weak long-range interaction

$$H_{\rm rp}=\sum_{\substack{ij\\k<\beta k_0}}\frac{\mathbf{k}\cdot(\mathbf{p}_i-\frac{1}{2}\hbar\mathbf{k})\mathbf{k}\cdot(\mathbf{p}_j+\frac{1}{2}\hbar\mathbf{k})}{2k^2mn}\times\exp[i\mathbf{k}\cdot(\mathbf{x}_i-\mathbf{x}_j)]. \quad (34)$$

The results (33) and (34) confirm the identification of g_k^2 as the coupling constant. The relative shift in plasmon frequency is of order g_k^2; $H_{\rm rp}$ represents an interaction between electrons which is reduced in effectiveness by one order of g_k^2 from the original Coulomb interaction.

The terms d, e', f', and g' in the series expansion (23) arise from the interaction $H_{\rm rp}$ and from the correction to the plasmon zero-point energy, $\hbar(\omega-\omega_p)/2$. The latter yields

$$e=0.70\text{ ry};\quad g'=0.21\text{ ry}. \quad (35)$$

The contribution made by $H_{\rm rp}$ to the correlation energy was estimated in SSP by calculating the expectation value, $\langle H_{\rm rp}\rangle_{\rm Av}$, that is, the exchange energy associated with $H_{\rm rp}$. One finds then

$$d=-\frac{9}{64\alpha^2}\approx0.517\text{ ry};\quad f'=0.058\text{ ry}. \quad (36)$$

There are, however, contributions of order β^4/r_s^2 and β^6/r_s^2 arising from all higher-order terms in a perturbation series expansion of the energy associated with $H_{\rm rp}$. The contribution these make to d is simply calculable by means of the GB summation technique. We consider only the correction to d; the calculation is given in Appendix I. The result, combined with (36), yields for d the value

$$d=0.98\text{ ry}. \quad (37)$$

We see that the series expansion (23) contains two kinds of terms. The first, corresponding to odd powers of β, and involving inverse powers of $r_s^{\frac{1}{2}}$, arise from the zero-point energy of the plasmons. The plasmon frequency ω may in general be expanded as

$$\omega=\omega_p+\frac{k^2}{m}E_1+k^4\left(\frac{E_2}{m}\right)^2+\cdots. \quad (38)$$

The β^3 term arises from ω_p; the β^5, β^7, etc. terms arise

[22] The coupling constant introduced in BP III and SSP is simply the average of g_k^2 over plasmon and electron momenta, $g^2\cong\beta^2/2r_s$.

from the k dependence of ω. The second class of terms, those even in β and involving inverse powers of r_s, arise from the interaction between the individual particles. The β^2 term comes from the long-range part of the exchange energy, as does part of the β^4/r_s term; there are contributions of order $\beta^4/r_s^2+\beta^6/r_s^2$ from H_{rp}; there will be a β^6/r_s^3 term when one considers the screened particle interaction to one higher order in g_k^2, etc.

The power series expansion, (23), may also be obtained from the GB and Hubbard expressions for the correlation energy. The derivation of the coefficients through β^5 within the RPA (and of the β^6/r_s^3 term) using the GB method may be found in Appendix I, where the odd and even terms in β are also seen to possess a different origin. The coefficient d obtained by the GB method is shown there to be identical with that calculated using the BP collective description. There has been little doubt as to the validity of the BP approach for plasmons; this identity verifies the validity of that approach for the interaction between individual particles, and the contribution arising thereby to the correlation energy: we thus obtain an explicit justification within the RPA of the BP neglect of subsidiary conditions in the calculation of the ground-state energy.

What is the radius of convergence of the series expansion (23)? An upper limit may be obtained by setting the coupling constant for the most strongly coupled plasmon equal to unity. If we neglect the k^4 term in (27), a proper procedure at both high and metallic densities, we then find

$$\frac{3\pi}{4\alpha r_s}\beta_{\max}^2=1;\quad \beta_{\max}=0.47r_s^{\frac{1}{2}}. \tag{39}$$

The choice (39) represents the wave-vector beyond which it is certainly not proper to regard the plasmon as an elementary excitation of the system. It likewise indicates the maximum wave-vector for which it is useful to use the BP approach, since the latter has as its basis the simplicity introduced in the problem by working explicitly with the plasmons in the region of wavelengths for which they are a well-defined elementary excitation. This cutoff for the plasmons essentially agrees with that proposed by SB and by Ferrell,[23] that the cutoff occurs for that value of k for which the plasmon spectrum merges with the individual electron continuum.[24] If one seeks to represent the long-range correlation energy accurately by the first few terms of the series (23), it may well be advisable to choose a somewhat smaller value of β. In any event, it is clear from these considerations that at high densities the long-range part of the correlation energy will not approximate the total correlation energy well, since the latter contains important contributions arising from momentum transfers of the order of the Fermi-Thomas wave-vector, $0.814r_s^{\frac{1}{2}}$.

We now consider the corrections to (23) arising from exchange diagrams. In the collective approach these arise from two kinds of terms; those involving U, which correspond to the coupling between different long-wavelength density fluctuations, and those involving the coupling between $H_{\rm int}$ and $H_{\rm sr}$ which is a coupling between the low and high momentum transfer part of the Coulomb interaction. The latter appear at a lower order in β, and we calculate them first.

The lowest order in β influence of exchange diagrams on the individual particle contribution to the correlation energy is of order β^4/r_s. We may estimate this term by using the Hamiltonian (31); there will be a contribution arising from the second-order exchange diagram, involving H_{rp} once and H_{sr} once. This is

$$\frac{9}{64\pi^4}\frac{1}{\alpha r_s}\frac{\int d^3k\int d^3p\int d^3q(\mathbf{k}\cdot\mathbf{p})(\mathbf{k}\cdot\mathbf{q})}{[\mathbf{k}\cdot(\mathbf{k}+\mathbf{p}+\mathbf{q})](\mathbf{k}+\mathbf{p}+\mathbf{q})^2k^2}\text{ ry}, \tag{40}$$

where the regions of integration for $\mathbf{p}$ and $\mathbf{q}$ are subject to the usual limits imposed by the Pauli principle, as for Eq. (10), and that for $\mathbf{k}$ is given by $k<\beta$.[25] On dropping terms which contribute to higher order than β^4, we have

$$\Delta d'=\frac{9}{64\pi^2}\frac{1}{\alpha r_s}\frac{\int d^3k\int d^3p\int d^3q(\mathbf{k}\cdot\mathbf{p})(\mathbf{k}\cdot\mathbf{q})}{k^2\mathbf{k}\cdot(\mathbf{p}+\mathbf{q})(\mathbf{p}+\mathbf{q})^2}\text{ ry}, \tag{41}$$

which may be estimated rather well by[26]

$$\approx\frac{9}{64\pi^2}\frac{1}{\alpha r_s}\frac{\int d^3k\int d^3p\int d^3q(\mathbf{k}\cdot\mathbf{p})(\mathbf{k}\cdot\mathbf{q})}{2k^2\mathbf{k}\cdot(\mathbf{p}+\mathbf{q})}\text{ ry}.$$

We thus find

$$\Delta d'=\frac{9}{160\pi}\frac{\beta^4}{\alpha r_s}\{2\ln 2-1\}\approx 0.0136\frac{\beta^4}{r_s}\text{ ry}. \tag{42}$$

The exchange terms (40) and (42) represent a "screened" version of the long-range part of $E_b^{(2)}$. The latter is given by

$$E_b^{(2)}=\frac{3}{16\pi^5}\frac{\int d^3k\int d^3p\int d^3q}{k^2(\mathbf{k}+\mathbf{p}+\mathbf{q})^2\mathbf{k}\cdot(\mathbf{k}+\mathbf{p}+\mathbf{q})}\cong 0.015\beta^2\text{ ry}, \tag{43}$$

where the limits are the same as in (40) and the nu-

[23] R. Ferrell, Phys. Rev. **107**, 450 (1957).

[24] In fact, we would get just their criterion if we had defined $g_k^2=\omega_{n0}^2(kp)/\omega^2$, $\omega_{n0}(kp)$ being the free-electron excitation frequency, instead of our choice, $g_k^2=\omega_{n0}^2(p,k)/\omega_p^2$. We believe the latter choice is better suited to the determination of the radius of convergence of the power series expansion of $E_c^{\rm lr}(\beta)$, and also represents a realistic choice for the maximum plasmon wave-vector.

[25] In the remainder of this section we measure momenta in units of k_0.

[26] A similar estimate for the low momentum part of $E_b^{(2)}$ may be shown to be accurate to $\sim 5\%$.

merical result has been obtained from the Monte-Carlo calculations carried out for Gell-Mann and Brueckner.[27] In (40) we have taken into account the screening of the long-range interactions between the electrons, which reduces the exchange contribution (43) by a factor of order g_k^2.

The lowest order exchange correction to the plasmon frequency is of order k^2, and hence contributes to the β^5 coefficient in the long-range correlation energy. It appears when we take into account the cross-terms between H_{int} and H_{sr} which arise, for instance, when we decouple the electrons from the plasmons to order g_k^2 by means of a canonical transformation. We estimate one such term, which involves only one power of H_{sr}, in Appendix II. We find that the e' of Eq. (24) is given approximately by

$$e'=0.039 \text{ ry}. \tag{44}$$

The contribution to the correlation energy from U is simply estimated using second-order perturbation theory, and is approximately[28]

$$g''\frac{\beta^7}{r_s^{\frac{3}{2}}}=-\frac{\sqrt{3}}{96}\frac{\beta^7}{r_s^{\frac{3}{2}}}\sim-0.018\frac{\beta^7}{r_s^{\frac{3}{2}}} \text{ ry}. \tag{45}$$

U thus first gives rise to a correction to the plasmon frequency of order k^4; its importance may be assessed by comparing (39) with the contribution from the two k^4 terms which arise within the random phase approximation[23]

$$g\frac{\beta^7}{r_s^{7/2}}+g'\frac{\beta^7}{r_s^{\frac{3}{2}}}=\left(0.21\frac{\beta^7}{r_s^{\frac{3}{2}}}-0.084\frac{\beta^7}{r_s^{7/2}}\right) \text{ ry}. \tag{46}$$

We may make the following remarks on the basis of our investigation of the role of the exchange diagrams:

(1) There are no corrections to the β^2 and β^3 terms of the series expansion (23). Hence for sufficiently long wavelengths, the RPA is valid *for any density* in the calculation of the long-range correlation energy.

(2) The β^4 and β^5 terms in the expansion of the long-range correlation energy may be written, approximately, as

$$-\beta^4\left(+\frac{0.98}{r_s^2}-\frac{0.033}{r_s}+\cdots\right) \text{ ry}, \tag{47a}$$

$$\frac{\beta^5}{r_s^{\frac{1}{2}}}\left(\frac{0.70}{r_s^2}-\frac{0.039}{r_s}+\cdots\right) \text{ ry}. \tag{47b}$$

The terms in parentheses of order $1/r_s$ represent the contribution from exchange diagrams involving H_{sr} once as well as the β^4/r_s term from (30). They are negligible in the high-density limit, as they should be. The corrections are numerically small for actual metallic densities. We do not at present know whether the higher order terms in the series (in β) display a similar behavior.

It is clear that where E_{lr}^c is accurately calculated by keeping terms only through β^4 or β^5, say, the use of the RPA result for these terms should provide a quite adequate approximation.

(3) On the basis of the structure of the exchange terms involving one power of H_{sr} in (47), it is tempting to conjecture that the result of taking into account all RPA corrections would be to yield an expression β^4 which is

$$\beta^4\left(\frac{s}{r_s^2}+\frac{t}{r_s}+u\ln r_s+v+wr_s\ln r_s+\cdots\right).$$

Thus the coefficients of β^4, $\beta^5/r_s^{\frac{1}{2}}$, etc., themselves may possess a power series expansion in r_s analogous to that for the electron energy in the high-density limit. We have not verified this conjecture in detail. If it is true, it is clear already in (47) that the coefficients of that expansion will not be the same as in the system energy; hence it might be possible to use the β^4, β^5, etc., contributions to the correlation energy with some confidence, even though the series expansion (2) does not apply. However, the question is by no means settled, and merely points up the difficulty of drawing definitive conclusions from our calculation of the first set of exchange corrections.

We conclude this section by considering $E^{\text{sr}}(\beta)$ briefly. We have given physical arguments in Sec. II that for a sufficiently large β this part of the correlation energy arises almost entirely from the interaction between electrons of antiparallel spin. We may obtain an idea of how well this hypothesis is born out by studying the contribution to $E_c^{\text{sr}}(\beta)$ arising from electrons of parallel spin, as calculated in second-order perturbation theory. This is

$$E_{11}^{(2)}=-\frac{3}{16\pi^5}\frac{\int d\mathbf{k}\int d\mathbf{p}\int d\mathbf{q}}{\mathbf{k}\cdot(\mathbf{k}+\mathbf{p}+\mathbf{q})}\times\left\{\frac{1}{k^4}-\frac{1}{k^2(\mathbf{k}+\mathbf{p}+\mathbf{q})^2}\right\} \text{ ry}. \tag{48}$$

The momenta here are measured in units of k_0 and the limits of integration are as in (10); the first term in the parenthesis is the direct contribution, the second is the exchange contribution. For large momentum transfers, ($k>1$), the regions of $\mathbf{p}$ and $\mathbf{q}$ which contribute to $E_{11}^{(2)}$ become small compared to $\mathbf{k}$, and the exchange contribution tends to cancel the direct term. The integration (48) may be carried out explicitly for $k>\beta$; one finds, with the aid of (43) and Eq. (6.22) of SSP,

$$E_{11}^{(2)}\approx(+0.021+0.062\ln\beta-0.021\beta^2) \text{ ry}. \tag{49}$$

The result (49) is valid for $\beta\lesssim\frac{3}{4}$, the limitation being imposed by the accuracy of (43); for larger values of β

[27] We should like to thank Professor M. Gell-Mann for making this result available to us.

[28] We have corrected the value given in BP III by a factor of 2, and estimated the reduction in the sum over states due to the Pauli principle by the approximate value β.

it is an underestimate. The values of $E_{11}^{(2)}(\beta)$ calculated from (49) tend to support our physical argument concerning the short-range correlation energy associated with electrons of parallel spin.

We have been discussing the breakdown of the RPA for $E_c^{sr}(\beta)$ due to the coupling between different short wavelength momentum transfers. We have also considered the breakdown of the RPA for $E_c^{lr}(\beta)$ due to the contributions arising from H_{sr}. We may mention that there will also be contributions to $E_c^{sr}(\beta)$ arising from the long-wavelength momentum transfers. We do not consider these corrections here because $E_c^{sr}(\beta)$ is itself so imperfectly known once one leaves the high-density limit.

V. INTERPOLATION PROCEDURES FOR ACTUAL METALLIC DENSITIES

On the basis of our qualitative and quantitative discussion of the preceding section we have reached the following conclusions.

(1) It does not appear feasible to extend systematically the GB scheme into the region of actual metallic densities. Therefore it is necessary to consider the possibility of developing suitable interpolation procedures to calculate the correlation energy at metallic densities.

(2) The behavior of the correlation energy at a given r_s as a function of interaction momentum, $E_c(k)$, provides a natural basis for the development of interpolation procedures. In the low momentum region, the RPA calculation of $E_c(k)$ is rigorous for the k and k^2 terms, and provides quite a good approximation for the k^3 and k^4 terms throughout the region of actual metallic densities. The maximum value of β for which $E_c(k)$ may be reliably calculated in the RPA is $\beta=0.47r_s^{\frac{1}{2}}$, for which $g_k^2\cong 1$. For sufficiently large k, the exchange diagrams cancel one-half the direct diagrams, so that one need consider only interactions between electrons of antiparallel spin. This region begins at approximately $\beta=1$.

One possible interpolation scheme would involve plotting $E_c(k)$ for all k according to the following prescription: (1) for $k<\beta_1$, take the value given by the RPA; β_1 should then be $\lesssim 0.47r_s^{\frac{1}{2}}$; (2) for $k>\beta_2$, take the value given by second-order perturbation theory, in which only interactions between electrons of antiparallel spin are considered; β_2 should be $\gtrsim 1.5$; and (3) for $\beta_1\lesssim k\lesssim\beta_2$, draw a smooth curve between the portions defined by (1) and (2). To get the correlation energy for a given value of r_s, one would then carry out a numerical integration of $E_c(k)$ over all interaction momenta k.

A somewhat simpler scheme, which yields an explicit formula for the correlation energy as a function of r_s, and very nearly the same numerical values as the above procedure, is based on the following expressions for the long-range and short-range parts of the correlation energy:

$$E_c^{lr}(\beta_1)\cong\left(-0.458\frac{\beta_1^2}{r_s}+0.866\frac{\beta_1^3}{r_s^{\frac{3}{2}}}-0.98\frac{\beta_1^4}{r_s^2}+0.706\frac{\beta_1^5}{r_s^{\frac{5}{2}}}\right)\text{ ry}, \tag{50}$$

$$E_c^{sr}(\beta_2)\cong(-0.025+0.062\ln\beta_2-0.006\beta_2^2)\text{ ry}. \tag{51}$$

The long-range part of the correlation energy, $E_c^{lr}(\beta_1)$, is calculated in the RPA. Only terms through β_1^5 are kept, because for actual metallic densities the β_1^6 terms are comparable with the exchange corrections to the β_1^4 term, and (50) may be regarded as accurate only as long as both such terms are small. The short-range part of the correlation energy, $E_c^{sr}(\beta_2)$, is calculated in second-order perturbation theory, and only interactions between electrons of antiparallel spin are kept. If we now choose $\beta_1=\beta_2=\beta$, it is clear that there will be a discontinuity in $E_c(k)$ at $k=\beta$, because both expressions (50) and (51) are approximations; however for a suitable choice of β this will not introduce any appreciable error. It is desirable to choose β as large as possible, in order that (51) represent a tolerably accurate approximation; on the other hand, if β is taken as too large, the long-range correlation energy will no longer be well represented by (50). We accordingly take for β that maximum value for which (50) is still a rapidly converging series,

$$\beta=0.47r_s^{\frac{1}{2}}. \tag{52}$$

With this choice, we find

$$E_c^{lr}(\beta)\cong -0.043\text{ ry}, \tag{53}$$

$$E_c^{sr}(\beta)\cong(-0.072+0.031\ln r_s)\text{ ry}. \tag{54}$$

We have dropped the β^2 term in $E_c^{sr}(\beta)$ since it is of the same order as the β^4/r_s exchange term. The total correlation energy is given by the sum of (53) and (54),

$$E_c\cong(-0.115+0.031\ln r_s)\text{ ry}. \tag{55}$$

The foregoing interpolation scheme is very nearly the same as that proposed earlier by one of us in SSP. The present proposal differs in that we have used the correct RPA coefficient of the β^4 term in $E_c^{lr}(\beta)$, instead of the lowest order approximation to it utilized in SSP. This has enabled us to go to a larger value of β.

Hubbard has proposed an alternative interpolation procedure, which is equivalent to multiplying the RPA value of $E_c(k)$ by a phenomenological correction factor, which is 1 for small k and $\frac{1}{2}$ for large k. This procedure is obviously satisfactory for large k. Detailed investigation shows that his procedure does not yield the β^4/r_s exchange correction, so that for small k Hubbard's result cannot be regarded as any more accurate than the RPA result.

In Table I we compare our numerical results for the correlation energy with those obtained by Hubbard. In Fig. 1 we plot $E_c(k)$ for $r_s=4$ as calculated by us and by

TABLE I. A comparison between the revised BP calculation of the correlation energy and that of Hubbard; the energies are given in ry.

r_s	2	3	4	5
E_{BP}	−0.094	−0.081	−0.072	−0.065
E_{Hubbard}	−0.099	−0.086	−0.074	−0.067

Hubbard. The close agreement of the two interpolation procedures is not surprising, since they agree in both the high and low momentum limits.

It is difficult to set any exact limit on the accuracy of either interpolation procedure. We can estimate rather well the accuracy of $E_c^{\mathrm{lr}}(\beta)$ in our scheme, but we do not know the accuracy of our perturbation theory expression for $E_c^{\mathrm{sr}}(\beta)$. Hubbard's interpolation procedure is sufficiently close to ours that it suffers from the same defect, an inadequate knowledge of the behavior of $E_c(k)$ in the region between $\beta=0.47r_s^{\frac{1}{2}}$ and that value of β for which the perturbation theory calculation of $E_c^{\mathrm{sr}}(\beta)$ becomes accurate. It is perhaps encouraging that there is such good agreement between Hubbard's estimate of $E_c(k)$ and the perturbation theory calculation for $\beta\gtrsim 1$; however both could be somewhat inaccurate in the region of $\beta\cong 1$. If one is content with 15% accuracy in the correlation energy, it is quite likely that both schemes may be regarded as satisfactory. On the other hand, if one demands a value of the correlation energy which is accurate to within 5%, it is doubtful that either scheme is that accurate except, perhaps, through a fortunate cancellation of rather larger errors.

We conclude that the Hubbard scheme is mathematically more satisfying, since it yields a smooth curve for $E_c(k)$. On the other hand, our modified SSP interpolation procedure is simpler, and rather closely tied to the physical behavior of the electron gas. It may also be more flexible, in that it permits straightforward calculation of other metallic properties, such as the one-electron energies, specific heats, etc.[29]

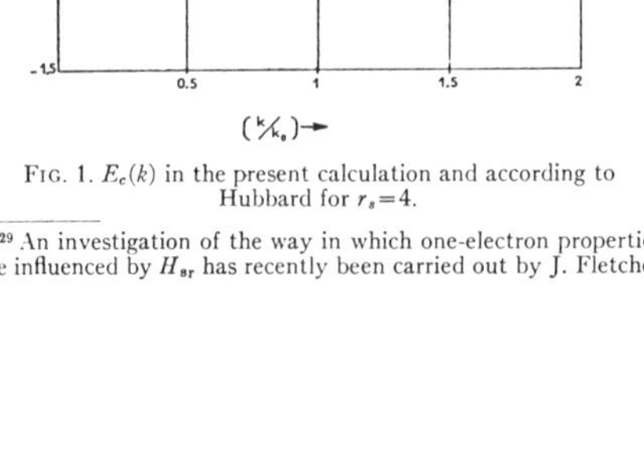

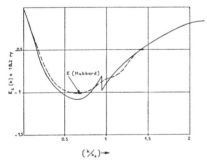

FIG. 1. $E_c(k)$ in the present calculation and according to Hubbard for $r_s=4$.

[29] An investigation of the way in which one-electron properties are influenced by H_{sr} has recently been carried out by J. Fletcher and D. C. Larson, following paper [Phys. Rev. **110**, 455 (1958)]. They have considered only the second-order interaction between electrons of antiparallel spin, and obtain thereby not unreasonable values for the specific heat, etc.

APPENDIX I

The GB result for the correlation energy may be written as

$$E_{\mathrm{corr}}=-\frac{3}{4\pi\alpha^2 r_s^2}\int k^3 dk \times\int_{-\infty}^{+\infty}du\sum_{n=}^{n=\infty}\left\{\frac{(-1)^n}{n}[Q_k(u)]^n\right\},\quad \text{(A1)}$$

where the function $Q_k(u)$ is given by

$$Q_k(u)=\frac{\alpha r_s}{\pi^2 k^2}\int_{\substack{|p|<1\\|\mathbf{p}+\mathbf{k}|>1}} d^3p \times\int_{-\infty}^{+\infty}\exp\{ituk-|t|(\tfrac{1}{2}k^2+\mathbf{k}\cdot\mathbf{p})\}dt.\quad \text{(A2)}$$

The total change in energy due to the Coulomb interaction, $E_{\mathrm{tot}}=E_{\mathrm{corr}}+E_{\mathrm{exch}}$, is given by

$$E_{\mathrm{tot}}=-\frac{3}{4\pi\alpha^2 r_s^2}\int_0^\infty dk\, k^3\int_{-\infty}^{+\infty}du \times\sum_{n=1}^{\infty}\frac{(-1)^n}{n}[Q_k(u)]^n-\sum_k\left(\frac{2\pi e^2}{k^2}\right).\quad \text{(A3)}$$

The first term on the right-hand side of (A3) may therefore be viewed as the energy change brought about by a perturbation

$$\sum_k\left(\frac{2\pi e^2}{k^2}\rho_k\rho_{-k}\right).\quad \text{(A4)}$$

The GB result is now in a form well suited for its application to the calculation of the long-range correlation energy arising from H_{rp}, E_{rp}. To lowest order in k^2, we may write

$$H_{\mathrm{rp}}=\sum_k\frac{2\pi e^2}{k^2}\rho_k{}^s\rho_{-k}{}^s,\quad \text{(A5)}$$

where $\rho_k{}^s$ is a screened density fluctuation defined by

$$\rho_k{}^s=\sum_i\left(\frac{-\mathbf{k}\cdot\mathbf{p}_i}{m\omega_p}\exp(-i\mathbf{k}\cdot\mathbf{x}_i)\right).$$

Thus (A5) and (A4) differ only in the replacement of ρ_k by $\rho_k{}^s$. E_{rp} is therefore given by

$$E_{\mathrm{rp}}=-\frac{3}{4\pi\alpha^2 r_s^2}\int_0^\beta dk\,k^3 \times\int_{-\infty}^{+\infty}du\sum_{n=1}^{\infty}\frac{(-1)^n}{n}[Q_k{}^s(u)]^n,\quad \text{(A6)}$$

where $Q_k{}^s(u)$ is the screened version of $Q_k(u)$ arising from the replacement of ρ_k in (A4) by ρ_k^s in (A5). Inspection of the matrix elements arising from (A4) and (A5) shows that the change from ρ_k to ρ_k^s introduces a multiplicative factor in the integration over p in (A2) which is simply

$$-(3\pi/4\alpha r_s)(\mathbf{k}\cdot\mathbf{p})^2,$$

that is, $-g_k^2(p)$ (to lowest order in k^2). The minus sign corresponds to the fact that H_{rp} induces attractive electron correlations.

We have therefore

$$Q_{k^s}(u)=-\frac{3}{4\pi}\int_{\substack{p<1\\|\mathbf{p}+\mathbf{k}|>1}} d^3p\,\frac{(\mathbf{k}\cdot\mathbf{p})^2}{k^2}\times\int_{-\infty}^{+\infty}dt\,\exp\left\{ituk-|t|\left(\frac{k^2}{2}+\mathbf{k}\cdot\mathbf{p}\right)\right\}$$

$$=3u^2R(u)-1, \tag{A7}$$

where

$$R(u)=1-u\arctan(1/u),$$

and we have carried out the integration in the first of Eqs. (A7) to lowest order in k. On substituting (A7) into (A6) and summing the series there, we find

$$E_{\mathrm{rp}}=\frac{3}{4\pi\alpha^2r_s^2}\int_0^\beta dk\,k^3\int_{-\infty}^{\infty}du\,\ln[3u^2R(u)]. \tag{A8}$$

The latter integral may be computed numerically, and yields the result

$$E_{\mathrm{rp}}=-0.98(\beta^4/r_s^2)\ \mathrm{ry}. \tag{A9}$$

We now show that the GB result, (A1), when restricted to sufficiently long wavelength momentum transfers β, may be expanded as a power series in β. Let us return to the expression (A2) for $Q_k(u)$. A careful examination of (A2) shows that $Q_k(u)$ may be expanded in powers of k, and contains only even powers of k. The leading contribution is

$$Q_k(u)=(4\alpha r_s/\pi k^2)R(u).$$

On substituting this value in (A1) and carrying out the sum there, we get

$$E_c^{\mathrm{lr}}(\beta)=-\frac{3}{4\pi\alpha^2r_s^2}\int_0^\beta dk\,k^3\int_{-\infty}^{\infty}du\left\{\left[\frac{4\alpha r_s}{\pi k^2}R(u)\right]-\ln\left[1+\frac{4\alpha r_s}{\pi k^2}R(u)\right]\right\}. \tag{A10}$$

In (A10) the first term in brackets yields the β^2 term of (23); we integrate the remaining term by parts, obtaining

$$-\frac{3}{4\pi\alpha^2r_s^2}\int_0^\beta k^3dk\,I(k), \tag{A11}$$

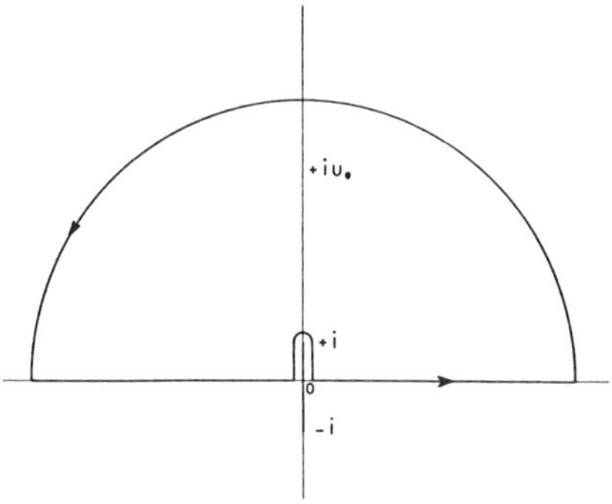

FIG. 2. The contour on which one must integrate $uR'(u)/[R(u)+\pi k^2/4\alpha r_s]$ in order to expand $I(k)$ in powers of k.

where

$$I(k)=\int_{-\infty}^{+\infty}du\,\frac{uR'(u)}{(\pi k^2/4\alpha r_s)+R(u)}. \tag{A12}$$

We wish to expand $I(k)$ in powers of k.

The function $R(u)$ is even and analytic. It has two branch points at $u=\pm i$, which we must join by a cut in the complex plane. Its only zero is at $u=\infty$, near which it behaves as

$$R(u)\sim 1/3u^2-1/5u^4\cdots.$$

For small enough k, the only poles of

$$\frac{uR'}{(\pi k^2/4\alpha r_s)+R(u)}$$

arise therefore at $\pm iu_0$, where u_0 is given by

$$u_0=i(4\alpha r_s/3\pi k^2)^{\frac{1}{2}}[1+O(k^2)].$$

The residue of the pole iu_0 is simply iu_0.

$I(k)$ may then be calculated by integrating over the contour shown on Fig. 2. We find

$$I(k)=2\pi i(iu_0)-\oint_c\frac{uR'(u)}{R(u)+\pi q^2/4\alpha r_s}, \tag{A13}$$

in which the contour c goes from -0 to i and then back to $+0$.

The first term of (A13) corresponds to the plasmon ground-state energy. It gives rise to the $\beta^3/r_s^{\frac{1}{2}}, \beta^5/r_s^{\frac{1}{2}},\cdots$ terms. The coefficients determined with (A11) agree perfectly with those obtained by the plasma theory. Remark that we have kept only the first term in the expansion of $Q_k(u)$. Since $Q_k(u)$ contains only even powers of k, such corrections will always lead to corrections to the energy odd in β, such as $\beta^7/r_s^{\frac{1}{2}},\cdots$.

The second term of (A13) corresponds to the energy of the individual particles. We can simplify it considerably by remarking that on the contour c, $R(u)$ is never small, being of order 1 or more. We may therefore expand the fraction in powers of k. The correlation energy arising from the individual particles is then given by

$$-\frac{3}{16\pi\alpha^2 r_s^2}\beta^4 \oint_c \frac{uR'(u)}{R}du + \frac{1}{32\alpha^3 r_s^3}\beta^6 \oint_c \frac{uR'}{R^2}du\cdots. \quad \text{(A14)}$$

It is easily shown that the inclusion of k^2 correction terms to $Q_k(u)$ will only introduce even powers of β.

In order to establish the equivalence of the plasma and the GB calculation of the β^4 term, we have to show that

$$\int_{-\infty}^{+\infty} du \ln[3u^2R(u)] = -\oint_c \frac{uR'(u)}{R}du. \quad \text{(A15)}$$

The integral on the left-hand side may be integrated by parts, yielding

$$-\int_{-\infty}^{+\infty}\left[2+\frac{uR'(u)}{R}\right]du.$$

Since the function $\{2+[uR'(u)/R]\}$ goes to zero as $1/u^2$ at infinity, we can again integrate on the contour of Fig. 2. The only contribution comes from the integration on the contour c. The term 2, being single valued, gives obviously zero, which establishes (A15).

We remark that it is simple to calculate the higher order terms in β by using (A14). In (A14) we give the explicit coefficient of β^6, which has been computed numerically. The contribution to the correlation energy from (A14) is thus

$$-0.98\frac{\beta^4}{r_s^2}-0.23\frac{\beta^6}{r_s^3}+\cdots \text{ ry}. \quad \text{(A16)}$$

APPENDIX II

In this appendix we discuss briefly the question of non-RPA corrections to the plasma frequency. In the BP approach such corrections occur for the following reason: when one eliminates the linear plasmon-electron coupling by a canonical transformation generated by S, the transformation acts also on H_{sr} and yields new terms coupling the plasmons with the individual particles. These new terms give rise to the extra "exchange" frequency shift of the plasmons. We shall restrict ourselves to a very limited class of such corrections, namely those of order k^2 which involve only one power of H_{sr}. Each power of H_{sr} introduces a factor e^2, i.e., a factor r_s: this discussion is therefore limited to the terms in $\beta^5/r_s^{\frac{1}{2}}$ in the correlation energy.

The canonical transformation gives rise to new terms of the form

$$\frac{i}{\hbar}[H_{sr},S]+\tfrac{1}{2}\left(\frac{i}{\hbar}\right)^2[[H_{sr},S],S]+\cdots. \quad \text{(A17)}$$

The first term of (A17) is an extra coupling term. It can be eliminated by a further canonical transformation: this would yield a frequency shift involving two powers of H_{sr}, and is therefore beyond the range of the study. The terms in which we are interested are in fact the second-order terms, or, more precisely, their expectation value with respect to the electronic wave function. Such terms are quadratic in the field coordinates and therefore represent a shift in the plasmon frequency.

If we limit ourselves to the leading order in k, the generating function of the canonical transformation performed in BP III can be written

$$S=\left(\frac{4\pi e^2}{k^2}\right)^{\frac{1}{2}}\sum_i \frac{\mathbf{k}\cdot\mathbf{p}_i}{m\omega_p^2}\exp(-i\mathbf{k}\cdot\mathbf{x}_i)\pi_{-k}.$$

It is a straightforward matter to calculate the double commutator of (A17), and to take the corresponding expectation value for a Fermi distribution of independent particles. The final result is to add to the Hamiltonian a correction term given for small k by

$$-\sum_k \frac{3}{40}\frac{k^2}{k_0^2}\pi_k^*\pi_k.$$

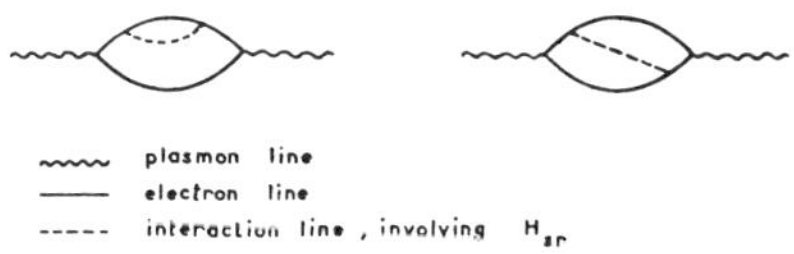

FIG. 3. The diagrams taken into account in the calculation of Appendix II.

This results in a shift of the frequency

$$\frac{\Delta\omega}{\omega_p}=-\frac{3}{40}\frac{k^2}{k_0^2}. \quad \text{(A18)}$$

The contribution to the long-range energy of the system may then be written as

$$-0.039(\beta^5/r_s^{\frac{1}{2}}) \text{ ry},$$

which is to be compared to the β^5 term occuring within the RPA

$$0.70(\beta^5/r_s^{\frac{1}{2}}) \text{ ry}.$$

It may be seen that, for actual metallic densities, the exchange correction to the plasma frequency is appreciably smaller than the RPA correction.

It is interesting to see what the foregoing procedure means in the language of diagrams. The diagrams which we take into account are those shown on Fig. 3. They involve first-order corrections to the excitation energy of the individual particles, *and* to the matrix elements for such excitations, $(\rho_k)_{0n'}$. Remark that in keeping only some of these diagrams, one can be led to a shift of the plasma frequency independent of k: such a result is obviously spurious, since it disappears when one takes consistently into account all diagrams of a given order.

A Dielectric Formulation of the Many Body Problem: Application to the Free Electron Gas.

P. NOZIÈRES and D. PINES (*)

Laboratoire de Physique, Ecole Normale Supérieure - Paris

(ricevuto il 19 Maggio 1958)

Summary. — Under suitable conditions it is shown that a generalized dielectric constant $\varepsilon(k, \Omega)$ for an arbitrary many-body problem may be defined by analogy with the macroscopic laws of electrostatics. The ground state energy may then be simply expressed in terms of $\varepsilon(k, \Omega)$ without resort to perturbation theoretic expansions. Within the random phase approximation (RPA), $\varepsilon(k, \Omega) = 1 + 4\pi\alpha(k, \Omega)$, where $\alpha(k, \Omega)$ is the complex polarizability obtained by a generalized Kramers-Heisenberg formula in the plane wave representation. For the free electron gas, this value of $\varepsilon(k, \Omega)$ leads directly to the expression for the ground state energy obtained by GELL-MANN and BRUECKNER and by HUBBARD in the high density limit. It is shown that the treatment of electron interaction within the RPA is equivalent to taking into account only the field of surface charges at the external boundary of the system; it thus corresponds to neglecting all potential and local field corrections. $\varepsilon(k, \Omega)$ may be inferred from the inelastic scattering of fast electrons or from the Compton scattering of X-rays; the circumstances are discussed under which such experiments yield information about the ground state energy of the electron gas.

---

1. – The aim of this paper is the derivation of an expression for the ground state energy and stopping power of a uniform gas of interacting particles without resort to a perturbation series expansion. This program is realized by the introduction of a generalized dielectric constant at arbitrary frequency Ω and wave vector k, $\varepsilon(k, \Omega)$, in terms of which the above quantities may be simply

(*) National Science Foundation Senior Post Doctoral Fellow on leave of absence from Princeton University, 1957-58.

expressed. The quantity $\varepsilon(k, \Omega)$ is closely related to the Fourier transform of the time dependent pair correlation function introduced by VAN HOVE (1). (In fact, our derivation of the electron and X-ray scattering cross-sections is essentially equivalent with that given in VAN HOVE's paper). For simplicity, we consider throughout this paper the case of the free electron gas; the methods employed are more general, and apply to any uniform system, of bosons or fermions which displays a linear response to a generalized test charge.

It is particularly simple to evaluate $\varepsilon(k, \Omega)$ within the random phase approximation (RPA). In the RPA, the coupling between different Fourier components of the Coulomb interaction is completely neglected. This approximation, first used for long wave lengths by BOHM and PINES (2), has recently been shown to be valid for all momentum transfers in the case of a very high density electron gas. The calculation of the ground state energy within the RPA has been performed by GELL-MANN and BRUECKNER (3), SAWADA (4), and HUBBARD (5). The present method enables us to calculate that result in very simple fashion without resort to field-theory techniques or perturbation-theory diagrams (6). We also calculate the stopping-power in the Born approximation, and establish the connection with the macroscopic approach of HUBBARD (7) and FRÖHLICH (8).

In Sect. **2**, we establish a rigorous expression for the ground state energy and stopping power in terms of the *true* dielectric constant of the system defined by analogy with the macroscopic laws of electrostatics. In Sect. **3**, we proceed to use the RPA to calculate this dielectric constant, which turns out to be simply given by the usual Kramers-Heisenberg formula (using a plane wave representation). We show how this result may arise from a perturbation series expansion. The corresponding ground state energy is identical with that given by HUBBARD. We then discuss the general behaviour of the results, and show how they may be interpreted as arising from a charge renormalization. In Sect. **4**, we discuss the physical meaning of the RPA, along the lines given by us previously. We conclude that the use of RPA is equi-

---

(1) L. VAN HOVE: *Phys. Rev.*, **95**, 249 (1954). Our approach is closely related to the treatment of the scattering and self energy of charged particles in a free electron gas given by LINDHARD: *Kgl. Danske Mat.-fys. Medd.*, **28**, 8 (1954).

(2) D. BOHM and D. PINES: *Phys. Rev.*, **92**, 609 (1953).

(3) K. A. BRUECKNER and M. GELL-MANN: *Phys. Rev.*, **106**, 364 (1957).

(4) K. SAWADA: *Phys. Rev.*, **106**, 372 (1957); K. A. BRUECKNER, N. FUKUDA, K. SAWADA and R. BROUT: *Phys. Rev.*, **108**, 507 (1957).

(5) J. HUBBARD: *Proc. Roy. Soc.*, A **240**, 539 (1957); A **243**, 336 (1958).

(6) A preliminary report on these results has been given by the authors: *Phys. Rev.*, **109**, 1009 (1958).

(7) J. HUBBARD: *Proc. Phys. Soc.*, A **68**, 441 (1955).

(8) H. FRÖLICH and H. PELZER: *Proc. Phys. Soc.*, A **68**, 525 (1955).

valent to treating only the field of surface charges at the boundary of the system. We then outline various possible corrections to the RPA to take into account potential effects and local field corrections. In Sect. **5**, we discuss the extent to which the quantities involved in our treatment may be obtained experimentally.

2. – Let us consider a gas of free electrons, N per unit volume, from a macroscopic point of view. Their density fluctuation, ϱ_k, is defined as

$$\varrho_k = \sum_i \exp\left[-ik\cdot x_i\right].$$

We now introduce an oscillating test charge of wave vector k and frequency Ω; its charge density is assumed to be

$$e\left\{r_k \exp\left[-i(\Omega t + \mathrm{k}\cdot\mathrm{r})\right] + \mathrm{c.\,c.}\right\}.$$

In the absence of the test charge, the average value of $\varrho_{k'}$, $\langle\varrho_{k'}\rangle$, is zero. When we switch on the test charge, $\langle\varrho_{k'}\rangle$ will stay zero for $k' \neq \pm k$, for reasons of translational invariance (this is no longer true if the electrons are imbedded in a periodic potential). On the contrary, $\langle\varrho_{\pm k}\rangle$ will be $\neq 0$. The ratio $\varrho_k/r_k \exp[-i\Omega t]$ may be related to the dielectric constant $\varepsilon(k, \Omega)$, with the aid of the Fourier expansion of the macroscopic Poisson equations

$$\left\{\begin{aligned} &\mathrm{k}.\, \varepsilon E_k = 4\pi e r_k \exp\left[-i\Omega t\right], \\ &\mathrm{k}.\, E_k = 4\pi e(r_k \exp\left[-i\Omega t\right] + \langle\varrho_k\rangle). \end{aligned}\right. \tag{1}$$

We have

$$\frac{\langle\varrho_k\rangle}{r_k \exp\left[-i\Omega t\right]} = \frac{1}{\varepsilon(k, \Omega)} - 1. \tag{2}$$

Obviously, macroscopic arguments will not have much physical meaning for very large k: we then may consider (2) as *a definition* of the *true* dielectric constant $\varepsilon(k, \Omega)$, which, for long wave lengths, agrees with the usual macroscopic definition.

We may write the Hamiltonian of the electrons as

$$H_0 = \sum_i \frac{p_i^2}{2m} + \sum_k \frac{2\pi e^2}{k^2}\left(\varrho_k \varrho_{-k} - N\right). \tag{3}$$

The test charges then act as a perturbation H_1 on H_0. Since there is an energy transfer, we must switch on the test charges adiabatically in order to avoid heating up the electrons. We therefore write H_1 as

$$H_1 = \frac{4\pi e^2}{k^2} (\varrho_{-k} r_k \exp[-i\Omega t] + \text{c. c.}) \exp[\eta t] , \tag{4}$$

where η is small, and will be set equal to zero at the end of our calculations. We calculate $\varepsilon(k\Omega)$ with the aid of the following basic assumption: *the response of the system to the test charge is linear.* In other words, the test charge is sufficiently weak that we may treat H_1 by second-order perturbation theory. For a Coulomb interaction, this is true for small enough r_k (one can then always construct a *weak* test charge). For highly singular interactions (such as a hard sphere repulsion) the response may not be linear, and the following development fails.

We calculate the response of the electron system to the test charge under the assumption that the system is initially in its ground state. Thus if Ψ_n and E_n are the eigenstates and eigenvalues of H_0 (the Ψ_n are many-body wave functions which we do not know how to calculate exactly), we may write the wave function $\Psi(t)$ of the electron gas as

$$\Psi(t) = \sum_n \Psi_n \exp\left[-i\frac{E_n}{\hbar} t\right] a_n(t) ,$$

with the boundary conditions: $a_0(-\infty) = 1$, $a_n(-\infty) = 0$, for $n \neq 0$. If we limit ourselves to the first order in r_k, the time dependent Schrödinger equation for $H_0 + H_1$ is trivially solved and yields:

$$a_n(t) = -\frac{4\pi e^2}{\hbar k^2}\left[\frac{r_k(\varrho_{-k})_{n0} \exp[i(-\Omega + \omega_{n0})t + \eta t]}{-\Omega + \omega_{n0} - i\eta} + \right.$$
$$\left. + \frac{r_{-k}(\varrho_k)_{n0} \exp[i(\omega_{n0} + \Omega)t + \eta t]}{\omega_{n0} + \Omega - i\eta}\right] . \tag{5}$$

From (5), we calculate the perturbed expectation value of ϱ_k, $\langle\varrho_k\rangle$. The result is considerably simplified if we notice that the translational invariance implies that $(\varrho_{-k})_{n0} = 0$ if $(\varrho_k)_{n0} \neq 0$. We find that, to lowest order in r_k, $\langle\varrho_k\rangle$ is given by [10]

$$\langle\varrho_k\rangle = -\frac{4\pi e^2}{\hbar k^2} r_k \exp[-i\Omega t + \eta t] \cdot \tag{6}$$
$$\cdot \left[\sum_n |(\varrho_k)_{n0}|^2 \left\{\frac{1}{-\Omega + \omega_{n0} - i\eta} + \frac{1}{\omega_{n0} + \Omega + i\eta}\right\}\right] .$$

---

[9] P. Nozières and D. Pines: *Phys. Rev.*, **109**, 762 (1958).

[10] To obtain (6), we have assumed that the system was invariant by reflexion. A detailed discussion of this point is given by the authors in *Phys. Rev.*, **109**, 741 (1958).

Comparing equations (2) and (6), we obtain ε:

$$(7)\qquad \frac{1}{\varepsilon(k,\Omega)} - 1 = -\frac{4\pi e^2}{\hbar k^2}\sum_n |(\varrho_k)_{n0}|^2 \left\{\frac{1}{-\Omega+\omega_{n0}-i\eta} + \frac{1}{\omega_{n0}+\Omega+i\eta}\right\}.$$

Remark that (7) looks very much like a Kramers-Heisenberg formula. However, the matrix elements and energy differences refer to the true eigenstates of H_0, and not these quantities calculated in a plane wave representation.

We now consider the imaginary part of (7) (letting η go to zero). We find:

$$(8)\qquad \operatorname{Im}\left(\frac{1}{\varepsilon}\right) = \frac{4\pi^2 e^2}{\hbar k^2}\sum_n |(\varrho_k)_{n0}|^2\{\delta(\omega_{n0}+\Omega) - \delta(\omega_{n0}-\Omega)\}.$$

As has been noticed by FANO ([11]), the quantity Im $(1/\varepsilon)$ is closely related to the time dependent pair correlation function $G(r, t)$ introduced by VAN HOVE ([1]). In fact, if we denote by $S_0(k, \Omega)$ the Fourier transform of $G(r, t)$ at zero temperature (defined by Equation (4) of VAN HOVE's paper), one has simply

$$(9)\qquad \operatorname{Im}\frac{1}{\varepsilon} = \frac{4\pi^2 e^2}{\hbar k^2}\,[S_0(k, -\Omega) - S_0(k, +\Omega)]\,.$$

Equation (9) shows clearly that, in general, the correlation function $S(k, \Omega)$ contains more information that Im $1/\varepsilon$. However, if we limit ourselves to the ground state, all the ω_{n0} are positive, and $S_0(k, -\Omega) = 0$. We can drop the first δ-function in (8), and it is possible to infer $S(k, \Omega)$ from Im $1/\varepsilon$. Such a procedure fails for an excited state.

We now integrate (8) over all frequencies, Ω, and obtain the following basic formula (valid only for the ground state)

$$(10a)\qquad -\int_0^\infty \operatorname{Im}\frac{1}{\varepsilon(k,\Omega)}\,d\Omega = \frac{4\pi e^2}{k^2}\langle 0|\varrho_k\varrho_{-k}|0\rangle\,.$$

The relation (10a) enables us to obtain the interaction energy in the ground state, E_{int}, and the static pair correlation function $S(k)$, from a knowledge of the dielectric constant at all wave vectors and frequencies. We have:

$$(10b)\qquad E_{\text{int}} = \langle 0|H_{\text{coul}}|0\rangle = -\sum_k \frac{\hbar}{2\pi}\int_0^\infty \operatorname{Im}\frac{1}{\varepsilon(k,\Omega)}\,d\Omega - \frac{2\pi N e^2}{k^2}$$

and

$$S(k) = \int_{-\infty}^{\infty} S_0(k,\Omega)\,d\Omega = \left\langle 0\left|\frac{\varrho_k^*\varrho_k}{N}\right|0\right\rangle = -\frac{\hbar k^2}{4\pi e^2}\int_0^\infty \operatorname{Im}\frac{1}{\varepsilon}\,d\Omega\,.$$

([11]) U. FANO: *Phys. Rev.*, **103**, 1202 (1956).

The form factor $S(k)$ yields directly the cross-section within the Born approximation for the inelastic scattering of photons or electrons.

In order to obtain the ground state energy E_0 from E_{int}, we use a relation frequently cited in the literature, which is apparently due to PAULI ([12])

$$E_0(g) = E_0(0) + \int_0^g \frac{dg'}{g'} E_{int}(g') , \tag{11}$$

where g is the coupling constant, in our case e^2. This relation can be demonstrated simply by switching on the Coulomb interaction adiabatically and requiring that the system « follow » the ground state. From (10) and (11), we see that, in the general case, E_0 can be inferred from a knowledge of ε at all k, Ω *and* e^2. The latter requirement is of course difficult to meet experimentally. We shall see however in the next section that, within the RPA, the integration over e^2 can be performed trivially. In any event, we see that the calculation of E_0 may be rigorously reduced to that of $\varepsilon(k, \Omega, e^2)$.

Within this framework, we can evaluate easily the energy transfer of the test charges to the electron gas. Using elementary lowest order time dependent perturbation theory for a harmonic perturbation, we find ([13])

$$\frac{dW}{dt} = \frac{2\pi}{\hbar}\left(\frac{4\pi e^2}{k^2}\right)^2 |r_k|^2 |(\varrho_k)_{n0}|^2 \hbar\Omega[\delta(\omega_{n0} - \Omega) - \delta(\omega_{n0} + \Omega)] . \tag{12}$$

Comparing (12) with (8), we see at once that

$$\frac{dW}{dt} = -\frac{8\pi e^2}{k^2} \Omega |r_k|^2 \operatorname{Im} \frac{1}{\varepsilon(k, \Omega)} . \tag{13}$$

This result is identical with that obtained by a direct macroscopic calculation using Ohm's law, in agreement with the correspondence principle. We may also obtain the rate of energy loss of a fast particle, say an electron, with velocity V_0. If V_0 is large, the Born approximation is satisfactory. Each Fourier component of the incoming particle density then behaves like a test charge of density fluctuation $r_k = 1$, with a frequency $\Omega = k \cdot V_0$. The energy loss suffered by the fast electron is therefore

$$\frac{dE}{dt} = -\frac{1}{2} \sum_k \frac{8\pi e^2}{k^2} (k \cdot V_0) \operatorname{Im} \frac{1}{\varepsilon(k, k \cdot V_0)} .$$

([12]) Private communication from V. WEISSKOPF. This relation is demonstrated in the paper of K. SAWADA (ref. ([4])).

([13]) See footnote ([10]).

If we perform the angular integration over k, assuming V_0 to be very large, we find the well known result (7,8)

$$\frac{\mathrm{d}E}{\mathrm{d}t} = -\frac{2e^2}{\pi V_0}\int \frac{\mathrm{d}k}{k}\int_0^\infty \Omega\,\mathrm{d}\Omega\,\mathrm{Im}\left(\frac{1}{\varepsilon}\right). \tag{14}$$

Here again, all the information needed for $\mathrm{d}E/\mathrm{d}t$ is contained in the dielectric constant $\varepsilon(k, \Omega)$. Remark that (14) offers a way to measure $\varepsilon(k, \Omega)$ experimentally: in the energy loss experiment first performed by WATANABE (14) one measures the absorption of an electron beam as a function of both the energy transfer $\hbar\Omega$ and the momentum transfer k: this yields directly $\varepsilon(k, \Omega)$. (In fact, one even obtains directly the correlation function $S(k, \Omega)$, since, in the scattering of a running wave, there is no mixing of the energy transfers $+\hbar\Omega$ and $-\hbar\Omega$).

As was first emphasized by HUBBARD (5), the above formulation may be used to establish in simple fashion the existence of the collective modes of the electron gas, the plasmons. Let us write the complex as $\varepsilon_1 + i\varepsilon_2$. Then $-\mathrm{Im}\,(1/\varepsilon) = \varepsilon_2/\varepsilon_1^2 + \varepsilon_2^2$. For small enough k, $\varepsilon_2(k, \Omega)$ displays the following approximate behaviour: for $\Omega \lesssim kv_0$ (v_0 is the Fermi velocity), ε_2 is non-zero, while for $\Omega \gg kv_0$, ε_2 is very small and in first approximation may be chosen to be zero. Hence if ε_1 happens to go through zero at a certain frequency $\Omega = \omega_{pl}$, ($\Omega \gg kv_0$), $\mathrm{Im}(1/\varepsilon)$ peaks sharply at this frequency. (In the limit $\varepsilon_2(\omega_{pl}, k) = 0$, $\mathrm{Im}\,(1/\varepsilon)$ is simply equal to $\pi\delta(\varepsilon_1)$). The contribution of this peak to the ground state energy and to the stopping power corresponds simply to the ground state energy of the plasmons and to the excitation of plasma modes by the incoming electron. If $\varepsilon_2(\omega_{pl}, k) = 0$, the plasma line is discrete, and there is no damping of the plasmons. In general, $\varepsilon_2(\omega_{pl}, k)$ is small, but $\neq 0$, and the plasmons are damped, giving rise to a finite line width. We remark that it is not necessary to introduce plasmon variables explicitly in order to describe such resonances in the properties of the system; however their explicit introduction often makes possible a simple description of the system properties.

We here note several important properties of $\varepsilon(k, \Omega)$. First of all, for very large Ω, we can expand (7) in powers of $1/\Omega$. We thus find the asymptotic form of ε

$$\varepsilon(k, \Omega) \underset{\Omega\to\infty}{\longrightarrow} 1 - \frac{8\pi e^2}{\hbar k^2 \Omega^2}\sum_n \omega_{n0}\,|(\varrho_k)_{n0}|^2\,. \tag{15}$$

The summation occurring in (15) is just that involved in the f-sum rule, which has been shown to be rigorous for the representation in terms of the eigen-

(14) H. WATANABE: *Journ. Phys. Soc. Japan*, **11**, 112 (1956). A discussion of this problem is given by D. PINES: *Rev. Mod. Phys.*, **28**, 184 (1956).

states of H_0 ([15]). Therefore (15) reduces to

$$\varepsilon(k, \Omega) \underset{\Omega\to\infty}{\longrightarrow} 1 - \frac{4\pi N e^2}{m\Omega^2} = 1 - \frac{\omega_P^2}{\Omega^2}, \tag{16}$$

where ω_P is the usual free electron plasma frequency.

It is well known that causality requires that ε be analytic with respect to Ω in the upper half complex Ω plane. Furthermore, the system is passive and must always dissipate energy. Therefore ε_2 must have a constant sign in the upper half plane and can only be zero on the real axis. Therefore $1/\varepsilon$ is also analytic in the upper half plane, although it may have poles on the real axis. A simple contour integration then leads to the following relations

$$\left\{\begin{aligned} &\int_0^\infty \varepsilon_2 \Omega \, \mathrm{d}\Omega = \frac{\pi}{2}\,\omega_P^2, \\ &\int_0^\infty \frac{\varepsilon_2}{\varepsilon_1^2 + \varepsilon_2^2}\,\Omega \, \mathrm{d}\Omega = \frac{\pi}{2}\,\omega_P^2, \end{aligned}\right. \tag{17}$$

(in the second integral, the contour must pass *above* any pole of the integrand on the real axis). Remark that (17*b*) may be deduced directly from (8) and the *f*-sum rule; if we compare it with (14), we find the well known result, due to Bethe, that the total energy loss of a fast particle depends only on the total number of scatterers.

We have defined unambiguously a dielectric constant in terms of the true eigenstates of the system; we have shown that both the ground state energy and stopping power may be rigorously deduced from a knowledge of $\varepsilon(k, \Omega)$ at various coupling constants. In fact, this result is of a purely academic interest as long as we do not have a way to calculate ε. The rigorous results cannot even be used to infer E_0 from the energy loss experiments, since the calculation of E_0 in principle requires a knowledge of $\varepsilon(k, \Omega)$ for different values of the electron charge e. Therefore, approximations are undoubtedly needed, and will, in fact, form the basis for the next section.

Before taking up the calculation of $\varepsilon(k, \Omega)$, we should like to emphasize again the limitations of these results. First, one must be able to define a weak test charge in order to treat H_1 as a perturbation: this requirement imposes certain conditions on the law of interaction between the particles. Second, the foregoing results are only valid for the ground state of the system. For

([15]) See for instance a paper by the authors: *Phys. Rev.*, **109**, 741 (1958). The *f* sum rule can be easily established by calculating the quantity $[[H_0, \varrho_k], \varrho_{-k}]_{00}$, first directly, and then in the representation in terms of the Ψ_n.

excited states, HUBBARD ([5]) has shown that essentially the same approach could be used, provided one replaces ε by a slightly modified expression, in which Im $(1/\varepsilon)$ involves the sum of two δ-functions rather than the difference (see Equation (8)). In such a case, in fact, it is easier to work directly with the time dependent pair correlation function $S(k, \Omega)$ of VAN HOVE, which contains all the information needed to get both the dielectric constant and Hubbard's $\varepsilon(k, \Omega)$.

3. – The simplest approximation for ε involves the complete neglect of the Coulomb interaction between the electrons. The eigenstates of H_0 are then Slater determinants of plane waves: we denote such eigenstates as Φ_μ to distinguish them from the true eigenstates Ψ_n. Equation (7) gives ε directly:

(18)
$$\frac{1}{\varepsilon} = 1 - 4\pi\alpha\,,$$

where α is the complex polarizability given by the Kramers-Heisenberg formula in this plane wave representation

(19)
$$4\pi\alpha(k, \Omega) = \frac{4\pi e^2}{\hbar k^2}\sum_\mu |(\varrho_k)_{0\mu}|^2 \left\{\frac{1}{\omega_{\mu 0} - \Omega - i\eta} + \frac{1}{\omega_{\mu 0} + \Omega + i\eta}\right\}.$$

For free electrons, α is easily computed, and the result is given by HUBBARD ([4]) With this value of ε, the calculation of the interaction energy E_{int}, (10) is straightforward. We find

(20)
$$E_{\text{int}} = \sum_k \frac{2\pi e^2}{k^2}\left\{\sum_u |(\varrho_k)_{0\mu}|^2 - N\right\}.$$

This is just the usual Hartree-Fock exchange energy. Furthermore the integration over de^2 does not introduce any correction to E_0 since E_{int} is linear in e^2. Therefore, this crude approximation is simply the Hartree-Fock approximation, and corresponds to treating the Coulomb interaction by first order perturbation theory.

We may improve our calculation of the energy by taking the Coulomb interaction into account within the RPA (*i.e.* neglecting all coupling between different Fourier components of the interaction). If k is the wave vector of the test charge we would thus only take into account the k-th Fourier component of the electron interaction in computing the system response. The ground state Ψ_0 of the reduced electron system will then consist of an admixture of the plane wave ground state Φ_0 with various excited configurations involving an even number of electron hole pairs each with a total momentum $+k$ or $-k$. The average number of excited electrons will be only of order 1, because we keep only *one* component of the interaction. If we consider an operator which is diagonal in the plane wave representation, the Coulomb interaction will modify its matrix elements to order $1/N$; such a modification is not negligible if we consider the kinetic energy, since a contribution of order $1/N$ arising from each of the N Fourier components leads to a finite correction. On the contrary, for a diagonal operator A specifically related to the wave vector k, we may replace the expectation value $\langle\Psi_0|A|\Psi_0\rangle$ by its plane wave value $\langle\Phi_0|A|\Phi_0\rangle$. The matrix elements $(\varrho_k)_{0n}$ will be very different from the imperturbed $(\varrho_k)_{0n}$, since ϱ_k leads to creation or destruction of an electron hole pair with momentum k, and therefore connects a large number of configurations involved in Ψ_0 and Ψ_n. We can therefore predict that the expectation value of an operator like $\varrho_k\varrho_{-k}$ will be strongly modified by the Coulomb interaction.

A simple way to calculate the dielectric constant ε is to perform a time dependent canonical transformation to eliminate H_1. Let $\Psi_0(r_k)$ and $\Psi_0(0)$ be the ground state wave function of the system with and without test charge. We may write

$$\Psi_0(r_k) = \exp[iS/\hbar]\Psi_0(0)\,, \tag{21}$$

where $S(t)$ is a Hermitian operator, proportional to r_k. If $\Psi_0(r_k)$ satisfies the Schrödinger equation

$$-i\hbar\,\frac{\partial\Psi_0(r_k)}{\partial t} = H\Psi_0(r_k)\,,$$

then $\Psi_0(0)$ must obey the following equation

$$-i\hbar\,\frac{\partial\Psi_0(0)}{\partial t} = \left\{\exp[-iS/\hbar]H\exp[iS/\hbar] + i\hbar\exp[-iS/\hbar]\frac{\partial}{\partial t}(\exp[iS/\hbar])\right\}\Psi_0(0)\,.$$

The solution of the Schrödinger equation is therefore equivalent to finding an operator such that

$$\exp[-iS/\hbar]H\exp[iS/\hbar] + i\hbar\exp[-iS/\hbar]\frac{\partial}{\partial t}(\exp[iS/\hbar])\,, \tag{22}$$

does not contain any linear term in r_k. Elementary algebra shows that (22) may be expressed as a series

$$H + \left\{\frac{i}{\hbar}[H, S] - \frac{\partial S}{\partial t}\right\} + \frac{1}{2}\frac{i}{\hbar}\left[\left\{\frac{i}{\hbar}[H, S] - \frac{\partial S}{\partial t}\right\}, S\right] + \\ + \dots + \frac{1}{n!}\left(\frac{i}{\hbar}\right)^n\left[\left[\dots\left[\left\{\frac{i}{\hbar}[H, S] - \frac{\partial S}{\partial t}\right\}, S\right]\dots, S\right], S\right]. \tag{23}$$

To lowest order, S therefore satisfies the following equation

$$\frac{i}{\hbar}[H_0, S] - \frac{\partial S}{\partial t} = -H_1 . \tag{24}$$

The operator S will turn out to be a « one electron » operator (*i.e.*, able only to excite *one* electron hole pair of momentum $\pm k$). Let us write its matrix elements in the plane wave representation in the following way

$$S_{\mu\nu} = g(\omega_{\mu\nu}, \Omega)(\varrho_{-k})_{\mu\nu} r_k \exp[-i\Omega t + \eta t] + \\ + g(\omega_{\mu\nu}, -\Omega)(\varrho_k)_{\mu\nu} r_{-k} \exp[i\Omega t + \eta t] . \tag{25}$$

We must find $g(\omega_{\mu\nu}, \Omega)$. The commutator of S with the kinetic energy is trivially calculated (its matrix elements are simply $\hbar\omega_{\mu\nu} S_{\mu\nu}$). The commutator with the Coulomb interaction may be written

$$\frac{4\pi e^2}{k^2}\{[\varrho_k, S]\varrho_{-k} + \varrho_k[\varrho_{-k}, S]\}.$$

An operator like $[\varrho_k, S]$, being a commutator, is necessarily a one electron operator. The first part of S yields a diagonal contribution: according to the discussion given earlier in this section, this term may be replaced by its expectation value in the state Φ_0. The second term of S gives rise to an operator creating or destroying an electron hole pair with momentum $2k$: this must be neglected within RPA. Therefore, the commutator of the Coulomb interaction with S reduces to:

$$\frac{i}{\hbar}[H_{\text{coul}}, S] = \frac{i}{\hbar}\frac{4\pi e^2}{k^2}\sum_\mu |(\varrho_k)_{0\mu}|^2 [g(\omega_{\mu 0}, \Omega) - g(\omega_{0\mu}, \Omega)] \cdot \\ \cdot \varrho_{-k} r_k \exp[-i\Omega t + \eta t] + \text{c. c.} \tag{26}$$

With the help of (26), the correct value of $g(\omega_{\mu\nu}, \Omega)$ may be shown to be given by

$$g(\omega_{\mu\nu}, \Omega) = -\frac{1}{1+4\pi\alpha}\frac{4\pi e^2}{ik^2}\frac{1}{\omega_{\mu\nu} + \Omega + i\eta}, \tag{27}$$

where $\alpha(k, \Omega)$ is the polarizability defined in equation (19). We next use (27) to calculate the perturbed expectation value of ϱ_k

$$\langle\varrho_k\rangle = \langle\Psi_0(r_k)|\varrho_k|\Psi_0(r_k)\rangle.$$

From (21), we see at once that, to first order in r_k, $\langle\varrho_k\rangle$ is given by

$$\langle\varrho_k\rangle = \frac{i}{\hbar}[\varrho_k, S]_{00} = -\frac{4\pi\alpha}{1+4\pi\alpha} r_k \exp[-i\Omega t + \eta t] . \tag{28}$$

On comparing (28) with the equation (2) defining ε, we conclude that, within the RPA, the complex dielectric constant is given by

$$\varepsilon(k, \Omega) = 1 + 4\pi\alpha(k, \Omega) \,. \tag{29}$$

The result (29) is in agreement with an earlier treatment ([9]) in which the effect of real transitions, *i.e.* the imaginary part of ε, was neglected. In the next section, we shall discuss its physical significance.

Although it is not needed for the present reasoning, it is interesting to calculate the second order terms, which are given by

$$\frac{1}{2}\frac{i}{\hbar}[H_1, S] \,.$$

Using (28), we find

$$\frac{i}{2\hbar}[H_1\,, S] = \frac{2\pi e^2}{k^2} r_k r_{-k}\left[\frac{1}{\varepsilon(k, \Omega)} + \frac{1}{\varepsilon(-k, -\Omega)}\right] \exp\,[2\eta t] \,. \tag{30}$$

If we write $\varepsilon = \varepsilon_1 + i\varepsilon_2$, we see that the Coulomb interaction between the test charges is screened by a factor $\varepsilon_1/(\varepsilon_1^2 + \varepsilon_2^2)$. When $\varepsilon_2 = 0$ this reduces to the usual macroscopic result of electrostatics; furthermore, in such a case, there is no energy transfer of the test charges to the system, and the concept of independent test charges interacting through a screened interaction is physically sensible. On the contrary, if $\varepsilon_2 \neq 0$, the test charges give up energy to the system. (This feature did not appear in the previous calculation because we switched on the test charges adiabatically in such a way that no appreciable. energy transfer could take place.) The test charges then have a finite « life time » τ. If $\Omega\tau \gg 1$. it makes sense to consider the effective interaction (30). On the other hand, if $\Omega\tau \lesssim 1$, the concept of a test charge at frequency Ω loses its meaning. Finally, (30) shows that, although the interaction has a pole at the plasma frequency ($\varepsilon_1 = \varepsilon_2 = 0$), it is regular at frequencies such that $\varepsilon_1 = 0$, $\varepsilon_2 \neq 0$, a definite improvement over the time independent result $(2\pi e^2/k^2\varepsilon_1) r_k r_{-k}$.

Remark that result (29) could have been obtained by a perturbation treatment of the Coulomb interaction, limited to those diagrams appearing within RPA. The first order contribution to $1/\varepsilon$ is that given in (18), and the following orders form a geometrical series which may be trivially summed:

$$\frac{1}{\varepsilon} = 1 - 4\pi\alpha + (4\pi\alpha)^2 + \ldots = \frac{1}{1 + 4\pi\alpha} \,.$$

Such an expansion only converges when $4\pi\alpha < 1$, while the present calculation does not suffer from that limitation.

If we substitute the value (29) for ε in the expressions for the ground state energy and stopping power given in the preceeding section, we obtain exactly the result of HUBBARD ([5]). The latter is easily seen to be the same as that of SAWADA and hence of GELL-MANN and BRUECKNER. The expression for the ground state energy E_0 may be simplified if we remark that $4\pi\alpha$ is proportional to e^2: the integration over $\mathrm{d}e^2$ then yields the following result (cf. HUBBARD)

$$E_0 = E_0(0) + \sum_k \left\{\frac{\hbar}{2\pi}\int_0^\infty \operatorname{arctg}\frac{\varepsilon_2}{\varepsilon_1}\,\mathrm{d}\Omega - \frac{2\pi N e^2}{k^2}\right\}, \tag{31}$$

where $E_0(0) = \frac{3}{5} N E_F$ is the ground state energy of a gas of non-interacting electrons. We wish to emphasize the fact that the result is approximate; its validity depends on the validity of the RPA. In cases in which the RPA is valid, we see that E_0 can be obtained from the experimental data on energy loss of fast electrons, through the following procedure:

$$\begin{cases} 1)\quad \text{Absorption spectrum} \rightarrow \operatorname{Im}\left(\dfrac{1}{\varepsilon}\right), \\ 2)\quad \text{Kramers-Kronig relation} + \operatorname{Im}\left(\dfrac{1}{\varepsilon}\right) \rightarrow \operatorname{Re}\left(\dfrac{1}{\varepsilon}\right), \\ 3)\quad \operatorname{Im}\left(\dfrac{1}{\varepsilon}\right) + \operatorname{Re}\left(\dfrac{1}{\varepsilon}\right) \rightarrow \varepsilon \rightarrow E_0\,. \end{cases}$$

Such a calculation has not been attempted. Remark that, if performed carefully, it might constitute a check of the validity of the RPA. In fact, it is probable that the crystalline periodic field gives rise to important corrections which will tend to mask those arising from RPA, so that this discussion is somewhat academic.

The ground state energy may be written as

$$\begin{cases} E_0 = E_0(0) + \sum_k \Delta E_0(k)\,, \\ \Delta E_0(k) = \dfrac{\hbar}{2\pi}\displaystyle\int_0^\infty \operatorname{arctg}\frac{\varepsilon_2}{\varepsilon_1}\,\mathrm{d}\Omega - \frac{2\pi N e^2}{k^2}\,. \end{cases} \tag{32}$$

It is interesting to see how $\Delta E_0(k)$ behaves for k much smaller than the Fermi momentum k_0. The arctg has the behaviour shown on Fig. 1: it goes from 0 to π between $\Omega = 0$ and $\Omega = kv_0$ (where v_0 is the Fermi velocity). It then stays equal to π until Ω reaches the plasma frequency ω_{pl} at which ε_1 is zero. For $\Omega > \omega_{\text{pl}}$, it is zero. Remark that for small k, ω_{pl} is very close to

$\omega_p = \sqrt{4\pi N e^2/m}$. The integral appearing in (32) may be written as

$$\pi\omega_{\text{pl}} - A \; ,$$

where A is the shaded area of Fig. 1. The first term corresponds to the zero-point energy of the plasmons, while the second corresponds to the screened long range electron interaction energy (that arising from H_{rp} in the Bohm-Pines formalism). One sees at once that the plasmon energy is k independent, while the individual particle energy is of order k. It is easy to expand E_0 in powers of k along these lines, and one finds exactly the results quoted by the authors in a previous paper ([16]), to which the interested reader is referred. For very large k, on the contrary, arctg $\varepsilon_2/\varepsilon_1$ is only different from 0 in the region $\Omega \sim \hbar k^2/2m$: the first term of $\Delta E_0(k)$ then tends towards $2\pi N e^2/k^2$, so that the result is convergent.

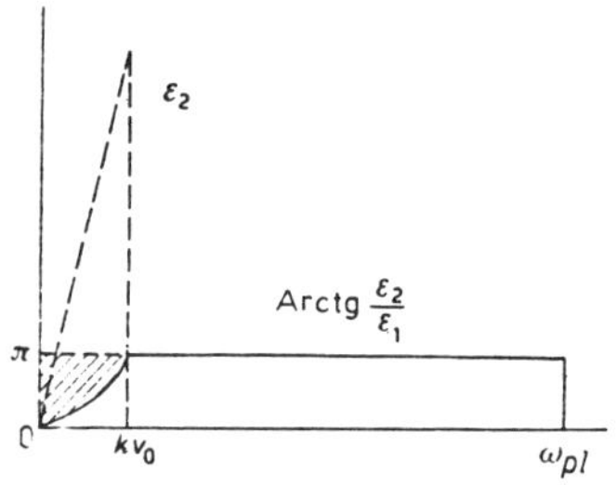

Fig. 1. – arctg $\varepsilon_2/\varepsilon_1$ as a function of Ω. Note the discontinuity at ω_{pl}. In the Hartree-Fock approximation (first order perturbation theory, arctg $\varepsilon_2/\varepsilon_1$ would be replaced by ε_2 (dotted curve).

It should be emphasized that the inclusion of all higher order terms, within the RPA, corresponds to a radical change in the description of the electron system. In the Hartree-Fock approximation, arctg $\varepsilon_2/\varepsilon_1$ shoud be replaced by ε_2 in Equation (32): the energy of the system then arises only from individual particles. Within the RPA, one sees from Fig. 1 that this individual particle contribution is heavily screened (by a factor of order k^2, with even its sign changed). Furthermore, collective effects appear, which extend considerably the frequency range contributing to E_0 (from kv_0 to ω_{pl}). Such a basic change is a consequence of the summation of an infinite perturbation series. The same effect appears in a still clearer fashion in the energy loss of fast electrons. There, the RPA result $\varepsilon_2/(\varepsilon_1^2+\varepsilon_2^2)$ is to be compared with the Hartree-Fock result ε_2. The effect of the electron correlations is to screen the low frequency part, and to concentrate most of the long wave length oscillator strengths at the collective plasma frequency ω_{pl}.

We shall not discuss in detail the numerical details of the calculation for free electrons, since they may be found in HUBBARD's paper. We shall, however, make a few remarks concerning the interpretation of these results. First, we note that the expectation value E_{int} of the Coulomb interaction in the ground state Ψ_0 is equal to the expectation value in the plane wave state Φ_0 of a

([16]) P. NOZIÈRES and D. PINES: submitted to the *Phys. Rev.*

renormalized interaction

$$H^S_{\text{coul}} = \sum_k \frac{2\pi e^2}{k^2} \varrho^S_k \varrho^{S*}_k$$

in which ϱ^S_k is a screened density fluctuation, defined by its matrix elements

$$(\varrho^S_k)_{\mu\nu} = \frac{(\varrho_k)_{\mu\nu}}{\varepsilon(k, \omega_{\mu\nu})} . \tag{33}$$

This may be interpreted as the replacement of the electric charge $e\varrho_k$ by a renormalized charge $e\varrho^S_k$. (Remark that the screening is frequency dependent, a definite improvement over the Fermi-Thomas method.) Physically, this substitution means that each electron carries a polarization cloud which modifies its effective electric charge.

To conclude this section, we compare the present results with the minority carrier technique described in a previous paper ([9]). There, it was shown that if one neglected resonant transitions (taking principal parts at any pole), it was possible to isolate a group of independent minority carriers, (with density fluctuation $\tilde{\varrho}_k$) which then interact amongst each other through a screened interaction

$$\sum_k \frac{2\pi e^2}{k^2} \tilde{\varrho}_{-k} \tilde{\varrho}^{S'}_{+k} .$$

in which $\tilde{\varrho}^{S'}_k$ was a screened density fluctuation (different from the above ϱ^S_k), defined by

$$(\tilde{\varrho}^{S'}_k)_{\mu\nu} = \frac{(\tilde{\varrho}_k)_{\mu\nu}}{\varepsilon_1(k, \omega_{\mu\nu})} .$$

Such minority carriers do not undergo any further correlations, and their contribution to the total energy of the system involves only their kinetic and exchange energy. The latter may be written as

$$\sum_k \left\{ \sum_\nu \frac{2\pi e^2}{k^2} \frac{|(\tilde{\varrho}_k)_{0\nu}|^2}{\varepsilon_1(k, \omega_{\nu 0})} - \frac{2\pi \tilde{N} e^2}{k^2} \right\} , \tag{34}$$

(where $\tilde{N}$ is the number of minority carriers). If we assume that this approach can be extended to all carriers, we find as the ground state energy of the system the following quantity

$$E_0 = E_0(0) + \sum_k \left\{ \frac{\hbar}{2\pi} \int_0^\infty \frac{\varepsilon_2}{\varepsilon_1} \, \mathrm{d}\Omega - \frac{2\pi N e^2}{k^2} \right\} . \tag{35}$$

(35) is the same as (31) but for the replacement of $\operatorname{arctg} \varepsilon_2/\varepsilon_1$ by $\varepsilon_2/\varepsilon_1$. This is not surprising, since ε_2 arises only from the real resonant transitions; taking

principal parts amounts to the neglect of higher powers of ε_2, therefore to the replacement of $\operatorname{arctg} \varepsilon_2/\varepsilon_1$ by $\varepsilon_2/\varepsilon_1$. This discussion shows that the minority carrier treatment presented in ref. ([9]) is only valid when one deals with frequencies such that $\varepsilon_2 \ll \varepsilon_1$ (*i.e.* for Ω much smaller *or* much larger than kv_0). This is the case for many physical problems, particularly for all problems dealing with transitions in an energy shell of width kT around the Fermi surface. On the other hand, the minority carrier approach, although it is suggested by the correspondence principle, does not lead to an accurate calculation of the cohesive energy.

4. – In discussing the meaning of the RPA, it is easier to consider the calculation of the meaning and validity of the RPA in the calculation of the dielectric constant than the ground state energy. A detailed discussion of the meaning and validity of the RPA in the calculation of the dielectric constant may be found in ref. ([9]); we summarize the arguments briefly here. Consider a slab of the electron gas between two condenser plates. Let E be the electric field in the gas arising from the charge on the condenser plates (oscillating at frequency Ω). Consider now a given electron at point M (Fig. 2), and let E_M be the effective electric field at point M, and P the polarization. In order to calculate E_M, we use the method of Lorentz, and draw a sphere around point M: E_M is then the sum of four terms

— the applied field E,
— the field of the polarized matter outside the sphere, which is equivalent to

- the field $E_1 = -4\pi P$ of surface charges at the boundary of the specimen
- the field $E_2 = \dfrac{4\pi P}{3}$ of surface charges at the boundary of the sphere

— the field E_3 of the polarized matter inside the sphere (which we do not know how to calculate).

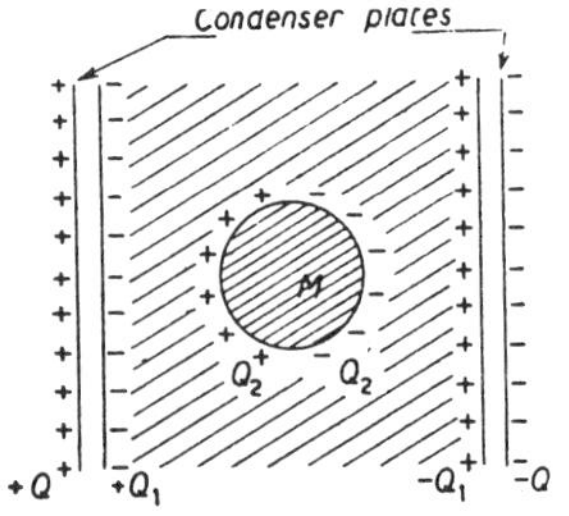

Fig. 2. – The local fields corrections to the dielectric constant. The fields E, E_1, E_2 arise respectively from the charges $\pm Q$, $\pm Q_1$, $\pm Q_2$.

Let α be the microscopic polarizability: we then have $P = \alpha E_M$. The problem is to calculate ε, which is defined by

$$\varepsilon = \frac{E}{E - 4\pi P}\,.$$

We therefore need to know both E_3 and α.

Let us now consider from a qualitative point of view the effects of the Coulomb interaction on these quantities. Let $\varrho(0)$ be the electron charge distribution in the absence of applied electric field ($E=0$), and $\delta\varrho(E)$ the distortion of this distribution due to E. The effect of $\varrho(0)$ is to modify the restoring forces acting on the electron M, therefore to change the polarizability α. This is typically a « potential » effect, which may be described by an appropriate modification of the electron mass. On the other hand, the effect of $\delta\varrho(E)$ is to modify the electric field at point M, giving rise to the corrections E_1, E_2 and E_3. Of these, E_1 arises from surface charges at the outer boundary of the specimen, and in fact is a direct consequence of the boundary condition of electrostatics, while E_2 and E_3 correspond to the field of polarized electrons *inside* the specimen; the latter fields give rise to the so-called « local field » corrections.

Suppose we neglect the potential correction and thereby assume the polarizability α to be that of free electrons given in the preceeding section: there remains the determination of E_M. If we further neglect completely the electron interaction, thus dropping E_1, E_2 and E_3: we find

$$\frac{1}{\varepsilon} = 1 - 4\pi\alpha\,,$$

which is just equation (8), as might be expected. We might attempt to improve this calculation by keeping E_1, that is, by taking into account the Coulomb interaction, but neglecting the local field corrections: we then find

$$\varepsilon = 1 + 4\pi\alpha\,,$$

which is just (29): the RPA is therefore equivalent to *neglecting potential and local field corrections* taking only into account surface charges at the outer boundary. In fact, this could have been predicted easily: since the RPA does not mix different Fourier components, it is impossible to build wave packets corresponding to the field of localized charges. Therefore the static Coulomb potential and the local field are automatically neglected within RPA. The field E_1 is included because it arises from charges at the boundary of the specimen (we always can take an infinite sample: E_1 then arises from charges at infinity and does not require building a wave packet). Note that we could have made this remark at the beginning, and that, therefore, the Brueckner-Gell-Mann result can be obtained with no calculation at all.

Having thus emphasized the physical meaning of the RPA, we are in a position to discuss its validity. Except in the case of high density, we know that the corrections are quite appreciable (for low density, they change the entire character of the results). It is convenient to discuss the corrections in the language of diagrams, introduced by GOLDSTONE ([17]), HUGENHOLTZ ([18]) and others. The so-called «polarization» diagrams belong to the general class shown on Fig. 3*a*. As shown by HUGENHOLTZ, the potential effects arise from the replacement of particle propagation lines by any « diagonal » subdiagram, of the general type shown on Fig. 3*b*. Of these diagonal diagrams, the simplest is certainly that shown on Fig. 3*c* which corresponds to the Hartree-Fock potential. Such a potential may be refined: in a further publication, we shall for instance discuss a screened Hartree-Fock potential including all subdiagrams of the type shown on Fig. 3*d*. Of course, this does not exhaust all possible potential corrections, and should be considered as an approximation whose validity must be discussed.

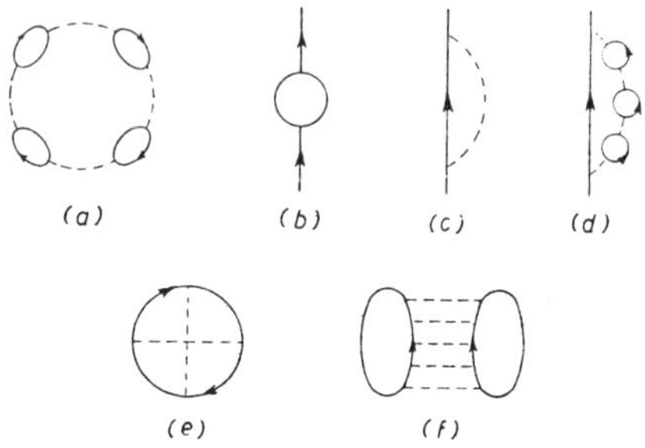

Fig. 3. – The diagrammatic representation of a perturbation expansion. Full lines are particle propagators, dotted lines are interaction lines. 3*a*, polarization diagram; 3*b*, general « diagonal » subdiagram; 3*c*, Hartree-Fock subdiagram; 3*d*, screened Hartree-Fock subdiagram; 3*e*, the second order exchange « cross diagram »; 3*f*, a « ladder » diagram, included in a Brueckner treatment.

In principle, all the other diagrams give rise to local field corrections. From the study of BRUECKNER and GELL-MANN, it appears that among the most important are the « cross diagrams » involving exchange of electrons. In Fig. 3*e* we draw such a diagram in lowest order; it is this diagram which in fact gives rise to the second order exchange energy (the E_b^2 of BRUECKNER and GELL-MANN). As emphasized by HUBBARD, this class of diagrams is important, since for high momentum transfer k, they cancel one half of the RPA diagrams. Unfortunately, it is not possible at present to compute their contribution to all orders, and one must rely on interpolation procedures between the low k and high k limits. Such procedures have been extensively discussed by HUBBARD ([5]) and the authors ([16]).

In the case of the electron gas, these seem to be the leading corrections. Of course, it is possible to consider yet other approaches; for instance one might try to use a Brueckner type calculation to correct the effective interaction and

([17]) J. GOLDSTONE: *Proc. Roy. Soc.*, A **239**, 267 (1957).
([18]) N. HUGENHOLTZ: *Physica*, **23**, 481, 533 (1957).

the one electron potential (including « ladder » diagrams of the type shown on Fig. 3*f*). Such a correction does not seem important in the electron case and nothing has yet been attempted in this direction.

5. – The discussion given in the preceeding section demonstrates the necessity of taking electron correlations into account in a description of any longitudinal phenomenon in the electron gas. The concentration of oscillator strengths at the collective plasma pole, and its counter part, the screening of the low frequency components, are basic features of the system, which are not obtainable from second-order perturbation theory. The RPA represents a considerable improvement over the Hartree-Fock method for the electron gas; it yields a qualitatively sensible physical picture of the system properties, even though it is not quantitatively correct at actual metallic densities. (Thus, as the authors have shown elsewhere ([16]), the RPA is only valid in the long wave length limit ($k \lesssim 0.47 k_0 r_s^{\frac{1}{2}}$); for the short wave length interactions, a perturbation theoretical treatment in which only interactions between electrons of anti-parallel spin are included represents a tolerable approximation)..

We should like to emphasize that the results presented in Sect. **3** are not new. The resolution of the many-body problem for free electrons within the RPA is now well understood, and the present approach is but one way to obtain the by now well known result. In our minds, however, the present approach has several advantages.

— It involves quite simple calculations, especially if one realizes at the outset that the RPA is equivalent to neglecting all potential and local field corrections; the calculation of ε is then trivial.

— It does not involve a perturbation series expansion, so that, as in Sawada's approach, there are no accompanying convergence difficulties.

— It affords a direct link between the dielectric properties of the system (*i.e.* the scattering cross-sections for fast electrons), and the ground state energy. (This relation bears some resemblance to the dispersion relations which relate the imaginary and reals parts of the scattering matrix.)

— It yields a very simple discussion of the physical meaning of the approximations.

We have already mentioned that $\varepsilon(k, \Omega)$ is a quantity which may be inferred directly from the energy loss experiments. As has been pointed out by VAN HOVE, the correlation function, *i.e.* Im $1/\varepsilon$ may also be obtained from the Compton scattering of X-rays by the gas of free electrons ([19]). In the

([19]) The authors wish to thank Prof. H. CURIEN, for drawing their attention to the interpretation of Compton scattering data.

latter case, in fact, the sensitivity of the experimental apparatus does not allow a spectral analysis of the scattered light, and one only measures with Geiger counters the numbers of photons scattered as a function of angle, which is proportional to the form factor $S(k)$ given in (10*b*) (the calculations are the same as those for the scattering of neutrons by liquid helium). The dielectric constant is therefore an easily measurable quantity. As we already pointed out, where the RPA is valid, one may then compute the ground state energy E_0 from the observed $\varepsilon(k, \Omega)$. (This does not appear to be feasible outside the RPA.)

We remark that, although we have only applied our approach to electrons in the present paper, the method is quite general, and may be extended to other types of systems for which it provides a simple recipe for the calculation of the system properties within RPA. For instance, the method may be applied to the dilute hard sphere boson gas, for which it yields the results of BOGOLJUBOV ([20]) and HUANG, LEE and YANG ([21]) on the ground state energy and excitation spectrum. Details of the calculation will be published elsewhere

To conclude, we see that the RPA yields results which, for actual metallic densities, are qualitatively correct, but that a detailed quantitative analysis requires more insight into both potential and local field corrections. The latter have been already treated by qualitative interpolation procedures, while the former are still an open problem.

* * *

The authors wish to thank Prof. P. AIGRAIN and Prof. J. LUTTINGER for many enlightening discussions on these and related subjects.

([20]) N. N. BOGOLJUBOV: *Journ. Phys. USSR*, **11**, 23 (1947).

([21]) K. HUANG, T. D. LEE and C. N. YANG: *Phys. Rev.*, **106**, 1135 (1957).

PHYSICAL REVIEW VOLUME 115, NUMBER 4 AUGUST 15, 1959

Self-Consistent Field Approach to the Many-Electron Problem*

H. EHRENREICH, *General Electric Research Laboratory, Schenectady, New York*

AND

M. H. COHEN, *Institute for the Study of Metals, University of Chicago, Chicago, Illinois*

(Received March 30, 1959)

The self-consistent field method in which a many-electron system is described by a time-dependent interaction of a single electron with a self-consistent electromagnetic field is shown to be equivalent for many purposes to the treatment given by Sawada and Brout. Starting with the correct many-electron Hamiltonian, it is found, when the approximations characteristic of the Sawada-Brout scheme are made, that the equation of motion for the pair creation operators is the same as that for the one-particle density matrix in the self-consistent field framework. These approximations are seen to correspond to (1) factorization of the two-particle density matrix, and (2) linearization with respect to off-diagonal components of the one-particle density matrix. The complex, frequency-dependent dielectric constant is obtained straightforwardly from the self-consistent field approach both for a free-electron gas and a real solid. It is found to be the same as that obtained by Noziéres and Pines in the random phase approximation. The resulting plasma dispersion relation for the solid in the limit of long wavelengths is discussed.

THE electromagnetic properties of crystals have long been studied by considering the time-dependent interaction of a single particle with a self-consistent electromagnetic field.[1] This procedure seems plausible for studying the response of electrons to any external perturbation, and Bardeen,[2] Wolff,[3] Lindhard,[4] Frölich and Pelzer,[5] Ferrell,[6] and others[7] have used this or a closely related approach with considerable success in discussing such phenomena as the electron-phonon interaction, the frequency and wave-number dependence of the dielectric constant, plasma oscillations, and characteristic energy losses in solids. These, and similar phenomena, have also been studied on the basis of more sophisticated treatments of the many-body problem[8–13] with largely identical results. The explicit relationship of the self-consistent field approach (e.g., Lindhard[4]) to the many-body approach (e.g., Sawada and Brout[10,11]) has not been stated. It is the purpose of this note to examine this relationship and to show that for many problems the two approaches may be regarded as rigorously equivalent. We do so by showing that the approximations introduced by Sawada and Brout are in fact sufficient to deduce the equation of the self-consistent field approach.

1. SELF-CONSISTENT FIELD, OR SCF, METHOD

We begin with a convenient formulation of the SCF method. We consider the single-particle Liouville equation

$$i\hbar(\partial\rho/\partial t)=[H,\rho], \tag{1}$$

as describing the response of any particle of the system to the self-consistent potential $V(x,t)$, where ρ is the operator represented by the single-particle density matrix. The single-particle Hamiltonian in (1) is

$$H=H_0+V(x,t), \tag{2}$$

where $H_0=p^2/2m$ is the Hamiltonian of a free electron satisfying Schrödinger's equation $H_0|\mathbf{k}\rangle=E_\mathbf{k}|\mathbf{k}\rangle$, and $|\mathbf{k}\rangle=\Omega^{-\frac{1}{2}}e^{i\mathbf{k}\cdot\mathbf{x}}$, Ω being the volume of the system. We expand the operator ρ in the form $\rho=\rho^{(0)}+\rho^{(1)}$. The unperturbed (Dirac or von Neumann) density matrix has the property $\rho^{(0)}|\mathbf{k}\rangle=f_0(E_\mathbf{k})|\mathbf{k}\rangle$, where $f_0(E_\mathbf{k})$ is the distribution function. Use of the von Neumann density matrix permits us to treat systems at finite temperatures. We now Fourier-analyze $V(x,t)$ in the form

$$V(x,t)=\sum_{q'}V(q',t)e^{-i\mathbf{q}'\cdot\mathbf{x}}, \tag{3}$$

and linearize Eq. (1) by neglecting products of the type $V\rho^{(1)}$. This approximation is equivalent to first-order self-consistent perturbation theory. Taking

* An account of this work has been presented at the 1959 March meeting of the American Physical Society [Bull. Am. Phys. Soc. Ser. II, 4, 129 (1959)].

[1] E.g., F. Seitz, *Modern Theory of Solids* (McGraw-Hill Book Company, New York, 1940), Chap. 17 (optical properties); G. E. H. Reuter and E. H. Sondheimer, Proc. Roy. Soc. (London) **A195**, 336 (1948) (anomalous skin effect).

[2] J. Bardeen, Phys. Rev. **52**, 688 (1937).

[3] P. Wolff, Phys. Rev. **92**, 18 (1953).

[4] J. Lindhard, Kgl. Danske Videnskab. Selskab, Mat.-fys. Medd. **28**, 8 (1954).

[5] H. Fröhlich and H. Pelzer, Proc. Phys. Soc. (London) **A68**, 525 (1955).

[6] R. A. Ferrell, Phys. Rev. **107**, 450 (1957).

[7] R. H. Ritchie, Phys. Rev. **106**, 874 (1957); H. Ehrenreich and M. H. Cohen, Bull. Am. Phys. Soc. Ser. II, **3**, 271 (1958); H. Ehrenreich, J. Phys. Chem. Solids **8**, 130 (1959). Since the present paper was submitted for publication, a manuscript by J. Goldstone and K. Gottfried which takes a similar point of view was called to our attention. We should like to thank Drs. Goldstone and Gottfried for sending us a copy of their work in advance of publication.

[8] D. Pines, in *Solid State Physics* edited by F. Seitz and D. Turnbull (Academic Press, Inc., New York, 1955), Vol. 1.

[9] M. Gell-Mann and K. A. Brueckner, Phys. Rev. **106**, 364 (1957).

[10] K. Sawada, Phys. Rev. **106**, 372 (1957); Sawada, Brueckner, Fukuda, and Brout, Phys. Rev. **108**, 507 (1957).

[11] R. Brout, Phys. Rev. **108**, 515 (1957).

[12] J. Hubbard, Proc. Roy. Soc. (London) **A240**, 539 (1957); **A243**, 336 (1958).

[13] P. Nozières and D. Pines, Phys. Rev. **109**, 741, 762, 1062 (1958), Nuovo cimento **9**, 470 (1958); Phys. Rev. **111**, 442 (1958), and work to be published.

matrix elements between states $\mathbf{k}$ and $\mathbf{k+q}$, we thus obtain

$$i\hbar(\partial/\partial t)\langle\mathbf{k}|\rho^{(1)}|\mathbf{k+q}\rangle$$
$$=\langle\mathbf{k}|[H_0,\rho^{(1)}]|\mathbf{k+q}\rangle+\langle\mathbf{k}|[V,\rho^{(0)}]|\mathbf{k+q}\rangle$$
$$=(E_{\mathbf{k}}-E_{\mathbf{k+q}})\langle\mathbf{k}|\rho^{(1)}|\mathbf{k+q}\rangle$$
$$+[f_0(E_{\mathbf{k+q}})-f_0(E_{\mathbf{k}})]V(q,t), \quad (4)$$

where $\langle\mathbf{k}|V|\mathbf{k+q}\rangle=V(q,t)$. The potential V consists of an external potential V_0 plus the screening potential V_s, which is related to the induced change in electron density,

$$n=\mathrm{Tr}\{\delta(\mathbf{x}_e-\mathbf{x})\rho^{(1)}\}$$
$$=\Omega^{-1}\sum_{\mathbf{q}}e^{-i\mathbf{q}\cdot\mathbf{x}}\sum_{\mathbf{k}'}\langle\mathbf{k}'|\rho^{(1)}|\mathbf{k}'+\mathbf{q}\rangle, \quad (5)$$

by Poisson's equation:

$$\nabla^2V_s=-4\pi e^2n. \quad (6)$$

Here $\delta(\mathbf{x}_e-\mathbf{x})$ is the charge density operator, $\mathbf{x}_e$ being the position operator and $\mathbf{x}$ referring to a specific point in space. We thus find

$$V_s(q,t)=v_q\sum_{\mathbf{k}'}\langle\mathbf{k}'|\rho^{(1)}|\mathbf{k}'+\mathbf{q}\rangle, \quad (7)$$

where $v_q=4\pi e^2/q^2\Omega$. By substituting the above expression giving V_s for V in Eq. (4) we obtain the Loiuville-Poisson equation determining $\langle\mathbf{k}|\rho^{(1)}|\mathbf{k+q}\rangle$ in the absence of an external perturbation:

$$i\hbar(\partial/\partial t)\langle\mathbf{k}|\rho^{(1)}|\mathbf{k+q}\rangle$$
$$=(E_{\mathbf{k}}-E_{\mathbf{k+q}})\langle\mathbf{k}|\rho^{(1)}|\mathbf{k+q}\rangle$$
$$+v_q[f_0(E_{\mathbf{k+q}})-f_0(E_{\mathbf{k}})]\sum_{\mathbf{k}'}\langle\mathbf{k}'|\rho^{(1)}|\mathbf{k}'+\mathbf{q}\rangle. \quad (8)$$

We have derived Eq. (8) in order to provide an explicit basis for comparison of the SCF method with the Sawada-Brout scheme. In solving problems by the SCF method, however, one can usually avoid the explicit expression of V_s in terms of $\rho^{(1)}$ within the equation of motion by making an *Ansatz* concerning the time dependence of $V(q,t)$. To illustrate this point, we calculate the frequency and wave-number dependence of the longitudinal dielectric constant $\epsilon(\omega,q)$. We imagine that the external potential $V_0(q,t)$ acts on the system with time dependence $e^{\alpha t}e^{i\omega t}$, where $\alpha\rightarrow 0$ corresponds to an adiabatic turning on of the perturbation. This potential polarizes the system. It follows from the definition of the dielectric constant and the Fourier analysis prescribed by Eq. (3) that

$$P(q,t)=(4\pi)^{-1}[\epsilon(\omega,q)-1]\mathcal{E}(q,t). \quad (9a)$$

The polarization $P(q,t)$ is related to the induced change in electron density by $\nabla\cdot\mathbf{P}=en$ or

$$-iqP(q,t)=en(q,t), \quad (9b)$$

and the electric field $\mathcal{E}(q,t)$ is given by

$$e\mathcal{E}(q,t)=-iqV(q,t). \quad (9c)$$

Equation (4) is readily solved for $\langle\mathbf{k}|\rho^{(1)}|\mathbf{k+q}\rangle$ by assuming that $\langle\mathbf{k}|\rho^{(1)}|\mathbf{k+q}\rangle$ and $V_s(q,t)$ have the same time dependence as $V_0(q,t)$. The induced change in electron density $n(q,t)$ may then be calculated from (5) and $\epsilon(\omega,q)$ deduced from the field equations (9a, b, c). We find

$$\epsilon(\omega,q)=1-\lim_{\alpha\rightarrow 0}v_q\sum_{\mathbf{k}}\frac{f_0(E_{\mathbf{k+q}})-f_0(E_{\mathbf{k}})}{E_{\mathbf{k+q}}-E_{\mathbf{k}}-\hbar\omega+i\hbar\alpha}. \quad (10)$$

This result was first obtained by Lindhard[4] with the SCF method and later by Nozières and Pines[13] using a many-particle approach based on the random-phase approximation for a Fermi gas at zero temperature.

2. PROOF OF EQUIVALENCE

In this section, we show that the approximations required to obtain (1) or (8) from the many-electron Hamiltonian are just those characteristic of the Sawada-Brout scheme. It proves useful to consider the total second-quantized Hamiltonian in the coordinate representation:

$$\mathcal{H}=-\left(\frac{\hbar^2}{2m}\right)\int\psi^\dagger(\mathbf{x})\nabla^2\psi(\mathbf{x})d\mathbf{x}$$
$$+\frac{1}{2}\int\left(\frac{e^2}{|\mathbf{x}-\mathbf{y}|}\right)\psi^\dagger(\mathbf{y})\psi^\dagger(\mathbf{x})\psi(\mathbf{x})\psi(\mathbf{y})d\mathbf{x}d\mathbf{y}. \quad (11)$$

The operator $\psi(\mathbf{x})=\sum_{\mathbf{k}}a_{\mathbf{k}}e^{i\mathbf{k}\cdot\mathbf{x}}$ develops in time according to

$$i\hbar(\partial/\partial t)\psi(\mathbf{x})=[\psi,\mathcal{H}], \quad (12)$$

where $a_{\mathbf{k}},a_{\mathbf{k}}^\dagger$ are, respectively, annihilation and creation operators referring to the state k and satisfy the usual commutation rules for Fermi particles. Performing the commutation indicated in (12), we find the Schrödinger-like equation

$$i\hbar(\partial/\partial t)\psi(\mathbf{x})=H_{\mathrm{op}}(\mathbf{x})\psi(\mathbf{x}), \quad (13)$$

with

$$H_{\mathrm{op}}(\mathbf{x})=-\left(\frac{\hbar^2}{2m}\right)\nabla_x^2+\int d\mathbf{y}\,\rho(\mathbf{y},\mathbf{y})\frac{e^2}{|\mathbf{x}-\mathbf{y}|}$$
$$\equiv H_0(\mathbf{x})+V_{\mathrm{op}}(\mathbf{x}). \quad (14)$$

The density matrix operators $\rho_{\mathrm{op}}(\mathbf{x},\mathbf{x}')$ and $\rho_{\mathrm{op}}(\mathbf{k},\mathbf{k}')$ in the x and k representations, respectively, are defined by

$$\rho_{\mathrm{op}}(\mathbf{x},\mathbf{x}')=\psi^\dagger(\mathbf{x}')\psi(\mathbf{x})$$
$$=\Omega^{-1}\sum_{\mathbf{kk}'}\rho_{\mathrm{op}}(\mathbf{k},\mathbf{k}')\exp[i(\mathbf{k}\cdot\mathbf{x}-\mathbf{k}'\cdot\mathbf{x}')], \quad (15)$$

and satisfy the equations of motion

$$i\hbar(\partial/\partial t)\rho_{\mathrm{op}}(\mathbf{x},\mathbf{x}')=H_{\mathrm{op}}(\mathbf{x})\rho_{\mathrm{op}}(\mathbf{x},\mathbf{x}')$$
$$-\rho_{\mathrm{op}}(\mathbf{x},\mathbf{x}')H_{\mathrm{op}}(\mathbf{x}'), \quad (16)$$

$$i\hbar(\partial/\partial t)\rho_{\mathrm{op}}(\mathbf{k},\mathbf{k+q})=(E_{\mathbf{k}}-E_{\mathbf{k+q}})\rho_{\mathrm{op}}(\mathbf{k},\mathbf{k+q})$$
$$+\sum_{\mathbf{k}'\mathbf{q}'}v_{q'}[\rho_{\mathrm{op}}(\mathbf{k}',\mathbf{k}'+\mathbf{q}')$$
$$\times\rho_{\mathrm{op}}(\mathbf{k}+\mathbf{q}',\mathbf{k+q})$$
$$-\rho_{\mathrm{op}}(\mathbf{k},\mathbf{k+q}-\mathbf{q}')\rho_{\mathrm{op}}(\mathbf{k}',\mathbf{k}'+\mathbf{q}')]. \quad (17)$$

In Eq. (16), $H_{op}(\mathbf{x})$ operates only on x, and $H_{op}(\mathbf{x}')$ operates only on x'. Equations (16) and (17) are exact.

We observe that the right-hand member of Eq. (16) is just the commutator of $H_{op}(\mathbf{x})$ and $\rho_{op}(\mathbf{x},\mathbf{x}')$ in a mixed representation. In complete operator form, (16) becomes

$$i\hbar\dot{\rho}_{op}=[H_{op},\rho_{op}]. \tag{18}$$

Equation (18) is just the formal analog of Eq. (1) in second quantization. For a single system in the state Ψ at time t, we have

$$\rho=(\Psi,\rho_{op}\Psi), \tag{19}$$

for the value of ρ at time t. For an ensemble of systems, (19) is replaced by its ensemble average. We denote either average by $\langle\rho_{op}\rangle_{Av}$. Without loss of generality we may average (18) and obtain

$$i\hbar\langle\dot{\rho}_{op}\rangle_{Av}=\langle[H_{op},\rho_{op}]\rangle_{Av}. \tag{20}$$

On the other hand, (1) is evidently

$$i\hbar\langle\dot{\rho}_{op}\rangle_{Av}=[\langle H_{op}\rangle_{Av},\langle\rho_{op}\rangle_{Av}], \tag{21}$$

because $V_s=\langle V_{op}\rangle_{Av}$ and therefore $H=\langle H_{op}\rangle_{Av}$ according to (2) and (14). Thus the approximation required for obtaining (1) from (18) is the replacement of the expectation value of a product of density matrices by the product of the expectation values, e.g.,

$$\langle\rho_{op}(\mathbf{y},\mathbf{y})\rho_{op}(\mathbf{x},\mathbf{x}')\rangle_{Av}\rightarrow\langle\rho_{op}(\mathbf{y},\mathbf{y})\rangle_{Av}\langle\rho_{op}(\mathbf{x},\mathbf{x}')\rangle_{Av}. \tag{22}$$

The approximation in (22) will be designated as "Hartree factorization." In order to arrive at the equation corresponding to (1) which contains exchange, we would have to replace the right side of (22) by the appropriately antisymmetrized combination (Hartree-Fock factorization).

In deriving Eqs. (4) and (8) the further approximation of linearization has been made. Without linearization, Eq. (8) would have been the factorized form of Eq. (17):

$$\begin{aligned}i\hbar(\partial/\partial t)\langle\mathbf{k}|\rho|\mathbf{k}+\mathbf{q}\rangle&=(E_{\mathbf{k}}-E_{\mathbf{k}+\mathbf{q}})\langle\mathbf{k}|\rho|\mathbf{k}+\mathbf{q}\rangle\\&\quad+\textstyle\sum_{\mathbf{k}'\mathbf{q}'}v_{q'}[\langle\mathbf{k}'|\rho|\mathbf{k}'+\mathbf{q}'\rangle\langle\mathbf{k}+\mathbf{q}'|\rho|\mathbf{k}+\mathbf{q}\rangle\\&\qquad-\langle\mathbf{k}|\rho|\mathbf{k}+\mathbf{q}-\mathbf{q}'\rangle\langle\mathbf{k}'|\rho|\mathbf{k}'+\mathbf{q}'\rangle],\end{aligned} \tag{23}$$

where

$$\langle\mathbf{k}|\rho|\mathbf{k}+\mathbf{q}\rangle=\langle\rho_{op}(\mathbf{k},\,\mathbf{k}+\mathbf{q})\rangle_{Av}. \tag{24}$$

By neglecting all terms $\mathbf{q}'\neq\mathbf{q}$ in (23), which corresponds to the random phase approximation[8] (RPA), we find

$$\begin{aligned}i\hbar(\partial/\partial t)\langle\mathbf{k}|\rho|\mathbf{k}+\mathbf{q}\rangle&=(E_{\mathbf{k}}-E_{\mathbf{k}+\mathbf{q}})\langle\mathbf{k}|\rho|\mathbf{k}+\mathbf{q}\rangle\\&\quad+v_q[\langle\mathbf{k}+\mathbf{q}|\rho|\mathbf{k}+\mathbf{q}\rangle\\&\qquad-\langle\mathbf{k}|\rho|\mathbf{k}\rangle]\textstyle\sum_{\mathbf{k}'}\langle\mathbf{k}'|\rho|\mathbf{k}'+\mathbf{q}\rangle.\end{aligned} \tag{25}$$

Equations (25) and (8) are seen to be almost identical. The correspondence with the formulation of the SCF method given in Sec. 1 is established completely by identifying $\langle\mathbf{k}|\rho|\mathbf{k}\rangle$ in (25) with $f_0(E_{\mathbf{k}})$ in (8) and $\langle\mathbf{k}|\rho|\mathbf{k}+\mathbf{q}\rangle$ with $\langle\mathbf{k}|\rho^{(1)}|\mathbf{k}+\mathbf{q}\rangle$. The RPA is therefore seen to be equivalent to treatment of the off diagonal elements of ρ as perturbations with respect to the diagonal components.

Sawada and Brout have derived an approximate equation of motion of the electron-hole pair creation operator $a_{\mathbf{k}+\mathbf{q}}{}^{\dagger}a_{\mathbf{k}}=\rho_{op}(\mathbf{k},\,\mathbf{k}+\mathbf{q})$. The approximations they employ are (1) the RPA and (2) the replacement of diagonal components $\rho_{op}(\mathbf{k},\mathbf{k})$ by their expectation values for the free-electron gas. The resulting equation for $\rho_{op}(\mathbf{k},\,\mathbf{k}+\mathbf{q})$,

$$\begin{aligned}i\hbar(\partial/\partial t)\rho_{op}(\mathbf{k},\,\mathbf{k}+\mathbf{q})&=(E_{\mathbf{k}}-E_{\mathbf{k}+\mathbf{q}})\rho_{op}(\mathbf{k},\,\mathbf{k}+\mathbf{q})+v_q[f_0(E_{\mathbf{k}+\mathbf{q}})-f_0(E_{\mathbf{k}})]\\&\quad\times\textstyle\sum_{\mathbf{k}'}\rho_{op}(\mathbf{k}',\,\mathbf{k}'+\mathbf{q}),\end{aligned} \tag{26}$$

has the same form as Eq. (25) for $\langle\mathbf{k}|\rho|\mathbf{k}+\mathbf{q}\rangle$ which corresponds to Eq. (8) in the SCF method. *Thus the* SCF *equation is simply the average of the Sawada-Brout equation.* Both schemes involve the RPA as well as a form of factorization. The extra generality of the scheme of Sawada and Brout resulting from their partial averaging is unnecessary for calculating properties of the system associated with one-electron operators only. Such a one-electron operator O has the form $O_{op}=\mathrm{tr}\{O_{op}\rho_{op}\}$ in second quantization. Its expectation value $\langle O_{op}\rangle_{Av}=(\Psi,O_{op}\Psi)=\mathrm{tr}\{O\rho\}$ involves only $\langle\rho_{op}\rangle_{Av}=\rho$. The same value of $\langle O_{op}\rangle_{Av}$ clearly obtains whether we solve the SCF equation for ρ or use the Sawada-Brout scheme. Therefore, the simplicity and ease of interpretation of the SCF method commend it for such problems as the calculation of the dielectric constant (Sec. 1) and the response of the system to a general external perturbation.

The SCF method does not yield by itself results depending on two-electron operators such as those entering the correlation energy. The more general Sawada-Brout equation (26), however, can be used in such problems. Nevertheless, the SCF calculation of the dielectric constant can be used in conjunction with an elegant formula due to Nozières and Pines[13] which exactly relates the correlation energy of the electron gas to its dielectric constant. Nozières and Pines obtain in this way the result of Sawada[10] and Hubbard[12] for the correlation energy.

A normal-mode analysis of the SCF equation, (8) or (25), can be performed following the procedure used by Brout[11] in connection with the Sawada-Brout equation, (26). The formal similarity of the two equations ensures that the condition for the existence of plasma oscillations and the corresponding dispersion relation is the same in both cases, namely the vanishing of the dielectric constant of Eq. (10). This normal mode analysis also affords a rigorous justification of the *Ansatz* for the time dependence of V_s used in obtaining (10).

Particular advantages of the present formulation of the SCF method are (1) the Liouville equation reduces in the classical limit to the ordinary Boltzmann equation used in transport theory and (2) the method applies to systems at finite temperatures. These features are not shared by formulations using the Hartree equation as the point of departure.[3,4] As a direct consequence of the normal-mode analysis already mentioned, one finds that at finite temperature the imaginary part of the dielectric constant does not vanish. Thus plasma oscillations do not exist as independent excitations of the system at finite temperature, and the plasma resonance is broadened. Further, the method is readily generalized to real crystals, as described in the next section.

3. APPLICATIONS TO REAL SOLIDS

To illustrate the simplicity and utility of the SCF method, we calculate the dielectric constant for a real solid and examine the plasma dispersion relation for plasma oscillations characterized by $q\cong 0$. We make the simplifying assumption that the core states of the atoms composing the solid are sufficiently tightly bound and the valence or conduction bands are sufficiently broad that local field corrections and hence umklapp processes may be neglected. The results obtained in elementary fashion by the SCF method agree with those previously obtained by Adams,[14] Wolff,[3] Nozières and Pines,[13] and others.

The generalization of the SCF method as presented in Sec. 1 is obtained by replacing the Hamiltonian H_0 by that for an electron in the unperturbed periodic lattice and the wave functions $|\mathbf{k}\rangle$ by $|\mathbf{k}l\rangle=\Omega^{-\frac{1}{2}}u_{\mathbf{k}l}(\mathbf{x})e^{i\mathbf{k}\cdot\mathbf{x}}$. Here $u_{\mathbf{k}l}(\mathbf{x})$ is the spatially periodic part of the wave function corresponding to wave vector $\mathbf{k}$ and band l. In terms of the integral

$$(\mathbf{k}l|\mathbf{k}+\mathbf{q},l')\equiv v_a^{-1}\int_0 u_{\mathbf{k}l}^*(\mathbf{x})u_{\mathbf{k}+\mathbf{q},l'}(\mathbf{x})d\mathbf{x},\qquad(27)$$

which extends over the unit cell, we find the generalizations of Eqs. (4), (5), and (8) to be, respectively,

$$\begin{aligned}i\hbar(\partial/\partial t)\langle\mathbf{k}l|\rho^{(1)}|\mathbf{k}+\mathbf{q},l'\rangle\\ =(E_{\mathbf{k}l}-E_{\mathbf{k}+\mathbf{q},l'})\langle\mathbf{k}l|\rho^{(1)}|\mathbf{k}+\mathbf{q},l'\rangle\\ +[f_0(E_{\mathbf{k}+\mathbf{q},l'})-f_0(E_{\mathbf{k}l})]V(q,t)(\mathbf{k}l|\mathbf{k}+\mathbf{q},l'),\qquad(28)\end{aligned}$$

$$\begin{aligned}n=\Omega^{-1}\textstyle\sum_{\mathbf{q}}e^{-i\mathbf{q}\cdot\mathbf{x}}\sum_{\mathbf{k},l,l'}(\mathbf{k}+\mathbf{q},l'|\mathbf{k}l)\\ \times\langle\mathbf{k}l|\rho^{(1)}|\mathbf{k}+\mathbf{q},l'\rangle,\qquad(29)\end{aligned}$$

and

$$\begin{aligned}i\hbar(\partial/\partial t)\langle\mathbf{k}l|R|\mathbf{k}+\mathbf{q},l'\rangle\\ =(E_{\mathbf{k}l}-E_{\mathbf{k}+\mathbf{q},l'})\langle\mathbf{k}l|R|\mathbf{k}+\mathbf{q},l'\rangle\\ +v_q|(\mathbf{k}l|\mathbf{k}+\mathbf{q},l')|^2[f_0(E_{\mathbf{k}+\mathbf{q},l'})-f_0(E_{\mathbf{k}l})]\\ \times\textstyle\sum_{\mathbf{k}'}\sum_{nn'}\langle\mathbf{k}'n|R|\mathbf{k}'+\mathbf{q},n'\rangle,\qquad(30)\end{aligned}$$

where

$$\langle\mathbf{k}l|R|\mathbf{k}+\mathbf{q},l'\rangle=(\mathbf{k}+\mathbf{q},l'|\mathbf{k}l)\langle\mathbf{k}l|\rho^{(1)}|\mathbf{k}+\mathbf{q},l'\rangle.\qquad(31)$$

[14] E. N. Adams, Phys. Rev. 85, 41 (1952).

The longitudinal frequency and wave-number dependent dielectric constant again is obtained straightforwardly from the relationship $\nabla\cdot\mathbf{P}=en$. We find

$$\epsilon(\omega,q)=1-\lim_{\alpha\to 0}v_q\sum_{\mathbf{k},l,l'}|(\mathbf{k}l|\mathbf{k}+\mathbf{q},l')|^2\times\frac{f_0(E_{k+q,l'})-f_0(E_{kl})}{E_{k+q,l'}-E_{kl}-\hbar\omega+i\hbar\alpha}.\qquad(32)$$

The Liouville-Poisson equation (30) and the expression for the dielectric constant (32) differ from that for the free electron case in that v_q is replaced by $v_q|(\mathbf{k}|\mathbf{k}+\mathbf{q},l')|^2$ and the summation over wave number is replaced by one extending over the band indices as well.

It is interesting to exhibit explicit expressions for the real and imaginary parts, $\epsilon_1(\omega,q)$ and $\epsilon_2(\omega,q)$, of the dielectric constant, which are easily given in the limit $q\to 0$. Perturbation theory in this limit yields

$$|(\mathbf{k}l|\mathbf{k}+\mathbf{q},l')|^2=\delta_{ll'}+(1-\delta_{ll'})(q/m\omega_{ll'})^2|P_{ll'}{}^{\mu}|^2,\qquad(33)$$

where

$$P_{ll'}{}^{\mu}=v_a^{-1}\int_0 u_{\mathbf{k}l'}{}^*p^{\mu}u_{\mathbf{k}l}d^3x,$$

p^{μ} being the momentum operator associated with the direction of propagation of the wave $\mathbf{q}$, and $\hbar\omega_{l'l}=E_{\mathbf{k}l'}-E_{\mathbf{k}l}$. After expanding other quantities depending on q about $q=0$, we find to lowest order

$$\begin{aligned}\epsilon_1(\omega,0)=1-(e/\pi\hbar\omega)^2\sum_l\int d^3k\,f_0(E_{\mathbf{k}l})\partial^2E_{\mathbf{k}l}/\partial k_\mu{}^2\\ +m^{-1}(e/\pi)^2\sum_{ll'}{}'\,\mathcal{P}\int d^3k\\ \times f_0(E_{\mathbf{k}l})f_{l'l}{}^{\mu}(\omega_{l'l}{}^2-\omega^2)^{-1},\qquad(34)\end{aligned}$$

$$\begin{aligned}\epsilon_2(\omega,0)=(e^2/q^2\pi)\sum_l\int d^3k\,f_0(E_{\mathbf{k}l})\\ \times[\delta(E_{\mathbf{k}+\mathbf{q},l}-E_{\mathbf{k}l}-\hbar\omega)-\delta(E_{\mathbf{k}-\mathbf{q},l}-E_{\mathbf{k}l}+\hbar\omega)]\\ +\pi^{-1}(e/m\omega)^2\sum_{ll'}{}'\int d^3k[f_0(E_{\mathbf{k}l})-f_0(E_{\mathbf{k}l'})]\\ \times|P_{ll'}{}^{\mu}|^2\delta(E_{\mathbf{k}l'}-E_{\mathbf{k}l}-\hbar\omega).\qquad(35)\end{aligned}$$

Here $\mathcal{P}$ denotes that principal parts of the corresponding integrals are to be taken and the prime on the summations that terms $l=l'$ are to be excluded. Further the oscillator strength

$$f_{l'l}{}^{\mu}=(2/\hbar\omega_{l'l}m)|P_{l'l}{}^{\mu}|^2$$

has been introduced.

We note that the terms in ϵ_1 and ϵ_2 consist of intraband and interband contributions. In ϵ_2 the first term, corresponding to one-electron intraband excitations, has not been approximated in order to exhibit its close relationship to the free electron case. The second term is associated with interband optical absorption. This is not surprising since for $q=0$ the transverse dielectric constant, which describes the interaction of the solid with electromagnetic waves, can be shown to equal the longitudinal dielectric constant which is associated with collective motion of the electrons. This leads to the result first pointed out by Wolff[3] and by Fröhlich and Pelzer[5] that the plasma frequency of infinite wavelength and the damping of these oscillations in the solid may be deduced from optical data. Further it is seen that even for q very small, with the exception of semiconductors or semimetals, ϵ_2 does not vanish for solids having any reasonable structure of the conduction bands so that plasma oscillations are always damped and cannot strictly exist as normal modes of the system. The plasma frequency in the limit of small q is obtained from $\epsilon_1(\omega,0)=0$.

In the case of insulators, for which all bands $l\leqslant L$ are filled and those for which $l>L$ are empty, the second term of Eq. (34) vanishes. Because of the relationship $f_{l'l}{}^{\mu}=-f_{ll'}{}^{\mu}$, the double sum in the third term of (34) may be written in the following two alternative forms: (i) $\sum_{l\leqslant L,\, l'>L}$ and (ii) $\sum_{l\leqslant L,\, l'}$. The first form immediately leads to the theorem that the plasma frequency of an insulator cannot fall inside the band gap. The second form permits the deduction of the plasma frequency in the limiting case that $\omega\ll\omega_{cv}$, where $\hbar\omega_{cv}$ is the energy difference between the valence and core bands, and that $\omega\gg\omega_{l'v}$ for all bands l' that contribute appreciably to the f-sum rule

$$\sum_l f_{l'v}{}^{\mu}=1-(m/\hbar^2)\partial^2E_{\mathbf{k}v}/\partial k_{\mu}{}^2.$$

In that case the plasma frequency is given by $\omega_p{}^2=4\pi n_v e^2/\epsilon_c m$ where n_v is the density of valence electrons, ϵ_c is the dielectric constant due to the core, and m is the free electron mass. We note also that the condition $\omega\gg\omega_{l'v}$ insures that ϵ_2 is small so that the plasma oscillations are only weakly damped.

For a "free-electron" metal with spherical surfaces of constant energy and a single partly occupied conduction band l, the second term of (34) becomes $-4\pi n_l e^2/\omega^2 m^*$ where m^* is the isotropic effective mass of the conduction band and n_l is the electron density in the band. The third term gives rise to a shift of the plasma frequency due to the presence of interband transitions. In the limit that this term is small, the shift agrees with that calculated by Adams.[14] The shift may be neglected in the limit that $\omega_p\ll\omega_{l'l}$ for all bands l' contributing appreciably to the f-sum rule. This is properly true only in semiconductors and possibly semimetals. On the other hand, in the case of a metal having a number of close-lying valence bands v in addition to the conduction band l, well separated from the core states, for which $\omega_p\gg\omega_{l'v}$ for all bands l' contributing appreciably to the f-sum rule, we find again that $\omega_p{}^2=4\pi n_{vl}e^2/\epsilon_c m$. Here n_{vl} is the density of valence plus conduction electrons. This shows that the plasma frequency is the same for metals and insulators provided that the plasma frequency is sufficiently large.

Finally, in the case of simple intrinsic semiconductors or semimetals, where the plasma frequency is sufficiently small, the interband term in Eq. (34) gives rise to a real dielectric constant ϵ due to the filled core and almost completely filled valence band. Collective motions can therefore exist as independent normal modes. Here, however, we have the simultaneous presence of an electron and a hole plasma. As first pointed out by Pines,[15] the normal modes of these interacting plasmas correspond to "optical" and "acoustical" vibrations just as in the case of lattice vibrations of a solid containing two atoms per unit cell. From Eq. (34) it is immediately seen that the optical mode frequency for infinite wavelength is $\omega_p{}^2=(4\pi ne^2/\epsilon)(m_n{}^{*-1}+m_p{}^{*-1})$, where $m_n{}^*$ and $m_p{}^*$ are the effective masses associated, respectively, with the conduction and valence band. The acoustic frequencies vanish at infinite wavelengths and are obtained only if the intraband terms in Eq. (32) are expanded to a higher order in q. If this is done, the results agree with those of Nozières and Pines.[13]

ACKNOWLEDGMENTS

Helpful comments from E. O. Kane and R. Brout are gratefully acknowledged.

[15] D. Pines, Can. J. Phys. 34, 1379 (1956).

The Theory of a Fermi Liquid

L. D. LANDAU
Institute for Physical Problems, Academy of Sciences, USSR
(Submitted to JETP editor March 7, 1956)
J. Exptl. Theoret. Phys. (U.S.S.R.) **30**, 1058-1064 (June, 1956)

A theory of the Fermi liquid is constructed, based on the representation of the perturbation theory as a functional of the distribution function. The effective mass of the excitation is found, along with the compressibility and the magnetic susceptibility of the Fermi liquid. Expressions are obtained for the momentum and energy flow.

As is well known, the model of a Fermi gas has been employed in a whole series of cases for the consideration of a system of Fermi particles, in spite of the fact that the interaction among such particles is not weak. Electrons in a metal serve as a classic example, Such a state of the theory is unsatisfactory, since it leaves unclear what properties of the gas model correspond to reality and what are intrinsic to such a gas.

For this purpose we must keep in mind that the problem is concerned with definite properties of the energy spectrum ("Fermi type spectrum"), for whose existence it is necessary, but not sufficient that the particles which compose the system obey Fermi statistics, i.e., that they possess half-integer spin. For example, the atoms of deuterium interact in such a manner that they form molecules. As a result, liquid deuterium possesses an energy spectrum of the Bose type. Thus the presence of a Fermi energy spectrum is connected not only with the properties of the particles, but also with the properties of their interaction.

A liquid of the Bose type was first considered by the author of the present article in application to the properties of He II. It follows from the character of the spectrum of such a liquid that a viscous liquid of Bose particles necessarily possesses superfluid properties. The converse theorem that a liquid consisting of Fermi particles cannot be superfluid, in accord with the above, is in general form not true.

1. THE ENERGY AS A FUNCTIONAL OF THE DISTRIBUTION ENERGY

If we consider a Fermi gas at temperatures which are low in comparison with the temperature of degeneration, and introduce some weak interaction between the atoms of this gas, then, as is known, the collision probability for a given atom, which is found in the diffuse Fermi zone, is proportional not only to the intensity of the interaction, but also to the square of the temperature. This shows that for a given intensity of interaction, the "indeterminacy of the momenta", associated with the finite path length, is also small for low temperatures, not only in comparison with the size of the momentum itself, but also in comparison with the width of the Fermi zone, proportional to the first power of the temperature.

As a basis for the construction of the type of spectrum under consideration, is the assumption that, as we gradually "turn on" the interaction between the atoms, i.e., in the transition from the gas to the liquid, classification of the levels remains invariant. The role of the gas particles in this classification is assumed by the "elementary excitations" (quasi-particles), each of which possesses a definite momentum. They obey Fermi statistics, and their number always coincides with the number of particles in a liquid. The quasi-particle can, in a well-known sense, be considered as a particle in a self-consistent field of surrounding particles. In the presence of a self-consistent field, the energy of the particle depends on the state of the surrounding particles, but the energy of the whole system is no longer equal to the sum of the energies of the individual particles, and is a functional of the distribution function.

We consider an infinitely small change in the distribution function of quasi-particles n Then we can write down the change in the energy density of the system in the form

$$\delta E = \int \varepsilon \delta n d\tau, \tag{1}$$

where $d\tau = dp_x dp_y dp_z / (2\pi\hbar)^3$. The quantity $\epsilon(p)$ is a function of the derivative of the energy with respect to the distribution function. It corresponds to a change in the energy of the system upon the addition of a single quasi-particle with momentum p, and it can be regarded as the Hamiltonian function of the added quasi-particle with given momentum in the self-consistent field.

However, we have not taken it into account in Eq. (1) that the particles possess spin. Since the spin is a quantum mechanical quantity, it cannot be considered by classical means. We must therefore consider the distribution function of the statistical matrices in regard to spin, and replace Eq. (1) by the following:

$$\delta E = \mathrm{Sp}_\sigma \int \varepsilon \delta n d\tau, \tag{2}$$

where Sp_σ is the spur over the spin states. The quantity ϵ in the general case is also an operator which depends on the spin operators. If we have an equilibrium liquid, which is not in an external magnetic field, then, because of isotropy, the energy cannot depend on the spin operators. We limit ourselves to the consideration of particles with $s = \frac{1}{2}$.

We can show that just this energy ϵ enters into the formula for the Fermi distribution of the quasi-particles. Actually, it is reasonable to determine the entropy of the liquid by the following way:

$$S = -\mathrm{Sp}_\sigma \int \{n \ln n + (1-n) \ln (1-n)\} d\tau. \tag{3}$$

By means of a variation, subject to the additional conditions

$$\delta N = \mathrm{Sp}_\sigma \int \delta n d\tau = 0, \quad \delta E = \mathrm{Sp}_\sigma \int \varepsilon \delta n \, d\tau = 0,$$

we can obtain the Fermi distribution

$$n(\varepsilon) = [e^{(\varepsilon-\mu)/\Theta} + 1]^{-1}. \tag{4}$$

from this equation. We note that ϵ, being a functional of n, naturally depends on the temperature also.

In correspondence with (4) the heat capacity of a Fermi liquid at low temperatures will be proportional to the temperature. It is determined by the same formula as for the Fermi gas, with one exception, that in place of the real mass m of the particles therein, we place the effective mass of the quasi-particles

$$m^* = \frac{p}{\partial \varepsilon \, \partial p}\bigg|_{p=p_0}, \tag{5}$$

where p_0 is the limiting momentum of the Fermi distribution of quasi-particles at absolute zero.

Not only $\epsilon(p)$ for a given distribution, but also the change in e produced by a change in n, is of essential importance for the theory of the Fermi liquid:

$$\delta\varepsilon(p) = \mathrm{Sp}_{\sigma'} \int f(\mathbf{p}, \mathbf{p}') \delta n' \, d\tau'. \tag{6}$$

Being a second variational derivative, the function f is a symmetric relative to p and p′; moreover, it depends on the spins.

If the principal distribution n is isotropy, then the function f in the general case contains terms of the form $\varphi_{ik}(\mathbf{p}, \mathbf{p}')\sigma_i\sigma_k'$, where σ_i is the spin operator, and if the interaction is exchange, only terms of the form

$$\varphi(\mathbf{p}, \mathbf{p}')(\sigma\sigma').$$

will appear.

We can consider the function f from the following point of view. The number of acts of scattering of quasi-particles per unit volume per unit time can be written in the form

$$dW = \frac{2\pi}{\hbar}|F(\mathbf{p}_1, \mathbf{p}_2; \mathbf{p}_1', \mathbf{p}_2')|^2 \delta (\varepsilon_1 + \varepsilon_2 - \varepsilon_1' - \varepsilon_2') n_1 n_2 (1 - n_1')(1 - n_2') d\tau_1 d\tau_2 d\tau_1', \tag{7}$$

where conservation of momentum is assumed: $\mathbf{p}_1 + \mathbf{p}_2 = \mathbf{p}_1' + \mathbf{p}_2'$. The quantity f is nothing else but $-F(\mathbf{p}_1, \mathbf{p}_2; \mathbf{p}_1, \mathbf{p}_2)$, i.e., the amplitude of the scattering on 0^0 (with opposite sign). Generally speaking, this amplitude is complex, its imaginary part being determined by the total effective scattering cross section. Inasmuch as we assume that the real acts of scattering are highly improbable, we can neglect the imaginary part.

2. RELATIONS WHICH FOLLOW FROM THE PRINCIPLE OF GALILEAN RELATIVITY

If we deal with a liquid which is not in an external field, then it follows from the principle of Galilean relativity that the momentum arriving at a unit volume must be equal to the density of mass flow*. Inasmuch as the velocity of the quasi-particle is $\partial\epsilon/\partial\mathbf{p}$, and the number of quasi-particles coincides with the number of real particles, we have

$$\mathrm{Sp}_\sigma \int \mathbf{p} n\, d\tau = \mathrm{Sp}_\sigma \int m \frac{\partial \varepsilon}{\partial \mathbf{p}} n d\tau. \tag{8}$$

Therefore, the variational derivatives with respect to n ought to be the same on both sides of this equation. Then

$$\frac{1}{m}\mathrm{Sp}_\sigma \int \mathbf{p}\delta n d\tau = \mathrm{Sp}_\sigma \int \frac{\partial \varepsilon}{\partial \mathbf{p}} \delta n d\tau + \mathrm{Sp}_\sigma \mathrm{Sp}_{\sigma'} \int \frac{\partial}{\partial \mathbf{p}} f(\mathbf{p}, \mathbf{p}')\delta n' n d\tau\, d\tau'.$$

Since the quantity δn is arbitrary, we obtain

$$\frac{\mathbf{p}}{m} = \frac{\partial \varepsilon}{\partial \mathbf{p}} + \mathrm{Sp}_{\sigma'} \int \frac{\partial f}{\partial \mathbf{p}'} n' d\tau' = \frac{\partial \varepsilon}{\partial \mathbf{p}} - \mathrm{Sp}_{\sigma'} \int f \frac{\partial n'}{\partial \mathbf{p}'} d\tau' \tag{9}$$

(the left side is understood as the unit matrix in the spins).

If we deal with the isotropic case, then it is sufficient that the Eq. (9) hold for the spurs, i.e.,

$$\frac{\mathbf{p}}{m} = \frac{\partial \varepsilon}{\partial \mathbf{p}} - \frac{1}{2}\mathrm{Sp}_\sigma \mathrm{Sp}_{\sigma'} \int f \frac{\partial n'}{\partial \mathbf{p}'} d\tau'. \tag{10}$$

We note that this formula determines the function ϵ through the quantity f with accuracy to within a constant.

Let us consider Eq. (10) for momenta close to the boundary of the Fermi distribution. For low temperatures, the function $\partial n/\partial p$ will differ slightly from the δ-function. For this reason, we can carry out the integration in Eq. (10) over the absolute value of the momentum, leaving only the integration over the angle. This gives the following relation between the real and the effective masses:

$$\frac{1}{m} = \frac{1}{m^*} + \frac{p_0}{2(2\pi\hbar)^3}\mathrm{Sp}_\sigma \mathrm{Sp}_{\sigma'} \int f \cos\theta\, d\Omega. \tag{11}$$

Inasmuch as, in this formula, both of the vector arguments in f correspond to the Fermi surface, the function f depends only on the angle between them.

3. COMPRESSIBILITY OF THE FERMI LIQUID

Let us express the compressibility (at absolute zero) by the more appropriate quantity for us, $\partial\mu/\partial N$. For this purpose, we note that as a consequence of homogeneity, the chemical potential μ depends only on the ratio N/V. Consequently, we have

$$\frac{\partial \mu}{\partial N} = -\frac{V \partial\mu\, \partial V}{N} = -\frac{V^2}{N}\frac{\partial p}{\partial V}. \tag{12}$$

* This conclusion does not apply in particular to electrons in a metal. For them, p is not the momentum, but the quasi-momentum.

For the square of the velocity of sound, we have

$$c^2 = \frac{\partial p}{\partial (mN/V)} = \frac{1}{m}\left(N\frac{\partial \mu}{\partial N}\right). \quad (13)$$

Thus the problem reduces to the calculation of the derivative $\partial\mu/\partial N$. Inasmuch as $\mu = \epsilon(p_0) = \epsilon_0$, the change in the chemical potential $\delta\mu$ which is brought about as a result of the change in the total number of particles δN, will be equal to*

$$\delta\mu = \frac{1}{2}\,\mathrm{Sp}_\sigma \mathrm{Sp}_{\sigma'}\int f\delta n'\,d\tau' + \frac{\partial \epsilon_0}{\partial p_0}\delta p_0. \quad (14)$$

The second term is connected with the fact that for a change δN the limiting momentum p_0 changes by an amount δp_0.

For the case of spin ½, δN and δp_0 are connected by the relation

$$\delta N = 8\pi p_0^2 \delta p_0 V/(2\pi\hbar)^3. \quad (15)$$

The value of the function under the integral in Eq. (14) is appreciable only for values of momentum close to p_0. Therefore, we can carry out integration over the absolute value of p, obtaining

$$\mathrm{Sp}_\sigma \mathrm{Sp}_{\sigma'}\int f\delta n'd\tau' = \frac{1}{8\pi V}\mathrm{Sp}_\sigma \mathrm{Sp}_{\sigma'}\int f do\,\delta N. \quad (16)$$

We get from Eq. (14), with the help of Eqs. (15) and (16):

$$\partial\mu/\partial N = \mathrm{Sp}_\sigma \mathrm{Sp}_{\sigma'}\int f do/16\pi V + (2\pi\hbar)^3/8\pi p_0 m^* V. \quad (17)$$

Now let us make use of Eq. (11) and express the effective mass m^* in the expression that has been obtained by the mass of the particles, m. We have

$$\frac{\partial\mu}{\partial N} = \frac{1}{16\pi V}\int \mathrm{Sp}_\sigma \mathrm{Sp}_{\sigma'} f(1-\cos\theta)\,do + \frac{(2\pi\hbar)^3}{8\pi p_0 m V}.$$

Furthermore, multiplying the resultant equation by $N/m = (1/m)8\pi p_0^3 V/3(2\pi\hbar)^3$, we find an expression for the square of the sound velocity:

$$c^2 = \frac{p_0^2}{3m^2} + \frac{1}{6m}\left(\frac{p_0}{2\pi\hbar}\right)^3 \int \mathrm{Sp}_{\sigma,\sigma'} f(1-\cos\theta)\,do. \quad (18)$$

* Equation (14) is obtained as a result of taking the spur of the analogous expression which contains the spin operators.

4. MAGNETIC SUSCEPTIBILITY

We calculate the magnetic susceptibility of a Fermi liquid. If the system is located in a magnetic field H, then the additional energy of a free particle in this field is equal to $\beta\sigma$H. Moreover, we must also consider the fact that the form of the distribution function also changes in the presence of a magnetic field. Consequently, in calculating the magnetic susceptibility, we must keep in mind that

$$\delta\varepsilon = -\beta(\sigma\mathbf{H}) + \mathrm{Sp}_{\sigma'}\int f\delta n'\,d\tau', \quad (19)$$

i.e., it is impossible to neglect the effect of the term containing f. We write f in the form

$$f = \varphi + \psi(\sigma\sigma'), \quad (20)$$

where the second term takes into account the exchange interaction between the particle s. Furthermore, in calculating δn, which depends on the field, the change in the chemical potential $\delta\mu$ does not have to be considered. This change appears as a quantity of second order of smallness relative to the field H, while $\delta\epsilon$ is of first order with respect to the field. Therefore, we can substitute $\delta n = (\partial n/\partial\epsilon)\,\delta\epsilon$ in Eq. (19). We then have

$$\delta\varepsilon = -\beta(\sigma\mathbf{H}) + \mathrm{Sp}_{\sigma'}\int f\frac{\partial n'}{\partial\varepsilon'}\delta\varepsilon'\,d\tau'. \quad (21)$$

We shall look for $\delta\epsilon$ in the form

$$\delta\varepsilon = -\gamma(\sigma\mathbf{H}). \quad (22)$$

The quantity γ is defined by Eq. (21)*

$$\gamma = \beta + \frac{1}{2}\int\psi\frac{\partial n}{\partial\varepsilon'}\gamma' d\tau'. \quad (23)$$

Remembering the δ-character of $\partial n/\partial\epsilon$, we than obtain

$$\gamma = \beta - {}^1\!/_2\overline{\psi}_0\gamma\,(\partial\tau/\partial\varepsilon)_0. \quad (24)$$

Here the index zero indicates that the values of all functions are taken at $p = p_0$; the bar over the symbol indicates averaging over the angles. On the other hand, the susceptibility is defined by the relation

$$\chi\mathbf{H} = \beta\,\mathrm{Sp}\int n\sigma d\tau$$

* Here we make use of the relation $\mathrm{Sp}_\sigma(\sigma\sigma')\sigma' = 1/3\,\sigma\mathrm{Sp}_{\sigma'}(\sigma'\sigma') = 1/2\,\sigma$.

or

$$\chi \mathbf{H} = -\beta \, \mathrm{Sp} \int \frac{\partial \tau}{\partial \varepsilon} \gamma (\mathbf{H}\sigma) \sigma \, d\tau = \frac{\mathbf{H}}{2} \beta \gamma \left(\frac{d\tau}{d\varepsilon}\right)_0 \quad (25)$$

Hence, we get finally,

$$\frac{1}{\chi} = \frac{2}{\beta \gamma_0 (\partial \tau / \partial \varepsilon)_0} \quad (26)$$

$$= \frac{2}{\beta^2 (d\tau/d\varepsilon)_0} \left(1 + \frac{1}{2} \overline{\psi}_0 \left(\frac{d\tau}{d\varepsilon}\right)_0\right).$$

Further, we can replace $(d\tau/d\epsilon)_0$ by the coefficient α in the linear heat capacity law. Then

$$1/\chi = \beta^{-2} \{2\pi^2 k^2 / 3\alpha + \overline{\psi}_0\}. \quad (27)$$

It is then evident that there does not exist in the liquid the relation between the heat capacity and the susceptibility that exists in gases. The term with $\overline{\psi}_0$ takes the exchange interaction into account and is large for liquids. Thus, for He^3, analysis of the experimental data[1] shows that $\overline{\psi}_0$ is negative and amounts to about 2/3 of the first term.

5. THE KINETIC EQUATION

In the absence of a magnetic field and for a neglect of the magnetic spin-orbit interaction, ϵ does not depend on the operator σ and the kinetic equation in the quasi-classical approximation takes the form

$$\frac{\partial n}{\partial t} + \frac{\partial n}{\partial \mathbf{r}} \frac{\partial \varepsilon}{\partial \mathbf{p}} - \frac{\partial n}{\partial \mathbf{p}} \frac{\partial \varepsilon}{\partial \mathbf{r}} = I(n). \quad (28)$$

The necessity of calculation of dervatives of the energy ϵ with respect to the coordinates in the absence of an external field is connected with the fact that ϵ is a functional of n, and the distribution function n depends on the coordinates.

We find the expression for the momentum flux. For This purpose, we multiply the left and right sides of the equation above by the momentum p_i and integrate over all phase space. We have

$$\frac{\partial}{\partial t} \mathrm{Sp} \int p_i n \, d\tau + \mathrm{Sp} \int p_i \left(\frac{\partial n}{\partial x_k} \frac{\partial \varepsilon}{\partial p_k} - \frac{\partial n}{\partial p_k} \frac{\partial \varepsilon}{\partial x_k}\right) d\tau \quad (29)$$

$$= \mathrm{Sp} \int p_i I(n) \, d\tau.$$

As a consequence of the conservation of momentum for collisions, the right side of the equation is zero, while the left side yields, after simple transformations,

$$\frac{\partial}{\partial t} \int p_i n \, d\tau + \frac{\partial}{\partial x_k} \int p_i \frac{\partial \varepsilon}{\partial p_k} n \, d\tau$$

$$- \int p_i \frac{\partial}{\partial p_k} \left(n \frac{\partial \varepsilon}{\partial x_k}\right) d\tau = 0. \quad (30)$$

Finally, integrating the three integrals by parts, we get

$$\frac{\partial}{\partial t} \int p_i n \, d\tau + \frac{\partial}{\partial x_k} \int p_i \frac{\partial \varepsilon}{\partial p_k} n \, d\tau \quad (31)$$

$$+ \int n \frac{\partial \varepsilon}{\partial x_i} d\tau = 0.$$

The integral $\mathrm{Sp} \int n (\partial \epsilon / \partial x_i) d\tau$ can be represented in the form [see (2)]

$$\mathrm{Sp} \int n \frac{\partial \varepsilon}{\partial x_i} d\tau = \mathrm{Sp} \frac{\partial}{\partial x_i} \int n \varepsilon \, d\tau - \mathrm{Sp} \int \varepsilon \frac{\partial n}{\partial x_i} d\tau$$

$$= \frac{\partial}{\partial x_i} [\mathrm{Sp} \int n \varepsilon \, d\tau - E].$$

Thus we finally obtain the law of conservation of momentum:

$$\frac{\partial}{\partial t} \mathrm{Sp} \int p_i n \, d\tau + \frac{\partial \Pi_{ik}}{\partial x_k} = 0, \quad (32)$$

where the tensor of momentum flux is

$$\Pi_{ik} = \mathrm{Sp} \int p_i \frac{\partial \varepsilon}{\partial p_k} n \, d\tau + \delta_{ik} \left[\mathrm{Sp} \int n \varepsilon \, d\tau - E\right]. \quad (33)$$

In a similar way we obtain the expression for the energy flow. We multiply the left and right sides of the kinetic equation (28) by ϵ and integrate over all phase space. We have

$$\mathrm{Sp} \int \varepsilon \frac{\partial n}{\partial t} d\tau + \mathrm{Sp} \int \varepsilon \left(\frac{\partial n}{\partial \mathbf{r}} \frac{\partial \varepsilon}{\partial \mathbf{p}} - \frac{\partial n}{\partial \mathbf{p}} \frac{\partial \varepsilon}{\partial \mathbf{r}}\right) d\tau$$

$$= \mathrm{Sp} \int \varepsilon I(n) \, d\tau.$$

As a consequence of the conservation of energy under collisions, the right side is zero while the left side reduces without difficulty to the form

$$\int \varepsilon \frac{\partial n}{\partial t} d\tau + \frac{\partial}{\partial \mathbf{r}} \int n \varepsilon \frac{\partial \varepsilon}{\partial \mathbf{p}} d\tau = 0.$$

Taking Eq. (2) into account, we have finally,

$$\frac{\partial E}{\partial t} + \mathrm{div} \, \mathbf{Q} = 0, \quad (34)$$

where the energy flow is

$$Q = \mathrm{Sp}\int ns\frac{\partial\varepsilon}{\partial\mathbf{p}}d\tau. \qquad (35)$$

In the solution of concrete kinetic problems it is necessary to keep in mind the following circumstances. For such a solution we usually write down the function n in the form of a sum of equilibrium functions n_0 and correction δn. In this case the departure of the tensor of momentum flow Π_{ik} and the vector of energy flow Q from their equilibrium values will result as a consequence of the direct change of the function n by the quantity δn, as well as from the change in ϵ which comes about as a result of the functional dependence of ϵ on n [Eq. (2)].

In conclusion, I express my gratitude to I. M. Khalatnikov and A. A. Abrikosov for fruitful discussions.

---

[1] Fairbank, Ard and Walters, Phys. Rev. **95**, 566 (1954).

Translated by R. T. Beyer
221

Oscillations in a Fermi Liquid

L. D. LANDAU
Institute for Physical Problems, Academy of Sciences, USSR

(Submitted to JETP editor September 15,1956)
J. Exptl. Theoret.Phys. (U.S.S.R.) 32, 59–66 (January,1957)

Different types of waves that can be propagated in a Fermi liquid, both at absolute zero and at non–zero temperatures, are investigated. Absorption of these waves is also considered.

THE present paper is devoted to the study of the propagation of waves in a Fermi liquid, and proceeds from the general theory of such liquids developed by the author.[1] These phenomena in a Fermi liquid should be distinguished by a large singularity, connected primarily with the impossibility of propagation in it of ordinary hydrodynamic sound waves at absolute zero. The latter circumstance is already evident from the fact that the path length, and therefore the viscosity of a Fermi liquid, tends to infinity for $T \to 0$, as a result of which the sound absorption coefficient increases without limit.

It is shown, however, that in a Fermi liquid at

absolute zero other waves can be propagated; these differ in nature from ordinary sound, and we shall call them waves of "zero sound".

Initially, the problem of vibrations in a Fermi liquid was considered by Gol'dman[2] in application to an electron gas with Coulomb interaction between the particles. The problem of a gas with uncharged particles, considered in detail here for liquids, was first considered in the research of Klimontovich and Silin,[3] and later in a series of works of Silin.[4-6] There, the gas was considered to be slightly non-ideal, with an interaction satisfying the conditions of applicability of perturbation theory.

1. VIBRATIONS IN A FERMI LIQUID AT ABSOLUTE ZERO

We begin with the investigation of those vibrations at absolute zero which do not involve the spin characteristics of the liquid. This means that not only the equilibrium distribution function n_0, but also the "perturbing" function

$$n = n_0 + \delta n(\mathbf{p}) \qquad (1)$$

is independent of the spin variables. At absolute zero, n_0 is a step function which is broken off at the limiting momentum $p = p_0$. *

The energy of the quasi-particles (elementary excitations) is a function of n, i.e., the form of the function $\epsilon(p)$ depends on the form of $n(p)$. By analogy to (1), we write it in the form

$$\varepsilon = \varepsilon_0(p) + \delta\varepsilon(\mathbf{p}), \qquad (2)$$

where the function $\epsilon_0(p)$ corresponds to the distribution $n_0(p)$. The value of $\delta\epsilon$ itself is connected with δn by a formula of the form (see Ref. 1):

$$\delta\varepsilon(\mathbf{p}) = \mathrm{Sp}_{\sigma'}\int f(\mathbf{p}, \mathbf{p}')\,\delta n' d\tau', \qquad (3)$$

$$d\tau = d^3\mathbf{p}/(2\pi\hbar)^3.$$

Inasmuch as δn is assumed to be independent of the spin variable, the operation Sp is applied only to the scattering amplitude f. But the scalar function $\mathrm{Sp}_{\sigma'}f$ can contain the spin operator σ only in the form of the product $\sigma\,[\mathbf{pp}']$ of two axial vectors: σ and $[\mathbf{pp}']$ (we do not consider expressions containing two products of components of σ, since for spin 1/2, as is well known, they reduce to expressions containing σ in the zeroth or first degree). But this product is not invariant to a time reversal and therefore cannot enter into the invariant quantity $\delta\epsilon$. Thus σ drops out completely and $\delta\epsilon$ is shown to be independent of the spin variable.

The kinetic equation for a Fermi liquid has the form:

$$\frac{\partial n}{\partial t} + \frac{\partial n}{\partial \mathbf{r}}\frac{\partial \varepsilon}{\partial \mathbf{p}} - \frac{\partial n}{\partial \mathbf{p}}\frac{\partial \varepsilon}{\partial \mathbf{r}} = I(n), \qquad (4)$$

where $I(n)$ is the integral of collisions between quasi-particles. The number of collisions is proportional to the square of the width of the diffusion zone, so that at absolute zero, $I(n) = 0$. Substituting (1) and (2) in (4), and considering that n_0 and ϵ_0 do not depend on r, we get

$$\frac{\partial \delta n}{\partial t} + \frac{\partial \delta n}{\partial \mathbf{r}}\frac{\partial \varepsilon_0}{\partial \mathbf{p}} - \frac{\partial \delta\varepsilon}{\partial \mathbf{r}}\frac{\partial n_0}{\partial \mathbf{p}} = 0,$$

and assuming δn and $\delta\epsilon$ to be proportional to

$$e^{-i\omega t + i\mathbf{kr}},$$

$$(\mathbf{kv} - \omega)\,\delta n = \mathbf{kv}\frac{\partial n_0}{\partial \varepsilon}\delta\varepsilon, \qquad (5)$$

where we have introduced the velocity of the quasi-particles $\mathbf{v} = \partial\epsilon_0/\partial\mathbf{p}$. In view of the absence of the δ-function $\partial n_0/\partial\epsilon$ from the right hand side of this equation, there actually enter in them only the values of all quantities taken at the limit $p = p_0$ of the (unperturbed) Fermi distribution. We introduce a new notation for what follows:

$$F = \mathrm{Sp}_{\sigma'}f(\mathbf{p}, \mathbf{p}')\,4\pi p^2 dp/(2\pi\hbar)^3 d\varepsilon. \qquad (6)$$

Then we can write Eq. (3) in the form:

$$\delta\varepsilon = \iint F\,\delta n' d\varepsilon' do'/4\pi.$$

Here only the $\delta n'$ are functions changing rapidly with ϵ'. Therefore, we can rewrite this expression in the form:

$$\delta\varepsilon = \int F\nu' do'/4\pi, \qquad (7)$$

where the function

---

*To avoid excessive complication of our study, we limited ourselves to the simplest and most important case of an energy spectrum with an occupied region represented by a uniform sphere of radius p_0.

$$\nu(\mathbf{n}) = \int \delta n(\mathbf{p})\, d\varepsilon \qquad (8)$$

has been introduced which depends only on the direction n of the vector p, and the function $F(\mathbf{p},\mathbf{p}')$ is taken on the boundary of the (unperturbed) Fermi distribution; here F depends only on the angle χ between p and p′.

We note for what follows that the relation found in Ref. 1, which connects the actual mass m of the particles with the effective mass m^* of the quasi–particles, can, with the help of the function $F(\chi)$ be written in the form

$$\overline{F \cos \chi} = (m^*/m) - 1, \qquad (9)$$

where the bar denotes averaging over the directions (in the derivation of this relation, we assume in (6) that $\epsilon = p^2 / 2m^*$). The equation for the velocity of ordinary sound c can be put in the form

$$\overline{F} = 3\, n m m^* c^2 / p_0^2 - 1. \qquad (10)$$

Let us substitute (7) in Eq. (5) and integrate the latter over $d\epsilon$. This gives

$$(\mathbf{kv} - \omega)\nu = -\mathbf{kv} \int F\nu\, do'/4\pi.$$

Let us take the direction of k as the polar axis, and let the angles θ, φ define the direction of the momentum p (and the direction of v coinciding with it) relative to this axis. Also, we introduce the propagation velocity $u = \omega/k$ of this wave, and the notation $\eta = u/v$, so that we can finally write the resultant equation in the form

$$(\eta - \cos\theta)\,\nu(\theta, \varphi) \qquad (11)$$

$$= \cos\theta \int F(\chi)\,\nu(\theta', \varphi')\, do'/4\pi.$$

This integral equation defines the principal velocity of propagation of the waves and the form of the function $\nu(\theta, \varphi)$ in them. The latter has the following graphic meaning. The fact that δn is proportional [as is evident from Eq. (5)] to the derivative $\partial n_0/\partial\epsilon$ means that the change of the distribution function for vibrations reduces to the deformation of the boundary of the Fermi surface (a sphere in the undisturbed distribution). The integral of (8) represents the magnitude of the displacement (in energy units) of this surface in the given direction n.

We at once note that it follows from the form of Eq. (11) that the real (only the undamped vibrations are of interest to us) value of η ought to exceed 1, i.e., the propagation velocity of the waves satisfies the inequality

$$u > v. \qquad (12)$$

As an example, let us investigate the case in which the function $F(\chi)$ reduces to a constant (we denote it by F_0). The integral on the right hand side of Eq. (11) does not depend on the angles θ, φ in this case. Therefore the desired function ν has the form (we omit the exponential factor):

$$\nu = \text{const} \cdot \cos\theta / (\eta - \cos\theta). \qquad (13)$$

The limiting Fermi surface has the form of a surface of revolution, elongated in the forward direction of the propagation of the wave, and flattened in the opposite direction. For comparison, let us point out that the ordinary sound wave corresponds to a function ν of the form $\nu = \text{const} \cdot \cos\theta$, which represents the displacement of the Fermi surface as a whole, without a change in shape.

For the determination of the velocity u, we substitute Eq. (13) in (11) and get

$$\frac{F_0}{4\pi} \int_0^{\pi} \frac{\cos\theta}{\eta - \cos\theta}\, 2\pi \sin\theta\, d\theta = 1.$$

Carrying out the integration, we find the following equation, which determines in implicit form the velocity of the wave for a given value of F_0:

$$\varphi(\eta) \equiv \frac{\eta}{2} \ln \frac{\eta + 1}{\eta - 1} - 1 = \frac{1}{F_0}. \qquad (14)$$

The function $\varphi(\eta)$ decreases monotonically from $+\infty$ to 0 for a change of η from 1 to ∞, always remaining positive. It then follows that the waves under consideration can exist only for $F_0 > 0$. Inasmuch as the function F is proportional to the scattering amplitude, taken with opposite sign (at the angle 0°), of the quasi–particles with one another [(see Ref. 1)], then the latter must be negative, which corresponds to the mutual collision of quasi–particles. However, it must be emphasized that this conclusion applies only to the case F = const. If the function $F(\chi)$ is not constant (and at the same time is not small compared with unity; see below), then propagation of zero sound is in general possible, bor both attractive and repulsive interactions of the quasi–particles.

For $\eta \to \infty$: $\varphi(\eta) \approx 1/3\eta^2$. Therefore, large F_0 corresponds to $\eta = \sqrt{F_0/3}$. In the opposite case of $F_0 \to 0$, we find that η tends toward unity according to the relation

$$\eta - 1 \sim e^{-2F_0}. \quad (15)$$

The latter case has much more general value. It corresponds to zero sound in an almost ideal Fermi gas for arbitrary form of the function $F(\chi)$. Actually, an almost ideal gas corresponds to a function F which is small in absolute magnitude. It is seen from Eq. (11) that in this case η will be close to unity and the function ν will be significantly different from zero only for small angles θ. On this basis, and being concerned only with this range of angles, we can replace the function F in the integral on the right side of Eq. (11) by its value for $\chi = 0$ (for $\theta \to 0$ and $\theta' \to 0$, $\chi \to 0$ also). As a result, we again recover Eqs. (13) and (15) with the constant F_0 replaced by $F(0)$ (this result coincides with that obtained earlier by Silin[4]).

We note that in a weakly non–ideal Fermi gas, the velocity of zero sound exceeds the velocity of ordinary sound by a factor of $\sqrt{3}$. Actually, for the former, we have $\eta \approx 1$, i.e., $u \approx v$. For the velocity of ordinary sound we get from Eq. (10) (neglecting the term $\bar{F}$ in it and setting $m^* \approx m$): $c^2 \approx p_0{}^2/3m^2 = v^2/3$

In the general case of an arbitrary dependence of $F(\chi)$, the solution of Eq. (11) is not well defined. In principle, it permits the existence of different types of zero sound, which are distinguished from one another by the angular dependence of their amplitude $\nu(\theta, \varphi)$, and which are propagated with different velocities. Along with the axially symmetric solutions of $\nu(\theta)$, asymmetric solutions can also exist. In these ν has an azimuthal factor $e^{\pm im\varphi}$ (m = integer)

Thus, for a function $F(\chi)$ of the form

$$F = F_0 + F_1 \cos\chi \quad (16)$$
$$= F_0 + F_1(\cos\theta\cos\theta' + \sin\theta\sin\theta'\cos(\varphi - \varphi'))$$

solutions can exist with

$$\nu \sim e^{\pm i\varphi}.$$

Actually, substituting Eq. (16) in (11) and carrying out the integration over $d\varphi'$ (assuming in this case that $\nu = f(\theta)e^{i\varphi}$), we obtain

$$(\eta - \cos\theta) f = \frac{F_1}{4}\cos\theta\sin\theta\int_0^\pi \sin^2\theta' f' d\theta'.$$

Thence,

$$\nu = \text{const}\cdot\frac{\sin\theta\cos\theta}{\eta - \cos\theta}e^{i\varphi}. \quad (17)$$

Conversely, substituting this expression in the equation, we obtain the relation

$$\int_0^\pi \frac{\sin^3\theta\cos\theta}{\eta - \cos\theta}d\theta = \frac{4}{F_1}, \quad (18)$$

which determines the dependence of the propagation velocity on F_1. The integral on the left side of the equation falls off monotonically with increase in the function η. Therefore its maximum possible value is achieved for $\eta = 1$. Computing the integral, we find that the corresponding (the least achieved) value of F_1 is 6. Thus, propagation of the asymmetric wave of the form (17) is possible only for $F_1 > 6$.

Turning to a real Fermi liquid—the liquid He^3—it is reasonable to attempt to approximate the unknown function $F(\chi)$ by the two term expression (16). We can determine the coefficients F_0 and F_1 entering into it by means of the relations

$$F_0 = 3mm^*c^2/p_0^2 - 1, \quad F_1/3 = m^*/m - 1$$

[see Eqs. (9) and (10)], knowing the values of the effective mass m^* and the velocity of ordinary sound c. We can derive the first from experimental data on the temperature dependence of the entropy (in the lowest temperature region). From the data available at present,[7] we get $m^* = 1.43\,m$ (m is the mass of the He^3 atom). For the velocity c, we get 195 m/sec from the data of Walters and Fairbank[8] on the compressibility of liquid He^3. Finally, p_0 is obtained directly from the density of the liquid:

$$p_0/\hbar = 0.76\times10^8 \text{ cm}^{-1}.$$

On the basis of these data, we obtain

$$F_0 = 5.4; \quad F_1 = 1.3. \quad (19)$$

From these values, we can draw a conclusion about the fact that in liquid He^3 the propagation of asymmetric zero sound is impossible. For symmetric zero sound, the solution of the equation with the function $F(\chi)$ from (16) and (19)* leads to the value $\eta = 1.83$, when we obtain $u = v = 1.83 p_0/m^* = 206$ m/sec.

The possibility of the propagation of waves in a Fermi liquid at absolute zero means that its energy spectrum can automatically possess a "Bose branch" in the form of phonons with energy $\epsilon = up$. However, one must say that it would be incorrect to introduce corrections corresponding to this branch in the thermodynamic quantities of the Fermi liquid, inasmuch as it has a much higher power of the temperature (T^3 in the heat capacity) than the departures from the approximate theory developed in Ref. 1.

2. VIBRATIONS OF A FERMI LIQUID AT TEMPERATURES ABOVE ZERO

For low, but non–zero, temperatures, mutual collisions of quasi–particles take place in the Fermi liquid. The number of these collisions is proportional to T^2. The corresponding relaxation time (the free path time) is $\tau \sim 1/T^2$. The character of the waves propagated in the liquid naturally depends fundamentally on the relations between their frequency and the reciprocal of the relaxation time.

For $\omega\tau \ll 1$ (which is actually equivalent to the condition of the shortness of the free path length of the quasi–particles in comparison with the wave–length λ), the collisions succeed in establishing thermodynamic equilibrium in each (small in comparison with λ) element of volume of the liquid. This means that we are dealing with ordinary hydrodynamical sound waves, propagated with a velocity c.

If $\omega\tau \gg 1$, then, on the contrary, the collisions do not play essential roles in the process of the propagation of the vibrations, and we will have the waves of zero sound considered in the preceding section.

In both these limiting cases, the propagation of waves is accompanied by a comparatively weak absorption. In the intermediate region, $\omega\tau \sim 1$, the absorption is very strong and isolation of the different types of waves as undamped processes is not possible here.

*These computations were carried out by A. A. Abrikosov and I. M. Khalatnikov.

One can easily obtain the temperature and frequency dependence of the absorption coefficient γ in the region of ordinary sound with the aid of the known formula for the absorption of sound (see Ref. 9, for example), according to which γ is proportional to the square of the frequency and to the viscosity coefficient*. Inasmuch as the viscosity of a Fermi liquid is proportional to $1/T^2$ [10], then we find that

$$\gamma \sim \omega^2/T^2 \quad \text{for} \quad \omega \ll 1/\tau. \tag{20}$$

Absorption in the region of zero sound differs essentially in its character from absorption of ordinary sound. In the latter, the collisions cannot lead to a dissipation of the energy "into the noise" of the distribution, which is changed only by the sound vibrations as such. This is connected with the circumstance already mentioned, that a distribution changed in this fashion remains in thermodynamic equilibrium in each element of the volume. Therefore, the absorption of ordinary sound is connected with the effect of the collisions on the distribution function itself.

In the region of zero sound the collisions lead to absorption "into the background" of the distribution which is changed only by the vibrations themselves, which in this case are not in thermodynamic equilibrium (inasmuch as the form of the limiting Fermi surface is deformed). This change in the distribution function does not depend on the frequency, and therefore the absorption coefficient will not depend on the frequency either. The dependence of γ on the temperature is determined by its proportionality to the number of collisions, i.e.,

$$\gamma \sim T^2 \quad \text{for} \quad \varkappa T/\hbar \gg \omega \gg 1/\tau. \tag{21}$$

The upper limit of the region of applicability of this formula is determined by the inequality $\hbar\omega \ll \varkappa T$ ($\varkappa$ is Boltzmann's constant), which allows a classical consideration of collisions. We recall that the inequality assumed here,

$$\varkappa T/\hbar \gg 1/\tau,$$

i.e.,

$$\hbar/\tau \ll \varkappa T$$

(smallness of the quantum uncertainty of the energy of quasi–particles in comparison with $\varkappa T$), must

*The contribution to γ from second viscosity and thermal conductivity is proportional to a much higher power of T and is therefore inconsiderable.

hold since it is the condition of applicability of everything generally developed in the theory of the Fermi liquid.[1]

The determination of the absorption coefficient of zero sound in the frequency range $\hbar\omega \gtrsim \varkappa T$ requires quantum consideration. The corresponding calculations can be simplified if we develop them in such a way that we express the desired "quantum" absorption coefficient in terms of the "classical" from Eq. (21).

The absorption of sound quanta takes place in the collisions of quasi-particles. If we denote by ϵ_1 and ϵ_2 the energies of the quasi-particles before and after collisions, then at a given frequency ω, they are connected by the law of conservation of energy

$$\varepsilon_1 + \varepsilon_2 + \hbar\omega = \varepsilon_1' + \varepsilon_2'.$$

In addition to the collisions, we must also consider the inverse collisions, which are accompanied by the emission of sound quanta. Taking into consideration the well known properties of the collision probabilities of Fermi particles, we find that the total rate of decrease of the number of sound quanta as a result of collisions is given by the expression

$$\iiiint w(\mathbf{p}_1, \mathbf{p}_2; \mathbf{p}_1', \mathbf{p}_2')\{n_1 n_2(1-n_1')(1-n_2') \quad (22)$$
$$- n_1' n_2'(1-n_1)(1-n_2)\}$$
$$\times \delta(\mathbf{p}_1' + \mathbf{p}_2' - \mathbf{p}_1 - \mathbf{p}_2 - \hbar\mathbf{k})$$
$$\times \delta(\varepsilon_1' + \varepsilon_2' - \varepsilon_1 - \varepsilon_2 - \hbar\omega)\, d\tau_1 d\tau_2 d\tau_1' d\tau_2'.$$

The delta functions in the integrand allow the satisfaction of the laws of conservation of energy and momentum.

In the integral (22), the essential values of the energy are only those in the region of diffuseness of the Fermi distribution. In this region, the expressions under the integral sign are changed strongly only by multipliers which contain $n(\epsilon)$. Furthermore, it should be noted that the angular integrals in (22) are practically unchanged in the transition from the "classical" region

$$\hbar\omega \ll \varkappa T\,.$$

to the "quantum" region

$$\hbar\omega \gg \varkappa T.$$

In view of this fact, it will be sufficient for us to calculate the integral

$$J = \iiiint \{n_1 n_2(1-n_1')(1-n_2') - n_1' n_2'(1-n_1)$$
$$\times (1-n_2)\,\delta(\varepsilon_1' + \varepsilon_2' - \varepsilon_1 - \varepsilon_2 - \hbar\omega)\, d\varepsilon_1 d\varepsilon_2 d\varepsilon_1' d\varepsilon_2',$$

taken only over the energy. Then, substituting

$$n(\varepsilon) = [e^{(\varepsilon-\mu)/\varkappa T} + 1]^{-1}$$

and introducing the notation

$$x = (\varepsilon - \mu)/\varkappa T, \quad \xi = \hbar\omega/\varkappa T,$$

we get (omitting the factor T^3)

$$J = \iiiint_{-\infty}^{+\infty} \frac{(1-e^{-\xi})\,\delta(x_1' + x_2' - x_1 - x_2 - \xi)\, dx_1 dx_2 dx_1' dx_2'}{(e^{x_1}+1)(e^{x_2}+1)(1+e^{-x_1'})(1+e^{-x_2'})}\,.$$

In view of the rapid convergence of the integral, the region of integration can be extended from $-\infty$ to $+\infty$.

For integration purposes, we transform to the variables x_1, x_2, y_1, y_2, where $y = x - x'$. Integration over x_1 and x_2 is elementary and gives

$$J = (1-e^{-\xi})$$
$$\times \iiiint_{-\infty}^{+\infty} \frac{\delta(y_1 + y_2 + \xi)\, dx_1\, dx_2\, dy_1\, dy_2}{(e^{x_1}+1)(e^{x_2}+1)(1+e^{-x_1+y_1})(1+e^{-x_2+y_2})}$$
$$= (1-e^{-\xi}) \iint_{-\infty}^{+\infty} \frac{y_1 y_2 \delta(y_1 + y_2 + \xi)\, dy_1\, dy_2}{(1-e^{y_1})(1-e^{y_2})}$$
$$= -(1-e^{-\xi}) \int_{-\infty}^{+\infty} \frac{y(\xi + y)\, dy}{(e^y - 1)(e^{-y-\xi} - 1)}$$
$$= \int_{-\infty}^{+\infty} y(\xi + y)\left\{\frac{1}{e^y - 1} - \frac{1}{e^{y+\xi} - 1}\right\} dx.$$

For calculation of the resulting difference of two diverging integrals, we introduce as an intermediate the finite lower limit $-\Lambda$ and write:

$$J = \int_{-\Lambda}^{+\infty} \frac{y(\xi + y)}{e^y - 1}\, dy - \int_{-\Lambda+\xi}^{+\infty} \frac{y(y-\xi)\, dy}{e^y - 1}$$
$$= 2\xi \int_{-\Lambda}^{\infty} \frac{y\,dy}{e^y - 1} - \int_{-\Lambda+\xi}^{-\Lambda} \frac{y(v-\xi)\, dy}{e^y - 1}.$$

Keeping in mind that we shall transform to the limit $\Lambda \to \infty$, we neglect e^y in the denominator of the second of the integrals. The first we rewrite in the form

$$\int_{-\Lambda}^{\infty} \frac{y\,dy}{e^y-1} = \int_{0}^{\infty} \frac{y\,dy}{e^y-1} + \int_{-\Lambda}^{0} \frac{y\,dy}{e^y-1}$$

$$= \frac{\pi^2}{6} + \int_{-\Lambda}^{0} \left(\frac{y}{1-e^{-y}} - y \right) dy$$

$$= \frac{\pi^2}{6} + \int_{0}^{\Lambda} \frac{y\,dy}{e^y-1} + \frac{\Lambda^2}{2}.$$

Carrying out reductions and then transforming to $\Lambda \to \infty$, we finally obtain

$$J = (2\xi\pi^2/3)(1 + \xi^2/4\pi^2).$$

The desired absorption coefficient γ is proportional to J. The coefficient of proportionality between them is so determined that for $\xi \ll 1$, $\gamma = \gamma_{cl}$. We then obtain:

$$\gamma = \gamma_{cl}\ [1 + (\hbar\omega/2\pi\varkappa T)^2] \quad \text{for} \quad \hbar\omega \ngtr \varkappa T. \tag{23}$$

Considering that $\gamma_{cl} \sim T^2$, we find that in the limit of high frequencies:

$$\gamma \sim \omega^2 \cdot \text{ for } \hbar\omega \ggg \varkappa T, \tag{24}$$

i.e., the absorption coefficient remains proportional to the square of the frequency, but does not depend on the temperature. We note that the transition from the formula for "low" to the formula for "high" frequencies takes place at

$$\hbar\omega \sim 2\pi\varkappa T,$$

(and not $\hbar\omega \sim \varkappa T$).* The result of (24 refers, in particular, to the zero sound of all frequencies at the absolute zero of temperature.

3. SPIN WAVES IN A FERMI LIQUID

In addition to a consideration of zero sound in Sec. 1, which does not involve the distribution of spins, in a Fermi liquid at absolute zero, waves of other types can also be propagated. These we call spin waves.*

In this section, we denote by K the function

$$K = f(\mathbf{p}, \mathbf{p}')\, 4\pi p^2\, dp\, / \,(2\pi\hbar)^3\, d\varepsilon, \tag{25}$$

in which the operator Sp is not used. In the calculation of exchange interaction between the quasi–particles, this function contains terms which are proportional to the product $\sigma\sigma'$, i.e., it has the form:[1]

$$K = {}^1/_2 F(\chi) + {}^1/_2 G(\chi)\, \sigma\sigma' \tag{26}$$

[F coincides with the function (6) used above].

In place of Eq. (11) we have now

$$(\eta - \cos\theta)\,\nu = \cos\theta\, \mathrm{Sp}_{\sigma'} \int F\nu'\, do' / 4\pi. \tag{27}$$

In addition to the solutions $\nu(n)$ considered earlier, which do not depend on the spin, this equation also has a solution of the form

$$\nu = \mu(\mathbf{n})\,\sigma. \tag{28}$$

Substituting (26) and (28) in (27), completing the operation Sp and dividing both sides of the equation by σ, we get

$$(\eta - \cos\theta)\,\mu = \cos\theta \int G\mu'\, do' / 16\pi. \tag{29}$$

We see that for each of the components of the vector μ, we obtain an equation which differs from (11) only by the replacement of F by $G/4$. Therefore, all the further calculations of Sec. 1 can immediately be applied to the spin waves.

In the real liquid He^3, we can determine from available experimental data on its magnetic susceptibility only the mean value of $\overline{G}$, which was pointed out previously—1.9. Inasmuch as this quantity is negative, then (in view of the results of Sec. 2) it is most probable that the propagation of spin waves in liquid He^3 is not possible. Such a conclusion, however, is in no sense categorical.

In conclusion, I wish to express my thanks to A. A. Abrikosov, E. M. Lifshitz and I. M. Khalatnikov for useful discussions.

*Considering the frequencies $\omega \gg \varkappa T/\hbar$, we at the same time assume satisfaction of the inequality

$$\hbar\omega \ll \varkappa T_0$$

(T_0 is the temperature of degeneration of the Fermi distribution). In the opposite case, particles from the "depth" of the Fermi distribution take part in the absorption and all the theory developed here would become inapplicable.

*The equation for spin waves in weakly non–ideal Fermi gas was considered by Silin.[6]

1 L. D. Landau, J. Exptl. Theoret. Phys. (U.S.S.R.) 30, 1058 (1956); Soviet Phys. JETP 3, 920 (1957).
2 I. I. Gol'dman, J. Exptl. Theoret. Phys. (U.S.S.R.) 17, 681 (1947).
3 Iu. L. Klimontovich and V. P. Silin, J. Exptl. Theoret. Phys. (U.S.S.R.) 23, 151 (1952).
4 V. P. Silin, J. Exptl. Theoret. Phys. (U.S.S.R.) 23, 641 (1952).
5 V. P. Silin, J. Exptl. Theoret. Phys. (U.S.S.R.) 27, 269 (1954).
6 V. P. Silin, J. Exptl. Theoret. Phys. (U.S.S.R.) 28, 749 (1955); Soviet Phys. JETP 1, 607 (1955).
7 Abraham, Osborne and Weisntock, Phys. Rev. 98, 551 (1955).
8 G. K. Walters and W. M. Fairbank, Phys. Rev. 103, 263 (1956).
9 L. D. Landau and E. M. Lifshitz, *Mechanics of Continuous Media*, 2nd edition; Moscow, 1954, Sec. 77.
10 I. Ia. Pomeranchuk, J. Exptl. Theoret. Phys. (U.S.S.R.) 20, 919 (1950).

Translated by R.T. Beyer
8

SOVIET PHYSICS JETP VOLUME 35 (8), NUMBER 1 JANUARY, 1959

ON THE THEORY OF THE FERMI LIQUID

L. D. LANDAU

Institute of Physical Problems, Academy of Sciences, U.S.S.R.

Submitted to JETP editor February 5, 1958

J. Exptl. Theoret. Phys. (U.S.S.R.) **35**, 97-103 (July, 1958)

A study is made of the zero-angle scattering in collisions of quasiparticles in a Fermi liquid. It is shown that the scattering amplitude for zero angle depends on the limit approached by the ratio of the momentum and energy transfers in the collision as both these quantities go to zero. It is ascertained which of these limits is connected with the inter: ction energy of the quasiparticles that occurs in the general theory of the Fermi liquid developed earlier by the writer.

A general theory of the Fermi liquid* has been developed in previous papers by the writer.[1,2] One of the quantities that plays an important part in this theory in characterizing the properties of the liquid is the function $f(p, p')$ which determines the interaction energy of the quasiparticles, i.e., the variation of the energy $\epsilon(p)$ of the quasiparticles arising from a variation of their distribution function:

$$\delta\varepsilon(p) = \mathrm{Sp}_{\sigma'} \int f(\mathbf{p}, \mathbf{p}')\, \delta n(\mathbf{p}')\, d\tau' \tag{1}$$

(where $d\ = d^3\mathbf{p}/(2\pi)^3$; here and below we take $\hbar = 1$).

In reference 1 it was shown that the function $f(\mathbf{p}, \mathbf{p}')$ is related in a definite way to the scattering amplitude of the quasi-particles in the liquid for their mutual collisions. The formulation given in reference 1 for this connection is not, however, quite accurate, as will be shown in the present paper.

We used below methods borrowed from quantum field theory; as is well known, these methods have recently been used with success by various authors in the study of the properties of quantum many-particle systems.

The main part in these methods is played by the Green's function G and the "vertex part" Γ. Let us recall the definitions and basic properties of these functions.

The function G is defined as the average value in the ground state of the system of the chronological product of two ψ operators:

$$G_{12} = -i\langle T(\psi_1 \overset{+}{\psi_2})\rangle. \tag{2}$$

The indices 1, 2 denote sets of values of the three coordinates and the time, and also of the spin index. As usual, we shall use below instead of the space-time representation (2) the Fourier expansion of this function. The only components different from zero are those with identical values of the two momenta and the two energies (that is, of the wave vectors and frequencies): $P_1 = P_2 \equiv P$; we denote by P the "four-momentum", i.e., the combination of the momentum $\mathbf{p}$ and the energy ϵ. In respcot to the spin indices (which we denote by Greek letters) the Fourier components $G_{\alpha\beta}(P) = \int G_{\alpha\beta}(X_1 - X_2)\, e^{-iP(X_1-X_2)} d^4(X_1 - X_2)$ are proportional to $\delta_{\alpha\beta}$; we shall write

$$G_{\alpha\beta}(P) = G(P)\,\delta_{\alpha\beta}. \tag{3}$$

As is well known, the poles of the function G(P) give the energies of the quasiparticles (the elementary excitations). In accordance with this, for p close to the boundary momentum p_0 and ϵ close to the boundary energy μ, G(P) has the form

$$G(P) \to \frac{a}{\varepsilon - \mu - v(p - p_0) + i\delta} \tag{4}$$

(μ is the chemical potential of the gas, and v is the speed of the quasiparticles at the Fermi boundary). This expression has a pole at

$$\varepsilon - \mu = v(p - p_0), \tag{5}$$

and the small constant δ is introduced in the usual way to specify the rule for going around the singularity in integrating; the sign of δ agrees with the sign of $\epsilon - \mu$ (or, what is the same thing near the pole, with the sign of $p - p_0$). The "renormalization" factor a is positive and, as has been shown by Migdal,[3] is smaller than unity:

$$a < 1. \tag{6}$$

*To avoid misunderstanding we emphasize that we are concerned not with simply a liquid composed of Fermi particles; it is also postulated that this liquid has an energy spectrum of the Fermi type, i.e., that it is not a superfluid.

The vertex part Γ is defined by means of the four-particle average value

$$\Phi_{1234} = \langle T(\psi_1\psi_2\psi_3^+\psi_4^+)\rangle. \tag{7}$$

The Fourier components of this function contain a part that is expressed in terms of functions $G(P)$ only, and a remainder that gives the definition of the Fourier component of the vertex part by the following formula:

$$\begin{aligned}\Phi_{\alpha\beta,\gamma\delta}(P_1, P_2; P_3, P_4) &= (2\pi)^8 G(P_1)G(P_2)[\delta(P_1-P_3)\delta(P_2-P_4)\delta_{\alpha\gamma}\delta_{\beta\delta} \\ &\quad - \delta(P_1-P_4)\delta(P_2-P_3)\delta_{\alpha\delta}\delta_{\beta\gamma}] \\ &\quad + iG(P_1)G(P_2)G(P_3)G(P_4) \\ &\quad \times\Gamma_{\alpha\beta,\gamma\delta}(P_1, P_2; P_3, P_4)(2\pi)^4\delta(P_1+P_2-P_3-P_4).\end{aligned} \tag{8}$$

Here the values of the arguments are connected by the relation

$$P_1 + P_2 = P_3 + P_4. \tag{9}$$

On interchange of the indices 1 and 2 (or 3 and 4) the function (7) changes sign; thus it follows from the definition (8) that Γ has the symmetry property:

$$\Gamma_{\alpha\beta,\gamma\delta}(P_1, P_2; P_3, P_4) = -\Gamma_{\beta\alpha,\gamma\delta}(P_2, P_1; P_3, P_4). \tag{10}$$

In the formation of the vertex part intermediate states occur that correspond to different values of the total number of particles in the system: the unchanged number N and the numbers $N \pm 2$. The latter arise from such arrangements of the ψ operators in the T-product as, for example, $\psi_1\psi_2\psi_3^+\psi_4^+$; the former correspond to arrangements such as, for example, $\psi_1\psi_3^+\psi_2\psi_4^+$. In accordance with this the contributions to the function Γ connected with these intermediate states have different characters in regard to their singularities. Namely, terms due to states that appear with the addition or removal of two particles have singularities with respect to the variables $P_1 + P_2$; terms corresponding to intermediate states with unchanged number of particles have singularities with respect to the variables $P_1 - P_3$ or $P_2 - P_4$.

The probability of scattering of quasi-particles with the transition

$$P_1\alpha,\ P_2\beta \rightarrow P_3\gamma,\ P_4\delta \tag{11}$$

is given in terms of the function Γ by the formula

$$\begin{aligned}dW_{\alpha\beta,\gamma\delta}(P_1, P_2; P_3, P_4) &= 2\pi|a^2\Gamma_{\alpha\beta,\gamma\delta}(P_1, P_2; P_3, P_4)|^2\delta(\varepsilon_1+\varepsilon_2-\varepsilon_3-\varepsilon_4) \\ &\quad\times n_1n_2(1-n_3)(1-n_4)\,d\tau_1 d\tau_2 d\tau_3\end{aligned} \tag{12}$$

[where $n_1, n_2, \ldots$ are the values of the distribution function for $P_1\alpha$, $P_2\beta$, and so on, and a is the renormalization constant from Eq. (4)]. The sign of Γ is defined in such a way that it corresponds to a positive scattering amplitude for repulsion and a negative amplitude for attraction.

Below we shall consider the function Γ for nearly equal values of the pairs of variables P_1, P_3 and P_2, P_4, i.e., we set $P_3 = P_1 + K$, $P_4 = P_2 - K$ with small K, and agree to write

$$\Gamma(P_1, P_2; P_1+K, P_2-K) \equiv \Gamma(P_1\, P_2; K). \tag{13}$$

In terms of the scattering process (11) this means that we are considering collisions of quasiparticles giving nearly "forward scattering."

In the lowest order of perturbation theory contributions to the function $\Gamma(P_1, P_2; K)$ are made by the diagrams shown in the figure (a, b, c).

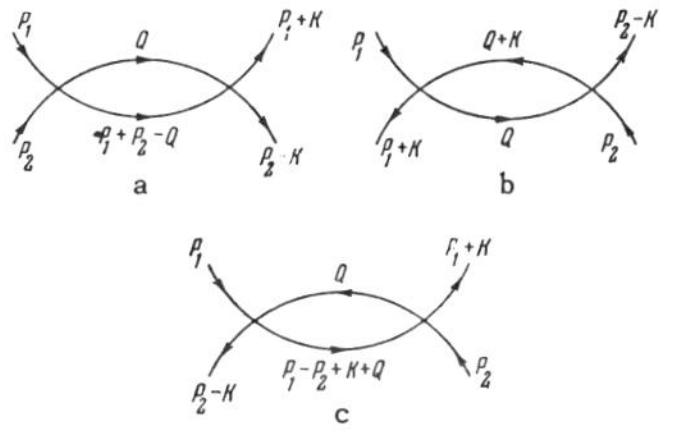

The internal parts of these diagrams correspond to the following propagation functions:

$$\text{(a) } G(Q)G(P_1+P_2-Q),\ \text{(b) } G(Q)G(K+Q),$$
$$\text{(c) } G(Q)G(P_1-P_2+K+Q),$$

where Q is the intermediate four-momentum over which one integrates. With arbitrary P_1 and P_2 there is nothing to distinguish the value $K = 0$ for the functions (a) and (c), and for small K we can put $K = 0$. In the case (b), on the other hand, for $K \rightarrow 0$ the poles of the two factors come together, so that diagrams of this type require special consideration.

To calculate Γ one must sum the entire series of perturbation theory. Since in doing this our purpose is to separate out the parts having a singularity at $K = 0$, we must first single out the contribution from all the diagrams that do not have any parallel pairs of lines with nearly equal (differing by K) values of the four-momentum. We denote by $\Gamma^{(1)}$ this part of the function Γ, which has no singularity at $K = 0$; in it we can simply put $K = 0$, so that $\Gamma^{(1)}$ will be a function of the variables P_1 and P_2 only: $\Gamma^{(1)} = \Gamma^{(1)}(P_1, P_2)$. The entire series that has to be summed can be written symbolically in the form

$$(:\Gamma:) = (:\Gamma_1:) + (:\Gamma_1:\Gamma_1:) + (:\Gamma_1:\Gamma_1:\Gamma_1:) + \cdots, \tag{14}$$

where the colons replace pairs of lines in the diagram with nearly equal values of the four-momentum, and Γ_1 denotes the set of all possible diagram elements that do not have such pairs.

The problem of summing this series (so-called "ladder" summation) reduces to the solution of an integral equation, to obtain which we "multiply" the series (14) by Γ_1, i.e., replace it by the series

$$(:\Gamma_1 : \Gamma :) = (:\Gamma_1 : \Gamma_1 :) + (:\Gamma_1 : \Gamma_1 : \Gamma_1 :) + \ldots .$$

Comparison of this with Eq. (14) leads to the equation

$$(:\Gamma :) - (:\Gamma_1 :) = (:\Gamma_1 : \Gamma :),$$

which, when written out in explicit form, is the desired integral equation

$$\Gamma_{\alpha\beta,\gamma\delta}(P_1, P_2; K) = \Gamma^{(1)}_{\alpha\beta,\gamma\delta}(P_1, P_2) - \frac{i}{(2\pi)^4}\int \Gamma^{(1)}_{\alpha\epsilon,\gamma\zeta}(P_1, Q)\, G(Q)\, G(Q+K)\, \Gamma_{\zeta\beta,\epsilon\delta}(Q, P_2; K)\, d^4Q \tag{15}$$

(in the first factor of the integrand we should, strictly speaking, have Q + K instead of the argument Q; but in view of the absence of singularities in $\Gamma^{(1)}$ we can here set K = 0). To investigate this equation we examine the product G(Q) G(Q + K) that occurs in the integrand. On substituting here G(P) in the form (4) we get

$$a^2/[\varepsilon - \mu - v(q - p_0) + i\delta_1] \times [\varepsilon + \omega - \mu - v(|\mathbf{q} + \mathbf{k}| - p_0) + i\delta_2]. \tag{16}$$

Here ϵ and **q** are the energy and momentum corresponding to the four-momentum Q, and $\epsilon + \omega$ and **q** + **k** are those corresponding to Q + K.

For small **k** and ω the expression (16), as a function of ϵ and q, behaves like δ functions of the arguments $\epsilon - \mu$ and $q - p_0$; that is, it has the form

$$A\delta(\varepsilon - \mu)\,\delta(q - p_0), \tag{17}$$

where the coefficient A depends on the angle θ between the vectors **k** and **q**. Comparing Eqs. (16) and (17), we see that this coefficient is given by the integral

$$A = \iint \frac{a^2 d\varepsilon\, dq}{[\varepsilon - \mu - v(q - p_0) + i\delta_1]\,[\varepsilon + \omega - \mu - v(|\mathbf{q} + \mathbf{k}| - p_0) + i\delta_2]}. \tag{18}$$

Let us first carry out the integration with respect to $d\epsilon$. The result of the integration depends essentially on the value of q. If the two differences $q - p_0$ and $|\mathbf{q} + \mathbf{k}| - p_0$ have the same sign, then we must also assign like signs to the quantities δ_1 and δ_2. The poles of the integrand then lie in one half-plane of the complex variable ϵ, and by closing the path of integration through the other half-plane we can see that the integral vanishes. Thus the integral is nonvanishing only for opposite signs of the differences $q - p_0$ and $|\mathbf{q} + \mathbf{k}| - p_0$. Let us first suppose that $\mathbf{qk} > 0$, i.e., $\cos\theta > 0$. Then the integral is nonvanishing for $q < p_0$, $|\mathbf{q} + \mathbf{k}| > p_0$, which, because of the smallness of **k**, is equivalent to the condition

$$p_0 - k\cos\theta < q < p_0. \tag{19}$$

In addition we must have for the quantities δ that $\delta_1 < 0$, $\delta_2 > 0$, so that the poles of the integrand lie in different half-planes. Closing the path of integration through one half-plane and calculating the integral from the residue at the corresponding pole, we find

$$A = \int \frac{2\pi i a^2 dq}{\omega - v(|\mathbf{q} + \mathbf{k}| - q)}.$$

Since by Eq. (19) q is nearly equal to p_0 and varies over a range $k\cos\theta$, we can put $|\mathbf{q} + \mathbf{k}| - q = k\cos\theta$, so that

$$A = \frac{2\pi i a^2 k\cos\theta}{\omega - vk\cos\theta}.$$

Let us note the peculiar character of this expression: its limit for $k \to 0$, $\omega \to 0$ depends on the limit approached by the ratio ω/k.

It is easy to show in the same way that for $\cos\theta < 0$ (in which case the integration must be taken over the region $q > p_0$, $|\mathbf{q} + \mathbf{k}| < p_0$) one gets the same expression for $A(\theta)$. Thus we have

$$G(Q)\,G(Q+K) = \frac{2\pi i a^2\, \mathbf{l}\cdot\mathbf{k}}{\omega - v\,\mathbf{l}\cdot\mathbf{k}}\,\delta(\varepsilon - \mu)\,\delta(q - p_0) + g(Q), \tag{20}$$

where $\boldsymbol{l}\mathbf{k}$ has been written instead of $k\cos\theta$ ($\boldsymbol{l}$ is the unit vector in the direction of **q**), and g(Q) does not contain any δ-function part (for small K), so that in it we can put K = 0.

Substituting Eq. (20) into Eq. (15), we get the fundamental integral equation in the form

$$\Gamma_{\alpha\beta,\gamma\delta}(P_1, P_2; K) = \Gamma^{(1)}_{\alpha\beta,\gamma\delta}(P_1, P_2) - \frac{i}{(2\pi)^4}\int \Gamma^{(1)}_{\alpha\epsilon,\gamma\zeta}(P_1, Q)\, g(Q)\, \Gamma_{\zeta\beta,\epsilon\delta}(Q, P_2; K)\, d^4Q + \frac{a^2 p_0^2}{(2\pi)^3}\int \Gamma^{(1)}_{\alpha\epsilon,\gamma\zeta}(P_1, Q)\, \Gamma_{\zeta\beta,\epsilon\delta}(Q, P_2, K)\, \frac{\mathbf{l}\cdot\mathbf{k}}{\omega - v\,\mathbf{l}\cdot\mathbf{k}}\, do. \tag{21}$$

In the last term we have put $d^4Q = q^2\, dq\, do\, d\epsilon$, where do is an element of solid angle in the direction of $\boldsymbol{l}$, and have carried out the integration of the δ functions in the integrand with respect to $dq\, d\epsilon$. In the arguments of the functions $\Gamma^{(1)}$ and Γ in this term Q is taken on the Fermi surface, i.e., it consists of the momentum $\mathbf{q} = p_0\boldsymbol{l}$

and the constant energy μ.

Because of the special character of the kernel of the integral equation as noted above, its solution also has just the same character: the limit of the function $\Gamma(P_1, P_2; K)$ for $K \to 0$ depends on the way in which **k** and ω go to zero, i.e., on the limit of the ratio ω/k.

Let us denote by $\Gamma^{\omega}(P_1, P_2)$ the limit

$$\Gamma^{\omega}_{\alpha\beta,\gamma\delta}(P_1, P_2) = \lim_{K\to 0} \Gamma_{\alpha\beta,\gamma\delta}(P_1, P_2; K) \quad \text{for } k/\omega \to 0 \tag{22}$$

[we shall see below that it is just this quantity with which the function $f(\mathbf{p}, \mathbf{p}')$ of Eq. (1) is related]. With this way of approaching the limit the kernel of the last term in Eq. (21) goes to zero, so that Γ^{ω} satisfies the equation

$$\Gamma^{\omega}_{\alpha\beta,\gamma\delta}(P_1, P_2) = \Gamma^{(1)}_{\alpha\beta,\gamma\delta}(P_1, P_2) - \frac{i}{(2\pi)^4}\int \Gamma^{(1)}_{\alpha\varepsilon,\gamma\zeta}(P_1, Q)\, g(Q)\, \Gamma^{\omega}_{\zeta\beta,\varepsilon\delta}(Q, P_2)\, d^4Q. \tag{23}$$

We can eliminate $\Gamma^{(1)}$ from (21) and (23). The result of the elimination is

$$\Gamma_{\alpha\beta,\gamma\delta}(P_1, P_2; K) = \Gamma^{\omega}_{\alpha\beta,\gamma\delta}(P_1, P_2) + \frac{a^2 p_0^2}{(2\pi)^3}\int \Gamma^{\omega}_{\alpha\varepsilon,\gamma\zeta}(P_1, Q)\, \Gamma_{\zeta\beta,\varepsilon\delta}(Q, P_2; K)\frac{\mathbf{lk}}{\omega - v\mathbf{lk}}\, do. \tag{24}$$

In fact, if we formally write Eq. (23) in the form

$$\Gamma^{(1)}_{\alpha\beta,\gamma\delta}(P_1, P_2) = \hat{L}\Gamma^{\omega}_{\alpha\beta,\gamma\delta}(P_1, P_2), \tag{25}$$

then Eq. (21) is written

$$\hat{L}\Gamma_{\alpha\beta,\gamma\delta}(P_1, P_2; K) = \Gamma^{(1)}_{\alpha\beta,\gamma\delta}(P_1, P_2) + \frac{a^2 p_0^2}{(2\pi)^3}\int \Gamma^{(1)}_{\alpha\varepsilon,\gamma\zeta}(P_1, Q)\, \Gamma_{\zeta\beta,\varepsilon\delta}(Q, P_2; K)\frac{\mathbf{lk}}{\omega - v\mathbf{lk}}\, do;$$

and substituting Eq. (25) and then applying the operator $\hat{L}^{-1}$ to both sides, we get Eq. (24).

Let us now introduce the function Γ^k defined by

$$\Gamma^{k}_{\alpha\beta,\gamma\delta}(P_1, P_2) = \lim_{K\to 0} \Gamma^{k}_{\alpha\beta,\gamma\delta}(P_1, P_2; K) \quad \text{for } \omega/k \to 0. \tag{26}$$

This function (multiplied by the renormalization constant a^2) is the "forward" scattering amplitude (i.e., that for the transition $P_1, P_2 \to P_1, P_2$), corresponding to actual physical processes occurring with quasi-particles on the Fermi surface: collisions leaving the quasiparticles on this surface involve changes of momentum without change of energy, so that the passage to the limit of zero momentum transfer **k** must be made for energy transfer ω strictly equal to zero. On the other hand the function Γ^{ω} introduced above corresponds to the nonphysical limiting case of "scattering" with small energy transfer and momentum transfer strictly equal to zero.

Setting $\omega = 0$ in Eq. (24) going to the limit $K \to 0$, and multiplying both sides of the equation by a^2, we get

$$a^2\Gamma^{k}_{\alpha\beta,\gamma\delta}(P_1, P_2) = a^2\Gamma^{\omega}_{\alpha\beta,\gamma\delta}(P_1, P_2) - \frac{p_0^2}{v(2\pi)^3}\int a^2\Gamma^{\omega}_{\alpha\varepsilon,\gamma\zeta}(P_1, Q)\cdot a^2\Gamma^{k}_{\zeta\beta,\varepsilon\delta}(Q, P_2)\, do. \tag{27}$$

Thus there exists a general relation connecting the two limiting forms of the forward scattering amplitude.

Let us now turn to the study of the poles of $\Gamma(P_1, P_2; K)$ as function of K. As was already pointed out at the beginning of this paper, the poles with respect to the variable $K = P_3 - P_1$ are due to contributions to Γ associated with intermediate states in which the number of particles in the system is not changed. Therefore these poles correspond to elementary excitations of the liquid without change of the number of quasiparticles in it. It is obvious that these are the excitations which can be described as sonic excitations in the gas of quasiparticles (phonons of the "zeroth sound").

Near a pole of the function $\Gamma(P_1, P_2; K)$ the left side and the integral on the right side of the equation (24) are arbitrarily large; the term $\Gamma^{\omega}(P_1, P_2)$, on the other hand, remains finite and therefore can be dropped. We note further that the variables P_2 and also the indices β and δ are not affected by the operations applied to the function Γ in Eq. (24), i.e., they here play the role of parameters. Finally, we shall consider Γ close to the Fermi surface, i.e., we shall consider the energy of the quasiparticle, which is one of the variables P_1, to be equal to μ, and the momentum to be equal to p_0, so that we write it in the form $p_0\mathbf{n}$, where **n** is a variable unit vector. Keeping all this in mind, we conclude that the determination of the sonic excitations in the liquid reduces to the problem of the eigenvalues of the integral equation

$$\chi_{\alpha\gamma}(\mathbf{n}) = \frac{a^2 p_0^2}{(2\pi)^3}\int \Gamma^{\omega}_{\alpha\varepsilon,\gamma\zeta}(\mathbf{n},\mathbf{l})\, \chi_{\zeta\varepsilon}(\mathbf{l})\frac{\mathbf{lk}}{\omega - v\mathbf{lk}}\, do, \tag{28}$$

where $\chi_{\alpha\gamma}(\mathbf{n})$ is an auxiliary function.

We transform this equation, introducing instead of χ a new function, by the substitution

$$\nu_{\alpha\gamma}(\mathbf{n}) = \frac{\mathbf{nk}}{\omega - v\mathbf{nk}}\, \chi_{\alpha\gamma}(\mathbf{n}). \tag{29}$$

Then Eq. (28) takes the form

$$(\omega - v\mathbf{nk})\, \nu_{\alpha\gamma}(\mathbf{n}) = (\mathbf{k}\cdot\mathbf{n})\frac{p_0^2 a^2}{(2\pi)^3}\int \Gamma^{\omega}_{\alpha\varepsilon,\gamma\zeta}(\mathbf{n},\mathbf{l})\, \nu_{\zeta\varepsilon}(\mathbf{l})\, do. \tag{30}$$

This equation agrees precisely in form with

equation (11) found in reference 2 for the distribution function ν in the zeroth sound, and moreover a comparison of the two equations (using the definition of F by Eq. (6) of reference 2) leads to the following correspondence between the function f (p, p')* and the function Γ^{ω}:

$$f_{\alpha\beta,\gamma\delta}(\mathbf{n},\mathbf{l}) = a^2\Gamma^{\omega}_{\alpha\beta,\gamma\delta}(\mathbf{n},\mathbf{l}) \tag{31}$$

This is the desired relation between f and the properties of the scattering of the quasiparticles. For clarity we point out that the four spin indices on this function correspond to the fact that f (p, p'), or more explicitly f (p, σ; p', σ'), depends on the spin operators (two-row matrices) σ and σ' of the two particles; thus to the two particles (momenta $p_0\mathbf{n}$ and $p_0\mathbf{l}$) there correspond the pairs of indices α, γ and β, δ (in the function $\Gamma_{\alpha\beta,\gamma\delta}(P_1, P_2; P_3, P_4)$ these pairs correspond to the pairs of nearly equal four-momenta P_1, P_3 and P_2, P_4).

Having thus found the connection of the function f with the properties of the scattering of the quasiparticles, let us return to the formula (27) and obtain with its aid explicit relations between the function f and the "physical" amplitude for zero-angle scattering on the Fermi surface, which we write in the form

$$A(\mathbf{n}_1,\sigma_1;\ \mathbf{n}_2,\sigma_2) = a^2\Gamma^k(\mathbf{n}_1,\sigma_1;\ \mathbf{n}_2,\sigma_2). \tag{32}$$

On the Fermi surface the relation (27) takes the form

$$A(\mathbf{n}_1,\sigma_1;\ \mathbf{n}_2,\sigma_2) = f(\mathbf{n}_1,\sigma_1;\ \mathbf{n}_2,\sigma_2) - \frac{1}{4\pi}\frac{d\tau}{d\varepsilon}\,\mathrm{Sp}_{\sigma'}\int f(\mathbf{n}_1,\sigma_1;\ \mathbf{n}',\sigma')\,A(\mathbf{n}',\sigma';\ \mathbf{n}_2,\sigma_2)\,do' \tag{33}$$

(where $d\tau/d\epsilon = 4\pi p_0^2/v(2\pi)^3$). The scalar functions A and f depend on all scalar combinations of the four vectors $\mathbf{n}_1$, $\mathbf{n}_2$, σ_1, σ_2. If, however, the interaction between the particles is an exchange interaction, then the only admissible scalar products are $\mathbf{n}_1\mathbf{n}_2$ and $\sigma_1\sigma_2$. Then we can expand A and f as functions of $\cos\theta$ in terms of Legendre polynomials:

$$A(\cos\theta) = \sum_l A_l P_l(\cos\theta), \quad f(\cos\theta) = \sum_l f_l P_l(\cos\theta). \tag{34}$$

Substituting this into Eq. (33) and performing the integration with respect to do', we get

$$A_l(\sigma_1,\sigma_2) = f_l(\sigma_1\sigma_2) - \frac{1}{2l+1}\frac{d\tau}{d\varepsilon}\,\mathrm{Sp}_{\sigma'}\, f_l(\sigma_1\sigma')\,A_l(\sigma'\sigma_2). \tag{35}$$

In the case of an exchange interaction the spin dependence of the function reduces to a term proportional to $\sigma_1\sigma_2$ (cf. reference 1*), so that

$$f_l = \varphi_l + \psi_l\sigma_1\sigma_2, \tag{36}$$

where φ_l, ψ_l do not depend on the spins. Corresponding to this we also set

$$A_l = B_l + C_l\sigma_1\sigma_2. \tag{37}$$

Substituting Eqs. (36) and (37) into Eq. (35), we get without difficulty

$$B_l = \varphi_l - \frac{2}{2l+1}\frac{d\tau}{d\varepsilon}B_l\varphi_l, \qquad C_l = \psi_l - \frac{1}{2(2l+1)}\frac{d\tau}{d\varepsilon}C_l\psi_l. \tag{38}$$

These formulas give a simple algebraic connection between the coefficients of the expansions of f and A in spherical harmonics. We note that only terms of the same l are related to each other, and that B is related only to the φ's and C only to the ψ's.

In conclusion, I would like to thank A. B. Migdal, who called my attention to the dependence of the forward scattering amplitude on the ratio ω/k, and also E. M. Lifshitz and L. P. Gor'kov for a discussion of this work.

[1] L. D. Landau, J. Exptl. Theoret. Phys. (U.S.S.R.) 30, 1058 (1956), Soviet Phys. JETP 3, 920 (1956).
[2] L. D. Landau, J. Exptl. Theoret. Phys. (U.S.S.R.) 32, 59 (1957), Soviet Phys. JETP 5, 101 (1957).
[3] A. B. Migdal, J. Exptl. Theoret. Phys. (U.S.S.R.) 32, 399 (1957), Soviet Phys. JETP 5, 333 (1957).

Translated by W. H. Furry
13

*In references 1 and 2 we did not write the spin indices explicitly.

*We take occasion to correct a mistake which got into reference 1: Equation (27) should be

$$1/\chi = \beta^{-2}\{4\pi^2k^2/3\alpha + \bar{\psi}_0\}.$$

SOVIET PHYSICS JETP VOLUME 34(7), NUMBER 1 JULY, 1958

THE ENERGY SPECTRUM OF A NON-IDEAL FERMI GAS

V. M. GALITSKII

Moscow Engineering-Physics Institute

Submitted to JETP editor July 12, 1957

J. Exptl. Theoret. Phys. (U.S.S.R.) 34, 151-162 (January, 1958)

We have evaluated the energy spectrum and ground state energy of a non-ideal Fermi gas with repulsive interactions, using an expansion in powers of the ratio of the range of the potential to the mean distance apart of the particles (gas approximation). We have obtained the first two terms of the expansion.

INTRODUCTION

It is well known that in many cases one can consider the excited states of a system of interacting Fermi particles as a gas of elementary excitations — quasiparticles. The energy of a quasiparticle is determined by its momentum in such a way that the energy of the excitation of the system ϵ_s is equal to $\epsilon(p_1) - \epsilon(p_2)$, where $p_1 > p_0 > p_2$ with p_0 the momentum at the Fermi surface. Such a spectrum is called a spectrum of the "Fermi type." A description of a system by means of the method of quasiparticles is exact only in the case of an ideal gas. If there are interactions between the particles, the excited states of the "Fermi type" do not represent the exact stationary states of the systems. This leads to the damping of the quasiparticles.

It was shown in Ref. 1 that it is convenient to apply the methods of quantum field theory to determine the energy spectrum of a system. The energy $\epsilon(p)$ and attenuation $\gamma(p)$ of the quasiparticles can be found as the poles of the analytical continuation of the single-particle Green function $G(p)$. In the present paper we shall apply the methods of quantum field theory to the problem of a non-ideal Fermi gas in which the interaction between the particles is short range $na^3 \ll 1$ (n is the density of the particles in the system and a the range of the potential), but not necessarily weak. We assume that the radially symmetrical potential $V(r)$ is positive and that the interaction between the particles is not retarded. We expand in powers of the parameter $p_0 f_0$, where f_0 is the real part of the scattering amplitude for small momenta. We shall find the energy spectrum of the system and the ground state energy up to quadratic terms in this parameter. Terms corresponding to higher powers than the cubic can not be expressed by means of two-particle parameters which makes it difficult to obtain them in a general form.* This fact was first remarked on in Ref. 2 in connection with the evaluation of the ground state energy.

1. SINGLE PARTICLE GREEN FUNCTION. THE METHOD OF GRAPHS

It is well known that the single particle Green

*The author is obliged to E. M. Lifshitz for this comment.

function of a system is given by the equation

$$iG(x-x') = \langle T\{\psi(x)\psi^+(x')S\}\rangle / S_{00}, \quad (1)$$

where ψ and ψ^+ are taken in the interaction representation and where the averaging is performed over the ground state of the non-interacting particles. The S-matrix of the system is in our case given by the equation

$$S \equiv T\left\{\exp\left(-i\int H'dt\right)\right\} = T\left\{\exp\left[-\frac{i}{2}\int V(\mathbf{r}_1-\mathbf{r}_2)\psi^+(\mathbf{r}_1 t)\psi^+(\mathbf{r}_2 t)\psi(\mathbf{r}_2 t)\psi(\mathbf{r}_1 t)\,dv_1 dv_2 dt\right]\right\} \quad (2)$$

(we choose our units in such a way that $\hbar = m = 1$). It is convenient to change this definition slightly by adding a dependence on a time t_2 to the dependence on the variable $\mathbf{r}_2$ and by introducing an extra integration over time,

$$S = T\left\{\exp\left[-\frac{i}{2}\int dx_1 dx_2 U(x_1-x_2)\right.\right.$$
$$\left.\left.\psi^+(x_1)\psi^+(x_2)\psi(x_2)\psi(x_1)\right]\right\}, \quad (2')$$

where

$$U(x_1-x_2) = V(\mathbf{r}_1-\mathbf{r}_2)\,\delta(t_1-t_2). \quad (3)$$

In (2′) the integration over the variables x_1 and x_2 is taken over the whole of the infinite four-dimensional space.

If we want to expand the S-matrix of (1) in powers of the interaction U, we must know the average value of the T product of the ψ operators. According to Wick's theorem this T product can be written in the form of a sum of normal products and different connections between operators. To apply the methods of quantum field theory it is necessary to put the average values of the normal products equal to zero. This condition will be fulfilled, if we take ψ in the form

$$\psi(\mathbf{r}) = u(\mathbf{r}) + v^+(\mathbf{r}); \quad u(\mathbf{r}) = V^{-1/2}\sum_{p>p_0} a_p e^{i\mathbf{pr}},$$
$$v^+(\mathbf{r}) = V^{-1/2}\sum_{p<p_0} a_\mathbf{p} e^{i\mathbf{pr}} \quad (4)$$

and if we define the normal products as those products in which all operators u and v are on the right, and all operators u^+ and v^+ on the left. In this representation the operators u and v play the role of annihilation operators and the operators u^+ and v^+ the role of creation operators for particles and holes. The average of the N products is equal to zero and the connection of two operators is equal to the Green function of non-interacting particles,

$$\dot\psi(x)\dot\psi^+(x') = iG_0(x-x') = \langle T\{\psi(x)\psi^+(x')\}\rangle. \quad (5)$$

In the momentum representation the Green function of non-interacting particles is of the form

$$G_0^{-1}(p) \equiv G_0^{-1}(\mathbf{p},\varepsilon) = \varepsilon - \varepsilon_p^0 + i\delta\theta(\mathbf{p}), \quad (6)$$

where $\epsilon_p^0 = \frac{1}{2}\mathbf{p}^2$ and

$$\theta(\mathbf{p}) = 1 - 2n_\mathbf{p} = \begin{cases} 1 & |\mathbf{p}| > p_0 \\ -1 & |\mathbf{p}| < p_0 \end{cases},$$

with $n_\mathbf{p}$ is the occupation number of the non-interacting particles in the ground state. We can use the corresponding graphs to assign a definite arrangement of T products to an assembly of connections. The graphs consist of full drawn and dotted lines; each full drawn line corresponds to a particle propagation function $iG_0(x-x')$, or in momentum representation $iG_0(p)$, and each dotted line to an interaction, $iU(x_1-x_2)$ or in momentum representation $iU(q)$, and at each vertex the law of conservation of the four-dimensional momentum is obeyed, $p_1 - p_2 + q = 0$. Since the interaction does not enter into the normal products, we can restrict ourselves in calculating them to the connections of operators which enter in the same H′, that is, to simultaneous operators. To determine those connections we note that in the interaction H′ the operators ψ^+ are to the left of the operators ψ so that $G_0(p,\tau)$ for $\tau = 0$ must necessarily be taken to be $G_0(p_1, -0) = in_\mathbf{p}$.

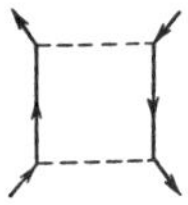

FIG. 1

For our further discussion it is very important that the interaction is not retarded [Eq. (3)]. It is convenient to indicate this absence of retardation by drawing the dotted lines in the graphs horizontally. This representation makes it possible to judge from the graphs the number of particles and holes taking part in the process. The process corresponding to the graph of Fig. 1, for instance, involves one particle and one hole since for each arrangement of vertices one of the lines of 1 or 2 is directed downwards on account of Eq. (3).

2. ESTIMATE OF THE GRAPHS. GAS APPROXIMATION

We introduce now the main part of the energy eigenvalue of the particle $\Sigma(p)$,

$$G^{-1}(p) = G_0^{-1}(p) - \Sigma(p) = \varepsilon - \varepsilon_p^0 - \Sigma(p)\,. \tag{7}$$

In the first approximation of perturbation theory $\Sigma(p)$ is determined by the two graphs of Fig. 2. Graph 2a corresponds to a non-exchange scattering by a particle of the Fermi sea (background particle) and graph 2b to an exchange scattering. In the case in which we are interested where $p_0a \ll 1$ both graphs are equal and are of the order $nV \sim p_0^2 \cdot p_0 a \cdot V_0 a^2$, where V is the Fourier component of the potential for small values of the momentum, and V_0 is the value of the potential inside its range.

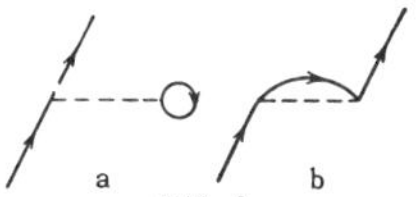

FIG. 2

We can consider now graphs which are more complicated in three different ways.

1. We can increase the number of dotted lines which connect solid lines which are in the same direction (graphs a and b in Fig. 3).

2. We can increase not only the dotted lines of the first kind, but also those which connect solid lines which are in opposite directions (graph c in Fig. 3).

3. We can increase the number of closed loops which are connected by dotted lines to the basic graph [it is well known that unconnected closed loops are eliminated by the denominator in Eq. (1)].

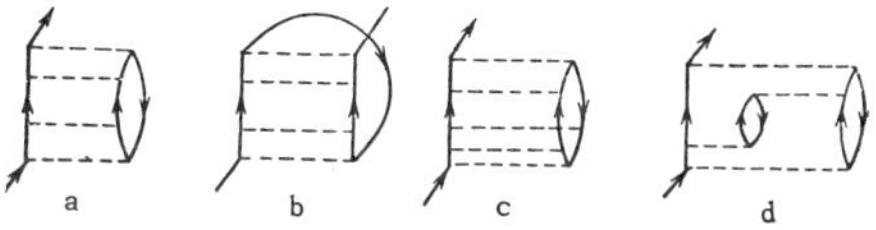

FIG. 3

Let us now estimate the value of these graphs. For graphs of the first kind each additional dotted line adds to Σ a factor $G_0^2U(q)$ and one integration over the four-dimensional momentum q. As $U(q)$ only depends on **q** we get after integrating over the fourth component of q an integral of the form

$$\int \frac{dq}{q^2} V(\mathbf{q}), \tag{8}$$

the convergence of which for large values of q is determined by the function $V(\mathbf{q})$. An estimate of this integral leads to Va^{-1} or V_0a^2, i.e., the parameter of the perturbation theory. The collection of graphs of the first kind gives thus the perturbation theory series. For graphs of the second kind, the additional part has the form pictured in Fig. 1. Since one of the two lines corresponds to the occurrence of a hole, the momentum of which does not exceed the momentum at the Fermi surface, p_0, the integration over **q** in Eq. (8) will in that case be taken over a limited region of dimensions p_0. We obtain as a result p_0V or $p_0a \cdot V_0a^2$, i.e., graphs of the second kind contain apart from the perturbation theory parameter also an additional "gaseousness" parameter p_0a. The difference between the magnitudes of graphs of the first and of the second kind can be given a simple physical interpretation. Indeed, graphs of the first kind correspond to a further approximation of the perturbation theory in terms of the interaction between two particles, while graphs of the second kind correspond to a further approximation in terms of an interaction between a particle and a hole. The interaction with a hole, however, is essentially an interaction with a background particle so that in the processes corresponding to the graphs 3a and 3b one background particle is taking place, but in the process 3c two background particles. This result can also be extended to graphs of the third kind (3d). Graphs 3c and 3d will be discarded. An exact estimate of these graphs which is not based upon perturbation theory is given in Sec. 5. The result of this estimate shows that graphs of the kind 3c and 3d can necessarily only be considered in the third approximation in terms of the small parameter p_0f.

The graphs of the kind 3a and 3b which determine the energy eigenvalues of the particles in the first and second gas approximation can be given in the form of block diagrams as given in Fig. 4. The

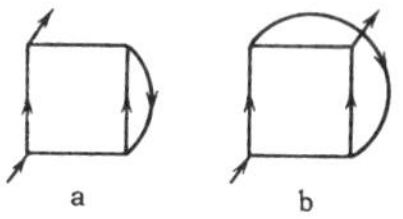

FIG. 4

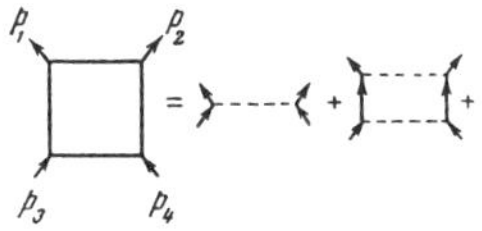

FIG. 5

square denotes as usual the Feynman diagram assembly describing the interaction of two particles in the "ladder" approximation (Fig. 5). This quantity we shall denote by

$$-i\Gamma(p_1p_2,\ p_3p_4)$$

and we shall call it the effective interaction potential. Graphs 4a and 4b have the same structure as the graphs 2a and 2b corresponding to the first approximation of perturbation theory with one difference, however, namely that the dotted lines corresponding to the Born approximation have been taken into account in our effective potential. The values of the particle energy eigenvalues determined by these graphs is of the form

$$\Sigma(p) = -i\int dp' G_0(p')\Gamma(pp', pp') + i\int dp' G_0(p')\Gamma(pp', p'p). \quad (9)$$

Here and henceforth we shall use the following notation

$$dp = d\mathbf{p}d\varepsilon/2\pi, \quad d\mathbf{p} = dp_x dp_y dp_z/(2\pi)^3;$$
$$\delta(p) = 2\pi\delta(\mathbf{p})\delta(\varepsilon), \quad \delta(\mathbf{p}) = (2\pi)^3\delta(p_x)\delta(p_y)\delta(p_z).$$

Equation (9) does not take the presence of particle spin into account. To do that we note first of all that the Green function $G_0(p)$ contains the delta function $\delta_{s,s'}$ (where s and s' are the projections of the spin at the points x and x') which we have omitted, and this means that the particles propagate without change in spin. This result is also still valid in the case where the particles interact during their propagation with other particles, since the potential does not depend on the spin variable. The presence of spin leads thus to the appearance of a factor $\delta_{s_1 s_2}\delta_{s_3 s_4}$ in $\Gamma(p_1 p_2, p_3 p_4)$. The result of summing over the spin variable s' in Eq. (9) is a factor 2s + 1 for the first term and a factor 1 for the second one. This result has a simple physical meaning. The graph 4b corresponds to an exchange scattering in which one must take into account only those background particles which have a spin the projection of which coincides with the projection of the spin of the impinging particle. In contradistinction to this, graph 4a corresponds a non-exchange scattering process in which all particles in the Fermi sea must be taken into account. If we restrict ourselves to the case of particles of spin ½ we get finally

$$\Sigma(p) = -2i\int dp' G_0(p')\Gamma(pp', pp') + i\int dp' G_0(p')\Gamma(pp', p'p). \quad (9')$$

3. THE EFFECTIVE INTERACTION POTENTIAL

To determine the effective interaction potential we introduce a function Q which is connected with Γ by the following equation

$$\Gamma(p_1 p_2, p_3 p_4) = \int dq U(q) Q(p_1 - q, p_2 + q, p_3 p_4). \quad (10)$$

The function Q differs from the two-particle Green function K in the "ladder" approximation by the absence of initial outside lines with moment p_3 and p_4. We can thus obtain the equation for Q from the corresponding equation for the function K by dividing by the product $iG_0(p_3)iG_0(p_4)$. We have thus finally for $Q' = \delta(p_1 + p_2 - p_3 - p_4)Q$,

$$Q'(p_1 p_2, p_3 p_4) = \delta(p_1 - p_3)\delta(p_2 - p_4) + iG_0(p_1)G_0(p_2)\int dq U(q) Q'(p_1 - q, p_2 + q, p_3 p_4). \quad (11)$$

If we go over to the relative momenta p and p' and the moment of the center of mass g,

$$p = (p_1 - p_2)/2, \quad p' = (p_3 - p_4)/2, \quad g = p_1 + p_2 = p_3 + p_4, \quad (12)$$

we get the equation

$$Q(p, p', g) \equiv Q\left(\frac{g}{2} + p, \frac{g}{2} - p; \frac{g}{2} + p', \frac{g}{2} - p'\right) = \delta(p - p') + iG_0\left(\frac{g}{2} + p\right)G_0\left(\frac{g}{2} - p\right)\int dq U(q) Q(p - q, p', g). \quad (13)$$

The potential $U(q)$ in Eq. (10) does not depend on the fourth component of q so that it is sufficient to know the function Q integrated over the fourth component of the relative momentum p, ϵ, in order to determine the effective potential. If we denote that function by $\chi(\mathbf{p}, \mathbf{p}', g)$ and integrate Eq. (13) over ϵ we find

$$\chi(\mathbf{p}, \mathbf{p}', g) - \frac{N(\mathbf{p})}{E - \mathbf{p}^2 + i\delta N(\mathbf{p})}\int d\mathbf{q} V(\mathbf{q})\chi(\mathbf{p} - \mathbf{q}, \mathbf{p}', g) = \delta(\mathbf{p} - \mathbf{p}'). \quad (14)$$

In Eq. (14) we have $E = g_0 - \frac{1}{4}\mathbf{g}^2$ (where g_0 is the fourth component of g) and the factor $N(\mathbf{p})$ takes the Pauli exclusion principle into account inasfar as it applies to the initial background,

$$N(\mathbf{p}) = 1 - n_{\mathbf{g}/2+\mathbf{p}} - n_{\mathbf{g}/2-\mathbf{p}}. \quad (15)$$

For further calculations it is convenient to consider the effective potential as a function of the relative and total momenta which are defined by Eqs. (12). If we write

$$\Gamma(\mathbf{p}, \mathbf{p}', g) \equiv \Gamma\left(\frac{g}{2} + p, \frac{g}{2} - p; \frac{g}{2} + p', \frac{g}{2} - p'\right),$$

we obtain from (10) the following connection between the effective potential and the function χ,

$$\Gamma(\mathbf{p}, \mathbf{p}', g) = \int d\mathbf{q} V(\mathbf{q})\chi(\mathbf{p} - \mathbf{q}, \mathbf{p}', g). \quad (16)$$

Equation (14) can not be solved in its general form; our problem is to express the solution of this equation in terms of the scattering amplitudes of the particles. We consider first of all the problem of the scattering in vacuo. In that case $N(\mathbf{p}) \equiv 1$ and Eq. (14) is of the form

$$\chi_0(\mathbf{p},\mathbf{p}',g)-\frac{1}{E-\mathbf{p}^2+i\delta}\int dqV(\mathbf{q})\chi_0(\mathbf{p},\mathbf{p}',g)=\delta(\mathbf{p}-\mathbf{p}'). \tag{14'}$$

Equation (14′), multiplied by $E-\mathbf{p}^2+i\delta$, is the same as the inhomogeneous Schrödinger equation for the relative motion of two particles. It is therefore easy to express its solution in terms of $\psi_{\mathbf{k}}(\mathbf{p})$, the wave function of the relative motion of particles which are scattered ($\mathbf{k}$ is the relative momentum at infinity),

$$\chi_0(\mathbf{p},\mathbf{p}',g)=(E-p'^2+i\delta)\int d\mathbf{k}\,\frac{\psi_{\mathbf{k}}(\mathbf{p})\,\psi^*_{\mathbf{k}}(\mathbf{p}')}{E-k^2+i\delta}. \tag{17}$$

Let us introduce the scattering amplitude $f(\mathbf{p},\mathbf{k})$ of the particles, by the equation

$$f(\mathbf{p},\mathbf{k})=\int dqV(\mathbf{q})\psi_{\mathbf{k}}(\mathbf{p}-\mathbf{q}). \tag{18}$$

This amplitude differs from the usual one by a factor -4π, that is, the usual amplitude is equal to $-f(\mathbf{p},\mathbf{k})/4\pi$. The wave function $\psi_{\mathbf{k}}(\mathbf{p})$ is connected with the amplitude by the relation

$$\psi_{\mathbf{k}}(\mathbf{p})=\delta(\mathbf{p}-\mathbf{k})+\frac{f(\mathbf{p},\mathbf{k})}{k^2-p^2+i\delta}. \tag{19}$$

Using this relation for $\psi^*_{\mathbf{k}}(\mathbf{p}')$ and substituting it into (17) we get

$$\chi_0(\mathbf{p},\mathbf{p}',g)=\psi_{\mathbf{p}'}(\mathbf{p})$$
$$+\int d\mathbf{k}\psi_{\mathbf{k}}(\mathbf{p})f^*(\mathbf{p}',\mathbf{k})\left\{\frac{1}{E-k^2+i\delta}+\frac{1}{k^2-p'^2-i\delta}\right\}, \tag{17'}$$

and we get for the effective interaction potential of particles in vacuo* which is defined by an equation which is analogous to equation (16)

$$\Gamma_0(\mathbf{p},\mathbf{p}',g)=f(\mathbf{p},\mathbf{p}')$$
$$+\int d\mathbf{k}f(\mathbf{p},\mathbf{k})f^*(\mathbf{p}',\mathbf{k})\left\{\frac{1}{E-k^2+i\delta}+\frac{1}{k^2-p'^2-i\delta}\right\}. \tag{20}$$

If we use Eq. (II) of the Appendix, we can express Γ_0 also in a slightly different, equivalent form,

$$\Gamma_0(\mathbf{p},\mathbf{p}',g)=f^*(\mathbf{p}',\mathbf{p})$$
$$+\int d\mathbf{k}f(\mathbf{p},\mathbf{k})f^*(\mathbf{p}',\mathbf{k})\left\{\frac{1}{E-k^2+i\delta}+\frac{1}{k^2-p^2+i\delta}\right\}. \tag{20'}$$

The effective interaction potential of particles in vacuo is thus equal in first approximation to the scattering amplitude $f(\mathbf{p},\mathbf{p}')$ or to $f\ (\mathbf{p},\mathbf{p}')$.

Going on to the solution of (14), we write it in the form

*The problems connected with Γ_0 were solved jointly with S. T. Beliaev, who was working simultaneously on the analogous problem of a Bose gas.

$$\chi(\mathbf{p},\mathbf{p}',g)-\frac{1}{E-p^2+i\delta}\int dqV(\mathbf{q})\chi(\mathbf{p}-\mathbf{q},\mathbf{p}',\mathbf{q})$$
$$=\delta(\mathbf{p}-\mathbf{p}')+\left\{\frac{N(\mathbf{p})}{E-p^2+i\delta N(\mathbf{p})}-\frac{1}{E-p^2+i\delta}\right\}\Gamma(\mathbf{p},\mathbf{p}',g), \tag{21}$$

where χ_0 is the Green function of the left hand side of this equation. Equation (21) admits thus of the following formal solution,

$$\chi(\mathbf{p},\mathbf{p}',g)=\chi_0(\mathbf{p},\mathbf{p}',g)$$
$$+\int d\mathbf{k}\chi_0(\mathbf{p},\mathbf{k},g)\left\{\frac{N(\mathbf{k})}{E-k^2+i\delta N(\mathbf{k})}-\frac{1}{E-k^2+i\delta}\right\}\Gamma(\mathbf{k},\mathbf{p}',g).$$

If we apply (16) we can go over to an integral equation for the effective potential Γ,

$$\Gamma(\mathbf{p},\mathbf{p}',g)=\Gamma_0(\mathbf{p},\mathbf{p}',g)+$$
$$+\int d\mathbf{k}\Gamma_0(\mathbf{p},\mathbf{k},g)\left\{\frac{N(\mathbf{k})}{E-k^2+i\delta N(\mathbf{k})}-\frac{1}{E-k^2+i\delta}\right\}\Gamma(\mathbf{k},\mathbf{p}',g). \tag{22}$$

Equation (22) can be solved by iteration methods since the term involving the integral is small. Indeed, the difference within the braces is different from zero only if $N\neq 1$, i.e., if the variation of k is of the order of p_0. To a first approximation we can substitute the scattering amplitude f for Γ_0. We then get, for $E\sim p_0^2$, the following order-of-magnitude estimate for the term involving the integral in (22):

$$p_0^3f\cdot p_0^{-2}\Gamma=p_0f\cdot\Gamma\ll\Gamma.$$

If we now take second-order terms $\ll\Gamma$ into account we can substitute in the integral in (22) $f(\mathbf{p},\mathbf{k})$ for $\Gamma_0(\mathbf{p},\mathbf{k},g)$ and $\Gamma_0(\mathbf{k},\mathbf{p}',g)\approx f(\mathbf{p}',\mathbf{k})$ for $\Gamma(\mathbf{k},\mathbf{p}',g)$. We get the following result,

$$\Gamma(\mathbf{p},\mathbf{p}',g)=\Gamma_0(\mathbf{p},\mathbf{p}',g)$$
$$+\int d\mathbf{k}f(\mathbf{p},\mathbf{k})f^*(\mathbf{p}',\mathbf{k})\left\{\frac{N(\mathbf{k})}{E-k^2+i\delta N(\mathbf{k})}-\frac{1}{E-k^2+i\delta}\right\} \tag{23}$$
$$=f(\mathbf{p},\mathbf{p}')+\int d\mathbf{k}f(\mathbf{p},\mathbf{k})f^*(\mathbf{p}',\mathbf{k})\left\{\frac{N(\mathbf{k})}{E-k^2+i\delta N(\mathbf{k})}+\frac{1}{k^2-p'^2+i\delta}\right\}$$

As can be seen from Eq. (23), the effective interaction potential Γ, like Γ_0, is equal in first approximation to the scattering amplitude $f(\mathbf{p},\mathbf{p}')$. In the second part, the integral converges for values of k^2 of the order of the larger of the quantities E or p'^2. For the energy eigenvalues of the particles $\Sigma(\mathbf{p},\epsilon)$ in which we are interested, and for values of ϵ near $\frac{1}{2}p^2$, both quantities E and p'^2 are of the order of magnitude p_0^2. The second term in (23) is thus p_0f times smaller than the first one.

If the momenta $\mathbf{p}$ and $\mathbf{p}'$ have the same absolute magnitude we can write the effective potential in a slightly different form. Using the equation

$$\frac{1}{k^2-p'^2+i\delta}=P\frac{1}{k^2-p^2}-i\pi\delta(k^2-p'^2)$$

and Eq. (IV) of the Appendix, we get

$$\Gamma(\mathbf{p}, \mathbf{p}', g) = \operatorname{Re} f(\mathbf{p}, \mathbf{p}') + \int d\mathbf{k} f(\mathbf{p}, \mathbf{k}) f(\mathbf{p}', \mathbf{k})\, \mathrm{P}\frac{1}{k^2 - p'^2}$$
$$+ \int d\mathbf{k} f(\mathbf{p}, \mathbf{k}) f(\mathbf{p}', \mathbf{k}) \frac{N(\mathbf{k})}{E - k^2 + i\delta N(\mathbf{k})}, \qquad (23')$$

where the symbol P indicates the principal part of the integral.

4. THE ENERGY SPECTRUM OF THE SYSTEM. GROUND STATE ENERGY

The main part of the energy eigenvalues of the particles can be written in the form

$$\Sigma(p) = -2i \int \frac{d\varepsilon'}{2\pi} d\mathbf{p}' G_0(p') \Gamma(\mathbf{q}, \mathbf{q}, g)$$
$$+ i \int \frac{d\varepsilon'}{2\pi} d\mathbf{p}' G_0(p') \Gamma(\mathbf{q}, -\mathbf{q}, g),$$

where $\mathbf{q} = (\mathbf{p} - \mathbf{p}')/2$, $g = \mathbf{p} + \mathbf{p}'$. The first two terms for Γ in (23′) do not depend on the fourth component of the momentum so that integrating them over ϵ' reduces to calculating

$$\int \frac{d\varepsilon'}{2\pi} G_0(\mathbf{p}', \varepsilon').$$

This integral is equal to $G_0(\mathbf{p}', \tau)$ with $\tau = 0$, i.e., to $G_0(\mathbf{p}', -0) = in_{\mathbf{p}'}$, according to the considerations of Sec. 1. The integral over ϵ' of the last term of (23′) is elementary. We get the following results

$$\Sigma(p) = \Sigma_1(p) + \Sigma_2(p), \qquad (24)$$

$$\Sigma_1(p) = 2\int d\mathbf{p}' n_{p'} \operatorname{Re} f(\mathbf{q}, \mathbf{q}) - \int d\mathbf{p}' n_{\mathbf{p}'} \operatorname{Re} f(\mathbf{q}, -\mathbf{q}), \qquad (25)$$

$$\Sigma_2(p) = \int d\mathbf{p}' d\mathbf{k} \{2|f(\mathbf{q}, \mathbf{k})|^2 - f(\mathbf{q}, \mathbf{k}) f^*(-\mathbf{q}, \mathbf{k})\}$$
$$\times \Big\{ n_{\mathbf{p}'} \mathrm{P} \frac{1}{k^2 - q^2} \qquad (26)$$
$$- N(\mathbf{k}) \frac{N(\mathbf{k}) - \theta(\mathbf{p}')}{2} \frac{1}{k^2 - q^2 - \varepsilon + \varepsilon_p^0 + i\delta[\theta(\mathbf{p}') - N(\mathbf{k})]} \Big\}.$$

For wavelengths that are considerably longer than the range a of the potential, the real part of the amplitude will not depend on the momenta. If we therefore restrict our considerations to excitations with momenta p which satisfy the condition

$$pf \ll 1, \qquad (27)$$

we can calculate the real part of the constant amplitude. Assuming that

$$\operatorname{Re} f = 4\pi f_0, \qquad (28)$$

where f_0 is the real part of the usual scattering amplitude with the opposite sign, and taking f_0 outside the integral sign in (25), we get for the first approximation to the energy eigenvalue

$$\Sigma_1 = 2\pi n f_0. \qquad (29)$$

To the first approximation in the gas parameter Σ is thus a real constant quantity. The Green function has in the first approximation the following form,

$$G_1(\mathbf{p}, \varepsilon) = 1/[\varepsilon - \varepsilon_p^0 - 2\pi n f_0 + i\delta\theta(\mathbf{p})]. \qquad (30)$$

If we introduce instead of ϵ a new variable ϵ',

$$\varepsilon' = \varepsilon - 2\pi n f_0, \qquad (31)$$

we can verify that $G_1(\mathbf{p}, \epsilon')$ is the same as the Green function of non-interacting particles. We can thus make our calculation more precise by assuming that everywhere instead of $G_0(\mathbf{p}, \epsilon)$ the Green function of the first approximation $G_1(\mathbf{p}, \epsilon)$ was used. If we make our considerations more precise in this way we must in all equations replace ϵ by ϵ' so that the Green function of the second approximation has the form

$$G_2^{-1}(\mathbf{p}, \varepsilon) = \varepsilon - \varepsilon_p^0 - 2\pi n f_0 - \Sigma_2(\mathbf{p}, \varepsilon') = \varepsilon' - \varepsilon_p^0 - \Sigma_2(\mathbf{p}, \varepsilon'), \qquad (32)$$

where $\Sigma_2(\mathbf{p}, \epsilon')$ is obtained from (26) by replacing ϵ by ϵ'.

We shall now calculate the energy eigenvalues of the second approximation, Σ_2. In the integral over the variable k in (26), the only appreciable contributions to the integral come from values of k in a region of the order of magnitude q. For excitations with momenta that satisfy condition (27), both arguments of the amplitude are small. The imaginary part of the amplitude can thus be neglected and the real part can be considered to be constant and can be taken outside the integral sign. Using the notation of Eq. (28), we get

$$\operatorname{Re}\Sigma_2 = 16\pi^2 f_0^2 \int d\mathbf{p}' d\mathbf{k} \Big\{ n_{\mathbf{p}'} \mathrm{P} \frac{1}{k^2 - q^2}$$
$$- N(\mathbf{k}) \frac{N(\mathbf{k}) - \theta(\mathbf{p}')}{2} \mathrm{P} \frac{1}{k^2 - q^2 - \varepsilon' + \varepsilon_p^0} \Big\}, \qquad (26')$$

$$\operatorname{Im}\Sigma_2 = -4\pi^3 f_0^2 \int d\mathbf{p}' d\mathbf{k} N(\mathbf{k}) [N(\mathbf{k})$$
$$- \theta(\mathbf{p}')]^2 \delta(k^2 - q^2 - \varepsilon' + \varepsilon_p^0). \qquad (26'')$$

The energy spectrum of the system is determined by the poles of the analytical continuation of the Green function (1), or, in our approximation, by the equation

$$\varepsilon_p - \varepsilon_p^0 - 2\pi n f_0 - \Sigma_2(\mathbf{p}, \varepsilon_p^0) = 0. \qquad (33)$$

After some simple calculations we get for the energy ϵ_p and the damping γ of the quasiparticles the following expressions

$$\frac{\varepsilon_p}{p_0^2} = \frac{1}{2}x^2 + \frac{2}{3\pi}p_0f_0 + \frac{2}{15\pi^2}p_0^2f_0^2\frac{1}{x}\left\{11x + 2x^5\ln\frac{x^2}{|x^2-1|} - 10(x^2-1)\ln\left|\frac{x+1}{x-1}\right| - (2-x^2)^{3/2}\ln\left|\frac{1+x\sqrt{2-x^2}}{1-x\sqrt{2-x^2}}\right|\right\},\quad x \leqslant \sqrt{2},$$

$$\frac{\varepsilon_p}{p_0^2} = \frac{1}{2}x^2 + \frac{2}{3\pi}p_0f + \frac{2}{15\pi^2}p_0^2f_0^2\frac{1}{x}\left\{11x + 2x^5\ln\frac{x^2}{x^2-1} - 10(x^2-1)\ln\left|\frac{x+1}{x-1}\right| - 2(x^2-2)^{3/2}\cot^{-1}\sqrt{x^2-2}\right\},\quad x \geqslant \sqrt{2}, \qquad (34)$$

$$\frac{\gamma}{p_0^2} = -\frac{1}{4\pi}p_0^2f_0^2(1-x^2)^2,\ x \leqslant 1;\quad \frac{\gamma}{p_0^2} = \frac{1}{15\pi}p_0^2f_0^2\frac{1}{x}\{5x^2-7+2(2-x^2)^{3/2}\},\ 1 \leqslant x \leqslant \sqrt{2},$$

$$\frac{\gamma}{p_0^2} = \frac{1}{15\pi}p_0^2f_0^2\frac{5x^2-7}{x},\quad \sqrt{2} \leqslant x, \qquad (35)$$

where $x = p/p_0$. The expansion (34) in a series in the momenta has for momenta near the Fermi surface the form

$$\frac{\varepsilon_p}{p_0^2} = \frac{1}{2}x^2 + \frac{2}{3\pi}p_0f_0 + \frac{2}{15\pi^2}p_0^2f_0^2(11-2\ln 2) - \frac{8}{15\pi^2}p_0^2f_0^2(7\ln 2 - 1)(x-1) + \ldots \qquad (34')$$

From Eq. (34′) we can obtain the effective mass, m^*, of particles on the Fermi surface,

$$\frac{m^*}{m} = 1 + \frac{8}{15\pi^2}(7\ln 2 - 1)p_0^2f_0^2 \qquad (36)$$

and the chemical potential of the system which, as is well known[1] is equal to the energy of the quasiparticles for $p = p_0$,

$$\mu = \frac{1}{2}p_0^2\left\{1 + \frac{4}{3\pi}p_0f_0 + \frac{4}{15\pi^2}(11-2\ln 2)p_0^2f_0^2\right\}. \qquad (37)$$

These expressions are the same as the ones obtained by Abrikosov and Khalatnikov.[3] Using the formula $\mu = (\partial E_0/\partial N)_V$, we find for the ground state energy of the system,[4]

$$\frac{E_0}{N} = \frac{3}{10}p_0^2\left\{1 + \frac{10}{9\pi}p_0f_0 + \frac{4}{21\pi^2}(11-2\ln 2)p_0^2f_0^2\right\}. \qquad (38)$$

The damping of the quasiparticles γ for momenta p which are nearly equal to p_0 is of the form

$$\gamma = -\frac{1}{\pi}p_0^2f_0^2(p_0-p)^2,\quad p < p_0,$$
$$\gamma = \frac{1}{\pi}p_0^2f_0^2(p-p_0)^2,\quad p > p_0, \qquad (35')$$

i.e., proportional to the square of the deviation from the Fermi surface.

Let us now consider excitations with large momenta,

$$p_0 \ll p \ll 1/f_0.$$

The quadratic correction term of the quasiparticle energy ϵ_p is small for such momenta. The imaginary part has the form

$$\gamma = \frac{1}{3\pi}p_0^3f_0^2p = \frac{1}{4}np\sigma.$$

If we use the connection between the imaginary part of the scattering amplitude and the total cross section σ we find

$$\varepsilon_p - i\gamma \sim \varepsilon_p^0 - 2\pi nf(\mathbf{p}/2, \mathbf{p}/2), \qquad (39)$$

where f is the usual scattering amplitude.

The evaluation of Σ_2 as a function of the variables $\mathbf{p}$ and ϵ leads to a very cumbersome expression. The expansion of its real part for the case when $|x-1| \ll 1$ and $|y| \ll 1$, where

$$y = (\varepsilon' - \varepsilon_p^0)/p_0^2,$$

is of the form

$$\mathrm{Re}\,\Sigma_2(\mathbf{p}, \varepsilon) = \frac{1}{\pi^2}p_0^2f_0^2\left\{\frac{2}{15}(11-2\ln 2) - \frac{8}{15}(7\ln 2 - 1)(x-1) - 4\ln 2\cdot y\right\}. \qquad (40)$$

From Eq. (40) we can get the renormalization constant of the Green function Z which is connected, as is well known,[5] with the discontinuity of the momentum distribution function of the particles,

$$Z = n(p_0 - 0) - n(p_0 + 0) = 1 - 2p_0^2f_0^2\frac{1}{\pi^2}4\ln 2 \qquad (41)$$

As far as the imaginary part of Σ_2 as a function of the variables p and ϵ is concerned, there are a number of regions in which $\mathrm{Im}\,\Sigma$ has a different analytical form. We shall show that this function tends to zero at the value $\epsilon' = p_0^2$, independent of the value of the momentum. We write Eq. (26″) in the following form

$$\mathrm{Im}\,\Sigma(\mathbf{p}, \varepsilon) = -4\pi^3f_0^2\int np'd\mathbf{p}_1d\mathbf{p}_2\{n_{\mathbf{p}'}(1-n_{\mathbf{p}_1})(1-n_{\mathbf{p}_2}) - (1-n_{\mathbf{p}'})n_{\mathbf{p}_1}n_{\mathbf{p}_2}\}$$
$$\times\,\delta\left(\frac{1}{2}p_1^2 + \frac{1}{2}p_2^2 - \frac{1}{2}p'^2 - \varepsilon'\right)\delta(\mathbf{p}_1 + \mathbf{p}_2 - \mathbf{p}' - \mathbf{p}).$$

The first term within the braces determines the attenuation of the quasiparticles, and the second one the attenuation of the holes. For $\epsilon' = \frac{1}{2}p_0^2$ each of those terms tends to zero since from the first one it follows that $p'^2 < p_0^2$, $\frac{1}{2}p_1^2 + \frac{1}{2}p_2^2 > p_0^2$, and from the second one that $p'^2 > p_0^2$, $\frac{1}{2}p_1^2 + \frac{1}{2}p_2^2 < p_0^2$, both in violation of the equation $p_1^2 + p_2^2 = p'^2 + p_0^2$. According to the general theory of Green functions of many-body systems, the value for which the imaginary part of the energy eigenvalue tends to zero determines the chemical potential of the system μ

(Ref. 1). The equation

$$\operatorname{Im} \Sigma (\mathbf{p}, p_0^2/2) = 0$$

shows that the expression which we have found for Im Σ makes it possible to determine the chemical potential only to a first approximation. We can make the result more accurate by including in ϵ' besides $2\pi m f_0$ also the second-order energy correction on the Fermi surface. This method of calculation is used by Beliaev.[6]

We shall now discuss the connection between the results obtained by us and the general theory of a Fermi liquid developed by Landau.[7]

Owing to the monotonic dependence of the quasiparticle energy ϵ_p on the momentum, the quasiparticles fill the Fermi sea up to a limiting momentum p_0. The occupation numbers of the quasiparticles are thus the same as the occupation numbers n_p of the non-interacting particles. This equality is also maintained for states near the ground state. It is easily seen that expressions (25), (26), and (26′) for the extra energy of the particles are correct for any distribution of non-interacting particles. To see this it is sufficient to define the operators u and v^+ in Eq. (4) by

$$u = V^{-1/2} \sum_{\mathbf{p}} (1 - n_p) a_p e^{i\mathbf{pr}}, \quad v^+ = V^{-1/2} \sum_{\mathbf{p}} n_p a_p e^{i\mathbf{pr}}, \qquad (4')$$

where n_p is the occupation number of the non-interacting particles. In the particular case of the ground state, Eq. (4′) is the same as Eq. (4). The sum of expressions (25) and (26′) at $\epsilon' = \epsilon_p^0$ gives the energy of the quasiparticles as a functional of the distribution function n_p of the quasiparticles. The variational derivative $\delta\epsilon_p/\delta n_{p'}$ determines the function $f_L(\mathbf{p}, \mathbf{p}')$ which was introduced by Landau (after a summation over the spins s and s′),

$$f_L(\mathbf{p}, \mathbf{p}') = 4\pi f_0 + 32\pi^2 f_0^2 \int d\mathbf{k} n_k \Big\{ \mathrm{P} \frac{1}{k^2 + \mathbf{pp}' - \mathbf{k}(\mathbf{p} + \mathbf{p}')} + \mathrm{P} \frac{1}{(\mathbf{p}' - \mathbf{k})(\mathbf{p}' - \mathbf{p})} - \mathrm{P} \frac{1}{(\mathbf{p} - \mathbf{k})(\mathbf{p}' - \mathbf{p})} \Big\}. \qquad (42)$$

Using Eq. (11) of Ref. 7 we obtain an expression for the effective mass of the particles at the Fermi surface which is the same as expression (36).

5. ESTIMATE OF THE GRAPHS OMITTED. HIGHER APPROXIMATIONS

When we estimated in Sec. 2 the importance of the omitted graphs, we considered an additional interaction with the background particles, corresponding to a single action of the potential V(q) (one additional dotted line). In the case where perturbation theory can not be used this estimate is incorrect. The Born approximation is the first term, which is large, of a series, the sum of which (the scattering amplitude) is small. To obtain a correct estimate it is necessary to sum the graphs corresponding to all orders of interaction of real gas particles, that is, all ladders of dotted lines connecting two full drawn lines going in the same direction. This sum is equal to the effective interaction potential Γ. To construct the graphs it is thus convenient to use the effective potential Γ (square), and not the potential U (dotted line).

Figure 6 illustrates a construction consisting of graphs omitted by us. To estimate the value of

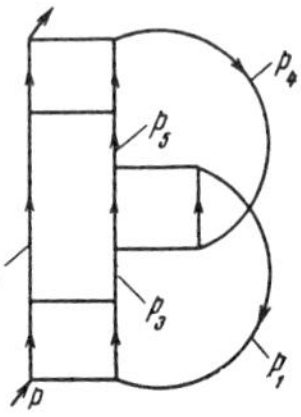

FIG. 6

Σ_3 defined by those graphs it is sufficient to consider Γ in the first gas approximation, $\Gamma \approx f$. We get then

$$\Sigma_3 = i f^3 \int dp_1 dp_2 dp_4 G_0(p_1) G_0(p_2) G_0(p + p_1 - p_2) \times G_0(p_4) G_0(p + p_4 - p_2) \qquad (43)$$

It is important for us to show that the integral in (43) converges since in that case its value can only depend on the momentum p_0 (we assume that the momentum of the particles is in the neighborhood of p_0) and from dimensional considerations it follows that Σ_3 must be of the order $p_0^5 f^3$. After integrating over the fourth momentum component, there are only two energy denominators left in (43). If we then integrate over the momentum $\mathbf{p}_2$ we get a final quantity which is of the order of a reciprocal momentum. It is essential that the two remaining integrals over $\mathbf{p}_1$ and $\mathbf{p}_4$ be taken over a domain that is bounded by the Fermi surface. Indeed, in our approximation Γ does not depend on the fourth momentum component, that is, the interaction takes place instantaneously. The lines corresponding to the momenta p_1 and $p_3 = p + p_1 - p_2$ thus form a closed loop and one of these lines must correspond to the propagation of a hole (compare the graph of Fig. 1). Since the momentum of a hole is less than the Fermi momentum p_0, one or other of the two conditions, $p_1 < p_0$, or $|\mathbf{p} + \mathbf{p}_1 - \mathbf{p}_2| < p_0$, must be fulfilled. In each of these cases p_1

varies only in a bounded region. The same argument applied to the lines with the momenta p_4 and $p_5 = p + p_4 - p_2$ shows that the domain of integration of p_4 is bounded. We have thus proved the convergence of the integral, and Σ_3 is of the order of magnitude

$$\Sigma_3 \sim p_0^5 f^3 \sim p_0^2 n f^3. \tag{44}$$

In conclusion the author wants to express his sincere thanks to A. B. Migdal and S. T. Beliaev for fruitful discussions and to A. F. Goriunov for help with the calculations.

APPENDIX

We shall derive certain relations to be satisfied by the scattering amplitude f. If we multiply Eq. (18) of the main text by $\psi_k^*(p')$ and integrate over k we find if we take into account that the ψ form a complete set of functions, that

$$V(p - p') = \int dk f(p, k) \psi_k^*(p'),$$

or, if we substitute (19) for $\psi_k^*(p')$,

$$V(p - p') = f(p, p') + \int dk \frac{f(p, k) f^*(p', k)}{k^2 - p'^2 - i\delta}. \tag{I}$$

If we use the condition that the potential be Hermitian we can obtain from (I) the following formula,

$$f(p, p') - f^*(p', p) = \int dk f(p, k) f^*(p', k) \left\{ \frac{1}{k^2 - p^2 + i\delta} - \frac{1}{k^2 - p'^2 - i\delta} \right\}. \tag{II}$$

In the case where the two vectors p and p' have the same absolute magnitude, the principal values of the integral cancel each other and (II) becomes

$$f(p, p') - f^*(p', p) = -2\pi i \int dk f(p, k) f^*(p', k) \delta(p^2 - k^2). \tag{III}$$

If we consider scattering in a central field of force, f(p, p') can depend only on p^2 and $(p \cdot p')$, if $p = p'$, so that $f(p, p') = f(p', p)$ and we get from (III)

$$\operatorname{Im} f(p, p') = -\frac{1}{16\pi^2} p \int dn f(p, pn) f^*(p', pn) \tag{IV}$$

where $|n| = 1$. Equation (IV) contains as a special case the well known relation between the imaginary part of the scattering amplitude at sero angle and the total cross section.

[1] V. M. Galitskii and A. B. Migdal, J. Exptl. Theoret. Phys. (U.S.S.R.) **34**, 139 (1958), Soviet Phys. JETP **7**, 96 (1958) (this issue).

[2] K. Huang and C. N. Yang, Phys. Rev. **105**, 767 (1957).

[3] A. A. Abrikosov and I. M. Khalatnikov, J. Exptl. Theoret. Phys. (U.S.S.R.) **33**, 1154 (1957), Soviet Phys. JETP **6**, 888 (1958).

[4] T. D. Lee and C. N. Yang, Phys. Rev. **105**, 1119 (1957).

[5] A. B. Migdal, J. Exptl. Theoret. Phys. (U.S.S.R.) **32**, 399 (1957), Soviet Phys. JETP **5**, 333 (1957).

[6] L. D. Landau, J. Exptl. Theoret. Phys. (U.S.S.R.) **30**, 1058 (1956), Soviet Phys. JETP **3**, 920 (1956).

Translated by D. ter Haar
23

Collective Excitations of Fermi Gases.

J. GOLDSTONE (*) and K. GOTTFRIED (**)
Universitetets Institut for Teoretisk Fysik - København

(ricevuto il 14 Maggio 1959)

In recent years a rather formidable array of successful but apparently distinct theories of collective motions in Fermi gases have been proposed (1). That most of these methods are actually equivalent within the approximations used can be surmised from the fact that they usually yield identical answers. It is our purpose here to point out that many of the results previously obtained can be retrieved from an approach which is completely elementary (2). The price which must be paid for this simplicity is lack of generality; in particular, the damping of the collective motion is completely beyond the power of this method. We should like to warn the reader at the outset that we have no new results to report, and that the following is, at the very most, a pedagogic contribution to the subject.

Our treatment is based on the time-dependent Hartree-Fock equation. Let $\varrho(t)$ be the density matrix, and $H[\varrho]$ the Hartree-Fock Hamiltonian,

$$\langle q|H[\varrho]|q'\rangle = \langle q|T|q'\rangle + \int \langle qq''|v|q'q'''\rangle\langle q'''|\varrho(t)|q''\rangle \,\mathrm{d}q''\,\mathrm{d}q'''$$

with T the kinetic energy, $\langle qq''|v|q'q'''\rangle$ the antisymmetrized matrix element of the interparticle potential v, and q, q', ... complete sets of single-particle quantum numbers. The equation of motion reads

$$i\frac{\partial}{\partial t}\varrho(t) = \left[H[\varrho], \varrho(t)\right]. \tag{1}$$

(*) Present address: CERN, Geneva.

(**) Present address: Physics Dept., Harvard University, Cambridge, Mass.

(1) D. BOHM and D. PINES: *Phys. Rev.*, **92**, 609 (1953); K. SAWADA, K. A. BRUECKNER, N. FUKUDA and R. BROUT: *Phys. Rev.*, **108**, 507 (1957); J. HUBBARD: *Proc. Roy. Soc.*, A **240**, 539 (1957); **243**, 336 (1957); L. D. LANDAU: *Žurn. Ėksp. Teor. Fiz.*, **32**, 59 (1957); V. M. GALITSKIJ and A. B. MIGDAL: *Žurn. Ėksp. Teor. Fiz.*, **34**, 139 (1958).

(2) After this manuscript was prepared we were shown a preprint by H. EHRENREICH and M. H. COHEN (submitted to *Phys. Rev.*) which also discusses this method and its application to solid-state physics. We should also point out that many of the concepts used by us were previously discussed by J. LINDHARD: *Dan. Mat. Fys. Medd.*, **28**, no. 8 (1954).

The ground state of the system Ψ_0 has the stationary density matrix

$$\varrho_0 = \sum_{n,\,\text{occ.}} |n\rangle\langle n| , \qquad H[\varrho_0]|n\rangle = E_n|n\rangle .$$

Suppose that the system is subjected to an arbitrarily weak external field $W(t)$ which results in the perturbation $\varrho_0 \to \varrho_0 + \varrho_1(t)$, with $\varrho_1(t)$ extremely small. Eq. (1) can therefore be linearized with respect to $\varrho_1(t)$ and $W(t)$. We immediately find that $\varrho_1(t)$ obeys

$$\text{(2)} \quad \left[i\frac{\partial}{\partial t} - (E_n - E_{n'})\right]\langle n|\varrho_1(t)|n'\rangle =$$
$$= [\theta(n') - \theta(n)][\sum_{mm'}\langle nm|v|n'm'\rangle\langle m'|\varrho_1(t)|m\rangle + \langle n|W(t)|n'\rangle] ,$$

with $\theta(n) = 1$ if $|n\rangle$ is occupied in Ψ_0, and zero otherwise.

If the system is extremely large the $|n\rangle$'s are momentum eigenstates $|\boldsymbol{k}\rangle$, and $E_k = (k^2/2m) + V_{\text{ex}}(k)$, with $V_{\text{ex}}(k)$ the one-particle potential arising from the exchange energy. The assumption that v and W are spin-independent local potentials, together with the neglect of the exchange term, reduces (2) to

$$\text{(3)} \quad \left[i\frac{\partial}{\partial t} - (E_{\boldsymbol{k}} - E_{\boldsymbol{k}+\boldsymbol{q}})\right]\langle \boldsymbol{k}|\varrho_1(t)|\boldsymbol{k}+\boldsymbol{q}\rangle =$$
$$= [\theta(\boldsymbol{k}) - \theta(\boldsymbol{k}+\boldsymbol{q})]\{W_{\boldsymbol{q}}(t) + g v(q)\sum_{\boldsymbol{k}'}\langle \boldsymbol{k}'|\varrho_1(t)|\boldsymbol{k}'+\boldsymbol{q}\rangle\} ,$$

where g is the spin-degeneracy.

Putting $W(t) = W\exp[-i\omega t] + \text{h.c.}$, $\varrho_1(t) = \varrho_1\exp[-i\omega t] + \text{h.c.}$, $\omega > 0$, in (3), we find that

$$\text{(4)} \qquad \operatorname{Tr}(W^+\varrho_1) = |W_q|^2 \frac{S(q\omega)}{1 - g v(q) S(q\omega)} ,$$

with

$$S(q\omega) = 2\sum_{\boldsymbol{k}} \frac{\theta(\boldsymbol{k}+\boldsymbol{q})[1-\theta(\boldsymbol{k})](E_{\boldsymbol{k}} - E_{\boldsymbol{k}+\boldsymbol{q}})}{\omega^2 - (E_{\boldsymbol{k}} - E_{\boldsymbol{k}+\boldsymbol{q}})^2} .$$

The physical content of eq. (4) can be understood from the following considerations which are independent of the Hartree-Fock approximation. To first order in W the state we are dealing with is

$$\Psi(t) = |0\rangle + \sum_{a\neq 0}|a\rangle\left\{\frac{\langle a|W|0\rangle}{\omega - E_a + i\varepsilon}\exp[-i\omega t] - \frac{\langle a|W^+|0\rangle}{\omega + E_a}\exp[i\omega t]\right\} ,$$

where $|0\rangle$, $|a\rangle$, are the *exact* eigenstates of the system with excitation energies E_a. Again writing

$$|\Psi(t)\rangle\langle\Psi(t)| = \varrho_0' + (\varrho_1'\exp[-i\omega t] + \text{h.c.}) + 0(W^2) ,$$

we have

$$\text{(4')}\qquad \operatorname{Tr}(W^+\varrho_1') = \sum_{a\neq 0}\left\{\frac{|\langle a|W|0\rangle|^2}{\omega - E_a + i\varepsilon} - \frac{|\langle 0|W|a\rangle|^2}{\omega + E_a}\right\} = R(\omega)\,.$$

Therefore the poles of $R(\omega)$ give the excitation energies of the system, while

$$-2\,\operatorname{Im} R(\omega) = 2\pi\sum_{a\neq 0}|\langle a|W|0\rangle|^2\,\delta(\omega - E_a)\,,$$

is the response, i.e. the total transition probability out of $|0\rangle$ induced by W. As is well known (3), the integral of the response over ω gives the pair correlation function in $|0\rangle$, and therefore this method enables us to find the total energy in the ground state. In fact, (4') is simply the Feynman propagator

$$-i\int_{-\infty}^{\infty}\langle 0|T(\exp[iHt]\,W^+\exp[-iHt],\,W)|0\rangle\exp[i\omega t]\,dt\,.$$

Comparing (4) and (4') we now find that the resonant frequencies of the system are characterized by the momentum $\boldsymbol{q}$ and are the solutions of the secular equation

$$\text{(5)}\qquad 1 - g\,v(q)\,S(q\omega) = 0\,.$$

This is precisely the eigenvalue condition discovered by HUBBARD and SAWADA. The spectrum consists of a continuum of single-particle excitations for $|\omega| < (qv_F + q^2/2m)$, v_F being the Fermi velocity, and provided $v(q)$ is repulsive and q not too large, a further isolated root $\omega_{\text{coll}}(q)$ above this continuum. If $v(q)\to q^{-2}$ as $q\to 0$ the isolated root is just the plasma frequency. On the other hand, if $v(q)\to$ const. as $q\to 0$, (short range forces)

$$\text{(6)}\qquad \omega_{\text{coll}}(q)\sim qv_F(1 + 2\exp[-1/\lambda - 2])\,,\qquad \lambda = 2\pi mk_F\,gv(0)\,,$$

in the weak coupling limit ($\lambda\to 0$). This latter solution, which was discovered by LANDAU and which he calls zeroth sound, remains experimentally undetected. If the forces are spin- (or iso-spin)-dependent, the situation is somewhat more complex, because then the separation into direct and exchange terms becomes artificial. Since the collective roots are isolated (i.e., like bound states in ordinary scattering theory) a variational attack using the solutions without exchange as trial functions is feasible.

The connection with Landau's Boltzmann equation can be seen as follows. Rewrite (2) in the coordinate representation:

$$\text{(7)}\qquad i\frac{\partial}{\partial t}\langle \boldsymbol{x}|\varrho_1(t)|\boldsymbol{x}'\rangle = \left\{-\frac{1}{2m}\left(\frac{\partial^2}{\partial \boldsymbol{x}^2} - \frac{\partial^2}{\partial \boldsymbol{x}'^2}\right) + V(\boldsymbol{x}) - V(\boldsymbol{x}')\right\}\langle \boldsymbol{x}|\varrho_1(t)|\boldsymbol{x}'\rangle +$$
$$+\langle \boldsymbol{x}|\varrho_0|\boldsymbol{x}'\rangle\int\{v(\boldsymbol{x}-\boldsymbol{y}) - v(\boldsymbol{x}'-\boldsymbol{y})\}\langle \boldsymbol{y}|\varrho_1(t)|\boldsymbol{y}\rangle\,d^3y\,,$$

---

(3) W. HEISENBERG: *Phys. Zeits.*, **32**, 737 (1931); A. AHIEZER and I. POMERANČUK: *Journ. Phys. USSR*, **11**, 167 (1947); G. C. WICK: *Phys. Rev.*, **94**, 1228 (1954); L. VAN HOVE: *Phys. Rev.*, **95**, 249 (1954); D. PINES and P. NOZIÈRES: *Nuovo Cimento*, **9**, 470 (1958).

where $V(x)$ is the ground state's self-consistent potential, and exchange terms are again deleted. Let $\boldsymbol{x}-\boldsymbol{x}'=\boldsymbol{\xi}$, $2\boldsymbol{r}=\boldsymbol{x}+\boldsymbol{x}'$, $\langle\boldsymbol{x}|\varrho|\boldsymbol{x}'\rangle=\varrho(\boldsymbol{r},\boldsymbol{\xi})$, and $\varrho(\boldsymbol{r},\boldsymbol{k})=\int\exp[-i\boldsymbol{k}\cdot\boldsymbol{\xi}]\cdot\varrho(\boldsymbol{r},\boldsymbol{\xi})\,\mathrm{d}^3\xi$. The Fourier transform of the last term of (7) then reads

$$(8)\qquad (2\pi)^{-3}\int\exp[i\boldsymbol{p}\cdot\boldsymbol{r}-\boldsymbol{r}']v(p)\{\varrho_0(\boldsymbol{r},\boldsymbol{k}-\tfrac{1}{2}\boldsymbol{p})-\varrho_0(\boldsymbol{r},\boldsymbol{k}+\tfrac{1}{2}\boldsymbol{p})\}\,\varrho_1(\boldsymbol{r}',\boldsymbol{p}',t)\,\mathrm{d}^3p\,\mathrm{d}^3r'\,\mathrm{d}^3p'\simeq$$

$$\simeq i(2\pi)^{-3}\int\exp[i\boldsymbol{p}\cdot\boldsymbol{r}-\boldsymbol{r}']v(p)\frac{\partial}{\partial\boldsymbol{r}'}\varrho_1(\boldsymbol{r}',\boldsymbol{p}',t)\cdot\frac{\partial}{\partial\boldsymbol{k}}\varrho_0(\boldsymbol{r},\boldsymbol{k})\,\mathrm{d}^3p\,\mathrm{d}^3r'\,\mathrm{d}^3p'\,.$$

If the motion is such that only wavelengths large compared to the range of v and the interparticle distance enter, and if $V(x)$ is assumed to be slowly varying, (7) and (8) may be simplified to the transport equation

$$(9)\qquad \frac{\partial}{\partial t}\varrho_1(\boldsymbol{r},\boldsymbol{k},t)=\left(-\frac{\boldsymbol{k}}{m}\cdot\frac{\partial}{\partial\boldsymbol{r}}+\frac{\partial V}{\partial\boldsymbol{r}}\cdot\frac{\partial}{\partial\boldsymbol{k}}\right)\varrho_1(\boldsymbol{r},\boldsymbol{k},t)+$$

$$+v(0)\frac{\partial}{\partial\boldsymbol{k}}\varrho_0(\boldsymbol{r},\boldsymbol{k})\cdot\frac{\partial}{\partial\boldsymbol{r}}\int\varrho_1(\boldsymbol{r},\boldsymbol{k}',t)\,\mathrm{d}^3k'\,.$$

Note that for the infinite system in the long wavelength limit (9) is an exact consequence of (3), and therefore Landau's apparently semi-classical treatment is equivalent to the other methods referred to. The solution of (9) in the infinite case is

$$(10)\qquad \varrho_1(\boldsymbol{r},\boldsymbol{k},t)=\exp[i(\boldsymbol{q}\cdot\boldsymbol{r}-\omega t)]\frac{\boldsymbol{k}\cdot\boldsymbol{q}\,\delta(k-k_F)}{\omega-\boldsymbol{k}\cdot\boldsymbol{q}/m}\,;$$

interpretation of (10) as a classical distribution function immediately reveals Landau's picture of the motion as a wave on the surface of the Fermi sphere.

In a recent article GLASSGOLD, HECKROTTE and WATSON (4) have employed the Sawada techniques to study collective excitations in nuclear matter. It is shown that when spin and isospin dependent forces are present collective modes can still arise even though the forces are attractive. One of these modes is related to the giant resonance in the photo-effect. It is apparent from the above that these modes are of the zeroth sound type, and are not sound waves in the ordinary sense. Unfortunately it is by no means clear that the simple results found in the infinite case retain an approximate validity in systems as small as atomic nuclei. For in a finite system the single-particle excitation spectrum does not have a sharp upper limit and so, loosely speaking, there will be « damping » of the collective motion even in this lowest order approximation. A study of the finite system can perhaps be based on the transport equation (9), which can be further simplified by taking a Thomas-Fermi distribution for ϱ_0.

* * *

We should like to thank A. BOHR and B. R. MOTTELSON for a number of helpful conversations, and Professor NIELS BOHR for the hospitality of his Institute.

---

(4) A. E. GLASSGOLD, W. HECKROTTE and K. M. WATSON: *Ann. Phys.*, **6**, 1 (1959).

Vol. XI, No. 1 JOURNAL of PHYSICS 1947

ON THE THEORY OF SUPERFLUIDITY *

By N. BOGOLUBOV

Mathematical Institute, Academy of Sciences of the Ukrainian SSR and Moscow State University

(Received October 12, 1946)

This paper presents an attempt of explaining the phenomenon of superfluidity on the basis of the theory of degeneracy of a non-perfect Bose-Einstein gas.

By using the method of the second quantization together with an approximation procedure we show that in the case of the small interaction between molecules the low excited states of the gas can be described as a perfect Bose-Einstein gas of certain "quasi-particles" representing the elementary excitations, which cannot be identified with the individual molecules.

The special form of the energy of a quasi-particle as a function of its momentum is shown to be connected with the superfluidity.

The object of this paper is an attempt to construct a consistent molecular theory explaining the phenomenon of superfluidity without assumptions concerning the structure of the energy spectrum.

The most natural starting point for such a theory seems to be the scheme of a non-perfect Bose-Einstein gas with a weak interaction between its particles.

It should be noted that similar attempts were done some time ago by Tisza and London to explain the phenomenon of superfluidity on the basis of the degeneracy of a perfect Bose-Einstein gas, but these attempts raised a counterblast of objections.

It has been pointed out, for example, that helium II has nothing to do with a perfect gas, because of the strong interaction between its molecules. However, this objection cannot be regarded as an essential one. Indeed, it is clear that a rigorous theoretical computation of the properties of a real liquid is hopelessly beyond the reach of a pure molecular theory based on usual "microscopic" equations of quantum mechanics. All we can require from a molecular theory of superfluidity, at least at the first stage of investigation, is to be able to account for the qualitative picture of this phenomenon being based on a certain simplified scheme.

A really essential objection one can make against this idea is the following one. The particles of a degenerate perfect Bose-Einstein gas in the ground state cannot possess the property of superfluidity, since nothing prevents them from exchanging their momenta with excited particles colliding with them, and, therefore, from friction in their movement through the liquid.

In the present paper we try to overcome this difficulty and to show that under certain conditions the "degenerate condensate" of a "nearly perfect" Bose-Einstein gas can move without any friction with respect to the elementary excitations, with an arbitrary, sufficiently small velocity. It is to be pointed out that the necessity of considering the collective elementary excitations rather than individual molecules was suggested by L. Landau in his well known paper "Theory of Superfluidity of Helium II" where he, by postulating their existence in form of phonons and rotons, was enabled to explain the property of the superfluidity.

In our theory the existence and the properties of the elementary excitations follow directly from the basic equations describing the Bose-Einstein condensation of non perfect gases.

* Presented to the Session of the Physical Mathematical Department of the Academy of Sciences of the USSR on October 21, 1946.

1. Let us consider a system of N identical monoatomic molecules enclosed in a certain macroscopic volume V and subjected to Bose statistics.

Suppose the Hamiltonian of our system to be, as usually, of the form:

$$H = \sum_{(1\leq i\leq N)} T(p_i) + \sum_{(1\leq i<j\leq N)} \Phi(|q_i - q_j|),$$

where

$$T(p_i) = \frac{|p_i|^2}{2m} = \sum_{(1\leq\alpha\leq 3)} \frac{(p_i^\alpha)^2}{2m}$$

represents the kinetic energy of the ith molecule and $\Phi(|q_i - q_j|)$ — the mutual potential energy of the pair (i, j).

Applying then the method of second quantization, let us present the basic equation in the form:

$$i\hbar\frac{\partial\Psi}{\partial t} = -\frac{\hbar^2}{2m}\Delta\Psi + \\ + \int \Phi(|q-q'|)\Psi^+(q')\Psi(q')dq' \cdot \Psi, \quad (1)$$

whereby

$$\Psi = \sum_f a_f\varphi_f(q); \quad \Psi^+ = \sum_f a_f^+\varphi_f^+(q).$$

Here a_f, a_f^+ are conjugated operators with commutation rules of the well-known type:

$$a_f a_{f'} - a_{f'} a_f = 0;$$

$$a_f a_{f'}^+ - a_{f'}^+ a_f = \Delta_{f,f'} = \begin{cases} 1, & f = f' \\ 0, & f \neq f' \end{cases}$$

and $\{\varphi_f(q)\}$ is a complete orthonormal set of functions:

$$\int \varphi_f^+(q)\varphi_{f'}(q)dq = \Delta_{f,f'}.$$

For the sake of simplicity, we shall further employ the system of eigenfunctions of the momentum operator of a single particle:

$$\varphi_f(q) = \frac{1}{V^{1/2}} e^{i\frac{(f\cdot q)}{\hbar}}; \quad (f\cdot q) = \sum_{(1\leq\alpha\leq 3)} f^\alpha q^\alpha,$$

the operator $N_f = a_f^+ a_f$ then representing the number of molecules with momentum f. For finite V vector f is obviously quantized. For example, under usual boundary conditions of the periodicity type,

$$f^\alpha = \frac{2\pi n^\alpha \hbar}{l},$$

where (n^1, n^2, n^3) are integers, and l denotes the side of a cube of the volume V.

However, since we are going to deal with thermodynamic volume properties of the system, we have to bear in mind a limiting process, such that when $N\to\infty$, the boundary of the container tends to infinity, $V\to\infty$, but the volume per molecule $v = V/N$ remains constant. Therefore, we shall pass over in the final results to a continuous spectrum, replacing the sums $\sum_f F(f)$ by integrals

$$\frac{V}{(2\pi\hbar)^3}\int F(f)\,df.$$

Equations (1) are the exact equations of the problem of N bodies. It becomes, therefore, a necessity in order to push forward the investigation of the motion of the considered system, to apply some approximation method based on a supposition that the energy of interaction is sufficiently small. According to this supposition, the potential function $\Phi(r)$ is assumed to be proportional to a certain small parameter ε.

Later on we shall see what dimensionless quantity can be chosen for ε. For the present it suffices to notice that the above supposition, strictly speaking, corresponds to a neglection of the finiteness of molecular radius, since we do not take into account the intensive increase of $\Phi(r)$ for small r, which causes the impenetrability of molecules. We shall see, however, that the results, which will be obtained, can be generalized so as to include the case of a finite molecular radius.

Now let us turn to the formulation of an approximate method. If there be no interaction at all, *i. e.* if ε be exactly zero, we could put at zero temperature: $N_0 = N$, $N_f = 0$ $(f\neq 0)$.

But in the case under consideration, when ε is small and the gas is in a weakly excited state, these relations are valid approximately, which means that the momenta of the overwhelming majority of molecules approach zero. Of course, the fact, that zero momentum is the limiting one for particles in the ground state, is due to a specific choice of the coordinate system: namely, we choose the system with respect to which our "condensate" is at rest.

Our approximation method, based on the above considerations, runs as follows.

1) Since $N_0 = a_0^+ a_0$ is very large as compared with unity, the expression:

$$a_0 a_0^+ - a_0^+ a_0 = 1,$$

is small as compared with a_0^+, a_0 themselves, which enables us to treat them as ordinary numbers* neglecting their non-commutability.

2) Let us put

$$\Psi = \frac{a_0}{\sqrt{V}} + \vartheta; \quad \vartheta = \frac{1}{\sqrt{V}} \sum_{(f\neq 0)} a_f e^{i\frac{(f\cdot q)}{\hbar}}$$

and consider ϑ as a "correction term of the first order". Neglecting all the terms in equation (1), involving the second and higher powers of ϑ, this being permissible since the excitations are supposed to be weak, we obtain the following basic approximate equations:

$$i\hbar\frac{\partial\vartheta}{\partial t} = -\frac{\hbar^2}{2m}\Delta\vartheta + \frac{N_0}{V}\Phi_0\vartheta +$$
$$+\frac{N_0}{V}\int \Phi(|q-q'|)\vartheta(q')dq' +$$
$$+\frac{a_0^2}{V}\int \Phi(|q-q'|)\vartheta^+(q')dq', \qquad (2)$$
$$i\hbar\frac{\partial a_0}{\partial t} = \frac{N_0}{V}\Phi_0 a_0,$$

where

$$\Phi_0 = \int \Phi(|q|)dq.$$

In order to pass from the operator wave function ϑ to operator amplitudes let us apply Fourier's expansion;

$$\Phi(|q-q'|) = \sum_f \frac{1}{V} e^{i\frac{(f\cdot q-q')}{\hbar}} \nu(f). \qquad (3)$$

From the radial symmetry of the potential function it follows that the amplitudes of this expansion:

$$\nu(f) = \int \Phi(|q|) e^{-i\frac{(f\cdot q)}{\hbar}} dq,$$

depend upon the length $|f|$ of vector f only.

The substitution of (3) into equation (2) gives:

$$i\hbar\frac{\partial a_f}{\partial t} = \left\{T(f) + E_0 + \frac{N_0}{V}\nu(f)\right\} a_f +$$
$$+\frac{a_0^2}{V}\nu(f)a_{-f}^+;$$

$$E_0 = \frac{N_0}{V}\Phi_0,$$

whence on setting

$$a_f = e^{i\frac{E_0}{\hbar}t} b_f; \quad a_0 = e^{i\frac{E_0}{\hbar}t} b \qquad (4)$$

we get:

$$i\hbar\frac{\partial b_f}{\partial t} = \left\{T(f) + \frac{N_0}{V}\nu(f)\right\} b_f + \frac{b^2}{V}\nu(f)b_{-f}^+, \qquad (5)$$
$$-i\hbar\frac{\partial b_{-f}^+}{\partial t} = \frac{(b^+)^2}{V}\nu(f)b_f + \left\{T(f) + \frac{N_0}{V}\nu(f)\right\} b_{-f}^+.$$

Solving this system of two differential equations with constant coefficients, we find that the operators b_f, b_f^+ depend upon the time by means of a linear combination of exponentials possessing the form

$$e^{\pm i\frac{E(f)}{\hbar}t},$$

where *

$$E(f) = \sqrt{2T(f)\frac{N_0}{V}\nu(f) + T^2(f)}\,. \qquad (6)$$

Now let us observe that the inequality:

$$\nu(0) = \int \Phi(|q|)dq > 0, \qquad (7)$$

implies the positiveness of the expression under the sign of the radical (6) since ε is considered to be sufficiently small; thus b_f, b_f^+ prove to be periodical functions of time. On the contrary, if $\nu(0) < 0$, this expression is negative for small momenta and, therefore, $E(f)$ receives complex values. As a consequence, b_f, b_f^+ will involve a real exponential increasing with time, whence it follows that the states with small $N_f = b_f^+ b_f$ are unstable.

In order to be sure in the stability of the excited states, let us restrict the class of possible types of interaction forces, supposing inequality (7) to be satisfied for all types we shall consider. It is interesting to note that the inequality (7) just represents the condition of thermodynamic stability of a gas at absolute zero.

* A similar procedure was used by Dirac in his book "The Principles of Quantum Mechanics" (second edition), *cf.* the end of § 63: Waves and Bose-Einstein particles.

* We may remark that if we form the corresponding frequency $\omega = \hbar E(f)$ and put: $\hbar \to 0$, $f/\hbar = k =$ const we obtain the classical formula for the dispersion of the frequency deduced by A. Vlasov, *cf.* Journal of Physics, **9**, 25 (1945).

Indeed, at absolute zero the free energy coincides with the mean energy. The main term of the latter has the form

$$E = \frac{N^2}{2V} \int \Phi(|q|)\, dq,$$

since the correction terms (for instance, the mean kinetic energy) are proportional to higher powers of ε.

Therefore, the pressure P is expressed by the following formula:

$$P = -\frac{\partial E}{\partial V} = \frac{N^2}{2V^2} \int \Phi(|q|)\, dq = \frac{\rho^2}{2m^2} \nu(0),$$

where $\rho = Nm/V$ represents the density of the gas.

This proves that the inequality (7) is equivalent to the condition of thermodynamic stability:

$$\frac{\partial P}{\partial \rho} > 0.$$

Finally, let us note that we can write instead of (6), with the same degree of accuracy:

$$E(f) = \sqrt{2T(f)\frac{N}{V}\nu(f) + T^2(f)} = \\ = \sqrt{\frac{|f|^2 \nu(f)}{mv} + \frac{|f|^4}{4m^2}}, \qquad (6')$$

since we take into account the main terms only; it follows from (6′) for small momenta:

$$E(f) = \sqrt{\frac{\nu(0)}{mv}}\, |f|(1 + \ldots) = \\ = \sqrt{\frac{\partial P}{\partial \rho}}\, |f|(1 + \ldots),$$

where by dots are denoted the terms vanishing together with f.

In what follows we shall attribute the positive sign to each square root we shall deal with. We thus have for small momenta:

$$E(f) = c\,|f|(1 + \ldots), \qquad (8)$$

where c denotes the velocity of sound at absolute zero.

On the contrary, for sufficiently large momenta $E(f)$ can be expanded in powers of ε:

$$E(f) = \frac{|f|^2}{2m} + \frac{\nu(f)}{v} + \ldots \qquad (9)$$

Since $\nu(f)$ tends to zero with increasing $|f|$, $E(f)$ is seen to approach the kinetic energy $T(f)$ of a single molecule for sufficiently large momenta.

Returning now to equations (5), let us introduce new mutually conjugated operators ξ_f, ξ_f^+ instead of b_f, b_f^+ by means of the following relations

$$\xi_f = \frac{b_f - L_f b_{-f}^+}{\sqrt{1 - |L_f|^2}}; \quad \xi_f^+ = \frac{b_f^+ - L_f^+ b_{-f}}{\sqrt{1 - |L_f|^2}}, \qquad (10)$$

where L_f are numbers determined by the equalities:

$$L_f = \frac{Vb^2}{N_0^2 \nu(f)} \left\{ E(f) - T(f) - \frac{N_0}{V}\nu(f) \right\},$$

so that

$$|L_f|^2 = \left(\frac{(N_0/V)\nu(f)}{E(f) + T(f) + (N_0/V)\nu(f)} \right)^2; \\ 1 - |L_f|^2 = \frac{2E(f)}{E(f) + T(f) + (N_0/V)\nu(f)}. \qquad (11)$$

Reversing (10) we obtain:

$$b_f = \frac{\xi_f + L_f \xi_{-f}^+}{\sqrt{1 - |L_f|^2}}; \quad b_f^+ = \frac{\xi_f^+ + L_f^+}{\sqrt{1 - |L_f|^2}}. \qquad (12)$$

The substitution of these expressions into equations (5) gives:

$$i\hbar \frac{\partial \xi_f}{\partial t} = E(f)\xi_f; \quad -i\hbar \frac{\partial \xi_f^+}{\partial t} = E(f)\xi_f^+. \qquad (13)$$

It can be immediately verified that the new operators satisfy the same commutation relations as the operators a_f, a_f^+:

$$\xi_f \xi_{f'} - \xi_{f'} \xi_f = 0; \quad \xi_f \xi_{f'}^+ - \xi_{f'}^+ \xi_f = \Delta_{f,f'}. \qquad (14)$$

This alone suffices to conclude that the excited states of the given assemblage of molecules can be treated as a perfect gas composed of "elementary excitations" — "quasi-particles" with energy depending on the momentum by means of the relation: $E = E(f)$. These quasi-particles are described by the operators ξ_f, ξ_f^+ in the same way as molecules were described by the operators a_f, a_f^+, and, therefore, they are also subjected to Bose statistics. The operator

$$n_f = \xi_f^+ \xi_f$$

represents the number of quasi-particles with the momentum f.

The above conclusion becomes quite clear if we consider the total energy

$$H = H_{\text{kin}} + H_{\text{pot}}$$

where

$$H_{\text{kin}} = -\frac{\hbar^2}{2m}\int \Psi^+(q)\Delta\Psi(q)\,dq,$$

$$H_{\text{pot}} = \frac{1}{2}\int \Phi(|q-q'|)\,\Psi^+(q)\,\Psi^+(q')\,\Psi(q)\,\Psi(q')\,dq\,dq' =$$

$$= \frac{1}{2V}\sum_f \nu(f)\int e^{i\frac{(f\cdot q-q')}{\hbar}}\,\Psi^+(q)\,\Psi^+(q')\,\Psi(q)\,\Psi(q')\,dq\,dq'.$$

For the kinetic energy we obtain

$$H_{\text{kin}} = \sum_f T(f)\,a_f^+a_f = \sum_f T(f)\,b_f^+b_f.$$

In order to compute the potential energy with the assumed degree of accuracy, we shall disregard in the expression:

$$\Psi^+(q)\,\Psi^+(q')\,\Psi(q)\,\Psi(q') = \left(\frac{a_0^+}{\sqrt{V}}+\vartheta^+(q)\right)\left(\frac{a_0^+}{\sqrt{V}}+\vartheta^+(q')\right)\left(\frac{a_0}{\sqrt{V}}+\vartheta(q)\right)\left(\frac{a_0}{\sqrt{V}}+\vartheta(q')\right)$$

all terms of the third and higher order with respect to ϑ, ϑ^+. This gives

$$H_{\text{pot}} = \Phi_0\left\{\frac{1}{2}\frac{N_0^2}{V}+\frac{N_0}{V}\sum_{f\neq 0} b_f^+b_f\right\}+\frac{b^2}{2V}\sum_{f\neq 0}\nu(f)\,b_f^+b_{-f}^+ + \frac{(b^+)^2}{2V}\sum_{f\neq 0}\nu(f)\,b_f b_{-f} + \frac{N_0}{V}\sum_{f\neq 0}\nu(f)\,b_f^+b_f.$$

Noticing here that

$$\sum_{f\neq 0} b_f^+b_f = \sum_{f\neq 0} N_f = N - N_0,$$

we can write with the same degree of accuracy:

$$\frac{1}{2}\frac{N_0^2}{V}+\frac{N_0}{V}\sum_{f\neq 0} b_f^+b_f = \frac{1}{2}\frac{N^2}{V}.$$

Hence

$$H = \frac{N^2}{2V}\Phi_0 + \frac{b^2}{2V}\sum_{f\neq 0}\nu(f)\,b_f^+b_{-f}^+ + \frac{(b^+)^2}{2V}\sum_{f\neq 0}\nu(f)\,b_f b_{-f} + \frac{N_0}{V}\sum_{f\neq 0}\nu(f)\,b_f^+b_f + \sum_f T(f)\,b_f^+b_f.$$

Replacing here the operators b_f, b_f^+ by the operators ξ_f, ξ_f^+ with the aid of relations (12), we obtain finally

$$H = H_0 + \sum_{f\neq 0} E(f)\,n_f; \qquad n_f = \xi_f^+\xi_f, \tag{15}$$

where

$$H_0 = \frac{1}{2}\frac{N^2}{V}\Phi_0 + \sum_{f\neq 0}\frac{E(f)-T(f)-(N_0/V)\,\nu(f)}{2} =$$

$$= \frac{1}{2}\frac{N^2}{V}\Phi_0 + \frac{V}{2(2\pi\hbar)^3}\int\left\{E(f)-T(f)-\frac{N_0}{V}\nu(f)\right\}df. \tag{16}$$

Thus we see that the total energy of the considered non-ideal gas consists of the energy of the ground state and the individual energies of each of the quasi-particles. It means that the quasi-particles do not interact with each other and thus form a perfect Bose-Einstein gas. The absence of interaction between the quasi-particles is evidently caused by the admitted approximation, namely by neglecting the terms of the third and higher order with

*

respect to ξ_f, ξ_f^+ involved in the expression for energy. Therefore, the above results are valid for weakly excited states only.

On having taken into account the disregarded third-order terms in the expression for energy or the second-order terms in equations (13) which are to be considered as a small perturbation, we could discover a weak interaction between the quasi-particles. This interaction enables the assemblage of quasi-particles to attain the state of statistical equilibrium. Proceeding to the study of this state, let us first observe that the total momentum of quasi-particles: $\sum_f f n_f$, is conserved. To prove this consider the components of the total momentum. From

$$\sum_{(1\leq i\leq N)} p_i^\alpha = \int \Psi^+(q)\left\{-i\hbar\frac{\partial\Psi(q)}{\partial q^\alpha}\right\}dq = \sum_f f^\alpha a_f^+ a_f = \sum_f f^\alpha b_f^+ b_f$$

it follows, according to the transformation relations (12), that

$$\sum_{(1\leq i\leq N)} p_i^\alpha = \sum_f f^\alpha \frac{(\xi_f^+ + L_f^+\xi_{-f})(\xi_f + L_f\xi_{-f}^+)}{1-|L_f|^2}\,.$$

The invariance of L_f, L_f^+ with respect to the replacement of f by $-f$ implies:

$$\sum_f f^\alpha \frac{L_f^+\xi_{-f}\xi_f}{1-|L_f|^2} = \sum_f f^\alpha \frac{L_f\xi_f^+\xi_{-f}^+}{1-|L_f|^2} = 0,$$

$$\sum_f f^\alpha \frac{|L_f|^2\xi_{-f}\xi_{-f}^+}{1-|L_f|^2} = \sum_f f^\alpha \frac{|L_f|^2(\xi_{-f}\xi_{-f}^+ - 1)}{1-|L_f|^2} = -\sum_f f^\alpha \frac{|L_f|^2\xi_f^+\xi_f}{1-|L_f|^2}\,.$$

Hence we conclude that

$$\sum_{(1\leq i\leq N)} p_i^\alpha = \sum_f f^\alpha n_f,$$

i. e. that the total momentum of the assemblage of molecules is equal to that of the assemblage of quasi-particles. Since the former is conserved, the sum $\sum_f f n_f$ proves to be conserved too.

It is also easy to see that the total number of quasi-particles $\sum_f n_f$ is not invariant; quasi-particles can appear and disappear.

For this reason we obtain in the usual way the following formula for average occupation numbers $\bar{n}_f$ $(f\neq 0)$ in the state of statistical equilibrium:

$$\bar{n}_f = \left\{A\exp\frac{E(f)-(f\cdot u)}{\Theta} - 1\right\}^{-1};\ A=1, \quad (17)$$

where Θ is the temperature modulus, while u denotes an arbitrary vector. However, the length of this vector must have an upper limit. In fact, since all average occupation numbers have to be positive, the inequality

$$E(f) > (f\cdot u),$$

and, therefore, the inequality

$$E(f) > |f|\cdot|u|$$

is satisfied for all $f\neq 0$.

But from the above properties of $E(f)$ it follows that the ratio

$$\frac{E(f)}{|f|}$$

is a continuous positive function of $|f|$. This function is equal to $c>0$ at $|f|=0$ and is increasing as $|f|^2/2m$, for $|f|\to\infty$; therefore, the minimum value of the considered ratio is essentially positive. Hence the condition for the positiveness of $\bar{n}_f$ is equivalent to inequality

$$|u| \leq \min\frac{E(f)}{|f|}\,. \quad (18)$$

If the decrease of $E(f)$ for small momenta be proportional not to the momentum itself, but to its square (as it is the case for the kinetic energy of a molecule), the right-hand side of the obtained inequality would be equal to zero, this value being the only possible one for u. But in the case under consideration the vector u can be chosen arbitrarily, provided that its length is sufficiently small.

Now let us note that from the momentum distribution over the gas, composed of quasi-particles, given by formula (17), it follows that this gas is moving as a whole with the velocity u. At first, we have chosen such a coordinate system with respect to which the condensate (*i. e.* the assemblage of molecules in the ground state) is at rest. Inversely, by transition to a coordinate system with respect to which the quasi-particle gas, as a whole, is at rest, we can discover the motion of the condensate with the velocity u.

This relative motion goes on stationarily in the state of statistical equilibrium without any external forces. Hence we see that it is not accompanied by friction and thus represents the property of superfluidity*.

As we have seen, the energy of a quasi-particle is asymptotically equal to $c|f|$ for small momenta, c being the velocity of sound. Therefore, a quasi-particle for small momenta is just a phonon. When the momentum is increasing, the kinetic energy becomes large as compared with the binding energy of a molecule, and the energy of a quasi-particle tends continuously toward the individual energy of a molecule $T(f)$.

Thus we see that no division of quasi-particles into two different types, phonons and rotons, can even be spoken of.

2. Now let us consider the distribution of momenta over an assemblage of molecules, in the state of statistical equilibrium. We introduce a function $W(f)$ defined in such a way that $NW(f)\,df$ represents the average number of molecules whose momenta belong to an elementary volume df of the momentum space. This function is seen to be normalized by means of

$$\int W(f)\,df = 1. \tag{19}$$

* By using the argument pointed out by Landau in § 4 of his paper "Theory of Superfluidity of Helium II" it may be noted that the existence of superfluidity is directly evident from the above mentioned properties of the function $E(f)$, following from the inequality: $\nu(0) > 0$. This inequality can thus be considered as the condition of superfluidity.

Suppose, further, $F(f)$ to be an arbitrary continuous function of the momentum. For the mean value of a dynamical variable

$$\sum_{(1\le i\le N)} F(p_i),$$

we then obtain obviously:

$$N\int F(f)\,W(f)\,df. \tag{20}$$

On the other hand, we have for the same mean value:

$$\sum_f F(f)\,\overline{N}_f = \frac{V}{(2\pi\hbar)^3}\int F(f)\,\overline{N}_f\,df. \tag{21}$$

Thus the comparison of expressions (20), (21) gives:

$$NW(f) = \frac{V}{(2\pi\hbar)^3}\overline{N}_f = N\frac{v}{(2\pi\hbar)^3}\overline{b_f^+ b_f},$$

whence we obtain, by expressing b_f, b_f^+ through ξ_f, ξ_f^+:

$$W(f) =$$

$$= \frac{v}{(2\pi\hbar)^3}(1-|L_f|^2)^{-1}\,\overline{(\xi_f^+ + L_f^+\xi_{-f})(\xi_f + L_f\xi_f^+)} =$$

$$= \frac{v}{(2\pi\hbar)^3}\,\frac{\bar{n}_f + |L_f|^2(\bar{n}_{-f}+1)}{1-|L_f|^2}, \tag{22}$$

where, in virtue of (17)

$$\bar{n}_f = \left\{\exp\left(\frac{E(f)-(f\cdot u)}{\Theta}\right)-1\right\}^{-1}. \tag{23}$$

Expression (22), obtained for the distribution function, holds evidently only for $f \neq 0$. Therefore, the complete expression for the momentum distribution function is seen to be, in virtue of the normalization condition (19):

$$W(f) = C\delta(f) + \frac{v}{(2\pi h)^3}\,\frac{\bar{n}_f + |L_f|^2(n_{-f}+1)}{1-|L_f|^2}, \tag{24}$$

where $\delta(f)$ is Dirac's δ-function, and the number C is determined by the equation:

$$C = 1 - \frac{v}{(2\pi\hbar)^3}\int\frac{\bar{n}_f + |L_f|^2(\bar{n}_{-f}+1)}{1-|L_f|^2}\,df. \tag{25}$$

C is obviously equal to N_0/N since CN represents the average number of molecules with zero momentum.

It follows from (11) that

$$\frac{|L_f|^2}{1-|L_f|^2}=\frac{\left(\frac{N_0}{V}\nu(f)\right)^2}{2E(f)\left\{E(f)+T(f)+\frac{N_0}{V}\nu(f)\right\}};$$
$$\frac{1}{1-|L_f|^2}=\frac{E(f)+T(f)+\frac{N_0}{V}\nu(f)}{2E(f)}. \qquad (26)$$

Hence the momentum distribution function at absolute zero is of the form:

$$W(f)=C\delta(f)+$$
$$+\frac{v}{(2\pi\hbar)^3}\frac{\left(\frac{N_0}{V}\nu(f)\right)^2}{2E(f)\left\{E(f)+T(f)+\frac{N_0}{V}\nu(f)\right\}},$$

whereby

$$1-C=$$
$$=\frac{v}{(2\pi\hbar)^3}\int\frac{\left(\frac{N_0}{V}\nu(f)\right)^2}{2E(f)\left\{E(f)+T(f)+\frac{N_0}{V}\nu(f)\right\}}\,df. \qquad (27)$$

We thus see that even at $\Theta=0$ only a fraction of molecules possesses momenta which are exactly zero, the rest being continuously distributed over the whole momentum spectrum.

As we have pointed out, our approximation method holds only when $(N-N_0)/N=1-C\ll 1$, and, therefore, the interaction of molecules must be sufficiently small in order to secure the smallness of the integral (27).

We can now make clear the meaning of the assumed smallness of the interaction.

Les us put $\Phi(r)=\Phi_m F(r/r_0)$, where $F(\rho)$ is a function assuming together with its derivatives the values of the order of unity for $\rho\sim 1$ and rapidly approaching zero as $\rho\longrightarrow\infty$. Then

$$\nu(f)=\Phi_m r_0^3\omega\left(\frac{|f|\,r_0}{\hbar}\right),$$

$\omega(x)$ being a function assuming the values ~ 1 for $x\sim 1$, rapidly approaching zero as $x\longrightarrow\infty$.

Transforming (27) to the dimensionless variables and reducing the three-dimensional integral to the one-dimensional one, we get:

$$\frac{N-N_0}{N}=\frac{v}{r_0^3}\eta\frac{1}{(2\pi)^2}\int_0^\infty\frac{\eta\omega^2(x)\,x\,dx}{\alpha(x)\{x\alpha(x)+x^2+\eta\omega(x)\}}, \qquad (28)$$

where

$$\alpha(x)=\sqrt{x^2+2\eta\omega(x)};$$
$$\eta=\frac{(r_0^3N_0/V)\,\Phi_m}{\hbar^2/2mr_0^2}\sim\frac{(r_0^3/v)\,\Phi_m}{\hbar^2/2mr_0^2}.$$

Now it is be easy to see that for small values of η the integral in (28) is of the order of $\sqrt{\eta}$, and hence the condition of the validity of our method may be represented by the inequalities:

$$\eta\ll 1;\quad \frac{v}{r_0^3}\eta^{3/2}\ll 1,$$

i. e.

$$\frac{r_0^3}{v}\Phi_m\ll\frac{\hbar^2}{2mr_0^2};\quad \sqrt[3]{\frac{r_0^3}{v}}\Phi_m\ll\frac{\hbar^2}{2mr^2}. \qquad (29)$$

For $\Theta>0$ an analogous consideration of the general formula (24) leads us to a supplementary condition of the weakness of the excitation which requires the temperature to be small as compared with the λ-point temperature.

It is to be pointed out that the inequalities (29) automatically exclude the possibility of accounting for the short-range repulsion forces, since this would require to admit the strong increment of $\Phi(r)$ in the vicinity of $r=0$.

It seems nevertheless possible to modify the obtained results in the way to get them extended to the more real case of a low density gas of molecules possessing a finite radius.

To this aim we observe that the potential function $\Phi(r)$ appears in our final formula only in the form of the expression:

$$\nu(f)=\int\Phi(|q|)\,e^{i\frac{(f\cdot q)}{\hbar}}\,dq, \qquad (30)$$

proportional to the amplitude of Born's collision probability for binary collision.

Hence, as the interaction of molecules for the low density gas reveals itself principally by means of these binary collisions, it seems that expression (30) is to be replaced* by the corresponding expression proportional to the amplitude of the exact probability of the binary collisions, calculated for the limiting case of zero density, *i. e.* we have to put:

$$\nu(f) = \int \Phi(|q|)\varphi(q, f)\, d \qquad (31)$$

where $\varphi(q, f)$ is the solution of the Schrödinger equation for the relative movement of an isolated pair of molecules:

$$-\frac{\hbar^2}{m}\Delta\varphi + \{\Phi(|q|) - E\}\varphi = 0,$$

going over into $e^{i\frac{(f\cdot q)}{\hbar}}$ at infinity. The replacement of (31) instead of (30) in the formula for $E(f)$ will lead us to the results, referring to low density gases.

This being admitted we see, *e. g.*, that the condition of the superfluidity $\nu(0) > 0$ may be written in the form:

$$\int \Phi(|q|)\varphi(|q|)\, dq > 0, \qquad (32)$$

where $\varphi(|q|)$ is the radially symmetric solution of the equation:

$$-\frac{\hbar^2}{m}\Delta\varphi + \Phi(|q|)\varphi = 0$$

going over into unity at infinity.

In order to connect, as before, inequality (32) with the condition of the thermodynamic stability let us compute the principal term in the expansion of the free energy in powers of density at the absolute zero of temperature. The free energy at absolute zero being equal to the mean energy, we have the following expression for this energy per one molecule:

$$\mathscr{E} = \overline{T} + \frac{1}{2v}\int \Phi(|q|)G(|q|)\, dq, \qquad (33)$$

where $\overline{T}$ is the mean kinetic energy of one molecule and $G(r)$ is the molecular distribution function normalized in the way that $G(r) \to 1$ as $r \to \infty$. On the other hand, by using the virial theorem we see that the pressure P can be determined by the formula

$$Pv = \frac{2}{3}T - \frac{1}{6v}\int \Phi'(|q|)|q|G(|q|)\, dq. \qquad (34)$$

Let us now remark that for $\Theta = 0$ the principal term in the expansion of the molecular distribution function in powers of density is obviously equal to $\varphi^2(|q|)$. Therefore, by neglecting in (33), (34) the terms of the second order in density one gets:

$$\mathscr{E} = \overline{T} + \frac{1}{2v}\int \Phi(|q|)\varphi^2(|q|)\, dq;$$

$$Pv = \frac{2}{3}\overline{T} - \frac{1}{6v}\int \Phi'(|q|)|q|\varphi^2(|q|)\, dq.$$

Hence, taking into account that

$$Pv = -v\frac{\partial\mathscr{E}}{\partial v},$$

we obtain the equation for the evaluation of the principal term in the expression for $\overline{T}$. In this way one gets:

$$\mathscr{E} = \frac{1}{2v}\int \Phi(|q|)\varphi(|q|)\, dq = \frac{\nu(0)}{2v}, \quad P = \frac{\nu(0)}{2v^2},$$

and thus the condition of superfluidity in the considered case of low density gas is also equivalent to the usual condition of the thermodynamic stability at absolute zero:

$$\frac{\partial P}{\partial v} < 0.$$

It can also be seen that the energy of a quasi-particle goes over again into $c|f|$ for small f.

Consider now, for instance, the model of hard impenetrable spheres of the radius $r_0/2$ and put:

$$\Phi(r) = +\infty, \qquad r < r_0;$$

$$\Phi(r) = 0, \qquad r < r_0.$$

* I am indebted to L. D. Landau for this important remark.

Then by simple calculation we get:

$$\nu(0) = 2\pi \frac{\hbar^2 r_0}{m}.$$

If a weak attraction between molecules is admitted here in the way that

$$\Phi(r) = +\infty, \qquad r < r_0;$$

$$\Phi(r) = \varepsilon\Phi_0(r) < 0, \qquad r > r_0.$$

ε being a small parameter, we obtain, up to the terms of the order of ε^2:

$$\nu(0) = 2\pi \frac{\hbar^2 r_0}{m} + 2\pi \int\limits_{r_0}^{\infty} r^2 \Phi(r)\, dr.$$

Thus, the superfluidity in the considered gas model is conditioned by the play of repulsion and attraction forces, the first "encouraging" and the latter "hindering" it.

Let us note in conclusion that the extension of the present theory to the case of the real liquid seems possible if we be permitted to use such semi-phenomenological conceptions as that of the free energy of slightly non-equilibrium states.

Eigenvalues and Eigenfunctions of a Bose System of Hard Spheres and Its Low-Temperature Properties

T. D. LEE, *Columbia University, New York, New York*

AND

KERSON HUANG AND C. N. YANG, *Institute for Advanced Study, Princeton, New Jersey*

(Received March 19, 1957)

It is shown that the pseudopotential method can be used for an explicit calculation of the first few terms in an expansion in power of $(\rho a^3)^{\frac{1}{2}}$ of the eigenvalues and the corresponding eigenfunctions of a system of Bose particles with hard-sphere interaction. The low-temperature properties of the system are discussed.

THIS paper is concerned with the low-temperature properties of a dilute system of Bose particles with hard-sphere interactions, at a low but finite density. An explicit mathematical calculation is made of the energies and wave functions of the ground state and the low-lying excited states. The results confirm the usual notion of phonon waves as the only low-lying excitation, and the idea of momentum space ordering. One concludes from the calculation that such a system does show superfluidity and exhibit the two-fluid behavior at low temperatures.

It may be appropriate here to describe the motivation underlying the study of a system of hard spheres. One would like, of course, to study the general many-body problem with any potential of interaction between the particles. Such a program can be formalistically carried out. It is, however, generally recognized that to draw any definite physical conclusions from such a general program is very difficult. If one makes approximations on the general problem in order to arrive at concrete results, one usually encounters the great difficulty of defining and justifying the validity of the approximation made. We therefore start instead from the concrete model of hard-sphere interactions, which is sufficiently simple so that one might hope to be able to discuss the validity of the method of approach.

The interaction between real He atoms contains besides a hard repulsive core, also an attractive interaction outside of the core. This attractive interaction is responsible for many properties of the He liquid. For example, the ground state of a system of He atoms is known to have a negative energy corresponding to a binding energy per He atom of $(k\times 7°)$, as determined from the experimental vapor pressure curve near the absolute zero of temperature. Such a bound system owes its origin, of course, to the attractive force. The strength of the attractive force also determines the density of the He atoms in the ground state. Now at this density the total attractive potential that a He atom experiences from its neighbors is expected not to fluctuate very much. This fact suggests the following approximate picture: One replaces the attractive interparticle forces by a constant uniform negative external potential that acts on the individual particles, the repulsive core is retained, and the system is kept by an external pressure at a density equal to that of the ground state of He. Many qualitative features of the behavior of this hypothetical model may then be expected to resemble those of real He. Since the uniform external potential does not influence the system except to give it a negative total energy, one may consider simply a system of hard spheres at a given density and in the end add the external potential separately. This kind of reasoning is essentially contained in the work of London[1] on the density and the energy of liquid He in the ground state.

In Secs. 1 and 2 the method of the pseudopotential[2,3] is applied to the problem. It is seen that the energy per particle in the ground state and the energy level spectrum near the ground state can be very easily obtained as power series expansions in the parameter $(\rho a^3)^{\frac{1}{2}}$, where ρ is the particle density and a the hard-sphere diameter. That the expansion parameter should be $(\rho a^3)^{\frac{1}{2}}$ was already pointed out before.[4] The ground state energy per particle calculated with the present method agrees with that given in reference 4. The excited levels immediately above the ground state represent "phonon" states. The excitation spectrum is the same as that of Bogoliubov's.[5]

In Sec. 3 the same method is used to calculate the wave functions for the ground state, and the pair distribution function for the ground state. The results are compared with the work of Feynman[6] and of Penrose and Onsager.[7] It emerges from these results that one can define a "correlation length" which characterizes the spatial extension of the correlation introduced by the hard-sphere interactions.

Section 4 is devoted to a critical discussion of the validity of the method of the pseudopotential in the present problem. The order of magnitude of the expected corrections to the present calculation is analyzed.

In Sec. 5 the physical properties of a dilute system of a gas of hard spheres are discussed briefly on the basis of the energy spectrum obtained in Sec. 2. The energy spectrum near the ground state is shown to be that of a collection of "phonons." The properties of the system, such as the existence of a normal fluid and a superfluid component, can therefore be inferred immediately from the work of Landau,[8] Kramers,[9] and others.[9]

In Sec. 6 the concept of a "correlation length" introduced in Sec. 3 is further emphasized, and related to London's idea[10] of an order in momentum space. The question of the flow of the superfluid is discussed by the method of Sec. 1. It is indicated that the superfluid flow is irrotational, as was pointed out by Onsager and Feynman.[11]

[1] F. London, *Superfluids* (John Wiley and Sons, Inc., New York, 1954), Chap. B.
[2] K. Huang and C. N. Yang, Phys. Rev. **105**, 767 (1957).
[3] Huang, Yang, and Luttinger, Phys. Rev. **105**, 776 (1957).
[4] T. D. Lee and C. N. Yang, Phys. Rev. **105**, 1119 (1957).
[5] N. N. Bogoliubov, J. Phys. U.S.S.R. **II**, 23 (1947).
[6] R. P. Feynman, Phys. Rev. **94**, 262 (1954).
[7] O. Penrose and L. Onsager, Phys. Rev. **104**, 576 (1956).
[8] L. D. Landau, J. Phys. U.S.S.R. **5**, 71 (1940).
[9] H. A. Kramers, Physica **18**, 653 (1952). R. B. Dingle, *Advances in Physics* (Taylor and Francis, Ltd., London, 1952), Vol. 1, p. 112.
[10] F. London, *Superfluids* (John Wiley and Sons, Inc., New York, 1954), pp. 142–144 and pp. 199–201.
[11] L. Onsager, Suppl. Nuovo cimento **6**, 249 (1949). R. P. Feynman, in *Progress in Low Temperature Physics*, edited by C. J. Gorter (North Holland Publishing Company, Amsterdam, 1955), Vol. 1, p. 17.

1. GROUND STATE ENERGY

We use mostly the same notation as that of reference 2 but choose units so that $\hbar=1$, $2m=1$, and recall that the Hamiltonian of a system of hard spheres can be replaced in certain approximations by the pseudopotential Hamiltonian [see Eqs. (32) and (33) of reference 2]:

$$H=-\sum_{i=1}^{N}\nabla_i^2+V,$$
$$V=8\pi a\sum_{i<j}\delta(\mathbf{r}_i-\mathbf{r}_j)\frac{\partial}{\partial r_{ij}}r_{ij}. \qquad (1)$$

By using the language of quantized fields, the pseudopotential V can be recast in the form [see Eq. (38) of reference 2]:

$$V=4\pi a\int d^3\mathbf{r}_1 d^3\mathbf{r}_2\psi^*(\mathbf{r}_1)\psi^*(\mathbf{r}_2)\delta(\mathbf{r}_1-\mathbf{r}_2)\frac{\partial}{\partial r_{12}}\times[r_{12}\psi(\mathbf{r}_1)\psi(\mathbf{r}_2)]. \qquad (2)$$

We shall not enter here into a discussion of the region of validity of the use of the pseudopotential, a subject that we shall come back to in Sec. 4. In the present section and the next section it will be shown that the pseudopotential (2) leads directly and simply to an expression of the ground state energy per particle of the Bose gas and to the energy spectrum near the ground state.

It was already observed and emphasized in reference 2 that the pseudopotential V, when operating on a wave function that is not singular at $r_{ij}=0$, is equivalent to the operator

$$V'=4\pi a\int d^3\mathbf{r}_1 d^3\mathbf{r}_2\psi^*(\mathbf{r}_1)\psi^*(\mathbf{r}_2)\delta(\mathbf{r}_1-\mathbf{r}_2)\psi(\mathbf{r}_1)\psi(\mathbf{r}_2). \qquad (3)$$

It was further observed that using the potential (3) leads to divergences which arise from the singularities of the correct wave function. The use of the correct pseudopotential V, however, does not lead to any divergencies. For clarity we shall adopt the following procedure in the present paper. The potential V' will first be used to compute the ground state energy per particle. It will be found that the expression obtained is divergent, as expected. It will then be easy to see that substituting the correct pseudopotential V, [Eq. (2)], for the potential V', [Eq. (3)], in the calculation leads very simply to a subtraction procedure which yields a correct finite result.

By expanding ψ into free-particle waves as was done in reference 2, we obtain

$$V'=\Omega^{-1}4\pi a\sum_{\alpha,\beta,\mu,\nu}a_\alpha^* a_\beta^* a_\mu a_\nu\delta(\mathbf{k}_\alpha+\mathbf{k}_\beta-\mathbf{k}_\mu-\mathbf{k}_\nu), \qquad (4)$$

where a_α^* and a_α are, respectively, the creation and annihilation operators of the free-particle states with momentum $\mathbf{k}_\alpha$, and $\Omega=L^3$ is the volume of the cube in which the N particles move. The delta symbol $\delta(\mathbf{k}_\alpha+\mathbf{k}_\beta-\mathbf{k}_\mu-\mathbf{k}_\nu)$ appearing in (4) is a Kronecker delta function. It is essential that the boundary condition at the edge of the box be taken to be the usual periodicity condition [compare reference 17]. The diagonal elements of (4) are

$$\langle n|V'|n\rangle=\Omega^{-1}4\pi a(2N^2-N-\textstyle\sum_\alpha n_\alpha{}^2), \tag{5}$$

where n_α is the occupation number $a_\alpha{}^*a_\alpha$. Equation (5) has already been obtained in reference 2. Subtracting a constant term $4\pi a\rho(N-1)$ from expression (5), one obtains

$$\langle n|V'|n\rangle-4\pi a\rho(N-1)$$

$$=8\pi a\rho\sum_{\alpha\neq0} n_\alpha-\frac{4\pi a}{\Omega}(\sum_{\alpha\neq0} n_\alpha)^2-\frac{4\pi a}{\Omega}\sum_{\alpha\neq0} n_\alpha, \tag{6}$$

with $\rho=N/\Omega$. If one takes a system for which the density ρ is fixed and for which N and Ω both approach infinity, Eq. (6) reduces to[12]

$$\langle n|V'|n\rangle-4\pi a\rho N=8\pi a\rho\sum_{\alpha\neq0} n_\alpha. \tag{7}$$

The off-diagonal matrix elements of the potential V' cause transitions in which two particles of momenta $\mathbf{k}_\alpha$ and $\mathbf{k}_\beta$ collide and go into the states $\mathbf{k}_\mu$ and $\mathbf{k}_\nu$. The periodicity boundary condition that we took insures that the matrix element is nonvanishing only if momentum is conserved: $\mathbf{k}_\alpha+\mathbf{k}_\beta=\mathbf{k}_\mu+\mathbf{k}_\nu$. The value of such an off-diagonal matrix element is equal to

$$(4\pi a/\Omega)[n_\alpha n_\beta(n_\mu+1)(n_\nu+1)]^{\frac{1}{2}}. \tag{8}$$

The crucial point is now to observe that as the total number of particles N approaches infinity, each of the n_α's is finite except n_0, which is $N-\sum_{\alpha\neq0}n_\alpha$. For large values of N, the off-diagonal matrix elements fall into three categories in magnitude:

(1) Those in which two of the four momenta $\mathbf{k}_\alpha$, $\mathbf{k}_\beta$' $\mathbf{k}_\mu$, $\mathbf{k}_\nu$ are equal to 0. Such matrix elements are proportional to $8\pi a\rho$.

(2) Those for which only one of the four momenta $\mathbf{k}_\alpha$, $\mathbf{k}_\beta$, $\mathbf{k}_\mu$, $\mathbf{k}_\nu$ is equal to 0. Such matrix elements are smaller than those of the first category by a factor $N^{-\frac{1}{2}}$.

(3) Those for which none of the four momenta $\mathbf{k}_\alpha$, $\mathbf{k}_\beta$, $\mathbf{k}_\mu$, $\mathbf{k}_\nu$ is 0. Such matrix elements are smaller than those of the category (1) by a factor N^{-1}.

[12] The neglect of the second and third terms of the right hand side of (6) as compared to the first term is consistent with the power series expansion of the energy in the parameter $(\rho a^3)^{\frac{1}{2}}$. It is shown later [see (41)] that

$$N^{-1}\langle\sum_{\alpha\neq0} n_\alpha\rangle\sim(\rho a^3)^{\frac{1}{2}},$$

where the expectation value is taken with respect to the *perturbed* ground state of the total system.

Starting from the free-particle ground state, by first considering only matrix elements of category (1), we would obtain the dominant term of the energy of the system. The matrix elements of categories (2) and (3) will later be shown in Sec. 4 to give rise to higher order corrections. To calculate the dominant term of energy, we thus need *only consider those free-particle states S which are connected to the free-particle ground state, directly or indirectly, through off-diagonal matrix elements of category* (1), i.e., matrix elements that represent the scattering of two particles of momenta $\mathbf{k}$ and $-\mathbf{k}$ into the ground state or vice versa. Evidently a state in S is specified by l_1 pairs of particles each with momenta $\mathbf{k}_1$ and $-\mathbf{k}_1$, l_2 pairs of particles each with momenta $\mathbf{k}_2$ and $-\mathbf{k}_2$, etc., and $N-2\sum l_i$ particles with momentum zero. We denote such a state by

$$|l_1, l_2, \cdots\rangle. \tag{9}$$

In terms of the annihilation operators $a_\mathbf{k}$, where $\mathbf{k}\neq0$ ranges over *half* of the momentum space, we can write down the diagonal matrix elements (7) for the pseudopotential V' between the states of S:

$$4\pi a\rho N+16\pi a\rho\textstyle\sum' a_\mathbf{k}{}^*a_\mathbf{k}, \tag{10}$$

where $\sum'$ represents a summation over *half* of the $\mathbf{k}$ space with $\mathbf{k}\neq0$. The off-diagonal matrix elements of V' are given by those of

$$8\pi a\rho\textstyle\sum' B_0(k), \tag{11}$$

where

$$B_0(\mathbf{k})=\begin{bmatrix}0&1&0&0&\\1&0&2&0&\\0&2&0&3&\\0&0&3&0&\\&&&&\cdots\end{bmatrix}, \tag{12}$$

in the standard representation in which $a_\mathbf{k}{}^*a_\mathbf{k}$ is diagonal:

$$a_\mathbf{k}{}^*a_\mathbf{k}=\begin{bmatrix}0&0&0&0&\\0&1&0&0&\\0&0&2&0&\\0&0&0&3&\\&&&&\cdots\end{bmatrix}. \tag{13}$$

One has evidently the commutation relations

$$0=[B_0(\mathbf{k}),a_{\mathbf{k}'}]=[B_0(\mathbf{k}),a_{\mathbf{k}'}{}^*] \quad \text{if} \quad \mathbf{k}\neq\mathbf{k}'. \tag{14}$$

The Hamiltonian $H'=-\sum\nabla_i{}^2+V'$ between the states of S is then

$$H'=4\pi a\rho N+2\textstyle\sum'(k^2+k_0{}^2)[a_\mathbf{k}{}^*a_\mathbf{k}+y_\mathbf{k}B_0(\mathbf{k})], \tag{15}$$

where

$$k_0{}^2=8\pi a\rho, \tag{16}$$

$$y_\mathbf{k}=\tfrac{1}{2}k_0{}^2(k^2+k_0{}^2)^{-1}. \tag{17}$$

The summation $\sum'$ in (15) is a sum of mutually commuting operators. Its lowest eigenvalue is therefore the sum of the lowest eigenvalues of the individual terms. It will be shown in Appendix I that the eigenvalues of

$$a^*a+yB_0$$

are

$$\lambda_m=-\tfrac{1}{2}+(m+\tfrac{1}{2})(1-4y^2)^{\frac{1}{2}}, \tag{18}$$

with $m=0, 1, 2, \cdots$. One thus obtains the lowest eigenvalue of the Hamiltonian (15):

$$\begin{aligned} E_0' &= 4\pi a\rho N+\sum{}'(k^2+k_0^2)[-1+(1-4y_{\mathbf{k}}^2)^{\frac{1}{2}}] \\ &= 4\pi a\rho N-\sum{}'[k^2+k_0^2-k(k^2+2k_0^2)^{\frac{1}{2}}]. \end{aligned} \tag{19}$$

The above expression contains a spurious term which makes the sum divergent. This is because we have used V' instead of the correct pseudopotential V. The situation is easily remedied by identifying the spurious term and subtracting it.

The correct interaction V, Eq. (2), expressed in momentum space, reads

$$V=\lim_{r\to 0}4\pi a\Omega^{-1}\frac{\partial}{\partial r}\{r\sum_{\mu,\nu}\exp[\tfrac{1}{2}i(\mathbf{k}_\mu-\mathbf{k}_\nu)\cdot\mathbf{r}]$$
$$\times\sum_{\alpha,\beta}a_\alpha{}^*a_\beta{}^*a_\mu a_\nu\delta(\mathbf{k}_\alpha+\mathbf{k}_\beta-\mathbf{k}_\mu-\mathbf{k}_\nu)\}. \tag{20}$$

It can be seen that the replacement of (4) by (20) does not affect in any essential way the general arguments that led to the Hamiltonian (15), which is now replaced by

$$H=4\pi a\rho N+2\sum{}'(k^2+k_0^2)a_{\mathbf{k}}{}^*a_{\mathbf{k}}$$
$$+\tfrac{1}{2}k_0^2\operatorname*{Lim}_{r\to 0}\frac{\partial}{\partial r}\{r\sum_{\mathbf{k}\neq 0}e^{i\mathbf{k}\cdot\mathbf{r}}B_0(\mathbf{k})\}. \tag{21}$$

Using this Hamiltonian, the calculation of E_0 proceeds in the same way as before except that in the final expression (19), the simple sum over $\mathbf{k}$ is replaced by a limiting process, namely

$$E_0=4\pi a\rho N-\tfrac{1}{2}\operatorname*{Lim}_{r\to 0}\frac{\partial}{\partial r}\{r\sum_{\mathbf{k}\neq 0}e^{i\mathbf{k}\cdot\mathbf{r}}[k^2+k_0^2$$
$$-k(k^2+2k_0^2)^{\frac{1}{2}}]\}. \tag{22}$$

The mathematical problem of evaluating this expression is similar to the corresponding problems encountered in reference 2. It can be shown without difficulty that

$$E_0=4\pi a\rho N-\sum{}'\left[k^2+k_0^2-k(k^2+2k_0^2)^{\frac{1}{2}}-\frac{k_0^4}{2k^2}\right]. \tag{23}$$

The sum can easily be evaluated in the limit $\Omega\to\infty$:

$$E_0=4\pi a\rho N+\frac{\Omega k_0^5}{4\pi^2}\int_0^\infty dy y^2\left[-1-y^2 +y(y+2)^{\frac{1}{2}}+\frac{1}{2y^2}\right], \tag{24}$$

or

$$E_0=4\pi aN\rho\left[1+\frac{128}{15\sqrt{\pi}}(a^3\rho)^{\frac{1}{2}}\right], \tag{25}$$

a result which was first obtained in reference 4 by the "binary collision expansion method."

Another way of proving that the correct pseudopotential V of Eq. (1) leads to the convergent expression (23) while V' leads to the divergent one [Eq. (19)] is the following: Treating the pseudopotential V or V' as a perturbation, one can calculate the ground state energy E_0 as a power series expansion in a. This was the procedure followed in reference 2. In the order a^2, using the potential V', one obtains a divergent expression. Using the correct pseudopotential V, however, one obtains zero in the order a^2. [See Eq. (53) below. Notice that $a/L=0$ in the limit $L\to\infty$.] Except for the order a^2, V and V' give the same results. [We stay here within the approximation of neglecting small off-diagonal matrix elements. As will be discussed in Sec. 4, this approximation is equivalent to retaining the maximum power of N to each order of a.] To obtain the energy expression when V is used, one therefore need only take the divergent expression (19) for the case of V' and expand it in powers of a and strike out the term a^2. Now

$$\sum{}'[k^2+k_0^2-k(k^2+2k_0^2)^{\frac{1}{2}}]=\sum{}'\left[\frac{k_0^4}{2k^2}-\frac{k_0^6}{2k^4}+\cdots\right],$$
$$k_0^2=8\pi a\rho.$$

Striking out the term a^2 therefore means subtracting from the summand $k_0^4/2k^2$, leading immediately to (23). We shall return to this discussion in Sec. 4.

2. ENERGY LEVELS NEAR THE GROUND STATE; PHONON SPECTRUM

The method of the last section can also be applied to discuss the energy of a state with a nonvanishing momentum. We start from an unperturbed state $|\mathbf{q}\rangle$ in which all particles have momentum zero except one, which has momentum $\mathbf{q}$. The set of unperturbed states, denoted by S', connected to $|\mathbf{q}\rangle$ by large off diagonal matrix elements are all of the form

$$|\mathbf{q}; l_{\mathbf{q}}; l_1, l_2, \cdots\rangle, \tag{26}$$

which means that there is a particle of momentum $\mathbf{q}$, and in addition, there are $l_{\mathbf{q}}$ pairs of particles $\mathbf{q}$, $-\mathbf{q}$; l_1 pairs $\mathbf{k}_1$, $-\mathbf{k}_1$; l_2 pairs $\mathbf{k}_2$, $-\mathbf{k}_2$; etc., with $\mathbf{k}_i\neq\mathbf{q}$. The rest of the particles, $(N-2\sum l_i-1)$ in number, have momentum $\mathbf{k}=0$. The total momentum of every state in S' is $\mathbf{q}$. The Hamiltonian H' for the states S' is very similar to that for the states S given before by Eq. (15). It is

$$H'=4\pi a\rho N+2\sum_{\mathbf{k}\neq\mathbf{q}}{}'(k^2+k_0^2)[a_{\mathbf{k}}{}^*a_{\mathbf{k}}+y_{\mathbf{k}}B_0(\mathbf{k})]$$
$$+2(q^2+k_0^2)[N_{\mathbf{q}}+y_{\mathbf{q}}B_1(\mathbf{q})]+8\pi a\rho+q^2, \tag{27}$$

where

$$N_{\mathbf{q}}=\begin{bmatrix} 0 & 0 & 0 & 0 & \\ 0 & 1 & 0 & 0 & \\ 0 & 0 & 2 & 0 & \\ 0 & 0 & 0 & 3 & \\ & & & & \cdots \end{bmatrix} \tag{28}$$

has diagonal values equal to $l_{\mathbf{q}}$, and

$$B_1(\mathbf{q})=\begin{bmatrix} 0 & (1\times2)^{\frac{1}{2}} & 0 & \\ (1\times2)^{\frac{1}{2}} & 0 & (2\times3)^{\frac{1}{2}} & \\ 0 & (2\times3)^{\frac{1}{2}} & 0 & \\ & & & \cdots \end{bmatrix}. \tag{29}$$

The matrix $B_0(\mathbf{k})$ is given by Eq. (12). The eigenvalue of $N+yB_1$ is discussed in Appendix I. The lowest eigenvalue is

$$-1+[1-4y^2]^{\frac{1}{2}}. \tag{30}$$

The difference of the lowest eigenvalue of (27) and that of (15) is the energy of excitation into a state of momentum $\mathbf{q}$. From (30) and (18) it is evidently equal to[13]

$$E_{\mathbf{q}}-E_0=q(q^2+2k_0^2)^{\frac{1}{2}}=q(q^2+16\pi a\rho)^{\frac{1}{2}}. \tag{31}$$

It will be shown in Appendix II that the wave function in coordinate space for the state we just discussed, i.e., for the lowest excited state with momentum $\mathbf{q}$, is to the order of approximation considered equal to

$$\sum_{j=1}^{N} e^{i\mathbf{q}\cdot\mathbf{r}_j}\Psi_0,$$

where Ψ_0 is the wave function of the ground state. This means that these excitations are density fluctuations (i.e., sound waves, or phonons), as has been discussed by Bijl[14] and Feynman.[6]

The velocity v of sound waves of infinite wavelength is directly related to the macroscopic compressibility, which can in turn be computed from the energy expression Eq. (25) for the ground state. In fact, remembering that in our units $m=\frac{1}{2}$, one has

$$v=\left(2\frac{dp}{d\rho}\right)^{\frac{1}{2}}, \quad \text{and} \quad p=\rho^2\frac{d}{d\rho}(E_0/N). \tag{32}$$

Equations (25) and (32) together give

$$v=(16\pi a\rho)^{\frac{1}{2}}[1+16\pi^{-\frac{1}{2}}(a^3\rho)^{\frac{1}{2}}]. \tag{33}$$

The first term of (33) agrees with the velocity that one computes from (31) for the sound waves with momentum $\mathbf{k}=0$, as it should. The second term in (33) represents a correction term that is beyond the accuracy of (31).

In an entirely similar way, one can solve other eigenvalues and eigenstates of the Hamiltonian (1), by considering the excited states of (15) and (27) and also considering the states connected to an unperturbed state that contains more than one particle having nonvanishing momentum. This is discussed in detail in Appendix I. The eigenvalues for these states can be shown to be

$$E=E_0+\sum_{\mathbf{k}\neq0} m_{\mathbf{k}}k(k^2+16\pi a\rho)^{\frac{1}{2}}, \tag{34}$$

with the corresponding total momentum

$$\mathbf{P}=\sum m_{\mathbf{k}}\mathbf{k}, \quad m_{\mathbf{k}}=0,1,2,\cdots. \tag{35}$$

They represent therefore states with $m_{\mathbf{k}}$ phonons of momentum $\mathbf{k}$.

3. WAVE FUNCTIONS AND THE PAIR DISTRIBUTION FUNCTION

The ground state wave function Ψ_0 of the Hamiltonian (15) can be written in terms of the free-particle states $|l_1,l_2,\cdots\rangle$ [Eq. (9)] as

$$\Psi_0=\sum_{l_i=0}^{\infty} A(l_1,l_2,\cdots)|l_1,l_2,\cdots\rangle, \tag{36}$$

with $A(l_1,l_2,\cdots)$ representing the probability amplitudes. The value of $A(l_1,l_2,\cdots)$ is found to be (see Appendix I)

$$A(l_1,l_2,\cdots)=C\prod_i{}'[-\alpha(\mathbf{k}_i)]^{l_i}, \tag{37}$$

where

$$\alpha(\mathbf{k})=(8\pi a\rho)^{-1}[k^2+8\pi a\rho-k(k^2+16\pi a\rho)^{\frac{1}{2}}], \tag{38}$$

and C is a normalization constant given by

$$C=\prod_i{}'[1-\alpha^2(\mathbf{k}_i)]^{\frac{1}{2}}. \tag{39}$$

In Eqs. (37) and (39) the product $\prod_i{}'$ extends over half of the $\mathbf{k}$ space with $\mathbf{k}_i\neq0$.

Upon using Eq. (37), it is easy to compute the average occupation number $\langle n_{\mathbf{k}}\rangle$ of the free-particle states with momentum $\mathbf{k}$ for the ground state wave function Ψ_0. One finds

$$\langle n_{\mathbf{k}}\rangle=\frac{\alpha^2(\mathbf{k})}{1-\alpha^2(\mathbf{k})} \quad \text{for} \quad \mathbf{k}\neq0, \tag{40a}$$

and

$$\langle n_{\mathbf{k}=0}\rangle=N\left[1-\frac{8}{3\sqrt{\pi}}(a^3\rho)^{\frac{1}{2}}\right], \tag{40b}$$

where N is the total number of particles and $\langle\ \rangle$ means taking the average over the ground state of the system. For an ideal Bose system the ground state of the system is characterized by the fact that all particles are in the free-particle ground state. In the present case, owing to the interactions, particles are excited from the state, $\mathbf{k}=0$, into various free-particle states

[13] To calculate the excitation energy $(E_{\mathbf{q}}-E_0)$, the identical result is obtained by using either V' [Eq. (3)] or the correct pseudopotential V [Eq. (2)].

[14] A. Bijl, Physica **7**, 869 (1940).

with $\mathbf{k}\neq 0$. Let f be the total fractional number of particles excited. We find for the ground state of the entire system, this fraction is

$$f \equiv N^{-1} \sum_{\mathbf{k}\neq 0} \langle n_{\mathbf{k}} \rangle = \frac{8}{3\sqrt{\pi}}(\rho a^3)^{\frac{1}{2}}. \quad (41)$$

It is important to note that the occupation number of the free-particle ground state $\langle n_{\mathbf{k}=0}\rangle$ is proportional to N while all the other free-particle states have finite occupation numbers as $N\rightarrow\infty$. The significance of these free-particle state occupation numbers in the discussion of a Bose system with interactions has recently been pointed out and emphasized by Penrose and Onsager.[7]

Another important quantity is the pair distribution function $D(r_{12})$, defined by

$$D(r_{12}) \equiv \rho^{-2} \langle \psi^*(\mathbf{r}_1)\psi^*(\mathbf{r}_2)\psi(\mathbf{r}_2)\psi(\mathbf{r}_1)\rangle. \quad (42)$$

The pair distribution function $D(r)$ describes the relative probability for finding two particles at a distance r apart. The normalization of the function is so chosen that $D(r)\rightarrow 1$ as $r\rightarrow\infty$. By using Eqs. (36)–(39), the function $D(r)$ can be readily evaluated. It is

$$D(r) = [1+G(r)]^2 + [1+F(r)]^2 - 1 - 4f[G(r)+F(r)], \quad (43)$$

where

$$F(r) = \frac{1}{8\pi^3\rho}\int \frac{\alpha^2(\mathbf{k})}{1-\alpha^2(\mathbf{k})} e^{i\mathbf{k}\cdot\mathbf{r}} d^3\mathbf{k},$$
$$G(r) = -\frac{1}{8\pi^3\rho}\int \frac{\alpha(\mathbf{k})}{1-\alpha^2(\mathbf{k})} e^{i\mathbf{k}\cdot\mathbf{r}} d^3\mathbf{k}, \quad (44)$$

with f and $\alpha(\mathbf{k})$ given by Eq. (41) and Eq. (38). To study the behavior of these two functions F and G, it is convenient to introduce a "correlation length" r_0, defined as

$$r_0 \equiv (8\pi a\rho)^{-\frac{1}{2}}. \quad (45)$$

r_0 is the inverse of k_0 introduced in Eq. (16). For $r\gg r_0$, the functions F and G approach, respectively,

$$F(r)\rightarrow +\frac{1}{\pi^2\rho r_0 r^2}$$

and

$$G(r)\rightarrow -\frac{1}{\pi^2\rho r_0 r^2}, \quad (46)$$

while for small distances $r\ll r_0$,

$$F(r)\rightarrow f = \frac{8}{3\sqrt{\pi}}(\rho a^3)^{\frac{1}{2}}$$

and

$$G(r)\rightarrow -\frac{a}{r} + \frac{8}{\sqrt{\pi}}(\rho a^3)^{\frac{1}{2}}. \quad (47)$$

Correspondingly, we see that for $r\ll r_0$,

$$D(r)\rightarrow \left(1-\frac{a}{r}\right)^2 + O\left(\frac{a}{r_0}\right)$$

and for $r\gg r_0$,

$$D(r)\rightarrow 1 + O\left(\frac{1}{r^4}\right). \quad (48)$$

Thus the correlation length r_0 characterizes the extension of the correlation between particles introduced by the hard-sphere interaction. Qualitative discussion of the physical implications of this correlation length will be given in Sec. 6.

It is of interest to compare the present result with the work of Feynman.[6] The function $S(\mathbf{k})$ in Feynman's paper can be defined in terms of the Fourier transform of $D(r)$ as

$$S(\mathbf{k}) \equiv 1 + \rho\int D(r) e^{i\mathbf{k}\cdot\mathbf{r}} d^3\mathbf{k}. \quad (49)$$

From Eq. (44), one finds

$$S(\mathbf{k}) = k(k^2+16\pi a\rho)^{-\frac{1}{2}}[1+O(\rho a^3)^{\frac{1}{2}}], \quad (\mathbf{k}\neq 0). \quad (50)$$

Substitution into the Feynman-Bijl relation[6,14] for the phonon energy,

$$E_{\mathbf{k}} - E_0 = k^2/S(\mathbf{k}) \quad (51)$$

leads to

$$E_{\mathbf{k}} - E_0 = k(k^2+16\pi a\rho)^{-\frac{1}{2}}, \quad (52)$$

in agreement with Eq. (31). This is not surprising since we shall see in Appendix II that the wave functions of the excited states have the form used by Feynman and Bijl from which Eq. (51) was derived.[6]

4. CRITICAL DISCUSSION OF THE VALIDITY OF THE PSEUDOPOTENTIAL METHOD FOR THE PRESENT PROBLEM

The method used in the present paper evokes many questions concerning its validity. In particular the following points need be analyzed:

(1) It has been emphasized in reference 2 that the pseudopotential (1) is in general accurate only to the order a^2, and that as applied to the ground state energy it is only accurate to the order a^3. The approximations involved include the neglect of the D-wave scattering and the genuine triple collisions as explained in Fig. 2 of reference 2. In the present paper we have used the pseudopotential (1) to calculate quantities which certainly involve contributions from infinitely high powers of a. How could one then be sure that such use of the pseudopotential is justified? Also, in reference 2 the energy per particle for the ground state was calculated

up to a^3. The result was

$$\frac{E_0}{N}=\frac{4\pi a(N-1)}{L^3}\left\{1+2.37\frac{a}{L}+\frac{a^2}{L^2}\left[(2.37)^2+\frac{\xi}{\pi^2}(2N-5)\right]\right\}, \quad (53)$$

$$\xi=\sum_{l,m,n=-\infty}^{\infty}\frac{1}{(l^2+m^2+n^2)^2}; \quad (l,m,n)\neq(0,0,0).$$

If one keeps $\rho=N/\Omega$ constant and allows $\Omega=L^3$ to approach ∞, expression (53) diverges as $N^{\frac{1}{3}}$. How does one reconcile this divergence with the finite result obtained in Sec. 1 of the present paper?

(2) Even assuming the validity of the use of the pseudopotential (1), how can one justify the neglect of the small off-diagonal matrix elements (8)?

(3) What is the nature of the series expansion of which (25) gives the first two terms? What is the limit of validity of the phonon spectrum (31)?

We start with a discussion of point (1) by examining the divergence of formula (53). If the expansion is carried out to higher orders of a, one can express the energy per particle E_0/N as a power series in a/L. The coefficient of $(a/L)^m$, $m\geqq 3$, is a polynomial in N:

$$\frac{1}{NL^2}\left(\frac{a}{L}\right)^m[AN^\nu+BN^{\nu-1}+\cdots+Z],$$

where A, B, $\cdots Z$ are numerical constants independent of a, L, or N, and ν is an integer depending on m, giving the maximum power of N that occurs in the coefficient of $(a/L)^m$. Of the terms in the polynomial, the most divergent one in the limit $N\rightarrow\infty$ at constant ρ is

$$\frac{1}{NL^2}\left(\frac{a}{L}\right)^m AN^\nu. \quad (A,\ \nu=\text{functions of } m). \quad (54)$$

Now in the discussion of Sec. 1 the guiding principle was that to each order of a, only the term with the maximum power for N be retained. The calculation that leads to (25) is therefore a calculation of the sum of the terms (54). This calculation shows that for the order $(a/L)^m$, the maximum exponent of N is

$$\nu=m \quad (m\geqq 3),$$

as one verifies immediately by expanding (23) in powers of a. The power series for E_0/N can therefore be written in the following way:

$$\frac{E_0}{N}-4\pi a\rho=\frac{1}{NL^2}\left[A\left(\frac{aN}{L}\right)^3+A'\left(\frac{aN}{L}\right)^4+A''\left(\frac{aN}{L}\right)^5\right.$$
$$+\cdots+B\left(\frac{aN}{L}\right)^3\frac{1}{N}+B'\left(\frac{aN}{L}\right)^4\frac{1}{N}+B''\left(\frac{aN}{L}\right)^5\frac{1}{N}$$
$$\left.+\cdots+C\left(\frac{aN}{L}\right)^3\frac{1}{N^2}+\cdots+\cdots\right], \quad (55)$$

where terms of the form (54) are written in the first line. The calculation that leads to (25) consists of summing the first line of the foregoing expression, and the result shows that this series, namely

$$\frac{1}{NL^2}\left[A\left(\frac{aN}{L}\right)^3+A'\left(\frac{aN}{L}\right)^4+\cdots\right],$$

approaches the finite limit

$$\frac{1}{NL^2}\frac{4\pi\times 128}{15\sqrt{\pi}}\left(\frac{Na}{L}\right)^{\frac{5}{2}}=4\pi a\rho\frac{128}{15\sqrt{\pi}}(a^3\rho)^{\frac{1}{2}},$$

as $Na/L\rightarrow\infty$.

It is clear that D-wave scattering introduces terms that contain higher powers of a for given powers of N. Triple collisions give rise to terms also of such nature. Therefore, their inclusion does not affect the first line of (55), but only subsequent lines.

It seems reasonable to expect that the sum of the terms in the second line of (55), i.e.,

$$\frac{1}{NL^2}\frac{1}{N}\left[B\left(\frac{aN}{L}\right)^3+B'\left(\frac{aN}{L}\right)^4+\cdots\right]$$

would also converge to a finite number of the limit $aN/L\rightarrow\infty$. This can happen only if the series in the square bracket approaches $(aN/L)^4$ as $aN/L\rightarrow\infty$. In that case the second line of (55) reduces to an expression of the form

$$(\text{constant})\rho^2a^4,$$

indicating that the expansion (25) is in powers of $(a^3\rho)^{\frac{1}{2}}$.

One arrives at the same conclusion in discussing question (2) mentioned at the beginning of this section. If one attempts to include the next dominant off-diagonal matrix elements, the additional perturbation energy is of the form

$$\Delta E=\sum(\text{matrix element})^2/(\text{energy difference}).$$

The matrix elements are of the order $N^{-\frac{1}{2}}a\rho$ and connects the ground state with states in which three phonons $\mathbf{k}_1$, $\mathbf{k}_2$, $\mathbf{k}_3$ are present, where $\mathbf{k}_1+\mathbf{k}_2+\mathbf{k}_3=0$. One therefore has a sum of the form

$$\Delta E=\sum\delta(\mathbf{k}_1+\mathbf{k}_2+\mathbf{k}_3)\frac{(a\rho)^2N^{-1}}{E(\mathbf{k}_1,\mathbf{k}_2,\mathbf{k}_3)}.$$

Using the energy spectrum for the phonons calculated in Sec. 2, one obtains

$$\Delta E=(a\rho)^2N^{-1}L^6\int d\mathbf{k}_1d\mathbf{k}_2F((a\rho)^{\frac{1}{2}},\mathbf{k}_1,\mathbf{k}_2).$$

By a dimensional argument one obtains

$$\Delta E=(a\rho)^2N^{-1}L^6(a\rho)^2(\text{constant})=(\text{constant})Na^4\rho^2,$$

indicating again that the expansion (25) is in powers of $(a^3\rho)^{\frac{1}{2}}$.

The surmise that the expansion (25) is in powers of $(a^3\rho)^{\frac{1}{2}}$ is in agreement with a conclusion already drawn[4] from the "binary collision expansion method."

We now come to the third point raised at the beginning of this section: the limit of validity of the formulas (25) and (31). The above discussions indicate that they represent the first terms of expansions in $(a^3\rho)^{\frac{1}{2}}$. As has been pointed out before,[4] such expansions are probably asymptotic expansions which even may not converge. For the phonon spectrum (31) the limit of validity,

$$ka \ll 1, \tag{56}$$

has to be imposed in addition to the condition

$$(a^3\rho)^{\frac{1}{2}} \ll 1.$$

Condition (56) is necessary for the validity of the pseudopotential (1).

We conclude this section by stating that to develop a systematic expansion method starting from the pseudopotential method of the present paper seems difficult, because the inclusion of triple collision terms presents grave obstacles. On the other hand, in the "binary collision expansion method"[4] triple and higher order collision terms can be automatically included. A systematic approach starting from the "binary collision expansion method" appears hopeful.

5. "TWO-FLUID MODEL" AND THE LOW-TEMPERATURE PROPERTIES OF THE HARD-SPHERE SYSTEM

In Sec. 2 we obtained the low-lying energy levels of a Bose system of hard spheres. The levels can be described as those of a collection of phonons with a spectrum given by (34). If one examines, by a method similar to the one already used, the low-lying energy levels of a corresponding Fermi-Dirac system, one finds that the energy level density near the ground state is infinitely greater than in the Bose case. The scarcity of low-lying energy levels in the Bose case has long been recognized[15] as the reason for the superfluid behavior of liquid helium. Feynman[15] has given arguments to show that for a Bose system of interacting particles such scarcity is to be expected. The results of Secs. 1 and 2 of the present paper confirms this conclusion in the case of a dilute hard sphere gas by an explicit mathematical treatment.

Knowing the spectrum of the phonons (i.e., of the low-lying states), one can easily obtain the specific heat of the system at low temperatures. Furthermore, by the reasoning developed by Landau,[8] Kramers,[9] and others[9] one can conclude that the system shows a two-fluid[16] behavior. According to these authors the ground state of the system is looked upon as a pure "superfluid." The low-lying excited states are looked upon as a mixture of "superfluid" and "normal fluid" components, with the collection of phonons constituting the "normal fluid" component. The "normal fluid" thus can be said to be moving against a "background superfluid." With such an identification of the two fluids, one can use all the formulas which the previously mentioned authors have established for the two-fluid model, and one can compute the density of the normal fluid, the velocity of second sound, and the magnitude of the fountain effect at very low temperatures. We shall not go into these discussions in detail as we have nothing new to add to the reasonings already developed in the literature quoted. It is to be noticed, however, that the present explicit mathematical treatment of a definite model allows one to visualize very clearly the fact that a phonon does carry a momentum equal to $\hbar\mathbf{k}$, where $\mathbf{k}$ is its wave number, and that by a superposition of phonon waves one does obtain a mass transport of the Bose particles.

6. MOMENTUM SPACE ORDER, CORRELATION LENGTH, AND SUPERFLUID FLOW

The method of Secs. 1 and 2 can be applied easily to the case where one starts from an unperturbed state in which almost all particles are in a given state of momentum $\mathbf{k}_0 \neq 0$. The lowest perturbed eigenstate there describes a background superfluid flow with velocity $2\mathbf{k}_0$ (notice that the mass per particle is $\frac{1}{2}$). The excited states represent various phonon states in such a background superfluid.

Is it possible to start from an unperturbed state in which a finite fraction of the particles occupy each of two different momentum states? In other words, is it possible to have an interpenetration of two superfluid velocities? The answer is no, because the method of Sec. 1 leads in this case to very large perturbations, indicating[17] that the unperturbed state is very far from an eigenstate.

The condensation of nearly all particles into a single free-particle momentum state is what London[10] called momentum space ordering. The foregoing discussion and the wave function and eigenvalues found in Secs. 1, 2, and 3 give explicit demonstrations of this concept for the special model of a dilute Bose system of hard spheres.

The influence of the order in momentum space does not, however, extend over infinite spatial distances. If it did, there would not be the possibility of superfluid flow, but only uniform motion of the superfluid as a whole. We shall in the following give a qualitative

[15] See, e.g., R. P. Feynman, in *Progress in Low Temperature Physics*, edited by C. J. Gorter (North Holland Publishing Company, Amsterdam, 1955), Vol. 1, p. 17.

[16] L. Tisza, J. phys. radium **1**, 164 (1940).

[17] For the same reason it is important to take periodic boundary conditions, as we remarked in Sec. 1. If one had chosen, e.g., the boundary condition $\Psi=0$ on the surface of the box, the unperturbed ground state would have an unphysical density variation across the box, so that the hard-sphere interaction would not be a small perturbation.

discussion[18] of the superfluid flow in the present model and of the stability of the flow. *The discussion is to be regarded as suggestive, rather than mathematically conclusive.*

We first notice that the number of particles within one correlation distance $r_0=k_0^{-1}=(8\pi a\rho)^{-\frac{1}{2}}$ is

$$\sim\rho r_0^3\sim(\rho a^3)^{-\frac{1}{2}}\gg 1.$$

The number of excited particles among these is computable from the fraction (41), and is a finite number of the order of 1. The correlation distance is therefore the distance within which the momentum space ordering is strongly effective.

In order to allow for a variation of the superfluid velocity, we divide the system into small boxes each of which is of the dimension of the correlation length, within which the ordering in momentum space forces practically all the particles to have the same momentum. The correlation between two different boxes is, however, not so strong, with the result that the superfluid velocity may vary from one small box to the other. This suggests that one makes use of the method of Secs. 1, 2, and 3, but takes the individual particle wave functions to be

$$e^{i\varphi+i\mathbf{k}\cdot\mathbf{r}}, \tag{57}$$

which form a complete set. Here, φ is a function of $\mathbf{r}$ (independent of $\mathbf{k}$) and $\nabla\varphi$ varies little within each small box. Expanding the second quantized wave function into these individual particle waves,

$$\psi(\mathbf{r})=\sum_{\mathbf{k}} a_{\mathbf{k}} e^{i\varphi+i\mathbf{k}\cdot\mathbf{r}},$$

one can calculate the matrix elements of the kinetic energy and the pseudopotential for the various eigenstates of the occupation numbers $a_{\mathbf{k}}{}^*a_{\mathbf{k}}$. It is then seen that the pseudopotential has the same matrix elements as in Sec. 1, and that the diagonal matrix elements of the kinetic energy is also the same as in Sec. 1 except for a uniform increment of the amount

$$\rho\int(\nabla\phi)^2 d\tau. \tag{58}$$

To give a physical meaning to $\nabla\varphi$, we notice that in each small box $\nabla\varphi$ may be taken as a constant vector. It is then evident that for the ground state in each small box the momentum of the superfluid is equal to $\nabla\varphi$ per particle. In other words,

$$\mathbf{v}_s=2\nabla\phi. \tag{59}$$

The expression (58) then gives simply the kinetic energy of the superfluid flow, which according to (59) is irrotational.

Neglecting the off-diagonal matrix elements of the kinetic energy, one could solve for the excited states too. The excited states are again describable as the states of phonon waves. The off-diagonal matrix elements of the kinetic energy then give rise to a possible transfer of momentum and energy from the superfluid background flow into the phonon waves.

The above discussion leads to the conclusion that the superfluid flow is described by a condensation of almost all particles [i.e., other than a fraction $\sim(\rho a^3)^{\frac{1}{2}}$] into the single-particle state (57). This is clearly exactly what London[10] meant by a macroscopic quantum state. It is clear that from the single-valuedness of φ one would obtain a quantization of the vortices, an interesting conclusion that has been discussed in detail by Onsager and by Feynman.[11]

One of us (K. Huang) would like to thank Dr. J. Robert Oppenheimer for the hospitality extended him during his stay at the Institute for Advanced Study.

APPENDIX I

In this Appendix, we discuss the eigenvalues and eigenfunctions of the matrix

$$M_s=N+yB_s, \tag{A1}$$

where

$$N=\begin{bmatrix} 0 & 0 & 0 & 0 & \cdot \\ 0 & 1 & 0 & 0 & \cdot \\ 0 & 0 & 2 & 0 & \cdot \\ 0 & 0 & 0 & 3 & \cdot \\ \cdot & \cdot & \cdot & \cdot & \cdots \end{bmatrix}, \tag{A2}$$

and

$$B_s=\begin{bmatrix} 0 & [1\times(s+1)]^{\frac{1}{2}} & 0 & \cdot \\ [1\times(s+1)]^{\frac{1}{2}} & 0 & [2(s+2)]^{\frac{1}{2}} & \cdot \\ 0 & [2(s+2)]^{\frac{1}{2}} & 0 & \cdot \\ \cdot & \cdot & \cdot & \cdots \end{bmatrix}. \tag{A3}$$

Let ψ be an eigenstate, with

$$M_s\psi=\lambda\psi, \tag{A4}$$

and

$$\psi=\begin{bmatrix} A_0 \\ A_1 \\ A_2 \\ \cdot \\ \cdot \\ \cdot \end{bmatrix}. \tag{A5}$$

By substituting ψ into (A4), we have

$$nA_n+y\{A_{n-1}[n(n+s)]^{\frac{1}{2}} +A_{n+1}[(n+1)(n+s+1)]^{\frac{1}{2}}\}=\lambda A_n.$$

It is convenient to introduce A_n' defined by

$$A_n'=\left[\frac{n!}{(n+s)!}\right]^{\frac{1}{2}}A_n. \tag{A6}$$

The difference equation for the A_n' becomes

$$(n-\lambda)A_n'+y[nA_{n-1}'+(n+s+1)A_{n+1}]=0, \tag{A7}$$

[18] See similar discussions by Onsager and Feynman, reference 11.

which can be readily solved by defining a generating function

$$H(z)\equiv\sum_{n=0}^{\infty}A_n'z^n. \tag{A8}$$

From (A7), we obtain the differential equation for H as

$$\frac{dH}{dz}[z+yz^2+y]=H\left[\lambda-yz-\frac{sy}{z}\right]. \tag{A9}$$

In order that ψ be normalizable we must have

$$\sum_{n=0}^{\infty}|A_n|^2=\text{finite},$$

which in turn implies that in the complex z plane except for $z=0$, $H(z)$ has no singularity inside the unit circle $|z|<1$. Thus, the eigenvalues of M_s are immediately determined. They are

$$\lambda_m=-\tfrac{1}{2}(1+s)+(\tfrac{1}{2}+m+\tfrac{1}{2}s)(1-4y^2)^{\frac{1}{2}}, \tag{A10}$$

with $m=0, 1, 2, \cdots$.

The corresponding eigenstates are given by Eqs. (A5) and (A6), with

$$A_n'=\text{coefficient of } z^n \text{ in } H_m(z), \quad (n\geq 0).$$

The generating function $H_m(z)$ is

$$H_m(z)=z^{-s}(z+\alpha)^{m+s}(1+\alpha z)^{-(m+1)}, \tag{A11}$$

with

$$\alpha=(2y)^{-1}[1-(1-4y^2)^{\frac{1}{2}}]. \tag{A12}$$

In particular, for $s=0$ and $m=0$

$$\lambda=-\tfrac{1}{2}+\tfrac{1}{2}(1-4y^2)^{\frac{1}{2}}, \tag{A13}$$

and the corresponding unnormalized A_n are

$$A_n=(-\alpha)^n, \quad (n=0, 1, 2\cdots), \tag{A14}$$

which yields Eq. (37).

The Hamiltonians (15) and (27) are related to the matrices M_s with $s=0$ and $s=1$. Consider now the more general case of starting with any unperturbed state which has $s_{\mathbf{k}}$ free particles with momentum $\mathbf{k}$. (Without loss of generality we can restrict the momentum $\mathbf{k}$ to range over only half of the $\mathbf{k}$ space.) Using the same arguments as that of Sec. 1, it is easy to see that the dominant part of the Hamiltonian H, Eq. (1), connects this state with other states which has in addition to these $s_{\mathbf{k}}$ particles also $l_{\mathbf{k}}$ pairs of particles each of momentum $\mathbf{k}$ and $-\mathbf{k}$, etc. Thus the Hamiltonian reduces to

$$H'=4\pi a\rho N+2\sum_{\mathbf{k}\neq 0}{}'(k^2+k_0^2)[N_{\mathbf{k}}+y_{\mathbf{k}}B_s(\mathbf{k})+\tfrac{1}{2}s_{\mathbf{k}}], \tag{A15}$$

where k_0^2 and $y_{\mathbf{k}}$ are given by Eqs. (16) and (17). The sum $\sum'$ extends over half of the $\mathbf{k}$ space with $\mathbf{k}\neq 0$. From the solution (A.10), we obtain immediately the complete phonon spectrums which are listed in Eqs. (34) and (35).

APPENDIX II

In this Appendix, we discuss the properties of the wave functions in the configuration space. From Eqs. (36) and (37), the ground state wave function Ψ_0 can be written in the configuration space as

$$\Psi_0=C\sum_{n=0}^{N/2}\chi_n, \tag{A16}$$

where

$$\chi_n=\Omega^{-N/2}\left[\frac{(N-2n)!}{N!}\right]^{\frac{1}{2}}N^n\sum f(r_{12})f(r_{34})\cdots, \quad (n\neq 0) \tag{A17}$$

and C is the normalization constant. The functions χ_n represent the part in which n pairs of particles are excited. In (A17), the sum extends over all different combinations of selecting n pairs made of $2n$ different particles among a total of N particles. Each term in the sum is a product of n functions $f(r_{ij})$ with the distances between these n pairs as arguments. Altogether there are

$$\frac{N!}{(N-2n)!n!2^n}$$

terms in the sum. The function $f(r)$ is

$$f(r)=-\frac{1}{8\pi^3\rho}\int\alpha_{\mathbf{k}}e^{i\mathbf{k}\cdot\mathbf{r}}d^3\mathbf{k}, \tag{A18}$$

with

$$\alpha_{\mathbf{k}}=(8\pi a\rho)^{-1}[k^2+8\pi a\rho-k(k^2+16\pi a\rho)^{\frac{1}{2}}].$$

Its behaviors at large and small distances are as follows:

$$f(r)\to -a/r \quad \text{as} \quad r\to 0, \tag{A19}$$

and

$$f(r)\to -(2\pi^{\frac{3}{2}}a^{\frac{1}{2}}\rho^{\frac{3}{2}}r^4)^{-1} \quad \text{as} \quad r\to\infty.$$

Using the ground state wave function Ψ_0 in the configuration space, it is also possible to obtain directly the pair distribution function $D(r_{12})$ [Eq. (43)] by integrating over the remaining spatial coordinates $\mathbf{r}_3, \cdots, \mathbf{r}_n$.

Our ground state wave function Ψ_0 satisfies the boundary condition,

$$\Psi_0=0 \quad \text{at} \quad r_{ij}=a, \tag{A20}$$

only *approximately*. Its violation of this boundary condition, however, has an effect on the energy spectrum only in higher orders of $(\rho a^3)^{\frac{1}{2}}$. To see this more clearly, let us consider the wave function

$$\Psi_0'=C'\Omega^{-N/2}\prod_{i<j}[1+f(r_{ij})], \tag{A21}$$

which satisfies the required boundary conditions. We

can obtain Ψ_0 from the above wave function by expanding the above product in powers of f and then omitting all terms in which the coordinate of any particle, say $\mathbf{r}_i$, occurs more than once. For example, a term like

$$f(r_{12})f(r_{13}) \tag{A22}$$

must be omitted. The difference between Ψ_0' and Ψ_0, therefore, consists of terms like (A22), which expresses a correlation among more than two particles. Such terms belong to a higher order of $(a^3\rho)^{\frac{1}{2}}$ than we have considered. For example, upon Fourier-analyzing (A22), we find that it is of the form of a sum over three momenta $\mathbf{k}_1$, $\mathbf{k}_2$, $\mathbf{k}_3$, subject to $\mathbf{k}_1+\mathbf{k}_2+\mathbf{k}_3=0$. Such terms arise from a calculation of order a^4, as shown in Sec. 4.

The wave function for the one-phonon state can be obtained directly from (A11). By an argument similar to the above one, it can be shown that upon neglecting terms of higher orders in $(\rho a^3)^{\frac{1}{2}}$ the wave function $\Psi_{\mathbf{q}}$ of one phonon with momentum $\mathbf{q}$ in the configuration space is

$$\Psi_{\mathbf{q}}=\sum_{i=1}^{N} e^{i\mathbf{q}\cdot\mathbf{r}_i}\Psi_0, \tag{A23}$$

where Ψ_0 is the ground state wave function [Eq. (A16) or Eq. (A21)]. Thus it is to be expected that the Feynman-Bijl relations [Eq. (52)] correlating the excitation energy of a phonon with the pair distribution function is satisfied for a dilute system of hard spheres with Bose statistics.

SOVIET PHYSICS JETP VOLUME 34(7), NUMBER 2 AUGUST, 1958

APPLICATION OF THE METHODS OF QUANTUM FIELD THEORY TO A SYSTEM OF BOSONS

S. T. BELIAEV

Academy of Sciences, U.S.S.R.

Submitted to JETP editor August 2, 1957

J. Exptl. Theoret. Phys. (U.S.S.R.), **34**, 417-432 (February, 1958)

It is shown that the techniques of quantum field theory can be applied to a system of many bosons. The Dyson equation for the one-particle Green's function is derived. Properties of the condensed phase in a system of interacting bosons are investigated.

1. INTRODUCTION

In recent years Green's functions have been widely used[1] in quantum field theory, and in particular in quantum electrodynamics. This has made possible the development of methods[2] which escape from ordinary perturbation theory. The method of Green's functions has also been shown* to be applicable to many-body problems. In such problems the one-particle Green's function determines the essential characteristics of the system, the energy spectrum, the momentum distribution of particles in the ground state, etc.[3]

The present paper develops the method of Green's functions for a system consisting of a large number N of interacting bosons. The special feature of this system is the presence in the ground state of a large number of particles with momentum $\mathbf{p} = 0$ (condensed phase), which prevent the usual methods of quantum field theory from being applied. We find that for large N the usual technique of Feynman graphs can be used for the particles with $\mathbf{p} \neq 0$, while the condensed phase (we show that it does not disappear when interactions are introduced) can be considered as a kind of external field.

The Green's function is expressed in terms of three effective potentials Σ_{ik}, describing pair-production, pair-annihilation and scattering, and in terms of a chemical potential μ. This is the analog of Dyson's equation in electrodynamics.[4,1] Some approximation must be made in the calculation of Σ_{ik} and μ. If these quantities are computed by perturbation theory, the quasi-particle spectrum of Bogoliubov[5] is obtained. In the following paper[6] we evaluate Σ_{ik} and μ in the limit of low density.

*Private communication from A. B. Migdal.

2. STATEMENT OF THE PROBLEM. FEYNMAN GRAPHS

We consider a system of N spinless bosons with mass $m = 1$, enclosed in a volume V. We suppose N and V become infinite, the density $N/V = n$ remaining finite. A summation over discrete momenta is then replaced by an integral according to the rule

$$\sum_{\mathbf{p}} \rightarrow (2\pi)^{-3} V \int d\mathbf{p}.$$

The Hamiltonian of the system is $H = H_0 + H_1$, where

$$H_0 = \frac{1}{2}\int \nabla \Psi^+(\mathbf{x}) \nabla \Psi(\mathbf{x})\, d\mathbf{x} = \sum_{\mathbf{p}} \varepsilon^0_{\mathbf{p}} a^+_{\mathbf{p}} a_{\mathbf{p}}; \quad \varepsilon^0_{\mathbf{p}} = \frac{p^2}{2}, \quad (2.1)$$

$$H_1 = \frac{1}{2}\int \Psi^+(\mathbf{x}) \Psi^+(\mathbf{x}') U(\mathbf{x}-\mathbf{x}') \Psi(\mathbf{x}') \Psi(\mathbf{x})\, d\mathbf{x}\, d\mathbf{x}' = \frac{1}{2V}\sum_{\mathbf{p}\mathbf{p}'\mathbf{q}} U_{\mathbf{q}} a^+_{\mathbf{p}} a^+_{\mathbf{p}'} a_{\mathbf{p}'-\mathbf{q}} a_{\mathbf{p}+\mathbf{q}}. \quad (2.2)$$

The units are chosen so that $\hbar = 1$. $U(\mathbf{x}-\mathbf{x}')$ is the interaction between a pair of particles, $U_{\mathbf{q}} = \int e^{-i\mathbf{qx}} U(\mathbf{x})\,d\mathbf{x}$ is its Fourier transform, and

$$\Psi = V^{-1/2} \sum_{\mathbf{p}} e^{ipx} a_p, \quad \Psi^+ = V^{-1/2} \sum_{\mathbf{p}} e^{-ipx} a_p^+,$$

where $a_{\mathbf{p}}$ and $a_{\mathbf{p}}^+$ are the usual boson operators with the commutation law $[a_{\mathbf{p}}, a_{\mathbf{p}'}^+] = \delta_{\mathbf{pp}'}$.

The one-particle Green's function may be defined in two equivalent ways. In terms of Heisenberg-representation operators we may write

$$iG(x-x') = \langle \Phi_0^N,\ T\{\Psi(x)\Psi^+(x')\}\,\Phi_0^N\rangle, \tag{2.3}$$

with the expectation value taken in the ground-state of the N interacting particles. In terms of interaction-representation operators we may write

$$iG(x-x') = \langle T\{\Psi(x)\Psi^+(x')S\}\rangle / \langle S\rangle, \tag{2.4}$$

with the expectation value taken in the ground-state of the non-interacting particles, which has all the particles in the condensed phase so that $N_{\mathbf{p}\neq 0} = 0$, $N_0 = N$. The S-matrix for this system has the form

$$S = T\Big\{\exp\Big(-\frac{i}{2}\int d^4x_1 d^4x_2 U (1-2)\,\Psi^+(1)\Psi^+(2)\Psi(2)\Psi(1)\Big)\Big\}, \tag{2.5}$$

where we have written for convenience $U(1-2) = U(\mathbf{x}_1 - \mathbf{x}_2)\delta(t_1 - t_2)$. Here and henceforth $x, \ldots, p$ are four-vectors, and $px = \mathbf{px} - p_0x_0$. The definition (2.3) is convenient for relating G to physical quantities, while Eq. (2.4) is convenient for calculations.

In the numerator of Eq. (2.4) we expand the S-matrix in a series, each term of which is a T-product of a certain number of factors Ψ and Ψ^+. A T-product can be expressed by standard methods[7] as a sum of normal products in which some of the factors Ψ and Ψ^+ have been paired. In quantum electrodynamics the vacuum expectation value of every term which contains an unpaired annihilation operator vanishes from this sum. The surviving terms, which contain only pairs of Ψ and Ψ^+, are represented by certain Feynman graphs. In our case the expectation value is taken in a state containing N particles with momentum $\mathbf{p} = 0$. The expectation value of an N-product containing a_0 does not vanish, and the usual method of constructing graphs is not applicable.

Because of the special role of the state with $\mathbf{p} = 0$, it is convenient to separate the operators a_0 and a_0^+ from Ψ and Ψ^+. Thus we write

$$\Psi = \Psi' + a_0/\sqrt{V}; \quad \Psi^+ = \Psi'^+ + a_0^+/\sqrt{V}. \tag{2.6}$$

The Green's function (2.4) is also divided into two parts. The uncondensed particles give

$$iG'(x-x') = \langle T\{\Psi'(x)\Psi'^+(x')S\}\rangle / \langle S\rangle \tag{2.7}$$

while the Green's function of the condensed phase, a function of $(t-t')$ only, is

$$iG_0(t-t') = \langle T\{a_0(t)a_0^+(t')S\}\rangle / V\langle S\rangle. \tag{2.8}$$

The two functions are not independently determined, since the S-matrix appears in the definition of both and itself contains both Ψ' and a_0 operators. We shall prove later that when N is large the usual method of Feynman graphs can be adapted to the calculation of G′, the condensed phase behaving just like an external field.

We divide the operations T and $\langle\ldots\rangle$ into two successive operations, the first acting only upon Ψ' and Ψ'^+, the second acting only upon a_0 and a_0^+. Thus

$$T = T^0 T', \quad \langle\ldots\rangle = \langle\langle\ldots\rangle'\rangle^0,$$

where T^0 and $\langle\ldots\rangle^0$ act on a_0 and a_0^+.

We now drop the prime from G′ and write Eq. (2.7) in the form

$$iG(x-x') = \langle T^0\{\mathfrak{G}(x-x')\}\rangle^0 / \langle S\rangle, \tag{2.9}$$

with

$$\mathfrak{G}(x-x') = \langle T'\{\Psi'(x)\Psi'^+(x')S\}\rangle'. \tag{2.10}$$

Eq. (2.10) has the same structure as the numerator of Eq. (2.7), but the operators a_0, a_0^+ occurring in S are now to be treated as parameters. The expectation value in Eq. (2.10) is taken in the ground state of the operators Ψ', Ψ'^+. This equivalent to a vacuum expectation value, and so the usual formalism of Feynman graphs can be used for calculating $\mathfrak{G}$.

We represent the potential $-iU(1-2)$ by a dotted line joining the points 1 and 2. The pair of operators $\dot{\Psi}'(1)\ \dot{\Psi}'^+(2) = iG^{(0)}(1-2)$ is represented by a continuous line directed from 2 to 1. From the form of the interaction Hamiltonian (2.2) it follows that every graph contributing to Eq. (2.10) is a combination of the eight elementary graphs shown in Fig. 1. These correspond to the various terms which appear in Eq. (2.2) after the substitution (2.6). A missing continuous line (incomplete vertex) corresponds to a factor $(a_0/\sqrt{V})$ or $(a_0^+/\sqrt{V})$. Fig. 2 shows an example of one graph which appears in $\mathfrak{G}(x_1 - x_2)$, corresponding to the integral

$$\mathfrak{M}_2(x_1;\ x_2) = i^2\int G^{(0)}(1-3)\,U(3-4)\,G^{(0)}(3-5)\,G^{(0)}(4-6)\,U(5-6) \times G^{(0)}(6-2)\,V^{-1}a_0^+(t_4)\,a_0(t_5)\,d^4x_3 d^4x_4 d^4x_5 d^4x_6. \tag{2.11}$$

Let $\mathfrak{M}(x; x')$ be any graph contributing to Eq. (2.10) and not containing disconnected parts or vacuum loops. Together with $\mathfrak{M}$ we may consider all graphs differing from $\mathfrak{M}$ by the addition of vacuum loops. The totality of such graphs gives $\mathfrak{M}$ multiplied by a factor which is just the vacuum expectation value of the S matrix, namely $\langle S\rangle'$ in this case, since we are taking matrix elements only of Ψ' and Ψ'^{+}. Thus the inclusion of vacuum loops changes $\mathfrak{M}$ into

$$\mathfrak{M}(x; x')\langle S\rangle'. \tag{2.12}$$

In quantum electrodynamics the factor $\langle S\rangle$ cancels the denominator of Eq. (2.4), so that we can ignore the vacuum loops and merely omit this denominator. In our case, as we shall see later, the factor $\langle S\rangle'$ has a real significance.

Eq. (2.12) substituted into Eq. (2.9) gives

$$\langle T^0\{\mathfrak{M}(x; x')\langle S\rangle'\}\rangle^0/\langle S\rangle, \tag{2.13}$$

where the operation T^0 acts on the factors a_0, a_0^+ occurring in $\mathfrak{M}$ and in $\langle S\rangle'$. Suppose that $\mathfrak{M}$ contains m pairs of operators a_0, a_0^+. Then

$$\mathfrak{M}(x; x') = V^{-m}\int M(x; x'; t_1\ldots t_m; t_1'\ldots t_m')\, a_0(t_1)\ldots$$
$$\ldots a_0(t_m)\, a_0^+(t_1')\ldots a_0^+(t_m')\,(dt)\,(dt'),$$

and Eq. (2.13) becomes

$$\int MiG_0(t_1\ldots t_m; t_1'\ldots t_m')\,(dt)\,(dt'), \tag{2.14}$$

where

$$iG_0(t_1\ldots t_m; t_1'\ldots t_m')$$
$$= \langle T\{a_0(t_1)\ldots a_0^+(t_m')S\}\rangle/V^m\langle S\rangle \tag{2.15}$$

is the m-particle Green's function of the condensed phase, Eq. (2.8) being the special case m = 1.

The graphs for the Green's function (2.9) thus coincide with the graphs for $\mathfrak{G}$, only the factors $(a_0a_0^+/V)$ in the integrals are replaced by the corresponding Green's function of the condensed phase. For example, in tne integral (2.11), the factor $(a_0^+(t_4)\,a_0(t_5)/V)$ is replaced by $iG_0(t_5 - t_4)$. We need not consider graphs with disconnected parts, since these are already included in G_0. The problem is therefore reduced to the determination of the Green's functions G_0 of the condensed phase.

3. THE GREEN'S FUNCTIONS OF THE CONDENSED PHASE

We write the m-particle Green's function (2.15) of the condensed phase in the form

$$iG_0(t_1\ldots t_m; t_1'\ldots t_m')$$
$$= \frac{1}{V^m\langle S\rangle}\langle T^0\{a_0(t_1)\ldots a_0(t_m)\,a_0^+(t_1')\ldots$$
$$\ldots a_0^+(t'_m)\langle S\rangle'\}\rangle^0. \tag{3.1}$$

The quantity $\langle S\rangle'$ is the sum of contributions from all vacuum loops. If λ is the sum of contributions from all connected vacuum loops, then[8] the sum of contributions from all pairs of connected loops is $(\lambda^2/2!)$, the sum of contributions from all triples is $(\lambda^3/3!)$, and so on. Therefore $\langle S\rangle' = e^{\lambda}$. In our case λ is a functional of $a_0a_0^+$, and is proportional to the volume V if we take $(a_0a_0^+/V)$ to be finite. This can be seen by considering any vacuum loop as obtained from a graph with two free ends, carrying momenta $\mathbf{p}$ and $\mathbf{p}'$, by setting $\mathbf{p} = \mathbf{p}' = 0$. The graph with free ends gives a contribution proportional to $\delta(\mathbf{p}-\mathbf{p}') \sim (2\pi)^{-3}V\delta_{\mathbf{pp}'}$, and so the vacuum loop becomes proportional to V. Therefore $\lambda = V\sigma$, where σ is a finite functional of $(a_0a_0^+/V)$, and

$$\langle S\rangle' = e^{V\sigma}. \tag{3.2}$$

The commutator of a_0 and a_0^+ is unity, and is small compared with their product which is of order N. At first glance it would seem that the order of factors a_0, a_0^+ was unimportant, and that the T-product in Eq. (3.1) could be omitted. But one must remember that the T^0 in Eq. (3.1) links the product $a_0\ldots a_0^+$ with the quantity $e^{V\sigma}$, which contains all powers of the volume and hence may compensate for the smallness of the commutator of a_0 with a_0^+. Only after disentangling $a_0\ldots a_0^+$ from the T-product may we neglect the commutators. We observe that a_0 and a_0^+ commute with H_0 given by Eq. (2.1), so these operators are independent of time in the interaction representation. The arguments of the $a_0(t)$ and $a_0^+(t')$ in Eq. (3.1) are only ordering symbols for the operation of T^0. After carrying out the T-ordering we may consider a_0 and a_0^+ as time-independent.

The disentangling of $a_0\ldots a_0^+$ from the T-product is done by means of the following theorem. Let $B(a_0a_0^+/V)$ and $\sigma(a_0a_0^+/V)$ be any functionals of $(a_0a_0^+/V)$, which is considered as a finite quantity. The "disentangling rule"

$$T^0\{B(a_0a_0^+/V)\,e^{V\sigma}\} = B(AA^+)\,T^0\{e^{V\sigma}\}. \tag{3.3}$$

holds with an error of order $(1/V)$. The quantities A and A^+ are defined by the integral equations

$$A(t) = C(AA^+) + \int dt'\,\theta(t-t')\,\delta\sigma(AA^+)/\delta A^+(t'),$$
$$A^+(t) = C^+(AA^+) + \int dt'\,\theta(t'-t)\,\delta\sigma(AA^+)/\delta A(t'), \tag{3.4}$$

where $\theta(t-t')$ is the contribution from a factor-pair (a_0, a_0^+),

$$\theta(t-t') = \dot{a}_0(t)\dot{a}_0^+(t') = \begin{cases} 1 & \text{for } t > t' \\ 0 & \text{for } t < t', \end{cases} \quad (3.5)$$

and C, C^+ are time-independent functionals defined by the quadratic equations

$$C^2 + C\int \frac{\delta\sigma(AA^+)}{\delta A^+(t)}dt = \frac{a_0^2}{V};$$

$$C^{+2} + C^+\int \frac{\delta\sigma(AA^+)}{\delta A(t)}dt = \frac{a_0^{+2}}{V}. \quad (3.6)$$

A proof of this theorem is given in the Appendix.

Applying Eq. (3.3) to (3.1), we obtain

$$iG_0(t_1 \ldots; \ldots t_m) = \langle A(t_1) \ldots A^+(t'_m)\rangle^0.$$

The denominator of Eq. (3.1) cancels against $\langle T^0 \langle S \rangle' \rangle^0 = \langle S \rangle$. When the expectation value of the product $(A \ldots A^+)$ is taken, we may with an error of order $(1/V)$ replace all factors a_0, a_0^+ by $\sqrt{N}$. Let K and K^+ denote the result of making this replacement in A and A^+. Then

$$iG_0(t_1 \ldots t_m; t'_1 \ldots t'_m)$$
$$= K(t_1) \ldots K(t_m) K^+(t'_1) \ldots K(t'_m), \quad (3.7)$$

holds, with K and K′ given according to Eq. (3.4) by the integral equations

$$K(t) = \bar{C} + \int dt' \theta(t-t') \frac{\delta\sigma(KK^+)}{\delta K^+(t')},$$

$$K^+(t) = \bar{C}^+ + \int dt' \theta(t'-t) \frac{\delta\sigma(KK^+)}{\delta K(t')}, \quad (3.8)$$

and with $\bar{C}$ and $\bar{C}^+$ defined by

$$\bar{C}^2 + \bar{C}\int \frac{\delta\sigma(KK^+)}{\delta K^+(t)}dt = \frac{N}{V};$$

$$\bar{C}^{+2} + \bar{C}^+\int \frac{\delta\sigma(KK^+)}{\delta K(t)}dt = \frac{N}{V}. \quad (3.9)$$

Eq. (3.7) shows that the Green's functions of the condensed phase are products of factors, each factor being a function of one time variable. The physical meaning of this result may be clarified by the following qualitative argument. For simplicity we consider the one-particle function for non-interacting particles $iG_0^{(0)}(t-t') = \langle a_0(t)\, a_0^+(t') \rangle / V$. It describes the propagation of a particle from t′ to t. If there was originally a vacuum, then this process can proceed only by creating a particle at time t′ and annihilating it at the later time t, which is represented by the factor-pairing $\mathring{a}_0 \mathring{a}_0^+ = \theta$. In this case $G^{(0)}$ coincides with the factor-pairing, as is the case in electrodynamics. But if the process occurs in the presence of N particles of the same type, the created and absorbed particles may be different. In this case the propagation of a particle from t′ to t is composed of two processes, the creation of an extra particle in the condensed phase at time t′, and the absorption of one particle from the condensed phase at time t. The time sequence of these two events is immaterial, to order N^{-1}, if N is large. The processes are therefore independent. These arguments are valid also for the exact function G_0. It is also a product of two factors K(t) and $K^+(t')$, describing the two independent processes of emission and absorption of a particle at the two corresponding times.

We consider in greater detail the one-particle function of the condensed phase

$$iG_0(t-t') = K(t) K^+(t'). \quad (3.10)$$

The left side is a function of the difference $(t-t')$. The right side is a product of functions of t and t′. Therefore K(t) and $K^+(t')$ must be exponentials

$$K(t) = \sqrt{n_0}\, e^{-i\mu t}; \quad K^+(t) = \sqrt{n_0}\, e^{i\mu t} \quad (3.11)$$

so that

$$iG_0(t-t') = n_0 e^{-i\mu(t-t')}. \quad (3.12)$$

To understand the physical meaning of the quantities n_0 and μ, we go back to the definition (2.3) of G_0

$$iG_0(t-t') = \langle \Phi_0^N, T\{a_0(t)\, a_0^+(t')\} \Phi_0^N \rangle / V. \quad (3.13)$$

Putting t′ = t in Eq. (3.13) we find

$$iG_0(0) = \langle \Phi_0^N, a_0^+ a_0 \Phi_0^N \rangle / V = \bar{N}_0 / V. \quad (3.14)$$

Comparing Eq. (3.14) with (3.12), we see that $n_0 = (\bar{N}_0/V)$ is the mean density of particles in the condensed phase.

Next, suppose for definiteness t > t′, and write Eq. (3.13) in the form

$$iG_0(t-t') = \frac{1}{V}\langle \Phi_0^N a_0(t) \Phi_0^{N+1} \rangle \langle \Phi_0^{N+1} a_0^+(t') \Phi_0^N \rangle$$
$$+ \frac{1}{V}\sum_{s \neq 0} \langle \Phi_0^N a_0(t) \Phi_s^{N+1} \rangle \langle \Phi_s^{N+1} a_0^+(t') \Phi_0^N \rangle,$$

Separating out the time dependence of the Heisenberg operators, this expression becomes

$$iG_0(t-t') = \frac{1}{V} \exp\{-i(E_0^{N+1} - E_0^N)(t-t')\} \langle \Phi_0^N a_0 \Phi_0^{N+1} \rangle \langle \Phi_0^{N+1} a_0^+ \Phi_0^N \rangle$$

$$+ \frac{1}{V} \sum_{s \neq 0} \exp\{-i(E_s^{N+1} - E_0^N)(t-t')\} \langle \Phi_0^N a_0 \Phi_s^{N+1} \rangle \langle \Phi_s^{N+1} a_0^+ \Phi_0^N \rangle. \quad (3.15)$$

We compare the exact Eq. (3.15) with the approximation (3.12) which is valid as $N \to \infty$, and conclude that the second term in Eq. (3.15) must vanish as $N \to \infty$. Comparison of the time dependence of the first term in Eq. (3.15) with that of Eq. (3.12) then shows that μ is the chemical potential of the system,

$$\mu = E_0^{N+1} - E_0^N \approx \partial E_0^N / \partial N. \quad (3.16)$$

The parameters n_0 and μ which appear in K and K^+ can be calculated in principle by solving Eq. (3.8). In practice this is very difficult. The trouble is that, in calculating the vacuum loops which contribute to σ, one has first to integrate over a finite time interval $(-T, T)$, so that the parameter T appears in Eq. (3.8). One may pass to the limit $T \to \infty$ in the solutions, but not in the equations. Thus it is incorrect to use in $\sigma(KK^+)$ the limiting expressions (3.11) for K and K^+. One has instead to solve the nonlinear equations (3.8) directly.

We can obtain from Eq. (3.8) one relation between the quantities n_0 and μ. Differentiating Eq. (3.8) with respect to t, and remembering that $d\theta(t-t')/dt = \delta(t-t')$, we obtain the differential equations

$$dK/dt = \delta\sigma(KK^+)/\delta K^+(t);$$

$$dK^+/dt = -\delta\sigma(KK^+)/\delta K(t). \quad (3.17)$$

Let σ be expanded in a series

$$\sigma(KK^+) = -i \sum_{(W)} \frac{1}{m} \int W_m(t_1' \ldots t_m'; t_1 \ldots t_m) K^+(t_1') \ldots$$

$$\ldots K^+(t_m') K(t_1) \ldots K(t_m)(dt)(dt'), \quad (3.18)$$

in which each term corresponds to a certain vacuum loop with m pairs of incomplete vertices, and the sum is taken over all such loops. The "vacuum amplitudes" W_m are functions of only $(2m-1)$ variables (time differences), so that the limiting values of K and K^+ would give an infinite result when substituted into Eq. (3.18). If Eq. (3.18) is varied with respect to K^+, one integration disappears, and the result becomes finite. The Fourier transform of $W_m(t'; t)$ may be written

$$W_m(\omega_1' \ldots; \ldots \omega_m)\, \delta\left(\sum\omega - \sum\omega'\right)$$

$$= \int W_m(t'; t)\, e^{i(\omega' t') - i(\omega t)} (dt)(dt').$$

Then Eq. (3.11) and (3.18) give

$$\frac{\delta\sigma(KK^+)}{\delta K^+(t)}$$

$$= -i\sqrt{n_0} \sum_{(W)} n_0^{m-1} W_m(\mu \ldots; \ldots \mu)\, e^{-i\mu t}. \quad (3.19)$$

Substituting Eq. (3.19) into (3.17) and using Eq. (3.11), we obtain the desired relation

$$\mu = \sum_{(W)} n_0^{m-1} W_m(\mu \ldots \mu; \mu \ldots \mu). \quad (3.20)$$

The summation here extends over the various vacuum loops which contribute to the quantity $(\delta\sigma/\delta K^+(t))$ according to Eq. (3.19). Each such loop is to be taken with unit weight, remembering that there is one special incomplete vertex t, at which the variation with respect to K^+ was taken. Two loops are to be counted as different if they have the same geometrical structure and differ only in the position of the special vertex.

Equation (3.20) may be considered as an equation for $\mu(n_0)$. There is one free parameter in the problem, the total particle number N or the density n. Thus μ and n_0 ought to be expressible in terms of n. However n does not appear explicitly in the equation. It is thus convenient to consider n_0 instead of n as the free parameter, and to express all other quantities as functions of n_0. The connection between n_0 and n can be found after the problem is solved. From this standpoint, Eq. (3.20) completely determines K and K^+ and consequently all the Green's functions of the condensed phase. We might also solve the problem with two free parameters μ and n_0, and only consider the connection between them in the final result.[6] But this procedure would considerably increase the mathematical difficulties.

4. PROPERTIES OF THE CONDENSED PHASE

The form of the functions G_0 of the condensed phase leads to some deductions concerning the properties of the condensed phase in a system of interacting particles.

In the absence of interaction, the momentum distribution of particles in the ground state is $\delta(\mathbf{p}) = (2\pi)^{-3} V \delta_{\mathbf{p}0}$. When interactions are introduced the distribution is smeared out. In principle two possibilities are open. Either the term in $\delta(\mathbf{p})$ completely disappears and the distribution becomes continuous (there is no condensed phase), or a term in $\delta(\mathbf{p})$ remains and the state $\mathbf{p} = 0$ is still exceptional (there is a condensed phase). In the first case all average occupation numbers $\overline{N}_\mathbf{p}$ are finite, and $\overline{N}_\mathbf{p} \to \overline{N}_0$ as $\mathbf{p} \to 0$. In the second case $\overline{N}_{\mathbf{p}\neq 0}$ is finite but $\overline{N}_0 \sim V$.

The neglect of the second term in Eq. (3.15) is equivalent to the assumption that a_0, operating on the ground state Φ_0^N, does not excite the system, or in symbols

$$a_0\Phi_0^N \approx (N_0)^{1/2}\Phi_0^{N-1} \tag{4.1}$$

This assumption seems at first glance strange. A change in the number of particles with $\mathbf{p} = 0$, disturbing the stationary relation between the occupation numbers, must excite the system. If N_0 were finite, a change of it by one unit would change the state appreciably, but if $N_0 \sim V$ this change will practically not disturb the ground state. Equation (4.1) supports the second alternative. Therefore the introduction of interactions never causes the condensed phase to disappear entirely.

We next examine the problem of the fluctuation of the number of particles in the condensed phase. The quantity N_0 does not have an exact value in the state Φ_0^N. We expand Φ_0^N into eigenstates of the operator N_0. The expansion may be written

$$\Phi_0^N = \sum_{m=0}^{N} C_{N-m}^N \varphi_0^{N-m} \chi_m^N, \tag{4.2}$$

where $\varphi_0^{N_0}$ is a function only of the occupation number of the condensed phase, while χ_m^N depends on the other variables. χ_m^N describes a state of m particles with momenta distributed over all values $\mathbf{p} \neq 0$. It is a superposition of states with definite occupation numbers for the momenta $\mathbf{p} \neq 0$. The coefficients in this superposition depend on the upper index N. The normalization of χ_m^N is given by $\langle\chi_m^N\chi_{m'}^N\rangle\ \delta_{mm'}$.

Equation (4.1), with the orthogonality of $\varphi_0^{N_0}$ and $\varphi_0^{N_0'}$ for $N_0 \neq N_0'$, now gives the result

$$\langle\Phi_0^{N-1}a_0\Phi_0^N\rangle = \sum_{m=0}^{N-1} \sqrt{N-m}\,(C_{N-m-1}^{N-1})^* C_{N-m}^N \langle\chi_m^{N-1}\chi_m^N\rangle \tag{4.3}$$

We assume that C_{N-m}^N and χ_m^N are smooth functions of N, so that

$$C_{N-m-1}^{N-1} \approx C_{N-m}^N - \partial C_{N-m}^N/\partial N = C_{N-m}^N\{1 + O(N^{-1})\};$$
$$\chi_m^{N-1} \approx \chi_m^N - \partial\chi_m^N/\partial N = \chi_m^N\{1 + O(N^{-1})\}.$$

Then Eq. (4.3) becomes

$$\langle\Phi_0^{N-1}a_0\Phi_0^N\rangle = \sum_{m=0}^{N-1}\sqrt{N-m}\,|C_{N-m}^N|^2\{1 + O(N^{-1})\}.$$

The sum here is simply $\overline{N_0^{1/2}}$, so that

$$\langle\Phi_0^{N-1}a_0\Phi_0^N\rangle = \overline{N_0^{1/2}}\{1 + O(N^{-1})\}. \tag{4.4}$$

A similar expression naturally holds also for $\langle\Phi_0^{N+1}a_0^+\Phi_0^N\rangle$.

We estimate the sum in Eq. (3.15) after setting $t = t'$. Using Eq. (4.4) and (3.14), we find

$$\sum_{s\neq 0} \Phi_0^N a_0\Phi_s^{N+1}\rangle\langle\Phi_s^{N+1}a_0^+\Phi_0^N\rangle \approx \overline{N}_0 - (\overline{N_0^{1/2}})^2, \tag{4.5}$$

from which it is clear that the sum is connected with the magnitude of the fluctuations in the number of particles in the condensed phase. From the fact that the sum is negligible as $N \to \infty$, we conclude

$$[\overline{N}_0 - (\overline{N_0^{1/2}})^2]/\overline{N}_0 \to \infty \text{ for } N \to \infty, \tag{4.6}$$

Thus the fluctuations in the number of particles in the condensed phase are relatively small.

5. GREEN'S FUNCTION FOR A PARTICLE WITH $\mathbf{p} \neq 0$

The expressions obtained in Sec. 3 for the functions of the condensed phase allow us to reformulate the rules which were described in Sec. 2 for the construction of graphs.

Every graph is a combination of eight elementary graphs (Fig. 1). Every incomplete vertex carries a factor $K(t) = \sqrt{n_0}\,e^{-i\mu t}$ corresponding to a missing incoming continuous line, or a factor $K^+(t) = \sqrt{n_0}\,e^{i\mu t}$ corresponding to a missing outgoing line. These factors mean that the interaction

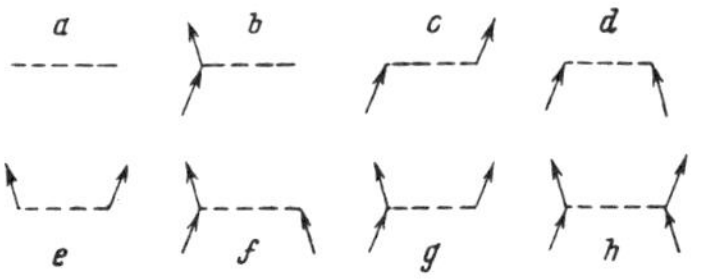

FIG. 1

involves the absorption or the emission of a particle of energy μ in the condensed phase. We may draw a wavy line corresponding to every incoming or outgoing particle of the condensed phase. All such lines have free ends. In analogy with quantum electrodynamics, we may say that the condensed phase behaves like an external field with frequency μ.

Consider the general structure of a graph which contributes to the Green's function (2.7). Every

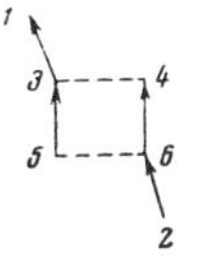

FIG. 2

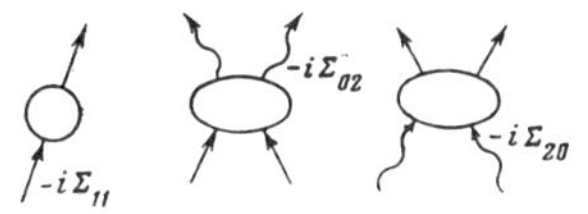

FIG. 3

graph contributing to G has the form of a chain consisting of separate irreducible parts connected to each other by only one continuous line. There are only three types of irreducible parts (i.e., parts which cannot be separated into pieces joined by only one continuous line). The three types differ in the number of outgoing and incoming continuous lines (Fig. 3). The sums of the contributions from all irreducible parts of each type we call respectively $-i\Sigma_{11}$, $-i\Sigma_{02}$, $-i\Sigma_{20}$. Σ_{11} describes processes in which the number of particles out of the condensed phase is conserved. Σ_{02} and Σ_{20} describe the absorption and emission of two particles out of the condensed phase; in these processes two particles in the condensed phase must be simultaneously emitted or absorbed, and their energy 2μ must be taken into account. In the momentum representation, $\Sigma_{11}(p_1; p_2)$ contains a factor $\delta(p_1 - p_2)$, while $\Sigma_{02}(p_1p_2)$ and $\Sigma_{20}(p_1p_2)$ contain* $\delta(p_1 + p_2 - 2\mu)$. Henceforth we shall assume momentum conservation in the arguments of the functions Σ_{ik}, representing the quantities which multiply the δ-functions by the notations

$$\Sigma_{11}(p;\ p) \equiv \Sigma_{11}(p);\ \Sigma_{02}(p + \mu, -p + \mu) \equiv \Sigma_{02}(p + \mu);$$
$$\Sigma_{20}(p + \mu, -p + \mu) \equiv \Sigma_{20}(p + \mu). \qquad (5.1)$$

The functions Σ_{ik} are characteristic of the particle interactions, and we may call them the effective potentials of the pair interaction.

Besides the Green's function G we introduce an auxiliary quantity $\hat{G}$, consisting of the sum of contributions from graphs with two ingoing lines. Graphs contributing to G have one ingoing and one outgoing. Figure 4 shows some of the graphs which contribute to $\hat{G}$. The quantity $\hat{G}$ describes the transition of two particles into the condensed phase. In momentum representation we write $\hat{G}(p + \mu)$ when the ingoing lines carry momenta $(p + \mu)$ and $(-p + \mu)$.

There are two equations, analogous to the Dyson equation in electrodynamics,[5,1] for the functions G and $\hat{G}$,

$$G(p + \mu) = G^{(0)}(p + \mu) + G^{(0)}(p + \mu)\Sigma_{11}(p + \mu)G(p + \mu)$$
$$+ G^{(0)}(p + \mu)\Sigma_{20}(p + \mu)\hat{G}(p + \mu),$$
$$\hat{G}(p+\mu) = G^{(0)}(-p+\mu)\Sigma_{11}(-p+\mu)\hat{G}(p + \mu)$$
$$+ G^{(0)}(-p + \mu)\Sigma_{02}(p+\mu)G(p + \mu). \qquad (5.2)$$

The structure of these equations is illustrated graphically by Fig. 5 and does not need any further explanation. Solving the system (5.2) for G and $\hat{G}$, we find

$$G(p + \mu)$$
$$= (G^{(0)^{-1}} - \Sigma_{11})^{-}\{(G^{(0)^{-1}} - \Sigma_{11})^{+}(G^{(0)^{-1}} - \Sigma_{11})^{-} - \Sigma_{20}\Sigma_{02}\}^{-1},$$
$$\hat{G}(p + \mu) = \Sigma_{02}\{(G^{(0)^{-1}} - \Sigma_{11})^{+}(G^{(0)^{-1}} - \Sigma_{11})^{-} - \Sigma_{20}\Sigma_{02}\}^{-1}, \qquad (5.3)$$

where the suffixes $\pm$ indicate the values $(\pm p + \mu)$ of the arguments. Equation (5.3) for G may be written in the usual form of a Dyson equation

$$G^{-1} = G^{(0)^{-1}} - \Sigma,$$
$$\Sigma(p + \mu) = \Sigma_{11}(p + \mu)$$
$$+ \Sigma_{20}\Sigma_{02}/[G^{(0)^{-1}}(-p + \mu) - \Sigma_{11}(-p + \mu)]. \qquad (5.4)$$

Into Eq. (5.3) we substitute the explicit form of the free-particle Green's function,

$$G^{(0)^{-1}}(p) = p^0 - \varepsilon_p^0 + i\delta; \qquad (\varepsilon_p^0 = p^2/2;\ \delta \to +0), \quad (5.5)$$

and obtain for G and $\hat{G}$ the expressions

$$G(p + \mu)$$
$$= \frac{p^0 + \varepsilon_p^0 + \Sigma_{11}^{-} - \mu}{[p^0 - (\Sigma_{11}^{+} - \Sigma_{11}^{-})/2]^2 - [\varepsilon_p^0 + (\Sigma_{11}^{+} + \Sigma_{11}^{-})/2 - \mu]^2 + \Sigma_{20}\Sigma_{02}}, \qquad (5.6)$$

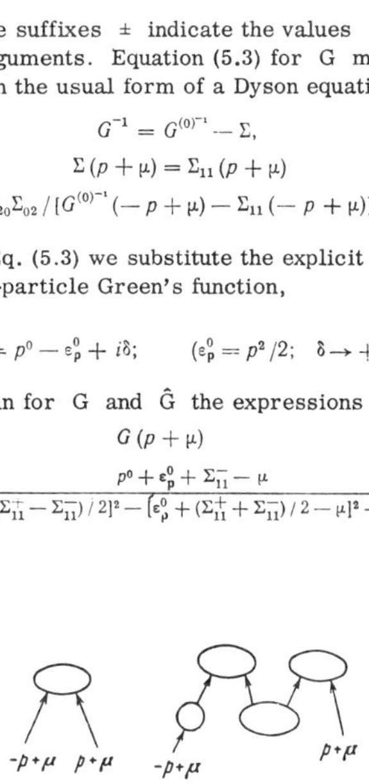

FIG. 4

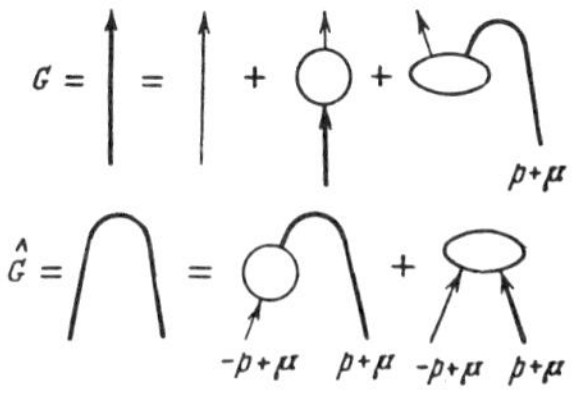

FIG. 5

*Here μ represents a 4-vector having only its fourth component non-zero.

$$\hat{G}(p+\mu) = \frac{-\Sigma_{02}}{[p^0 - (\Sigma_{11}^+ - \Sigma_{11}^-)/2]^2 - [\varepsilon_p^0 + (\Sigma_{11}^+ + \Sigma_{11}^-)/2 - \mu]^2 + \Sigma_{20}\Sigma_{02}}. \tag{5.7}$$

Equation (5.6) determines the Green's function in terms of the effective potentials Σ_{ik} and the chemical potential μ of the system. Equations for Σ_{ik} and μ cannot be obtained in so general a form. To calculate these quantities we have to use approximate methods to sum over series of graphs. In Sec. 7 we shall calculate Σ_{ik} and μ by perturbation theory. In the following paper[6] we develop an approximation in which the density is considered as a small parameter.

6. CONNECTION BETWEEN THE GREEN'S FUNCTION AND PROPERTIES OF THE SYSTEM

The energy E_0 of the ground state is the expectation value of the Hamiltonian (2.1), (2.2) in the state Φ_0^N,

$$E_0 = \langle \Phi_0^N H \Phi_0^N \rangle = \sum_p \varepsilon_p^0 \langle a_p^+ a_p \rangle + \frac{1}{2V} \sum_{pp'q} U_q \langle a_p^+ a_{p'}^+ a_{p'-q} a_{p+q} \rangle. \tag{6.1}$$

The last term in Eq. (6.1) is connected with the Green's function G. Consider G in the (p, t) representation, i.e., in the momentum representation for the space-components only. Taking the expectation value in the state Φ_0^N, Eq. (2.3) becomes

$$iG(p;\, t-t') = \langle T\{a_p(t)\, a_p^+(t')\}\rangle, \qquad (p \neq 0), \tag{6.2}$$

from which it is easy to deduce

$$(i\partial/\partial t - \varepsilon_p^0)\, G(p;\, t-t') = \delta(t-t') + R(p;\, t-t'), \tag{6.3}$$

with

$$R(p;\, t-t') = -\frac{i}{V} \sum_{p'q} U_q \langle T\{a_{p'}^+(t)\, a_{p'-q}(t)\, a_{p+q}(t)\, a_p^+(t')\}\rangle. \tag{6.4}$$

We multiply Eq. (6.3) by $e^{ip^0(t-t')}$ and integrate with respect to t. Then using Eq. (5.5) we obtain

$$G^{(0)^{-1}}(p)\, G(p) = 1 + R(p),$$

This, with the definition (5.4) of Σ, gives immediately

$$R(p) = \Sigma(p)\, G(p), \qquad (p \neq 0). \tag{6.5}$$

On the other hand, Eq. (6.4) shows that $R(p; -0)$ is related to the last sum in Eq. (6.1), namely

$$\frac{1}{2V} \sum_{pp'q} U_q \langle a_p^+ a_{p'}^+ a_{p'-q} a_{p+q} \rangle = \frac{i}{2} \sum_p R(p;\, -0). \tag{6.6}$$

The expression (6.5) for R holds only when $p \neq 0$. For $p = 0$ the left side of Eq. (6.2) is $iVG_0(t-t')$ according to Eq. (3.13). Instead of Eq. (6.3) we have in this case

$$i\frac{\partial}{\partial t} G_0(t-t') = \frac{1}{V}\delta(t-t') + \frac{1}{V} R(0;\, t-t'). \tag{6.7}$$

We neglect the δ-function since it is of order V^{-1}, and use Eq. (3.12) for G_0. This gives

$$R(0;\, -0) = i\left[\frac{\partial}{\partial \tau} G_0(\tau)\right]_{\tau=0} = -i\mu n_0. \tag{6.8}$$

Equations (6.6) and (6.8) bring the expression (6.1) for E_0 into the form

$$E_0 = \sum_p \varepsilon_p^0 \langle a_p^+ a_p \rangle + \frac{i}{2} \sum_{p \neq 0} R(p;\, -0) + \frac{1}{2}\mu n_0. \tag{6.9}$$

From Eq. (6.2) we find

$$\bar{N}_p = \langle a_p^+ a_p \rangle = iG(p;\, -0) = i\int G(p)\, dp^0/2\pi \tag{6.10}$$

Using Eq. (6.5) and (6.10), and passing from summation to integration in Eq. (6.9), we obtain the following expression* for the ground-state energy E_0,

$$E_0/V = i\int [\varepsilon_p^0 + {}^1\!/_2 \Sigma(p)]\, G(p)\, d^4p/(2\pi)^4 + \mu n_0/2 \tag{6.11}$$

The p_0-integration is to be taken with a small detour into the upper half-plane.

The quantities Σ and G depend parametrically upon μ and n_0, supposing that Eq. (3.20) has not been used in order to eliminate one of these parameters. Therefore Eq. (6.11) gives a relation between (E_0/V), μ, and n_0. There are two further relations between these quantities. First there is the definition of the chemical potential μ,

$$\mu = \frac{\partial E_0}{\partial N} = \frac{\partial}{\partial n}\left(\frac{E_0}{V}\right), \tag{6.12}$$

and second there is the condition that the total number of particles is conserved, which by Eq. (6.10) can be written in the form

$$n = n_0 + i\int G(p)\, d^4p/(2\pi)^4. \tag{6.13}$$

Equations (6.11), (6.12), and (6.13) determine (E_0/V), μ, and n_0 in terms of the density n, or determine any three of these quantities in terms of the fourth. The relation (3.20) which we found earlier does not give any new information; it is

*V. M. Galitskii informed me that a similar relation exists for Fermi systems.

satisfied identically when Eq. (6.11), (6.12), and (6.13) hold.

7. PERTURBATION THEORY APPROXIMATION TO Σ_{ik} AND μ

In the first order of perturbation theory, the graphs which contribute to Σ_{ik} are the elementary graphs shown in Fig. 1. Graphs b and c refer to Σ_{11}, d to Σ_{02} and e to Σ_{20}. These graphs give the contributions

$$\Sigma_{02} = \Sigma_{20} = n_0 U_{\mathbf{p}}; \qquad \Sigma_{11}^{+} = n_0 (U_0 + U_{\mathbf{p}}). \tag{7.1}$$

In first approximation the only vacuum loop is the elementary graph a of Fig. 1. Thus Eq. (3.20) gives for μ the value

$$\mu = n_0 U_0. \tag{7.2}$$

Inserting Eq. (7.1) and (7.2) into the expression (5.6) for the Green's function, we find

$$G(p+\mu) = p^0 + \varepsilon_{\mathbf{p}}^0 + n_0 U_{\mathbf{p}} / (p^{0\,2} - \varepsilon_{\mathbf{p}}^{0\,2} - 2n_0 U_{\mathbf{p}} \varepsilon_{\mathbf{p}}^0 + i\delta). \tag{7.3}$$

The Green's function $G(p+\mu)$ has a pole at a value $p_0(\mathbf{p})$ which defines the energy of an elementary excitation of the system[3] (quasi-particle). Equation (7.3) gives for the quasi-particle energy

$$\varepsilon_{\mathbf{p}} = \sqrt{\varepsilon_{\mathbf{p}}^{0\,2} + 2n_0 U_{\mathbf{p}} \varepsilon_{\mathbf{p}}^0}. \tag{7.4}$$

Substituting Eq. (7.3) into (6.10), we obtain the mean occupation number of the ground state,

$$\begin{aligned} \overline{N}_p &= (-\varepsilon_{\mathbf{p}} + \varepsilon_{\mathbf{p}}^0 + n_0 U_{\mathbf{p}}) / 2\varepsilon_p \\ &= (n_0 U_{\mathbf{p}})^2 / 2\varepsilon_p (\varepsilon_{\mathbf{p}} + \varepsilon_{\mathbf{p}}^0 + n_0 U_{\mathbf{p}}). \end{aligned} \tag{7.5}$$

Equations (7.4) and (7.5) coincide with the results of the well-known work of Bogoliubov.[5]

APPENDIX. PROOF OF THEOREM (3.3)

We use the method of Wick[9] to transform the T-product in Eq. (3.3), which may be written symbolically

$$T\{Be^{V\sigma}\} = N\{e^{\Delta} B e^{V\sigma}\}, \tag{A.1}$$

where Δ is an operator which changes a pair $a_0 a_0^+$ into its replacement (3.5),

$$\Delta = \frac{1}{V} \int dt dt' \theta(t-t') \frac{\delta^2}{\delta\alpha(t)\,\delta\alpha^+(t')}; \qquad (\alpha = a_0/\sqrt{V}; \quad \alpha^+ = a_0^+/\sqrt{V}). \tag{A.2}$$

The proof of the theorem proceeds in two stages: (1) pulling B out across the operator e^{Δ}, and (2) a final disentangling of the N-product.

(1) The result of the first stage can be formulated as follows. With an error of order V^{-1} we have

$$e^{\Delta}\{B(\alpha;\,\alpha^+)e^{V\sigma}\} = B(\beta;\,\beta^+)e^{\Delta}\{e^{V\sigma}\}, \tag{A.3}$$

where β and β^+ are defined by the equations

$$\begin{aligned} \beta(t) &= \alpha + \int dt' \theta(t-t')\, \delta\sigma(\beta\beta^+)/\delta\beta^+(t'); \\ \beta^+(t) &= \alpha^+ + \int dt' \theta(t'-t)\, \delta\sigma(\beta\beta^+)/\delta\beta(t'). \end{aligned} \tag{A.4}$$

Proof. The factor V^{-1} in Δ can be compensated on the left side of Eq. (A.3) only if $e^{V\sigma}$ is involved in at least one operation of Δ. We write $\Delta = \Delta_{\sigma\sigma} + \Delta_{B\sigma} + \Delta_{\sigma B}$, where the first suffix indicates the object upon which the variation with respect to α operates, and the second suffix refers to the variation with respect to α^+. The result of operating with $e^{\Delta\sigma\sigma}$ can be written

$$e^{\Delta_{\sigma\sigma}}\{e^{V\sigma}\} = e^{V\sigma'} \tag{A.5}$$

It will be shown later that σ' is independent of V. Equation (A.5) gives

$$e^{\Delta}\{Be^{V\sigma}\} = \exp(\Delta_{B\sigma} + \Delta_{\sigma B})\{Be^{V\sigma'}\}. \tag{A.6}$$

We let $e^{\Delta_{B\sigma}}$ operate first on $e^{V\sigma'}$. From Eq. (A.2) we obtain

$$\Delta_{B\sigma} e^{V\sigma'} = e^{V\sigma'} \int dt dt' \theta(t-t') \frac{\delta\sigma'}{\delta\alpha^+(t')} \left[\frac{\delta}{\delta\alpha(t)}\right]_B \equiv e^{V\sigma'} D_B. \tag{A.7}$$

Successive application of Eq. (A.7) gives $(\Delta_{B\sigma})^k \times e^{V\sigma'} = e^{V\sigma'}(D_B)^k$, since the operators $\Delta_{B\sigma}$ produce variations only in the exponential. Therefore

$$e^{\Delta_{B\sigma}} e^{V\sigma'} = e^{V\sigma'} e^{D_B}. \tag{A.8}$$

From (A.7) it is clear that e^{D_B} is a displacement operator, displacing $\alpha(t)$ by the quantity

$$\alpha_1(t) = \int dt' \theta(t-t')\, \delta\sigma' / \delta\alpha^+(t'),$$

Therefore

$$\begin{aligned} e^{\Delta_{B\sigma}}\{B(\alpha;\,\alpha^+)e^{V\sigma'}\} &= e^{V\sigma'} e^{D_B} B(\alpha;\,\alpha^+) \\ &= e^{V\sigma'} B(\alpha+\alpha_1;\,\alpha^+). \end{aligned} \tag{A.9}$$

An analogous result holds for the operator $e^{\Delta_{\sigma B}}$, which displaces $\alpha^+(t)$. We obtain finally from Eq. (A.6)

$$e^{\Delta}\{B(\alpha;\,\alpha^+)e^{V\sigma}\} = B(\beta;\,\beta^+)e^{V\sigma'} = B(\beta;\,\beta^+)e^{\Delta}\{e^{V\sigma}\}, \tag{A.10}$$

where

$$\begin{aligned} \beta(t) &= \alpha + \int dt' \theta(t-t')\, \delta\sigma' / \delta\alpha^+(t'); \quad \beta^+(t) \\ &= \alpha^+ + \int dt' \theta(t'-t)\, \delta\sigma' / \delta\alpha(t'); \end{aligned} \tag{A.11}$$

Equation (A.10) is identical with (A.3). It remains to show that Eq. (A.11) and (A.4) are identical. Varying both sides of Eq. (A.5) with respect to

$\alpha(t)$ and $\alpha^+(t)$, and using Eq. (A.10), we obtain

$$\delta\sigma'(\alpha\alpha^+)/\delta\alpha(t) = \delta\sigma(\beta\beta^+)/\delta\beta(t);$$
$$\delta\sigma'(\alpha\alpha^+)/\delta\alpha^+(t) = \delta\sigma(\beta\beta^+)/\delta\beta^+(t); \quad \text{(A.12)}$$

Equations (A.12) and (A.11) define σ'. When Eq. (A.12) is substituted into (A.11), the result is Eq. (A.4).

(2) In the second stage of the proof we may consider α and α^+ to be constant operators (see the beginning of Sec. 3). Then σ' and $B(\beta\beta^+) = B'(\alpha\alpha^+)$ are functions of α, α^+ instead of functionals. By integrating Eq. (A.12) with respect to time we obtain the connection between the functions $\sigma'(\alpha\alpha^+)$ and σ,

$$\int \frac{\delta\sigma(\beta\beta^+)}{\delta\beta(t)}dt = \int \frac{\delta\sigma'(\alpha\alpha^+)}{\delta\alpha(t)}dt = \frac{\partial\sigma'}{\partial\alpha}; \quad \int \frac{\delta\sigma(\beta\beta^+)}{\delta\beta^+(t)}dt = \frac{\partial\sigma'}{\partial\alpha^+}. \quad \text{(A.13)}$$

For the following argument it is important that σ' and β' depend only on $\nu = \alpha^+\alpha$. This being so, the disentangling proceeds according to the rule

$$N\{B'(\nu)e^{V\sigma'(\nu)}\} = B'(\bar{\nu})N\{e^{V\sigma'}\}, \quad \text{(A.14)}$$

with $\bar{\nu}$ obtained from ν by the relation

$$\nu = \bar{\nu}[1 + \partial\sigma'(\bar{\nu})/\partial\bar{\nu}] \equiv \bar{\nu}X^2 \quad \text{(A.15)}$$

We defer the proof of Eq. (A.14), and show first that Eq. (A.3) and (A.14) imply the truth of the theorem (3.3). After the infinite factor $e^{V\sigma'}$ is removed, the lack of commutativity of α and α^+ can be neglected. The substitution $\nu \to \bar{\nu}$ which appears in Eq. (A.14), can therefore be divided into the two substitutions $\alpha \to \bar{\alpha}$, $\alpha^+ \to \bar{\alpha}^+$, where $\alpha = \bar{\alpha}X$ and $\alpha^+ = \bar{\alpha}^+X$ according to Eq. (A.15). After some algebra we find

$$\alpha^2 = \bar{\alpha}^2 + \bar{\alpha}\partial\sigma'(\bar{\nu})/\partial\bar{\alpha}^+; \quad \alpha^{+2} = \bar{\alpha}^{+2} + \bar{\alpha}^+\partial\sigma'(\bar{\nu})/\partial\bar{\alpha}. \quad \text{(A.16)}$$

We denote by A, A^+ the quantities into which β, β^+ are transformed under the substitution $\alpha, \alpha^+ \to \bar{\alpha}, \bar{\alpha}^+$. The equations for A and A^+ are obtained from Eq. (A.4) by changing the terms outside the integrals into $\bar{\alpha}$ and $\bar{\alpha}^+$, thus

$$A(t) = \bar{\alpha} + \int dt'\theta(t-t')\delta\sigma(AA^+)/\delta A^+(t');$$
$$A^+(t) = \bar{\alpha}^+ + \int dt'\theta(t'-t)\delta\sigma(AA^+)/\delta A(t'), \quad \text{(A.17)}$$

By the definition of $B'(\nu)$, $B'(\bar{\nu}) = B(AA^+)$, and so Eq. (A.3) and (A.14) imply (3.3). To complete the proof of the theorem it remains to show the equivalence of Eq. (A.17) and (3.4). To do this, we substitute $\alpha, \alpha^+ \to \bar{\alpha}, \bar{\alpha}^+$ in Eq. (A.13), and find the result

$$\frac{\partial\sigma'(\bar{\nu})}{\partial\bar{\alpha}} = \int \frac{\delta\sigma(AA^+)}{\delta A(t)}dt; \quad \frac{\partial\sigma'(\bar{\nu})}{\partial\bar{\alpha}^+} = \int \frac{\delta\sigma(AA^+)}{\delta A^+(t)}dt. \quad \text{(A.18)}$$

Equation (A.16) for $\bar{\alpha}$ and $\bar{\alpha}^+$ are identical with Eq. (3.6) by virtue of Eq. (A.18). Therefore $\bar{\alpha}$, $\bar{\alpha}^+$ are identical with C, C^+, and the theorem is proved.

We now return to the proof of Eq. (A.14). Let L be a quantity related to σ' by the equation

$$N\{e^{V\sigma'}\} = e^{VL}. \quad \text{(A.19)}$$

We shall later express L explicitly in terms of σ' and shall verify that L is independent of V. Suppose that $B'(\nu)$ has the form

$$B'(\nu) = \sum_k b_k\nu^k, \quad \text{(A.20)}$$

Then Eq. (A.19) implies

$$N\{B'e^{V\sigma'}\} = \sum_k b_k\alpha^{+k}N\{e^{V\sigma'}\}\alpha^k = \sum_k b_k\alpha^{+k}e^{VL}\alpha^k. \quad \text{(A.21)}$$

The commutation relation $\alpha y(\nu) = \left(y + \frac{1}{V}\frac{\partial y}{\partial\nu}\right)\alpha$ holds for any function $y(\nu)$. Applying it repeatedly, we find

$$\alpha^k e^y = \exp\left\{\left(1 + \frac{1}{V}\frac{\partial}{\partial\nu}\right)^k y\right\}\alpha^k. \quad \text{(A.22)}$$

We choose y to satisfy $\left(1 + \frac{1}{V}\frac{\partial}{\partial\nu}\right)^k y = VL$. Then Eq. (A.22) gives the rule for pulling α^k through e^{VL},

$$\exp\{VL\}\alpha^k = \alpha^k\exp\left\{\left(1 + \frac{1}{V}\frac{\partial}{\partial\nu}\right)^{-k}VL\right\}$$
$$\approx \alpha^k\exp\left(-k\frac{\partial L}{\partial\nu}\right)\exp\{VL\}. \quad \text{(A.23)}$$

Applying Eq. (A.23) to Eq. (A.21), we obtain

$$N\{B'e^{V\sigma'}\} = \sum_k b_k(\nu e^{-\partial L/\partial\nu})^k e^{VL} = B'(\bar{\nu})e^{VL}, \quad \text{(A.24)}$$

with

$$\bar{\nu} = \nu e^{-\partial L/\partial\nu} \quad \text{(A.25)}$$

L can be determined by differentiating both sides of Eq. (A.19) with respect to α. On the left side we find

$$\frac{1}{V}\frac{\partial}{\partial\alpha}N\{e^{V\sigma'}\} = N\left\{\frac{\partial\sigma'}{\partial\alpha}e^{V\sigma'}\right\} = \alpha^+N\left\{\frac{\partial\sigma'}{\partial\nu}e^{V\sigma'}\right\}$$

which with Eq. (A.24) and (A.19) gives

$$\frac{1}{V}\frac{\partial}{\partial\alpha}N\{e^{V\sigma'}\} = \alpha^+\frac{\partial\sigma'(\bar{\nu})}{\partial\bar{\nu}}e^{VL}. \quad \text{(A.26)}$$

In differentiating the right side of Eq. (A.19) with respect to α, we must remember that $(\partial L/\partial\alpha)$ and L do not commute. Since $(\partial L/\partial\alpha) = \alpha^+(\partial L/\partial\nu)$, the commutation rule $L\alpha^+ = \alpha^+\left(L + \frac{1}{V}\frac{\partial L}{\partial\nu}\right)$ implies

$$\frac{1}{V}\frac{\partial}{\partial\alpha}L^k = \alpha^+\left\{\left(L + \frac{1}{V}\frac{\partial L}{\partial\nu}\right)^k - L^k\right\}, \quad \text{(A.27)}$$

and hence

$$\frac{1}{V}\frac{\partial}{\partial\alpha}e^{VL}=\alpha^{+}(e^{\partial L/\partial\nu}-1)e^{VL}. \qquad (A.28)$$

By comparing Eq. (A.28) with (A.26), we obtain the desired relation between L and σ',

$$\partial\sigma'(\bar{\nu})/\partial\bar{\nu}=e^{\partial L/\partial\nu}-1. \qquad (A.29)$$

Equations (A.29) and (A.25) imply Eq. (A.15), and by Eq. (A.24) this proves Eq. (A.14).

[1] V. B. Berestetskii and A. D. Galanin, Проблемы современной физики (Problems of Modern Physics), No. 3 (1955), (introductory paper).

[2] Abrikosov, Landau, and Khalatnikov, Dokl. Akad. Akad. Nauk SSSR **95**, 497, 773 and 1177; **96**, 261 (1954).

[3] V. M. Galitskii and A. B. Migdal, J. Exptl. Theoret. Phys. (U.S.S.R.) **34**, 139 (1958); Soviet Phys. JETP **7**, 96 (1958).

[4] F. J. Dyson, Phys. Rev. **75**, 1736 (1949).

[5] N. N. Bogoliubov, Izv. Akad. Nauk SSSR, ser. fiz. **11**, 77 (1947).

[6] S. T. Beliaev, J. Exptl. Theoret. Phys. (U.S.S.R.) **34**, 433 (1958); Soviet Phys. JETP **7**, 289 (1958) (this issue).

[7] G. C. Wick, Phys. Rev. **80**, 268 (1950).

[8] R. P. Feynman, Phys. Rev. **76**, 749 (1949).

[9] S. Hori, Prog. Theoret. Phys. **7**, 578 (1952).

Translated by F. J. Dyson
77

SOVIET PHYSICS JETP VOLUME 34(7), NUMBER 2 AUGUST, 1958

ENERGY-SPECTRUM OF A NON-IDEAL BOSE GAS

S. T. BELIAEV

Academy of Sciences, U.S.S.R.

Submitted to JETP editor August 2, 1957

J. Exptl. Theoret. Phys. (U.S.S.R.) **34**, 433-446 (February, 1958)

The one-particle Green's function is calculated in a low-density approximation for a system of interacting bosons. The energy spectrum of states near to the ground state (quasi-particle spectrum is derived.

1. INTRODUCTION

IN the preceding paper[1] the method of Green's functions was developed for a system consisting of a large number of bosons. The one-particle Green's function was expressed in terms of the effective potentials Σ_{ik} of pair interactions and the chemical potential μ of the system. Approximate methods must be used to determine Σ_{ik} and μ. In the present paper we study a "gaseous" approximation, in which the density n, or the ratio between the volume occupied by particles and the total volume, is treated as a small parameter. The interaction between particles is assumed to be central and short-range, but not necessarily weak. The first two orders of approximation involve only the scattering amplitude f of a two-particle system. But in the next order (proportional to $(\sqrt{nf^3})^2$) the effects of three-particle interaction amplitudes appear, which means that practical calculations to this order are hardly possible.

From the Green's function which we calculate, we derive the energy spectrum of excitations or quasi-particles, the energy of the ground state, and also the momentum distribution of particles in the ground state.

2. ESTIMATE OF THE GRAPHS CONTRIBUTING TO THE EFFECTIVE POTENTIALS

The definition of the potentials Σ_{ik}, and the rules for constructing Feynman graphs, were described in our earlier paper,[1] which we shall call I.

We shall estimate by perturbation theory the various graphs contributing to Σ_{ik} and μ. For the Fourier transform of the potential $U(\mathbf{p}) = U_{\mathbf{p}}$, we assume for simplicity* $U_p = U_0$ for $p < 1/a$, and $U_p = 0$ for $p > 1/a$. Then a is of the order of magnitude of the particle radius.

For definiteness we examine Σ_{20}. The graphs for Σ_{02}, Σ_{11} and μ are essentially similar. The first order of perturbation theory, as we saw in Sec. (I, 7), gives $\Sigma_{20}^{(1)} = n_0 U_p$; $\mu = n_0 U_0$.

In the estimate of any graph there may appear three parameters — U_0 and a, characterizing the interaction, and n_0, characterizing the density of particles in the condensed phase. The three parameters can be combined into two dimensionless ratios,

$$\xi = U_0/a; \qquad \beta = \sqrt{n_0 a^3}. \tag{2.1}$$

The quantity ξ is the usual parameter which appears in perturbation theory (in ordinary units $\xi \sim mU(r)a^2/\hbar^2$), while β is a parameter of gas-density.

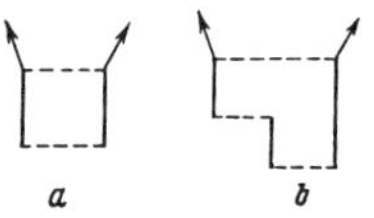

FIG. 1

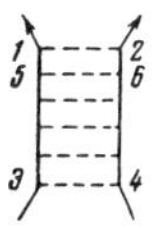

FIG. 2

The only non-vanishing graph in second order is the one shown in Fig. 1a. This gives a contribution

$$M_a \sim n_0 \int G^0(q+\mu)\, G^0(-q+\mu)\, U_0 U_{\mathbf{p}+\mathbf{q}} d^4q$$

Substituting for G^0 from

$$G^0(p) = (p^0 - \varepsilon_p^0 + i\delta)^{-1}, \qquad \varepsilon_p^0 = p^2/2, \qquad \delta \to +0 \tag{2.2}$$

and carrying out the q^0-integration, we find

$$M_a \sim n_0 U_0^2 \int_{qa<1} d\mathbf{q} \int \frac{dq^0}{(q^0 + \mu - \varepsilon_q^0 + i\delta)(-q^0 + \mu - \varepsilon_q^0 + i\delta)}$$

$$\sim n_0 U_0^2 \int_{qa<1} d\mathbf{q} \frac{1}{\mu - \varepsilon_q^0 + i\delta}$$

In the last integral the main contribution comes from $q \sim 1/a$, where $\mu/\epsilon^0 \sim n_0 U_0 a^2 = \xi\beta^2 \ll 1$. Therefore

$$M_a \sim n_0 U_0^2 / a = \Sigma_{20}^{(1)} \xi. \tag{2.3}$$

We consider next the third-order graph (1b). This gives a contribution

$$M_b \sim n_0^2 \int G^0(q+\mu)\, [G^0(-q+\mu)]^2\, U_0 U_{\mathbf{q}} U_{\mathbf{p}+\mathbf{q}} d^4q$$

$$\sim n_0^2 U_0^3 \int d\mathbf{q} / (\mu - \varepsilon_{\mathbf{q}}^0 + i\delta)^2.$$

The last integral, unlike the previous one, converges at the upper limit, and the main contribution now comes from the range $q \sim \sqrt{\mu} = \sqrt{n_0 U_0}$. Therefore

$$M_b \sim n_0^2 U_0^3 / \sqrt{\mu} = \Sigma_{20}^{(1)} \xi^{3/2} \beta. \tag{2.4}$$

From Eqs. (2.3) and (2.4) we see that $M_b/M_a \sim \xi^{1/2}\beta$. This is a consequence of the fact that M_a contains an integral of a product of two factors G^0, formally diverging at the upper limit, while M_b contains an integral of a product of three factors G^0 and converges without any cut-off. In the graphs this difference is indicated by the number of continuous lines in the closed circuit formed by the continuous and dotted lines. The same result holds when the circuits form part of a more complicated graph.

Thus every circuit containing more than two continuous lines introduces the small parameter β, while circuits with two continuous lines do not involve β. In the lowest order we need consider only graphs whose circuits are all of the two-line type. All such graphs are of the "ladder" construction shown in Fig. 2. We denote by $-i\Gamma(12; 34)$ the total contribution from all such graphs. The first-order approximation in β then differs from the first-order approximation in perturbation theory by changing the potential U (arising from a ladder with one rung) into Γ (arising from ladders of all lengths). Similar conclusions hold also for the higher approximations. A summation over a set of graphs, differing only by the insertion of ladder circuits into a fixed skeleton, produces a change of U into Γ. If we represent Γ by a rectangle, all graphs can be constructed by means of rectangles and continuous lines only. In this way the potential U is eliminated from the problem. The effective potential is Γ.

3.* EQUATION FOR THE EFFECTIVE POTENTIAL Γ

We can write down an integral equation

$$\Gamma(12;34) = U(1-2)\,\delta(1-3)\,\delta(2-4)$$
$$+ i \int U(1-2)\, G^0(1-5)\, G^0(2-6)\, \Gamma(56;34)\, d^4x_5 d^4x_6. \tag{3.1}$$

for the sum of contributions from all graphs of the ladder type (see Fig. 2). The notations are the

*The letters p, q, ... are used to denote the lengths of 3-vectors, or to denote 4-vectors. There can be no confusion, because they denote 4-vectors only when they appear as arguments in G(p), Σ(p), etc.

*The problems connected with Γ were solved in collaboration with V. M. Galitskii, who was working simultaneously on the analogous problems in Fermion systems.

same as in I. We next transform Eq. (3.1) into momentum representation. In order to relieve the equations of factors of 2π, we shall use the conventions

$$d^4p = (2\pi)^{-4} dp^1 dp^2 dp^3 dp^0;$$

$$\delta(p) = (2\pi)^4 \delta(p^1)\delta(p^2)\delta(p^3)\delta(p^0),$$

and similarly we understand $\mathbf{dp}$ and $\delta(\mathbf{p})$ to carry factors of $(2\pi)^3$. We write

$$\begin{aligned}\Gamma(p_1p_2;\ p_3p_4)\,\delta(p_1+p_2-p_3-p_4) &= \int \exp\{-ip_1x_1 - ip_2x_2 + ip_3x_3 \\ &\quad + ip_4x_4\}\,\Gamma(12;34)\,d^4x_1d^4x_2d^4x_3d^4x_4,\end{aligned} \tag{3.2}$$

and introduce the relative and total momenta by

$$\begin{aligned}p_1 + p_2 = P';\ p_3 + p_4 = P;\\ p_1 - p_2 = 2p',\ p_3 - p_4 = 2p,\end{aligned} \tag{3.3}$$

Then, by Eq. (3.1), $\Gamma(p';\ p;\ P) \equiv \Gamma(p_1p_2;\ p_3p_4)$ satisfies the equation

$$\begin{aligned}\Gamma(p';\ p;\ P) = U(p'-p) + i\int d^4q U(p'-q)\,G^0(P/2+q)\\ \times G^0(P/2-q)\,\Gamma(q;\ p;\ P).\end{aligned} \tag{3.4}$$

Since the interaction U is instantaneous, $U(1-2) = U(\mathbf{x}_1 - \mathbf{x}_2)\,\delta(t_1 - t_2)$, and therefore the points 1, 2 and 3, 4 in $\Gamma(12;\ 34)$ must be simultaneous. In momentum representation this means that $\Gamma(p_1p_2;\ p_3p_4)$ depends on the fourth components only in the combination $p_1^0 + p_2^0 = p_3^0 + p_4^0 = P^0$. Therefore $\Gamma(p';\ p;\ P)$ is independent of the fourth components of its first two arguments (the relative momenta). The q^0-integration in Eq. (3.4) can thus be carried out, giving

$$\begin{aligned}\int dq^0 G^0\left(\tfrac{1}{2}P + q\right) G^0\left(\tfrac{1}{2}P - q\right)\\ = -i\left(P^0 - \tfrac{1}{4}\mathbf{P}^2 - \mathbf{q}^2 + i\delta\right)^{-1},\end{aligned} \tag{3.5}$$

and then Eq. (3.4) takes the form

$$\begin{aligned}\Gamma(\mathbf{p}';\ \mathbf{p};\ P) = U(\mathbf{p}'-\mathbf{p}) + \int d\mathbf{q}\,\frac{U(\mathbf{p}'-\mathbf{q})\,\Gamma(\mathbf{q};\ \mathbf{p};\ P)}{k_0^2 - q^2 + i\delta}\ ;\\ k_0^2 = P^0 - \tfrac{1}{4}\mathbf{P}^2.\end{aligned} \tag{3.6}$$

Equation (3.6) cannot be solved explicitly, but its solution can be expressed in terms of the scattering amplitude of two particles in a vacuum. We write $\chi(\mathbf{q}) = (k_0^2 - q^2 + i\delta)^{-1}\,\Gamma(\mathbf{q};\ \mathbf{p};\ P)$. Then Eq. (3.6) becomes

$$(k_0^2 - p'^2)\,\chi(\mathbf{p}') - \int U(\mathbf{p}'-\mathbf{q})\,\chi(\mathbf{q})\,d\mathbf{q} = U(\mathbf{p}'-\mathbf{p}). \tag{3.7}$$

Let $\Psi_{\mathbf{k}}(\mathbf{p}')$ be the normalized wave-function which satisfies the equation

$$(k^2 - p'^2)\,\Psi_{\mathbf{k}}(\mathbf{p}') - \int U(\mathbf{p}'-\mathbf{q})\,\Psi_{\mathbf{k}}(\mathbf{q})\,d\mathbf{q} = 0, \tag{3.8}$$

Then the solution of Eq. (3.7) may be written

$$\chi(\mathbf{p}') = \int \frac{\Psi_{\mathbf{k}}(\mathbf{p}')\,\Psi_{\mathbf{k}}^*(\mathbf{q})}{k_0^2 - k^2 + i\delta}\,U(\mathbf{q}-\mathbf{p})\,d\mathbf{q},$$

and so $\Gamma(\mathbf{p}';\ \mathbf{p};\ P)$ becomes

$$\Gamma(\mathbf{p}';\ \mathbf{p};\ P) = (k_0^2 - p'^2)\int \frac{\Psi_{\mathbf{k}}(\mathbf{p}')\,\Psi_{\mathbf{k}}^*(\mathbf{q})}{k_0^2 - k^2 + i\delta}\,U(\mathbf{q}-\mathbf{p})\,d\mathbf{q}. \tag{3.9}$$

We observe now that Eq. (3.8) is the Schrödinger equation in momentum representation. Thus $\Psi_{\mathbf{k}}(\mathbf{p})$ is the wave-function for a scattering problem with potential U. The scattering amplitude* $f(\mathbf{p}'\,\mathbf{p})$ is related to the Ψ-function by

$$f(\mathbf{p}'\mathbf{p}) = \int e^{-i\mathbf{p}'\mathbf{r}}\,U(\mathbf{r})\,\Psi_{\mathbf{p}}(\mathbf{r})\,d\mathbf{r} = \int U(\mathbf{p}'-\mathbf{q})\,\Psi_{\mathbf{p}}(\mathbf{q})\,d\mathbf{q}, \tag{3.10}$$

or by

$$\Psi_{\mathbf{p}}(\mathbf{p}') = \delta(\mathbf{p}-\mathbf{p}') + f(\mathbf{p}'\mathbf{p})/(p^2 - p'^2 + i\delta). \tag{3.11}$$

In the first Eq. (3.10), $\Psi_{\mathbf{p}}(\mathbf{r})$ is the wave-function in coordinate space which behaves at infinity like a plane wave with momentum $\mathbf{p}$ and an outgoing spherical wave. The usual scattering amplitude is the value of $f(\mathbf{p}'\,\mathbf{p})$ at $p' = p$. We consider arbitrary values of the arguments, so that $f(\mathbf{p}'\,\mathbf{p})$ is in general defined by Eq. (3.10).

Because $\Psi_{\mathbf{p}}(\mathbf{p}')$ satisfies orthogonality conditions in both its arguments, $f(\mathbf{p}'\,\mathbf{p})$ satisfies the unitarity conditions

$$\begin{aligned}f(\mathbf{p}'\mathbf{p}) - f^*(\mathbf{p}'\mathbf{p})\\ = \int d\mathbf{q} f(\mathbf{p}'\mathbf{q})\,f^*(\mathbf{p}\mathbf{q})\left[\frac{1}{q^2 - p'^2 + i\delta} - \frac{1}{q^2 - p^2 - i\delta}\right]\\ = \int d\mathbf{q} f^*(\mathbf{q}\mathbf{p}')\,f(\mathbf{q}\mathbf{p})\left[\frac{1}{q^2 - p'^2 + i\delta} - \frac{1}{q^2 - p^2 - i\delta}\right].\end{aligned} \tag{3.12}$$

When $\mathbf{p}' = \pm\mathbf{p}$, Eq. (3.12) gives the imaginary part of the forward and backward scattering amplitudes. Since $f(-\mathbf{p}' - \mathbf{p}) = f(\mathbf{p}'\,\mathbf{p})$, Eq. (3.12) implies

$$\operatorname{Im} f(\pm\mathbf{p}\mathbf{p}) = -i\pi\int d\mathbf{q} f(\mathbf{p}\mathbf{q})\,f^*(\pm\mathbf{p}\mathbf{q})\,\delta(q^2 - p^2). \tag{3.13}$$

For the forward scattering amplitude, Eq. (3.13) gives just the well-known relation between the imaginary part of the amplitude and the total cross-section σ, $\operatorname{Im} f(\mathbf{p}\,\mathbf{p}) = -ip\sigma$.

We substitute Eq. (3.11) into (3.9) and use Eq. (3.12). This gives two equivalent expressions for $\Gamma(\mathbf{p}';\ \mathbf{p};\ P)$,

$$\begin{aligned}\Gamma(\mathbf{p}';\ \mathbf{p};\ P) = f(\mathbf{p}'\mathbf{p})\\ + \int d\mathbf{q} f(\mathbf{p}'\mathbf{q})\,f^*(\mathbf{p}\mathbf{q})\left[\frac{1}{k_0^2 - q^2 + i\delta} + \frac{1}{q^2 - p^2 - i\delta}\right]\\ = f^*(\mathbf{p}\mathbf{p}') + \int d\mathbf{q} f(\mathbf{p}'\mathbf{q})\,f^*(\mathbf{p}\mathbf{q})\left[\frac{1}{k_0^2 - q^2 + i\delta} + \frac{1}{q^2 - p'^2 + i\delta}\right].\end{aligned} \tag{3.14}$$

*The quantity $f(\mathbf{p}'\,\mathbf{p})$ differs by a numerical factor from the usual amplitude $a(\mathbf{p}'\,\mathbf{p})$, in fact $f = -4\pi a$.

expressing the effective potential $\Gamma(\mathbf{p}'; \mathbf{p}; P)$ in terms of the scattering amplitudes of a two-particle system.

4. FIRST-ORDER GREEN'S FUNCTION

The effective potentials Σ_{ik} are determined by special values which Γ takes when two out of the four particles involved in a process belong to the condensed phase. Thus two of the four particles must have $\mathbf{p} = 0$, $p^0 = \mu$. Each particle of the condensed phase also carries a factor $\sqrt{n_0}$. Therefore we find

$$\Sigma_{20}(p+\mu) = n_0\Gamma(\mathbf{p}; 0; 2\mu); \quad \Sigma_{02}(p+\mu) = n_0\Gamma(0; \mathbf{p}; 2\mu),$$

$$\Sigma_{11}(p+\mu) = n_0\Gamma(\mathbf{p}/2; \mathbf{p}/2; p+2\mu) + n_0\Gamma(-\mathbf{p}/2; \mathbf{p}/2; p+2\mu). \tag{4.1}$$

To obtain the chemical potential we must let all four particles in Γ belong to the condensed phase, and divide by one power of n_0 [see Eq. (I, 3.20)]. We then have

$$\mu = n_0\Gamma(0; 0; 2\mu). \tag{4.2}$$

Substituting into Eqs. (4.1) and (4.2) the value of Γ from Eq. (3.14), we find

$$\mu = n_0 f(00) + n_0\int d\mathbf{q}\, |f(0\mathbf{q})|^2 \left[\frac{1}{2\mu - q^2 + i\delta} + \frac{1}{q^2}\right],$$

$$\Sigma_{20}(p+\mu) = n_0 f(\mathbf{p}0) + n_0\int d\mathbf{q}\, f(\mathbf{pq}) f^*(0\mathbf{q}) \left[\frac{1}{2\mu - q^2 + i\delta} + \frac{1}{q^2}\right],$$

$$\Sigma_{02}(p+\mu) = n_0 f^*(\mathbf{p}0) + n_0\int d\mathbf{q}\, f(0\mathbf{q}) f^*(\mathbf{pq}) \left[\frac{1}{2\mu - q^2 + i\delta} + \frac{1}{q^2}\right],$$

$$\Sigma_{11}(p+\mu) = 2n_0 f_s\left(\frac{\mathbf{p}}{2}\,\frac{\mathbf{p}}{2}\right) + 2n_0\int d\mathbf{q} \left|f_s\left(\frac{\mathbf{p}}{2}\,\mathbf{q}\right)\right|^2 \left[\frac{1}{p^0 + 2\mu - p^2/4 - q^2 + i\delta} + \frac{1}{q^2 - p^2/4 - i\delta}\right]. \tag{4.3}$$

In the last equation we have introduced the symmetrized amplitude

$$f_s(\mathbf{p}'\mathbf{p}) = [f(\mathbf{p}'\mathbf{p}) + f(-\mathbf{p}'\mathbf{p})]/2.$$

All the integrals in Eq. (4.3) converge at high momentum, even if the amplitudes are taken to be constant. For dimensional reasons these terms are of order $n_0 f^2 \sqrt{\mu}$. Compared with the first terms in Eq. (4.3), these terms contain an extra factor $\sqrt{n_0 f^3}$, which is just the gas-density parameter (2.1) obtained by substituting the amplitude f for the particle radius. In first approximation we neglect the integral terms in Eq. (4.3) and obtain

$$\mu = n_0 f(00); \quad \Sigma_{20}(p+\mu) = \Sigma_{02}^*(p+\mu) = n_0 f(\mathbf{p}0);$$
$$\Sigma_{11}^{\pm} \equiv \Sigma_{11}(\pm p + \mu) = 2n_0 f_s\left(\frac{\mathbf{p}}{2}\,\frac{\mathbf{p}}{2}\right). \tag{4.4}$$

The Green's function G is given by Eq. (I, 5.6),

$$G(p+\mu) = \frac{p^0 + \varepsilon_0^0 + \Sigma_{11}^- - \mu}{[p^0 - (\Sigma_{11}^+ - \Sigma_{11}^-)/2]^2 - [\varepsilon_\mathbf{p}^0 + (\Sigma_{11}^+ + \Sigma_{11}^-)/2 - \mu]^2 + \Sigma_{20}\Sigma_{02} + i\delta} \tag{4.5}$$

and after substituting from Eq. (4.4) this becomes

$$G(p+\mu) = \frac{p^0 + \varepsilon^0 + 2n_0 f_s\left(\frac{\mathbf{p}}{2}\,\frac{\mathbf{p}}{2}\right) - n_0 f(00)}{p^{02} - \varepsilon_\mathbf{p}^2 + i\delta}, \tag{4.6}$$

with

$$\varepsilon_\mathbf{p} = \sqrt{\left[\varepsilon_\mathbf{p}^0 + 2n_0 f_s\left(\frac{\mathbf{p}}{2}\,\frac{\mathbf{p}}{2}\right) - n_0 f(00)\right]^2 - n_0^2 |f(\mathbf{p}0)|^2}. \tag{4.7}$$

The point $p_0(\mathbf{p})$, at which the Green's function $G(p+\mu)$ has a pole, determines the energy $\epsilon_\mathbf{p}$ of elementary excitations or quasi-particles[2] carrying momentum $\mathbf{p}$. To calculate $\epsilon_\mathbf{p}$ we must know three distinct amplitudes. $f_s(\mathbf{p}/2\ \mathbf{p}/2)$ is the ordinary symmetrized amplitude for forward scattering, and $f(0\ 0)$ is a special value of the same amplitude. However, $f(\mathbf{p}\ 0)$ does not have any obvious meaning in the two-particle problem, since it refers to a process which is forbidden for two particles in a vacuum.

At small momenta, we may neglect the momentum dependence of $f_s(\mathbf{p}/2\ \mathbf{p}/2)$ and of $f(\mathbf{p}\ 0)$, setting $f_s(\mathbf{p}/2\ \mathbf{p}/2) \approx f(\mathbf{p}\ 0) \approx f(0\ 0) \equiv f_0$. This approximation is allowed when the wavelength is long compared with the characteristic size of the interaction region, which has an order of magnitude given by the scattering amplitude f_0. Therefore when $p < f_0^{-1}$ we may consider all the amplitudes in Eq. (4.6) and (4.7) to be constant. For higher excitations with $p \gtrsim f_0^{-1}$, the momentum dependence of the amplitude becomes important, and the problem cannot be treated in full generality. We shall examine the higher excitations (in Sec. 8) for the special example of a hard-sphere gas.

Confining ourselves to the case $pf_0 < 1$, we deduce from Eq. (4.6) and (4.7)

$$G(p+\mu) = (p^0 + \varepsilon_\mathbf{p}^0 + n_0 f_0)/(p^{02} - \varepsilon_\mathbf{p}^2 + i\delta), \tag{4.8}$$

with

$$\varepsilon_\mathbf{p} = \sqrt{\varepsilon_\mathbf{p}^{02} + 2n_0 f_0 \varepsilon_p^0}. \tag{4.9}$$

Equations (4.8) and (4.9) are formally identical with the results obtained from perturbation theory in Eq. (I, 7.3) and (I, 7.4). Only the scattering amplitude f_0 now appears instead of the Fourier transform $U_\mathbf{p}$ of the potential.

Equation (4.9) shows that quasi-particles with $p \ll \sqrt{n_0 f_0}$ have a sound-wave type of dispersion law $\epsilon_\mathbf{p} \approx p\sqrt{n_0 f_0}$. When $p \gg \sqrt{n_0 f_0}$ they go over into almost free particles with $\epsilon_\mathbf{p} \approx \epsilon_\mathbf{p}^0 + n_0 f_0$.

This sort of energy spectrum appears also when one considers particles moving in a continuous medium with a refractive index. The transition from phonon to free particle behavior occurs at $p \sim \sqrt{n_0 f_0} \ll 1/f_0$, so that the approximation of constant amplitudes is valid in both ranges.

The conditions $\sqrt{n_0 f_0^3} \ll 1$ and $pf_0 \ll 1$ are not independent. If we look at momenta p not greatly exceeding $\sqrt{n_0 f_0}$, then the second condition is a consequence of the first. If we are then neglecting quantities of order $\sqrt{n_0 f_0^3}$, we must also treat the amplitudes as constant.

In Sec. (I, 5) we introduced the quantity $\hat{G}(p + \mu)$, the analog of the Green's function G but constructed from graphs with two ingoing ends instead of one ingoing and one outgoing. The analogous quantity with two outgoing ends will be denoted by $\check{G}(p + \mu)$. It is obtained from $\hat{G}(p + \mu)$ when Σ_{02} is replaced by Σ_{20}. In the constant-amplitude approximation, Eq. (I, 5.7) and (4.4) give

$$\hat{G}(p+\mu) = \check{G}(p+\mu) = -n_0 f_0/(p^{02} - \varepsilon_{\mathbf{p}}^2 + i\delta). \quad (4.10)$$

5. SECOND APPROXIMATION FOR THE GREEN'S FUNCTION

For the second approximation to Σ_{ik} and μ, we must retain quantities of order $\sqrt{n_0 f_0^3}$. As we saw at the end of the preceding section, we must then also retain terms of order pf_0 in the amplitudes. The real part of the amplitude involves only even powers of p, and the imaginary part only odd powers. Terms of order pf_0 arise only from the lowest approximation to the imaginary part of the amplitudes. The imaginary part of $f_s(\mathbf{p}/2\ \mathbf{p}/2)$ is given by Eq. (3.13), and from Eq. (3.10) we see that the amplitude $f(\mathbf{p}\ 0)$ is real [and anyway in this approximation we need only the square of the modulus of $f(\mathbf{p}\ 0)$].

The graphs of the first approximation give terms of order $\sqrt{n_0 f_0^3}$, namely the integral terms in Eq. (4.3). In these terms, as in Eq. (3.13), we may take the amplitudes to be constant. We have seen in Sec. 2 that graphs containing one circuit with three or more continuous lines give contributions of the same order. The summation over sets of graphs, which differ only in the number of continuous lines in a circuit, is automatically performed if one replaces the zero-order Green's function G^0 by the first-order functions G, $\hat{G}$ and $\check{G}$. We therefore consider immediately the circuits which can be built out of G, $\hat{G}$, $\check{G}$ and Γ. There are altogether ten essentially different circuits (see Fig. 3). A rectangle with a cross denotes a sum of two rectangles, one being a direct interaction and the other an exchange interaction. The two differ only by an

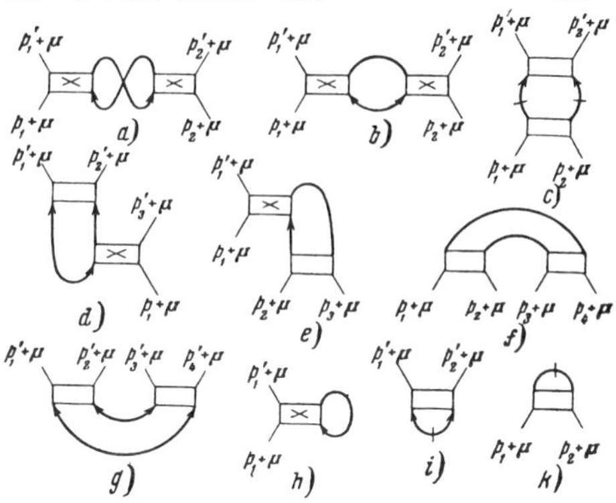

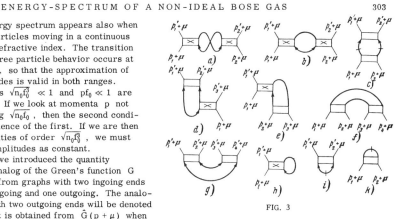

FIG. 3

interchange of the upper or the lower ends. The sum of the two rectangles introduces a factor $-i[\Gamma(12;34) + \Gamma(12;43)]$, or in momentum representation $-i[\Gamma(\mathbf{p}';\mathbf{p};P) + \Gamma(-\mathbf{p}';\mathbf{p};P)]$. If G, $\hat{G}$ and $\check{G}$ are expanded in powers of the effective potential Γ, then in the lowest approximation the graphs (3c, 3i, 3k) become circuits with two continuous lines. But all such circuits are already included in Γ and must therefore be omitted. This omission is represented in Fig. 3 by the strokes across the continuous lines. Let $-iF_{a,b\ldots}(p_1',\ldots,p_1\ldots)$ denote the contributions from the graphs of Fig. 3. In the constant-amplitude approximation these contributions are:

$$\begin{aligned}
F_a(p_1'p_2';\ p_1p_2) &= i4f_0^2\int G(q+\mu)\,G(p_1-p_1'+q+\mu)\,d^4q;\\
F_b &= i4f_0^2\int \hat{G}(q+\mu)\,\check{G}(p_1-p_1'+q+\mu)\,d^4q;\\
F_c &= if_0^2\int\{G(q+\mu)\,G(p_1+p_2-q+\mu)\\
&\quad - G^0(q+\mu)\,G^0(p_1+p_2-q+\mu)\}\,d^4q;\\
F_d &= i2f_0^2\int \hat{G}(q+\mu)\,G(p_2'+p_3'-q+\mu)\,d^4q,\\
F_e &= i2f_0^2\int \check{G}(q+\mu)\,G(p_1'+p_2'-q+\mu)\,d^4q;\\
F_f &= if_0^2\int \hat{G}(q+\mu)\,\hat{G}(p_1+p_2-q+\mu)\,d^4q;\\
F_g &= if_0^2\int \check{G}(q+\mu)\,\check{G}(p_1'+p_2'-q+\mu)\,d^4q;\\
F_h &= i2f_0\int G(q+\mu)\,d^4q;\\
F_i &= if_0\int\{\check{G}(q+\mu)-n_0f_0G^0(q+\mu)\,G^0(-q+\mu)\}\,d^4q;\\
F_k &= if_0\int\{\hat{G}(q+\mu)-n_0f_0G^0(q+\mu)\,G^0(-q+\mu)\}\,d^4q.
\end{aligned} \quad (5.1)$$

Momentum conservation $\Sigma p = \Sigma p'$ is assumed to hold everywhere. The q^0-integration in F_h is performed with a detour into the upper half-plane, since this contribution must vanish as $G \to G^0$. Everywhere on the right of Eq. (5.1) the first-approximation value $\mu^{(1)} = n_0 f_0$ should be substituted for μ.

The Σ_{ik} involve special values of the F, together with a factor $\sqrt{n_0}$ for each particle of the condensed phase:

$$\Sigma'_{20}(p+\mu) = n_0 F_a(p-p;\ 00)$$
$$+ n_0 F_b(p-p;\ 00) + n_0 F_e(p0-p;\ 0)$$
$$+ n_0 F_e(0p-p;\ 0) + n_0 F_e(-p0p;\ 0)$$
$$+ n_0 F_e(0-pp;\ 0) + n_0 F_g(p0-p0;)$$
$$+ n_0 F_g(p00-p;) + F_i(p-p;);$$

$$\Sigma'_{02}(p+\mu) = n_0 F_a(00;\ p-p)$$
$$+ n_0 F_b(00;\ p-p) + n_0 F_d(0;\ p-p0)$$
$$+ n_0 F_d(0;\ p0-p)$$
$$+ n_0 F_d(0;\ -pp0) + n_0 F_d(0;\ -p0p)$$
$$+ n_0 F_f(;\ p0-p0)$$
$$+ n_0 F_f(;\ p00-p) + F_k(;\ p-p); \tag{5.2}$$

$$\Sigma'_{11}(p+\mu) = n_0 F_a(p0;\ 0p) + n_0 F_b(p0;\ 0p)$$
$$+ n_0 F_c(p0;\ p0) + n_0 F_c(p0;\ 0p)$$
$$+ n_0 F_d(p;\ 0p0) + n_0 F_d(p;\ 00p)$$
$$+ n_0 F_e(p00;\ p) + n_0 F_e(0p0;\ p) + F_h(p;\ p).$$

To enumerate the vacuum loops which contribute to μ, we must first distinguish one incoming or outgoing particle of the condensed phase (see Section (I, 4)). After this we must sum the loops, counting separately all possible geometric structures and all possible positions of the distinguished particle. The vacuum loops include three types of rectangle, differing in the numbers of incoming and outgoing continuous lines, and corresponding to factors $\Sigma_{11}^{(1)}$, $\Sigma_{02}^{(1)}$ and $\Sigma_{20}^{(1)}$. The distinguished particle of the condensed phase may come out from $\Sigma_{11}^{(1)}$ or from $\Sigma_{02}^{(1)}$. The sums of contributions from graphs of these two types are respectively $-iF_h(0;0)$ and $-iF_i(00;)$. The term in μ arising from all these vacuum loops is thus

$$\mu' = F_h(0;\ 0) + F_i(00;). \tag{5.3}$$

To carry out the q^0-integration in Eq. (5.1), it is convenient to represent G and $\hat{G} = \check{G}$ in the following form,

$$G(q+\mu) = \frac{A_{\mathbf{q}}}{q^0 - \varepsilon_{\mathbf{q}} + i\delta} - \frac{B_{\mathbf{q}}}{q^0 + \varepsilon_{\mathbf{q}} - i\delta};$$
$$\hat{G} = \check{G}(q+\mu)$$
$$= -C_{\mathbf{q}}\left[\frac{1}{q^0 - \varepsilon_{\mathbf{q}} + i\delta} - \frac{1}{q^0 + \varepsilon_{\mathbf{q}} - i\delta}\right], \tag{5.4}$$

with

$$A_{\mathbf{q}} = (\varepsilon_{\mathbf{q}} + \varepsilon_{\mathbf{q}}^0 + n_0 f_0)/2\varepsilon_{\mathbf{q}};$$
$$B_{\mathbf{q}} = (-\varepsilon_{\mathbf{q}} + \varepsilon_{\mathbf{q}}^0 + n_0 f_0)/2\varepsilon_{\mathbf{q}}$$
$$= n_0^2 f_0^2 / 2\varepsilon_{\mathbf{q}}(\varepsilon_{\mathbf{q}} + \varepsilon_{\mathbf{q}}^0 + n_0 f_0);\ C_{\mathbf{q}} = n_0 f_0 / 2\varepsilon_{\mathbf{q}} \tag{5.5}$$

depending only on $|\mathbf{q}|$. The q^0-integrations are now performed and the results substituted into Eq. (5.2) and (5.3). After some manipulations we obtain

$$\Sigma'_{02\,(20)}(p+\mu) = 2n_0 f_0^2 \int d\mathbf{q}\,[(A_{\mathbf{q}};\ B_{\mathbf{k}})$$
$$- (A_{\mathbf{q}} + B_{\mathbf{q}};\ C_{\mathbf{k}}) + 3C_{\mathbf{q}}C_{\mathbf{k}}]$$
$$\times\left(\frac{1}{p^0 - \varepsilon_{\mathbf{q}} - \varepsilon_{\mathbf{k}} + i\delta} - \frac{1}{p^0 + \varepsilon_{\mathbf{q}} + \varepsilon_{\mathbf{k}} - i\delta}\right)$$
$$- f_0 \int d\mathbf{q}\left\{C_{\mathbf{q}} + \frac{n_0 f_0}{2n_0 f_0 - 2\varepsilon_{\mathbf{q}}^0 + i\delta}\right\},$$

$$\Sigma'_{11}(p+\mu) = 2n_0 f_0^2 \int d\mathbf{q}\left\{\frac{(A_{\mathbf{q}};\ B_{\mathbf{k}}) + 2C_{\mathbf{q}}C_{\mathbf{k}} + A_{\mathbf{q}}A_{\mathbf{k}} - 2(A_{\mathbf{q}};\ C_{\mathbf{k}})}{p^0 - \varepsilon_{\mathbf{q}} - \varepsilon_{\mathbf{k}} + i\delta}\right.$$
$$- \frac{(A_{\mathbf{q}};\ B_{\mathbf{k}}) + 2C_{\mathbf{q}}C_{\mathbf{k}} + B_{\mathbf{q}}B_{\mathbf{k}} - 2(B_{\mathbf{q}};\ C_{\mathbf{k}})}{p^0 + \varepsilon_{\mathbf{q}} + \varepsilon_{\mathbf{k}} - i\delta} \tag{5.6}$$
$$\left. - \frac{1}{p^0 + 2n_0 f_0 - \varepsilon_{\mathbf{q}}^0 - \varepsilon_{\mathbf{k}}^0 + i\delta}\right\} + 2f_0 \int d\mathbf{q}\,B_{\mathbf{q}},$$

$$\mu' = 2f_0 \int d\mathbf{q}\,B_{\mathbf{q}} - f_0 \int d\mathbf{q}\left\{C_{\mathbf{q}} + \frac{n_0 f_0}{2n_0 f_0 - 2\varepsilon_{\mathbf{q}}^0 + i\delta}\right\},$$

Here $\mathbf{k} = \mathbf{p} - \mathbf{q}$, and the symbol (;) denotes a symmetrized product, $(A_{\mathbf{q}};\ B_{\mathbf{k}}) = A_{\mathbf{q}}B_{\mathbf{k}} + B_{\mathbf{q}}A_{\mathbf{k}}$. The integrands are all symmetrical in $\mathbf{q}$ and $\mathbf{k}$.

Before we add to Eq. (5.6) the second-order terms from Eq. (4.2), we transform the expression (4.3) for Σ_{11}. Remembering that

$$q^2 + p^2/4 = \varepsilon_{\mathbf{p}/2+\mathbf{q}}^0 + \varepsilon_{\mathbf{p}/2-\mathbf{q}}^0$$

and introducing the new integration variable $\mathbf{q}' = \mathbf{q} + \mathbf{p}/2$, we find that Eq. (4.3) gives to the required approximation

$$\Sigma_{11}(p+\mu) = 2n_0 f_0 + 2n_0 \operatorname{Im} f_s\left(\frac{\mathbf{p}}{2}\,\frac{\mathbf{p}}{2}\right)$$
$$+ 2n_0 f_0^2 \int d\mathbf{q}\left[\frac{1}{p^0 + 2n_0 f_0 - \varepsilon_{\mathbf{q}}^0 - \varepsilon_{\mathbf{k}}^0 + i\delta}\right.$$
$$\left. - \frac{1}{\varepsilon_p^0 - \varepsilon_{\mathbf{q}}^0 - \varepsilon_{\mathbf{k}}^0 + i\delta}\right]. \tag{5.7}$$

The total of all second-order terms in now obtained from Eq. (5.6), (4.3), and (5.7), and after some algebra becomes

$$\mu^{(2)} = 2f_0 \int d\mathbf{q}\, B_{\mathbf{q}} + \frac{1}{2} n_0 f_0^2 \int d\mathbf{q} \left(\frac{1}{\varepsilon_{\mathbf{q}}^0} - \frac{1}{\varepsilon_{\mathbf{q}}} \right),$$

$$\Sigma_{20(02)}^{(2)} (p + \mu) = 2n_0 f_0^2 \int d\mathbf{q}\, [(A_{\mathbf{q}};\ B_{\mathbf{k}}) - (A_{\mathbf{q}} + B_{\mathbf{q}};\ C_{\mathbf{k}}) + 3C_{\mathbf{q}} C_{\mathbf{k}}]$$
$$\times \left(\frac{1}{p^0 - \varepsilon_{\mathbf{q}} - \varepsilon_{\mathbf{k}} + i\delta} - \frac{1}{p^0 + \varepsilon_{\mathbf{q}} + \varepsilon_{\mathbf{k}} - i\delta} \right) + \frac{1}{2} n_0 f_0^2 \int d\mathbf{q} \left(\frac{1}{\varepsilon_{\mathbf{q}}^0} - \frac{1}{\varepsilon_{\mathbf{q}}} \right);$$

$$\Sigma_{11}^{(2)} (p + \mu) = 2n_0 f_0^2 \int d\mathbf{q} \left\{ \frac{(A_{\mathbf{q}};\ B_{\mathbf{k}}) + 2C_{\mathbf{q}} C_{\mathbf{k}} + A_{\mathbf{q}} A_{\mathbf{k}} - 2(A_{\mathbf{q}};\ C_{\mathbf{k}})}{p^0 - \varepsilon_{\mathbf{q}} - \varepsilon_{\mathbf{k}} + i\delta} - \frac{(A_{\mathbf{q}};\ B_{\mathbf{k}}) + 2C_{\mathbf{q}} C_{\mathbf{k}} + B_{\mathbf{q}} B_{\mathbf{k}} - 2(B_{\mathbf{q}};\ C_{\mathbf{k}})}{p^0 + \varepsilon_{\mathbf{q}} + \varepsilon_{\mathbf{k}} - i\delta} + \frac{1}{4} \left(\frac{1}{\varepsilon_{\mathbf{q}}} + \frac{1}{\varepsilon_{\mathbf{k}}} \right) \right\} + 2f_0 \int d\mathbf{q}\, B_{\mathbf{q}} \tag{5.8}$$
$$+ 2n_0 \operatorname{Im} f_s \left(\frac{\mathbf{p}}{2} \frac{\mathbf{p}}{2} \right) - 2n_0 f_0^2 \int d\mathbf{q} \left[\frac{1}{\varepsilon_p^0 - \varepsilon_{\mathbf{q}}^0 - \varepsilon_{\mathbf{k}}^0 + i\delta} + \frac{1}{4\varepsilon_{\mathbf{q}}} + \frac{1}{4\varepsilon_{\mathbf{k}}} \right].$$

The value of $\operatorname{Im} f_S(\mathbf{p}/2\ \mathbf{p}/2)$ can be obtained from Eq. (3.13), and the integrals not involving p^0 can be carried out exactly. In this way Eq. (5.8) becomes

$$\Sigma_{20(02)}^{(2)} (p + \mu) = \frac{1}{2} n_0 f_0^2 \int \frac{dq}{\varepsilon_{\mathbf{q}} \varepsilon_{\mathbf{k}}} R(qk) \left[\frac{1}{p^0 - \varepsilon_{\mathbf{q}} - \varepsilon_{\mathbf{k}} + i\delta} - \frac{1}{p^0 + \varepsilon_{\mathbf{q}} + \varepsilon_{\mathbf{k}} - i\delta} \right] + \frac{1}{\pi^2} \sqrt{n_0 f_0^3}\, n_0 f_0, \tag{5.9}$$

$$\Sigma_{11}^{(2)} (p + \mu) = \frac{1}{2} n_0 f_0^2 \int \frac{dq}{\varepsilon_{\mathbf{q}} \varepsilon_{\mathbf{k}}} \left[\frac{Q^-(qk)}{p^0 - \varepsilon_{\mathbf{q}} - \varepsilon_{\mathbf{k}} + i\delta} - \frac{Q^+(qk)}{p^0 + \varepsilon_{\mathbf{q}} + \varepsilon_{\mathbf{k}} - i\delta} + \varepsilon_{\mathbf{q}} + \varepsilon_{\mathbf{k}} \right] + \frac{8}{3\pi^2} \sqrt{n_0 f_0^3}\, n_0 f_0, \tag{5.10}$$

$$\mu^{(2)} = (5/3\pi^2) \sqrt{n_0 f_0^3} \cdot n_0 f_0, \tag{5.11}$$

with

$$R(qk) = 2\varepsilon_{\mathbf{q}}^0 \varepsilon_{\mathbf{k}}^0 - 2\varepsilon_{\mathbf{q}} \varepsilon_{\mathbf{k}} + n_0^2 f_0^2,$$
$$Q^{\mp}(qk) = 3\varepsilon_{\mathbf{q}}^0 \varepsilon_{\mathbf{k}}^0 - \varepsilon_{\mathbf{q}} \varepsilon_{\mathbf{k}} + n_0 f_0 (\varepsilon_{\mathbf{q}}^0 + \varepsilon_{\mathbf{k}}^0) + n_0^2 f_0^2 \mp [n_0 f_0 (\varepsilon_{\mathbf{q}} + \varepsilon_{\mathbf{k}}) - \varepsilon_{\mathbf{q}} \varepsilon_{\mathbf{k}}^0 - \varepsilon_{\mathbf{k}} \varepsilon_{\mathbf{q}}^0]. \tag{5.12}$$

It is convenient to express the Green's function (4.5) in a form analogous to Eq. (5.4). In this approximation we find

$$G(p + \mu) = \frac{A_p + \alpha_p}{p^0 - \varepsilon_p - \Lambda_p^-} - \frac{B_p + \alpha_p}{p^0 + \varepsilon_p + \Lambda_p^+}, \tag{5.13}$$

where α_p and $\Lambda_p^{\mp}$ are the second-order corrections

$$\alpha_p = \frac{n_0 f_0}{4\varepsilon_{\mathbf{p}}^3} \{ 2\varepsilon_{\mathbf{p}}^0 \Sigma_{20}^{(2)} - n_0 f_0 (\Sigma_{11}^+ + \Sigma_{11}^- - 2\mu - 2\Sigma_{20})^{(2)} \};$$

$$\Lambda_p^{\mp} = \frac{\varepsilon_{\mathbf{p}}^0}{2\varepsilon_{\mathbf{p}}} (\Sigma_{11}^+ + \Sigma_{11}^- - 2\mu)^{(2)} + \frac{n_0 f_0}{2\varepsilon_{\mathbf{p}}} (\Sigma_{11}^+ + \Sigma_{11}^- - 2\mu - 2\Sigma_{20})^{(2)} \pm \frac{1}{2} (\Sigma_{11}^+ - \Sigma_{11}^-)^{(2)}. \tag{5.14}$$

These α_p and $\Lambda_p^{\mp}$ are combinations of the integrals (5.9) and (5.10). In the limits of small and large momentum (compared with $\sqrt{n_0 f_0}$), explicit expressions can be obtained for the functions $\alpha_p = \alpha(p^0; \mathbf{p})$ and $\Lambda_p^{\mp} = \Lambda^{\mp}(p^0; \mathbf{p})$. When these are examined it is found that there are no new poles of the Green's function. We here exhibit the behavior of the Green's function near to the poles $p_0 \approx \pm\epsilon_{\mathbf{p}}$. In this region we may write $|p^0| = \epsilon_{\mathbf{p}}$ in α_p, and we need retain only terms of first order in the difference $(\epsilon_{\mathbf{p}} \mp p^0)$ in $\Lambda_p^{\mp}$. For small momenta ($p \ll \sqrt{n_0 f_0}$) we then find

$$\alpha_p = \sqrt{n_0 f_0^3} \left(\frac{2}{3\pi^2} \frac{n_0 f_0}{\varepsilon_p} + i \frac{1}{64\pi} \frac{\varepsilon_p}{n_0 f_0} \right)$$
$$(p \ll \sqrt{n_0 f_0};\ \varepsilon_p \approx p \sqrt{n_0 f_0}),$$
$$\Lambda_p^{\mp} \equiv \Omega_p + \lambda_p (\varepsilon_{\mathbf{p}} \mp p^0) = \sqrt{n_0 f_0^3} \left(\frac{7}{6\pi^2} \varepsilon_{\mathbf{p}} - i \frac{3}{640\pi} \frac{\varepsilon_p^5}{n_0^4 f_0^4} \right) + (\varepsilon_{\mathbf{p}} \mp p^0) \sqrt{n_0 f_0^3} \left(\frac{1}{2\pi^2} + i \frac{1}{32\pi} \frac{\varepsilon_{\mathbf{p}}^2}{n_0^2 f_0^2} \right). \tag{5.15}$$

For large momenta only the imaginary part of $\Lambda_p^{\mp}$ is important,

$$\Lambda_p^{\mp} = \Omega_p = -\frac{i}{4\pi} p f_0 n_0 f_0 \quad (p \gg \sqrt{n_0 f_0}). \tag{5.16}$$

For small momenta, in virtue of Eq. (5.13) and (5.15), the Green's function near to the poles may be written in the form

$$G(p + \mu) = (1 - \lambda_{\mathbf{p}}) \left[\frac{A_p + \alpha_p}{p^0 - \varepsilon_{\mathbf{p}} - \Omega_p} - \frac{B_p + \alpha_p}{p^0 + \varepsilon_p + \Omega_p} \right]. \tag{5.17}$$

For large momenta, α_p and λ_p may be neglected in Eq. (5.17).

6. QUASI-PARTICLE SPECTRUM AND GROUND-STATE ENERGY

We have already mentioned that the energy of a quasi particle is determined by the value of $p^0(\mathbf{p})$ at a pole of $G(p + \mu)$. Only those poles are to be considered for which the imaginary part of the energy is negative, so that the damping is positive. In the range $p \ll \sqrt{n_0 f_0}$, Eq. (5.15) and (5.17) give

$$\varepsilon = p \sqrt{n_0 f_0} \left(1 + \frac{7}{6\pi^2} \sqrt{n_0 f_0^3} \right) - i \frac{3}{640\pi} \sqrt{n_0 f_0^3} \frac{p^5}{(n_0 f_0)^{3/2}} \quad (p \ll \sqrt{n_0 f_0}), \tag{6.1}$$

In the high-momentum range, according to Eq. (5.16), we have

$$\varepsilon = \varepsilon_p^0 + n_0 f_0 \left(1 - \frac{i}{4\pi} p f_0\right) \approx \varepsilon_p^0$$

$$+ n_0 f\,(\mathbf{pp}) \quad (p \gg \sqrt{n_0 f_0}). \tag{6.2}$$

Equation (6.1) shows that for small p the quasi particles are phonons. The second approximation gives a correction to the sound velocity, and a damping proportional to p^5 which is connected with a process of decay of one phonon into two. In the high-momentum range, the second approximation gives a damping which is related to the imaginary part of the forward scattering amplitude, and so to the total cross section.

In Sec. (I, 7) we found connections between the Green's function and various physical properties of the system. The mean number of particles $\overline{N}_p$ with a given momentum **p** in the ground state of the system is related to the residue of the Green's function at its upper pole,

$$\overline{N}_{\mathbf{p}} = i \int G\, dp^0 / 2\pi = (B_{\mathbf{p}} + \alpha_p)(1 - \lambda_p). \tag{6.3}$$

When $p \ll \sqrt{n_0 f_0}$, Eqs. (5.15) and (5.5) give

$$\overline{N}_{\mathbf{p}} = \frac{n_0 f_0}{2\varepsilon_p}\left(1 + \frac{5}{6\pi^2}\sqrt{n_0 f_0^3}\right) \tag{6.4}$$

The imaginary parts of α_p and λ_p here cancel, as they should. To find the total number of particles with $\mathbf{p} \neq 0$, we need to know $\overline{N}_{\mathbf{p}}$ for all momenta. We therefore use only the first approximation formula for $\overline{N}_{\mathbf{p}}$, namely $\overline{N}_{\mathbf{p}} = B_{\mathbf{p}}$. For the density of particles with $\mathbf{p} \neq 0$ we find

$$n - n_0 = i \int G\,(p + \mu)\, d^4 p$$

$$= \int B_{\mathbf{p}}\, d\mathbf{p} = \sqrt{n_0 f_0^3}\, n_0 / 3\pi^2. \tag{6.5}$$

Equation (6.5) gives the relation between the density n_0 of particles in the condensed phase, which appeared as a parameter in all our equations, and the total number of particles in the system.

We note here one important point. It can be seen from the way the calculations were done that the validity of the "gaseous" approximation requires that n_0 be small. It is not directly required that the total density n be small, since n does not appear explicitly in the problem. But Eq. (6.5) shows that when n_0 is small n is necessarily small, too. This means that it is not possible to decrease significantly the density of the condensed phase by increasing the interaction or the total density, so long as $n_0 \ll f_0^{-3}$. This result confirms and strengthens the assertion made in I that the condensed phase does not disappear when interactions are introduced.

We can calculate the ground-state energy from the chemical potential μ. By Eq. (4.4) and (5.11),

$$\mu = n_0 f_0 \left(1 + \frac{5}{3\pi^2}\sqrt{n_0 f_0^3}\right), \tag{6.6}$$

Expressing n_0 in terms of n by means of Eq. (6.5), we have in the same approximation

$$\mu = n f_0 \left(1 + \frac{4}{3\pi^2}\sqrt{n f_0^3}\right). \tag{6.7}$$

By definition we have $\mu = \frac{\partial}{\partial n}\left(\frac{E_0}{V}\right)$. Therefore, integrating Eq. (6.7) with respect to n, we obtain the ground-state energy

$$\frac{E_0}{V} = \frac{1}{2} n^2 f_0 \left(1 + \frac{16}{15\pi^2}\sqrt{n f_0^3}\right), \tag{6.8}$$

This coincides with the result of Lee and Yang[3] for the hard-sphere gas, if we remember that in that case $f_0 = 4\pi a$.

The condition for the system to be thermodynamically stable is $\partial P/\partial V = -\partial^2 E/\partial V^2 < 0$. This condition reduces to $f_0 > 0$. Our results are only meaningful when this condition is satisfied.

7. POSSIBILITY OF HIGHER APPROXIMATIONS

In the first two approximations, all the results can be expressed in terms of the amplitudes f. Thus the problem of many interacting particles is reducible to the problem of two particles.

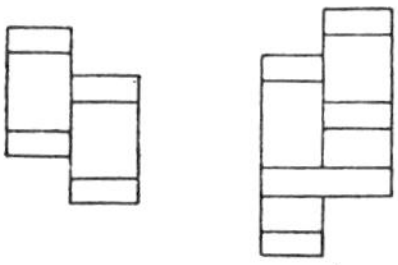

a b

FIG. 4

In the next approximation we must consider contributions to Σ_{ik} proportional to $n_0 f_0^3$. Among other graphs, we must include the "triple ladders" illustrated in Fig. 4. The integrals arising from graphs of this type diverge at high momenta and become finite only when the momentum dependence of f is taken into account. For an estimate we may cut the integrals off at a momentum $p \sim f_0^{-1}$. We see then that an increase in the number of "rungs" does not change the order of magnitude of the integral. In fact, each rung adds a factor $f_0 G^2$ and an integration over one momentum 4-vector. For a rough estimate we take $q^0 \sim q^2$, $G \sim q^{-2}$, and find

$$\int f_0 G^2 d^4q \sim f_0 \int dq \sim 1,$$

Therefore we have to consider simultaneously all such graphs with any number of rungs. The totality of these triple ladders describes completely the interaction of three particles. Therefore the sum of contributions from such graphs can be expressed only by means of three-particle amplitudes.

In the third approximation (terms proportional to $n_0 f_0^3$) we thus require a solution of the three-particle problem (see also Ref. 4). Since the problem of three strongly interacting particles is in general insoluble, the higher approximations to the many-particle problem are physically meaningless.

8. HIGH EXCITATIONS ($pf_0 \sim 1$) IN A HARD-SPHERE GAS

For the high-energy excitations, the momentum dependence of the amplitudes becomes important. We therefore consider as an example the case of a gas of hard spheres of radius ($a/2$). We also consider only the first approximation in the density expansion, i.e., we use Eq. (4.6). The amplitude $f(\mathbf{p}\,0)$ can be computed exactly from Eq. (3.10). For $f_s(\mathbf{p}/2\ \mathbf{p}/2)$ we consider only s-waves. The higher waves (the symmetrized amplitude involves only even values of ℓ) add a numerically unimportant contribution. For example the d-waves at $pa \sim 1$ contribute about 10 per cent. We substitute into Eq. (4.7) the values of the amplitudes

$$f(\mathbf{p}0) = 4\pi \frac{\sin pa}{p}; \quad f_s\left(\frac{\mathbf{p}}{2}\frac{\mathbf{p}}{2}\right) = \frac{8\pi}{p} \sin\frac{pa}{2}\, e^{-ipa/2}, \tag{8.1}$$

and obtain for the quasi-particle energy

$$\varepsilon = \left[\left(\frac{p^2}{2} + 8\pi n_0 \frac{\sin pa}{p} - 4\pi n_0 a\right)^2 - 16\pi^2 n_0^2 \frac{\sin^2 pa}{p^2}\right]^{1/2}, \tag{8.2}$$

At high momenta this becomes

$$\varepsilon \approx \frac{p^2}{2} + 4\pi n_0 a \left(2\frac{\sin pa}{pa} - 1\right). \tag{8.3}$$

The second term in Eq. (8.3) changes sign at $pa \approx 1.9$. An oscillating component is superimposed on the usual parabolic dependence. This oscillation will not be important since the magnitude of the term is small; when $pa \sim 1$ it is of relative order $n_0 a^3$. However, if one formally allows the parameter $n_0 a^3$ to become larger in Eqs. (8.3) or (8.2), the second term of Eq. (8.3) produces an increasing departure of the dispersion law from the parabolic form, until at sufficiently high densities there appears first a point of inflection and finally a maximum and a minimum in the curve. The spectrum then resembles qualitatively the spectrum postulated by L. D. Landau[5] to explain the properties of liquid helium II. This extrapolation is certainly unwarranted. But it allows one to suppose that the difference between liquid helium and a non-ideal Bose gas is only a quantitative one, and that no qualitatively new phenomena arise in the transition from gas to liquid.

9. CONCLUSION

We summarize the main features of the approximation which we have studied.

(1) The interaction between particles is specified not by a potential but by an exact scattering amplitude. This allows us to deal with strong interactions. After the potential has been replaced by the amplitude, it is possible to make a perturbation expansion in powers of the amplitude, or more precisely in powers of $\sqrt{n_0 f_0^3}$.

(2) We make a series expansion not of the quasiparticle energy (this appears as the denominator of the Green's function), but of the effective interaction potentials Σ_{ik} and the chemical potential μ. The formula giving the Green's function in terms of Σ_{ik} and μ is exact.

From Eq. (4.7) and (4.9) we see that ϵ_p can be expanded in powers of f only for high-momentum excitations with $p \gg \sqrt{n_0 f_0}$. The low-lying excitations of the system are in principle impossible to obtain by perturbation theory. For this reason, the expression obtained by Huang and Yang[4] for the energy of the low excitations of a Bose hard-sphere gas is incorrect. They used perturbation theory with a "pseudopotential," and their result agrees with a formal expansion of Eq. (4.7) in powers of f_0.

In conclusion I wish to thank A. B. Migdal and especially V. M. Galitskii for fruitful discussions, and also L. D. Landau for criticism of the results.

[1] S. T. Beliaev, J. Exptl. Theoret. Phys. (U.S.S.R.) 34, 417 (1958); Soviet Phys. JETP **7**, 289 (1958) (this issue).

[2] V. M. Galitskii and A. B. Migdal, J. Exptl. Theoret. Phys. (U.S.S.R.) **34**, 139 (1958); Soviet Phys. JETP **7**, 96 (1958).

[3] T. D. Lee and C. N. Yang, Phys. Rev. **105**, 1119 (1957).

[4] K. Huang and C. N. Yang, Phys. Rev. **105**, 767 (1957).

[5] E. M. Lifshitz, Usp. Fiz. Nauk **34**, 512 (1948).

Translated by F. J. Dyson
78

Ground-State Energy and Excitation Spectrum of a System of Interacting Bosons

N. M. HUGENHOLTZ AND D. PINES*
The Institute for Advanced Study, Princeton, New Jersey
(Received May 4, 1959)

In this paper properties of a boson gas at zero temperature are investigated by means of field-theoretic methods. Difficulties arising from the depletion of the ground state are resolved in a simple way by the elimination of the zero-momentum state. The result of this procedure when applied to the calculation of the Green's functions of the system is identical to that of Beliaev. It is then shown generally that for a repulsive interaction the energy $E(\mathbf{k})$ of a phonon of momentum $\mathbf{k}$, which is found as the pole of a one-particle Green's function, approaches zero for zero momentum, which means that the phonon spectrum does not exhibit an energy gap.

The Green's function method is applied to the calculation of the properties of a low-density boson gas. The next order term beyond that calculated by Lee and Yang, and Beliaev for the ground-state energy is obtained and the general form of the series expansion is found to be

$$(E_0/\Omega)=\tfrac{1}{2}n^2f_0[1+a(nf_0^3)^{\frac{1}{2}}+b(nf_0^3)\ln nf_0^3+c(nf_0^3)+d(nf_0^3)^{\frac{3}{2}}\ln(nf_0^3)+\cdots],$$

where n is the density and f_0 is the scattering length for the assumed two-body interaction between the bosons. The coefficients a and b are independent of the shape of the interaction, and are the only terms thus far calculated. The coefficient b is in agreement with the hard-sphere gas calculations of Wu and of Sawada.

A discussion is given of the intermediate-density calculation of Brueckner and Sawada, and certain possible improvements in the method of summing a selected set of higher-order terms are proposed.

1. INTRODUCTION

THE realization that there exists a great formal similarity between the quantum theory of a large number of interacting Fermi particles and quantum field theory has led in recent years to the development of new methods for the treatment of such a fermion gas,[1] in particular at zero temperature.

The application of similar methods to a system of particles obeying Bose statistics gives rise to two difficulties of a different nature. The first difficulty has to do with the particular role played by the large number of particles of momentum zero. In the noninteracting system all particles have zero momentum. In the interacting system the zero-momentum state likewise contains very many particles, since only a finite fraction of these is excited as a consequence of the interaction. The fraction of particles of nonzero momentum in the ground state of the interacting system is a function of the density and is very small for low density. Hence for low density this so-called "depletion" of the ground state can be neglected, as in the work of Bogoliubov[2] and in the pseudopotential method of Lee, Huang, and Yang.[3] However, for calculations of the energies of the ground state and of low-lying excited states going beyond the extreme low density case, the depletion effect must be taken into account.

Another difficulty, which also is absent in the fermion case, is the fact that even for a regular, repulsive interaction perturbation theory diverges Bogoliubov has shown how in the case of very weak interaction and low density the divergences can be removed by means of a canonical transformation. This same procedure was also used by Lee, Huang, and Yang in their pseudopotential method.

Recently Beliaev[4] developed a method which enables one to take into account the depletion effect rigorously and which furthermore leads to a formulation which to all orders is free of divergences. In this method essential use is made of the Green's functions, which are well-known in field theory. From these functions both the phonon spectrum and the energy E_0 of the ground state can be obtained.

In the present paper we present in the first place another, and to our opinion, simpler and more transparent treatment of the depletion effect, in which we do not make use of any form of perturbation theory. This forms the content of Sec. 3, which follows a rather extensive discussion of the difficulties in Sec. 2.

In Sec. 4 we introduce the one-particle Green's functions. We follow Beliaev and obtain a closed expression for the Green's functions in terms of two functions Σ_{11} and Σ_{20} which are the analog of the proper self-energy parts in field theory. This procedure involves a partial summation of the perturbation series expansion of the Green's functions and is sufficient to remove all low-momentum divergences. One then has a consistent scheme where both difficulties, mentioned above, have been resolved. We use this scheme in Sec. 6 to derive a quite general relationship between the chemical potential $\mu=dE_0/dN$ and the functions Σ_{11} and Σ_{20}, both for zero momentum and energy. This relation permits one to prove that the phonon energy is equal to zero for zero momentum. That some calculations give rise to an

* Present address: Department of Physics and Department of Electrical Engineering, University of Illinois, Urbana, Illinois.

[1] J. Goldstone, Proc. Roy. Soc. (London) **A239**, 267 (1957). N. M. Hugenholtz, Physica **23**, 481 (1957). V. M. Galitskii and A. B. Migdal, J. Exptl. Theoret. Phys. U.S.S.R. **34**, 139 (1958) [translation: Soviet Phys. JETP **7**, 96 (1958)]. See also A. Klein and R. Prange (to be published).

[2] N. N. Bogoliubov, J. Phys. U.S.S.R. **9**, 23 (1947).

[3] Lee, Huang, and Yang, Phys. Rev. **106**, 1135 (1957).

[4] S. T. Beliaev, J. Exptl. Theoret. Phys. U.S.S.R. **34**, 417 (1958) [translation: Soviet Phys. JETP **7**, 289 (1958)].

energy gap in the phonon spectrum[5] is due either to an incorrect treatment of the depletion effect, or to an inconsistent treatment of some of the terms in the interaction.

In Sec. 5 we give a short discussion of more general Green's functions and their significance for treating the scattering of neutrons by a boson gas.

Sections 7 and 8 are devoted to the calculation of the properties of the low-density boson gas. In Sec. 8 it is shown that, in agreement with Sawada[6] and Wu,[7] there exists a term in the expansion for E_0/N of the form $n^3 f_0^4 \ln(n f_0^3)$, the coefficient of which is calculated. Finally in Sec. 9, we discuss the general form of the series expansion, and, somewhat briefly, the intermediate-density theory of the hard-sphere gas due to Brueckner and Sawada.[9]

2. DIFFICULTIES WITH DEPLETION OF THE GROUND STATE

The system under consideration consists of N interacting particles enclosed in a cubic box of volume Ω. We assume these particles to obey Bose statistics. The Hamiltonian can be written as $H=H_0+V$, in which the kinetic energy H_0 and the interaction V have the usual form in second quantization[9]:

$$H_0=\sum_{\mathbf{k}} \tfrac{1}{2}k^2 a_{\mathbf{k}}^* a_{\mathbf{k}},$$

$$V=\tfrac{1}{4}\Omega^{-1} \sum_{\mathbf{k}_1\mathbf{k}_2\mathbf{k}_3\mathbf{k}_4} [v(\mathbf{k}_1-\mathbf{k}_3)+v(\mathbf{k}_1-\mathbf{k}_4)] \times \delta_{\mathrm{Kr}}(\mathbf{k}_1+\mathbf{k}_2-\mathbf{k}_3-\mathbf{k}_4) a_{\mathbf{k}_1}^* a_{\mathbf{k}_2}^* a_{\mathbf{k}_3} a_{\mathbf{k}_4}. \tag{2.1}$$

The operators $a_{\mathbf{k}}^*$ and $a_{\mathbf{k}}$ are creation and annihilation operators, satisfying the commutation relations

$$[a_{\mathbf{k}},a_{\mathbf{l}}]=[a_{\mathbf{k}}^*,a_{\mathbf{l}}^*]=0; \quad [a_{\mathbf{k}},a_{\mathbf{l}}^*]=\delta_{\mathbf{kl}}.$$

The function $v(\mathbf{k})$ is the Fourier transform of the central two-body interaction

$$v(\mathbf{k})=\int d^3x\, v(\mathbf{x}) e^{-i\mathbf{k}\cdot\mathbf{x}}.$$

The Kronecker symbol δ_{Kr} is equal to one if the argument is zero, and zero otherwise.

Since we are interested in the limiting case in which both N and Ω are infinite, with a finite particle density $n=N/\Omega$, we find it convenient to use another notation, which is more suitable for that case. We define

$$\int_{\mathbf{k}} = (2\pi)^3\Omega^{-1}\sum_{\mathbf{k}}, \quad \delta^3(\mathbf{k}-\mathbf{k}')=(2\pi)^{-3}\Omega\delta_{\mathrm{Kr}}(\mathbf{k}-\mathbf{k}'),$$

$$\xi_{\mathbf{k}}=\Omega^{\frac{1}{2}}(2\pi)^{-\frac{3}{2}}a_{\mathbf{k}}.$$

The commutation relations for the ξ-operators are

$$[\xi_{\mathbf{k}},\xi_{\mathbf{l}}]=[\xi_{\mathbf{k}}^*,\xi_{\mathbf{l}}^*]=0; \quad [\xi_{\mathbf{k}},\xi_{\mathbf{l}}^*]=\delta^3(\mathbf{k}-\mathbf{l}).$$

In the limit of an infinite system, $\delta^3(\mathbf{k}-\mathbf{l})$ is the Dirac δ-function and the symbol $\int_{\mathbf{k}}$ is replaced by the integration sign $\int d^3k$. In our new notation the Hamiltonian becomes

$$H_0=\int_{\mathbf{k}} \tfrac{1}{2}k^2 \xi_{\mathbf{k}}^*\xi_{\mathbf{k}},$$

and

$$V=\tfrac{1}{4}(2\pi)^{-3}\int_{\mathbf{k}_1\mathbf{k}_2\mathbf{k}_3\mathbf{k}_4} [v(\mathbf{k}_1-\mathbf{k}_3)+v(\mathbf{k}_1-\mathbf{k}_4)] \times\delta^3(\mathbf{k}_1+\mathbf{k}_2-\mathbf{k}_3-\mathbf{k}_4)\xi_{\mathbf{k}_1}^*\xi_{\mathbf{k}_2}^*\xi_{\mathbf{k}_3}\xi_{\mathbf{k}_4}.$$

It is our purpose to calculate the energies of the ground state and of low-lying excited states of this system, all at zero temperature. Since practically all available methods for the treatment of such problems are based on some form of perturbation theory (in which the interaction between the particles is considered as the perturbation) let us first consider the noninteracting system. Here we notice a marked difference with the fermion gas, a difference which, as we shall see, gives rise to definite complications in any treatment of the boson gas. In the ground state $|\phi_0\rangle$ of the noninteracting system all particles have zero momentum. Consequently in any form of perturbation theory this zero-momentum state will play a role different from the other single-particle states. It is therefore convenient to rewrite the interaction Hamiltonian in a form in which all terms with one or more $\mathbf{k}$'s equal to zero are written separately. When we do this, we find

$$H=H_0+V_a+V_b+V_c+V_d+V_e+V_f+V_g, \tag{2.2}$$

where

$$\begin{aligned}
V_a&=\tfrac{1}{4}(2\pi)^{-3}\int'_{\mathbf{k}_1\cdots\mathbf{k}_4}[v(\mathbf{k}_1-\mathbf{k}_3)+v(\mathbf{k}_1-\mathbf{k}_4)] \times\delta^3(\mathbf{k}_1+\mathbf{k}_2-\mathbf{k}_3-\mathbf{k}_4)\xi_{\mathbf{k}_1}^*\xi_{\mathbf{k}_2}^*\xi_{\mathbf{k}_3}\xi_{\mathbf{k}_4},\\
V_b&=\tfrac{1}{2}(2\pi)^{-3}a_0(2\pi)^{\frac{3}{2}}\Omega^{-\frac{1}{2}}\int'_{\mathbf{k}_1\mathbf{k}_2\mathbf{k}_3}[v(\mathbf{k}_1)+v(\mathbf{k}_2)] \times\delta^3(\mathbf{k}_1+\mathbf{k}_2-\mathbf{k}_3)\xi_{\mathbf{k}_1}^*\xi_{\mathbf{k}_2}^*\xi_{\mathbf{k}_3},\\
V_c&=\tfrac{1}{2}(2\pi)^{-3}a_0^*(2\pi)^{\frac{3}{2}}\Omega^{-\frac{1}{2}}\int'_{\mathbf{k}_2\mathbf{k}_3\mathbf{k}_4}[v(\mathbf{k}_3)+v(\mathbf{k}_4)] \times\delta^3(\mathbf{k}_2-\mathbf{k}_3-\mathbf{k}_4)\xi_{\mathbf{k}_2}^*\xi_{\mathbf{k}_3}\xi_{\mathbf{k}_4},\\
V_d&=\tfrac{1}{4}(2\pi)^{-3}a_0^2(2\pi)^3\Omega^{-1}\int'_{\mathbf{k}_1\mathbf{k}_2}[v(\mathbf{k}_1)+v(\mathbf{k}_2)] \times\delta^3(\mathbf{k}_1+\mathbf{k}_2)\xi_{\mathbf{k}_1}^*\xi_{\mathbf{k}_2}^*,\\
V_e&=\tfrac{1}{4}(2\pi)^{-3}a_0^{*2}(2\pi)^3\Omega^{-1}\int'_{\mathbf{k}_3\mathbf{k}_4}[v(\mathbf{k}_3)+v(\mathbf{k}_4)] \times\delta^3(\mathbf{k}_3+\mathbf{k}_4)\xi_{\mathbf{k}_3}\xi_{\mathbf{k}_4},\\
V_f&=(2\pi)^{-3}a_0^*a_0(2\pi)^3\Omega^{-1}\int'_{\mathbf{k}_2\mathbf{k}_3}[v(\mathbf{k}_2)+v(0)] \times\delta^3(\mathbf{k}_2-\mathbf{k}_3)\xi_{\mathbf{k}_2}^*\xi_{\mathbf{k}_3},\\
V_g&=\tfrac{1}{2}(2\pi)^{-3}a_0^{*2}a_0^2(2\pi)^6\Omega^{-2}\delta^3(0)v(0).
\end{aligned} \tag{2.3}$$

---

[5] M. Girardeau and R. Arnowitt, Phys. Rev. **113**, 755 (1959). See also S. Butler and J. Valatin, Nuovo cimento **10**, 37 (1958).
[6] K. Sawada (to be published).
[7] T. T. Wu (to be published).
[8] K. A. Brueckner and K. Sawada, Phys. Rev. **106**, 1117 (1957).
[9] We choose such units that $\hbar=M=1$.

Here the primed summation symbols mean that the summation is extended over all $\mathbf{k}\neq 0$.

The operators a_0^* and a_0 always appear together with a factor $\Omega^{-\frac{1}{2}}$ in the Hamiltonian (2.2). At first sight one might suppose therefore that the operators $a_0^*\Omega^{-\frac{1}{2}}$ and $a_0\Omega^{-\frac{1}{2}}$ could be neglected. However, because the state of zero momentum contains a large number of particles, this is not the case; thus $a_0^*a_0\Omega^{-1}|\phi_0\rangle = n|\phi_0\rangle$, if $|\phi_0\rangle$ is the unperturbed state of the system, corresponding to N particles with momentum zero.

Let us now see the way in which the operators a_0^* and a_0 prevent the immediate application to the boson problem of methods which have been very successful in the case of fermions. These methods possess the common feature that the ground state of the system is considered as the analog of the vacuum in field theory; that is, the ground state of the noninteracting system is then defined by the condition that all annihilation operators applied to that state give zero. In the case of the Fermi gas this means that neither holes nor additional particles are present. For the Bose gas a similar situation does not exist. Although in the unperturbed ground state there are no particles of momentum $\mathbf{k}\neq 0$, a large number N of particles has momentum zero. Hence the noninteracting ground state cannot be considered as vacuum with respect to the operators a_0^* and a_0.

This leads to serious difficulties, which become apparent as soon as one starts calculating the ground-state energy, using perturbation theory. The energy of a gas of interacting Fermi particles can be expressed as a power series in the interaction V, and the various contributions can be represented in terms of diagrams. As shown by Goldstone,[1] an expression for the ground-state energy may be derived, in which only connected diagrams appear. The derivation depends upon a rule which makes it possible to express the contributions of disconnected diagrams in terms of their connected parts.

For the boson gas one can likewise represent the various terms arising from the power series expansion of V in terms of diagrams. The difference lies in the fact that, when one calculates the contribution from a given diagram, the resulting expression is multiplied by the expectation value for the unperturbed ground state of products of $a_0\Omega^{-\frac{1}{2}}$ and $a_0^*\Omega^{-\frac{1}{2}}$, arising from the various terms in V. For instance, for the fourth-order diagram of Fig. 1(a), where the dashed lines represent particles of zero momentum, this factor is equal to $\Omega^{-3}\langle\phi_0|a_0^{*3}a_0^3|\phi_0\rangle = \Omega^{-3}N(N-1)(N-2)$. For any connected diagram the contribution for large systems is asymptotically proportional to Ω, so that in the foregoing expression one may replace the factors $N-1$ and $N-2$ by N, the neglected terms being of relative order N^{-1}. This approximation amounts to replacing the operators a_0^* and a_0 by the c-number $N^{\frac{1}{2}}$.

In contributions from disconnected diagrams this replacement would lead to incorrect results. Consider the disconnected fourth-order diagrams of Fig. 1(b) and (c).

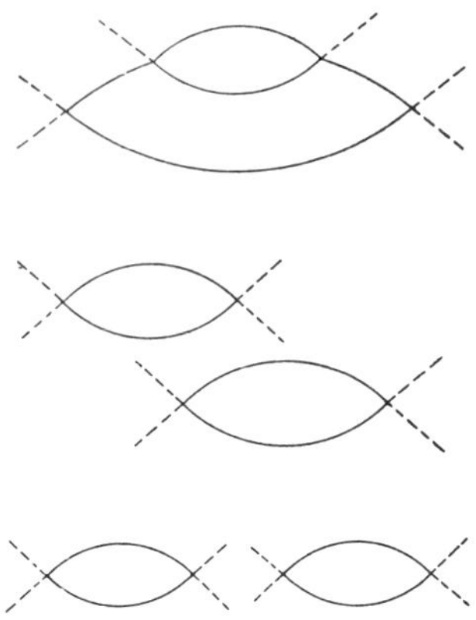

FIG. 1. Three fourth-order ground-state diagrams; diagram a is connected, b and c disconnected. The dashed lines refer to annihilation and creation of particles with zero momentum.

They give contributions of order Ω^2;[10] therefore a term of relative order N^{-1} can no longer be neglected. Thus the diagrams b and c differ not only in the fact that the energies of the intermediate states which appear are different (as is the case for analogous diagrams in the fermion problem) but also in that the operators a_0 and a_0^* appear in different orders, contributing in the one case a factor $N(N-1)(N-2)(N-3)$ and in the other $N^2(N-1)^2$. It is this latter difference which renders invalid the theorem on the contribution of disconnected diagrams and makes a linked cluster expansion impossible for the boson gas.

For the case of extreme low density, the correction terms we have discussed may be neglected, since they lead to higher powers of the density $n=N/\Omega$. In that case one is justified in replacing the operators a_0 and a_0^* by the c-number $N^{\frac{1}{2}}$, a procedure well known from the work of Bogoliubov.[2] Then also the theorem on the disconnected diagrams and hence the Goldstone formula are valid. But for cases where n is not small, the depletion of the ground state spoils the validity of both.

3. TREATMENT OF THE DEPLETION EFFECT

Two essentially different ways are open to resolve the difficulties related to the depletion of the zero-momentum state discussed in the preceding section. One possibility is to carry through all necessary calculations, treating the operators a_0^* and a_0 exactly, until a stage has been reached in which only Ω-independent expressions appear. In such expressions one is justified in neglecting terms of order Ω^{-1}, a procedure which leads to great simplifications. This is the basis for the Green's function approach of Beliaev.[4] Recently Sawada[6] has handled the problem along these lines, with the aid of ordinary time-independent perturbation theory. In both

[10] See, for instance, N. M. Hugenholtz, reference 1.

methods one runs into considerable complications in the derivation of Ω-independent expressions.

We propose another and simpler method of dealing with the depletion of the zero-momentum state; we use a Lagrangian multiplier technique to eliminate this state at the outset. Our method amounts to a generalization of the original argument of Bogoliubov[2] concerning the role played by a_0 and a_0^*, and forms the natural extension of his argument to finite densities.

We remark that, no matter what the density is, the commutator of the operators $a_0^*\Omega^{-\frac{1}{2}}$ and $a_0\Omega^{-\frac{1}{2}}$ is equal to Ω^{-1}, and therefore vanishes for an infinitely large system. Furthermore, these operators commute with all other operators in the problem, so that in this limit they can be considered as c-numbers. The operator $a_0^*a_0/\Omega$, the density n_0 of particles of zero momentum, is then likewise a c-number. Thus it seems natural to replace the operators $a_0\Omega^{-\frac{1}{2}}$ and $a_0^*\Omega^{-\frac{1}{2}}$ by a c-number $n_0^{\frac{1}{2}}$. The variable n_0 is to be determined by the properties of the interacting system, in a way which will be discussed below. It will turn out that for low densities, n_0 is approximately equal to n, the particle density, so that the foregoing procedure then reduces to that of Bogoliubov.

Let us see what happens when one replaces $a_0\Omega^{-\frac{1}{2}}$ and $a_0^*\Omega^{-\frac{1}{2}}$ by $n_0^{\frac{1}{2}}$. The zero-momentum state then simply disappears from the problem. The new Hamiltonian is

$$H(n_0)=H_0+V(n_0). \tag{3.1}$$

Now the number of particles is no longer conserved, since $V(n_0)$ contains terms which do not commute with the operator $N'=\Sigma' a_k^*a_k$. However N' is still approximately a good quantum number, in that it commutes with H to order Ω^{-1}. Thus

$$N'H=HN'[1+O(\Omega^{-1})].$$

Our original problem was to determine the ground state of the system of N interacting Bose particles, that is, that eigenstate of the total Hamiltonian H which has the lowest eigenvalue E_0, subject to the condition that the number of particles is equal to N. In the modified problem, in which the momentum-state zero has been eliminated, we must therefore impose the subsidiary condition that

$$\langle N'\rangle=N-n_0\Omega. \tag{3.2}$$

The variable n_0 must be determined in such a way that the energy we find is minimal.

In a theory in which N' commutes rigorously with H, the subsidiary condition (3.2) could be satisfied most easily by imposing it on the unperturbed wave functions. It would then be automatically satisfied for the true wave function, since N' would then also commute with the operator e^{-iHt}, which enters when one wants to describe the transition from the unperturbed to the perturbed wave function. In the present case, N' commutes with H only to order Ω^{-1}; since e^{-iHt} has matrix elements which contain arbitrarily high powers of the volume Ω (due to disconnected diagrams) a correction term of order Ω^{-1} cannot be neglected. Hence we cannot satisfy the subsidiary condition by imposing it on the unperturbed states and we must turn to another method for satisfying (3.2). To that purpose we use the method of the undetermined multiplier.

We first remark that the ground state of the Hamiltonian (3.1) with the subsidiary condition (3.2) is also the ground state of the Hamiltonian

$$H'=H(n_0)-\mu N', \tag{3.3}$$

without any subsidiary condition. Clearly now the ground-state wave function $|\psi_0(n_0,\mu)\rangle$, and thus also the expectation values $E_0'(n_0,\mu)$, $E_0(n_0,\mu)$ and $N'(n_0,\mu)$ of H', H, and N', respectively, depend on the parameter μ, which is determined by the condition

$$n'(n_0,\mu)=n-n_0. \tag{3.4}$$

This relation expresses, for fixed n, the parameter μ in terms of n_0. As said before n_0 is determined by the condition that, again for n fixed,

$$\frac{d}{dn_0}\left(\frac{E_0}{\Omega}\right)=0. \tag{3.5}$$

Using (3.4) and (3.5) we may derive two useful equations for μ. From the observation that $|\psi_0(n_0,\mu)\rangle$ is the ground-state wave function of H' in (3.3) we conclude that the expectation value of H' for the wave function $|\psi_0(n_0',\mu')\rangle$ has a minimum for $n_0'=n_0$ and $\mu'=\mu$, and hence

$$\frac{\partial}{\partial n_0}\left(\frac{E_0'}{\Omega}\right)=\left\langle\psi_0\left|\frac{d}{dn_0}\left(\frac{V}{\Omega}\right)\right|\psi_0\right\rangle, \tag{3.6}$$

and

$$\frac{\partial}{\partial \mu}\left(\frac{E_0}{\Omega}\right)-\mu\frac{\partial n'}{\partial \mu}=0. \tag{3.7}$$

Keeping n fixed and using (3.4) and (3.7), one finds easily

$$\frac{d}{dn_0}\left(\frac{E_0}{\Omega}\right)=\frac{\partial}{\partial n_0}\left(\frac{E_0'}{\Omega}\right)-\mu,$$

so that

$$\mu=\frac{\partial}{\partial n_0}\left(\frac{E_0'}{\Omega}\right). \tag{3.8}$$

The second equation we obtain by noticing that E_0/Ω is a function of n, since μ, which is a solution of (3.4), is a function of n_0 and n. By virtue of (3.5) we have

$$\frac{d}{dn}\left(\frac{E_0}{\Omega}\right)=\frac{\partial}{\partial \mu}\left(\frac{E_0}{\Omega}\right)\cdot\left(\frac{\partial \mu}{\partial n}\right)_{n_0}=-\frac{\partial}{\partial \mu}\left(\frac{E_0}{\Omega}\right)\bigg/\frac{\partial n'}{\partial \mu},$$

which with (3.7) reduces to

$$\frac{d}{dn}\left(\frac{E_0}{\Omega}\right)=\mu. \tag{3.9}$$

We thus arrive at the following procedure. Instead of the Hamiltonian $H(n_0)$ we consider H', given by Eq. (3.3), in which the kinetic energies of the particles $k^2/2$ are replaced by $k^2/2-\mu$, with $\mu=dE_0/dN$ playing the role of a potential. We then calculate the ground-state energy $E_0'(n_0,\mu)$ of H', the parameter n_0 being determined by (3.4).

Our original problem of calculating the ground-state energy of a system of N interacting boson is thereby reduced to the mathematically simpler problem of finding the smallest eigenvalue of the Hamiltonian $H'=H_0'+V(n_0)$, in which

$$H_0'=\int_{\mathbf{k}} \xi_{\mathbf{k}}^*\xi_{\mathbf{k}}(\tfrac{1}{2}k^2-\mu),$$

and $V(n_0)$ is a sum of terms, which are obtained from (2.3) by replacing the operators $a_0\Omega^{-\frac{1}{2}}$ and $a_0^*\Omega^{-\frac{1}{2}}$ by $n_0^{\frac{1}{2}}$.

It should be emphasized that here all difficulties connected with the zero-momentum state are absent, since this state has been eliminated. Hence, in this respect, we are now free to use the same methods, which are applied successfully to a gas of fermions, or to field theory. In particular, one can now use the linked cluster expansion of Goldstone[1]:

$$E_0'=\left\langle 0\left|\left[V-V\frac{1}{H_0'}V+V\frac{1}{H_0'}V\frac{1}{H_0'}V-\cdots\right]_C\right|0\right\rangle. \quad (3.10)$$

Here $|0\rangle$ is the vacuum state, which now replaces the state $|\phi_0\rangle$ with N particles of zero momentum in the original problem. The subscript C means that only connected diagrams contribute to the expression.

We will end this section by making a comment on the number of zero-momentum particles in the ground state of the interacting system. It is very easy to show that, as long as perturbation methods are applicable, a finite fraction of particles always occupies the zero-momentum state. Suppose namely that the interaction be such that n_0, the density of particles of zero momentum, is equal to zero. In that case the only nonvanishing term in the interaction (2.3) is V_a, which describes the interaction of two particles in excited states. With such an interaction as a perturbation, it is clearly impossible to get the perturbed ground state, if one starts from an unperturbed state in which all particles have zero momentum.

4. THE GREEN'S FUNCTIONS

There is a further complication in the boson problem which arises from the fact that the interaction allows for the creation of pairs of equal and opposite momenta. As a result one has terms in which two or more intermediate states consist of pairs of particles of the same energy, as shown in Fig. 2. This leads to divergent integrals for low momenta.

One way of handling this problem is to transform to a new set of variables, as proposed by Bogoliubov[2]:

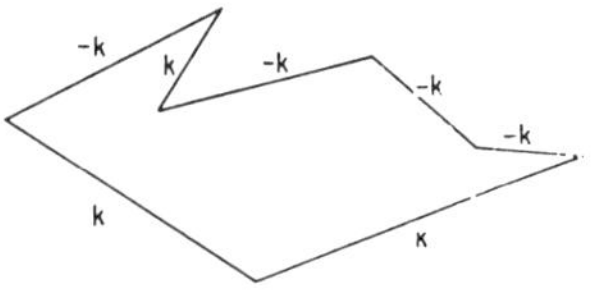

FIG. 2. A ground-state diagram leading to an integral which is highly divergent for small k.

$$a_{\mathbf{k}}=u_k'b_{\mathbf{k}}-v_k'b_{-\mathbf{k}}^*;\quad a_{\mathbf{k}}^*=u_k'b_{\mathbf{k}}^*-v_k'b_{-\mathbf{k}}, \quad (4.1)$$

where u_k' and v_k' are real, and satisfy $u_k'^2-v_k'^2=1$. One can then determine the coefficients of the transformation by imposing the principle of the compensation of dangerous diagrams,[11] and work with the linked cluster expansion (3.10) in terms of the new variables $b_{\mathbf{k}}$ and $b_{\mathbf{k}}^*$.

We shall here follow another method, due to Beliaev,[4] which is closely related to the Green's function methods of quantum field theory. The relationship between the Green's function method and the Bogoliubov transformation will be discussed briefly in Sec. 7. In the Green's function method the divergences are removed by making partial summations over classes of diagrams. Our presentation differs from Beliaev's mainly through the fact that in our work the zero-momentum state has already been removed. The results turn out to be completely equivalent to those obtained by Beliaev.

We define a one-particle Green's function by[12]

$$G(\mathbf{x}_1-\mathbf{x}_2,\,t_1-t_2)=-i\langle\psi_0|T\psi(\mathbf{x}_1t_1)\psi^*(\mathbf{x}_2t_2)|\psi_0\rangle, \quad (4.2)$$

in which $|\psi_0\rangle$ is the ground-state wave function of the interacting system and the $\psi(\mathbf{x}t)$ are field operators in Heisenberg representation. They are related to the creation and annihilation operators $\xi_{\mathbf{k}}^*$ and $\xi_{\mathbf{k}}$ by

$$\psi(\mathbf{x}t)=(2\pi)^{-\frac{3}{2}}\int_{\mathbf{k}} e^{i\mathbf{k}\cdot\mathbf{x}}\xi_{\mathbf{k}}(t).$$

Defining the Fourier transform of $G(\mathbf{x},t)$ by

$$G(\mathbf{x},t)=(2\pi)^{-3}\int d^3p\, e^{i\mathbf{p}\cdot\mathbf{x}}G(\mathbf{p},t),$$

one finds

$$G(\mathbf{p},\,t-t')\delta^3(\mathbf{p}-\mathbf{p}')=-i\langle\psi_0|T\xi_{\mathbf{p}}(t)\xi_{\mathbf{p}'}^*(t')|\psi_0\rangle. \quad (4.3)$$

For an extensive discussion of the use of Green's functions in the theory of many-particle systems we refer to the work of Beliaev and also to Migdal and Galitskii,[1] who studied the fermion problem in this way.

The one-particle Green's functions, as defined above, are appropriate tools to describe single-particle excitations. In particular, the analytical behavior near the

[11] N. N. Bogoliubov, Nuovo cimento **7**, 794 (1958).

[12] The sign agrees with that of reference 4. In field theory the opposite sign is customary.

real ϵ-axis of the function $G(\mathbf{p},\epsilon)$, defined by

$$G(\mathbf{p},\epsilon)=\int_{-\infty}^{+\infty} G(\mathbf{p},t)e^{i\epsilon t}dt, \tag{4.4}$$

tells us the energy and the life-time of a single-particle excitation of momentum $\mathbf{p}$.[13]

It is also possible to derive a formula expressing the total energy of the system in terms of the one-particle Green's function. One proceeds as follows. Using (4.3) for $t<t'$ and taking the derivative with respect to t, one has

$$\begin{aligned}\frac{d}{dt}G(\mathbf{p},\,t-t')\delta^3(\mathbf{p}-\mathbf{p}')&\\ =\langle\psi_0|\,\xi_{\mathbf{p}'}{}^*(t')[H'(t),\xi_{\mathbf{p}}(t)]\,|\psi_0\rangle&\\ =-i(\tfrac{1}{2}p^2-\mu)G(\mathbf{p},\,t-t')\delta^3(\mathbf{p}-\mathbf{p}')&\\ +\langle\psi_0|\,\xi_{\mathbf{p}'}{}^*(t')[V(t),\xi_{\mathbf{p}}(t)]\,|\psi_0\rangle.&\end{aligned}$$

Here $H'(t)$ and $V(t)$ are the total and interaction Hamiltonian in Heisenberg representation, and one should bear in mind the fact that the unperturbed energies are shifted from $\frac{1}{2}p^2$ to $\frac{1}{2}p^2-\mu$. Using (4.4), taking the limit $t\rightarrow t'$ (always keeping $t<t'$), and summing both over $\mathbf{p}$ and $\mathbf{p}'$, one gets

$$\Omega^{-1}\int d^3p\langle\psi_0|\,\xi_{\mathbf{p}}{}^*[V,\xi_{\mathbf{p}}]\,|\psi_0\rangle =\frac{-i}{(2\pi)^4}\int d^3p\int_C d\epsilon(\epsilon-\tfrac{1}{2}p^2+\mu)G(\mathbf{p},\epsilon). \tag{4.5}$$

The path of integration C is a contour consisting of the real axis from $-\infty$ to $+\infty$, together with a semicircle in the upper half plane. Making use of (2.3) for V, in which $a_0\Omega^{-\frac{1}{2}}$ and $a_0{}^*\Omega^{-\frac{1}{2}}$ must be replaced by $n_0{}^{\frac{1}{2}}$, one easily finds that the left-hand side of Eq. (4.5) can be written

$$\Omega^{-1}\left[-2\langle\psi_0|\,V\,|\psi_0\rangle+n_0\left\langle\psi_0\middle|\frac{d}{dn_0}V\middle|\psi_0\right\rangle\right]. \tag{4.6}$$

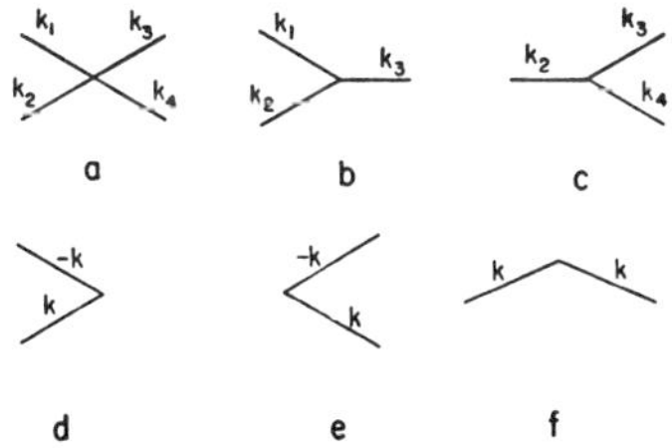

FIG. 3. The vertices a, b, $\cdots$, f correspond to the terms V_a, V_b, $\cdots$, V_f of Eq. (2.3).

[13] The function $G(p,\epsilon)$ plays a role very similar to that of the function $D_p(z)$ of Van Hove and Hugenholtz. See, e.g., Physica **24**, 363 (1958).

The ground-state expectation value of the kinetic energy H_0 is

$$\langle\psi_0|\,H_0\,|\psi_0\rangle=\Omega(2\pi)^{-3}\int d^3p\,\tfrac{1}{2}p^2\langle\psi_0|\,N_{\mathbf{p}}\,|\psi_0\rangle. \tag{4.7}$$

The expectation value $\langle\psi_0|N_{\mathbf{p}}|\psi_0\rangle$ for the number of particles of momentum $\mathbf{p}$ can be expressed in terms of the one-particle Green's function

$$\langle\psi_0|\,N_{\mathbf{p}}\,|\psi_0\rangle=iG(\mathbf{p},\,-0)=\frac{i}{2\pi}\int_C d\epsilon\,G(\mathbf{p},\epsilon). \tag{4.8}$$

Equations (4.5), (4.6), (4.7), and (4.8), together with (3.6), and (3.8) lead to

$$\frac{E_0}{\Omega}-\tfrac{1}{2}n_0\mu=\frac{i}{(2\pi)^4}\int d^3p\int_C d\epsilon\,\tfrac{1}{2}(\epsilon+\tfrac{1}{2}p^2+\mu)G(\mathbf{p},\epsilon),$$

or with

$$n'=i(2\pi)^{-4}\int d^3p\int_C d\epsilon\,G(\mathbf{p},\epsilon), \tag{4.9}$$

$$\frac{E_0}{\Omega}-\tfrac{1}{2}n\mu=\frac{i}{(2\pi)^4}\int d^3p\int_C d\epsilon\,\tfrac{1}{2}(\epsilon+\tfrac{1}{2}p^2)G(\mathbf{p},\epsilon). \tag{4.10}$$

This is actually a rather complicated differential equation in E_0/Ω as a function of n, since $\mu=(d/dn)(E_0/\Omega)$ appears not only at the left-hand side, but also in $G(\mathbf{p},\epsilon)$. In the low-density case the situation is, in fact, much less complicated. If one uses Eq. (4.10) to calculate E_0/Ω to a certain accuracy, one can use for μ on the right-hand side an expression of lower order. One therefore finds a simple linear first order differential equation for E_0/Ω which leaves the term $\sim n^2$ undetermined. This is however the first term in the expansion for E_0/Ω, which can easily be calculated by other methods; for example, by the formula (6.2) for μ derived in Sec. 6.

Having seen that both single particle properties and the ground-state energy can be derived from the one-particle Green's function, our task is to derive a useful expression for $G(\mathbf{p},\epsilon)$. Here we use the method of Beliaev.[4] Since we will use the same ideas in Sec. 8 to obtain a new result, we will have to repeat some of his arguments.

Using the interaction representation we have

$$G(\mathbf{p},\,t-t')\delta^3(\mathbf{p}-\mathbf{p}')=-i\langle 0|\,T\xi_{\mathbf{p}}(t)\xi_{\mathbf{p}'}{}^*(t')S\,|0\rangle, \tag{4.11}$$

where $|0\rangle$ is the vacuum state (no particles of momentum $|\mathbf{k}|\neq 0$) and $\xi_{\mathbf{p}}(t)=e^{iH_0t}\xi_{\mathbf{p}}e^{-iH_0t}$. The S-matrix can be expanded in powers of the interaction V:

$$S=\sum_{n=0}^{\infty}\frac{(-i)^n}{n!}\int_{-\infty}^{+\infty}dt_1\cdots\int_{-\infty}^{+\infty}dt_n\,TV(t_1)\cdots V(t_n).$$

If one substitutes this expression in (4.11), one obtains an infinite series, in which each term contains a vacuum expectation value of a time-ordered product of ξ and ξ^*

operators. Using the method of Wick, one can write such a vacuum expectation value as a sum of terms, which are found by forming pairs of creation and annihilation operators in all possible ways and taking the product of the vacuum expectation values of the time-ordered product of each such pair. From the definition (4.3) of the Green's function we see that each such pair carries a factor

$$\langle 0|T\xi_{\mathbf{k}}(t)\xi_{\mathbf{k}'}{}^*(t')|0\rangle = iG_0(\mathbf{k},\,t-t')\delta^3(\mathbf{k}-\mathbf{k}'),$$

G_0 being the Green's function of the noninteracting system.

As usual in field theory, we represent the various ways in which such pairs of creation and annihilation operators can be formed by Feynman diagrams. In that way one obtains a one-to-one correspondence between diagrams and terms of our Green's function. It is now a simple matter to find the following rules for calculating the contribution of each diagram to $G(\mathbf{p},\epsilon)$. (a) For each line, either internal or external, of momentum $\mathbf{k}$ and energy ϵ one has a factor $G_0(\mathbf{k},\epsilon)\equiv(\epsilon-\frac{1}{2}k^2+\mu+i\delta)^{-1}$. (b) For each vertex one has a factor $\delta^3(\sum_i \mathbf{k}_i)\,\delta(\sum_i \epsilon_i)$ for conservation of momentum and energy. (c) For each vertex of type a, b, c, d, e, f of Fig. 3, which correspond to the terms V_a, V_b, V_c, V_d, V_e, and V_f of Eq. (2.3) a factor $v(\mathbf{k}_1-\mathbf{k}_3)+v(\mathbf{k}_1-\mathbf{k}_4)$, $v(\mathbf{k}_1)+v(\mathbf{k}_2)$, $v(\mathbf{k}_3)+v(\mathbf{k}_4)$, $v(\mathbf{k})$, $v(\mathbf{k})$, $v(\mathbf{k})+v(0)$, respectively. (d) For each pair of equivalent lines (i.e., two lines connecting the same pair of vertices) a factor $\frac{1}{2}$. (e) For each incomplete vertex (i.e., a vertex with one or two missing lines) a factor $n_0^{\frac{1}{2}}$ or n_0 as the case may be. (f) A numerical factor $1/r(2\pi)^{-4m+4s}\cdot i^{m-s}$, where m is the order of the diagram and s is the number of factors n_0, due to the incomplete vertices; r is the number of ways in which the vertices can be permuted without changing the diagram. In addition to $G(\mathbf{p},\epsilon)$ it is advantageous to introduce two similar functions

$$\tilde{G}(\mathbf{p},\,t-t')\delta^3(\mathbf{p}+\mathbf{p}') = -i\langle\psi_0|T\xi_{\mathbf{p}}(t)\xi_{\mathbf{p}'}(t')|\psi_0\rangle,$$

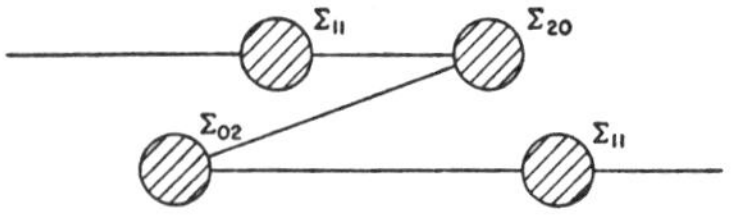

FIG. 4. The general form of the diagrams contributing to $G(p,\epsilon)$.

and

$$\bar{G}(\mathbf{p},\,t-t')\delta^3(\mathbf{p}+\mathbf{p}') = -i\langle\psi_0|T\xi_{\mathbf{p}}{}^*(t)\xi_{\mathbf{p}'}{}^*(t')|\psi_0\rangle,$$

which are represented by diagrams with two outgoing lines (i.e., external lines running to the left) and two ingoing lines (i.e., lines running to the right), respectively. These functions obviously have no counterpart in the unperturbed system.

The general structure of the diagrams contributing to $G(\mathbf{p},\epsilon)$ is shown in Fig. 4. It simply forms a chain consisting of three types of proper parts, connected by a single line. These proper parts will be called $\Sigma_{11}(\mathbf{p},\epsilon)$, $\Sigma_{02}(\mathbf{p},\epsilon)$, or $\Sigma_{20}(\mathbf{p},\epsilon)$ depending on whether they have one ingoing and one outgoing line, two ingoing lines or two outgoing lines, respectively. One sees easily that Σ_{02} and Σ_{20} are equal. Clearly $\tilde{G}$ and $\bar{G}$ have diagrams of the same general structure as G.

It is now very easy to express $G(p,\epsilon)$, $\tilde{G}(p,\epsilon)$, and $\bar{G}(p,\epsilon)$ in terms of these three quantities Σ_{11}, Σ_{02}, and Σ_{20}. One can immediately write down the equations

$$G(p,\epsilon)=G_0(p,\epsilon)+G(p,\epsilon)\Sigma_{11}(p,\epsilon)G_0(p,\epsilon)+\tilde{G}(p,\epsilon)\Sigma_{20}(p,\epsilon)G_0(p,\epsilon),$$
$$\tilde{G}(p,\epsilon)=\tilde{G}(p,\epsilon)\Sigma_{11}(p,\,-\epsilon)G_0(p,\,-\epsilon)+G(p,\epsilon)\Sigma_{02}(p,\epsilon)G_0(p,\,-\epsilon),$$
$$\bar{G}(p,\epsilon)=\bar{G}(p,\epsilon)\Sigma_{11}(p,\,-\epsilon)G_0(p,\,-\epsilon)+G(p,\epsilon)\Sigma_{02}(p,\epsilon)G_0(p,\,-\epsilon).$$

These equations are represented graphically by Fig. 5, where the thick lines are exact Green's functions and the thin lines the unperturbed Green's functions. These three algebraic equations can be solved and one finds the expressions

$$G(p,\epsilon)=\frac{\epsilon+\frac{1}{2}p^2-\mu+\Sigma_{11}^-}{[\epsilon-\frac{1}{2}(\Sigma_{11}^+-\Sigma_{11}^-)]^2-[\frac{1}{2}p^2-\mu+\frac{1}{2}(\Sigma_{11}^++\Sigma_{11}^-)]^2+\Sigma_{02}^2}, \tag{4.12}$$

$$\bar{G}(p,\epsilon)=\tilde{G}(p,\epsilon)=\frac{-\Sigma_{02}}{[\epsilon-\frac{1}{2}(\Sigma_{11}^+-\Sigma_{11}^-)]^2-[\frac{1}{2}p^2-\mu+\frac{1}{2}(\Sigma_{11}^++\Sigma_{11}^-)]^2+\Sigma_{02}^2}, \tag{4.13}$$

as derived by Beliaev. Here $\Sigma_{11}{}^+=\Sigma_{11}(p,\epsilon)$ and $\Sigma_{11}{}^-=\Sigma_{11}(p,\,-\epsilon)$. By calculating these explicit expressions of the Green's functions in terms of Σ_{11} and Σ_{02}, one has performed the partial summation necessary to remove the divergences from the theory.

5. THE CORRELATION FUNCTION

One is also interested in more general Green's functions than the one-particle Green's functions defined in the preceding sections, namely those which characterize the interaction of the many-boson system with external fields. For instance, the only way in which the elementary excitation spectrum of liquid helium may be directly measured is through the inelastic scattering of slow neutrons. As has been shown by Van Hove,[14] the probability per unit time that a slow neutron give up energy ω and momentum $\mathbf{k}$ to a boson gas in its ground

[14] L. Van Hove, Phys. Rev. 95, 249 (1954).

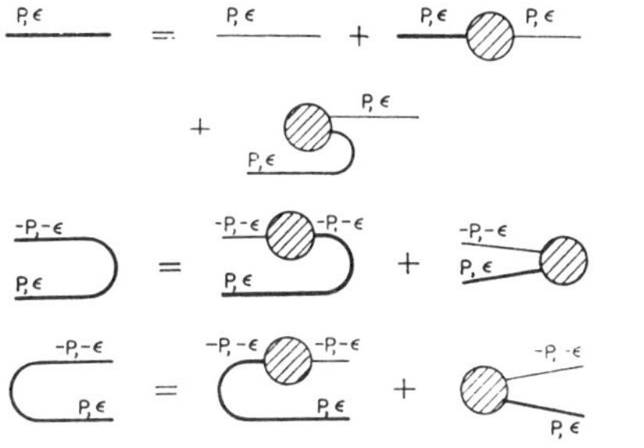

FIG. 5. The graphical representation of the algebraic relations between $G(p,\epsilon)$, $\tilde{G}(p,\epsilon)$ and $\bar{G}(p,\epsilon)$.

state may be written in the Born approximation as

$$w(\mathbf{k},\omega)=AS(\mathbf{k},\omega). \tag{5.1}$$

A is a constant which characterizes the neutron-boson interaction, and $S(k,\omega)$ characterizes the elementary excitation spectrum of the boson system according to

$$S(\mathbf{k},\omega)=\sum_n(\rho_\mathbf{k})_{n0}{}^2\delta(\omega-\omega_{n0}). \tag{5.2}$$

Here $\rho_\mathbf{k}$ is the density fluctuation of momentum $\mathbf{k}$,

$$\rho_\mathbf{k}=\int n(\mathbf{x})e^{-i\mathbf{k}\cdot\mathbf{x}}d^3x=\sum_i e^{-i\mathbf{k}\cdot\mathbf{x}_i},$$

which may easily be written in second quantization as $\rho_\mathbf{k}=\sum_\mathbf{q} a_\mathbf{q}{}^*a_{\mathbf{q}+\mathbf{k}}$. $(\rho_\mathbf{k})_{n0}$ denotes the matrix element between the exact wave functions $|\psi_0\rangle$ of the ground state and $|\psi_n\rangle$ of the excited state, which correspond to the exact eigenvalues E_0 and E_n, so that $\omega_{n0}=E_n-E_0$ is the exact energy of the excitation produced by the neutron.

$S(\mathbf{k},\omega)$ is the Fourier-transform in space and time of the pair distribution function. It is simply related to the following function:

$$iF_\mathbf{k}(t-t')=\langle 0|T\{\rho_\mathbf{k}(t)\rho_{-\mathbf{k}}(t')\}|0\rangle. \tag{5.3}$$

This relationship can be seen if we define the Fourier transform of $F_k(t-t')$ by

$$F_\mathbf{k}(t-t')=\frac{1}{2\pi}\int_{-\infty}^{\infty}d\omega\, F(\mathbf{k},\omega)e^{-i\omega(t-t')}. \tag{5.4}$$

It then follows that

$$F(\mathbf{k},\omega)=\sum_n(\rho_\mathbf{k})_{n0}{}^2\left\{\frac{1}{\omega-\omega_{n0}+i\delta}-\frac{1}{\omega+\omega_{n0}-i\delta}\right\}, \tag{5.5}$$

from which one immediately finds

$$\mathrm{Im}F(\mathbf{k},\omega)=\pi S(\mathbf{k},\omega). \tag{5.6}$$

We further remark that the structure factor, $S(\mathbf{k})$, which is defined by

$$S(\mathbf{k})=\langle 0|\rho_\mathbf{k}{}^*\rho_\mathbf{k}|0\rangle/n, \tag{5.7}$$

is given then by

$$S(\mathbf{k})=\frac{1}{n}\int_C d\omega\,\frac{i}{2\pi}F(\mathbf{k},\omega), \tag{5.8}$$

where the contour may be closed either above or below the real axis (since $\rho_\mathbf{k}$ and $\rho_{-\mathbf{k}}$ commute).

If we now eliminate the condensed state operators, a_0 and $a_0{}^*$, according to the prescription of the preceding sections, we see that there are three distinct contributions to $iF_\mathbf{k}(t-t')$, corresponding to diagrams with two, three, and four external lines, respectively. Thus we may write

$$iF_\mathbf{k}(t-t')=i\{F_\mathbf{k}{}^a(t-t')+F_\mathbf{k}{}^b(t-t')+F_\mathbf{k}{}^c(t-t')\}, \tag{5.9}$$

where

$$F_\mathbf{k}{}^a(t-t')=n_0\langle\psi_0|T\{[a_{-\mathbf{k}}{}^*(t)+a_\mathbf{k}(t)]\times[a_\mathbf{k}{}^*(t')+a_\mathbf{k}(t')]\}|\psi_0\rangle, \tag{5.10a}$$

$$F_\mathbf{k}{}^b(t-t')=2(n_0)^{\frac{1}{2}}\langle\psi_0|T\{[a_{-\mathbf{k}}{}^*(t)+a_\mathbf{k}(t)]\times[\textstyle\sum_\mathbf{q} a_{\mathbf{q}+\mathbf{k}}{}^*(t')a_\mathbf{q}(t')]\}|\psi_0\rangle, \tag{5.10b}$$

$$F_\mathbf{k}{}^c(t-t')=\langle\psi_0|T\{[\textstyle\sum_\mathbf{q} a_{\mathbf{q}-\mathbf{k}}{}^*(t)a_\mathbf{q}(t)]\times[\textstyle\sum_{\mathbf{q}'} a_{\mathbf{q}'+\mathbf{k}}{}^*(t')a_{\mathbf{q}'}(t')]\}|\psi_0\rangle. \tag{5.10c}$$

$F_\mathbf{k}{}^a(t-t')$ may be expressed in terms of G and $\bar{G}$ as

$$F_\mathbf{k}{}^a(t-t')=n_0\{G_\mathbf{k}(t-t')+G_\mathbf{k}(t'-t)+\bar{G}_\mathbf{k}(t-t')+\bar{G}_\mathbf{k}(t'-t)\}. \tag{5.11}$$

With the aid of (4.12) and (4.13), we may write its Fourier transform, $F^a(k,\omega)$, as

$$F^a(k,\omega)=n_0\frac{k^2+\Sigma_{11}{}^++\Sigma_{11}{}^--2\mu-2\Sigma_{02}}{[\epsilon-\frac{1}{2}(\Sigma_{11}{}^+-\Sigma_{11}{}^-)]^2-[\frac{1}{2}p^2-\mu+\frac{1}{2}(\Sigma_{11}{}^++\Sigma_{11}{}^-)]^2+\Sigma_{02}{}^2}. \tag{5.12}$$

6. THEOREM ON THE PHONON SPECTRUM

The expression for $G(p,\epsilon)$, derived in Sec. 4 can be used to calculate the energy $E(k)$ of a single particle excitation as a function of its momentum $\mathbf{k}$. In the low-density approximation it appears that for small momenta $E(k)$ is proportional to k, so that in particular $E(0)=0$.

It is the purpose of this section to show generally i.e., to all orders in the interaction, that the phonon energy is equal to zero for zero momentum. The proof is based on a simple relationship we shall establish between the chemical potential μ and the functions $\Sigma_{11}(0,0)$ and $\Sigma_{02}(0,0)$ for $\mathbf{p}$ and ϵ equal to zero.

In order to derive this relationship we shall start with

the well-known expansion

$$U(t-t')\equiv e^{-iH(t-t')}=e^{-iH_0t}\sum_{n=0}^{\infty}(-i)^n\int_{t'}^{t}dt_1$$
$$\times\int_{t'}^{t_1}dt_2\cdots\int_{t'}^{t_{n-1}}dt_n\,V(t_1)\cdots V(t_n)e^{iH_0t'},$$

where, as before, $V(t)$ is the interaction in interaction representation. Introducing the time-ordering operator T, one can easily write

$$U(t-t')=e^{-iH_0t}\sum_{n=0}^{\infty}\frac{(-i)^n}{n!}$$
$$\times\int_{t'}^{t}dt_1\cdots\int_{t'}^{t}dt_n\,TV(t_1)\cdots V(t_n)e^{iH_0t'}. \quad (6.1)$$

Since the asymptotic behavior of $\langle 0|U(t)|0\rangle$ for large t is of the form $N_0\exp(-iE_0t)$,[15] one can use the diagonal element $\langle 0|U(t)|0\rangle$ to derive a convenient expression for the total energy E_0 of the system in its ground state. Using diagrams to represent the various terms of the expansion (6.1), one finds that $\langle 0|U(t)|0\rangle$ can be expressed in terms of connected diagrams only, by the formula[15]

$$\langle 0|U(t)|0\rangle=\exp(\langle 0|\bar{U}(t)|0\rangle),$$

where $\langle 0|\bar{U}(t)|0\rangle$ is defined by

$$\langle 0|\bar{U}(t)|0\rangle=\sum_n\frac{(-i)^n}{n!}\int_{-\frac{1}{2}t}^{\frac{1}{2}t}dt_1\cdots$$
$$\times\int_{-\frac{1}{2}t}^{\frac{1}{2}t}dt_n\langle 0|[TV(t_1)\cdots V(t_n)]_C|0\rangle.$$

The subscript C means that only connected ground-state diagrams contribute to $\langle 0|\bar{U}(t)|0\rangle$. In exactly the same way as in the case of the Green's function, the time-ordered product can be expressed in terms of normal products. We now study this function in the limit $t\to\infty$. As before, the integrations over t_i lead to factors $\delta(\sum_j\epsilon_j)$ in each vertex, saying that the sum of the ϵ_j's, with appropriate signs, must be zero for each interaction. However, in the case without external lines, one of these relations is identically satisfied as soon as the others are fulfilled. This means that the last integration simply leads to a factor t, expressing the fact that $\langle 0|\bar{U}(t)|0\rangle$ is asymptotically proportional to t. The proportionality factor must clearly be $-iE_0$. It is now very easy to establish that E_0/Ω can be calculated as the sum of the contributions of all connected ground state diagrams. The contribution of each diagram must be calculated according to the rules given in Sec. 4, with the following modifications: (b'). For each vertex except one, there is a factor $\delta^3(\sum_i\mathbf{k}_i)\delta(\sum_i\epsilon_i)$. ($f'$). A numerical factor $1/r(2\pi)^{-4m+4s-4}i^{m-s+1}$.

[15] Cl. Bloch, Nuclear Phys. 7, 451 (1958).

We shall now use this expansion of E_0 to establish the following equation involving the chemical potential μ, Σ_{11}, and Σ_{02}:

$$\mu=\Sigma_{11}(0,0)-\Sigma_{02}(0,0). \quad (6.2)$$

To prove this equation we remark that all diagrams of $\Sigma_{11}(0,0)$ can be obtained in a unique fashion from the connected ground-state diagrams by attaching in all possible ways one ingoing and one outgoing line of momentum and energy zero to one or two incomplete vertices; to obtain $\Sigma_{02}(0,0)$ one attaches two ingoing lines in a similar fashion. Let us consider an arbitrary connected ground-state diagram, which is built up from n_a vertices of type a of Fig. 3, n_b of type b, etc., and denote its value by $\{n_a,n_b,n_c,n_d,n_e,n_f\}$. Obviously $n_b+2n_d=n_c+2n_e$. The number s of factors n_0 is given by

$$s=\tfrac{1}{2}n_b+\tfrac{1}{2}n_c+n_d+n_e+n_f=n_b+2n_d+n_f. \quad (6.3)$$

With the Eq. (3.8) derived in Sec. 3, the value of μ arising from this diagram is found to be

$$\mu^{(d)}=sn_0^{-1}\{n_a,\cdots,n_f\}. \quad (6.4)$$

We now calculate the sum of all terms of $\Sigma_{11}(0,0)$ and $\Sigma_{02}(0,0)$ which can be obtained from this ground-state diagram. The process of attaching two external lines diminishes the number of factors n_0 by one. Hence, in the numerical factor for Σ_{11} or Σ_{02}, as given in Sec. 4 under f, we must replace s by $s-1$, which makes this factor identical to the corresponding one for $n_0^{-1}E_0/\Omega$.

Let us now start adding the two external lines, one by one. An ingoing line can only be added to vertices b, d, and f, transforming them into a, b, and c, respectively. The transitions $b\to a$ and $f\to c$ do not change the value of the diagram, since the vertex-functions $v(\mathbf{k}_1)+v(\mathbf{k}_2)$ and $v(\mathbf{k}_1-\mathbf{p})+v(\mathbf{k}_2-\mathbf{p})$ are equal for $\mathbf{p}=0$, and similarly for $v(\mathbf{k})+v(0)$ and $v(\mathbf{k})+v(\mathbf{p})$. However the transition $d\to b$ leads to a factor of 2.

If we denote by $I\{n_a,n_b,n_c,n_d,n_e,n_f\}$ the sum of terms one gets by adding one ingoing line in all possible ways, one finds

$$\begin{aligned}I\{n_a,n_b,\cdots,n_f\}&=n_b\{n_a+1,\,n_b-1,\,n_c,\,n_d,\,n_e,\,n_f\}\\&\quad+n_d\{n_a,\,n_b+1,\,n_c,\,n_d-1,\,n_e,\,n_f\}\\&\quad+n_f\{n_a,\,n_b,\,n_c+1,\,n_d,\,n_e,\,n_f-1\},\end{aligned}$$

where one should remember that the different bracket expressions, representing different diagrams, have all equal values, except for a factor of 2 for each missing vertex of type d or e. Adding one more ingoing line in exactly the same manner, one finds

$$\begin{aligned}\Sigma_{02}^{(d)}(0,0)=II\{n_a\cdots n_f\}&=[n_b(n_b-1+2n_d+n_f)\\&\quad+2n_d(n_b+1+2n_d-2+n_f)\\&\quad+n_f(n_b+2n_d+n_f-1)]n_0^{-1}E_0/\Omega,\end{aligned}$$

which, by virtue of (6.3), gives

$$\Sigma_{02}^{(d)}(0,0)=(s^2-s)n_0^{-1}E_0/\Omega. \quad (6.5)$$

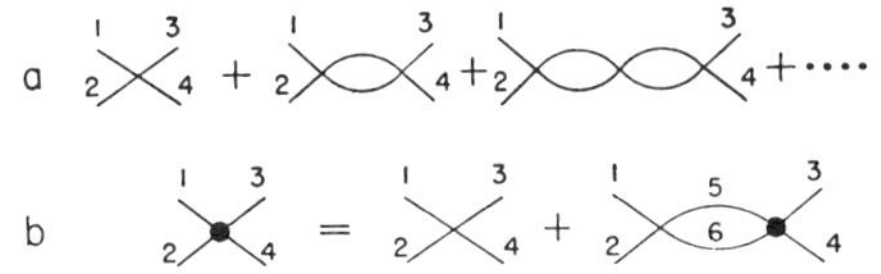

FIG. 6. (a) The multiple scattering terms which for low density are all of equal importance; (b) the graphical representation of Eq. (7.3) for the scattering matrix t.

Similarly

$$\Sigma_{11}^{(d)}(0,0)=OI\{n_a\cdots n_f\}=s^2n_0^{-1}E_0/\Omega, \qquad (6.6)$$

in which O is the operation of adding an outgoing line. Equations (6.4), (6.5), and (6.6) give immediately

$$\mu^{(d)}=\Sigma_{11}^{(d)}(0,0)-\Sigma_{02}^{(d)}(0,0).$$

Since we proved this relation for an arbitrary diagram of order $n>1$. and since for $n=1$ this relation is also satisfied, we have proved Eq. (6.2) generally.

This equation makes it very easy to prove our assertion concerning the phonon spectrum. Indeed, it follows immediately from (4.12) that for μ satisfying (6.2) both poles of $G(p,\epsilon)$ coincide at $\epsilon=0$. Hence there can be no energy gap in the phonon spectrum.

We therefore conclude that for those theories of the boson gas for which an energy gap has appeared in the elementary excitation spectrum, the cause of the apparent gap is to be found in an inconsistent treatment of the vertices Σ_{11} and Σ_{02}, or the depletion effect, in the higher order terms of the perturbation-theoretic expansion.[5]

A few remarks should still be made concerning the conditions under which Eq. (6.2) is valid. We made use of a power series expansion for E_0 which, as can be proved easily, is equivalent to the linked cluster expansion (3.10). We obtained E_0 from the limiting process

$$E_0=i\lim_{t\to\infty}\frac{d}{dt}\langle 0|\bar{U}(t)|0\rangle. \qquad (6.7)$$

It is clear that this expansion in powers of V does not converge in the boson case. In fact, many of the terms are infinite, due to the divergence of integrals for small momenta. For actual calculations this expansion is therefore not very useful and we prefer (4.10). However, in our proof we implicitly used a cutoff for small momenta, knowing that the result will not depend on the cutoff, provided the limit in (6.7) exists. A criterion for this existence is not known at present. We believe that, at least for repulsive interactions, this condition is fulfilled.

7. THE LOW-DENSITY LIMIT

In this and the succeeding section we will be concerned with the calculation of the properties of a dilute boson gas at zero temperature. Our goal is to calculate the first few terms in the expansion of these properties in an ascending series in n; for instance, the ground-state energy may be written as

$$E_0=E_0^{(1)}+E_0^{(2)}+E_0^{(3)}+\cdots,$$

where in the low-density limit $E_0^{(3)}\ll E_0^{(2)}\ll E_0^{(1)}$. As discussed in Sec. 4, the calculation of the properties of a given system in the present method begins with the calculation of the effective potentials, Σ_{11} and Σ_{02}. Once these are determined, in a given order, say, the Green's function $G(p,\epsilon)$ is obtained from (4.12). The poles of $G(p,\epsilon)$ then yield the low-lying elementary excitations, while the ground-state energy, E_0, may be obtained by a suitable integration over $\mathbf{p}$ and ϵ, according to (4.10).

It should be emphasized that the Green's function method differs markedly from a conventional perturbation-theoretic approach, in that a first-order determination of the G's (and the system properties deriving therefrom) already corresponds to the summation of an infinite sequence of terms in a perturbation-theory approach. As an example, calculating Σ_{11} and Σ_{02} in first order, one finds

$$\Sigma_{11}=n_0(V_0+V_p), \qquad (7.1)$$

$$\Sigma_{02}=n_0V_p, \qquad (7.2)$$

where Σ_{11} represents the sum of a direct and an exchange term. As Beliaev has remarked, the results (7.1), (7.2) when combined with (4.12) already contain the classic result of Bogoliubov for the excitation spectrum of a dilute gas of weakly interacting bosons.[2] We may further remark that a calculation of the ground-state energy based on (7.1) and (7.2) is formally equivalent to the high-density electron gas calculation of Gell-Mann and Brueckner.[16] It represents a sum of all ground-state diagrams which are topologically equivalent to a continuous line, punctuated by dots to represent the interactions, as illustrated in Fig. 2. The use, then, of (7.1) and (7.2) for the Σ's is likewise equivalent to the random-phase approximation introduced by Bohm and Pines for the electron gas.[17]

The first-order calculation above is not sufficiently accurate to describe the properties of the system in the low-density limit. The reason is that there is, in fact, a whole sequence of contributions to the Σ's which are of equal importance in this limit. These correspond to the repeated scatterings of a given pair of particles.[8,4] Consider the scattering of particles of momentum $\mathbf{p}_1$ and $\mathbf{p}_2$ to $\mathbf{p}_3$ and $\mathbf{p}_4$. Then, as shown in Fig. 6(a), not only is the first-order scattering of importance, but also all the additional multiple scatterings which are indicated there. Thus all the terms of Fig. 6(a) contribute to the Σ's in the same order of n_0. From Fig. 6(b), it is clear that this infinite sequence of terms may be summed with the aid of an integral equation which may be

[16] M. Gell-Mann and K. A. Brueckner, Phys. Rev. **106**, 364 (1957).
[17] D. Bohm and D. Pines, Phys. Rev. **92**, 608 (1953). See also P. Nozières and D. Pines, Nuovo cimento **9**, 470 (1958).

written symbolically as

$$t_{12;34}=v_{12;34}+v_{12;56}G_5^0G_6^0t_{56;34}. \tag{7.3}$$

Henceforth, in all diagrams, we shall assume the vertex to be given by t according to (7.3), so that the point representing a given vertex is in reality a sum over an infinite sequence of diagrams. The class of diagrams which must then be considered is correspondingly considerably reduced.

The solutions of (7.3) may be expressed in terms of the scattering amplitude for two particles in a vacuum. The particular vertices which are of interest to us here have been calculated by Beliaev, who finds

$$t(00\mathbf{p}-\mathbf{p})=f^*(\mathbf{p},0)+\int\frac{d^3q}{(2\pi)^3}f(0,\mathbf{q})f^*(\mathbf{p},\mathbf{q})\times\left\{\frac{1}{2\mu-q^2+i\delta}+\frac{1}{q^2}\right\}, \tag{7.4}$$

$$t(0\mathbf{pp}0)+t(0\mathbf{p}0\mathbf{p})=2f_s(\tfrac{1}{2}\mathbf{p},\tfrac{1}{2}\mathbf{p})+2\int\frac{d^3q}{(2\pi)^3}|f_s(\tfrac{1}{2}\mathbf{p},\mathbf{q})|^2\times\left\{\frac{1}{\epsilon+2\mu-\frac{1}{4}p^2-q^2+i\delta}+\frac{1}{q^2-\frac{1}{4}p^2-i\delta}\right\}, \tag{7.5}$$

where $f_s(\mathbf{p}',\mathbf{p})=\{f(\mathbf{p}',\mathbf{p})+f(-\mathbf{p}',\mathbf{p})\}/2$, $f(\mathbf{p}',\mathbf{p})$ is the scattering amplitude defined by

$$f(\mathbf{p}',\mathbf{p})=\int d^3x\, v(\mathbf{x})e^{-i\mathbf{p}'\cdot\mathbf{x}}\Psi_{\mathbf{p}}(\mathbf{x}), \tag{7.6}$$

and $\Psi_{\mathbf{p}}(\mathbf{x})$ is the eigenfunction for a particle moving in a potential $v(\mathbf{x})$ which behaves at infinity like a plane wave of momentum $\mathbf{p}$ plus an outgoing spherical wave.

It is not necessary to know the complete solutions of (7.4) and (7.5) in order to determine the leading terms in the low-density expansion of the ground-state energy and excitation spectrum. As we shall see these are completely determined by the properties of the Σ's for momentum transfers which are small compared to the inverse of the zero-momentum scattering amplitude, f_0. In this limit, we have

$$f_s(\tfrac{1}{2}\mathbf{p},\tfrac{1}{2}\mathbf{p})\cong f(\mathbf{p},0)\cong f(0,0)=f_0,\quad (p\ll f_0^{-1}). \tag{7.7}$$

Further, the integrals in (7.4) and (7.5) give rise to terms of order $f_0(n_0f_0^3)^{\frac{1}{2}}$. As we shall see, the expansion parameter for the low-density hard sphere gas is just $(n_0f_0^3)^{\frac{1}{2}}$, so that contributions arising from the integrals may properly be regarded as giving rise to second-order corrections to the first-order Σ's formed from the t's. The latter are, therefore,

$$\Sigma_{02}^{(1)}=n_0f_0, \tag{7.8a}$$

$$\Sigma_{11}^{(1)}=2n_0f_0; \tag{7.8b}$$

while the first-order chemical potential is, according to (6.2),

$$\mu^{(1)}=\Sigma_{11}(0)-\Sigma_{02}(0)=n_0f_0. \tag{7.8c}$$

Let us now consider the properties of the system in first and second orders. The first-order ground-state energy $E_0^{(1)}$ is determined from the "zeroth" order Green's function, $(\Sigma_{11}^{(0)}=\Sigma_{02}^{(0)}=0)$,

$$G^0(p,\epsilon)=1/(\epsilon-\tfrac{1}{2}p^2+i\delta),$$

and the first-order chemical potential, $\mu^{(1)}$. According to (4.10) we have

$$E_0^{(1)}/\Omega=\tfrac{1}{2}n\mu^{(1)}=\tfrac{1}{2}n^2f_0, \tag{7.9}$$

since there is no contribution from the integral in (4.10). The energy spectrum of the elementary excitations derived from $G^0(p,\epsilon)$ is of course that of a gas of free particles.

The first-order excitation spectrum and the second-order ground-state energy are obtained from the first-order Green's functions, which are, according to (4.12) and (7.8),

$$G^{(1)}(p,\epsilon)=(\epsilon+\tfrac{1}{2}p^2+n_0f_0)/(\epsilon^2-\omega_p^2+i\delta), \tag{7.10}$$

$$\tilde{G}^{(1)}(p,\epsilon)=\bar{G}^{(1)}(p,\epsilon)=-n_0f_0/(\epsilon^2-\omega_p^2+i\delta), \tag{7.11}$$

where the poles of G, and hence the energies of the low-lying elementary excitations, are given by

$$\omega_p^2=p^2n_0f_0+\tfrac{1}{4}p^4. \tag{7.12}$$

The dispersion relation, (7.12), for the excitation spectrum shows that in the low-momentum region $(p\ll(n_0f_0)^{\frac{1}{2}})$ the elementary excitations behave like sound waves with a constant velocity, $(n_0f_0)^{\frac{1}{2}}$. In the high-momentum region $(p\gg(n_0f_0)^{\frac{1}{2}})$, ω_p may be expanded in powers of n_0f_0;

$$\omega_p\cong\tfrac{1}{2}p^2+n_0f_0-(n_0^2f_0^2/p^2)+\cdots. \tag{7.13}$$

The elementary excitations then correspond to almost free particles moving in an "optical potential," n_0f_0.

It is convenient to re-express the first-order Green's functions in the following form:

$$G^{(1)}(p+\mu)=\frac{u_p^2}{\epsilon-\omega_p+i\delta}-\frac{v_p^2}{\epsilon+\omega_p-i\delta}, \tag{7.14}$$

$$\tilde{G}^{(1)}(p+\mu)=\bar{G}^{(1)}(p+\mu)=-u_pv_p\left\{\frac{1}{\epsilon-\omega_p+i\delta}-\frac{1}{\epsilon+\omega_p-i\delta}\right\}, \tag{7.15}$$

where

$$u_p^2=(\tfrac{1}{2}p^2+n_0f_0+\omega_p)/2\omega_p, \tag{7.16a}$$

$$v_p^2=(\tfrac{1}{2}p^2+n_0f_0-\omega_p)/2\omega_p, \tag{7.16b}$$

$$u_pv_p=n_0f_0/2\omega_p. \tag{7.16c}$$

We remark that the coefficients u_p and v_p may be regarded as coherence factors which, for a given momentum, measure the way in which the interaction between the particles influences the system properties. The role that these coherence factors play depends in turn on the relative size of the momentum **p**, and the momentum which characterizes the strength of the interaction, $(n_0 f_0)^{\frac{1}{2}}$. Two limiting cases are of interest:

Case 1: $p \ll (2n_0 f_0)^{\frac{1}{2}}$,

$$u_p^2 \cong v_p^2 \cong u_p v_p \cong (n_0 f_0)^{\frac{1}{2}}/2p,$$

Case 2: $p \gg (2n_0 f_0)^{\frac{1}{2}}$,

$$u_p^2 \cong 1; \quad v_p^2 \cong -n_0^2 f_0^2/p^4 \ll 1.$$

Thus for small momenta the coherence factors u_p and v_p are large and equal. The resulting Green's function differs markedly from its free-particle value, $G^{(0)}$, and the related properties of the system are determined by the sound-wave type excitations in a way which is not at all accessible to a conventional perturbation-theoretic treatment. On the other hand, for large momenta, the Green's function, $G^{(1)}$, approaches that of a free particle (as do the elementary excitations), $\bar{G}^{(1)}$ is negligible, and the interaction could easily be treated by ordinary perturbation theory, to which, in fact, the present treatment reduces.

The second-order ground-state energy is found from (4.10) and (7.14) to be

$$\frac{E_0^{(2)}}{\Omega} - \tfrac{1}{2} n\mu^{(2)} = \frac{1}{2}\int \frac{d^3p}{(2\pi)^3} \int_C \frac{id\epsilon}{2\pi}(\epsilon + \tfrac{1}{2}p^2) \times \left\{ \frac{u_p^2}{\epsilon - \omega_p + i\delta} - \frac{v_p^2}{\epsilon + \omega_p - i\delta} \right\}, \qquad (7.17)$$

where the contour integral is to be closed in the upper half plane. We find, on carrying out the integration over ϵ,

$$\frac{E_0^{(2)}}{\Omega} - \tfrac{1}{2} n\mu^{(2)} = \frac{1}{2}\int \frac{d^3p}{(2\pi)^3} \times \frac{(\frac{1}{2}p^2 + n_0 f_0 - \omega_p)(\frac{1}{2}p^2 - \omega_p)}{2\omega_p}. \qquad (7.18)$$

According to (7.13), for large momenta $(p \gtrsim (2n_0 f_0)^{\frac{1}{2}})$, the integrand on the right side of (7.18) is of order p^{-4}; the dominant contributions to the integral come from the low-momentum part $(p \lesssim (2n_0 f_0)^{\frac{1}{2}})$. The integration is straightforward, and one finds

$$(E_0^{(2)}/\Omega) - \tfrac{1}{2} n\mu^{(2)} = -2(n_0 f_0)^{\frac{5}{2}}/15\pi^2. \qquad (7.19)$$

We remark that we can now see that the neglect of the dispersion in $f(\mathbf{p}'\mathbf{p})$ is justified in the calculation of the first-order ground-state energy: the contributions to (7.19) come from

$$p \lesssim (n_0 f_0)^{\frac{1}{2}} \ll f_0^{-1},$$

since

$$(n_0 f_0^3)^{\frac{1}{2}} \ll 1.$$

To complete the calculation, we consider (7.19) as a differential equation in E_0/Ω as a function of n, since $\mu = (d/dn)(E_0/\Omega)$. Putting in this order,

$$E_0^{(2)}/\Omega = \alpha (n f_0)^{\frac{5}{2}}, \qquad (7.20)$$

one gets

$$\mu^{(2)} = \tfrac{5}{2}\alpha n^{\frac{3}{2}} f_0^{\frac{5}{2}}. \qquad (7.21)$$

If we substitute (7.20) and (7.21) in (7.19) and bear in mind that in this order we may replace n_0 by n in the right-hand side of (7.19), we find $\alpha = 8/15\pi^2$, and thus

$$E_0^{(2)}/\Omega = \frac{8}{15\pi^2} n^2 f_0 (n f_0^3)^{\frac{1}{2}}, \qquad (7.22)$$

in agreement with the results of Beliaev.[4] For the special case of a gas of hard spheres, the result (7.22) is at once seen to yield the result of Lee, Huang, and Yang,[3] since for this potential $f_0 = 4\pi a$, where a is the diameter of the spheres.

It is of interest to calculate the depletion of the ground state in this order. We have from (4.9) and (7.14)

$$n - n_0 = \int \frac{d^3p}{(2\pi)^3} \int_C d\epsilon \frac{i}{2\pi} \times \left\{ \frac{u_p^2}{\epsilon - \omega_p + i\delta} - \frac{v_p^2}{\epsilon + \omega_p - i\delta} \right\}, \qquad (7.23)$$

and hence

$$n = n_0 + \int \frac{d^3p}{(2\pi)^3} \frac{\frac{1}{2}p^2 + n_0 f_0 - \omega_p}{2\omega_p} = n_0\left(1 + \frac{(n_0 f_0^3)^{\frac{1}{2}}}{3\pi^2}\right). \qquad (7.24)$$

As in (7.18) the main contributions to the integral come again from low momenta $(p \lesssim (2n_0 f_0)^{\frac{1}{2}})$, since for large p, the integrand is of order p^{-4}. We see that the depletion of the ground state as a result of the interaction between the particles is of order $n_0^{\frac{3}{2}} f_0^{\frac{3}{2}}$.

We may also calculate the time-dependent correlation function $F(k,\omega)$ and the structure factor $S(k)$ defined in Sec. 5 in this order. It is straightforward to show that in lowest-order only the two-line part of $F(k,\omega)$ is of importance. One finds, on substituting Eqs. (7.8) in (5.12)

$$F^{(1)}(k,\omega) = \frac{nk^2}{(\epsilon - \omega_k + i\delta)(\epsilon + \omega_k - i\delta)}. \qquad (7.25)$$

We then have, using (5.8)

$$S^{(1)}(k) = \frac{k^2}{2\omega_k} = \frac{k}{2(n_0 f_0 + k^2/4)^{\frac{1}{2}}}. \qquad (7.26)$$

We thus find from (7.26) that in this order the phonon excitation spectrum takes the form proposed by Feynman[18]

$$\omega_k = k^2/2S(k).$$

[18] R. P. Feynman, Phys. Rev. 91, 1291 (1953).

We note that for large $k(k\gg(n_0f_0)^{\frac{1}{2}})$, $S(k)$ approaches unity, which is its free particle value. For small $k\,[k\ll(n_0f_0)^{\frac{1}{2}}]$ on the other hand, $S(k)$ differs greatly from unity, and in fact varies linearly with k. It is natural, then, to regard $(n_0f_0)^{-\frac{1}{2}}$ as the correlation length in the problem, i.e., the length over which correlations brought about by the particle interactions play an important role.

We see from (7.9), (7.22), and (7.24) that the parameter which characterizes the series expansion of the properties of the dilute boson gas is $(n_0f_0{}^3)^{\frac{1}{2}}$. That this parameter should enter (rather than, say, $f_0/n_0{}^{-\frac{1}{3}}$) is a direct consequence of the fact that it is the low momentum transfers which determine the properties of the system. Thus the interaction is weak when the scattering length f_0 is small compared to the correlation length, $(n_0f_0)^{-\frac{1}{2}}$, and it is this ratio which appears as the expansion parameter.

We conclude this section by remarking on the connection between the present method and that of Bogoliubov.[2] In the latter approach one obtains the first-order excitation spectrum and $E_0{}^{(2)}$ by keeping only the terms V_d, V_e, V_f, and V_g of (2.3) in the Hamiltonian (assuming one has first introduced an effective interaction by the pseudopotential method or by using [7.3]). The resulting Hamiltonian then may be diagonalized by means of a canonical transformation of the form (4.1). The condition that the new Hamiltonian be diagonal is simply

$$u_p'=u_p;\quad v_p'=v_p,$$

where u_p and v_p are defined by (7.16). In this fashion one may obtain a ground-state energy and excitation spectrum in accord with (7.22) and (7.12). It is interesting to note that the coefficient v_p which measures the admixture of the new creation operator in the old annihilation operator is likewise a measure of the strength of the negative frequency pole in our Green's function, (7.14), as might perhaps have been expected.

8. THE NEXT ORDER TERM IN THE GROUND-STATE ENERGY

We now carry out the calculation of the next term in the series expansion for the ground-state energy. To do this, we need to know the second-order effective potentials, $\Sigma_{11}{}^{(2)}$ and $\Sigma_{02}{}^{(2)}$. With the aid of these we may determine $G^{(2)}$, and then calculate $E_0{}^{(3)}$ and $\mu^{(3)}$ in a fashion directly analogous to our calculation of $E_0{}^{(2)}$ and $\mu^{(2)}$ in the preceding section.

We find it convenient to begin by obtaining an expression for $G^{(2)}$ which differs somewhat from that, one finds on direct application of (4.12). We do this by considering the algebraic equations for G and $\tilde{G}$ which obtain if one uses $G^{(1)}(p,\epsilon)$, $\tilde{G}^{(1)}(p,\epsilon)$, and $\bar{G}^{(1)}(p,\epsilon)$ as the "bare" propagation functions instead of $G^0(p,\epsilon)$; one must likewise introduce new effective potentials, $\Sigma_{11}'=\Sigma_{11}-\Sigma_{11}{}^{(1)}$, and $\Sigma_{02}'=\Sigma_{02}-\Sigma_{02}{}^{(1)}$, in place of Σ_{11} and Σ_{02}. The new coupled integral equations may be simply obtained by analysis of the appropriate new diagrams in a way directly analogous to the procedure of Sec. 4, as shown in Fig. 7. The resulting equations are[19]

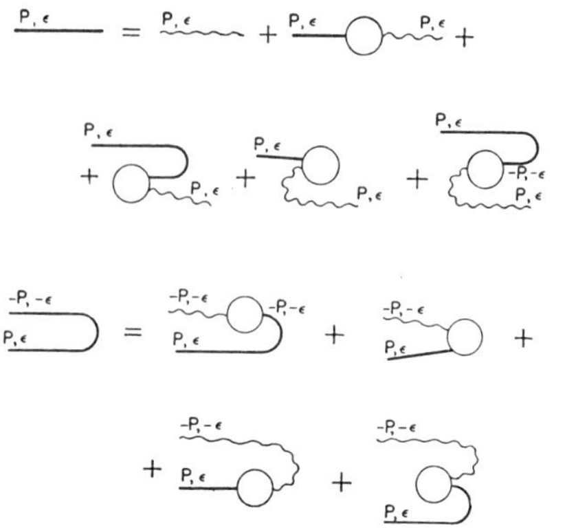

FIG. 7. The graphical representation of the equations (8.1) and (8.2). The wavy lines correspond to the Green's functions $G^{(1)}$, $\tilde{G}^{(1)}$, and $\bar{G}^{(1)}$.

$$\begin{aligned}G(p,\epsilon)=G^{(1)}(p,\epsilon)&+G(p,\epsilon)\Sigma_{11}'(p,\epsilon)G^{(1)}(p,\epsilon)\\&+\tilde{G}(p,\epsilon)\Sigma_{02}'(p,\epsilon)G^{(1)}(p,\epsilon)\\&+G(p,\epsilon)\Sigma_{20}'(p,\epsilon)\tilde{G}^{(1)}(p,\epsilon)\\&\qquad+\tilde{G}(p,\epsilon)\Sigma_{11}'(p,-\epsilon)\bar{G}^{(1)}(p,\epsilon),\end{aligned}\qquad(8.1)$$

and

$$\begin{aligned}\tilde{G}(p,\epsilon)=\tilde{G}(p,\epsilon)&\Sigma_{11}'(p,-\epsilon)G^{(1)}(p,-\epsilon)\\&+G(p,\epsilon)\Sigma_{20}'(p,\epsilon)G^{(1)}(p,-\epsilon)\\&+G(p,\epsilon)\Sigma_{11}'(p,\epsilon)\tilde{G}^{(1)}(p,\epsilon)\\&\qquad+\tilde{G}(p,\epsilon)\Sigma_{02}(p,\epsilon)\tilde{G}^{(1)}(p,\epsilon).\end{aligned}\qquad(8.2)$$

These equations may also be obtained by suitable algebraic manipulations from (4.12) and (4.13), using (7.10) and (7.11). We now remark that for the calculation of $E_0{}^{(3)}$ it will suffice to work directly with (8.1), substituting $G^{(1)}$, $\tilde{G}^{(1)}$, and $\bar{G}^{(1)}$ for G, $\tilde{G}$, and $\bar{G}$ on the right-hand side of the equation, and then using the resulting expression in (4.10). Thus we may write $G=G^{(1)}+G^{(2)}$, where

$$\begin{aligned}G^{(2)}(p,\epsilon)=G^{(1)}&(p,\epsilon)\Sigma_{11}{}^{(2)}(p,\epsilon)G^{(1)}(p,\epsilon)\\&+G^{(1)}(p,\epsilon)\Sigma_{02}{}^{(2)}(p,\epsilon)\tilde{G}^{(1)}(p,\epsilon)\\&+\tilde{G}^{(1)}(p,\epsilon)\Sigma_{11}{}^{(2)}(p,-\epsilon)\bar{G}^{(1)}(p,\epsilon)\\&\qquad+\tilde{G}^{(1)}(p,\epsilon)\Sigma_{02}{}^{(2)}(p,\epsilon)G^{(1)}(p,\epsilon),\end{aligned}\qquad(8.3)$$

and the $\Sigma^{(2)}$'s denote the appropriate expressions of lowest order in n_0 for Σ's.

[19] Actually, the $G^{(1)}(p,\epsilon)$, and $\bar{G}^{(1)}(p,\epsilon)$ of (8.1) and (8.2) differ from the expressions (7.10) and (7.11), in that one should include as well the higher terms in the expansion for μ. To the order of the calculations carried out in this section, these higher terms make no contribution, so we drop them at the outset.

There is a further very great simplification which results because $E_0^{(3)}$ turns out to depend logarithmically on $n_0 f_0^3$. As will presently become obvious, we can obtain the correct coefficient of this term by taking only the leading terms for small n_0 in (8.3). We therefore find

$$G^{(2)}(p,\epsilon)\cong\frac{\Sigma_{11}^{(2)}(p,\epsilon)}{(\epsilon-\frac{1}{2}p^2+i\delta)^2}-\frac{2n_0 f_0\Sigma_{02}^{(2)}(p,\epsilon)}{(\epsilon-\frac{1}{2}p^2+i\delta)^2(\epsilon+\frac{1}{2}p^2-i\delta)}+\frac{n_0^2 f_0^2\Sigma_{11}^{(2)}(p,-\epsilon)}{(\epsilon-\frac{1}{2}p^2+i\delta)^2(\epsilon+\frac{1}{2}p^2-i\delta)^2}. \quad (8.4)$$

The corresponding expression for $E_0^{(3)}$ is

$$\frac{E_0^{(3)}}{\Omega}-\tfrac{1}{2}n\mu^{(3)}=\frac{1}{2}\int\frac{d^3p}{(2\pi)^3}\int_C d\epsilon\frac{i}{(2\pi)}(\epsilon+\tfrac{1}{2}p^2)G^{(2)}(p,\epsilon),$$

which upon substitution of (8.4) becomes

$$\frac{E_0^{(3)}}{\Omega}-\tfrac{1}{2}n\mu^{(3)}=\frac{1}{2}\int\frac{d^3p}{(2\pi)^3}\int_C\frac{d\epsilon i}{(2\pi)}\left\{\frac{[\epsilon+\frac{1}{2}p^2]\Sigma_{11}^{(2)}(p,\epsilon)}{(\epsilon-\frac{1}{2}p^2+i\delta)^2}-\frac{2n_0 f_0\Sigma_{02}^{(2)}(p,\epsilon)}{(\epsilon-\frac{1}{2}p^2+i\delta)^2}+\frac{n_0^2 f_0^2\Sigma_{11}^{(2)}(p,-\epsilon)}{(\epsilon-\frac{1}{2}p^2+i\delta)^2(\epsilon+\frac{1}{2}p^2-i\delta)}\right\}. \quad (8.5)$$

In carrying out the integral over ϵ in (8.5), it is important to bear in mind the fact that any singularities of $G(p,\epsilon)$ must lie slightly above the negative real ϵ axis, or slightly below the positive real ϵ axis, as follows from the definition of G, (4.2). Therefore, because the contour of integration in (8.5) is closed above the real axis, the only contributions to $E_0^{(3)}$ arise from the singularities along the negative real ϵ axis. As a result we need consider only those parts of $\Sigma_{02}^{(2)}(p,\epsilon)$ which have singularities on the negative real ϵ-axis, while for the terms involving $\Sigma_{11}^{(2)}(p,\epsilon)$ and $\Sigma_{11}^{(2)}(p,-\epsilon)$ we shall need to consider the leading term in n_0 for $\Sigma_{11}^{(2)}(p,p^2/2)$, as well as the leading terms which possess singularities above the negative real ϵ axis.

Inspection of the terms of $\Sigma_{02}^{(2)}$ which might then contribute shows that the only diagram of importance (in the limit $n_0\to 0$) is that shown in Fig. 8(a). If we now apply the rules of Sec. 4 and make use of (7.10) and (7.11) to calculate the contribution from this diagram, we find for small n_0

$$\Sigma_{02}^{(2)}\cong 4n_0 f_0^2\int\frac{d^3q}{(2\pi)^3}\int d\epsilon'\frac{i}{2\pi}\times\left\{\frac{u_k^2}{\epsilon'+\epsilon-\frac{1}{2}k^2+i\delta}-\frac{v_k^2}{\epsilon'+\epsilon+\frac{1}{2}k^2-i\delta}\right\}\times\left\{-\frac{u_q v_q}{\epsilon-\frac{1}{2}q^2+i\delta}+\frac{u_q v_q}{\epsilon+\frac{1}{2}q^2-i\delta}\right\}, \quad (8.6)$$

where we have introduced $\mathbf{k}=\mathbf{q}+\mathbf{p}$, and approximated the vertices by $(n_0 f_0)^{\frac{1}{2}}$. On carrying out the integral over ϵ', we find for the relevant term,

$$\Sigma_{02}^{(2)}\cong 4n_0^2 f_0^3\int\frac{d^3q}{(2\pi)^3}\frac{1}{q^2[\epsilon+\frac{1}{2}q^2+\frac{1}{2}k^2-i\delta]}. \quad (8.7)$$

There are two kinds of contributions to $\Sigma_{11}^{(2)}(p,\epsilon)$ which are relevant to our purposes. Those of the first kind give rise to singularities above the negative real ϵ axis, and arise from the diagrams of Fig. 8(b) and (c). The contribution from (b) is

$$4n_0 f_0^2\int\frac{d^3q}{(2\pi)^3}\int d\epsilon'\frac{i}{2\pi}\left\{\frac{u_q v_q}{\epsilon'-\frac{1}{2}q^2+i\delta}-\frac{u_q v_q}{\epsilon'+\frac{1}{2}q^2-i\delta}\right\}\times\left\{\frac{u_k v_k}{\epsilon+\epsilon'-\frac{1}{2}k^2+i\delta}-\frac{u_k v_k}{\epsilon+\epsilon'+\frac{1}{2}k^2-i\delta}\right\}, \quad (8.8)$$

while that from (c) is

$$4n_0 f_0^2\int\frac{d^3q}{(2\pi)^3}\int d\epsilon'\frac{i}{2\pi}\times\left\{\frac{u_k^2}{\epsilon+\epsilon'-\frac{1}{2}k^2+i\delta}-\frac{v_k^2}{\epsilon+\epsilon'+\frac{1}{2}k^2-i\delta}\right\}\times\left\{\frac{u_q^2}{\epsilon'-\frac{1}{2}q^2+i\delta}-\frac{v_q^2}{\epsilon'+\frac{1}{2}q^2-i\delta}\right\}. \quad (8.9)$$

On carrying out the integrals over ϵ', and keeping only the relevant terms for small n_0, we find for the sum of (8.8) and (8.9)

$$-4n_0^3 f_0^4\int\frac{d^3q}{(2\pi)^3}\left(\frac{1}{q^2k^2}+\frac{1}{k^4}\right)\frac{1}{\epsilon+\frac{1}{2}q^2+\frac{1}{2}k^2-i\delta}. \quad (8.10)$$

The contribution to $\Sigma_{11}^{(2)}(p,\epsilon)$ to be substituted in the last term of (8.5) arises from the momentum dependence of $f_s(\frac{1}{2}\mathbf{p},\frac{1}{2}\mathbf{p})$ and the integral in (7.5). It is, to lowest order in n_0,

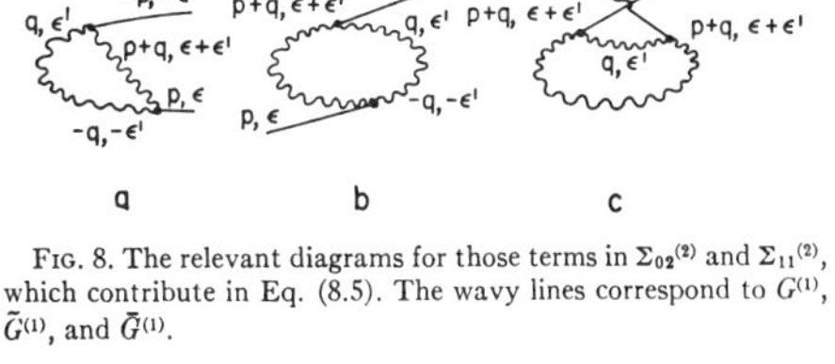

FIG. 8. The relevant diagrams for those terms in $\Sigma_{02}^{(2)}$ and $\Sigma_{11}^{(2)}$, which contribute in Eq. (8.5). The wavy lines correspond to $G^{(1)}$, $\tilde{G}^{(1)}$, and $\bar{G}^{(1)}$.

$$2n_0 \operatorname{Im} f_s(\tfrac{1}{2}\mathbf{p},\tfrac{1}{2}\mathbf{p}) + 2n_0 f_0^2 \int \frac{d^3q}{(2\pi)^3} \times \left\{ \frac{1}{\epsilon - \frac{1}{4}p^2 - q^2 + i\delta} + \frac{1}{q^2 - \frac{1}{4}p^2 - i\delta} \right\}, \quad (8.11)$$

where, as Beliaev has shown, one may apply (7.6) to obtain

$$\operatorname{Im} f_s(\tfrac{1}{2}\mathbf{p},\tfrac{1}{2}\mathbf{p}) = -\operatorname{Im} f_0^2 \int \frac{d^3q}{(2\pi)^3} \frac{1}{q^2 - \frac{1}{4}p^2 - i\delta}$$

$$= -f_0^2 \pi \int \frac{d^3q}{(2\pi)^3} \delta(q^2 - \tfrac{1}{4}p^2).$$

On summing (8.10) and (8.11) we find then:

$$\Sigma_{11}^{(2)}(p,\epsilon) = 2n_0 f_0^2 \int \frac{d^3q}{(2\pi)^3} \times \left\{ \frac{1}{\epsilon - \frac{1}{2}q^2 - \frac{1}{2}k^2 + i\delta} - \frac{2n_0^2 f_0^2 [k^2 + q^2]}{q^2 k^4 (\epsilon + \frac{1}{2}q^2 + \frac{1}{2}k^2 - i\delta)} \right\}$$

$$+ 2n_0 f_0^2 \, \text{P.P.} \int \frac{d^3\mathbf{q}}{(2\pi)^3} \frac{1}{\frac{1}{2}q^2 + \frac{1}{2}k^2 - \frac{1}{2}p^2}, \quad (8.12)$$

where we have shifted variables in order to write the contribution from (8.11) in a more symmetric way, and P.P. denotes the principle part. The results (8.7) and (8.12) are in agreement with the high momentum ($p \gg (n_0 f_0)^{\frac{1}{2}}$) expansion of the Beliaev effective potentials.

The singularities in (8.7) and (8.12), which give rise to an imaginary part of Σ_{02} and Σ_{11}, are associated with the fact that in this order it is possible for an excitation of momentum $\mathbf{p}$ to decay into two excitations of momentum $-\mathbf{q}$ and $\mathbf{p}+\mathbf{q}$.

If we now substitute (8.7) and (8.12) into (8.5) and carry out the integration over ϵ, we find

$$\frac{E_0^{(3)}}{\Omega} - \tfrac{1}{2} n \mu^{(3)} = 2n_0^3 f_0^4 \int \frac{d^3p}{(2\pi)^3} \times \left[\int \frac{d^3q}{(2\pi)^3} \{A(p,q) + B(p,q) + C(p,q)\} + \operatorname{Im} \int \frac{d^3q}{(2\pi)^3} \frac{1}{p^2(q^2 + k^2 - p^2 - i\delta)} \right], \quad (8.13)$$

where

$$A(p,q) = \frac{2(p^2 - q^2 - k^2)(1 + q^2/k^2)}{(p^2 + q^2 + k^2)^2 k^2 q^2}, \quad (8.14a)$$

$$B(p,q) = \frac{8}{q^2(p^2 + q^2 + k^2)^2}, \quad (8.14b)$$

and

$$C(p,q) = \frac{4}{(p^2 + q^2 + k^2)(p^2 - q^2 - k^2 + i\delta)}, \quad (8.14c)$$

are the contributions from the singularities of $\Sigma_{11}(p,\epsilon)$, $\Sigma_{02}(p,\epsilon)$, and $\Sigma_{11}(p,-\epsilon)$, respectively, and the last term in (8.13) arises from the singularity at $\epsilon = -\frac{1}{2}p^2$ in (8.5).

It is convenient to combine A and B. On doing this, and taking advantage of the symmetry between k and q, k and p, and of course p and q (when the expression appears as the integrand of [8.13]), one finds

$$A + B = \frac{2}{k^2 q^2 (k^2 + p^2 + q^2)} - \frac{2}{q^2 (k^2 + p^2 + q^2)^2}. \quad (8.15a)$$

We also have

$$C = \frac{2}{p^2 (p^2 + k^2 + q^2)^2} + \frac{1}{p^4} \left\{ \frac{1}{k^2 + p^2 + q^2} - \frac{1}{q^2 + k^2 - p^2 + i\delta} \right\}. \quad (8.15b)$$

We now note there is considerable cancellation amongst the terms arising from the different singularities. The resultant ground-state energy is real (as of course it must be) and is given by

$$\frac{E_0^{(3)}}{\Omega} - \tfrac{1}{2} n \mu^{(3)} = 2n_0^3 f_0^4 \int \frac{d^3p}{(2\pi)^3} \left\{ \int \frac{d^3\mathbf{q}}{(2\pi)^3} \frac{2}{p^2 q^2 (p^2 + q^2 + k^2)} + \frac{1}{p^4} \frac{1}{(k^2 + p^2 + q^2)} - \frac{\text{P.P.}}{p^4} \int \frac{d^3\mathbf{q}}{(2\pi)^3 (k^2 + q^2 - p^2)} \right\}. \quad (8.16)$$

On carrying out the integration over $\mathbf{q}$ and over the solid angle of $\mathbf{p}$, we find

$$\frac{E_0^{(3)}}{\Omega} - \tfrac{1}{2} n \mu^{(3)} = \frac{n_0^3 f_0^4}{16\pi^2} \left(\frac{4}{3} - \frac{\sqrt{3}}{\pi} \right) \int \frac{dp}{p}. \quad (8.17)$$

Our limiting procedure has led us to a logarithmically divergent expression. This need not concern us unduly however. We know that, had we kept the next-order in n_0 terms, the expression would be well-behaved in the low-momentum region, and possess a natural cutoff at $p_{\min} \sim (n_0 f_0)^{\frac{1}{2}}$. Further, the logarithmic divergence for large momenta is a consequence of the fact that we replaced the $f(\mathbf{p},\mathbf{q})$, which properly appear in the Σ's, by f_0; had we not done this, we would have found a natural cutoff occurring, in the case of hard spheres, for $p_{\max} \sim f_0^{-1}$. We are thus led to write

$$\int \frac{dp}{p} \approx \ln \frac{p_{\max}}{p_{\min}} = \ln \frac{g}{(n_0 f_0^3)^{\frac{1}{2}}}, \quad (8.18)$$

where g is a constant which will influence only the next order terms, $E_0^{(4)}$ and $\mu^{(4)}$. We have therefore

$$\frac{E_0^{(3)}}{\Omega} - \tfrac{1}{2} n \mu^{(3)} = \frac{n^3 f_0^4}{32\pi^2} \left(\frac{\sqrt{3}}{\pi} - \frac{4}{3} \right) \ln(n f_0^3), \quad (8.19)$$

in which we replaced n_0 by n.

To solve this equation we proceed as in Sec. 7 and write

$$E_0{}^{(3)}=\beta n^3 f_0{}^4 \ln(nf_0{}^3),$$

so that

$$\mu^{(3)}=3\beta n^2 f_0{}^4 \ln(nf_0{}^3),$$

in which β is determined by substituting this in (8.19). We find

$$\frac{E_0{}^{(3)}}{\Omega}=\frac{n^3 f_0{}^4}{16\pi^2}\left\{\frac{4}{3}-\frac{\sqrt{3}}{\pi}\right\}\ln(nf_0{}^3), \tag{8.20}$$

$$\mu^{(3)}=\frac{3n^3 f_0{}^4}{16\pi^2}\left\{\frac{4}{3}-\frac{\sqrt{3}}{\pi}\right\}\ln(nf_0{}^3). \tag{8.21}$$

Our result (8.20) is in agreement with the results (obtained by different methods) of Wu[7] and Sawada.[6]

The present calculations also permit us to determine the second-order excitation spectrum in the high-momentum region ($p\gg(n_0f_0)^{\frac{1}{2}}$). In this region, one has, from (4.12) and (8.11) a pole at

$$\epsilon(p)\cong\omega_p+\Sigma_{11}{}^{(2)}(p,\tfrac{1}{2}p^2)$$
$$=\tfrac{1}{2}p^2+n_0f_0\left(1-\frac{i}{4\pi}f_0p\right)\cong\tfrac{1}{2}p^2+f(\tfrac{1}{2}\mathbf{p},\tfrac{1}{2}\mathbf{p}), \tag{8.22}$$

in agreement with the results of Beliaev[4] and Lee and Yang.[20] In this momentum region, the imaginary part of $\epsilon(p)$ is seen to be simply related to the imaginary part of the forward scattering amplitude.

Beliaev has calculated the second-order excitation spectrum in the low-momentum region. For this calculation it is necessary to include many more terms in the calculation of $\Sigma_{11}{}^{(2)}$ and $\Sigma_{02}{}^{(2)}$, that is terms which are of the same order at low momentum as those we have considered. Beliaev finds[21]

$$\epsilon(p)\cong p(n_0f_0)^{\frac{1}{2}}[1+(7/6\pi^2)(n_0f_0{}^3)^{\frac{1}{2}}]$$
$$-i(3/640\pi)p^5/n_0. \tag{8.23}$$

Thus in second-order, for low momentum, there is a correction to the real part of the sound wave frequency, and an imaginary part, corresponding to the fact that a phonon of momentum $\mathbf{p}$ can decay into two phonons of momenta $\mathbf{q}$ and $\mathbf{p}-\mathbf{q}$. That this latter process goes as p^5 is a consequence of great cancellation amongst the coherence factors appearing in the $\Sigma^{(2)}$'s and in the low-momentum expansion of the denominator in (4.12).

It is interesting to note that the microscopic calculation of the sound velocity, $(n_0f_0)^{\frac{1}{2}}[1+(7/6\pi^2)(n_0f_0{}^3)^{\frac{1}{2}}]$, is in agreement with the macroscopic calculation. The latter derives from the fact that the velocity of sound waves of infinite wavelength is related to the compressibility, which may in turn be obtained from the ground-state energy. We calculate the first three terms in the expansion of the macroscopic sound velocity, s, with the aid of the following equations:

$$s=(dp/dn)^{\frac{1}{2}}, \tag{8.24}$$

$$p=n^2\frac{d}{dn}\left(\frac{E_0}{N}\right). \tag{8.25}$$

We find, from (7.9), (7.22), (7.24), and (8.20),

$$s=(nf_0)^{\frac{1}{2}}[1+(nf_0{}^3)^{\frac{1}{2}}/\pi^2$$
$$+(3/16\pi^2)nf_0{}^3\{\tfrac{4}{3}-\sqrt{3}/\pi\}\ln(nf_0{}^3)+\cdots. \tag{8.26}$$

9. DISCUSSION

In the preceding section we have seen that both E_0 and μ possess a series expansion of the form

$$E_0=\tfrac{1}{2}n^2f_0[1+a(nf_0{}^3)^{\frac{1}{2}}+bnf_0{}^3\ln(nf_0{}^3)+\cdots], \tag{9.1}$$

$$\mu=nf_0[1+a'(nf_0{}^3)^{\frac{1}{2}}+b'nf_0{}^3\ln(nf_0{}^3)+\cdots]. \tag{9.2}$$

It is not difficult to see that $n'=n-n_0$ possesses a similar series expansion

$$n-n_0=n'=\frac{1}{3\pi^2}(nf_0)^{\frac{3}{2}}[1+a''(nf_0{}^3)^{\frac{1}{2}}+\cdots]. \tag{9.3}$$

Depletion effects associated with the difference between n_0 and n only make their appearance in the higher order terms of the expansion (9.1) for the ground-state energy. To see this, recall Eq. (7.19), which states that

$$E_0{}^{(2)}-\tfrac{1}{2}n\mu^{(2)}\sim(n_0f_0)^{\frac{5}{2}}.$$

If one then substitutes the series expansion (9.3) for n_0 into (7.19), one finds contributions to higher order terms in the expansion for E_0, which have the form:

$$a_1n^3f_0{}^4+a_2n^{7/2}f_0{}^{11/2}+\cdots.$$

If one further remembers that $E_0{}^{(3)}-\frac{1}{2}n\mu^{(3)}\sim n_0{}^3f_0{}^4\times\ln(n_0f_0{}^3)$, one finds further contributions from the depletion effect which take the form

$$n^{7/2}f_0{}^{11/2}\ln(nf_0{}^3)+\cdots.$$

It thus appears likely that the series expansion for E_0 takes the form

$$E_0=\tfrac{1}{2}n^2f_0[1+a(nf_0{}^3)^{\frac{1}{2}}+bnf_0{}^3\ln(nf_0{}^3)+cnf_0{}^3$$
$$+dn^{\frac{3}{2}}f_0{}^{9/2}\ln(nf_0{}^3)+cn^{\frac{3}{2}}f_0{}^{9/2}+\cdots]. \tag{9.4}$$

Of these coefficients only a and b are known at present.

It should further be remarked that while the coefficients a and b are independent of the particular "shape" of the interaction potential, the coefficients of the higher-order terms will depend on the specific law of force. Thus, no matter what the interaction which gives rise to the scattering length f_0, the coefficients a and b remain the same, while c is a shape-dependent parameter [as may be seen directly from (8.18), for instance].

The coefficient c appears to be quite difficult to calculate. Consider the calculation of $\Sigma_{02}{}^{(3)}$, for instance. It

[20] T. D. Lee and C. N. Yang, Phys. Rev. **112**, 1419 (1958).

[21] An independent calculation of (8.23) has been carried out by Lee and Yang.[20]

may readily be seen that to calculate this quantity, one must include, amongst other terms, the entire sequence of diagrams shown in Fig. 9. As Beliaev has remarked, the summation of these diagrams cannot be expressed in terms of two-body scattering amplitudes, but requires that one obtain a solution to the three-body problem in closed form.

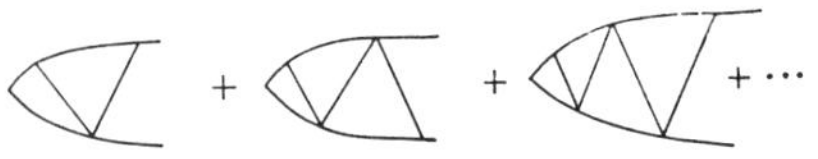

FIG. 9. An infinite series of terms which contribute to the $n_0{}^3f_0{}^4$ term in E_0/Ω.

Actually the incentive does not appear great for performing a calculation which yields c. The reason is that if one studies the relative size of the logarithmic term and the term immediately preceding it in the series expansion for E_0, one finds that when one gets to densities and scattering lengths which might characterize the behavior of liquid helium, the logarithmic term is much arger than its predecessor. For a model of hard spheres of diameter 2.2 A, with a density equal to that of liquid helium, $(nf_0{}^3)^{\frac{1}{2}} \cong 21.4$, and $E_0{}^{(3)} \sim 6E_0{}^{(2)}$.

The one hope, then, of carrying out a microscopic calculation of the properties of liquid helium would seem to lie in summing a selected class of higher order terms, as has been done by Brueckner and Sawada.[8] Their procedure is equivalent in the present formulation to the following approach. One takes, as in the extreme low-density limit, the effective potentials to be

$$\begin{aligned}\Sigma_{11}{}^{(a)} &= n[t^{(a)}(0\mathbf{q}0\mathbf{q}) + t^{(a)}(0\mathbf{q}\mathbf{q}0)],\\ \Sigma_{02}{}^{(a)} &= nt^{(a)}(00\mathbf{q}-\mathbf{q}),\\ \mu^{(a)} &= nt^{(a)}(0000).\end{aligned} \tag{9.5}$$

We have introduced the notation $t^{(a)}$ to reflect the fact that the $t^{(a)}$'s are now determined by a nonlinear integral equation, which is similar in form to (7.3), but in which the propagators are altered to include the effect of forward scattering via the effective potential, $\Sigma_{11}{}^{(a)}$. Thus (7.3) is replaced by

$$t_{12;34} = v_{12;34} + v_{12;56} G_5{}^{(a)} G_6{}^{(a)} t_{56;34}{}^{(a)}, \tag{9.6}$$

where

$$G^{(a)}(p,\epsilon) = 1/(\epsilon - \tfrac{1}{2}p^2 - \Sigma_{11}{}^{(a)} + \mu^{(a)} + i\delta). \tag{9.7}$$

In this fashion one is able to sum a selected class of higher-order diagrams.

It is always difficult to justify including some higher order terms and not others. For instance, certain terms which are of importance in the calculation of the logarithmic term, $E_0{}^{(3)}$, in the low-density expansion, are neglected in the Bruecker-Sawada approximation. Whether this neglect is justifiable remains to be seen. We believe the methods described in this paper offer a useful way to carry out such an investigation.

We may further remark that Brueckner and Sawada have neglected the depletion effect. This omission would appear to lead to nonnegligible corrections for a system of hard spheres of diameter 2.2 A at the density of liquid helium (this being the Brueckner-Sawada model for liquid helium). It is not difficult to see how to apply that correction in the present formulation. In the expressions, (9.5), one should properly have n_0 appearing in place of n, where n_0 is determined, in this order by

$$n - n_0 = \int \frac{d^3p}{(2\pi)^3} \int d\epsilon \frac{i}{2\pi} G'(p,\epsilon), \tag{9.8}$$

where $G'(p,\epsilon)$ differs from $G^{(1)}(p,\epsilon)$ only in that the modified effective potentials (with n_0) given by (9.5) are to be substituted in place of $\Sigma_{11}{}^{(1)}$ and $\Sigma_{02}{}^{(1)}$.

It is also not difficult, using the Green's function method, to formulate procedures which go well beyond Brueckner and Sawada in the number of higher-order terms which are summed. One approach, which has been proposed by one of us and Nozières,[22] consists in taking for the effective potentials

$$\begin{aligned}\Sigma_{11}{}^{(b)} &= n_0[t^{(b)}(0\mathbf{q}0\mathbf{q}) + t^{(b)}(0\mathbf{q}\mathbf{q}0)],\\ \Sigma_{02}{}^{(b)} &= n_0 t^{(b)}(00\mathbf{q}-\mathbf{q}),\\ \mu &= n_0 t^{(b)}(0000),\end{aligned} \tag{9.9}$$

where the $t^{(b)}$'s are determined by a nonlinear integral equation with the more accurate propagators, $G^{(b)}(p,\epsilon)$, defined in terms of the effective potentials (9.9), according to (4.12). Thus

$$t_{12;34}{}^{(b)} = v_{12;34} + v_{12;56} G_5{}^{(b)} G_6{}^{(b)} t_{56;34}{}^{(b)}, \tag{9.10}$$

where

$$G^{(b)}(p,\epsilon) = \frac{\epsilon + \frac{1}{2}p^2 + \Sigma_{11}{}^{(b)}(p,-\epsilon) - \mu^{(b)}}{\{\epsilon - \frac{1}{2}[\Sigma_{11}{}^{(b)}(p,\epsilon) - \Sigma_{11}{}^{(b)}(p,-\epsilon)]\}^2 - \{\frac{1}{2}p^2 + \frac{1}{2}[\Sigma_{11}{}^{(b)}(p,\epsilon) + \Sigma_{11}{}^{(b)}(p,-\epsilon)] - \mu^{(b)}\}^2 + [\Sigma_{02}{}^{(b)}(p,\epsilon)]^2},$$

and the relation between n_0 and n is to be determined according to (4.9) using the above value of $G^{(b)}(p,\epsilon)$. How successful such a procedure may be (it will certainly require extensive machine computations) remains to be seen.

10. CONCLUSION

We have shown that a simple treatment of the depletion of the ground state as a consequence of the interaction between particles permits one to apply the powerful methods of field theory to the many-boson problem. It is then possible to present a consistent divergence-free formulation of the problem to any order. This enables one to obtain certain general relationships between the quantities of interest in the theory. We were thus able to prove that the low-lying excitations of

[22] P. Nozières and D. Pines (private communication).

the system would not possess an energy gap. Another general relation, which perhaps can be proved by the methods of the present paper, is the equality which appears to obtain between the macroscopic and microscopic sound velocities.

The present approach also affords a straightforward way to calculate the series expansion of the properties of the dilute boson gas. We have calculated the next term beyond the results of Beliaev and Lee, Huang, and Yang, and find that the expansion is not a power series, but involves as well the logarithms of the expansion parameter $(nf_0^3)^{\frac{1}{2}}$. We have likewise seen that for the terms up to and including the $n^3 f_0^4 \ln(nf_0^3)$ term in the ground state energy, the character of the forces between the particles in a dilute boson gas is irrelevant; only the zero energy scattering amplitude enters. It is obvious that the series expansion has no meaning in the case of forces which are attractive, so that the scattering amplitude, f_0, is negative. In this case one would expect a complete breakdown of the perturbation-theoretic expansion in the low-density region (no condensed state, etc.) along with the appearance of two- or more-particle bound states.

11. ACKNOWLEDGMENTS

We should like to thank Professor C. N. Yang and Dr. K. Sawada for interesting discussions. We wish to thank Professor Robert Oppenheimer for the warm hospitality afforded us by the Institute for Advanced Study, and to acknowledge support from the Institute, from a National Science Foundation Grant (D.P.), and from a Fulbright Travel Grant (N.H.).

Electron-Phonon Interaction in Metals*

John Bardeen and David Pines†
Physics Department, University of Illinois, Urbana, Illinois
(Received April 4, 1955)

The role of electron-electron interactions in determining the electron-phonon interaction in metals is investigated by extending the Bohm-Pines collective description to take into account the ionic motion. Collective coordinates are introduced to describe the long-range electron-ionic correlations, and it is shown by a series of canonical transformations that these give rise to plasma waves and to coupled electron-ion waves which correspond to longitudinal sound waves. The dispersion relation for the sound waves is identical with that derived by Toya and Nakajima by self-consistent field methods. The velocity of these sound waves is calculated from first principles for sodium and is found to be in good agreement with experiment. The effective matrix element for the electron-phonon interaction is determined and is found to be identical for long wavelengths with that found earlier by Bardeen using a self-consistent field method which neglects exchange and correlation effects. The agreement with the earlier work is explained by the fact that the residual electron-electron interaction is of quite short range, so that an independent-particle treatment is rather well justified. The effects of Coulomb correlations on superconductivity are likewise shown to be small, so that the neglect of Coulomb interactions in the formulation of the superconductivity problem is justified.

1. INTRODUCTION

IN the coupled motion of electrons and ions in a metal, the tendency of the electrons to screen out the field of the ions is of primary importance. This screening greatly influences the effective matrix element for the electrical conductivity of the metal; it is also determining for the longitudinal sound wave velocity. The extent and effectiveness of this screening depend, in turn, on the electron-electron interaction in the metal. The usual theories of metallic electrical conduction treat these electron interactions on an independent-particle model in which each electron is assumed to move independently in a self-consistent field determined by the ions and the other electrons. Often no attempt is made to determine this self-consistent field; instead an empirical constant is introduced to describe the electron-lattice interaction. In 1937, one of us extended the self-consistent field method to take into account the motion of the ions, and thus to determine the matrix element for electron-lattice interactions.[1] The electron-electron interactions were treated in the Hartree approximation. More recently Nakajima[2] has given a simple field-theoretic treatment of this problem which also invokes the Hartree approximation and which leads to results essentially equivalent to *A*.

The independent-particle model has been successfully applied to a wide variety of problems. This success is somewhat surprising, because the marked correlations in electronic positions due to Coulomb interaction and exchange have been neglected. One of the main objects of this paper is to give a justification for this approach to the theory of conductivity.

The question of electronic correlations is of particular importance in the theory of superconductivity. Here one would expect that the large Coulomb correlation energy would play a more important role than the relatively small electron-lattice interaction energy in determining the transition from the normal to the superconducting state. However, experimentally this is not the case, as is shown by the isotope effect. We shall show why this is to be expected on theoretical grounds, and thereby justify the application of the independent-particle model to the formulation of this problem.

The physical basis for the unimportance of electron-electron correlations in conductivity and superconductivity arises from the tendency of the electrons to stay apart from one another in such a way that the field of a given electron is screened out within a distance of the order of the inter-particle spacing. Bohm and one of the authors have established this by showing that the long-range part of the Coulomb interaction leads only to coherent plasma oscillations of the electron gas and may be described in terms of these oscillations.[3] Energies of the plasma quanta are so high that the

* This research was supported in part by the Office of Ordnance Research, U. S. Army.
† Now at: Palmer Physical Laboratory, Princeton University, Princeton, New Jersey.
[1] J. Bardeen, Phys. Rev. 52, 688 (1937), hereafter referred to as A.
[2] S. Nakajima, Proceedings of the International Conference on Theoretical Physics, Kyoto and Tokyo, September, 1953 (Science Council of Japan, Tokyo, 1954).
[3] D. Bohm and D. Pines, Phys. Rev. 82, 625 (1951); 85, 338 (1952); 92, 609 (1953); and D. Pines, Phys. Rev. 92, 626 (1953), hereafter referred to as BP I, BP II, BP III, and P IV respectively.

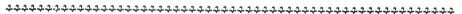

oscillations are not excited thermally. There remains the short-range screened interaction between the aforementioned electrons. This analysis shows why electron-electron collisions are not important for either thermal or electrical conduction in metals. Abrahams[4] has estimated the collision cross section and mean free path for electrons interacting via the screened Coulomb interaction in alkali metals. While the cross sections are of a reasonable order of magnitude ($\sim 10^{-15}$ cm^2) the scatterings possible are so greatly restricted by the exclusion principle that collisions are infrequent. The electron-electron mean free path is long compared with the electron-phonon mean free path at all temperatures of interest.

The BP collective description is extended here to treat the coupled system of moving electrons and ions. Plasma coordinates are introduced to take into account the effect of electron-electron interactions on the electron-ion system. It is found that the effect of the long-range interactions may then be described rather simply, and explicit expressions are derived for the effective electron-phonon matrix element and for the sound wave frequencies. There remains the short-range electron-electron interaction which, as we have mentioned, does not influence the motion of a given electron appreciably, so that it is not surprising that superconductivity is little affected by the electronic Coulomb correlations.

The results we obtain for the influence of the electronic response on the electron-phonon interaction are in good agreement with those obtained in A. This problem has also been treated from another point of view by Bohm and Staver,[5] who describe the ions and electrons as a set of coupled plasmas. They used a semi-classical approach closely related to BP II, and obtained the phonon dispersion relation. Our dispersion relation reduces to theirs in the appropriate limit.

The foregoing collective description is not applicable in the short wavelength region (corresponding to Fourier components of the electron-electron or electron-phonon interaction of wave vector greater than about the inverse particle spacing). Here the electron response to the ionic motion is probably best accounted for using the formulation of Nakajima. However, this formulation should be extended to include exchange effects, for it is precisely in this region (distances comparable or small compared to an electron de Broglie wavelength) that exchange begins to play an important role. We discuss the Nakajima method with exchange, but do not attempt to apply it to a detailed calculation because of the mathematical difficulties encountered in so doing.

2. DERIVATION OF HAMILTONIAN

We shall derive here for a monatomic crystal the Hamiltonian which will be used in the subsequent discussion. A number of approximations are made in order to simplify the calculations. It is assumed, as is usually done in the Bloch theory, that the matrix elements of the electron-lattice and Coulomb interactions depend only on the difference in wave vectors of the initial and final states. Another approximation is that lattice waves are either longitudinal or transverse, and that the electrons interact only with the longitudinal component. Both of these amount to neglect of anisotropic effects which greatly complicate the equations are probably important only for detailed quantitative calculations.

The general Hamiltonian is of the form

$$H=\sum_i (\mathbf{p}_i^2/2m)+\sum_{i,j} v(\mathbf{x}_i-\mathbf{X}_j)+H_{\text{ion-ion}}+H_{\text{Coul}}. \quad (2.1)$$

where i runs over the valence electrons, j over the ions, $v(\mathbf{x}_i-\mathbf{X}_j)$ represents the electron-ion interaction, $H_{\text{ion-ion}}$ includes the Coulomb and exchange repulsion of the ion cores and H_{Coul} the Coulomb interaction of the electrons. There are $n_0=ZN$ electrons/cm^3 where Z is the valency and N the ionic density. In order to eliminate infinities which appear in the separate terms of (2.1) (although not in the sum) we suppose there is subtracted from the electron-ion interaction the interaction of each electron with a uniform positive charge, from the electron-electron interaction the self-energy of a uniform negative charge, and from the ion-ion interaction the self-energy of a uniform positive charge. The sum of these adds to zero, so that the energy is unchanged. The ion-ion interaction energy less the self-energy of a uniform positive charge is equivalent to the energy of the ions in a uniform negative sea, including the self-energy of the negative charge.

We describe the electron wave function in second quantization by occupation numbers of a set of Bloch functions. It is assumed that the individual electrons move in a potential $V(\mathbf{x})$ which is the potential of the ions compensated by a uniform negative charge:

$$V(\mathbf{x})=\sum_j v(\mathbf{x}-\mathbf{X}_j)+\text{comp. charge}. \quad (2.2)$$

The Bloch equation is

$$[(\mathbf{p}^2/2m)+V(\mathbf{x})]\psi_\kappa=E_\kappa\psi_\kappa. \quad (2.3)$$

We have used the extended zone scheme, so that $\boldsymbol{\kappa}$ is not necessarily in the first Brillouin zone. A discrete set of $\boldsymbol{\kappa}$ values is obtained in the usual way by introduction of periodic boundary conditions, with N the number of atoms in a fundamental period of unit volume. Creation and destruction operators, $c_{\kappa s}^*$, $c_{\kappa s}$, are defined in the usual way and obey commutation laws for Fermi particles:

$$[c_{\kappa' s'}^*, c_{\kappa s}]^+=\delta_{\kappa\kappa'}\delta_{ss'}. \quad (2.4)$$

We shall omit the spin index s except where necessary for clarity.

The departure of the ions from equilibrium positions,

$$\delta\mathbf{X}_j=\mathbf{X}_j-\mathbf{X}_j^0, \quad (2.5)$$

[4] E. Abrahams, Phys. Rev. 95, 839 (1954).

[5] D. Bohm and T. Staver, Phys. Rev. 84, 836 (1952); T. Staver, Ph.D. thesis, Princeton University, 1952 (unpublished).

may be expressed in terms of normal coordinates, $q_{k\sigma}$ $(\sigma=1, 2, 3)$:

$$\delta \mathbf{X}_j = (NM)^{-\frac{1}{2}} \sum_{k\sigma} q_{k\sigma} \boldsymbol{\varepsilon}_{k\sigma} \exp[i\mathbf{k}\cdot\mathbf{X}_j^0], \qquad (2.6)$$

in which $\boldsymbol{\varepsilon}_{k\sigma}$ is a unit vector in the direction of polarization of the lattice waves, taken in the same sense for $\mathbf{k}$ and $-\mathbf{k}$ so that $q_{-k}=q_k^*$. We assume that one of these directions is longitudinal and two transverse, and that the transverse frequencies are determined entirely from ion-ion interactions. These assumptions are not really valid, particularly for short wavelengths for which the distinction between longitudinal and transverse waves is no longer sharp. While much of the analysis can be carried through without making these assumptions, the slight additional generality gained does not seem warranted by the increase in complexity of the equations. We shall use the symbol q_k without polarization subscript to denote the longitudinal mode. In some of the summations, $\mathbf{k}$ may run out of the first zone. In such a case, we always mean the q_k for the corresponding $\mathbf{k}$ in the first zone. The phonon Hamiltonian is

$$H_{\text{ph}} = \tfrac{1}{2} \sum_{\text{zone}} (p_k^* p_k + \Omega_k^2 q_k^* q_k), \qquad (2.7)$$

where Ω_k^2 is determined solely from ion-ion interactions. Calculation of the correct frequency, ω_k, which requires inclusion of the electron-lattice interactions, is discussed later.

The remaining parts of the Hamiltonian are expressed in terms of the c_κ's. The electron-lattice interaction may be written

$$\sum_{i,j} v(\mathbf{x}_i - \mathbf{X}_j) = \sum_{i,j} v(\mathbf{x}_i - \mathbf{X}_j^0) - (NM)^{-\frac{1}{2}} \sum_k \boldsymbol{\varepsilon}_k \cdot \nabla v(x - X_j^0) q_k \exp[i\mathbf{k}\cdot\mathbf{X}_j^0]. \qquad (2.8)$$

The first term may be combined with the kinetic energy of the electrons to give

$$H_{\text{el}} = \sum_i [(\mathbf{p}_i^2/2m) + V(\mathbf{x}_i)] = \sum_{\kappa s} E_\kappa c_{\kappa s}^* c_{\kappa s}. \qquad (2.9)$$

The second term of (2.8) may be expressed in terms of the matrix elements,

$$v_k^i = -(NM)^{-\frac{1}{2}} \int \psi_{\kappa+k}^* \{\sum_j \boldsymbol{\varepsilon}_k \cdot \nabla v(\mathbf{x}_i - \mathbf{X}_j^0) \times \exp[i\mathbf{k}\cdot\mathbf{X}_j^0]\} \psi_\kappa d\tau, \qquad (2.10)$$

assumed to be independent of $\boldsymbol{\kappa}$, to give

$$H_{\text{int}} = \sum_{\kappa k s} c_{\kappa+k,s}^* c_{\kappa,s} q_k v_k^i = \sum_k q_k v_k^i \rho_{-k}, \qquad (2.11)$$

where

$$\rho_k = \sum_{\kappa s} c_{\kappa-k,s}^* c_{\kappa,s}; \quad \rho_{-k} = \sum_{\kappa s} c_{\kappa,s}^* c_{\kappa-k,s}. \qquad (2.12)$$

The sum over $\mathbf{k}$ and $\boldsymbol{\kappa}$ extend over all values; q_k refers to the reduced vector in the first zone. Note that $(v_k)^* = v_{-k}^i$. The Coulomb interaction may be expressed in the form

$$H_{\text{Coul}} = \tfrac{1}{2} \sum_k M_k^2 \rho_{-k} \rho_k, \qquad (2.13)$$

where for free electrons, $M_k^2 = 4\pi e^2/k^2$. It has again been assumed that the matrix elements depend only on the vector difference between initial and final states.

The Hamiltonian is now expressed in the form

$$H = E_{\text{ion-ion}} + H_{\text{tr}} + H_1,$$

where

$$H_1 = H_{\text{el}} + H_{\text{ph}} + H_{\text{int}} + H_{\text{Coul}}, \qquad (2.14)$$

and $E_{\text{ion-ion}}$ is the ion-ion interaction energy for the ions in equilibrium positions, H_{tr} is the Hamiltonian for transverse phonons, and the terms in H_1 are given by (2.7), (2.9), (2.11), and (2.13). From now on, we shall be concerned with the Hamiltonian H_1. In Sec. III, we discuss earlier methods for calculation of the electron-lattice interaction by self-consistent field methods and in Sec. 4 we introduce plasma coordinates to treat the long-range part of the Coulomb interaction.

3. EARLIER METHODS FOR THE CALCULATION OF v_k AND ω_k

We are concerned in this section with a review of calculations which have been made of the effect of the shielding of the valence electrons on the interaction potential and on the sound-wave vibrational frequencies. For long wavelengths, the interaction of a given electron with the ionic motion is radically altered by the field due to the other electrons, which in the course of responding to the ionic motion move in just such a way as to produce a field which very nearly cancels the ionic field. The effective matrix element for the electron-lattice interaction, v_k, may be written as the sum of v_k^i due to the motion of the ions and v_k^p due to the motion of the electrons:

$$v_k = v_k^i + v_k^p. \qquad (3.1)$$

In A, v_k^p was computed by what amounts to a Hartree self-consistent field calculation. It was assumed that the wave functions of the individual electrons change adiabatically with the ion motion. Recently Toya has determined the longitudinal sound wave frequency, ω_k, using this same model.[6] We here give a simple derivation of the effect of the electron response on v_k and ω_k which agrees in the long wavelength limit with the foregoing calculations. Our treatment parallels in some respects that of Bohm and Staver.

According to (2.14) the equation of motion of the $\mathbf{k}$th sound wave amplitude is

$$\ddot{q}_k + \Omega_k^2 q_k = -v_{-k}^i \rho_k. \qquad (3.2)$$

The electronic density fluctuation ρ_k consists of two parts; one, ρ_k^0, associated with the motion of electrons in the absence of a sound wave, and another, $\delta\rho_k$ which represents the motion of the electrons associated with the sound wave. For pure Coulomb interactions between free electrons $\delta\rho_k$ is related to the Fourier component of that part of the effective one-electron interaction potential v_k associated with the electronic re-

[6] T. Toya, Busseiron Kenkyu **59**, 179 (1952).

sponse v_k^ρ, by Poisson's equation; $k^2 v_k^\rho q_k = 4\pi e^2 \delta\rho_k$, or more generally,

$$v_k^\rho q_k = M_k^2 \delta\rho_k. \quad (3.3)$$

Thus we may write for our sound-wave, q_k,

$$\ddot{q}_k + \omega_k^2 q_k = -v_{-k}^i \rho_k^0, \quad (3.4)$$

with ω_k specified by the dispersion relation

$$\omega_k^2 = \Omega_k^2 + M_k^{-2} v_{-k}^i v_k^\rho. \quad (3.5)$$

Equation (3.4) describes the interaction of the sound waves with the *independently* moving electrons (here described in terms of their density fluctuation ρ_k^0) and in principle the resistance may be calculated from it.

The matrix element v_k^ρ may be simply calculated in the Fermi-Thomas approximation.[7] In this approximation the electron density, $\rho(\mathbf{x})$, is proportional to $[E_0 - \delta V(\mathbf{x})]^{\frac{3}{2}}$, where E_0 is the energy of an electron at the top of the Fermi distribution and $\delta V(\mathbf{x})$ is the effective potential acting on the electrons due to the ionic motion. We then have for the density change $\delta\rho(\mathbf{x})$ associated with the sound-wave potential $\delta V(\mathbf{x})$,

$$\delta\rho(\mathbf{x}) = -\frac{3}{2}\frac{n_0}{E_0}\delta V(\mathbf{x}). \quad (3.6)$$

We Fourier-analyze (3.6) and apply (3.1) to obtain

$$\delta\rho_k = -\frac{3}{2}\frac{n_0}{E_0} v_k q_k = -\frac{3}{2}\frac{n_0}{E_0}(v_k^i + v_k^\rho) q_k. \quad (3.7)$$

If we now compare (3.7) with (3.3), we see that the requirement that our treatment be self-consistent yields

$$v_k^\rho = \frac{-v_k^i}{1 + (2E_0/3n_0 M_k^2)}. \quad (3.8)$$

For long wavelengths, M_k^2 is given by its free-electron value, $4\pi e^2/k^2$, and we have

$$v_k^\rho = \frac{-v_k^i}{1 + (k^2 v_0^2/3\omega_p^2)} \cong -v_k^i\left(1 - \frac{k^2 v_0^2}{3\omega_p^2}\right), \quad (3.9)$$

where v_0 is the velocity of an electron at the top of the Fermi distribution and

$$\omega_p^2 = 4\pi n e^2/m \quad (3.10)$$

is the square of the plasma frequency. Thus we see that the effect of the field of the other electrons is to screen out the "bare" electron-ion interaction within a distance of order ω_p/v_0. For long wavelengths the effective matrix element for electron-phonon interaction is thus drastically reduced to

$$v_k \cong (k^2 v_0^2/3\omega_p^2) v_k^i. \quad (3.11)$$

The corresponding expression for the sound-wave frequencies is

$$\omega_k^2 = \Omega_k^2 - \frac{k^2}{4\pi e^2} v_{-k}^i v_k^i \left(1 - \frac{k^2 v_0^2}{3\omega_p^2}\right) \quad (3.12)$$

Our expressions (3.9) for v_k and (3.12) for ω_k agree in this limit with the corresponding results of A and of Toya.

We may obtain a quite general expression for ω_k if we neglect the effect of lattice periodicity on the Ω_k and v_k^i and consider only Coulombic electron-ion and ion-ion interactions. This corresponds to treating the positive ions as a plasma; the "bare" phonon frequency is then

$$\Omega_k^2 \cong \Omega_P^2 = 4\pi N Z^2 e^2/M. \quad (3.13)$$

The v_k^i are given by

$$v_k^i \cong -(4\pi Z e^2 i/k)(N/M)^{\frac{1}{2}}. \quad (3.14)$$

We then have

$$\omega_k^2 = k^2 v_0^2 \Omega_P^2/3\omega_P^2 = m Z^2 v_0^2 k^2/3M. \quad (3.15)$$

Thus in this approximation the square of the phonon frequency has been reduced by the same factor as the electron-phonon interaction matrix element,[8] the phonon frequency is altered from a constant to one linear in $\mathbf{k}$, with a sound velocity c_l, which depends only on the density and which is given by

$$c_l = m Z v_0/\sqrt{3}M. \quad (3.16)$$

Equation (3.16) was first obtained by Bohm and Staver,[5] and as shown by Staver[5] is in fairly good agreement with the experimental observations. The extent of this agreement (~20% for the alkali metals, no worse than a factor of two for any metal) indicates that one can get a good order of magnitude estimate, by such simple Coulombic considerations. In Appendix A, we discuss the results obtained with our more accurate dispersion relation (3.12). It is interesting to note that this same simplified model leads to a Sommerfeld-Bethe interaction constant $C = E_F/\sqrt{2}$.

Nakajima's field-theoretic derivation for v_k^ρ and ω_k goes somewhat beyond the adiabatic approximation but gives nearly equivalent results. We sketch his derivation here because as we shall see, it offers the best framework for the treatment of the short wavelength electron-phonon interaction. Frohlich[9] has used a field-theoretic method which differs in some respects. He used an individual particle model which did not include Coulomb interactions between the electrons. His choice of creation and destruction operators and method of analyzing is such as to give the correct result for his problem when the change in the frequency from electron-phonon interactions is small. However, his dis-

[7] See also A. W. Overhauser, Phys. Rev. 89, 689 (1953).

[8] The same shielding factor occurs because in the limit we are considering we may write $\omega^2 = n k^2 v_k^i (v_k^i + v_k^\rho)/m$, so that the sound wave frequencies are proportional to the effective matrix element v_k.

[9] H. Frohlich, Proc. Roy. Soc. (London) A215, 291 (1952).

persion relation is not correct when the change in frequency is large.

Nakajima writes the Hamiltonian, H_1, in a form equivalent to the following:

$$H_1=\sum_{\kappa s} E_\kappa c_{\kappa s}^* c_{\kappa s}+\tfrac{1}{2}\sum_{\text{zone}} (p_k^* p_k+\omega_k^2 q_k^* q_k) +\sum_k v_k q_k \rho_{-k}+\tfrac{1}{2}\sum_k M_k^2 \rho_k\rho_{-k}+\sum_k (v_k^i - v_k) q_k\rho_{-k} +\tfrac{1}{2}\sum_{\text{zone}} (\Omega_k^2-\omega_k^2) q_k^* q_k. \quad (3.17)$$

A canonical transformation is made to eliminate the first linear term in q_k. The interaction v_k is determined so as to eliminate the second linear terms in q_k, the second line, and ω_k^2 is chosen so as to cancel diagonal terms in $q_k^* q_k$ in the last line.

The transformed Hamiltonian is

$$H_1'=e^{-iS/\hbar}H_1 e^{iS/\hbar}=H_1+i\hbar^{-1}[H_1,S] -\tfrac{1}{2}\hbar^{-2}[[H_1,S],S]+\text{etc.} \quad (3.18)$$

For S we take

$$S=i\sum_{k\kappa} c_\kappa^* c_{\kappa-k}[f(\mathbf{k},\boldsymbol{\kappa})q_k - ig(\mathbf{k},\boldsymbol{\kappa})p_{-k}]. \quad (3.19)$$

The required commutators are:

$$[\sum_\kappa E_\kappa c_\kappa^* c_\kappa, S]=i\sum (E_\kappa - E_{\kappa-k})c_\kappa^* c_{\kappa-k} \times[f(\mathbf{k},\boldsymbol{\kappa})q_k - ig(\mathbf{k},\boldsymbol{\kappa})p_{-k}]. \quad (3.20)$$

$$[\tfrac{1}{2}\sum (p_k^* p_k+\omega_k^2 q_k^* q_k),S]=i\sum c_\kappa^* c_{\kappa-k} \times\{-i\hbar p_k f(\mathbf{k},\boldsymbol{\kappa})+\hbar\omega_k^2 q_k g(\mathbf{k},\boldsymbol{\kappa})\}. \quad (3.21)$$

$$[\rho_{-k}\rho_k,S]=\rho_{-k}[\rho_k,S]+[\rho_{-k},S]\rho_k. \quad (3.22)$$

$$[\rho_k,S]=i\sum_{\kappa,k',\kappa'} \{\delta_{\kappa\kappa'} c_{\kappa-k}^* c_{\kappa'-k'} - \delta_{\kappa'-k',\kappa-k} c_{\kappa'}^* c_\kappa^*\} \times\{f(\mathbf{k}',\boldsymbol{\kappa}')q_{k'} - ig(\mathbf{k}',\boldsymbol{\kappa}')p_{-k'}\}. \quad (3.23)$$

Diagonal terms are obtained if $\mathbf{k}'=\mathbf{k}$. If we keep only these, we may treat $[\rho_k,S]$ as a c-number. This is the procedure which Nakajima used, and is essentially equivalent to the Hartree approximation since there are evidently diagonal exchange terms in $\rho_{-k}[\rho_k,S]$ which are neglected. In this approximation, then,

$$\tfrac{1}{2}\sum_{\mathbf{k}} M_k^2[\rho_{-k}\rho_k,S]=i\sum_{\kappa,\mathbf{k},s} M_k^2\rho_{-k}\{n(\boldsymbol{\kappa}-\mathbf{k})-n(\boldsymbol{\kappa})\} \times\{f(\mathbf{k},\boldsymbol{\kappa})g_k - ig(\mathbf{k},\boldsymbol{\kappa})p_{-k}\}. \quad (3.24)$$

Elimination of linear terms in q_k and p_{-k} in the first line gives the following for f and g:

$$g(\mathbf{k},\boldsymbol{\kappa})=-\frac{\hbar f(\mathbf{k},\boldsymbol{\kappa})}{E_\kappa - E_{\kappa-k}}, \quad (3.25)$$

$$f(\mathbf{k},\boldsymbol{\kappa})=\frac{\hbar(E_\kappa - E_{\kappa-k})v_k}{(E_\kappa - E_{\kappa-k})^2-\hbar^2\omega_k^2}. \quad (3.26)$$

Principal parts are to be taken in the sums over the energy denominators. Elimination of linear terms in q_k from the second line then gives

$$\sum_{k,s} M_k^2\{n(\boldsymbol{\kappa}-\mathbf{k})-n(\boldsymbol{\kappa})\}f(\mathbf{k},\boldsymbol{\kappa})=\hbar(v_k^i - v_k), \quad (3.27)$$

or, substituting for $f(\mathbf{k},\boldsymbol{\kappa})$,

$$v_k^i - v_k = -M_k^2 v_k \sum_{\kappa,s} \frac{(E_\kappa - E_{\kappa-k})(n(\boldsymbol{\kappa})-n(\boldsymbol{\kappa}-\mathbf{k}))}{(E_\kappa - E_{\kappa-k})^2-\hbar^2\omega_k^2} = M_k^2 v_k \sum_{\kappa,s} \frac{n(\boldsymbol{\kappa})-n(\boldsymbol{\kappa}-\mathbf{k})}{E_{\kappa-k}-E_\kappa+\hbar\omega_k}. \quad (3.28)$$

which can be solved readily to give v_k. Terms in p_{-k} vanish for a symmetrical distribution, such that $n(\boldsymbol{\kappa})=n(-\boldsymbol{\kappa})$.

Finally, elimination of diagonal terms in $q_k^* q_k$ gives the following equation for ω_k^2:

$$\Omega_k^2-\omega_k^2=-\mathcal{S} v_{-k}^i v_k \sum_{\kappa,s} \frac{n(\boldsymbol{\kappa}-\mathbf{k})-n(\boldsymbol{\kappa})}{E_{\kappa-k}-E_\kappa+\hbar\omega_k}. \quad (3.29)$$

The $\boldsymbol{\kappa}$ on the left refers to the reduced vector in the first Brillouin zone. The sum $\mathcal{S}$ on the right is over all $\mathbf{k}$ which correspond to this same reduced wave vector, so that transitions due to Umklapp processes may be included. Contributions of electrons from different bands, if treated separately should also be added to the right-hand side. An alternative expression is

$$\Omega_k^2-\omega_k^2=\mathcal{S} M_k^{-2} v_{-k}^i (v_k^i - v_k). \quad (3.30)$$

This is of the same form as Eq. (3.5). Equation (3.28) for the matrix element agrees with the 1937 result if $\hbar\omega_k$ in the denominator is neglected. Since $\hbar\omega_k \sim 10^{-4} E_F$, where E_F is the Fermi energy, the term will have practically no effect on the matrix element or frequency.

It should be noted that the Coulomb interaction between electrons still appears in the transformed Hamiltonian and we are still left with the problem of taking into account correlation effects on electron motion. The transformation does serve to introduce a set of oscillators which to second order are not coupled with the individual electrons. Scattering of electrons by oscillators is accounted for by interactions which conserve energy; i.e., those with zero energy denominators in the expansion. When summed by taking principal parts, these give a negligible contribution to the matrix element.

The exchange terms which Nakajima neglects are considered in Appendix B. Perhaps the most important effect of these terms is to add an exchange energy, W_κ, to the individual particle energy. It is known that such a term leads to an abnormally small density of states at the Fermi surface, and a correspondingly small low-temperature specific heat, which is contrary to observation.[10] This has been explained by the collective description in which the fields of the individual electrons are effectively screened by the other electrons. The exchange terms are drastically reduced, and one finds that the calculated specific heat is not far from the free-electron value neglecting exchange.[11] It is

[10] J. Bardeen, Phys. Rev. **50**, 1098 (1936).
[11] See P IV for a derivation of this result.

likely that inclusion of the unshielded exchange energy in the Nakajima calculation would lead to an incorrect electron-lattice interaction for long wavelengths. We shall show in Sec. 4 that the collective description is such as to introduce a shielded interaction into the calculation of the short-wavelength exchange terms, and no exchange effects at all appear for long wavelengths.

There remains the question of convergence of the expansion of the canonical transformation in a power series in v_k. The transformation is such as to introduce new individual particle wave functions which depend on the q_k's. The expansion coefficients are, neglecting $\hbar^2\omega_k^2$ in the denominator,

$$\hbar^{-1}f(\mathbf{k},\boldsymbol{\kappa})\cong\frac{v_k q_k}{E_{\kappa-k}-E_\kappa}. \tag{3.31}$$

One may expect that the expansion will converge rapidly if

$$\sum_k \frac{v_k^* v_k q_k^* q_k}{(E_{\kappa-k}-E_\kappa)^2}\ll 1. \tag{3.32}$$

The v_k are small and difficulty arises only from the terms for which the energy denominators are correspondingly small.

We shall now show that since principal parts are used, terms with small energy denominators contribute a negligible amount to the calculation of the matrix elements and frequency. Our analysis is similar to that of Frohlich[9] who suggested omitting from the transformation those $f(\mathbf{k},\boldsymbol{\kappa})$ which have small energy denominators, i.e., those for which

$$|E_{\kappa-k}-E_\kappa|<\Delta E. \tag{3.33}$$

When the same is done in our calculation, we find that we can choose ΔE large enough so that (3.32) is satisfied and small enough to have a negligible effect on the self-consistent field. To estimate the orders of magnitude involved, let us replace the matrix element by an average value, so that (3.32) becomes

$$|v_k q_k|_{\mathrm{Av}}^2\left\{\int_0^{E_\kappa-\Delta E}\frac{N(E)dE}{(E-E_\kappa)^2}+\int_{E_\kappa+\Delta E}^{E_F}\frac{N(E)dE}{(E-E_\kappa)^2}\right\}\ll 1, \tag{3.34}$$

where $N(E)$ is the density of states in energy. This is roughly equivalent to

$$N(E_\kappa)|v_k q_k|_{\mathrm{Av}}^2\ll\Delta E. \tag{3.35}$$

If terms which satisfy (3.33) are omitted, what is the effect on the principal parts summation? The integral is of order:

$$\int_0^{E_F}\frac{N(E)dE}{E-E_\kappa}\sim N(E). \tag{3.36}$$

The omitted terms contribute roughly

$$\int_{E_\kappa-\Delta E}^{E_\kappa+\Delta E}\frac{N(E)dE}{E-E_\kappa}\cong 2\frac{dN}{dE}\Delta E. \tag{3.37}$$

The relative error is thus of order

$$\frac{2}{N(E)}\frac{dN}{dE}\Delta E\sim\frac{\Delta E}{E_F}. \tag{3.38}$$

The left-hand side of (3.35) is of order $\hbar\omega_{\max}\sim 10^{-4}E_F$ for most metals. Thus if one takes $\Delta E\sim 10^{-2}E_F$, one may satisfy (3.32) and at the same time not affect appreciably the calculation of the matrix element and vibrational frequency.

It is, however, just these omitted terms which are all important in the theory of superconductivity.[12] In fact, the criterion for the occurrence of superconductivity is roughly that

$$N(E)|v_k q_k|_{\mathrm{Av}}^2>\hbar\omega_{\max}. \tag{3.39}$$

The conclusion is that even though the electron-lattice interaction is so large as to give superconductivity, one may still use the expansion in powers of v_k for those terms which give almost the entire contribution to the matrix element and vibrational frequency. This is true because the virtual transitions which give difficulty are only a very small fraction of the total and do not contribute abnormally.

4. COLLECTIVE DESCRIPTION OF ELECTRON-ION INTERACTION

In this section we wish to treat the electron response to the ionic motion by describing the electrons in terms of the appropriate plasma variables. As pointed out by BP, such a description of the electrons is only appropriate for long wavelengths, corresponding to $k<k_c$, where k_c is a critical wave vector (or the order of k_0) which is determined by minimizing the total system energy as discussed in BP III. In this section we shall therefore only apply the collective description to the long-wavelength phonons, their interaction with the electrons, and the long-range electron-electron interactions (all corresponding to Fourier components with wave vectors less than k_c). We follow a procedure closely related to that used by one of us in dealing with the electron-electron interaction problem in the absence of ionic motion.[13]

Our basic aim in the collective description is the introduction of a new set of field variables which describe the independent plasma oscillations of the system as a whole. We may do this by first introducing a new field into our basic Hamiltonian, and then carrying out a series of canonical transformations which enable us to relate this field variable to the plasma oscillations we

[12] H. Frohlich, Phys. Rev. 79, 845 (1950); J. Bardeen, Revs. Modern Phys. 23, 261 (1951).

[13] D. Pines, Report to the Tenth Solvay Conference, Brussels (1954).

wish to describe. As a first step, we add a field-energy term to our initial Hamiltonian, (2.14) so that our Hamiltonian is now

$$H=H_1+\tfrac{1}{2}\sum_{|\mathbf{k}|<k_c} P_k{}^*P_k. \tag{4.1}$$

The P_k are here a quite arbitrary set of field variables, as yet undefined, which commute with all operators appearing in H_1. In order that the energy and number of degrees of freedom of our system remain unchanged by this field, we must further impose a set of subsidiary conditions on our combined system wave function:

$$P_k\Psi=0 \quad (|\mathbf{k}|<k_c). \tag{4.2}$$

We now wish to transform to a representation in which the P_k will describe independent plasma oscillations. In the absence of ionic motion, the ρ_k in the long-wavelength limit describe almost free collective oscillation, and the appropriate transformation would relate the P_k to the ρ_k. In our case, since the electrons also interact with the phonons, we would expect that it is only some linear combination of density fluctuations and phonon field variables which would carry out uncoupled collective plasma oscillation. Thus we are led to try a canonical transformation which relates the P_k to both ρ_k and the q_k. The first transformation we consider is generated by

$$S=\sum_{|\mathbf{k}|<k_c}(-iM_k\rho_{-k}+u_kq_{-k})Q_k, \tag{4.3}$$

where Q_k is the coordinate conjugate to the plasma field momentum P_k, and u_k is a real constant to be determined. After this transformation our subsidiary condition, (4.2) becomes

$$e^{-iS/\hbar}P_ke^{iS/\hbar}\Psi=[P_k-iM_k\rho_{-k}+u_kq_{-k}]\Psi=0. \tag{4.4}$$

Our Hamiltonian (4.1) is transformed to

$$\begin{aligned}H=&\sum_{\kappa}E_\kappa\, c_\kappa{}^*c_\kappa+\tfrac{1}{2}\sum_{|\mathbf{k}|<k_c}\{p_k{}^*p_k+(\Omega_k{}^2-\omega_k{}^2)q_k{}^*q_k\}\\&+\tfrac{1}{2}\sum_{|\mathbf{k}|<k_c}\{P_k{}^*P_k+(\omega_p{}^2+u_k{}^2)Q_k{}^*Q_k\}\\&+\sum_{|\mathbf{k}|<k_c}\{v_k{}^i-iM_ku_k\}q_k\rho_k+\sum_{|\mathbf{k}|<k_c}u_k\rho_k{}^*Q_k\\&-\sum_{|\mathbf{k}|<k_c,\kappa}M_\kappa\frac{\hbar\mathbf{k}}{m}\cdot(\boldsymbol{\kappa}-\tfrac{1}{2}\mathbf{k})c_\kappa{}^*c_{\kappa-k}Q_k\\&+\tfrac{1}{2}\sum_{|\mathbf{k}|>k_c}\{p_k{}^*p_k+\Omega_k{}^2q_k{}^*q_k\}+\sum_{|\mathbf{k}|>k_c}v_k{}^iq_k\rho_{-k}\\&+\tfrac{1}{2}\sum_{|\mathbf{k}|>k_c}M_k{}^2\rho_{-k}\rho_k.\end{aligned} \tag{4.5}$$

In obtaining this expression we have made use of the new subsidiary condition, (4.4), and we have applied the "random phase" approximation of BP.[14] We have also applied the effective mass approximation, $E(\boldsymbol{\kappa})=\hbar^2\boldsymbol{\kappa}^2/2m$.[15]

The first three terms of our new Hamiltonian, (4.5), describe the one-electron energy levels, the phonon field, and the plasma field. The next three describe the electron-phonon interaction, plasma-phonon interaction, and plasma-electron interaction, while the remaining terms describe the short-wavelength phonons, their interaction with the electrons, and the short-range electron-electron interaction. We see that we have here redescribed the long-range electron-electron interactions in terms of the plasma oscillations, which however are not yet "isolated," in the sense that there remains a plasma-electron and plasma-phonon interaction, and the plasma field variable still appears in the subsidiary condition. The effect of the electron-electron interactions on the effective interaction matrix element and on the sound wave frequency is contained in the u_k, which as we see measure the coupling between the plasma waves and the phonons. The u_k are simply related to the $v_k{}^\bullet$ we introduced in the preceding section to describe the effect of electron shielding on the v_k; we have

$$v_k{}^\bullet=-iM_ku_k. \tag{4.6}$$

Our remaining problem is thus the determination of u_k and the new sound wave frequencies.

We may actually do this without explicitly solving for the effect of the electron-plasma interaction and plasma-phonon interaction on the problem at hand. For, as is the case in the absence of ionic motion, the coupling between the electrons and the plasma oscillations is weak. Furthermore, because of the great disparity between the plasma frequencies and phonon frequencies (a disparity of order m/M), the coupling between the plasma waves and the phonons is likewise weak. Both of these interactions may easily be taken into account by appropriate canonical transformations; however because of the weakness of the coupling, such transformations will leave the relevant terms in our Hamiltonian essentially unaltered, so we do not need to carry them out explicitly here.

How, then, do we choose the u_k? It is clear from the discussion of the preceding section that we must choose the u_k in such a way that our treatment is self-consistent. The requirement of self-consistency appears in the following guise. In our problem the electrons and phonons are coupled not only in the Hamiltonian (4.8) but in the subsidiary condition (4.9) as well. The requirement of self-consistency is then just the requirement that the coupling via the subsidiary condition be completely equivalent to the coupling via the Hamil-

[14] If the phonon terms in (4.5) are suppressed (and $u_k=0$), it may be seen that the form of (4.5) is identical with the Hamiltonian used in BP III as the starting point for the collective description.

[15] The use of the effective mass approximation for this problem is discussed by J. Hubbard, Proc. Phys. Soc. (London) **A67**, 1058 (1954); P. Wolff, Phys. Rev. **95**, 56 (1954). E. N. Adams II [Phys. Rev. **98**, 947 (1955)] has pointed out that interband transitions may have an important effect. N. F. Mott [Proceedings of the Tenth Solvay Conference, Brussels (1954)] has given reasons for expecting that the free electron mass rather than an effectual mass would appear in many cases.

tonian. This guarantees that we introduced the correct admixture of phonon coordinates (as determined by the electron-phonon interaction) in the plasma oscillation amplitude. We obtain an explicit solution for the u_k and the sound-wave frequencies ω_k by carrying out a canonical transformation which eliminates to a given order of accuracy the electron-phonon interaction terms in the Hamiltonian, (4.5). Our self-consistency requirement is then that to this same order of accuracy there be no coupling of the sound waves to the electrons in the transformed subsidiary condition; this will be the case if the phonon variable no longer appears in the transformed subsidiary condition in this order.

Our desired canonical transformation is just that generated by (3.19), since we have identified v_k with $v_k{}^i - iM_k\rho_{-k}$. As before $f(\mathbf{k},\boldsymbol{\kappa})$ and $g(\mathbf{k},\boldsymbol{\kappa})$ are defined through (3.25) and (3.26), and as discussed in the preceding section a perturbation theory expansion in powers of v_k is valid. Our transformed Hamiltonian is[16]

$$\begin{aligned} H=&\sum_{\kappa} E_\kappa c_\kappa{}^* c_\kappa + \tfrac{1}{2}\sum_{|\mathbf{k}|<k_c}\{p_k{}^*p_k+\omega_k{}^2 q_k{}^*q_k\} \\ &+\tfrac{1}{2}\sum_{|\mathbf{k}|<k_c}\{P_k{}^*P_k+(\omega_p{}^2+u_k{}^2)Q_k{}^*Q_k\} \\ &-\sum_{|\mathbf{k}|<k_c,\kappa} M_\kappa \frac{\hbar\mathbf{k}}{m}\cdot(\boldsymbol{\kappa}-\tfrac{1}{2}\mathbf{k})c_\kappa{}^*c_{\kappa-k}Q_k \\ &+\tfrac{1}{2}\sum_{|\mathbf{k}|>k_c}(p_k{}^*p_k+\Omega_k{}^2q_k{}^*q_k)+\sum_{|\mathbf{k}|>k_c} v_k{}^i q_k\rho_{-k} \\ &+\tfrac{1}{2}\sum_{|\mathbf{k}|>k_c} M_k{}^2\rho_k\rho_{-k} \\ &-\tfrac{1}{2}\sum_{|\mathbf{k}|<k_c,\kappa} v_{-k}\rho_k c_\kappa{}^*c_{\kappa-k}g(\mathbf{k},\boldsymbol{\kappa}). \end{aligned} \quad (4.7)$$

The dispersion relation for our "uncoupled" phonons is

$$\omega_k{}^2=\Omega_k{}^2-u_k{}^2+v_k{}^2\sum_\kappa \frac{n(\boldsymbol{\kappa}-\mathbf{k})-n(\boldsymbol{\kappa})}{\hbar\omega_k+E_{\kappa-k}-E_\kappa} \quad (4.8a)$$

We further find, using (3.23) that our self-consistency requirement that q_k no longer appear in the subsidiary condition yields:

$$u_k=-iM_k v_k\sum_\kappa \frac{n(\boldsymbol{\kappa}-\mathbf{k})-n(\boldsymbol{\kappa})}{\hbar\omega_k+E(\boldsymbol{\kappa}-\mathbf{k})-E(\boldsymbol{\kappa})}. \quad (4.8b)$$

On applying (4.6) we see that this choice of u_k (or v_k) is in complete agreement with the result of Nakajima (3.28). Furthermore, on combining (4.8a) with (4.8b) we obtain the by now familiar dispersion relation (3.5).

A transformation similar to that of Nakajima may be used to eliminate the terms linear in q_k for $|\mathbf{k}|>k_c$. The result of both transformations is to replace the electron-lattice interaction by an interaction between electrons:

$$\begin{aligned} H_2=&-\tfrac{1}{2}\sum_{\kappa,|\mathbf{k}|<k_c} v_{-k}\rho_k c_\kappa{}^*c_{\kappa-k}g(\mathbf{k},\boldsymbol{\kappa}) \\ &-\tfrac{1}{2}\sum_{\kappa,|\mathbf{k}|>k_c} v_{-k}{}^i\rho_k c_\kappa{}^*c_{\kappa-k}g(\mathbf{k},\boldsymbol{\kappa}). \end{aligned} \quad (4.9)$$

[16] We have omitted the weak plasma-phonon interaction term.

Note that the unshielded interaction, $v_{-k}{}^i$, appears from the Nakajima transformation. Since $v_{-k}{}^i$ becomes infinite while v_{-k} approaches zero as $\mathbf{k}\to 0$, the difference is particularly important when $\mathbf{k}$ is small.

The interaction term, H_2, is similar to one derived by Frohlich[9] without explicit introduction of Coulomb interactions. His interaction constant is to be identified with the shielded interaction, v_k. As he points out, the usual second-order perturbation theory expression for the electron-lattice interaction energy is the sum of the diagonal component of H_2 (which Frohlich calls E_2) and the change in zero-point energy of the oscillators. It is the true frequency, ω_k, rather than Ω_k, which enters into these expressions.

There are several other important differences between our treatment and that of Nakajima. In (4.7) the long-range Coulomb interactions no longer appear, but have been redescribed by the high-frequency plasma oscillations. There remains the weak plasma-electron interaction but this may be easily transformed away, as is done in BP III, and it may then be seen that only effective residual electron-electron interaction is that described by

$$H_{\text{s.r.}}=\tfrac{1}{2}\sum_{|\mathbf{k}|>k_c} M_k{}^2\rho_k\rho_{-k}.$$

Thus the long-range electron-electron correlations have been described in terms of the plasma oscillations, and are seen explicitly not to affect the electron response to the long wavelength phonons. We further note that no exchange effects are associated with the long-range part of the electron-electron interactions, so that the use of the Hartree approximation in determining v_k for $|\mathbf{k}|<k_c$ is completely justified.

There remains the complication introduced by the transformed subsidiary condition, which on applying (4.8b) agrees with that for the electron-plasma system in the absence of ionic motion:

$$(P_k-iM_k\rho_{-k})\Psi=0. \quad (4.10)$$

This complication has been discussed in BP III. The same transformation which eliminates the plasma-electron interaction in the Hamiltonian (4.7) also eliminates the plasma momentum from the subsidiary condition (4.10). There remains then a subsidiary condition acting on the electrons alone. As shown in BP III this subsidiary condition will be automatically satisfied for the lowest state of the system. It will influence the excited states in such a way as to reduce the effective number of electronic degrees of freedom slightly, but because the number of plasma degrees of freedom is relatively small (of order 5%–10%) this reduction should be correspondingly small.

We still have in our Hamiltonian (4.7) the short-wavelength terms corresponding to $|\mathbf{k}|>k_c$; the short-wavelength phonons, their interaction with the electrons, and the short-range part of the electron-electron interaction. As mentioned earlier, these terms may be treated by the method of Nakajima in order to determine the effective electron-lattice interaction and sound wave frequencies. As discussed in Appendix B, for this wavelength region exchange effects are of importance. However, they are in principle calculable and should not lead to anomalous results, since the effective electron interaction is screened. Because of the mathematical complexity of the system of coupled equations obtained in Appendix B, we have not been able to obtain an explicit solution for the short wavelength v_k and ω_k when exchange effects are included.

The short-range electron-electron interaction occurring in (4.7) will not have an appreciable effect on the electronic wave functions. For as is shown in P IV, the range of this interaction is of the order of the interparticle spacing, and it may consequently in fact be treated as a relatively small perturbation on the electronic motion, and thus on the electronic wave functions.

5. CONCLUSIONS

We are led to the following physical picture of the coupled motion of ions and electrons in a metal. As the ions move, the electrons tend to follow their motion so as to screen out the ion field in a distance comparable with the interparticle spacing. Thus the fluctuations in potential caused by the motion of the ions are greatly reduced. The effective matrix element for electron-lattice interaction is determined by a screened field which can be calculated rather accurately for long wavelengths by elementary considerations. The positions of the electrons are correlated so that the field of a given electron is also screened out in a distance of the order of the interparticle spacing. Long-range correlation effects, according to the collective description, give rise to plasma waves and to coupled ion-electron waves which correspond to longitudinal sound waves. The remaining short-range interactions can be taken into account in a satisfactory way by the individual particle model.

In the collective description, extra coordinates are introduced to describe the plasma oscillations, and introduction of these requires that the system wave function satisfy supplementary conditions. A series of canonical transformations is made in order to isolate the plasma oscillations and the longitudinal sound waves. The effective matrix element for electron-phonon interaction is thereby determined and found to be identical for long wavelengths with that found earlier by self-consistent field methods which neglect effects of exchange and correlation. There remain in the transformed Hamiltonian terms which describe individual particles which interact with each other with a screened Coulomb force and are coupled with the phonon field in the usual way. The screened interaction is sufficiently weak so that it can be treated by perturbation methods and does not have a large effect on the particle wave functions. In this way we see that the usual empirical individual-particle model is justified.

An expression is derived for phonon frequencies of long wavelength and is found to be identical with that derived by Toya and Nakajima by self-consistent field methods. An explicit calculation for monovalent metals, given in an appendix, yields, when applied to Na, good agreement with values calculated from observed elastic constants. Our calculation makes use of matrix elements derived in A by the self-consistent field method. The elastic constants may also be obtained by a direct calculation of the energy of a distorted crystal, as has been done by Fuchs. The close agreement both with the observed values and with the direct calculation makes one confident that the general method for calculating matrix elements and vibrational frequencies is correct.

The calculation of the matrix elements and vibrational frequencies for short wavelengths is less reliable, because we have not been able to take into account the influence of exchange on these quantities. Perhaps the best procedure in calculations of conductivity at room temperatures, where short wavelength phonons and Umklapp processes are important, is to introduce an empirical scattering factor, as has recently been suggested by Ziman.[17]

In the formulation of the superconductivity problem, it is probably a reasonably good approximation to use an empirical interaction constant, v_k, and to omit explicit introduction of Coulomb interaction terms, as has been done by Frohlich and by Bardeen. The fact that v_k^i rather than v_k appears in the interaction Hamiltonian, H_2, in (4.9) for $|\mathbf{k}|>k_c$ raises some doubt about this procedure, but the difference is not large for these short wavelengths.

Except for terms which correspond to transitions which nearly satisfy the conservation of energy,

$$|E_{\kappa'}-E_\kappa\pm\hbar\omega_k|<\Delta E, \tag{5.1}$$

the electron-lattice interaction can be eliminated by a canonical transformation such that in the final Hamiltonian the lattice oscillators are not coupled with the electrons. It is the remaining interaction terms which do satisfy (5.1) that are responsible for scattering of electrons and also presumably account for superconductivity. The value of ΔE always can be chosen sufficiently small so that these terms have a negligible effect on the electron-lattice matrix element and on the vibrational frequency. On the other hand, they cannot be treated by perturbation theory, and so can have a pronounced effect on the electron wave functions.

In the theory of superconductivity, then, one need only consider virtual transitions which satisfy an expression of the form (5.1). This approach was followed by one of the authors,[12] with ΔE chosen to be of the order of the electron-lattice interaction energy resulting

[17] J. M. Ziman, Proc. Roy Soc. (London) **A226**, 432 (1954).

from these virtual transitions. This gives [see Eq. (3.35)]

$$\Delta E \sim N(E)\langle |v_k q_k|^2\rangle_{Av}, \tag{5.2}$$

which is an order of magnitude or so larger than kT_c (T_c= transition temperature) for most superconductors If one assumes that these interactions contribute to the superconducting state, but not to the normal state, one would have far too large an energy difference between the two phases. About the correct order of magnitude for this energy difference is obtained if we arbitrarily take $\Delta E \sim kT_c$; that is if electrons with energies within $\sim kT_c$ of the Fermi surface have their energies lowered by $\sim kT_c$. Undoubtedly the virtual transitions determined by (5.2) contribute to both the normal and superconducting states, and only a very small fraction of the interaction energy is involved in the change of state. It is evident that better pictures of both the normal and superconducting phases are required. The equations we have presented here should provide a good basis for development of an adequate theory.

APPENDIX A: CALCULATION OF ω_k FOR A MONOVALENT METAL

We give here an explicit calculation of the sound-wave frequencies for the long wavelength limit, $\mathbf{k}\to 0$, by making use of the matrix elements of the electron-lattice interaction in Eq. (3.12). The Nakajima expression, Eq. (3.30), gives equivalent results in this limit. The matrix elements used are those derived originally for a calculation of the conductivity of monovalent metals.

Determining the frequencies for long wavelengths is equivalent to calculating the elastic constants, which are usually obtained by a direct calculation of the energy of a distorted crystal. Compressibilities of monovalent metals have been determined by a Wigner-Seitz calculation of the energy as a function of volume. Fuchs[18] has shown that the shear constants of the alkali metals are given quite closely by the Coulomb interactions of the ions in a uniform negative sea. Although important for the noble metals, repulsion between the closed shells is almost negligible for the alkalis.

It is most convenient to determine the sum of the squares of the frequencies for the three different directions of polarization, since thus sum is independent of the direction of propagation. Kohn[19] has shown that the sum for the Coulomb interactions of the ions alone is just the square of the plasma frequency for a gas of the same density:

$$\sum_{\sigma=1,2,3} \Omega_{k\sigma}^2 = \frac{4\pi N e^2}{M}. \tag{A1}$$

[18] K. Fuchs, Proc. Roy. Soc. (London) **153**, 662 (1936); **157**, 444 (1936).

[19] W. Kohn (private communication). We should like to thank Professor Kohn for communicating his results to us in advance of publication.

Only the longitudinal component is affected by the electron-lattice interaction, so that, from Eq. (3.12),

$$\sum_\sigma \omega_{k\sigma}^2 = \sum_\sigma \Omega_{k\sigma}^2 - \frac{k^2}{4\pi e^2}|v_k{}^i|^2\left(1-\frac{k^2E_F}{6\pi N e^2}\right), \tag{A2}$$

where

$$E_F = \hbar^2 k_0^2/2m \tag{A3}$$

is the Fermi energy. In terms of the elastic constants

$$\textstyle\sum_\sigma \omega_{k\sigma}^2 = NMk^2(c_{11}+2c_{44}). \tag{A4}$$

The expression for the matrix element $v_k{}^i$ is taken from reference 1,

$$v_k{}^i = -\frac{ik}{(NM)^{\frac{1}{2}}}\left\{\frac{4\pi N e^2}{k^2}+\gamma(V_0(r_s)-E_0)\right\} \times\left\{\frac{3(\sin kr_s - kr_s\cos kr_s)}{(kr_s)^3}\right\}, \tag{A5}$$

where r_s is the radius of a sphere of atomic volume, $V_0(r_s)\cong 0$ is the potential at the boundary of the s-sphere, E_0 is the energy of the lowest state, and $\gamma = |\psi_0(r_s)|^2/\langle\psi_0(r)^2\rangle_{Av}$. Both V_0 and E_0 refer to a potential in which the field of the ions is compensated by a uniform negative charge. Wigner and Seitz calculated the energy, $\mathcal{E}_0$, for a cell in which the field is that of the bare ion at the center. If $v_0(r_s)$ is the corresponding potential, there is the approximate relation

$$V_0(r_s)-E_0 = v_0(r_s)-\mathcal{E}_0-0.2e^2/r_s. \tag{A6}$$

There is also the following relation, originally derived by Frohlich:

$$\frac{r_s}{3}\frac{d\mathcal{E}_0}{dr_s} = \gamma(v_0(r_s)-\mathcal{E}_0). \tag{A7}$$

An expansion of (A5) in a power series in k is

$$\begin{aligned} v_k{}^i &= -i\frac{4\pi e^2}{k}\left(\frac{N}{M}\right)^{\frac{1}{2}}\left\{1+\frac{k^2\gamma}{4\pi N e^2}[V_0(r_s)-E_0] - \frac{1}{10}k^2r_s^2+\cdots\right\} \\ &= -i\frac{4\pi e^2}{k}\left(\frac{N}{M}\right)^{\frac{1}{2}}\left[1+\frac{k^2}{4\pi N e^2} \times\left[\gamma[V_0(r_s)-E_0]-\frac{3}{10}\frac{e^2}{r_s}\right]+\cdots\right\}, \end{aligned} \tag{A8}$$

where we have used $N=3/4\pi r_s^3$. Substitution of (A8) into (A2) gives after some reduction

$$\sum_\sigma \omega_{k\sigma}^2 = \frac{k^2}{M}\left\{\frac{2}{3}E_F+\frac{6e^2}{10r_s}-2\gamma[V_0(r_s)-E_0]+\cdots\right\},$$

so that, from (A4),

$$c_{11}+2c_{44}=N\left\{\frac{2}{3}E_F+\frac{6e^2}{10r_s}-2\gamma[V_0(r_s)-E_0]+\cdots\right\}.$$

Inserting appropriate values for sodium, with the free-electron value for E_F and $\gamma=1$, we find

$$2.6\times10^{22}\{3.3+6.6-0.6\}\times10^{-12}=2.4\times10^{11}\ \text{ergs/cm}^3,$$

while

$$c_{11}+2c_{44}=\{0.95+2(0.59)\}\times10^{11}=2.13\times10^{11}\ \text{ergs/cm}^3.$$

The difference between calculated and observed values is only about 10%. This agreement gives one confidence in the method of calculation and in the values of the matrix elements for electron-lattice scattering for small k.

APPENDIX B: EXCHANGE TERMS FROM COULOMB INTERACTION

There are a number of exchange terms in the commutator of the Coulomb interaction:

$$\tfrac{1}{2}\sum M_{k'}^2[\rho_{-k'}\rho_{k'},S]=\tfrac{1}{2}\sum M_{k'}^2 \times\{\rho_{-k'}[\rho_{k'},S]+[\rho_{-k'},S]\rho_{k'}\},\quad \text{(B1)}$$

which were omitted in Nakajima's treatment. Exchange terms appear only for electrons of parallel spins. Expansion of the commutators for these gives

$$\frac{i}{2}\sum M_{k'}^2\{c_{\kappa'}^*c_{\kappa'-k'}(c_{\kappa-k'}^*c_{\kappa-k}-c_\kappa^*c_{\kappa-k+k'}) + (c_{\kappa-k'}^*c_{\kappa-k}-c_\kappa^*c_{\kappa-k+k'})c_{\kappa'}^*c_{\kappa'-k'}\} \times\{f(\mathbf{k},\boldsymbol{\kappa})q_k-ig(\mathbf{k},\boldsymbol{\kappa})p_{-k}\}.\quad \text{(B2)}$$

In the second line we have changed the sign of $\mathbf{k}'$.

Our problem is to find diagonal parts of the coefficients of $c_\kappa^*c_{\kappa-k}q_k$ and corresponding terms. In addition to those for which $\mathbf{k}'=\mathbf{k}$, $\boldsymbol{\kappa}'$ arbitrary, used in obtaining (3.24), there are several others: (a) $\boldsymbol{\kappa}'=\boldsymbol{\kappa}$, (b) $\boldsymbol{\kappa}'=\boldsymbol{\kappa}-\mathbf{k}$, (c) $\boldsymbol{\kappa}=\boldsymbol{\kappa}'-\mathbf{k}'$, (d) $\boldsymbol{\kappa}'=\boldsymbol{\kappa}-\mathbf{k}+\mathbf{k}'$. The sum of these may be expressed in the form

$$\tfrac{1}{2}\sum M_{k'}^2[\rho_{-k'}\rho_{k'},S]=i\sum_{\kappa,k}\{\sum_{\kappa',s}M_k^2 \times\{n(\boldsymbol{\kappa}'-\mathbf{k})-n(\boldsymbol{\kappa}')\}f(\mathbf{k},\boldsymbol{\kappa}') +\sum_{k'}M_{k'}^2\{n(\boldsymbol{\kappa}-\mathbf{k}-\mathbf{k}')-n(\boldsymbol{\kappa}-\mathbf{k}')\}f(\mathbf{k},\boldsymbol{\kappa}) +\sum_{\kappa'}M_{\kappa'-\kappa}^2\{n(\boldsymbol{\kappa}')-n(\boldsymbol{\kappa}'-\mathbf{k})\}f(\mathbf{k},\boldsymbol{\kappa}')\}c_\kappa^*c_{\kappa-k}q_k +\text{corresponding terms in } g(\mathbf{k},\boldsymbol{\kappa}).\quad \text{(B3)}$$

The first sum in the curly brackets is over both spin states, the other two only over the spin which is parallel to that of $c_\kappa^*c_{\kappa-k}$.

When exchange terms are included, the effects of the other electrons on a given electron can no longer be expressed in terms of a potential, and it is no longer advantageous to introduce v_k. In order to eliminate linear terms in q_k and p_{-k} in the transformed Hamiltonian, $f(\mathbf{k},\boldsymbol{\kappa})$ and $g(\mathbf{k},\boldsymbol{\kappa})$ must satisfy the following equations:

$$\sum_{\kappa'}(2M_k^2-M_{\kappa-\kappa'}^2)[n(\boldsymbol{\kappa}'-\mathbf{k})-n(\boldsymbol{\kappa}')]f(\mathbf{k},\boldsymbol{\kappa}') -(W_{\kappa-k}-W_\kappa)f(\mathbf{k},\boldsymbol{\kappa})+(E_\kappa-E_{\kappa-k})f(\mathbf{k},\boldsymbol{\kappa}) +\hbar\omega_k^2g(\mathbf{k},\boldsymbol{\kappa})+v_k^i=0,$$

$$\sum_{\kappa'}(2M_k^2-M_{\kappa-\kappa'}^2)[n(\boldsymbol{\kappa}'-\mathbf{k})-n(\boldsymbol{\kappa}')]g(\mathbf{k},\boldsymbol{\kappa}') -(W_{\kappa-k}-W_\kappa)g(\mathbf{k},\boldsymbol{\kappa})+(E_\kappa-E_{\kappa-k})g(\mathbf{k},\boldsymbol{\kappa}) +\hbar f(\mathbf{k},\boldsymbol{\kappa})=0,\quad \text{(B4)}$$

where W_κ is the exchange energy of an electron in the state $\boldsymbol{\kappa}$:

$$W_\kappa=-\sum_{k'}M_{k'}^2n(\boldsymbol{\kappa}-\mathbf{k}').\quad \text{(B5)}$$

The factor two multiplying M_k^2 takes account of the sum over spins. The equations are such that for a symmetrical distribution of electrons in $\boldsymbol{\kappa}$ space, $n(\boldsymbol{\kappa})=n(-\boldsymbol{\kappa})$, the solutions satisfy the relations:

$$f(\mathbf{k},\mathbf{k}-\boldsymbol{\kappa})=-f(\mathbf{k},\boldsymbol{\kappa});\quad g(\mathbf{k},\mathbf{k}-\boldsymbol{\kappa})=g(\mathbf{k},\boldsymbol{\kappa}).\quad \text{(B6)}$$

The direct term in the equation for $g(\mathbf{k},\boldsymbol{\kappa})$ then vanishes:

$$\sum_{\kappa'}M_k^2[n(\boldsymbol{\kappa}'-\mathbf{k})-n(\boldsymbol{\kappa}')]g(\mathbf{k},\boldsymbol{\kappa}')=0.\quad \text{(B7)}$$

When the exchange terms are included, the equations cannot be solved algebraically.

There are two types of exchange terms. One simply adds the exchange energy W_κ to the individual particle energy E_κ. The other adds sums over $f(\mathbf{k},\boldsymbol{\kappa}')$ and $g(\mathbf{k},\boldsymbol{\kappa}')$ to the equations for f and g, and it is these which make the solution difficult. An estimate of the magnitude of the latter terms can be obtained by taking an average over $\boldsymbol{\kappa}$. We have

$$\sum_\kappa M_{\kappa-\kappa'}^2n(\boldsymbol{\kappa})=-W_{\kappa'}.\quad \text{(B8)}$$

An average of $W_{\kappa'}$ is the exchange energy, $-0.92e^2/r_s$ for a monovalent metal. This is to be compared with

$$2M_k^2\sum n(\boldsymbol{\kappa})=n_0M_k^2=4\pi n_0e^2/k^2=(3e^2/r_s)(kr_s)^{-2},\quad \text{(B9)}$$

where $n_0=3/4\pi r_s^3$ is the concentration of electrons. These exchange terms may be neglected when $kr_s\ll1$, i.e., for long wavelengths. However, as discussed in the text, the exchange energies of the individual electrons, W_κ, would make marked changes at long wavelengths. As the collective description shows, these should not be included.

SOVIET PHYSICS JETP VOLUME 34(7), NUMBER 6 DECEMBER, 1958

INTERACTION BETWEEN ELECTRONS AND LATTICE VIBRATIONS IN A NORMAL METAL

A. B. MIGDAL

Moscow Institute of Engineering Physics

Submitted to JETP editor July 12, 1957; resubmitted March 20, 1958

J. Exptl. Theoret. Phys. (U.S.S.R.) **34**, 1438-1446 (June, 1958)

A method is developed which enables one to obtain the electron-energy spectrum and dispersion of the lattice vibrations without assuming that the interaction between electrons and phonons is small.

1. INTRODUCTION

THE attraction between electrons due to the exchange of phonons leads in superconductors to the formation of a bound state of two electrons with opposite momenta. In the ground state of a superconductor a condensed component consisting of these bound electrons is formed and a gap results in the energy spectrum.[1]

In papers on the theory of superconductivity[1] the interaction between electrons and lattice vibrations has been assumed to be small, although we know that this condition is not fulfilled for all superconductors. It is therefore of interest to construct a theory which is not limited in this way.

In the present paper we develop a method which enables one to consider the interaction between electrons and lattice vibrations in a normal metal without assuming that the interaction is small. The method is based on the use of quantum field-theoretical equations.

The application of field theory to superconductors involves certain difficulties. The state which contains the "condensate" of bound electrons cannot be obtained from the ground state of noninteracting particles by applying the interaction adiabatically. The necessary condition for the use of ordinary field-theoretical methods is thus violated. The method developed below for a normal metal, where this difficulty does not occur, can therefore be extended to a superconductor only through a separate investigation.

The interaction between electrons and lattice vibrations in a normal metal is certainly of interest in itself. Fröhlich[2] used perturbation theory to investigate this interaction. He considered an isotropic model of a metal described by the Hamiltonian

$$H = H_0 + H_1, \quad H_0 = \sum_{\mathbf{p}} \varepsilon^0_{\mathbf{p}} a^+_{\mathbf{p}} a_{\mathbf{p}} + \sum_{q<q_m} \omega^0_{\mathbf{q}} b^+_{\mathbf{q}} b_{\mathbf{q}},$$

$$H_1 = \sum_{\mathbf{p},\, q<q_m} \alpha_{\mathbf{q}} a^+_{\mathbf{p}+\mathbf{q}} a_{\mathbf{p}} (b_{\mathbf{q}} + b^+_{-\mathbf{q}}), \tag{1}$$

where $a_{\mathbf{p}}$, $a^+_{\mathbf{p}}$ and $b_{\mathbf{q}}$, $b^+_{\mathbf{q}}$ are the annihilation and creation operators of electrons and phonons and q_m is the maximum phonon momentum. We know that $\alpha^2_{\mathbf{q}}$, which determines the interaction between electrons and phonons, is given for small q (in atomic units) by

$$\alpha^2_q = (\lambda_0 \pi^2 / p_0)\, \omega^0_{\mathbf{q}}, \quad \omega^0_{\mathbf{q}} = c_0 q, \tag{2}$$

where c_0 is the unrenormalized velocity of sound, $c_0 \sim M^{-1/2}$, M is the mass of an ion and λ_0 is a dimensionless parameter, introduced by Fröhlich,[2] which does not contain the ion mass; $\lambda_0 \lesssim 1$.

It will be shown below that the energy spectrum of the Hamiltonian (1) cannot be obtained by perturbation theory, despite the smallness of the parameter $M^{-1/2}$ in $\alpha^2_{\mathbf{q}}$. The criterion for the applicability of perturbation theory is the smallness of λ_0, which does not contain the ion mass. Field theoretical methods[3] enable us to obtain the energy spectrum, without assuming that λ_0 is small, as a power series in $M^{-1/2}$.

2. METHOD OF SOLUTION

We introduce the electron and phonon propagation functions G and D:

$$G = i \langle T \Psi(1) \Psi^+(2) \rangle, \quad D = i \langle T \varphi(1) \varphi(2) \rangle, \tag{3}$$

where the averaging is performed over the ground state of the system

$$\varphi = e^{iHt} \sum_{q<q_m} (b_{\mathbf{q}} + b^+_{-\mathbf{q}})\, e^{i\mathbf{q}\mathbf{r}} \alpha_{\mathbf{q}} e^{-iHt},$$

$$\Psi = e^{iHt} \sum_{\mathbf{p}} a_{\mathbf{p}} e^{i\mathbf{p}\mathbf{r}} e^{-iHt}.$$

Dyson's equations relate D and G to the vertex

part Γ, which is defined by the following set of diagrams:

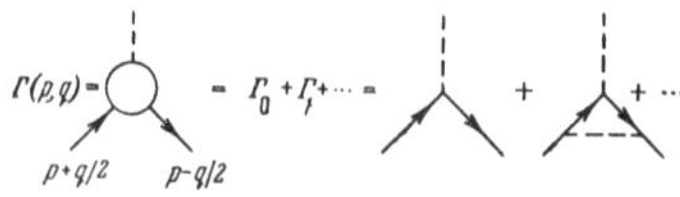

(4)

Here $q = (\mathbf{q}, \omega)$ and $p = (\mathbf{p}, \epsilon)$. The interaction energy in (1) can be written in the form

$$H_1 = \sum_{\substack{\mathbf{p} \\ q<q_m}} \Psi^+_{\mathbf{p}+\mathbf{q}} \Psi_{\mathbf{p}} \varphi_{\mathbf{q}},$$

where $\Psi_{\mathbf{p}}$ and $\varphi_{\mathbf{q}}$ are spatial Fourier components of the operators in the Green's functions. Therefore the first diagram in (4) corresponds to $\Gamma = \Gamma_0 = 1$. It will be shown below that the following terms in (4) are of the order of $M^{-1/2}$. Therefore Γ can be replaced by 1 in Dyson's equations, after which a closed system of equations is obtained for D and G. In the momentum representation Dyson's equations are

$$\begin{aligned} G(p) &= G_0(p) + G_0(p)\Sigma(p)G(p), \\ \Sigma(p) &= \frac{1}{i}\int G(p-q)D(q)\Gamma\left(p-\frac{q}{2},\, q\right)d^4q, \\ D(q) &= D_0(q) + D_0(q)\Pi(q)D(q), \\ \Pi(q) &= \frac{1}{i}\int G\left(p+\frac{q}{2}\right)G\left(p-\frac{q}{2}\right)\Gamma(p, q)\,d^4p, \end{aligned} \tag{5}$$

where

$$d^4p = d\mathbf{p}\,d\varepsilon/(2\pi)^4, \quad d^4q = d\mathbf{q}\,d\omega/(2\pi)^4,$$

Σ and Π are the irreducible parts of the electron and phonon self energies, and D_0 and G_0 are the electron and phonon Green's functions in the absence of interaction:[2]

$$G_0(p) = \frac{1}{\varepsilon^0_{\mathbf{p}} - \varepsilon - i\Delta(p)},$$

$$D_0(q) = \alpha_q^2\left\{\frac{1}{\omega_q^0 - \omega - i\delta} + \frac{1}{\omega_q^0 + \omega - i\delta}\right\}, \tag{6}$$

where

$$\Delta(p) \to \begin{cases} +0 & p > p_0 \\ -0 & p < p_0 \end{cases}, \quad \delta \to +0.$$

Assuming $\Gamma = 1$ in (5), we obtain G and D. As was shown in reference 3, the energy spectrum is determined by the poles of the analytic continuation of $G(\mathbf{p}, \epsilon)$ and $D(\mathbf{q}, \omega)$ in the complex plane.

3. THE VERTEX PART

We shall show that the vertex part differs from $\Gamma_0 = 1$ by a quantity of the order of $M^{-1/2}$. Let us consider a first-order perturbation correction to Γ.

We shall assume that

$$p \sim p_0, \quad \varepsilon \sim \mu_0, \quad \omega \lesssim \omega_q|_{q=2p_0} = \omega_0,$$

since for our further calculations this is the important range of values of p and ω. From our definition of G and D each internal line of the diagrams corresponds to a Green's function divided by i. We obtain

$$\Gamma_1(p, q) = i\int D_0(p-p_1)\,G_0\left(p_1 + \frac{q}{2}\right)G_0\left(p_1 - \frac{q}{2}\right)dp_1. \tag{7}$$

In accordance with (6), the function $D_0(\mathbf{p}-\mathbf{p}_1, \epsilon-\epsilon_1)$ possesses the following properties:

When $|\mathbf{p}-\mathbf{p}_1| \sim p_0$ and $\epsilon - \epsilon_1 \ll \omega_0$, D_0 can be replaced by

$$D_0 = 2\pi^2\lambda_0/p_0.$$

When $\epsilon - \epsilon_1 \gg \omega_0$, D_0 diminishes as $(\epsilon-\epsilon_1)^{-2}$. For $|\mathbf{p}_1 - \mathbf{p}| > q_m$, $D_0 = 0$. Using these properties of D_0, we obtain for Γ_1:

$$\Gamma_1 \sim \frac{\lambda_0 i}{8\pi^2 p_0}\int_{\varepsilon-\omega_0/2}^{\varepsilon+\omega_0/2} d\varepsilon_1 \int_{|\mathbf{p}-\mathbf{p}_1|<q_m} G_0\left(\mathbf{p}_1 + \frac{\mathbf{q}}{2},\, \varepsilon_1 + \frac{\omega}{2}\right) \times G_0\left(\mathbf{p}_1 - \frac{\mathbf{q}}{2},\, \varepsilon_1 - \frac{\omega}{2}\right)d\mathbf{p}_1. \tag{8}$$

Integration with respect to ϵ_1 gives the factor $\omega_0 \sim M^{-1/2}$ and leads to

$$\Gamma_1 \sim \lambda_0\omega_0/p_0^2 \sim \lambda_0/\sqrt{M},$$

if integration over p_1 does not introduce factors $\sim 1/\omega_0$. Such factors result only for small $q \lesssim \omega_0/p_0$ and $\omega < \omega_0$, when the two poles of the integrand approach each other. Then the integrand has a maximum near $p_1 = g$, where g is given by $\epsilon^0_g = \epsilon_1 \simeq \epsilon$ and thus $g \sim p_0$. Integration over regions far from this maximum does not introduce factors $\sim 1/\omega_0$. We can therefore limit ourselves to consideration of the integral over p_1 in the region $(p_1 - g)/g \ll 1$. Using (6) and the notation $\epsilon^0_{p_1} - \epsilon = E$, we obtain from (8)

$$\Gamma_1 \sim i\lambda_0 p_0 \int_{\varepsilon-\omega_0/2}^{\varepsilon+\omega_0/2} d\varepsilon_1 \int_{-1}^{1} dx \int_{-\infty}^{\infty} dE \Big/ \left[E + \frac{v_g qx}{2} - \frac{\omega}{2} - i\Delta\left(\varepsilon_1 + \frac{\omega}{2}\right)\right]\left[E - \frac{v_g qx}{2} + \frac{\omega}{2} - i\Delta\left(\varepsilon_1 - \frac{\omega}{2}\right)\right].$$

For simplicity it is assumed here that $q_m > p + g$, and the condition $|p - p_1| < q_m$ imposes no limitation on integration near $p_1 = g$.

Integration with respect to E gives

$$\Gamma_1 \sim \lambda_0 p_0 \int_{\varepsilon-\omega_0/2}^{\varepsilon+\omega_0/2} d\varepsilon_1 \int_{-1}^{1} dx \frac{\theta(\varepsilon_1 - \mu_0 + \omega/2) - \theta(\varepsilon_1 - \mu_0 - \omega/2)}{v_g qx - \omega + i\delta\omega/|\omega|},$$

where

$$\theta(y) = \begin{cases} 1 & y \geqslant 0 \\ 0 & y < 0 \end{cases}, \qquad v_g = \frac{\partial \varepsilon_g^0}{\partial g}.$$

Integrating with respect to x, we obtain

$$\Gamma_1 \sim \frac{\lambda_0}{p_0 q}\left[\ln\left|\frac{v_g q + \omega}{v_g q - \omega}\right| - i\pi\theta(v_g q - |\omega|)\right]$$
$$\times \int_{\varepsilon-\mu_0-\omega_0/2}^{\varepsilon+\omega_0/2-\mu_0} \left[\theta\left(t + \frac{\omega}{2}\right) - \theta\left(t - \frac{\omega}{2}\right)\right] dt. \qquad (9)$$

The last integral differs from zero in the region $|\epsilon - \mu_0| \lesssim \omega_0$ and is of the order of ω_1, where ω_1 is the smaller of the numbers ω_0 and ω. It follows from (9) that the largest value $\Gamma_1 \sim \lambda_0$ is reached for $\omega \sim \omega_0$ and $q \sim \omega_0/p_0$. These values of q and ω play no part in our subsequent calculations. Indeed, for the calculation of $\Sigma(p)$ according to (5) the essential values are $q \sim p_0$ and $\omega \sim \omega_0$, for which it follows from (8) that $\Gamma_1 \sim M^{-1/2}$. For obtaining $\Pi(q)$ the essential values are $\omega \sim \omega_q \sim qp_0/\sqrt{M} \ll p_0 q$ (ω_q is the frequency of a phonon of momentum q). From (9) we obtain

$$\Gamma_1 \sim \lambda_0 \frac{\omega}{p_0 q}\left[\frac{\omega}{2v_g q} - i\pi\right] \sim \frac{\lambda_0}{M} + i\frac{\lambda_0}{\sqrt{M}}.$$

We thus have $\Gamma = 1 + O(M^{-1/2})$. It can be shown that this estimate is not changed when diagrams of a higher order are taken into account.

4. THE PHONON GREEN'S FUNCTION

As will be seen from our subsequent calculations, $G(\mathbf{p}, \epsilon)$ differs essentially from $G_0(\mathbf{p}, \epsilon)$ only in a narrow range of values of p and ϵ: $|\mathbf{p} - \mathbf{p}_0| \sim \omega_0/p_0$; $\epsilon - \epsilon_0^0 \sim \omega_0$. In the calculation of $\Pi(q, \omega)$ according to (5) the integration is performed over a wide range of the variables, which permits us to replace G by G_0 accurately to terms $\sim M^{-1/2}$. For $\Pi(\mathbf{q}, \omega)$ we obtain from (5)

$$\Pi(\mathbf{q},\omega) \approx \frac{1}{i}\int G_0\left(p + \frac{q}{2}\right) G_0\left(p - \frac{q}{2}\right) d^4p. \qquad (10)$$

Integration of (10) with respect to ϵ gives

$$\Pi(\mathbf{q},\omega) = \frac{1}{(2\pi)^3}\int \frac{n(\mathbf{p} - \mathbf{q}/2) - n(\mathbf{p} + \mathbf{q}/2)}{\varepsilon^0_{\mathbf{p}+\mathbf{q}/2} - \varepsilon^0_{\mathbf{p}-\mathbf{q}/2} - \omega - i\delta\omega/|\omega|} d^3p, \qquad (11)$$

where

$$n(p) = \begin{cases} 1 & p > p_0 \\ 0 & p < p_0. \end{cases}$$

We see from (10) and (11) that $\Pi(\mathbf{q}, \omega)$ is an even function of ω. For subsequent calculations the important values of $\Pi(\mathbf{q}, \omega)$ are obtained for

$$\omega \sim \omega_q \sim p_0 q/\sqrt{M} \ll p_0 q.$$

With these values of ω we have from (11)

$$\Pi(\mathbf{q},\omega) = \frac{p_0}{(2\pi)^2}\left[g\left(\frac{q}{2p_0}\right) + \pi i \frac{|\omega|}{2p_0 q}\right], \qquad (12)$$

where

$$g(x) = \frac{1}{2}\left[1 + \frac{1 - x^2}{2x}\ln\left|\frac{1 + x}{1 - x}\right|\right]. \qquad (13)$$

$g(x)$ can be represented in the interval $0 < x < 1$ with sufficient accuracy by

$$g(x) \approx 1 - x^2/2. \qquad (13')$$

From (2), (5) and (12) we obtain

$$D(\mathbf{q},\omega) = \frac{1}{D_0^{-1}(\mathbf{q},\omega) - \Pi(\mathbf{q},\omega)}$$
$$= \frac{2\omega_q^0 \alpha_q^2}{\left[(\omega_q^0)^2 - \omega^2 - (\omega_q^0)^2 \lambda_0 \left(g\left(\frac{q}{2p_0}\right) + i\pi\frac{|\omega|}{2p_0 q}\right)\right]}.$$

The real part of the pole of $D(\mathbf{q}, \omega)$ gives the renormalized phonon frequency

$$\omega_q^2 = (\omega_q^0)^2\left[1 - \lambda_0 g\left(\frac{q}{2p_0}\right)\right] \approx (\omega_q^0)^2\left(1 - \lambda_0 + \lambda_0\frac{q^2}{8p_0^2}\right). \qquad (14)$$

Eq. (13′) was used in the derivation of this last equation. The imaginary part of the pole gives the phonon attenuation

$$\delta_1(q) = {}^1\!/_4\pi\lambda_0(\omega_q^0)^2/p_0 q. \qquad (15)$$

The relative attenuation is given by

$$\delta_1(q)/\omega_q = \lambda_0(\omega_q^0)^2\pi/4p_0 q\omega_q \sim \lambda_0/\sqrt{M} \ll 1.$$

From (14) and (15)

$$D(\mathbf{q},\omega) = \alpha_q^2\frac{\omega_q^0}{\omega_q}\left(\frac{1}{\omega_q - \omega - i\delta_1(q)\omega/|\omega|}\right.$$
$$\left. + \frac{1}{\omega_q + \omega - i\delta_1(q)\omega/|\omega|}\right). \qquad (16)$$

Thus the phonon Green's function D, which was obtained by taking the interaction with electrons into account, differs from D_0 through replacement of the frequencies ω_q^0 by $\omega_q - i\delta_1(q)$ and the occurrence of the renormalizing factor ω_q^0/ω_q.

5. THE ELECTRON GREEN'S FUNCTION

From (5) we have

$$G = 1/[\varepsilon_p^0 - \varepsilon - \Sigma(\mathbf{p},\varepsilon)], \tag{17}$$

$$\Sigma(\mathbf{p},\varepsilon) = \frac{1}{i(2\pi)^4} \int\limits_{|\mathbf{p}-\mathbf{p}_1|<q_m} D(\mathbf{p}-\mathbf{p}_1, \varepsilon-\varepsilon_1)\, d\mathbf{p}_1\, d\varepsilon_1/[\varepsilon_{p_1}^0 - \varepsilon_1 - \Sigma(\mathbf{p}_1, \varepsilon_1)]. \tag{18}$$

G will now be obtained by solving the integral equation (18).

It is easily seen that $\Sigma(\mathbf{p}, \epsilon) \sim \omega_0$, so that G differs essentially from G_0 only for $|\epsilon_p^0 - \epsilon| \sim \omega_0$. The relative change of the excitation energy is large only for $\epsilon_p^0 - \mu_0 \sim \omega_0$. Thus the electron excitation spectrum varies appreciably only close to the Fermi surface in the range $p - p_0 \sim \omega_0/p_0$.

We introduce the notation

$$(\varepsilon_p^0 - \mu_0)/\omega_0 = \xi,\ (\varepsilon - \mu)/\omega_0 = \eta,$$
$$\Sigma(\xi, \eta) = \Sigma(0, 0) + \omega_0 f(\xi, \eta), \tag{19}$$

where

$$\mu = \mu_0 + \Sigma(0,0),$$
$$\omega_0 = \omega_q|_{q=2p_0} = \omega_{2p_0}^0\left(1 - \frac{\lambda_0}{2}\right). \tag{19'}$$

As was shown in reference 3, the imaginary part of $\Sigma(\xi, \eta)$ must vanish for any value of ξ when ϵ equals the chemical potential.

As will be shown below $\Sigma(0, 0)$ is real. Therefore

$$\operatorname{Im} \Sigma(0, \eta) = \omega_0 \operatorname{Im} f(0, \eta);$$

since according to (19)

$$f(0, \eta)|_{\eta=0} = 0$$

and thus Im Σ vanishes for $\epsilon = \mu$ (μ is the chemical potential).

In the notation of (19) and (19′) G becomes

$$G(\xi, \eta) = \frac{1}{\omega_0} \frac{1}{\xi - \eta + f(\xi, \eta)}. \tag{20}$$

From the foregoing discussion we are interested in ξ and $\eta \sim 1$.

In (19) we pass from integration over the angles of the vector $\mathbf{p}_1$ to integration over $q = |\mathbf{p} - \mathbf{p}_1|$: $q\,dq = pp_1\,dx$, where x is the cosine of the angle between $\mathbf{p}$ and $\mathbf{p}_1$. We have

$$\Sigma(\mathbf{p},\varepsilon) = \frac{1}{i(2\pi)^3 p} \int_{-\infty}^{\infty} d\omega \int\limits_{|p-p_1|<q<p+p_1;\ q<q_m} q\,dq\,p_1\,dp_1$$
$$\times D(q,\omega)/[\varepsilon_{p_1}^0 - \varepsilon - \omega - \Sigma(p_1, \varepsilon+\omega)], \tag{21}$$

or

$$\Sigma(p, \varepsilon) = \frac{1}{(2\pi)^3 ip} \int_{-\infty}^{\infty} d\omega \int_0^{q_m} q\,dq D(q, \omega)$$
$$\times \int_{|p-q|}^{p+q} p_1\,dp_1/[\varepsilon_{p_1}^0 - \varepsilon - \omega - \Sigma(p_1, \varepsilon+\omega)]. \tag{21'}$$

As mentioned previously, values of p close to p_0 are of interest, so that $(p - p_0)/p_0 \sim 1/\sqrt{M}$. Therefore in the right-hand side of (21) p can be replaced by p_0 accurately to within $M^{-1/2}$. We divide the integration over p_1 into two regions defined by

$$1)\ |\xi_1| = |\varepsilon_{p_1}^0 - \mu_0|/\omega_0 \leqslant \gamma \text{ and } 2)\ |\xi_1| > \gamma,$$

where γ lies between the following limits:

$$1 \ll \gamma \ll 1/\nu,\quad \nu = \omega_0/p_0 \sim 1/\sqrt{M}. \tag{22}$$

In the integral over region 1 $\Sigma(p, \epsilon)$ in the integrand can be replaced by $\Sigma(p_0, \epsilon+\omega)$ accurately to within $\sim M^{-1/2}$.

We note that for $q_m > 2p_0$ region 1 exists only for $q < 2p_0$. Therefore for the integration over p_1 in (21), in the term corresponding to region 1 the integration over q is carried as far as the smaller of the quantities q_m and $2p_0$.

Since $D(q, \omega)$ for $\omega \gg \omega_0$ vanishes as ω^{-2}, in the integral of (21) the essential result is found for $\omega \sim \omega_0$, and for integration in region 2 we can neglect $\epsilon - \mu_0$ and $\Sigma(p_1, \epsilon+\omega)$ in the denominator of the integrand compared with $\epsilon_{p_1}^0 - \mu_0$, with accuracy $\sim 1/\gamma$. Therefore integration over the region $|\xi_1| > \gamma$ introduces into Σ a term which is independent of p and ϵ. This term is $\Sigma(0, 0) = \mu - \mu_0$, since integration over the region $|\xi_1| < \gamma$ yields an expression which vanishes for $\eta = 0$.

Therefore the change of the chemical potential is given by

$$\mu - \mu_0 = \Sigma(0, 0)$$

$$= \frac{1}{ip(2\pi)^3} \int_{-\infty}^{\infty} d\omega \int_0^{q_m} q\,dq D(q,\omega) \int_{|p_0-q|}^{p_0+q} \frac{p_1\,dp_1}{\varepsilon_{p_1}^0 - \mu_0}. \tag{23}$$

The integral over p_1 in (23) is taken in the sense of the principal value, which corresponds to dropping of the region $|\xi_1| < \gamma$, which is small compared with the essential region of integration.

Subtracting $\Sigma(0, 0)$ from the left and right members of (21) and dividing by ω_0, we obtain an integral equation for $f(0, \eta) = f(\eta)$:

$$f(\eta) = \frac{1}{ip(2\pi)^3} \int_{-\infty}^{\infty} d\eta' \int_{-\gamma}^{\gamma} d\xi_1 \int_0^{q_1} \frac{D(q, \omega_0\eta')\, q\,dq}{\xi_1 - \eta - \eta' - f(\eta+\eta')}, \tag{24}$$

where q_1 is the smaller of the numbers q_m and $2p_0$. Here terms $\nu\xi$, $\nu\xi_1 \ll 1$ have been dropped.

Since the essential values in (24) are $\eta' \ll \gamma$ the limits with respect to ξ_1 can be replaced by $\pm\infty$.

Integration over ξ_1 gives

$$\lim_{\gamma\to\infty}\int_{-\gamma}^{\gamma}\frac{d\xi_1}{\xi_1-\varphi(\eta,\eta')}=\pi i\,\mathrm{Sgn}\,\mathrm{Im}\,\varphi(\eta,\eta')=\pi i\,\mathrm{Sgn}\,f_1(\eta+\eta'),$$

where f_1 denotes the imaginary part of f. As was shown in reference 2, the imaginary part of the Green's function, and thus $f_1(\eta)$, reverses its sign for $\eta = 0$; moreover, $f_1(\eta) > 0$ for $\eta > 0$. Therefore

$$\mathrm{Sgn}\,f_1(\eta)=\mathrm{Sgn}\,\eta.$$

Inserting these results in (24), we obtain

$$f(\eta)=\frac{1}{8\pi^2p_0}\int_0^{q_1}q\,dq\int_0^{\infty}\mathrm{Sgn}(\eta+\eta')D(q,\omega_0\eta')\,d\eta'$$
$$=\frac{1}{8\pi^2p_0}\int_0^{q_1}q\,dq\int_{-\eta}^{\eta}D(q,\omega_0\eta')\,d\eta'.$$

We use the notation

$$f(\eta)=f_0(\eta)+if_1(\eta).$$

For f_0 and f_1 we obtain

$$f_0=\frac{\alpha_q^2}{8\pi^2p_0\omega_q^0}\int_0^{q_1}q\,dq\int_{-\eta}^{\eta}(\omega_q^0)^2\frac{2}{\omega_q^2-\omega_0^2\eta'^2}d\eta',\quad\text{(25a)}$$

$$f_1=\frac{\alpha_q^2}{8\pi^2p_0\omega_q^0}2\pi\int_0^{q_1}(\omega_q^0)^2q\,dq\int_{-\eta}^{\eta}\frac{2\delta_1(q)\,|\eta'|\,\omega_0}{(\omega_q^2-\omega_0^2\eta'^2)^2+4\delta_1^2\omega_0^2\eta'^2}\frac{d\eta'}{\pi}.\quad\text{(25b)}$$

It is easily seen that for $\eta \gg 1/\sqrt{M}$ the integrand with respect to η' in (25b) can be replaced by $\delta(\omega_q - \omega_0\eta')$. Using the notation $x = q/2p_0$ and integrating over η', we obtain

$$f_0=\lambda_0\int_0^{x_1}x\,dx\frac{(\omega_q^0)^2}{\omega_0\omega_q}\ln\left|\frac{\eta+\omega_q/\omega_0}{\eta-\omega_q/\omega_0}\right|,$$
$$f_1=\pi\lambda_0\int_0^{y}x\,dx\frac{(\omega_q^0)^2}{\omega_0\omega_q}.\quad\text{(26)}$$

Here $x_1 = q_1/2p_0$, $y = g/2p_0$, where g is given by the condition $\omega_g = \omega_0\eta$ for $|\eta| < 1$ and $y = x_1$ for $|\eta| > 1$.

We introduce the variable $t = \omega_q/\omega_0$. According to (14) and (19′), t is related to x by

$$t^2=\frac{1}{1-\lambda_0/2}x^2\left(1-\lambda_0+\frac{\lambda_0}{2}x^2\right).$$

For f_0 we obtain

$$f_0=\int_0^1 dt\ln\left|\frac{t+\eta}{t-\eta}\right|\left(1-\frac{a}{\sqrt{a^2+t^2}}\right),\ a^2=\frac{(1-\lambda_0)^2}{\lambda_0(2-\lambda_0)},\quad\text{(27)}$$

For $\eta \ll 1$ we have

$$f_0(\eta)=2\eta\ln\frac{1-\lambda_0/2}{1-\lambda_0}=\lambda\eta.\quad\text{(28)}$$

The imaginary part, $f_1(\eta)$, is

$$f_1(\eta)=\pi\int_0^{t_1}\left[1-\frac{a}{\sqrt{a^2+t^2}}\right]dt=\pi\left\{t_1-a\ln\frac{t_1+\sqrt{a^2+t_1^2}}{a}\right\},\quad\text{(29)}$$

where $t_1 = \eta$ for $|\eta| < 1$ and $t_1 = 1$ for $|\eta| > 1$; for $M^{-1/2} \ll \eta \ll 1$ we obtain

$$f_1(\eta)=\frac{\pi\eta^3}{6a^2}=\frac{\pi\lambda_0(2-\lambda_0)}{6(1-\lambda_0)^2}\eta^3.\quad\text{(30)}$$

For $\eta \ll 1/\sqrt{M}$ in the denominator of the integrand of (25b) η' can be neglected. This gives

$$f_1=\lambda_0^2\frac{\omega_0}{4p_0^2}\int_0^{x_1}\frac{dx}{[1-\lambda_0 g(x)]^2}\eta|\eta|.\quad\text{(31)}$$

The attenuation of the electron excitations given by (30) results from the emission of phonons. When the energy of a quasi-particle is very close to the Fermi surface ($\eta \ll M^{-1/2}$), a different attenuation mechanism is more important; this is attenuation due to the interaction between electrons, which results from phonon exchange. As mentioned above, interelectronic interaction leads to attenuation which is proportional to the square of the short distance from the Fermi surface, as follows from (31).

The electron energy spectrum is determined by the poles of G, that is, by the condition

$$\eta+f(\eta)=\xi.\quad\text{(32)}$$

For small η we have

$$\eta(\xi)\approx\xi/(1+\lambda).$$

Returning to the usual notation and subtracting the energy of a hole, we obtain for the excitation energy

$$E_{p_1p_2}=\varepsilon_{p_2}-\varepsilon_{p_1}=(\varepsilon_{p_2}^0-\varepsilon_{p_1}^0)/(1+\lambda)$$
$$=v_0^0(p_2-p_1)/(1+\lambda)=v_0(p_2-p_1),\quad\text{(33)}$$

where $p_2 > p_0$, $p_1 < p_0$ and v_0^0 is the unrenormalized velocity on the Fermi surface. Renormalization of the velocity on the Fermi surface is given by

$$v_0=\frac{v_0^0}{1+\lambda};\quad\lambda=2\ln\frac{1-\lambda_0/2}{1-\lambda_0}.\quad\text{(34)}$$

For $\lambda_0 \ll 1$ we obtain from (34)

$$v_0=v_0^0(1-\lambda_0).$$

This equation agrees with the result that Fröhlich obtained by using perturbation theory.

Equation (34) shows that $v_0 > 0$ for all values of λ, and the rearrangement of the Fermi distribution which Fröhlich predicted does not occur.

It follows from (30) that for $\lambda_0 \sim 1$ the excitation attenuation equals the excitation energy in order of magnitude for $\eta \sim 1$, i.e., for the excitation energy

$$E_{p_1 p_2} \sim \omega_0.$$

With further increase of the excitation energy, the attenuation ceases to increase and becomes smaller than the excitation energy. Thus for $\lambda_0 \sim 1$ electron excitations in the region $E_{p_1 p_2} \sim \omega_0$ cannot be described by means of quasi-particles.

[1] L. N. Cooper, Phys. Rev. **104**, 1189 (1956); Bardeen, Cooper and Schrieffer, Phys. Rev. **106**, 162 (1957); N. N. Bogoliubov, J. Exptl. Theoret. Phys. (U.S.S.R.) **34**, 58 (1958), Soviet Phys. JETP 7, 41 (1958); L. P. Gor'kov, J. Exptl. Theoret. Phys. (U.S.S.R.) **34**, 735 (1958), Soviet Phys. JETP 7, 505 (1958).

[2] H. Fröhlich, Phys. Rev. **79**, 845 (1950).

[3] V. M. Galitskii and A. B. Migdal, J. Exptl. Theoret. Phys. (U.S.S.R.) **34**, 139 (1958), Soviet Phys. JETP **7**, 96 (1958).

Translated by I. Emin
293

PHYSICAL REVIEW VOLUME 104, NUMBER 4 NOVEMBER 15, 1956

Letters to the Editor

PUBLICATION of brief reports of important discoveries in physics may be secured by addressing them to this department. The closing date for this department is five weeks prior to the date of issue. No proof will be sent to the authors. The Board of Editors does not hold itself responsible for the opinions expressed by the correspondents. Communications should not exceed 600 words in length and should be submitted in duplicate.

Bound Electron Pairs in a Degenerate Fermi Gas*

LEON N. COOPER
Physics Department, University of Illinois, Urbana, Illinois
(Received September 21, 1956)

IT has been proposed that a metal would display superconducting properties at low temperatures if the one-electron energy spectrum had a volume-independent energy gap of order $\Delta \simeq kT_c$, between the ground state and the first excited state.[1,2] We should like to point out how, primarily as a result of the exclusion principle, such a situation could arise.

Consider a pair of electrons which interact above a quiescent Fermi sphere with an interaction of the kind that might be expected due to the phonon and the screened Coulomb fields. If there is a net attraction between the electrons, it turns out that they can form a bound state, though their total energy is larger than zero. The properties of a noninteracting system of such bound pairs are very suggestive of those which could produce a superconducting state. To what extent the actual many-body system can be represented by such noninteracting pairs will be discussed in a forthcoming paper.

Because of the similarity of the superconducting transition in a wide variety of complicated and differing metals, it is plausible to assume that the details of metal structure do not affect the qualitative features of the superconducting state. Thus, we neglect band and crystal structure and replace the periodic ion potential by a box of volume V. The electrons in this box are free except for further interactions between them which may arise due to Coulomb repulsions or to the lattice vibrations.

In the presence of interaction between the electrons, we can imagine that under suitable circumstances there will exist a wave number q_0 below which the free states are unaffected by the interaction due to the large energy denominators required for excitation. They provide a floor (so to speak) for the possible transitions of electrons with wave number $k_i > q_0$. One can then consider the eigenstates of a pair of electrons with k_1, $k_2 > q_0$.

For a complete set of states of the two-electron system we take plane-wave product functions, $\varphi(\mathbf{k}_1,\mathbf{k}_2;\,\mathbf{r}_1,\mathbf{r}_2) = (1/V)\exp[i(\mathbf{k}_1\cdot\mathbf{r}_1+\mathbf{k}_2\cdot\mathbf{r}_2)]$ which satisfy periodic boundary conditions in a box of volume V, and where $\mathbf{r}_1$ and $\mathbf{r}_2$ are the coordinates of electron one and electron two. (One can use antisymmetric functions and obtain essentially the same results, but alternatively we can choose the electrons of opposite spin.) Defining relative and center-of-mass coordinates, $\mathbf{R}=\frac{1}{2}(\mathbf{r}_1+\mathbf{r}_2)$, $\mathbf{r}=(\mathbf{r}_2-\mathbf{r}_1)$, $\mathbf{K}=(\mathbf{k}_1+\mathbf{k}_2)$ and $\mathbf{k}=\frac{1}{2}(\mathbf{k}_2-\mathbf{k}_1)$, and letting $\mathcal{E}_K+\epsilon_k=(\hbar^2/m)(\frac{1}{4}K^2+k^2)$, the Schrödinger equation can be written

$$(\mathcal{E}_K+\epsilon_k-E)a_{\mathbf{k}}+\sum_{\mathbf{k}'} a_{\mathbf{k}'}(\mathbf{k}|H_1|\mathbf{k}')\times\delta(\mathbf{K}-\mathbf{K}')/\delta(0)=0 \quad (1)$$

where

$$\Psi(\mathbf{R},\mathbf{r})=(1/\sqrt{V})e^{i\mathbf{K}\cdot\mathbf{R}}\chi(\mathbf{r},K),\qquad \chi(\mathbf{r},K)=\sum_{\mathbf{k}}(a_{\mathbf{k}}/\sqrt{V})e^{i\mathbf{k}\cdot\mathbf{r}}, \quad (2)$$

and

$$(\mathbf{k}|H_1|\mathbf{k}')=\left(\frac{1}{V}\int d\mathbf{r}e^{-i\mathbf{k}\cdot\mathbf{r}}H_1e^{i\mathbf{k}'\cdot\mathbf{r}}\right)_{0\text{ phonons}}$$

We have assumed translational invariance in the metal. The summation over $\mathbf{k}'$ is limited by the exclusion principle to values of k_1 and k_2 larger than q_0, and by the delta function, which guarantees the conservation of the total momentum of the pair in a single scattering. The K dependence enters through the latter restriction.

Bardeen and Pines[3] and Fröhlich[4] have derived approximate formulas for the matrix element $(\mathbf{k}|H_1|\mathbf{k}')$; it is thought that the matrix elements for which the two electrons are confined to a thin energy shell near the Fermi surface, $\epsilon_1\simeq\epsilon_2\simeq\epsilon_F$, are the principal ones involved in producing the superconducting state.[2-4] With this in mind we shall approximate the expressions for $(\mathbf{k}|H_1|\mathbf{k}')$ derived by the above authors by

$$(\mathbf{k}|H_1|\mathbf{k}')=-|F| \quad \text{if} \quad k_0\leqslant k,\ k'\leqslant k_m$$
$$=0 \quad \text{otherwise}, \quad (3)$$

where F is a constant and $(\hbar^2/m)(k_m^2-k_0^2)\simeq 2\hbar\omega\simeq 0.2$ ev. Although it is not necessary to limit oneself so strongly, the degree of uncertainty about the precise form of $(\mathbf{k}|H_1|\mathbf{k}')$ makes it worthwhile to explore the consequences of reasonable but simple expressions.

With these matrix elements, the eigenvalue equation becomes

$$1=-|F|\int_{\epsilon_0}^{\epsilon_m}\frac{N(K,\epsilon)d\epsilon}{E-\epsilon-\mathcal{E}_K}, \quad (4)$$

where $N(K,\epsilon)$ is the density of two-electron states of total momentum K, and of energy $\epsilon=(\hbar^2/m)k^2$. To a very good approximation $N(K,\epsilon)\simeq N(K,\epsilon_0)$. The resulting spectrum has one eigenvalue smaller than $\epsilon_0+\mathcal{E}_K$, while the rest lie in the continuum. The lowest eigenvalue is $E_0=\epsilon_0+\mathcal{E}_K-\Delta$, where Δ is the binding energy of the pair

$$\Delta=(\epsilon_m-\epsilon_0)/(e^{1/\beta}-1), \quad (5)$$

where $\beta = N(K,\epsilon)|F|$. The binding energy, Δ, is independent of the volume of the box, but is strongly dependent on the parameter β.

Following a method of Bardeen,[5] by which the coupling constant for the electron-electron interaction, which is due to phonon exchange, is related to the high-temperature resistivity which is due to phonon absorption, one gets $\beta \simeq \rho n \times 10^{-6}$, where ρ is the high-temperature resistivity in esu and n is the number of valence electrons per unit volume. The binding energy displays a sharp change of behavior in the region $\beta \simeq 1$ and it is just this region which separates, in almost every case, the superconducting from the nonsuperconducting metals.[5] (Also it is just in this region where the attractive interaction between electrons, due to the phonon field, becomes about equal to the screened Coulomb repulsive interaction.)

The ground-state wave function,

$$\chi_0(r,K) = (\text{const}) \int \frac{e^{i\mathbf{k}\cdot\mathbf{r}} N(K,\epsilon(k))}{\mathcal{E}_K + \epsilon(k) - E} \left(\frac{d\epsilon}{dk}\right) d\mathbf{k}, \tag{6}$$

represents a true bound state which for large values of r decreases at least as rapidly as const$/r^2$. The average extension of the pair, $[\langle r^2 \rangle_{\text{Av}}]^{\frac{1}{2}}$, is of the order of 10^{-4} cm for $\Delta \simeq kT_c$. The existence of such a bound state with nonexponential dependence for large r is due to the exclusion of the states $k < k_0$ from the unperturbed spectrum, and the concomitant degeneracy of the lowest energy states of the unperturbed system. One would get no such state if the potential between the electrons were always repulsive. All of the excited states $\chi_{n>0}(\mathbf{r},K)$ are very nearly plane waves.

The pair described by $\chi_0(r)$ may be thought to have some Bose properties (to the extent that the binding energy of the pair is larger than the energy of interaction between pairs).[6] However, since $N(K,\epsilon)$ is strongly dependent on the total momentum of the pair, K, the binding energy Δ is a very sensitive function of K, being a maximum where $K=0$ and going very rapidly to zero where $K \simeq k_m - k_0$. Thus the elementary excitations of the pair might correspond to the splitting of the pair rather than to increasing the kinetic energy of the pair.

In either case the density of excited states (dN/dE) would be greatly reduced from the free-particle density and the elementary excitations would be removed from the ground state by what amounted to a small energy gap.

If the many-body system could be considered (at least to a lowest approximation) a collection of pairs of this kind above a Fermi sea, we would have (whether or not the pairs had significant Bose properties) a model similar to that proposed by Bardeen which would display many of the equilibrium properties of the superconducting state.

The author wishes to express his appreciation to Professor John Bardeen for his helpful instruction in many illuminating discussions.

* This work was supported in part by the Office of Ordnance Research, U. S. Army.

[1] J. Bardeen, Phys. Rev. **97**, 1724 (1955).
[2] See also, for further references and a general review, J. Bardeen, *Theory of Superconductivity Handbuch der Physik* (Springer-Verlag, Berlin, to be published), Vol. 15, p. 274.
[3] J. Bardeen and D. Pines, Phys. Rev. **99**, 1140 (1955).
[4] H. Fröhlich, Proc. Roy. Soc. (London) **A215**, 291 (1952).
[5] John Bardeen, Phys. Rev. **80**, 567 (1950).
[6] It has also been suggested that superconducting properties would result if electrons could combine in even groupings so that the resulting aggregates would obey Bose statistics. V. L. Ginzburg, Uspekhi Fiz. Nauk **48**, 25 (1952); M. R. Schafroth, Phys. Rev. **100**, 463 (1955).

Theory of Superconductivity*

J. Bardeen, L. N. Cooper,† and J. R. Schrieffer‡
Department of Physics, University of Illinois, Urbana, Illinois
(Received July 8, 1957)

A theory of superconductivity is presented, based on the fact that the interaction between electrons resulting from virtual exchange of phonons is attractive when the energy difference between the electrons states involved is less than the phonon energy, $\hbar\omega$. It is favorable to form a superconducting phase when this attractive interaction dominates the repulsive screened Coulomb interaction. The normal phase is described by the Bloch individual-particle model. The ground state of a superconductor, formed from a linear combination of normal state configurations in which electrons are virtually excited in pairs of opposite spin and momentum, is lower in energy than the normal state by amount proportional to an average $(\hbar\omega)^2$, consistent with the isotope effect. A mutually orthogonal set of excited states in one-to-one correspondence with those of the normal phase is obtained by specifying occupation of certain Bloch states and by using the rest to form a linear combination of virtual pair configurations. The theory yields a second-order phase transition and a Meissner effect in the form suggested by Pippard. Calculated values of specific heats and penetration depths and their temperature variation are in good agreement with experiment. There is an energy gap for individual-particle excitations which decreases from about $3.5kT_c$ at $T=0°K$ to zero at T_c. Tables of matrix elements of single-particle operators between the excited-state superconducting wave functions, useful for perturbation expansions and calculations of transition probabilities, are given.

I. INTRODUCTION

THE main facts which a theory of superconductivity must explain are (1) a second-order phase transition at the critical temperature, T_c, (2) an electronic specific heat varying as $\exp(-T_0/T)$ near $T=0°K$ and other evidence for an energy gap for individual particle-like excitations, (3) the Meissner-Ochsenfeld effect ($\mathbf{B}=0$), (4) effects associated with infinite conductivity ($\mathbf{E}=0$), and (5) the dependence of T_c on isotopic mass, $T_c\sqrt{M}=\text{const}$. We present here a theory which accounts for all of these, and in addition gives good quantitative agreement for specific heats and penetration depths and their variation with temperature when evaluated from experimentally determined parameters of the theory.

When superconductivity was discovered by Onnes[1] (1911), and for many years afterwards, it was thought to consist simply of a vanishing of all electrical resistance below the transition temperature. A major advance was the discovery of the Meissner effect[2] (1933), which showed that a superconductor is a perfect diamagnet; magnetic flux is excluded from all but a thin penetration region near the surface. Not very long afterwards (1935), London and London[3] proposed a phenomenological theory of the electromagnetic properties in which the diamagnetic aspects were assumed basic. F. London[4] suggested a quantum-theoretic approach to a theory in which it was assumed that there is somehow a coherence or rigidity in the superconducting state such that the wave functions are not modified very much when a magnetic field is applied. The concept of coherence has been emphasized by Pippard,[5] who, on the basis of experiments on penetration phenomena, proposed a nonlocal modification of the London equations in which a coherence distance, ξ_0, is introduced. One of the authors[6,7] pointed out that an energy-gap model would most likely lead to the Pippard version, and we have found this to be true of the present theory. Our theory of the diamagnetic aspects thus follows along the general lines suggested by London and by Pippard.[7]

The Sommerfeld-Bloch individual-particle model (1928) gives a fairly good description of normal metals, but fails to account for superconductivity. In this theory, it is assumed that in first approximation one may neglect correlations between the positions of the electrons and assume that each electron moves independently in some sort of self-consistent field determined by the other conduction electrons and the ions. Wave functions of the metal as a whole are designated by occupation of Bloch individual-particle states of energy $\epsilon(\mathbf{k})$ defined by wave vector $\mathbf{k}$ and spin σ; in the ground state all levels with energies below the Fermi energy, $\mathcal{E}_F$, are occupied; those above are unoccupied. Left out of the Bloch model are correlations between electrons brought about by Coulomb forces and interactions between electrons and lattice vibrations (or phonons).

* This work was supported in part by the Office of Ordnance Research, U. S. Army. One of the authors (J. R. Schrieffer) was aided by a Fellowship from the Corning Glass Works Foundation. Parts of the paper are based on a thesis submitted by Dr. Schrieffer in partial fulfillment of the requirements for a Ph.D. degree in Physics, University of Illinois, 1957.
† Present address: Department of Physics and Astronomy, The Ohio State University, Columbus, Ohio.
‡ Present address: Department of Theoretical Physics, University of Birmingham, Birmingham, England.

[1] H. K. Onnes, Comm. Phys. Lab. Univ. Leiden, Nos. 119, 120, 122 (1911).
[2] W. Meissner and R. Ochsenfeld, Naturwiss. **21**, 787 (1933).
[3] H. London and F. London, Proc. Roy. Soc. (London) **A149**, 71 (1935); Physica **2**, 341 (1935).
[4] F. London, Proc. Roy. Soc. (London) **A152**, 24 (1935); Phys. Rev. **74**, 562 (1948).
[5] A. B. Pippard, Proc. Roy. Soc. (London) **A216**, 547 (1953).
[6] J. Bardeen, Phys. Rev. **97**, 1724 (1955).
[7] For a recent review of the theory of superconductivity, which includes a discussion of the diamagnetic properties, see J. Bardeen, *Encyclopedia of Physics* (Springer-Verlag, Berlin, 1956), Vol. 15, p. 274.

Most of the relatively large energy associated with correlation effects occurs in both normal and superconducting phases and cancels out in the difference. One of the problems in constructing a satisfactory microscopic theory of superconductivity has been to isolate that part of the interaction which is responsible for the transition. Heisenberg[8] and Koppe[9] proposed a theory based on long-wavelength components of the Coulomb interaction, which were presumed to give fluctuations in electron density described roughly by wave packets localizing a small fraction of the electrons on lattices moving in different directions. A great break-through occurred with the discovery of the isotope effect,[10] which strongly indicated, as had been suggested independently by Fröhlich,[11] that electron-phonon interactions are primarily responsible for superconductivity.

Early theories based on electron-phonon interactions have not been successful. Fröhlich's theory, which makes use of a perturbation-theoretic approach, does give the correct isotopic mass dependence for H_0, the critical field at $T=0°$K, but does not yield a phase with superconducting properties and further, the energy difference between what is supposed to correspond to normal and superconducting phases is far too large. A variational approach by one of the authors[12] ran into similar difficulties. Both theories are based primarily on the self-energy of the electrons in the phonon field rather than on the true interaction between electrons, although it was recognized that the latter might be important.[13]

The electron-phonon interaction gives a scattering from a Bloch state defined by the wave vector $\mathbf{k}$ to $\mathbf{k}'=\mathbf{k}\pm\boldsymbol{\kappa}$ by absorption or emission of a phonon of wave vector $\boldsymbol{\kappa}$. It is this interaction which is responsible for thermal scattering. Its contribution to the energy can be estimated by making a canonical transformation which eliminates the linear electron-phonon interaction terms from the Hamiltonian. In second order, there is one term which gives a renormalization of the phonon frequencies, and another, H_2, which gives a true interaction between electrons, independent of the vibrational amplitudes. A transformation of this sort was given first by Fröhlich[14] in a formulation in which Coulomb interactions between electrons were disregarded. In a later treatment, Nakajima[15] showed how such interactions could be included. Particularly for the long-wavelength part of the interaction, it is important to take into account the screening of the Coulomb field of any one electron by other conduction electrons. Such effects are included in a more complete analysis by Bardeen and Pines,[16] based on the Bohm-Pines collective model, in which plasma modes are introduced for long wavelengths.

We shall call the interaction, H_2, between electrons resulting from the electron-phonon interaction the "phonon interaction." This interaction is attractive when the energy difference, $\Delta\epsilon$, between the electron states involved is less than $\hbar\omega$. Diagonal or self-energy terms of H_2 give an energy of order of $-N(\mathcal{E}_F)(\hbar\omega)^2$, where $N(\mathcal{E}_F)$ is the density of states per unit energy at the Fermi surface. The theories of Fröhlich and Bardeen mentioned above were based largely on this part of the energy. The observed energy differences between superconducting and normal states at $T=0°$K are much smaller, of the order of $-N(\mathcal{E}_F)(kT_c)^2$ or about 10^{-8} ev/atom. The present theory, based on the off-diagonal elements of H_2 and the screened Coulomb interaction, gives energies of the correct order of magnitude. While the self-energy terms do depend to some extent on the distribution of electrons in $\mathbf{k}$ space, it is now believed that this part of the energy is substantially the same in the normal and superconducting phases. The self-energy terms are also nearly the same for all of the various excited normal state configurations which make up the superconducting wave functions.

In a preliminary communication,[17] we gave as a criterion for the occurrence of a superconducting phase that for transitions such that $\Delta\epsilon<\hbar\omega$, the attractive H_2 dominate the repulsive short-range screened Coulomb interaction between electrons, so as to give a net attraction. We showed how an attractive interaction of this sort can give rise to a cooperative many-particle state which is lower in energy than the normal state by an amount proportional to $(\hbar\omega)^2$, consistent with the isotope effect. We have since extended the theory to higher temperatures, have shown that it gives both a second-order transition and a Meissner effect, and have calculated specific heats and penetration depths.

In the theory, the normal state is described by the Bloch individual-particle model. The ground-state wave function of a superconductor is formed by taking a linear combination of many low-lying normal state configurations in which the Bloch states are virtually occupied in pairs of opposite spin and momentum. If the state $\mathbf{k}\uparrow$ is occupied in any configuration, $-\mathbf{k}\downarrow$ is also occupied. The average excitation energy of the virtual pairs above the Fermi sea is of the order of kT_c. Excited states of the superconductor are formed by specifying occupation of certain Bloch states and by using all of the rest to form a linear combination of

[8] W. Heisenberg, *Two Lectures* (Cambridge University Press, Cambridge, 1948).

[9] H. Koppe, Ergeb. exakt. Naturw. **23**, 283 (1950); Z. Physik **148**, 135 (1957).

[10] E. Maxwell, Phys. Rev. **78**, 477 (1950); Reynolds, Serin, Wright, and Nesbitt, Phys. Rev. **78**, 487 (1950).

[11] H. Fröhlich, Phys. Rev. **79**, 845 (1950).

[12] J. Bardeen, Phys. Rev. **79**, 167 (1950); **80**, 567 (1950); **81**, 829 (1951).

[13] For a review of the early work, see J. Bardeen, Revs. Modern Phys. **23**, 261 (1951).

[14] H. Fröhlich, Proc. Roy. Soc. (London) **A215**, 291 (1952).

[15] S. Nakajima, *Proceedings of the International Conference on Theoretical Physics, Kyoto and Tokyo, September, 1953* (Science Council of Japan, Tokyo, 1954).

[16] J. Bardeen and D. Pines, Phys. Rev. **99**, 1140 (1955).

[17] Bardeen, Cooper, and Schrieffer, Phys. Rev. **106**, 162 (1957).

virtual pair configurations. There is thus a one-to-one correspondence between excited states of the normal and superconducting phases. The theory yields an energy gap for excitation of individual electrons from the superconducting ground state of about the observed order of magnitude.

The most important contribution to the interaction energy is given by short- rather than long-wavelength phonons. Our wave functions for the suerconducting phase give a coherence of short-wavelength components of the density matrix which extend over large distances in real space, so as to take maximum advantage of the attractive part of the interaction. The coherence distance, of the order of Pippard's ξ_0, can be estimated from uncertainty principle arguments.[5,7] If intervals of the order of $\Delta k\sim(kT_c/\mathcal{E}_F)k_F\sim10^4$ cm^{-1} are important in $\mathbf{k}$ space, wave functions in real space must extend over distances of at least $\Delta x\sim1/\Delta k\sim10^{-4}$ cm. The fraction of the total number of electrons which have energies within kT_c of the Fermi surface, so that they can interact effectively, is approximately $kT_c/\mathcal{E}_F\sim10^{-4}$. The number of these in an interaction region of volume $(\Delta x)^3$ is of the order of $10^{22}\times(10^{-4})^3\times10^{-4}=10^6$. Thus our wave functions must describe coherence of large numbers of electrons.[18]

In the absence of a satisfactory microscopic theory, there has been considerable development of phenomenological theories for both thermal and electromagnetic properties. Of the various two-fluid models used to describe the thermal properties, the first and best known is that of Gorter and Casimir,[19] which yields a parabolic critical field curve and an electronic specific heat varying as T^3. In this, as well as in subsequent theories of thermal properties, it is assumed that all of the entropy of the electrons comes from excitations of individual particles from the ground state. In recent years, there has been considerable experimental evidence[20] for an energy gap for such excitations, decreasing from $\sim3kT_c$ at $T=0$°K to zero at $T=T_c$. Two-fluid models which yield an energy gap and an exponential specific heat curve at low temperatures have been discussed by Ginsburg[21] and by Bernardes.[22] Koppe's theory may also be interpreted in terms of an energy-gap model.[7] Our theory yields an energy gap and specific heat curve consistent with the experimental observations.

The best known of the phenomenological theories for the electromagnetic properties is that of F. and H. London.[23] With an appropriate choice of gauge for the vector potential, $\mathbf{A}$, the London equation for the superconducting current density, $\mathbf{j}$, may be written

$$-c\Lambda\mathbf{j}=\mathbf{A}. \tag{1.1}$$

The London penetration depth is given by:

$$\lambda_L{}^2=\Lambda c^2/4\pi. \tag{1.2}$$

F. London has pointed out that (1.1) would follow from quantum theory if the superconducting wave functions are so rigid that they are not modified at all by the application of a magnetic field. For an electron density n/cm^3, this approach gives $\Lambda=m/ne^2$.

On the basis of empirical evidence, Pippard[5] has proposed a modification of the London equation in which the current density at a point is given by an integral of the vector potential over a region surrounding the point:

$$\mathbf{j}(\mathbf{r})=-\frac{3}{4\pi c\Lambda\xi_0}\int\frac{\mathbf{R}[\mathbf{R}\cdot\mathbf{A}(\mathbf{r}')]e^{-R/\xi_0}}{R^4}d\tau', \tag{1.3}$$

where $\mathbf{R}=\mathbf{r}-\mathbf{r}'$. The "coherence distance," ξ_0, is of the order of 10^{-4} cm in a pure metal. For a very slowly varying $\mathbf{A}$, the Pippard expression reduces to the London form (1.1).

The present theory indicates that the Meissner effect is intimately related to the existence of an energy gap, and we are led to a theory similar to, although not quite the same as, that proposed by Pippard. Our theoretical values for ξ_0 are close to those derived empirically by Pippard. We find that while the integrand is relatively independent of temperature, the coefficient in front of the integral (in effect Λ) varies with T in such a way as to account for the temperature variation of penetration depth.

Our theory also accounts in a qualitative way for those aspects of superconductivity associated with infinite conductivity and a persistent current flowing in a ring. When there is a net current flow, the paired states $(\mathbf{k}_1\uparrow,\mathbf{k}_2\downarrow)$ have a net momentum $\mathbf{k}_1+\mathbf{k}_2=\mathbf{q}$, where $\mathbf{q}$ is the same for all virtual pairs. For each value of $\mathbf{q}$, there is a metastable state with a minimum in free energy and a unique current density. Scattering of individual electrons will not change the value of $\mathbf{q}$ common to virtual pair states, and so can only produce fluctuations about the current determined by $\mathbf{q}$. Nearly all fluctuations will increase the free energy; only those which involve a majority of the electrons so as to change

[18] Our picture differs from that of Schafroth, Butler, and Blatt, Helv. Phys. Acta **30**, 93 (1957), who suggest that pseudomolecules of pairs of electrons of opposite spin are formed. They show if the size of the pseudomolecules is less than the average distance between them, and if other conditions are fulfilled, the system has properties similar to that of a charged Bose-Einstein gas, including a Meissner effect and a critical temperature of condensation. Our pairs are not localized in this sense, and our transition is not analogous to a Bose-Einstein condensation.

[19] C. J. Gorter and H. B. G. Casimir, Physik. Z. **35**, 963 (1934); Z. techn. Physik **15**, 539 (1934).

[20] For discussions of evidence for an energy gap, see Blevins, Gordy, and Fairbank, Phys. Rev. **100**, 1215 (1955); Corak, Goodman, Satterthwaite, and Wexler, Phys. Rev. **102**, 656 (1956); W. S. Corak and C. B. Satterthwaite, Phys. Rev. **102**, 662 (1956); R. E. Glover and M. Tinkham, Phys. Rev. **104**, 844 (1956), and to be published.

[21] W. L. Ginsburg, Fortschr. Physik **1**, 101 (1953); also see reference 7.

[22] N. Bernardes, Phys. Rev. **107**, 354 (1957).

[23] An excellent account may be found in F. London, *Superfluids* (John Wiley and Sons, Inc., New York, 1954), Vol. 1.

the common $\mathbf{q}$ can decrease the free energy. These latter are presumably extremely rare, so that the metastable current carrying state can persist indefinitely.[24]

It has long been recognized that there is a law of corresponding states for superconductors. The various properties can be expressed approximatly in terms of a small number of parameters. If the ratio of the electronic specific heat at T to that of the normal state at T_c, $C_s(T)/C_n(T_c)$, is plotted on a reduced temperature scale, $t=T/T_c$, most superconductors fall on nearly the same curve. There are two parameters involved: (1) the density of states in energy at the Fermi surface, $N(\mathcal{E}_F)$, determined from $C_n(T)=\gamma T$ and (2) one which depends on the phonon interaction, which can be estimated from T_c. A consequence of the similarity law is that $\gamma T_c^2/V_m H_0^2$ (where V_m is the molar volume and H_0 the critical field at $T=0°$K) is approximately the same for most superconductors.

A third parameter, the average velocity, v_0, of electrons at the Fermi surface,

$$v_0=\hbar^{-1}|\partial\mathcal{E}/\partial\mathbf{k}|_F \tag{1.4}$$

is required for penetration phenomena. As pointed out by Faber and Pippard,[25] this parameter is most conveniently determined from measurements of the anomalous skin effect in normal metals in the high-frequency limit. The expression, as given by Chambers[26] for the current density when the electric field varies over a mean free path, l, may be written in the form:

$$\mathbf{j}_n(r)=\frac{e^2N(\mathcal{E}_F)v_0}{2\pi}\int\frac{\mathbf{R}[\mathbf{R}\cdot\mathbf{A}(\mathbf{r}')]e^{-R/l}}{R^4}d\tau'. \tag{1.5}$$

The coefficient $N(\mathcal{E}_F)v_0$ has been determined empirically for tin and aluminum.

Pippard based his Eq. (1.3) on Chambers' expression. London's coefficient, Λ, for $T=0°$K may be expressed in the form:

$$\Lambda^{-1}=\tfrac{2}{3}e^2N(\mathcal{E}_F)v_0^2. \tag{1.6}$$

Faber and Pippard suggest that if ξ_0 is written:

$$\xi_0=a\hbar v_0/kT_c, \tag{1.7}$$

the dimensionless constant a has approximately the same value for all superconductors and they find it equal to about 0.15 for Sn and Al.[27]

Our theory is based on a rather idealized model in which anisotropic effects are neglected. It contains three parameters, two corresponding to $N(\mathcal{E}_F)$ and v_0, and one dependent on the electron-phonon interaction which determines T_c. The model appears to fit the law of corresponding states about as well as real metals do ($\sim$10% for most properties). We find a relation corresponding to (1.7) with $a=0.18$. It thus appears that superconducting properties are not dependent on the details of the band structure but only upon the gross features.

Section II is concerned with the nature of the ground state and the energy of excited states near $T=0°$K, Sec. III with excited states and thermal properties, Sec. IV with calculation of matrix elements for application to perturbation theory expansions and transition probabilities and Sec. V with electrodynamic and penetration phenomenon. Some of the computational details are given in Appendices.

We give a fairly complete account of the equilibrium properties of our model, but nothing on transport or boundary effects. Starting from matrix elements of single-particle scattering operators as given in Sec. IV, it should not be difficult to determine transport properties in the superconducting state from the corresponding properties of the normal state.

II. THE GROUND STATE

The interaction which produces the energy difference between the normal and superconducting phases in our theory arises from the virtual exchange of phonons and the screened Coulomb repulsion between electrons. Other interactions, such as those giving rise to the single-particle self-energies, are thought to be essentially the same in both states, their effects thus cancelling in the energy difference. The problem is therefore one of calculating the ground state and excited states of a dense system of fermions interacting via two-body potentials.

The Hamiltonian for the fermion system is most conveniently expressed in terms of creation and annihilation operators, based on the renormalized Bloch states specified by wave vector $\mathbf{k}$ and spin σ, which satisfy the usual Fermi commutation relations:

$$[c_{\mathbf{k}\sigma},c_{\mathbf{k}'\sigma'}^*]_+=\delta_{\mathbf{k}\mathbf{k}'}\delta_{\sigma\sigma'}, \tag{2.1}$$

$$[c_{\mathbf{k}\sigma},c_{\mathbf{k}'\sigma'}]_+=0. \tag{2.2}$$

The single-particle number operator $n_{\mathbf{k}\sigma}$ is defined as

$$n_{\mathbf{k}\sigma}=c_{\mathbf{k}\sigma}^*c_{\mathbf{k}\sigma}. \tag{2.3}$$

The Hamiltonian for the electrons may be expressed in

---

[24] Blatt, Butler, and Schafroth, Phys. Rev. **100**, 481 (1955) have introduced the concept of a "correlation length," roughly the distance over which the momenta of a pair of particles are correlated. M. R. Schafroth, Phys. Rev. **100**, 502 (1955), has argued that there is a true Meissner effect only if the correlation length is effectively infinite. In our theory, the correlation length (not to be confused with Pippard's coherence distance, ξ_0) is most reasonably interpreted as the distance over which the momentum of virtual pairs is the same. We believe that in this sense, the correlation length *is* effectively infinite. The value of $\mathbf{q}$ is exactly zero everywhere in a simply connected body in an external field. When there is current flow, as in a torus, there is a unique distribution of $\mathbf{q}$ values for minimum free energy.

[25] T. E. Faber and A. B. Pippard, Proc. Roy. Soc. (London) **A231**, 53 (1955).

[26] See A. P. Pippard, *Advances in Electronics* (Academic Press, Inc., New York, 1954), Vol. 6, p. 1.

[27] From analysis of data on transmission of microwave and far infrared radiation through superconducting films of tin and lead, Glover and Tinkham (reference 20) find $a=0.27$.

the form

$$H=\sum_{k>k_F}\epsilon_{\mathbf{k}}n_{\mathbf{k}\sigma}+\sum_{k<k_F}|\epsilon_{\mathbf{k}}|(1-n_{\mathbf{k}\sigma})+H_{\text{Coul}}+\tfrac{1}{2}\sum_{\mathbf{k},\mathbf{k}',\sigma,\sigma',\boldsymbol{\kappa}}\times\frac{2\hbar\omega_\kappa|M_\kappa|^2c^*(\mathbf{k}'-\boldsymbol{\kappa},\sigma')c(\mathbf{k}',\sigma')c^*(\mathbf{k}+\boldsymbol{\kappa},\sigma)c(\mathbf{k},\sigma)}{(\epsilon_{\mathbf{k}}-\epsilon_{\mathbf{k}+\boldsymbol{\kappa}})^2-(\hbar\omega_\kappa)^2}=H_0+H_I, \quad (2.4)$$

where $\epsilon_{\mathbf{k}}$ is the Bloch energy measured relative to the Fermi energy, $\mathcal{E}_F$. We denote by $k>k_F$ states above the Fermi surface, by $k<k_F$ those below. The fourth term on the right of (2.4) is H_2, the phonon interaction, which comes from virtual exchange of phonons between the electrons. The matrix element for phonon-electron interaction, M_κ, calculated for the zero-point amplitude of the lattice vibrations, is related to the v_κ introduced by Bardeen and Pines[16] by

$$|M_\kappa|^2=|v_\kappa|^2\langle q_\kappa^2\rangle_{\text{Av}}=|v_\kappa|^2(\hbar/2\omega_\kappa). \quad (2.5)$$

Since $|M_\kappa|^2$ varies with isotopic mass in the same way that ω_κ does, the ratio $|M_\kappa|^2/\hbar\omega_\kappa$ is independent of isotopic mass. We consider only the off-diagonal interaction terms of H_2, assuming that the diagonal terms are taken into account by appropriate renormalization of the Bloch energies, $\epsilon_{\mathbf{k}}$. The third term is the screened Coulomb interaction.

Following Bardeen and Pines,[16] the phonons are assumed to be decoupled from the electrons by a renormalization procedure and their frequencies are taken to be unaltered by the transition to the superconducting state. While this assumption is not strictly valid, the shift in self-energy can be taken into account after we have solved for the electronic part of the wave function. This separation is possible because the phonons depend only upon the average electron distribution in momentum space and the wave function for electrons at any temperature is formed from configurations with essentially the same distribution of particles. The Bloch energies are also assumed to be constant; however, their shift with temperature could be treated as in the phonon case.

The form of the phonon interaction shows that it is attractive (negative) for excitation energies $|\epsilon_{\mathbf{k}}-\epsilon_{\mathbf{k}+\boldsymbol{\kappa}}|<\hbar\omega_\kappa$. Opposed to this is the repulsive Coulomb interaction, which may be expressed in a form similar to H_2. For free electrons in a system of unit volume the interaction in momentum space is $4\pi e^2/\kappa^2$. In the Bohm-Pines theory, the long-wavelength components are expressed in the form of plasma oscillations, so that κ can be no smaller than minimum value κ_c, usually slightly less than the radius of the Fermi surface, k_F. One could also take screening into account by a Fermi-Thomas method, in which case κ^2 would be replaced by $\kappa^2+\kappa_s^2$, where κ_s depends on the electron density. Our criterion for superconductivity is that the attractive phonon interaction dominate the Coulomb interaction for those matrix elements which are of importance in the superconducting wave function:

$$-V\equiv\left\langle-\frac{2|M_\kappa|^2}{\hbar\omega_\kappa}+\frac{4\pi e^2}{\kappa^2}\right\rangle_{\text{Av}}<0. \quad (2.6)$$

The most important transitions are those for which $|\epsilon_{\mathbf{k}}-\epsilon_{\mathbf{k}+\boldsymbol{\kappa}}|\sim kT_c\ll\hbar\omega_\kappa$. A detailed discussion of the criterion (2.6) has been given by Pines,[28] who shows that it accounts in a reasonable way for the empirical rules of Matthias[29] for the occurrence of superconductivity. Numerically, the criterion is not much different from one given earlier by Fröhlich,[11] based on a different principle.

To obtain the ground state function, we observe that the interaction Hamiltonian connects a large number of nearly degenerate occupation number configurations with each other via nonzero matrix elements. If the matrix elements were all negative in sign, one could obtain a state with low energy by forming a linear combination of the basis functions with expansion coefficients of the same sign. The magnitude of the interaction energy obtained in this manner would be approximately given by the number of configurations which connect to a given typical configuration times an average matrix element. This was demonstrated by one of the authors[30] by solving a problem in which two electrons with zero total momentum interact via constant negative matrix elements in a small shell above the Fermi surface. It was shown that the ground state of this system is separated from the continuum by a volume independent energy. This type of coherent mixing of Bloch states produces a state with qualitatively different properties from the original states.

In the actual problem, the interaction which takes a pair from $(\mathbf{k}_1\sigma_1,\mathbf{k}_2\sigma_2)$ to $(\mathbf{k}_1'\sigma_1,\mathbf{k}_2'\sigma_2)$ contains the operators,

$$c^*(\mathbf{k}_2',\sigma_2)c(\mathbf{k}_2,\sigma_2)c^*(\mathbf{k}_1',\sigma_1)c(\mathbf{k}_1,\sigma_1). \quad (2.7)$$

Conservation of momentum requires that

$$\mathbf{k}_1+\mathbf{k}_2=\mathbf{k}_1'+\mathbf{k}_2'. \quad (2.8)$$

Because of Fermi-Dirac statistics, matrix elements of (2.7) between arbitrary many electron configurations alternate in sign so that if the configurations occur in the ground state with roughly equal weight, the net interaction energy would be small. We can, however, produce a coherent low state by choosing a subset of configurations between which the matrix elements are negative. Such a subset can be formed by those configurations in which the Bloch states are occupied in pairs, $(\mathbf{k}_1\sigma_1,\mathbf{k}_2\sigma_2)$; that is, if one member of the pair is occupied in any configuration in the subset, the other is also. Since the interaction conserves momentum, a

[28] D. Pines (to be published).

[29] B. Matthias, *Progress in Low Temperature Physics* (North-Holland Publishing Company, Amsterdam, 1957), Vol. 2.

[30] L. N. Cooper, Phys. Rev. **104**, 1189 (1956).

maximum number of matrix elements will be obtained if all pairs have the same net momentum, $\mathbf{k}_1+\mathbf{k}_2=\mathbf{q}$. It is further desirable to take pairs of opposite spin, because exchange terms reduce the interaction for parallel spins. The best choice for q for the ground state pairing is $\mathbf{q}=0$, $(\mathbf{k}\uparrow, -\mathbf{k}\downarrow)$.

We start then by considering a reduced problem in which we include only configurations in which the states are occupied in pairs such that if $\mathbf{k}\uparrow$ is occupied so is $-\mathbf{k}\downarrow$. A pair is designated by the wave vector $\mathbf{k}$, independent of spin. Creation and annihilation operators for pairs may be defined in terms of the single-particle operators as follows:

$$b_{\mathbf{k}}=c_{-\mathbf{k}\downarrow}c_{\mathbf{k}\uparrow}, \tag{2.9}$$

$$b_{\mathbf{k}}^*=c_{\mathbf{k}\uparrow}^*c_{-\mathbf{k}\downarrow}^*. \tag{2.10}$$

These operators satisfy the commutation relations

$$[b_{\mathbf{k}},b_{\mathbf{k}'}^*]_-=(1-n_{\mathbf{k}\uparrow}-n_{-\mathbf{k}\downarrow})\delta_{\mathbf{k}\mathbf{k}'}, \tag{2.11}$$

$$[b_{\mathbf{k}},b_{\mathbf{k}'}]_-=0, \tag{2.12}$$

$$[b_{\mathbf{k}},b_{\mathbf{k}'}]_+=2b_{\mathbf{k}}b_{\mathbf{k}'}(1-\delta_{\mathbf{k}\mathbf{k}'}), \tag{2.13}$$

where $n_{\mathbf{k}\sigma}$ is given by (2.3). While the commutation relation (2.12) is the same as for bosons, the commutators (2.11) and (2.13) are distinctly different from those for Bose particles. The factors $(1-n_{\mathbf{k}\uparrow}-n_{-\mathbf{k}\downarrow})$ and $(1-\delta_{\mathbf{k}\mathbf{k}'})$ arise from the effect of the exclusion principle on the single particles.

That part of the Hamiltonian which connects pairs with zero net momentum may be derived from the Hamiltonian (2.4) and expressed in terms of the b's. Measuring the energy relative to the Fermi sea, we obtain:

$$H_{\text{red}}=2\sum_{k>k_F}\epsilon_{\mathbf{k}}b_{\mathbf{k}}^*b_{\mathbf{k}}+2\sum_{k<k_F}|\epsilon_{\mathbf{k}}|b_{\mathbf{k}}b_{\mathbf{k}}^* - \sum_{\mathbf{k}\mathbf{k}'}V_{\mathbf{k}\mathbf{k}'}b_{\mathbf{k}'}^*b_{\mathbf{k}}. \tag{2.14}$$

We have defined the interaction terms with a negative sign so that $V_{\mathbf{k}\mathbf{k}'}$ will be predominantly positive for a superconductor. There are many other terms in the complete interaction which connect pairs with total momentum $\mathbf{q}\neq 0$. These have little effect on the energy, and can be treated as a perturbation. Although the interaction terms kept in H_{red} may appear to have a negligible weight, it is this part which contributes overwhelmingly to the interaction energy.

We have used a Hartree-like method to determine the expansion coefficients, which appears to give an excellent approximation, and may, indeed, even be correct in the limit of a large number of particles.[31] (See Appendix A.)

[31] Since (2.14) is quadratic in the b's, one might hope to get an exact solution for the ground state by an appropriate redefinition of the single-particle states, as can be done for either Fermi-Dirac or Einstein-Bose statistics. Our pairs obey neither of these, and no such simple solution appears possible.

Excited states are treated in much the same way as the ground state. One must distinguish between singly excited particles, in which one and only one of a pair $(\mathbf{k}\uparrow, -\mathbf{k}\downarrow)$ is occupied, and excited or "real" pair states. We treat singly excited particles in the Bloch scheme, as in the normal metal. They contribute a negligible amount to the interaction energy directly, but reduce the amount of phase space available for real and virtual pairs. Thus the interaction portion of H_{red} is modified by deleting from the sums over $\mathbf{k}$ and $\mathbf{k}'$ all singly occupied states, and the remainder is used to determine the interaction energy associated with the pairs.

One might expect to get some interaction energy from singly occupied states by associating them in various pairs with $\mathbf{q}\neq 0$. However, an appreciable energy is obtained only if a finite fraction of the pairs have the same $\mathbf{q}$, and this will not be true for randomly excited particles. States with a net current flow can be obtained by taking a pairing $(\mathbf{k}_1\uparrow,\mathbf{k}_2\downarrow)$, with $\mathbf{k}_1+\mathbf{k}_2=\mathbf{q}$, and $\mathbf{q}$ the same for all virtual pairs.

The most general wave function satisfying the pairing condition $(\mathbf{k}\uparrow, -\mathbf{k}\downarrow)$ is of the form

$$\sum_{\mathbf{k}_1\cdots\mathbf{k}_n}[h(\mathbf{k}_1\cdots\mathbf{k}_n)]^{\frac{1}{2}}f(\cdots 1(k_1)\cdots 1(k_n)\cdots), \tag{2.15}$$

where the sum extends over all distinct pair configurations. To construct our ground state function we make a Hartree-like approximation in which the probability that a specific configuration of pairs occurs in the wave function is given by a product of occupancy probabilities for the individual pair states. If for the moment we relax the requirement that the wave function describes a system with a fixed number of particles, then a function having this Hartree-like property is

$$\Psi=\prod_{\mathbf{k}}[(1-h_{\mathbf{k}})^{\frac{1}{2}}+h_{\mathbf{k}}^{\frac{1}{2}}b_{\mathbf{k}}^*]\Phi_0, \tag{2.16}$$

where Φ_0 is the vacuum. It follows from (2.16) that the probability of the n states $\mathbf{k}_1\cdots\mathbf{k}_n$ being occupied is $h(\mathbf{k}_1)\cdots h(\mathbf{k}_n)$, and since n is unrestricted we see that Ψ is closely related to the intermediate coupling approximation.

For any specified wave vector $\mathbf{k}'$, it is convenient to decompose Ψ into two components, in one of which, φ_1, the pair state designated by $\mathbf{k}'$ is certainly occupied and the other, φ_0, for which it is empty:

$$\Psi=h_{\mathbf{k}'}^{\frac{1}{2}}\varphi_1+(1-h_{\mathbf{k}'})^{\frac{1}{2}}\varphi_0. \tag{2.17}$$

The coefficient $h_{\mathbf{k}'}$ is the probability that state $\mathbf{k}'$ is occupied and the φ's are the normalized functions:

$$\varphi_1=b_{\mathbf{k}'}^*\varphi_0=b_{\mathbf{k}'}^*\prod_{\mathbf{k}(\neq\mathbf{k}')}[(1-h_{\mathbf{k}})^{\frac{1}{2}}+h_{\mathbf{k}}^{\frac{1}{2}}b_{\mathbf{k}}^*]\Phi_0. \tag{2.18}$$

In the limit of a large system, the weights of states with different total numbers of pairs in Ψ will be sharply peaked about the average number, N, which will be dependent on the choice of the h's. We take for our ground state function, Ψ_N, the projection of Ψ onto the space of exactly N pairs.[32] This function may also be decomposed as in (2.17), but since φ_{N1} and φ_{N0} now have the same number of pairs, φ_{N1} is not equal to $b_{\mathbf{k}'}^*\varphi_{N0}$. To decompose Ψ_N, we suppose that $\mathbf{k}$ space is divided into elements, $\Delta\mathbf{k}$, with $\mathfrak{N}_{\mathbf{k}}$ available states of which in a typical configuration $m_{\mathbf{k}}$ are occupied by pairs. The $m_{\mathbf{k}}$'s are restricted so that the total number of pairs is specified:

$$\sum_{\text{all }\Delta\mathbf{k}} m_{\mathbf{k}}=N=\sum_{\text{all }\Delta\mathbf{k}}\langle m_{\mathbf{k}}\rangle_{\text{Av}}. \tag{2.19}$$

The total weight of a given distribution of $m_{\mathbf{k}}$'s in Ψ is

$$W(m_{\mathbf{k}})=\prod_{\text{all }\Delta\mathbf{k}}\frac{\mathfrak{N}_k!}{m_{\mathbf{k}}!(\mathfrak{N}_{\mathbf{k}}-m_k)!}h_{\mathbf{k}}^{m_k}(1-h_{\mathbf{k}})^{\mathfrak{N}_{\mathbf{k}}-m_{\mathbf{k}}} \tag{2.20}$$

and the total weight of functions with specified N is

$$W_N=\sum_{(\Sigma m_{\mathbf{k}}=N)} W(m_{\mathbf{k}}), \tag{2.21}$$

where the sum is over-all distributions of the $m_{\mathbf{k}}$'s subject to the conditions (2.19).

The decomposition of Ψ_N into a part in which a specified pair state $\mathbf{k}'$ is occupied and one where it is not can be carried out by calculating from Ψ_N the total weight, $W_{N,\mathbf{k}'}$, corresponding to $\mathbf{k}'$ occupied with the restriction $\sum m_{\mathbf{k}}=N$. When $\mathbf{k}'$ is occupied, there are $\mathfrak{N}_{\mathbf{k}'}-1$ other states in the cell over which the remaining $m_{\mathbf{k}'}-1$ particles can be distributed and this cell will contribute a factor

$$\frac{(\mathfrak{N}_{\mathbf{k}'}-1)!h_{\mathbf{k}'}^{m_{\mathbf{k}'}}(1-h_{\mathbf{k}'})^{\mathfrak{N}_{\mathbf{k}'}-m_{\mathbf{k}'}}}{(m_{\mathbf{k}'}-1)!(\mathfrak{N}_{\mathbf{k}'}-m_{\mathbf{k}'})!} \tag{2.22}$$

to the weight for a given distribution of $m_{\mathbf{k}}$'s. It follows that

$$W_{N,\mathbf{k}'}=\sum_{(\Sigma m_{\mathbf{k}}=N)}\frac{m_{\mathbf{k}'}}{\mathfrak{N}_{\mathbf{k}'}}W(m_{\mathbf{k}})=h_{\mathbf{k}'}W_N. \tag{2.23}$$

The last equality holds except for terms which vanish in the limit of a large system because the state vector Ψ gives a probability $\langle m_{\mathbf{k}'}/\mathfrak{N}_{\mathbf{k}'}\rangle_{\text{Av}}=h_{\mathbf{k}'}$ that a given state in $\Delta\mathbf{k}'$ is occupied. Now the weights for different numbers of pairs in Ψ are strongly peaked about the most probable number N and therefore the average over distributions with exactly N pairs is essentially equal to the average over all distributions. Since all terms in the wave function come in with a positive sign, it follows that the normalized Ψ_N may be decomposed in the form

$$\Psi_N=h_{\mathbf{k}'}^{\frac{1}{2}}\varphi_{N1}+(1-h_{\mathbf{k}'})^{\frac{1}{2}}\varphi_{N0}, \tag{2.24}$$

where φ_{N1} and φ_{N0} are normalized functions.

For purposes of calculating matrix elements and interaction energies, a further decomposition into states in which occupancy of two pair states, $\mathbf{k}$ and $\mathbf{k}'$, is specified is often convenient. Thus we may write:

$$\Psi_N=(hh')^{\frac{1}{2}}\varphi_{N11}+[h(1-h')]^{\frac{1}{2}}\varphi_{N10}+[(1-h)h']^{\frac{1}{2}}\varphi_{N01}+[(1-h)(1-h')]^{\frac{1}{2}}\varphi_{N00}, \tag{2.25}$$

where the first index gives the occupancy of $\mathbf{k}$ and the second of $\mathbf{k}'$. It follows from the definition of the functions that

$$b_{\mathbf{k}'}^*b_{\mathbf{k}}\varphi_{N10}=\varphi_{N01}. \tag{2.26}$$

Thus the diagonal matrix element of $b_{\mathbf{k}'}^*b_{\mathbf{k}}$ is

$$(\Psi_N|b_{\mathbf{k}'}^*b_{\mathbf{k}}|\Psi_N)=[h_{\mathbf{k}}(1-h_{\mathbf{k}})h_{\mathbf{k}'}(1-h_{\mathbf{k}'})]^{\frac{1}{2}}. \tag{2.27}$$

Ground-State Energy

If the wave function (2.24) is used as a variational approximation to the true ground-state function, the ground-state energy relative to the energy of the Fermi sea is given by

$$W_0=(\Psi_0,H_{\text{red}}\Psi_0), \tag{2.28}$$

where Ψ_0 is the N-pair function Ψ_N, for the ground state. The Bloch energies, $\epsilon_{\mathbf{k}}$, are measured with respect to the Fermi energy and

$$-V_{\mathbf{kk}'}=(-\mathbf{k}'\downarrow,\mathbf{k}'\uparrow|H_I|-\mathbf{k}\downarrow,\mathbf{k}\uparrow)+(\mathbf{k}'\uparrow,-\mathbf{k}'\downarrow|H_I|\mathbf{k}\uparrow,-\mathbf{k}\downarrow). \tag{2.29}$$

The decomposition (2.23) leads to the Bloch energy contribution to W_0 of the form

$$W_{\text{KE}}=2\sum_{k>k_F}\epsilon_{\mathbf{k}}h_{\mathbf{k}}+2\sum_{k<k_F}|\epsilon_{\mathbf{k}}|(1-h_{\mathbf{k}}), \tag{2.30}$$

where "KE" stands for kinetic energy. The matrix elements of the interaction term in H_{red} are given by (2.27) and the interaction energy is

$$W_I=-\sum_{\mathbf{k},\mathbf{k}'}V_{\mathbf{kk}'}[h_{\mathbf{k}}(1-h_{\mathbf{k}})h_{\mathbf{k}'}(1-h_{\mathbf{k}'})]^{\frac{1}{2}}, \tag{2.31}$$

and therefore

$$W_0=W_{\text{KE}}+W_I=2\sum_{k>k_F}\epsilon_{\mathbf{k}}h_{\mathbf{k}}+2\sum_{k<k_F}|\epsilon_{\mathbf{k}}|(1-h_{\mathbf{k}})-\sum_{\mathbf{k},\mathbf{k}'}V_{\mathbf{kk}'}[h_{\mathbf{k}}(1-h_{\mathbf{k}})h_{\mathbf{k}'}(1-h_{\mathbf{k}'})]^{\frac{1}{2}}. \tag{2.32}$$

By minimizing W_0 with respect to $h_{\mathbf{k}}$, we are led to an integral equation determining the distribution function:

$$\frac{[h_{\mathbf{k}}(1-h_{\mathbf{k}})]^{\frac{1}{2}}}{1-2h_{\mathbf{k}}}=\frac{\sum_{\mathbf{k}'}V_{\mathbf{kk}'}[h_{\mathbf{k}'}(1-h_{\mathbf{k}'})]^{\frac{1}{2}}}{2\epsilon_{\mathbf{k}}}. \tag{2.33}$$

We shall neglect anisotropic effects and assume for simplicity that the matrix element $V_{\mathbf{kk}'}$ can be replaced by a constant average matrix element,

$$V=\langle V_{\mathbf{kk}'}\rangle_{\text{Av}}, \tag{2.34}$$

for pairs making transitions in the region $-\hbar\omega<\epsilon<\hbar\omega$

---

[32] It is easily seen that Ψ_N has zero total spin, corresponding to a singlet state.

and by zero outside this region, where ω is the average phonon frequency. This cutoff corresponds to forming our wave function from states in the region where the interaction is expected to be attractive and not mixing in states outside this region. The average is primarily one over directions of $\mathbf{k}$ and $\mathbf{k}'$ since the interaction is insensitive to the excitation energy for those transitions of importance in describing the superconducing phase. The average may also be viewed as choosing h to be a function of energy alone, thus neglecting the details of band structure. The laws of similarity indicate this to be a reasonable assumption and the good agreement of our theory with a wide class of superconductors supports this view.

Introduction of the average matrix element into (2.33) leads to

$$h_{\mathbf{k}}=\frac{1}{2}\left[1-\frac{\epsilon_{\mathbf{k}}}{(\epsilon_{\mathbf{k}}^2+\epsilon_0^2)^{\frac{1}{2}}}\right], \qquad (2.35)$$

and

$$[h_{\mathbf{k}}(1-h_{\mathbf{k}})]^{\frac{1}{2}}=\frac{\epsilon_0}{2(\epsilon_{\mathbf{k}}^2+\epsilon_0^2)^{\frac{1}{2}}}, \qquad (2.36)$$

where

$$\epsilon_0=V\sum_{\mathbf{k}'}[h_{\mathbf{k}'}(1-h_{\mathbf{k}'})]^{\frac{1}{2}}, \qquad (2.37)$$

the sum extending over states within the range $|\epsilon_{\mathbf{k}}|<\hbar\omega$. If (2.36) and (2.37) are combined, one obtains a condition on ϵ_0:

$$\frac{1}{V}=\sum_{\mathbf{k}}\frac{1}{2(\epsilon_{\mathbf{k}}^2+\epsilon_0^2)^{\frac{1}{2}}}. \qquad (2.38)$$

Replacing the sum by an integral and recalling that $V=0$ for $|\epsilon_{\mathbf{k}}|>\hbar\omega$, we may replace this condition by

$$\frac{1}{N(0)V}=\int_0^{\hbar\omega}\frac{d\epsilon}{(\epsilon^2+\epsilon_0^2)^{\frac{1}{2}}}. \qquad (2.39)$$

Solving for ϵ_0, we obtain

$$\epsilon_0=\hbar\omega\Big/\sinh\left[\frac{1}{N(0)V}\right]. \qquad (2.40)$$

where $N(0)$ is the density of Bloch states of one spin per unit energy at the Fermi surface.

The ground state energy is obtained by combining the expressions for $h_{\mathbf{k}}$ and ϵ_0, (2.35), (2.36), and (2.37), with (2.32). We find

$$W_0=4N(0)\int_0^{\hbar\omega}\epsilon h(\epsilon)d\epsilon-\frac{\epsilon_0^2}{V}$$

$$=2N(0)\int_0^{\hbar\omega}\left[\epsilon-\frac{\epsilon^2}{(\epsilon^2+\epsilon_0^2)^{\frac{1}{2}}}\right]d\epsilon-\frac{\epsilon_0^2}{V}, \qquad (2.41)$$

where we have used the fact that $[1-h(-\epsilon)]=h(\epsilon)$, that is, the distribution function is symmetric in electrons and holes with respect to the Fermi surface. Using the relations (2.40), we find that the difference in energy between the superconducting and normal states at the absolute zero becomes

$$W_0=N(0)(\hbar\omega)^2\left\{1-\left[1+\left(\frac{\epsilon_0}{\hbar\omega}\right)^2\right]^{\frac{1}{2}}\right\}=\frac{-2N(0)(\hbar\omega)^2}{e^{[2/N(0)V]}-1}. \qquad (2.42)$$

If there is a net negative interaction on the average, no matter how weak, there exists a coherent state which is lower in energy than the normal state. Thus our criterion for superconductivity is that $V>0$, as given in (2.6).

For excitations which are small compared to $\hbar\omega$, the phonon interaction is essentially independent of isotopic mass and therefore the total mass dependence of W_0 comes from $(\hbar\omega)^2$, in agreement with the isotope effect. Empirically, W_0 is of the order of $N(0)(kT_c)^2$ and in general kT_c is much less than $\hbar\omega$. According to (2.42), this will occur if $N(0)V<1$, that is, the weak coupling limit.

It should be noted that the ground state energy cannot be obtained in any finite-order perturbation theory. In the strong-coupling limit, (2.42) gives the correct result, $-N(0)(\hbar\omega)^2V$, for the average interaction approximation and it is possible that our solution is accurate in the statistical limit over the entire range of coupling. (See Appendix A.)

In the weak-coupling limit, the energy becomes

$$W_0=-2N(0)(\hbar\omega)^2\exp\left[-\frac{2}{N(0)V}\right]. \qquad (2.43)$$

which may be expressed in terms of the number of electrons, n_c, in pairs virtually excited above the Fermi surface as

$$W_0=-\tfrac{1}{2}n_c^2/N(0), \qquad (2.44)$$

where

$$n_c=2N(0)\hbar\omega\exp\left[-\frac{1}{N(0)V}\right]. \qquad (2.45)$$

In this form, the cooperative nature of the ground state is evident. Using the empirical order of magnitude relation between W_0 and kT_c, we might estimate

$$kT_c\sim\hbar\omega\exp\left[-\frac{1}{N(0)V}\right]. \qquad (2.46)$$

In the next chapter we shall see that the explicit calculation of kT_c from the free energy as a function of temperature leads to nearly this result.

Energy gap at $T=0°$K

An important feature of the reduced Hamiltonian is that there are no excitations from the ground state, analogous to single-particle excitations of the Bloch theory, with vanishing excitation energy. This is easily seen by considering a function

$$\Psi_{\text{exc}}=\{\prod_{\mathbf{k}\neq\mathbf{k}',\mathbf{k}''}[(1-h_{\mathbf{k}})^{\frac{1}{2}}+h_{\mathbf{k}}^{\frac{1}{2}}b_{\mathbf{k}}^*]\}c_{-\mathbf{k}'\downarrow}^*c_{\mathbf{k}''\uparrow}^*\Phi_0, \quad (2.47)$$

which is orthogonal to the ground state function and corresponds to breaking up a pair in $\mathbf{k}'$, the spin-up member going to $\mathbf{k}''\uparrow$. The projection of Ψ_{exc} onto the space with N pairs is also orthogonal to Ψ_0. The decomposition of $\Psi_{N\text{exc}}$ is the same as that of Ψ_0, except that $-\mathbf{k}'\downarrow$ and $\mathbf{k}''\uparrow$ are definitely known to be occupied and $\mathbf{k}'\uparrow$ and $-\mathbf{k}''\downarrow$ are unoccupied. This leads to the excitation energy

$$W_{\mathbf{k}',\mathbf{k}''}-W_0=\epsilon_{\mathbf{k}'}(1-2h_{\mathbf{k}'})+\epsilon_{\mathbf{k}''}(1-2h_{\mathbf{k}''}) +2V\sum_{\mathbf{k}}[h_{\mathbf{k}}(1-h_{\mathbf{k}})]^{\frac{1}{2}}\{[h_{\mathbf{k}'}(1-h_{\mathbf{k}'})]^{\frac{1}{2}} +[h_{\mathbf{k}''}(1-h_{\mathbf{k}''})]^{\frac{1}{2}}\}, \quad (2.48)$$

the decrease in interaction energy arising from the fact that pairs cannot make transitions into or out of pair states $\mathbf{k}'$ and $\mathbf{k}''$ in the excited function because these states are occupied by single particles. Combining (2.35), (2.36), and (2.37) with (2.48), we find

$$W_{\mathbf{k}',\mathbf{k}''}-W_0=\frac{\epsilon_{\mathbf{k}'}^2}{E_{\mathbf{k}'}}+\frac{\epsilon_{\mathbf{k}''}^2}{E_{\mathbf{k}''}}+\epsilon_0^2\left(\frac{1}{E_{\mathbf{k}'}}+\frac{1}{E_{\mathbf{k}''}}\right) =E_{\mathbf{k}'}+E_{\mathbf{k}''}, \quad (2.49)$$

where

$$E_{\mathbf{k}}=(\epsilon_{\mathbf{k}}^2+\epsilon_0^2)^{\frac{1}{2}}. \quad (2.50)$$

When $\epsilon_{\mathbf{k}}\to 0$, then $E_{\mathbf{k}}\to\epsilon_0$ and (2.49) shows that the minimum excitation energy is $2\epsilon_0$. These single-particle-like excitations have the new dispersion law (2.50) which goes over to the normal law when $\epsilon_{\mathbf{k}}\gg\epsilon_0$.

To obtain a complete set of excitations, we must include excited state pair functions generated by

$$[(1-h_{\mathbf{k}})^{\frac{1}{2}}b_{\mathbf{k}}^*-h_{\mathbf{k}}^{\frac{1}{2}}], \quad (2.51)$$

which by construction are orthogonal to the ground state pair functions generated by

$$[(1-h_{\mathbf{k}})^{\frac{1}{2}}+h_{\mathbf{k}}^{\frac{1}{2}}b_{\mathbf{k}}^*]. \quad (2.52)$$

The decomposition of an excited state with an excited pair in $\mathbf{k}'$ and a ground pair in $\mathbf{k}$ would be

$$\Psi_{\mathbf{k}'}=[h_{\mathbf{k}}(1-h_{\mathbf{k}'})]^{\frac{1}{2}}\varphi_{11}-[h_{\mathbf{k}}h_{\mathbf{k}'}]^{\frac{1}{2}}\varphi_{10} +[(1-h_{\mathbf{k}})(1-h_{\mathbf{k}'})]^{\frac{1}{2}}\varphi_{01}-[(1-h_{\mathbf{k}})h_{\mathbf{k}'}]^{\frac{1}{2}}\varphi_{00}, \quad (2.53)$$

where the functions φ are normalized and the second script denotes the occupancy of $\mathbf{k}'$. Taking the expectation value of H_{red} with respect to (2.53), we find the energy to form an excited pair in state $\mathbf{k}'$ is:

$$W_{\mathbf{k}'}-W_0=2\epsilon_{\mathbf{k}'}(1-2h_{\mathbf{k}'}) +4V\sum_{\mathbf{k}}[h_{\mathbf{k}}(1-h_{\mathbf{k}})h_{\mathbf{k}'}(1-h_{\mathbf{k}'})]^{\frac{1}{2}}=2E_{\mathbf{k}'}. \quad (2.54)$$

Again the minimum energy required to form an excitation is $2\epsilon_0$ and an energy gap of width $2\epsilon_0$ appears in the excitation spectrum in a natural way. It follows that in general the energy difference between two states, 1 and 2, is given by the difference in the sums of the excited-particle energies,

$$W_1-W_2=\sum\nolimits_1 E_k-\sum\nolimits_2 E_k, \quad (2.55)$$

and it is unnecessary to distinguish between single particles and two members of an excited pair in calculating the sums.

Collective excitations corresponding to long-range density fluctuations are suppressed by the subsidiary condition on the wave function resulting from the collective description of the electron-ion interaction.[16] The effect of the terms neglected in the Hamiltonian, $H-H_{\text{red}}=H'$, can be estimated by a perturbation expansion of H' in eigenfunctions of H_{red}. This expansion is carried out to second order in Appendix A and it is concluded that H' will contribute little to the condensation energy. The shift in the zero-point energy of the lattice associated with the transition at the absolute zero is estimated in Appendix B and it is shown that this effect contributes a small correction to W_0. The electron self-energy shift has not been calculated at the present time; however, it is also believed that the correction is small.

III. EXCITED STATES

An excited state of the system will be formed by specifying the set of states, $\mathcal{S}$, which are occupied by single particles and the set of states, $\mathcal{P}$, occupied by excited pairs. The rest of the states, $\mathcal{G}$, will be available for occupation by ground pairs. The term "single-particle" occupation means that either $\mathbf{k}\uparrow$ or $-\mathbf{k}\downarrow$ is occupied by an electron, but not both. "Excited pair" and "ground pair" occupation refers to pairs which are in functions generated by operators of the form (2.51) and (2.52) respectively. Wave functions with different distributions of single particles and excited pairs are orthogonal to each other and the totality of such functions constitutes a complete set of excited states which are in one-to-one correspondence with the Bloch-type excitations in the normal metal.

The energy of the excited states will be evaluated by using the reduced Hamiltonian plus the Bloch energy for single particles:

$$H_e=\sum_{k>k_F,\sigma}\epsilon_{\mathbf{k}}n_{\mathbf{k}\sigma}+\sum_{k<k_F,\sigma}|\epsilon_{\mathbf{k}}|(1-n_{\mathbf{k}\sigma}) -\sum_{\mathbf{k},\mathbf{k}'}V_{\mathbf{k}\mathbf{k}'}b_{\mathbf{k}'}^*b_{\mathbf{k}}, \quad (3.1)$$

where the second term gives the Bloch energy of the holes and the energies $\epsilon_{\mathbf{k}}$ are measured relative to the Fermi energy. Since H_I contains only terms for transitions by ground pairs and excited pairs, the single particles contribute only to the Bloch energy and hence they are treated as in the Bloch scheme.

The equilibrium condition of the system at a specified temperature will be determined by minimizing the free energy with respect to the distribution function for excited particles, f, and ground pairs, h.

It turns out that $h_{\mathbf{k}}$ is a function of temperature and therefore excited and ground pairs are not necessarily orthogonal to each other at different temperatures, although the excited states form a complete orthonormal set at each temperature. The most probable distribution of single particles also varies with temperature and thus the great majority of states contributing to the free energy at different temperatures will be orthogonal in any event. The situation is similar to taking the lattice constant temperature dependent as a result of thermal expansion. This freedom in choosing $h_{\mathbf{k}}$ allows us to work in that representation which minimizes the free energy at the specified temperature.

A typical excited state wave function can be written as the projection of

$$\Psi_{\text{exc}}=\prod_{\mathbf{k}(\mathcal{G})}[(1-h_{\mathbf{k}})^{\frac{1}{2}}+h_{\mathbf{k}}^{\frac{1}{2}}b_{\mathbf{k}}^{*}]\prod_{\mathbf{k}'(\mathcal{P})}[(1-h_{\mathbf{k}'})^{\frac{1}{2}}b_{\mathbf{k}'}^{*}-h_{\mathbf{k}'}^{\frac{1}{2}}]\prod_{\mathbf{k}''(\mathcal{S})}c_{(\mathbf{k}'')}^{*}\Phi_0, \quad (3.2)$$

onto the space with N pairs, where $\mathcal{G}$, $\mathcal{P}$, and $\mathcal{S}$ specify the states occupied by ground pairs, excited pairs, and single particles respectively and $c_{(\mathbf{k}'')}^{*}$ denotes either $\mathbf{k}''\uparrow$ or $-\mathbf{k}''\downarrow$ is included in the product. For any specified $\mathbf{k}$, this function can be decomposed into a portion with $\mathbf{k}$ occupied and a portion with $\mathbf{k}$ unoccupied. The decompositions for the three cases in which $\mathbf{k}$ is in the sets $\mathcal{G}$, $\mathcal{P}$, and $\mathcal{S}$ are

Ground:

$$\Psi=h_{\mathbf{k}}^{\frac{1}{2}}\varphi_1(\cdots 1_{\mathbf{k}}\cdots)+(1-h_{\mathbf{k}})^{\frac{1}{2}}\varphi_0(\cdots 0_{\mathbf{k}}\cdots), \quad (3.3)$$

Excited:

$$\Psi=(1-h_{\mathbf{k}})^{\frac{1}{2}}\varphi_1(\cdots 1_{\mathbf{k}}\cdots)-h_{\mathbf{k}}^{\frac{1}{2}}\varphi_0(\cdots 0_{\mathbf{k}}\cdots), \quad (3.4)$$

Single in $\mathbf{k}\uparrow$:

$$\Psi=c_{\mathbf{k}\uparrow}^{*}\varphi_0(\cdots 0_{\mathbf{k}}\cdots), \quad (3.5)$$

where the φ's are normalized functions with $1_{\mathbf{k}}$ representing pair state k being occupied and $0_{\mathbf{k}}$ unoccupied.

To determine the distribution functions, we need the free energy

$$F=W-TS, \quad (3.6)$$

where W is the energy calculated by an ensemble average over the wave functions of the form (3.2) and S is the entropy.

To enumerate the systems in the ensemble we divide k-space into cells $\Delta\mathbf{k}$ containing $\mathfrak{N}_{\mathbf{k}}$ pair states as before. Let there be $S_{\mathbf{k}}$ single particles and $P_{\mathbf{k}}$ excited pairs in $\Delta\mathbf{k}$, with the rest of the $\mathfrak{N}_{\mathbf{k}}$ states being occupied by ground pairs. The probability that either $\mathbf{k}\uparrow$ or $-\mathbf{k}\downarrow$ is occupied by a single particle is $s_{\mathbf{k}}=S_{\mathbf{k}}/\mathfrak{N}_{\mathbf{k}}$, while the probability for an excited pair in state $\mathbf{k}$ is $p_{\mathbf{k}}=P_{\mathbf{k}}/\mathfrak{N}_{\mathbf{k}}$ and therefore the probability for a ground pair is $(1-s_{\mathbf{k}}-p_{\mathbf{k}})$. Above the Fermi surface, $h<\frac{1}{2}$ and s and p refer to excited electrons; below the Fermi surface, $h>\frac{1}{2}$ and s and p refer to holes.

The diagonal element of $n_{\mathbf{k}\sigma}$ follows immediately from (3.3), (3.4), and (3.5). Including the factor giving fractional number of configurations for which each decomposition applies, we have

$$\begin{aligned}(\psi|n_{\mathbf{k}\sigma}|\psi)&=(1/2)s_{\mathbf{k}}+p_{\mathbf{k}}(1-h_{\mathbf{k}})\\&\quad+(1-s_{\mathbf{k}}-p_{\mathbf{k}})h_{\mathbf{k}}, \quad \epsilon>0;\\(\psi|1-n_{\mathbf{k}\sigma}|\psi)&=(1/2)s_{\mathbf{k}}+p_{\mathbf{k}}h_{\mathbf{k}}\\&\quad+(1-s_{\mathbf{k}}-p_{\mathbf{k}})(1-h_{\mathbf{k}}), \quad \epsilon<0.\end{aligned} \quad (3.7)$$

Upon using (3.7) and the fact that $1-h_{\mathbf{k}}(-\epsilon)=h_{\mathbf{k}}(\epsilon)$, the Bloch energy contribution to W,

$$(\psi|\sum_{\mathbf{k}>k_{F\sigma}}\epsilon_{\mathbf{k}}n_{\mathbf{k}\sigma}+\sum_{\mathbf{k}<k_{F\sigma}}|\epsilon_{\mathbf{k}}|(1-n_{\mathbf{k}\sigma})|\psi), \quad (3.8)$$

becomes

$$W_{\text{KE}}=\sum_{\mathbf{k}}|\epsilon_{\mathbf{k}}|[s_{\mathbf{k}}+2p_{\mathbf{k}}+2(1-s_{\mathbf{k}}-2p_{\mathbf{k}})h_{\mathbf{k}}(|\epsilon_{\mathbf{k}}|)], \quad (3.9)$$

where we have carried out the spin sum.

To calculate the matrix elements of the pair interaction operator $\sum V_{\mathbf{k}\mathbf{k}'}b_{\mathbf{k}'}^{*}b_{\mathbf{k}}$, we assume that $V_{\mathbf{k}\mathbf{k}'}$ varies continuously with $\mathbf{k}$ and $\mathbf{k}'$ so that $V_{\mathbf{k}\mathbf{k}'}$ may be considered to be the same for all transitions from states in $\Delta\mathbf{k}$ to states in $\Delta\mathbf{k}'$. Let $\mathbf{k}$ and $\mathbf{k}'$ represent two specified wave vectors in $\Delta\mathbf{k}$ and $\Delta\mathbf{k}'$. To obtain nonvanishing matrix elements for $b_{\mathbf{k}'}^{*}b_{\mathbf{k}}$, these states must be occupied by either excited $(-)$ or ground $(+)$ pairs, giving the four possibilities $++$, $--$, $+-$, $-+$ for $\mathbf{k}$ and $\mathbf{k}'$, respectively. For any one of these cases, a typical wave function may be decomposed into components in which the pair occupancy of $\mathbf{k}$ and $\mathbf{k}'$ is specified:

$$\begin{aligned}\Psi_{\mathbf{k},\mathbf{k}'}=&\alpha_{11}\varphi_{11}(\cdots 1_{\mathbf{k}}\cdots 1_{\mathbf{k}'}\cdots)\\&+\alpha_{10}\varphi_{10}(\cdots 1_{\mathbf{k}}\cdots 0_{\mathbf{k}'}\cdots)\\&+\alpha_{01}\varphi_{01}(\cdots 0_{\mathbf{k}}\cdots 1_{\mathbf{k}'}\cdots)\\&+\alpha_{00}\varphi_{00}(\cdots 0_{\mathbf{k}}\cdots 0_{\mathbf{k}'}\cdots),\end{aligned} \quad (3.10)$$

where the φ's are normalized functions. Table I gives the values of the α's for the different cases along with the fractional number of configurations for which they apply.

The diagonal elements of $b_{\mathbf{k}'}^{*}b_{\mathbf{k}}$ are given by $\alpha_{10}\alpha_{01}$ in each case. If we sum these, weighted according to the probability they occur in the ensemble, we obtain

$$\begin{aligned}&[h(1-h)h'(1-h')]^{\frac{1}{2}}\{(1-s-p)(1-s'-p')\\&\quad+pp'-(1-s-p)p'-p(1-s'-p')\}\\&\quad=[h(1-h)h'(1-h')]^{\frac{1}{2}}\\&\qquad\times\{(1-s-2p)(1-s'-2p')\}.\end{aligned} \quad (3.11)$$

Introducing these matrix elements into the ensemble average of the interaction Hamiltonian, we find

$$W_I=-\sum_{\mathbf{k},\mathbf{k}'}V_{\mathbf{k}\mathbf{k}'}[h_{\mathbf{k}}(1-h_{\mathbf{k}})h_{\mathbf{k}'}(1-h_{\mathbf{k}'})]^{\frac{1}{2}}\times\{(1-s_{\mathbf{k}}-2p_{\mathbf{k}})(1-s_{\mathbf{k}'}-2p_{\mathbf{k}'})\}. \quad (3.12)$$

TABLE I. Coefficients for the decomposition of Ψ according to Eq. (3.10).

Wave function k	k′	Fractional No. of cases	α_{11}	α_{10}	α_{01}	α_{00}
+	+	$(1-s-p)(1-s'-p')$	$[hh']^{\frac{1}{2}}$	$[h(1-h')]^{\frac{1}{2}}$	$[(1-h)h']^{\frac{1}{2}}$	$[(1-h)(1-h')]^{\frac{1}{2}}$
−	−	pp'	$[(1-h)(1-h')]^{\frac{1}{2}}$	$-[(1-h)h']^{\frac{1}{2}}$	$-[h(1-h')]^{\frac{1}{2}}$	$[hh']^{\frac{1}{2}}$
+	−	$(1-s-p)p'$	$[h(1-h')]^{\frac{1}{2}}$	$-[hh']^{\frac{1}{2}}$	$[(1-h)(1-h')]^{\frac{1}{2}}$	$-[(1-h)h']^{\frac{1}{2}}$
−	+	$p(1-s'-p')$	$[(1-h)h']^{\frac{1}{2}}$	$[(1-h)(1-h')]^{\frac{1}{2}}$	$-[hh']^{\frac{1}{2}}$	$-[h(1-h')]^{\frac{1}{2}}$

It should be noted that the Bloch energy (3.9) and the interaction energy (3.12) depend on $(s+2p)$ or the total occupancy probability. The energy does not depend upon the relative probability for single-particle and excited-pair occupation. Thus one may use a distribution function f which gives the over-all probability of occupancy, where

$$s_{\mathbf{k}}=2f_{\mathbf{k}}(1-f_{\mathbf{k}}), \quad (3.13)$$

and

$$p_{\mathbf{k}}=f_{\mathbf{k}}^2, \quad (3.14)$$

which follows from the fact that $s_{\mathbf{k}}$ is the probability that either $\mathbf{k}\uparrow$ is occupied and $-\mathbf{k}\downarrow$ is empty or the reverse and $p_{\mathbf{k}}$ is the probability that both $\mathbf{k}\uparrow$ and $-\mathbf{k}\downarrow$ are occupied. The free energy can be minimized directly with respect to $h_{\mathbf{k}}$, $p_{\mathbf{k}}$, and $s_{\mathbf{k}}$ without introducing $f_{\mathbf{k}}$ and one indeed finds that (3.13) and (3.14) hold.

Since the excited particles are specified independently for each system in the ensemble, the usual expression for the entropy in terms of f may be used:

$$-TS=2kT\sum_{\mathbf{k}'}\{f_{\mathbf{k}'}\ln f_{\mathbf{k}'}+(1-f_{\mathbf{k}'})\ln(1-f_{\mathbf{k}'})\}. \quad (3.15)$$

Minimization of the Free Energy

If the expressions (3.13) and (3.14) are introduced into (3.9) and (3.12), the free energy becomes

$$F=2\sum_{\mathbf{k}}|\epsilon_{\mathbf{k}}|[f_{\mathbf{k}}+(1-2f_{\mathbf{k}})h_{\mathbf{k}}(|\epsilon_{\mathbf{k}}|)] - \sum_{\mathbf{k},\mathbf{k}'}V_{\mathbf{k}\mathbf{k}'}[h_{\mathbf{k}}(1-h_{\mathbf{k}})h_{\mathbf{k}'}(1-h_{\mathbf{k}'})]^{\frac{1}{2}} \times\{(1-2f_{\mathbf{k}})(1-2f_{\mathbf{k}'})\}-TS. \quad (3.16)$$

When we minimize F with respect to $h_{\mathbf{k}}$, we find that

$$2\epsilon_{\mathbf{k}}-\sum_{\mathbf{k}'}V_{\mathbf{k}\mathbf{k}'}[h_{\mathbf{k}'}(1-h_{\mathbf{k}'})]^{\frac{1}{2}}(1-2f_{\mathbf{k}'}) \times\frac{(1-2h_{\mathbf{k}})}{[h_{\mathbf{k}}(1-h_{\mathbf{k}})]^{\frac{1}{2}}}=0, \quad (3.17)$$

or

$$\frac{[h_{\mathbf{k}}(1-h_{\mathbf{k}})]^{\frac{1}{2}}}{1-2h_{\mathbf{k}}}=\sum_{\mathbf{k}'}\frac{V_{\mathbf{k}\mathbf{k}'}[h_{\mathbf{k}'}(1-h_{\mathbf{k}'})]^{\frac{1}{2}}(1-2f_{\mathbf{k}'})}{2\epsilon_{\mathbf{k}}}, \quad (3.18)$$

where the energy $\epsilon_{\mathbf{k}}$ is measured relative to the Fermi energy and $\epsilon_{\mathbf{k}}<0$ for $k<k_F$. Assuming as before that the interaction can be replaced by a constant average matrix element $-V$, defined by (2.34) for $|\epsilon_{\mathbf{k}}|<\hbar\omega$ and by zero outside this region, it follows that $h_{\mathbf{k}}$ is again of the form

$$h_{\mathbf{k}}=\tfrac{1}{2}[1-(\epsilon_{\mathbf{k}}/E_{\mathbf{k}})], \quad (3.19)$$

and

$$[h_{\mathbf{k}}(1-h_{\mathbf{k}})]^{\frac{1}{2}}=\tfrac{1}{2}\epsilon_0/E_{\mathbf{k}}. \quad (3.20)$$

The energy $E_{\mathbf{k}}$, a positive definite quantity, is defined as

$$E_{\mathbf{k}}=+(\epsilon_{\mathbf{k}}^2+\epsilon_0^2)^{\frac{1}{2}}, \quad (3.21)$$

where

$$\epsilon_0=V\sum_{\mathbf{k}'}[h_{\mathbf{k}'}(1-h_{\mathbf{k}'})]^{\frac{1}{2}}(1-2f_{\mathbf{k}'}). \quad (3.22)$$

It will turn out that $2\epsilon_0$ is the magnitude of the energy gap in the single-particle density of states and therefore the distribution of ground pairs is determined by the magnitude of the gap at that temperature.

When we minimize F with respect to $f_{\mathbf{k}}$, we find that

$$2\epsilon_{\mathbf{k}}(1-2h_{\mathbf{k}})+4\sum_{\mathbf{k}'}V_{\mathbf{k}\mathbf{k}'}[h_{\mathbf{k}'}(1-h_{\mathbf{k}'})h_{\mathbf{k}}(1-h_{\mathbf{k}})]^{\frac{1}{2}} \times(1-2f_{\mathbf{k}'})+2kT\ln(f_{\mathbf{k}}/(1-f_{\mathbf{k}}))=0, \quad (3.23)$$

and using (3.19) through (3.22), we find that

$$-\ln(f_{\mathbf{k}}/(1-f_{\mathbf{k}}))=\beta\left[\frac{\epsilon_{\mathbf{k}}^2}{E_{\mathbf{k}}}+\frac{\epsilon_0^2}{E_{\mathbf{k}}}\right]=\beta E_{\mathbf{k}}, \quad (3.24)$$

where $\beta=1/kT$ and $E_{\mathbf{k}}$ is a positive quantity. The solution for f_k is

$$f_{\mathbf{k}}=\frac{1}{e^{\beta E_{\mathbf{k}}}+1}=f(E_{\mathbf{k}}). \quad (3.25)$$

Thus the single particles and excited pairs describe a set of independent fermions with the modified dispersion law (3.21). For $k>k_F$, $f_{\mathbf{k}}$ specifies electron occupation while for $k<k_F$, $f_{\mathbf{k}}$ specifies hole occupation. These electrons and holes are identified with the normal component of the two-fluid model.

When $\epsilon_{\mathbf{k}}\to0^+$, then $E_{\mathbf{k}}\to\epsilon_0$ for the electron and when $\epsilon_{\mathbf{k}}\to0^-$, then $E_{\mathbf{k}}\to\epsilon_0$ for the hole or the corresponding electron energy$\to-\epsilon_0$. Thus the new density of states has an energy gap of magnitude $2\epsilon_0$ centered about the Fermi energy. The modified density of states is given by

$$\frac{dN(E)}{dE}=\frac{dN(\epsilon)}{d\epsilon}\frac{d\epsilon}{dE}=N(0)\frac{E}{(E^2-\epsilon_0^2)^{\frac{1}{2}}}, \quad (3.26)$$

which is singular at the edges of the gap, $E=\epsilon_0$. The total number of states is of course unaltered by the interaction.

If the distribution functions (3.19) and (3.25) are introduced, the condition determining the energy gap

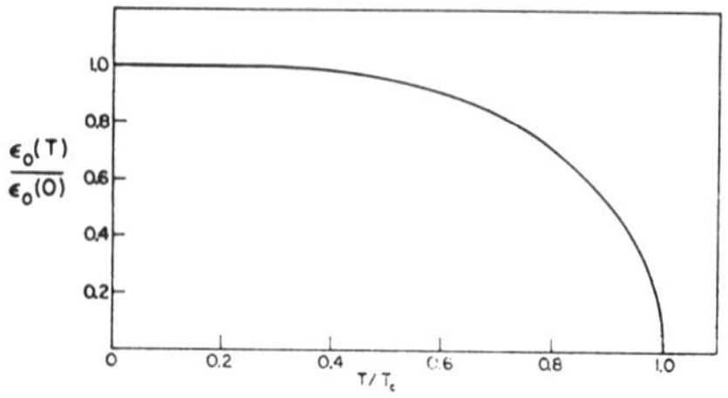

FIG. 1. Ratio of the energy gap for single-particle-like excitations to the gap at $T=0°K$ *vs* temperature.

(3.22) becomes (dividing by ϵ_0)

$$\frac{1}{N(0)V}=\int_0^{\hbar\omega}\frac{d\epsilon}{(\epsilon^2+\epsilon_0^2)^{\frac{1}{2}}}\tanh[\tfrac{1}{2}\beta(\epsilon^2+\epsilon_0^2)^{\frac{1}{2}}], \quad (3.27)$$

where we have replaced the sum by an integral and used the fact that the distribution functions are symmetric in holes and electrons with respect to the Fermi energy. The transition temperature, T_c, is defined as the boundary of the region beyond which there is no real, positive ϵ_0 which satisfies (3.27). Above T_c therefore, $\epsilon_0=0$ and $f(E_{\mathbf{k}})$ becomes $f(\epsilon_{\mathbf{k}})$, so that the metal returns to the normal state. Below T_c the solution of (3.27), $\epsilon_0\neq 0$, minimizes the free energy and we have the superconducting phase. Thus (3.26) can be used to determine the critical temperature and we find

$$\frac{1}{N(0)V}=\int_0^{\hbar\omega}\frac{d\epsilon}{\epsilon}\tanh(\tfrac{1}{2}\beta_c\epsilon), \quad (3.28)$$

or

$$kT_c=1.14\hbar\omega\exp\left[-\frac{1}{N(0)V}\right], \quad (3.29)$$

as long as $kT_c\ll\hbar\omega$, which corresponds to the weak-coupling case discussed in Sec. II. The transition temperature is proportional to $\hbar\omega$, which is consistent with the isotope effect. The small magnitude of T_c compared to the Debye temperature is presumably due to the cancellation of the phonon interaction and the screened Coulomb interaction for transitions of importance in describing the superconducting state, and the resulting effect of the exponential.

The transition temperature is a strong function of the electron concentration since the density of states enters exponentially. It should be possible to make estimates of the change in transition temperature with pressure, alloying, etc., from (3.29).

A plot of the energy gap as a function of temperature is given in Fig. 1. The ratio of the energy gap at $T=0°K$ to kT_c is given by combining (2.36) and (3.28):

$$2\epsilon_0/kT_c=3.50. \quad (3.30)$$

From the law of corresponding states, this ratio is predicted to be the same for all superconductors. Near T_c, the gap may be expressed as

$$\epsilon_0=3.2kT_c[1-(T/T_c)]^{\frac{1}{2}}, \quad (3.31)$$

which has the form suggested by Buckingham.[33]

It can be seen from the distribution functions that our theory goes over into the Bloch scheme above the transition temperature. As $T\rightarrow T_c$, $E_{\mathbf{k}}\rightarrow|\epsilon_{\mathbf{k}}|$, and $h_{\mathbf{k}}$ vanishes for $k>k_F$ and is unity for $k<k_F$. According to (3.4) the excited-pair function specifies complete electron occupancy for $k>k_F$ and complete hole occupancy for $k<k_F$ in this case. Thus, in the normal state, the ground pairs vanish above the Fermi surface and form the Fermi sea below, while the single particles and excited pairs combine to describe excited electrons for $k>k_F$ and excited holes for $k<k_F$.

Critical Field and Specific Heat

The critical field for a bulk specimen of unit volume is given by

$$H_c^2/8\pi=F_n-F_s, \quad (3.32)$$

where F_n is the free energy of the normal state:

$$F_n=-4N(0)kT\int_0^{\infty}d\epsilon\log(1+e^{-\beta\epsilon})$$

$$=-\tfrac{1}{3}\pi^2N(0)(kT)^2. \quad (3.33)$$

With the aid of (3.25) the entropy in the superconducting state, (3.15) may be expressed as

$$TS=4kT\sum_{k>k_F}[\ln(1+e^{-\beta E_{\mathbf{k}}})+\beta E_{\mathbf{k}}f_{\mathbf{k}}]. \quad (3.34)$$

Replacing the sum by an integral and performing a partial integration, we find

$$TS=4N(0)\int_0^{\infty}d\epsilon\left[\frac{\epsilon^2}{E}+E\right]f(\beta E), \quad (3.35)$$

where the upper limit has been extended to infinity because $f(\beta E)$ decreases rapidly for $\beta\epsilon>1$. If (3.34) and (3.16) are combined with the distribution functions (3.19) and (3.25), the free energy becomes

$$F_s=-4N(0)\int_0^{\infty}d\epsilon Ef(\beta E)$$

$$+2N(0)\int_0^{\hbar\omega}d\epsilon\left[\epsilon-\frac{\epsilon^2}{E}\right]-\frac{\epsilon_0^2}{V}, \quad (3.36)$$

which with the aid of (3.27) may be expressed as

$$F_s=-2N(0)\int_0^{\infty}d\epsilon\left[\frac{2\epsilon^2+\epsilon_0^2}{E}\right]f(\beta E)$$

$$-N(0)(\hbar\omega)^2\left\{\left[1+\left(\frac{\epsilon_0}{\hbar\omega}\right)^2\right]^{\frac{1}{2}}-1\right\}. \quad (3.37)$$

[33] M. J. Buckingham, Phys. Rev. **101**, 1431 (1956).

The critical field is given by combining (3.31), (3.32), and (3.37):

$$\frac{H_c^2}{8\pi}=N(0)(\hbar\omega)^2\left\{\left[1+\left(\frac{\epsilon_0}{\hbar\omega}\right)^2\right]^{\frac{1}{2}}-1\right\}-\frac{\pi^2}{3}N(0)(kT)^2\times\left\{1-\beta^2\int_0^\infty d\epsilon\left[\frac{2\epsilon^2+\epsilon_0^2}{E}\right]f(\beta E)\right\}. \quad (3.38)$$

A plot of the critical field as a function of $(T/T_c)^2$ is given in Fig. 2. The curve agrees fairly well with the $1-(T/T_c)^2$ law of the Gorter-Casimir two-fluid model,[19] the maximum deviation being about four percent. There is good experimental support for a similar deviation in vanadium, thallium, indium, and tin; however, our deviation appears to be somewhat too large to fit the experimental results.

The critical field at $T=0$ is

$$H_0=[4\pi N(0)]^{\frac{1}{2}}\epsilon_0(0)=1.75[4\pi N(0)]^{\frac{1}{2}}kT_c, \quad (3.39)$$

where $2\epsilon_0(0)$ is the energy gap at $T=0$ and the density of Bloch states $N(0)$ is taken for a system of unit volume.

A law of corresponding states follows from (3.39) and may be expressed as

$$\gamma T_c^2/H_0^2=\tfrac{1}{6}\pi[kT_c/\epsilon_0(0)]^2=0.170, \quad (3.40)$$

where the electronic specific heat in the normal state is given by

$$C_{en}=\gamma T(\text{ergs/°C cm}^3), \quad (3.41)$$

and

$$\gamma=\tfrac{2}{3}\pi^2N(0)k^2. \quad (3.42)$$

The Gorter-Casimir model gives the value of 0.159 for the ratio (3.40). The scatter of experimental data is too great to choose one value over the other at the present time.

Near $T=0$, the gap is practically independent of temperature and large compared to kT, and hence for

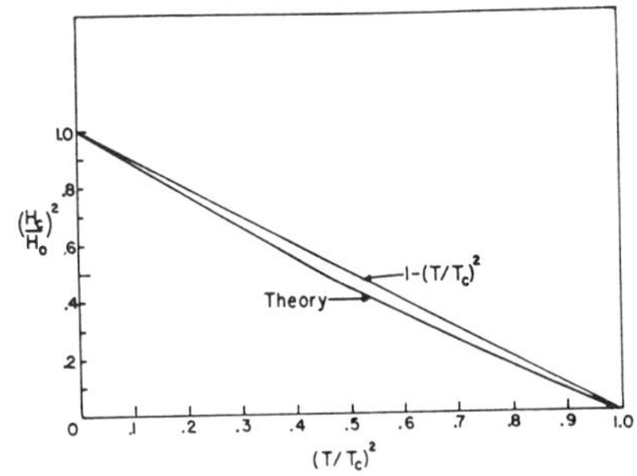

FIG. 2. Ratio of the critical field to its value at $T=0$°K *vs* $(T/T_c)^2$. The upper curve is the $1-(T/T_c)^2$ law of the Gorter-Casimir theory and the lower curve is the law predicted by the theory in the weak-coupling limit. Experimental values generally lie between the two curves.

$T/T_c\ll 1$ we have the relation

$$H_c^2=H_0^2[1-\tfrac{2}{3}\pi^2(kT/\epsilon_0)^2], \quad (3.43)$$

or

$$H_c\cong H_0[1-1.07(T/T_c)^2]. \quad (3.44)$$

This approximation corresponds to neglecting the free-energy change of the superconducting state, the total effect coming from F_n.

The electronic specific heat is most readily obtained from the entropy, (3.34):

$$C_{es}=T\frac{dS}{dT}=-\beta\frac{dS}{d\beta}=-4k\beta\sum_{k>k_F}\beta E_{\mathbf{k}}\frac{df_{\mathbf{k}}}{d\beta}, \quad (3.45)$$

or

$$C_{es}=4k\beta^2\sum_{k>k_F}f_{\mathbf{k}}(1-f_{\mathbf{k}})\left[E_{\mathbf{k}}^2+\frac{\beta}{2}\frac{d\epsilon_0^2}{d\beta}\right]. \quad (3.46)$$

The expression for C_{es} is simply interpreted as the specific heat due to electrons and holes with the modified spectrum (3.21) plus the change in condensation energy with temperature.

At the transition temperature, the energy gap vanishes and the jump in specific heat associated with the second order transition is given by

$$(C_{es}-C_{en})|_{T_c}=2k\beta^3\sum_{k>k_F}f_{\mathbf{k}n}(1-f_{\mathbf{k}n})\left[\frac{d\epsilon_0^2}{d\beta}\right]_{T_c}=kN(0)\beta_c^2\left[\frac{d\epsilon_0^2}{d\beta}\right]_{T_c}, \quad (3.47)$$

where

$$f_{\mathbf{k}n}=1/(e^{\beta\epsilon_{\mathbf{k}}}+1). \quad (3.48)$$

The derivative $d\epsilon_0^2/d\beta$ can be obtained from the relation between ϵ_0 and T, (3.27). After some calculation we find

$$\left.\frac{d\epsilon_0^2}{d\beta}\right|_{T_c}=\frac{10.2}{\beta_c^3}, \quad (3.49)$$

and the jump in specific heat becomes

$$\left.\frac{C_{es}-\gamma T_c}{\gamma T_c}\right|_{T_c}=1.52. \quad (3.50)$$

The Gorter-Casimir model gives 2.00 and the Koppe theory[9] gives 1.71 for this ratio. The experimental data in general range between our value and 2.00.

The initial slope of the critical-field curve at the transition temperature is given by the thermodynamic relation

$$\left.\frac{T_c}{4\pi}\left(\frac{dH_c}{dT}\right)^2\right|_{T_c}=(C_s-C_n)\Big|_{T_c}. \quad (3.51)$$

With use of (3.47) this becomes

$$\left.\frac{1}{\gamma}\left(\frac{dH_c}{dT}\right)^2\right|_{T_c}=19.4, \quad (3.52)$$

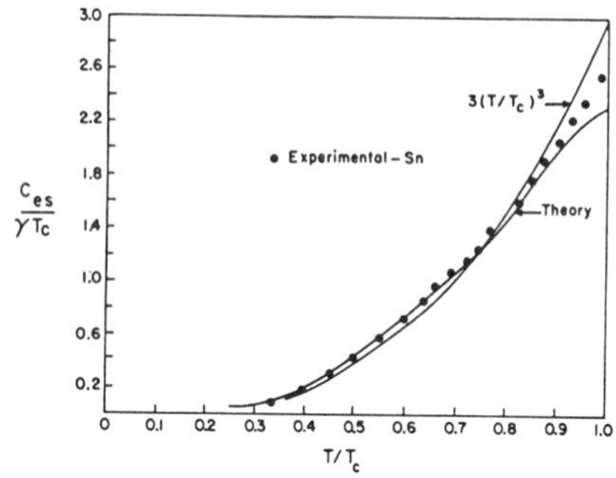

FIG. 3. Ratio of the electronic specific heat to its value in the normal state at T_c *vs* T/T_c for the Gorter-Casimir theory and for the present theory. Experimental values for tin are shown for comparison. *Note added in proof.*—The plotted theoretical curve is incorrect very near T_c; the intercept at T_c should be 2.52.

or with (3.39),

$$\left.\frac{dH_c}{dT}\right|_{T_c} = -\frac{1.82H_0}{T_c}. \tag{3.53}$$

When $\beta\epsilon_0 \gg 1$, the specific heat can be expressed in the form

$$\frac{C_{es}}{\gamma T_c} = \frac{3}{2\pi^2}\left(\frac{\epsilon_0}{kT_c}\right)^3\left(\frac{T_c}{T}\right)^2[3K_1(\beta\epsilon_0)+K_3(\beta\epsilon_0)] \cong 8.5e^{-1.44T_c/T}, \tag{3.54}$$

where K_n is the modified Bessel function of the second kind.

The ratio $C_{es}/(\gamma T_c)$ is plotted in Fig. 3 from (3.46) and compared with the T^3 law and the experimental values for tin. The agreement is rather good except near T_c where our specific heat is somewhat too small. The logarithm of the same ratio is plotted in Fig. 4 to bring out the experimental deviation from the T^3 law. The recent work of Goodman *et al.*[20] shows that the data for tin and vanadium fit the law:

$$C_{es}/(\gamma T_c) = ae^{-bT_c/T}, \tag{3.55}$$

with high accuracy for $T_c/T > 1.4$, where $a = 9.10$ and $b = 1.50$. These values are in good agreement with our results in this region, (3.54).

Thus we see that our theory predicts the thermodynamic properties of a superconductor quite accurately and in particular gives an exponential specific heat for $T/T_c \ll 1$ and explicitly exhibits a second-order phase transition in the absence of a magnetic field.

IV. CALCULATION OF MATRIX ELEMENTS

There are many problems for which one would like to determine matrix elements of a single-particle scattering operator of the form $U = \sum_j H_j$, where H_j involves only the coordinates of particle j. In terms of creation and destruction operators,

$$U = \sum_j H_j = \sum_{\mathbf{k},\mathbf{k}',\sigma,\sigma'} B_{\mathbf{k}\sigma\mathbf{k}'\sigma'} c_{\mathbf{k}'\sigma'}^* c_{\mathbf{k}\sigma}, \tag{4.1}$$

where

$$B_{\mathbf{k}\sigma\mathbf{k}'\sigma'} = \int \psi_{\mathbf{k}'\sigma'}^* H_j \psi_{\mathbf{k}\sigma} d\tau_j \tag{4.2}$$

is the matrix element for scattering of a single electron from $\mathbf{k}\sigma$ to $\mathbf{k}'\sigma'$. In this section we shall determine matrix elements of U between two of our many-particle excited-state wave functions for a superconductor and give tables which should be useful for application to perturbation theory and transport problems. We first give a brief review of the corresponding problems for the normal state.

The matrix element of $c_{\mathbf{k}'\sigma'}^* c_{\mathbf{k}\sigma}$ between two normal state configurations is zero unless the occupation numbers differ only in transfer of an electron from $\mathbf{k}\sigma$ in the initial to $\mathbf{k}'\sigma'$ in the final configuration, in which case it is unity. If one wishes to calculate the probability that at temperature T an electron be scattered from a state of spin σ in an element $\Delta\mathbf{k}$ to one of spin σ' in $\Delta\mathbf{k}'$, one must multiply the usual single-particle expression by $f(1-f')$, the probability that $\mathbf{k}\sigma$ be occupied and $\mathbf{k}'\sigma'$ unoccupied in a typical initial configuration. A similar factor occurs in the second-order perturbation theory expansion of U:

$$\sum_{\mathbf{k},\sigma,\mathbf{k}',\sigma'} \frac{|B_{\mathbf{k}\sigma\mathbf{k}'\sigma'}|^2 f(1-f')}{\epsilon-\epsilon'}. \tag{4.3}$$

If H_j is independent of spin, $\sigma'=\sigma$ and the sum reduces to

$$2\sum_{\mathbf{k},\mathbf{k}'} \frac{|B_{\mathbf{k}\mathbf{k}'}|^2 f(1-f')}{\epsilon-\epsilon'} = -\sum_{\mathbf{k},\mathbf{k}'} |B_{\mathbf{k}\mathbf{k}'}|^2 \left(\frac{f'-f}{\epsilon-\epsilon'}\right). \tag{4.4}$$

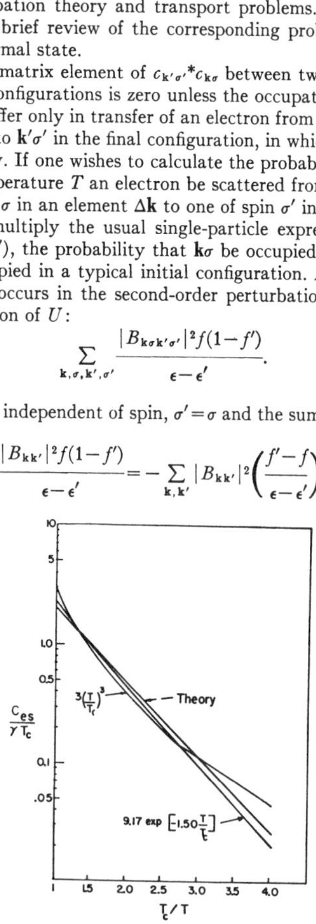

FIG. 4. A logarithmic plot of the ratio of the electronic specific heat to its value in the normal state at T_c *vs* T_c/T. The simple exponential fits the experimental data for tin and vanadium well for $T_c/T > 1.4$.

TABLE II. Matrix elements of single-particle scattering operator.

Wave functions[a] Initial, Ψ_i $(\mathbf{k}\uparrow, -\mathbf{k}\downarrow)$	$(\mathbf{k}'\uparrow, -\mathbf{k}'\downarrow)$	Final, Ψ_f $(\mathbf{k}\uparrow, -\mathbf{k}\downarrow)$	$(\mathbf{k}'\uparrow, -\mathbf{k}'\downarrow)$	Ground (+) or excited (−) $\mathbf{k}$	$\mathbf{k}'$	Energy difference $W_i - W_f$	Probability of initial state	Matrix elements $c_{\mathbf{k}'\uparrow}^* c_{\mathbf{k}\uparrow}$ or $c_{-\mathbf{k}'\downarrow}^* c_{\mathbf{k}\uparrow}$	$c_{-\mathbf{k}\downarrow}^* c_{-\mathbf{k}'\downarrow}$ or $-c_{-\mathbf{k}\downarrow}^* c_{\mathbf{k}'\uparrow}$
(a)				+	+	$E-E'$	$\frac{1}{2}s(1-s'-p')$	$[(1-h)(1-h')]^{\frac{1}{2}}$	$-(hh')^{\frac{1}{2}}$
$X0$	00	00	$X0$	−	−	$E'-E$	$\frac{1}{2}sp'$	$(hh')^{\frac{1}{2}}$	$-[(1-h)(1-h')]^{\frac{1}{2}}$
$X0$	XX	XX	$X0$	+	−	$E+E$	$\frac{1}{2}sp'$	$-[(1-h)h']^{\frac{1}{2}}$	$-[h(1-h')]^{\frac{1}{2}}$
				−	+	$-(E+E')$	$\frac{1}{2}s(1-s'-p')$	$-[h(1-h')]^{\frac{1}{2}}$	$-[(1-h)h']^{\frac{1}{2}}$
(b)				+	+	$E'-E$	$\frac{1}{2}s'(1-s-p)$	$(hh')^{\frac{1}{2}}$	$-[(1-h)(1-h')]^{\frac{1}{2}}$
XX	$0X$	$0X$	XX	−	−	$E-E'$	$\frac{1}{2}s'p$	$[(1-h)(1-h')]^{\frac{1}{2}}$	$-(hh')^{\frac{1}{2}}$
00	$0X$	$0X$	00	+	−	$-(E+E')$	$\frac{1}{2}s'(1-s-p)$	$[h(1-h')]^{\frac{1}{2}}$	$[h'(1-h)]^{\frac{1}{2}}$
				−	+	$E+E'$	$\frac{1}{2}s'p$	$[(1-h)h']^{\frac{1}{2}}$	$[h(1-h')]^{\frac{1}{2}}$
(c)				+	+	$E+E'$	$\frac{1}{4}ss'$	$[(1-h)h']^{\frac{1}{2}}$	$[h(1-h')]^{\frac{1}{2}}$
$X0$	$0X$	00	XX	−	−	$-(E+E')$	$\frac{1}{4}ss'$	$-[h(1-h')]^{\frac{1}{2}}$	$-[h'(1-h)]^{\frac{1}{2}}$
		XX	00	+	−	$E-E'$	$\frac{1}{4}ss'$	$[(1-h)(1-h')]^{\frac{1}{2}}$	$-(hh')^{\frac{1}{2}}$
				−	+	$E'-E$	$\frac{1}{4}ss'$	$-(hh')^{\frac{1}{2}}$	$[(1-h)(1-h')]^{\frac{1}{2}}$
(d)				+	+	$-(E+E')$	$(1-s-p)(1-s'-p')$	$[h(1-h')]^{\frac{1}{2}}$	$[(1-h)h']^{\frac{1}{2}}$
XX	00	$0X$	$X0$	−	−	$E+E'$	pp'	$-[(1-h)h']^{\frac{1}{2}}$	$-[h(1-h')]^{\frac{1}{2}}$
00	XX			+	−	$E'-E$	$(1-s-p)p'$	$-(hh')^{\frac{1}{2}}$	$[(1-h)(1-h')]^{\frac{1}{2}}$
				−	+	$E-E'$	$p(1-s'-p')$	$[(1-h)(1-h')]^{\frac{1}{2}}$	$-(hh')^{\frac{1}{2}}$

[a] For transitions which change spin, reverse designations of $(\mathbf{k}'\uparrow, -\mathbf{k}'\downarrow)$ in the initial and in the final states.

The factor of two comes from the sum over spins in the initial configuration and the second form from the fact that $|B_{\mathbf{kk}'}|^2$ is symmetric in $\mathbf{k}$ and $\mathbf{k}'$.

The calculation of the corresponding factors for the superconducting case is complicated by the fact that any given state $\mathbf{k}\sigma$ may be occupied singly or by either ground or excited pairs, and these possibilities must be weighted by the probability that they occur in a typical initial wave function. First consider matrix elements of an operator which does not give a spin change:

$$(\Psi_f | \textstyle\sum_{\mathbf{k},\mathbf{k}',\sigma} B_{\mathbf{kk}'} c_{\mathbf{k}'\sigma}^* c_{\mathbf{k}\sigma} | \Psi_i). \tag{4.5}$$

Nonvanishing matrix elements of $c_{\mathbf{k}'\uparrow}^* c_{\mathbf{k}\uparrow}$ are obtained only when the single and excited-pair occupancy of Ψ_i and Ψ_f is the same except for those designated by wave vectors $\mathbf{k}$ and $\mathbf{k}'$. Further, Ψ_i must contain a configuration in which $\mathbf{k}\uparrow$ is occupied and $\mathbf{k}'\uparrow$ unoccupied and Ψ_f one in which $\mathbf{k}'\uparrow$ is occupied and $\mathbf{k}\uparrow$ unoccupied. The various possible transitions along with the matrix elements are listed in Table II. While the individual matrix elements are complicated, fairly simple results are obtained when a sum is made over all transitions in which $\mathbf{k}$ is in a volume element $\Delta\mathbf{k}$ and $\mathbf{k}'$ in $\Delta\mathbf{k}'$, it being assumed that $B_{\mathbf{kk}'}$ is a continuous function of $\mathbf{k}$ and $\mathbf{k}'$.

The first type of transition, (a), listed in Table II corresponds to single occupancy of $\mathbf{k}\uparrow$ in the initial and of $\mathbf{k}'\uparrow$ in the final state. Pair occupancy of $\mathbf{k}'$ (i.e., the pair $\mathbf{k}'\uparrow, -\mathbf{k}'\downarrow$) in Ψ_i and of $\mathbf{k}$ in Ψ_f may be either excited or ground, giving the four possible combinations listed in the second column. These have components in which the pair states $\mathbf{k}'$ in Ψ_i and $\mathbf{k}$ in Ψ_f are unoccupied, designated by $X0\,00$ for Ψ_i and $00\,X0$ for Ψ_f. There is a nonvanishing matrix element of $c_{\mathbf{k}'\uparrow}^* c_{\mathbf{k}\uparrow}$ between these components. Other components of the same wave functions have the pair states $\mathbf{k}'$ in Ψ_i and $\mathbf{k}$ in Ψ_f occupied, as is indicated by the designations $X0\,XX$ and $XX\,X0$, respectively. While the matrix element of $c_{\mathbf{k}'\uparrow}^* c_{\mathbf{k}\uparrow}$ between these latter vanishes, that of $c_{-\mathbf{k}\downarrow}^* c_{-\mathbf{k}'\downarrow}$ does not. Since both $c_{\mathbf{k}'\uparrow}^* c_{\mathbf{k}\uparrow}$ and $c_{-\mathbf{k}\downarrow}^* c_{-\mathbf{k}'\downarrow}$ are included in the sum in (4.5), they will give coherent contributions and must be considered together. Matrix elements between these states of all other terms in the sum are zero.

Matrix elements of type (b) are for single-particle occupancy of $-\mathbf{k}'\downarrow$ in Ψ_i and of $-\mathbf{k}\downarrow$ in Ψ_f, while $\mathbf{k}$ in Ψ_i and $\mathbf{k}'$ in Ψ_f may be occupied by either excited or ground pairs. With interchange of spin and of $\mathbf{k}$ and $\mathbf{k}'$, they are similar to type (a). Type (c) represents single occupancy of $\mathbf{k}\uparrow$ and $-\mathbf{k}'\downarrow$ in Ψ_i and either excited or ground pair occupancy of both $\mathbf{k}$ and $\mathbf{k}'$ in Ψ_f. Transitions of $c_{\mathbf{k}'\uparrow}^* c_{\mathbf{k}\uparrow}$ are allowed for the component $00\,XX$ of Ψ_f and of $c_{-\mathbf{k}\downarrow}^* c_{-\mathbf{k}'\downarrow}$ for the component $XX\,00$. Again, these are coherent. Finally, type (d) represents excited or ground pair occupancy of Ψ_i and single particle occupancy of $-\mathbf{k}\downarrow$, $\mathbf{k}'\uparrow$ in Ψ_f.

The energy differences $W_i - W_f$ listed in the third column are obtained by taking an energy $2E$ for an excited pair, E for single occupancy, and zero for a ground pair. We have used the notation $E = E(\mathbf{k})$; $E' = E(\mathbf{k}')$, etc.

In column 4 are given the probabilities of the initial state designations, based on taking $\frac{1}{2}s$ for a specified single occupancy, p for an excited pair, and $(1-s-p)$ for a ground pair. For example, Ψ_i in the top row corresponds to $\mathbf{k}\uparrow$ occupied and a ground pair in $\mathbf{k}'$, and the fraction of the states $\mathbf{k}$ in the volume element $\Delta\mathbf{k}$ and $\mathbf{k}'$ in $\Delta\mathbf{k}'$ which have this designation is $\frac{1}{2}s(\mathbf{k})[1-s(\mathbf{k}')-p(\mathbf{k}')]$.

To calculate the matrix elements, it is convenient to decompose the wave functions into components corresponding to definite occupancy of excited and ground pairs as in (3.10). Thus, for the top row in which $\mathbf{k}'$ in Ψ_i and $\mathbf{k}$ in Ψ_f are both ground pairs,

$$\Psi_i = c_{\mathbf{k}\uparrow}^*[h'^{\frac{1}{2}}\varphi_{01}(0_{\mathbf{k}},1_{\mathbf{k}'})+(1-h')^{\frac{1}{2}}\varphi_{00}(0_{\mathbf{k}},0_{\mathbf{k}'})], \quad (4.6a)$$

$$\Psi_f = c_{\mathbf{k}'\uparrow}^*[h^{\frac{1}{2}}\varphi_{10}(1_{\mathbf{k}},0_{\mathbf{k}'})+(1-h)^{\frac{1}{2}}\varphi_{00}(0_{\mathbf{k}},0_{\mathbf{k}'})], \quad (4.6b)$$

where

$$\varphi_{10} = b_{\mathbf{k}}^* b_{\mathbf{k}'}\varphi_{01}.$$

The matrix element of $c_{\mathbf{k}'\uparrow}^* c_{\mathbf{k}\uparrow}$ is

$$(\Psi_f|c_{\mathbf{k}'\uparrow}^* c_{\mathbf{k}\uparrow}|\Psi_i) = [(1-h)(1-h')]^{\frac{1}{2}}(c_{\mathbf{k}'\uparrow}^*\varphi_{00}|c_{\mathbf{k}'\uparrow}^*\varphi_{00}) = [(1-h)(1-h')]^{\frac{1}{2}}. \quad (4.7)$$

The matrix element of $c_{-\mathbf{k}\downarrow}^* c_{-\mathbf{k}'\downarrow}$ is given by

$$(\Psi_f|c_{-\mathbf{k}\downarrow}^* c_{-\mathbf{k}'\downarrow}|\Psi_i) = (hh')^{\frac{1}{2}}(c_{\mathbf{k}'\uparrow}^* b_{\mathbf{k}}^* b_{\mathbf{k}'}\varphi_{01}|c_{-\mathbf{k}\downarrow}^* c_{-\mathbf{k}'\downarrow}c_{\mathbf{k}\uparrow}^*\varphi_{01}). \quad (4.8)$$

Since

$$c_{\mathbf{k}'\uparrow}^* b_{\mathbf{k}}^* b_{\mathbf{k}'}\varphi_{01} = -c_{-\mathbf{k}'\downarrow} b_{\mathbf{k}}^*\varphi_{01}, \quad (4.9)$$

$$c_{-\mathbf{k}\downarrow}^* c_{-\mathbf{k}'\downarrow} c_{\mathbf{k}\uparrow}^*\varphi_{01} = c_{-\mathbf{k}'\downarrow} b_{\mathbf{k}}^*\varphi_{01}, \quad (4.10)$$

the matrix element is $-(hh')^{\frac{1}{2}}$. The other matrix elements of types (a) and (b) may be calculated in a similar manner.

For types (c) and (d) we make use of the decomposition (3.10). For example, for type (d),

$$\Psi_i = \alpha_{11}\varphi_{11}(1_{\mathbf{k}},1_{\mathbf{k}'})+\alpha_{10}\varphi_{10}(1_{\mathbf{k}},0_{\mathbf{k}'}) +\alpha_{01}\varphi_{01}(0_{\mathbf{k}},1_{\mathbf{k}'})+\alpha_{00}\varphi_{00}(0_{\mathbf{k}},0_{\mathbf{k}'}), \quad (4.11)$$

where the α's are as listed in Table I. The final wave function is

$$\Psi_f = c_{\mathbf{k}'\uparrow}^* c_{\mathbf{k}\uparrow}\varphi_{10}. \quad (4.12)$$

Thus the matrix element of $c_{\mathbf{k}'\uparrow}^* c_{\mathbf{k}\uparrow}$ is just α_{10}. The matrix element of $c_{-\mathbf{k}\downarrow}^* c_{-\mathbf{k}'\downarrow}$ is found from

$$c_{-\mathbf{k}\downarrow}^* c_{-\mathbf{k}'\downarrow}\varphi_{01} = c_{-\mathbf{k}\downarrow}^* c_{-\mathbf{k}'\downarrow} b_{\mathbf{k}'}^* b_{\mathbf{k}}\varphi_{10} = c_{\mathbf{k}'\uparrow}^* c_{\mathbf{k}\uparrow}\varphi_{10}, \quad (4.13)$$

so that we find

$$(\Psi_f|c_{-\mathbf{k}\downarrow}^* c_{-\mathbf{k}'\downarrow}|\Psi_i) = \alpha_{01}. \quad (4.14)$$

Those for type (c) can be obtained by interchanging initial and final states and spin up and spin down.

We have so far assumed a spin-independent interaction. One involving a spin flip may be treated by exactly similar methods. Initial and final states differ from the parallel spin case by interchange of spin designation of $\mathbf{k}'$ in the initial and in the final state. There is a coherence between the matrix elements for $c_{-\mathbf{k}'\downarrow}^* c_{\mathbf{k}\uparrow}$ and $c_{-\mathbf{k}\downarrow}^* c_{\mathbf{k}'\uparrow}$. They are the same as the corresponding matrix elements for parallel spin, except for a *reversal of sign* of the reverse spin transitions. For example, for the type (a) transition of the top row, the final state is now

$$\Psi_f = c_{-\mathbf{k}'\downarrow}^*[h^{\frac{1}{2}}\varphi_{10}(1_{\mathbf{k}},0_{\mathbf{k}'})+(1-h)^{\frac{1}{2}}\varphi_{00}(0_{\mathbf{k}},0_{\mathbf{k}'})]. \quad (4.15)$$

The matrix element of $c_{-\mathbf{k}'\downarrow}^* c_{\mathbf{k}\uparrow}$ is $[(1-h)(1-h')]^{\frac{1}{2}}$ as before. To obtain matrix element of $c_{-\mathbf{k}\downarrow}^* c_{\mathbf{k}'\uparrow}$, we now have, corresponding to (4.9) and (4.10),

$$c_{-\mathbf{k}'\downarrow}^* b_{\mathbf{k}'} b_{\mathbf{k}}^*\varphi_{01} = c_{\mathbf{k}'\uparrow} b_{\mathbf{k}}^*\varphi_{01},$$

$$c_{-\mathbf{k}\downarrow}^* c_{\mathbf{k}'\uparrow} c_{\mathbf{k}\uparrow}^*\varphi_{01} = c_{\mathbf{k}'\uparrow} b_{\mathbf{k}}^*\varphi_{01},$$

giving $+(hh')^{\frac{1}{2}}$. We have indicated the change in sign in the table by listing $-c_{-\mathbf{k}\downarrow}^* c_{\mathbf{k}'\uparrow}$ at the top of column 6.

In a second-order perturbation theory calculation, one is interested in determining

$$\sum_f \frac{|(\Psi_f|\sum_{\mathbf{k},\mathbf{k}',\sigma} B_{\mathbf{k}\mathbf{k}'} c_{\mathbf{k}'\sigma}^* c_{\mathbf{k}\sigma}|\Psi_i)|^2}{W_i - W_f}, \quad (4.16)$$

where the sum is over all intermediate states, f. The initial state should be a typical one for a given temperature T. In general, one might have either

$$B_{\mathbf{k}\mathbf{k}'} = +B_{-\mathbf{k}',-\mathbf{k}}, \quad \text{(case I)} \quad (4.17)$$

or

$$B_{\mathbf{k}\mathbf{k}'} = -B_{-\mathbf{k}',-\mathbf{k}}. \quad \text{(case II)} \quad (4.18)$$

The latter applies to the magnetic interaction. To take the coherence into account, one may take the spin-independent sum over $\mathbf{k}$ and $\mathbf{k}'$, which designate initial and intermediate states:

$$+\sum_{\mathbf{k},\mathbf{k}'} \frac{|B_{\mathbf{k}\mathbf{k}'}|^2\langle|(\Psi_f|c_{\mathbf{k}'\uparrow}^* c_{\mathbf{k}\uparrow} \pm c_{-\mathbf{k}\downarrow}^* c_{-\mathbf{k}'\downarrow}|\Psi_i)|^2\rangle_{\text{Av}}}{W_i - W_f}, \quad (4.19)$$

where the average is taken over volume elements $\Delta\mathbf{k}$ and $\Delta\mathbf{k}'$ for the initial state.

For terms with an energy denominator $W_i - W_f = E - E'$, we have

$$\{[(1-h)(1-h')]^{\frac{1}{2}} \mp (hh')^{\frac{1}{2}}\}^2[\tfrac{1}{2}s(1-s'-p') +\tfrac{1}{2}ps'+\tfrac{1}{4}ss'+p(1-s'-p')] = \tfrac{1}{2}\left\{1+\frac{\epsilon\epsilon' \mp \epsilon_0^2}{EE'}\right\}f(1-f'). \quad (4.20)$$

Table III lists the average matrix elements for the various values of $W_i - W_f$.

The second-order perturbation theory sum may be written

$$-\sum_{\mathbf{k},\mathbf{k}'} |B_{\mathbf{k}\mathbf{k}'}|^2 L(\epsilon,\epsilon'), \quad (4.21)$$

TABLE III. Mean square matrix elements for possible values of $W_i - W_f$.

$W_i - W_f$	$\langle |(\Psi_f|c_{k'\uparrow}^* c_{k\uparrow} \pm c_{-k\downarrow}^* c_{-k'\downarrow}|\Psi_i)|^2\rangle_{Av}$
$E-E'$	$\frac{1}{2}\left(1+\frac{\epsilon\epsilon'\mp\epsilon_0^2}{EE'}\right)f(1-f')$
$E'-E$	$\frac{1}{2}\left(1+\frac{\epsilon\epsilon'\mp\epsilon_0^2}{EE'}\right)f'(1-f)$
$-(E+E')$	$\frac{1}{2}\left(1-\frac{\epsilon\epsilon'\mp\epsilon_0^2}{EE'}\right)(1-f)(1-f')$
$E+E'$	$\frac{1}{2}\left(1-\frac{\epsilon\epsilon'\mp\epsilon_0^2}{EE'}\right)ff'$

where

$$L(\epsilon,\epsilon')=\frac{1}{2}\left(1+\frac{\epsilon\epsilon'\mp\epsilon_0^2}{EE'}\right)\left(\frac{f'-f}{E-E'}\right)+\frac{1}{2}\left(1-\frac{\epsilon\epsilon'\mp\epsilon_0^2}{EE'}\right)\left(\frac{1-f-f'}{E+E'}\right)$$
$$=\frac{1}{2}\left(\frac{(1-2f)E-(1-2f')E'}{\epsilon^2-\epsilon'^2}\right)+\frac{1}{2}\left(\frac{\epsilon\epsilon'\mp\epsilon_0^2}{EE'}\right)\left(\frac{(1-2f)E'-(1-2f')E}{\epsilon^2-\epsilon'^2}\right). \quad (4.22)$$

The upper signs correspond to case I, the lower to case II.

To determine the probability of a transition in which an energy quantum $h\nu$ is absorbed, we have a sum of the form

$$\frac{2\pi}{\hbar}\sum_{\mathbf{k},\mathbf{k}'}|B_{\mathbf{kk}'}|^2\langle|(\Psi_f|c_{k'\uparrow}^*c_{k\uparrow}\pm c_{-k\downarrow}c_{-k'\downarrow}|\Psi_i)|^2\rangle_{Av}\times\delta(W_f-W_i-h\nu). \quad (4.23)$$

For the matrix elements for which $W_f-W_i=E-E'$, we may interchange $\mathbf{k}$ and $\mathbf{k}'$ in the sum and combine them with those for which $W_f-W_i=E'-E$. This just gives either one multiplied by a factor of two:

$$\frac{2\pi}{\hbar}\sum_{\mathbf{k},\mathbf{k}'}2|B_{\mathbf{kk}'}|^2\tfrac{1}{2}\left(1+\frac{\epsilon\epsilon'\mp\epsilon_0^2}{EE'}\right)\times f(1-f')\delta(W_f-W_i-h\nu). \quad (4.24)$$

One may interpret the factor of two as accounting for the sum over the two spin possibilities of the initial state. If $|B_{\mathbf{kk}'}|^2$ is symmetric with respect to the Fermi surface, so that we may sum over + and − values of ϵ and ϵ', terms odd in ϵ and ϵ' drop out, and we find

$$\frac{2\pi}{\hbar}\sum_{k,k'>k_F}4|B_{\mathbf{kk}'}|^2\left(1\mp\frac{\epsilon_0^2}{EE'}\right)\times f(1-f')\delta(E'-E-h\nu). \quad (4.25)$$

The corresponding expressions for $W_f-W_i=E+E'$ and $-(E+E')$ are:

$$\frac{2\pi}{\hbar}\sum_{k,k'>k_F}2|B_{\mathbf{kk}'}|^2\left(1\pm\frac{\epsilon_0^2}{EE'}\right)\times(1-f)(1-f')\delta(E+E'-h\nu), \quad (4.26)$$

$$\frac{2\pi}{\hbar}\sum_{k,k'>k_F}2|B_{\mathbf{kk}'}|^2\left(1\pm\frac{\epsilon_0^2}{EE'}\right)ff'\delta(E+E'+h\nu), \quad (4.27)$$

respectively, where again we have dropped terms odd in ϵ and ϵ'.

Hebel and Slichter[34] have used (4.25) to estimate the temperature dependence of the relaxation time for nuclear spin resonance in the superconducting state from the corresponding value in the normal state. They are able to account for an observed initial *decrease* in relaxation time in Al as the temperature is lowered below T_c. The increased density of states in energy in the superconducting phase more than makes up for the decrease in number of excited electrons at temperatures not too far below T_c. For this problem, the lower sign (+) is appropriate.

These expressions may also be used to determine transport properties, such as electrical conductivity in the microwave region and thermal conductivity.

Note added in proof.—The marked effect of coherence on the matrix elements is verified experimentally by comparing absorption of ultrasonic waves, which follows case I, with nuclear spin relaxation or electromagnetic absorption, both of which follow case II. For frequencies such that $h\nu\ll kT_c$, one expects for case II an initial *increase* in absorption just below T_c, followed by a decrease to values below that of the normal state as the temperature is lowered, as is observed experimentally. On the other hand, for case I one expects the absorption to drop with an infinite slope at T_c, such as is found for ultrasonic waves.

The expressions for the transition probabilities are simplified if we change our convention for the moment to give E the same sign as ϵ, so that $E=-(\epsilon^2+\epsilon_0^2)^{\frac{1}{2}}$ below the Fermi surface. One may then write (4.26) and (4.27) in the same form as (4.25), with E and E' now taking on both positive and negative values. Considering both direct absorption and induced emission, the net rate of absorption of energy in the superconducting state is proportional to

$$\alpha_s\propto\int\left(1\mp\frac{\epsilon_0^2}{EE'}\right)[f(1-f')-f'(1-f)]\rho(E)\rho(E')dE, \quad (4.28)$$

where $E'=E+h\nu$ and $\rho(E)=N(0)E/(E^2-\epsilon_0^2)^{\frac{1}{2}}$ is the density of states in energy.

With the upper sign (case I) and with $h\nu\ll kT$, the density of states terms are cancelled by the first factor, and the expression reduces to

$$\alpha_s\propto 2[N(0)]^2\int_{\epsilon_0}^{\infty}(f-f')dE\cong 2[N(0)]^2h\nu f(\epsilon_0),$$

The factor 2 comes from adding contributions above and below the Fermi surface. The corresponding expression for the normal

[34] L. C. Hebel and C. P. Slichter, Phys. Rev. **107**, 901 (1957). We are indebted to these authors for considerable help in working out the details of the calculation of matrix elements, particularly in regard to taking into account the coherence of matrix elements of opposite spin.

state is similar, except that $\epsilon_0=0$. We thus find

$$\alpha_s/\alpha_n=2f(\epsilon_0). \tag{4.29}$$

R. W. Morse and H. V. Bohm (to be published) have used (4.29) for analysis of data on ultrasonic attenuation in an indium specimen for which the electronic mean free path is large compared with the wavelength of the ultrasonic wave, so that one might expect the theory to apply. Values of $\epsilon_0(T)$ estimated from the data by use of (4.29) are in excellent agreement with our theoretical values (Fig. 1).

For case II, the integral may be expressed in the form:

$$\frac{\alpha_s}{\alpha_n}=\frac{1}{h\nu}\int\frac{(EE'+\epsilon_0^2)(f-f')dE}{\{(E^2-\epsilon_0^2)[(E+h\nu)^2-\epsilon_0^2]\}^{\frac{1}{2}}}. \tag{4.30}$$

The integral diverges at $E=\epsilon_0$ if $h\nu$ is set equal to zero in the denominator. Numerical evaluation of the integral indicates that for $h\nu\sim\frac{1}{2}kT_c$ or less, (4.30) gives an increase in absorption just below T_c as observed by Hebel and Slichter in nuclear magnetic resonance and by Tinkham and co-workers (private communication) for microwave absorption in thin superconducting films.

In order to have absorption at $T=0$, $h\nu$ must be greater than the energy gap, $2\epsilon_0$. We take E negative and E' positive, and find for this case:

$$\frac{\alpha_s}{\alpha_n}=\frac{1}{h\nu}\int_{\epsilon_0-h\nu}^{-\epsilon_0}\frac{[E(E+h\nu)+\epsilon_0^2]dE}{\{(E^2-\epsilon_0^2)[(E+h\nu)^2-\epsilon_0^2]\}^{\frac{1}{2}}}. \tag{4.31}$$

The integral may be evaluated in terms of the complete elliptic integrals, $E(\gamma)$ and $K(\gamma)$ as follows:

$$\frac{\alpha_s}{\alpha_n}=\left(1+\frac{2\epsilon_0}{h\nu}\right)E(\gamma)-2\left(\frac{2\epsilon_0}{h\nu}\right)K(\gamma), \tag{4.32}$$

where

$$\gamma=(h\nu-2\epsilon_0)/(h\nu+2\epsilon_0). \tag{4.33}$$

This expression is in excellent agreement with data of Glover and Tinkham (reference 20, Fig. 6) on infrared absorption in thin films.

V. ELECTRODYNAMIC PROPERTIES

The electrodynamic properties of our model are determined using a perturbation treatment in which the first order change in the wave function is used to calculate the current as a functional of the field. For such properties as the Meissner effect this approach is quite rigorous since we are interested in the limit as $\mathbf{A}(\mathbf{r})$ approaches zero. It is assumed that the medium is infinite and that the sources of the field may be introduced by inserting current sheets in the interior. This method has been applied previously to the calculation of the diamagnetic properties of an electron gas.[35]

We first derive an expression, valid for arbitrary temperatures, relating the current density to the total field (the field due to the sources and to the induced currents). The fact that the system displays a Meissner effect is established by investigating the Fourier transform of the current density in the limit that $q\to 0$. In this limit we obtain the equation,

$$\lim_{\mathbf{q}\to 0}\mathbf{j}(\mathbf{q})=-\frac{1}{c\Lambda_T}\mathbf{a}(\mathbf{q}), \tag{5.1}$$

where Λ_T is a function of temperature, increasing, in the free-electron approximation, from the London value $\Lambda=m/ne^2$ at $T=0$ to infinity at the transition temperature.

The limiting expression (5.1) is valid only for values of q smaller than those important for most penetration phenomena. In general we find the current density is a functional of the vector potential $\mathbf{A}$ which, with div $\mathbf{A}=0$, may be expressed in a form similar to that proposed by Pippard (1.3):

$$\mathbf{j}(\mathbf{r})=-\frac{3}{4\pi c\Lambda_T\xi_0}\int\frac{\mathbf{R}[\mathbf{R}\cdot\mathbf{A}(\mathbf{r}')]J(R,T)d\mathbf{r}'}{R^4}. \tag{5.2}$$

The kernel, $J(R,T)$, is a relatively slowly varying function of temperature, and at $T=0°$K is not far different from Pippard's $\exp(-R/\xi_0)$.

To calculate penetration depths, it is more convenient to use the Fourier transform of (5.2), which may be expressed in the form

$$\mathbf{j}(\mathbf{q})=-(c/4\pi)K(q)\mathbf{a}(\mathbf{q}), \tag{5.3}$$

where $K(q)$ is a scalar which approaches the constant value $4\pi/(\Lambda_T c^2)$ in the limit $q\to 0$. One may determine $K(q)$ directly from the perturbation expansion of the wave function, or one may first calculate $J(R,T)$ and then find the transform of (5.2). The latter procedure is followed in Appendix C, where an explicit expression for $K(q)$ valid for q not too small is derived. In this section we shall give a direct derivation of the transform which can be used to investigate the limit $q\to 0$, and then give the derivation of (5.2). A comparison of calculated and observed values of penetration depths is given at the end of the section.

In the absence of the electromagnetic field, the system at a given temperature is characterized by the complete orthonormal set of wave functions which we denote by

$$\Psi_0(T),\ \Psi_1(T),\ \cdots\Psi_n(T),\ \cdots,$$

with corresponding energies

$$W_0(T),\ W_1(T)\cdots W_n(T)\cdots. \tag{5.4}$$

We choose for $\Psi_0(T)$ a typical wave function of the type described in the previous sections, where the occupation of "single particles" and "excited pairs" is given by the s and p distributions, respectively, appropriate to the temperature T; the rest of the phase space is available for "ground pairs" whose distribution is specified by h which is also a function of T. The set of orthogonal states is obtained by varying s and p (in analogy with the normal metal) and not changing h. We thus are choosing a representative configuration of the most probable distribution and taking system averages with respect to this representative configuration.

The electromagnetic interaction term for an electron of charge $q=-e$, $e>0$, is, in second quantized form,

[35] This method was first applied to the calculation of the diamagnetic properties of an electron gas by O. Klein, Arkiv Mat. Astron Fysik, **A31**, No. 12 (1944). Our treatment follows that of one of the authors as given in reference 7, pp. 303–321, where further references to the literature may be found.

$$H_I=\int d\mathbf{r}\psi^*(\mathbf{r})\left[\frac{-ie\hbar}{2mc}(\mathbf{A}\cdot\nabla+\nabla\cdot\mathbf{A})+\frac{e^2}{2mc^2}\mathbf{A}^2(\mathbf{r})\right]\psi(\mathbf{r}). \quad (5.5)$$

We choose a gauge in which $\nabla\cdot\mathbf{A}=0$ and in which $\mathbf{A}=0$ if the magnetic field is zero.

We expand ψ and ψ^* in creation and annihilation operators[36]:

$$\psi(\mathbf{r})=\frac{1}{\Omega^{\frac{1}{2}}}\sum_{\mathbf{k},\sigma}c_{\mathbf{k},\sigma}u_\sigma e^{i\mathbf{k}\cdot\mathbf{r}},$$
$$\psi^*(\mathbf{r})=\frac{1}{\Omega^{\frac{1}{2}}}\sum_{\mathbf{k}',\sigma'}c_{\mathbf{k}',\sigma'}^*u_{\sigma'}^*e^{-i\mathbf{k}'\cdot\mathbf{r}}, \quad (5.6)$$

where the c's satisfy the usual fermion anticommutation relations, (2.1) and (2.2), u_σ is a two-component spinor, and Ω is the volume of the container. The interaction Hamiltonian becomes, when one neglects the term of higher order in $\mathbf{A}$,

$$H_I=\frac{e\hbar}{mc}\frac{(2\pi)^{\frac{3}{2}}}{\Omega}\sum_{\mathbf{k},\mathbf{q},\sigma}c_{\mathbf{k}+\mathbf{q},\sigma}^*c_{\mathbf{k},\sigma}\mathfrak{a}(\mathbf{q})\cdot\mathbf{k}, \quad (5.7)$$

where

$$\mathfrak{a}(\mathbf{q})=\left(\frac{1}{2\pi}\right)^{\frac{3}{2}}\int d\mathbf{r}\ \mathbf{A}(\mathbf{r})e^{-i\mathbf{q}\cdot\mathbf{r}}.$$

The current operator $\mathfrak{J}(\mathbf{r})$ is

$$\mathfrak{J}(\mathbf{r})=\frac{ie\hbar}{2m}(\psi^*\nabla\psi-\text{Herm. conj.})-\frac{e^2}{mc}\psi^*\mathbf{A}\psi=\mathfrak{J}_P(\mathbf{r})+\mathfrak{J}_D(\mathbf{r}). \quad (5.8)$$

Expanding ψ and ψ^* as in (5.6), we get

$$\mathfrak{J}_P(\mathbf{r})=\frac{e\hbar}{2m\Omega}\sum_{\mathbf{k},\mathbf{q},\sigma}c_{\mathbf{k}+\mathbf{q},\sigma}^*c_{\mathbf{k},\sigma}e^{-i\mathbf{q}\cdot\mathbf{r}}(2\mathbf{k}+\mathbf{q}),$$
$$\mathfrak{J}_D(\mathbf{r})=-\frac{e^2}{mc}\frac{1}{\Omega}\sum_{\mathbf{k},\mathbf{q},\sigma}c_{\mathbf{k}+\mathbf{q},\sigma}^*c_{\mathbf{k},\sigma}e^{-i\mathbf{q}\cdot\mathbf{r}}\mathbf{A}(\mathbf{r}). \quad (5.9)$$

In the presence of the electromagnetic field the wave function for the system may be written

$$\Phi(\mathbf{A})=\Phi_0+\Phi_1+(\text{terms of order }\mathbf{A}^2\cdots), \quad (5.10)$$

where the usual perturbation expression for Φ_1 is

$$\Phi_1=\sum_{i\neq0}\frac{(\Psi_i|H_I|\Psi_0)}{W_0-W_i}|\Psi_i). \quad (5.11)$$

[36] At this point we insert plane waves for the Bloch functions. It would be possible to carry through an analogous procedure formally with Bloch functions. The average matrix elements which enter cannot be evaluated explicitly, but can be expressed in terms of empirically determined parameters. The appropriate modifications of the free-electron expressions are as indicated in the introduction.

To the lowest order in $\mathbf{A}(\mathbf{r})$, the expectation value of the current operator is then

$$\mathbf{j}(\mathbf{r})=(\Phi|\mathfrak{J}(\mathbf{r})|\Phi)=(\Phi_1|\mathfrak{J}_P(\mathbf{r})|\Phi_0)+(\Phi_0|\mathfrak{J}_P(\mathbf{r})|\Phi_1)+(\Phi_0|\mathfrak{J}_D(\mathbf{r})|\Phi_0), \quad (5.12)$$

where the last equality follows if the current in the field free state, $\mathbf{j}_0(\mathbf{r})=(\Phi_0|\mathfrak{J}(\mathbf{r})|\Phi_0)$, vanishes.[36a]

If (5.7), (5.9), and (5.11) are combined with (5.12) and the condition $\mathbf{q}\cdot\mathfrak{a}(\mathbf{q})=0$ is used, we obtain for the paramagnetic part of the current density,

$$\mathbf{j}_P(\mathbf{r})=-\frac{e^2\hbar^2(2\pi)^{\frac{3}{2}}}{2m^2c\Omega^2}\sum_{i\neq0}\sum_{\mathbf{k},\mathbf{q},\sigma}\sum_{\mathbf{k}',\mathbf{q}',\sigma'}(2\mathbf{k}+\mathbf{q})\mathbf{k}'\cdot\Big[\mathfrak{a}(\mathbf{q}')e^{-i\mathbf{q}\cdot\mathbf{r}}(\Psi_0(T)|c_{\mathbf{k}'+\mathbf{q}',\sigma'}^*c_{\mathbf{k}'\sigma'}|\Psi_i(T))\times(\Psi_i(T)|c_{\mathbf{k}+\mathbf{q},\sigma}^*c_{\mathbf{k},\sigma}|\Psi_0(T))\frac{1}{W_0-W_i}+\text{complex conjugate}\Big], \quad (5.13)$$

while for the diamagnetic part we have

$$\mathbf{j}_D(\mathbf{r})=-(ne^2/mc)\mathbf{A}(\mathbf{r}), \quad (5.14)$$

where n is the number of conduction electrons, of both spin directions, per unit volume.

The spin sums and the calculation of the average matrix elements can be carried out as indicated in Sec. IV, (4.16) to (4.22). We then obtain

$$\mathbf{j}_P(\mathbf{r})=\frac{e^2\hbar^2}{2m^2c}\frac{(2\pi)^{\frac{3}{2}}}{\Omega^2}\sum_{\mathbf{k},\mathbf{q}}(2\mathbf{k}+\mathbf{q})\mathbf{k}\cdot\mathfrak{a}(-\mathbf{q})e^{-i\mathbf{q}\cdot\mathbf{r}}L(\epsilon_{\mathbf{k}},\epsilon_{\mathbf{k}+\mathbf{q}}), \quad (5.15)$$

where $L(\epsilon_{\mathbf{k}},\epsilon_{\mathbf{k}+\mathbf{q}})$ is given by (4.22) with the lower signs, corresponding to case II. Setting $\epsilon_{\mathbf{k}}=\epsilon$ and $\epsilon_{\mathbf{k}+\mathbf{q}}=\epsilon'$, the explicit expression for L is

$$L(\epsilon,\epsilon')=\frac{1}{2}\left(\frac{1-f-f'}{E+E'}\right)\left(1-\frac{\epsilon\epsilon'+\epsilon_0^2}{EE'}\right)+\frac{1}{2}\left(\frac{f'-f}{E-E'}\right)\left(1+\frac{\epsilon\epsilon'+\epsilon_0^2}{EE'}\right). \quad (5.16)$$

[36a] *Note added in proof.*—We neglect the effects of the momentum dependent cutoff on the expression for the current density; as shown explicitly by P. W. Anderson (private communication) errors introduced are negligible in the weak coupling limit. In a general gauge, $\mathbf{A}=\mathbf{A}_0+\text{grad}\varphi$, with div $\mathbf{A}_0=0$, it would be necessary to include in the perturbation expansion collective excitations of the electrons, the simplest of which corresponds to a uniform displacement of all of the electrons in momentum space. Energies and matrix elements of collective excitations are nearly the same in normal and superconducting phases. We assume the collective excitations make a negligible contribution to the current form $\mathbf{A}_0$ (see J. Bardeen, Nuovo Cimento 5, 1766 (1957)).

To show how (5.16) reduces to the usual expression for a free-electron gas in the limit $\epsilon_0 \to 0$, we note that E is defined so that it is intrinsically positive, while the Bloch energy ϵ may be either positive or negative. As $\epsilon_0 \to 0$, $E \to |\epsilon|$. If ϵ is negative,

$$f(\epsilon)=f(-|\epsilon|)=1-f(|\epsilon|)=1-f(E), \quad (5.17)$$

To get a nonvanishing contribution from the first line, ϵ and ϵ' must have opposite signs, while for the second line they must have the same signs. We thus find that $L(\epsilon,\epsilon') \to (f'-f)/(\epsilon-\epsilon')$, as it should [see (4.4)].

The Meissner Effect

To establish the existence of the Meissner effect, we investigate the Fourier transform of $\mathbf{j}(\mathbf{r})$ in the limit $\mathbf{q} \to 0$. If $\mathbf{j}(\mathbf{q})$ does not go to zero in this limit, and is opposite in sign to $\mathfrak{a}(\mathbf{q})$, then the system will expel the field from its interior and behave like a perfect diamagnet.

The Fourier transform of the paramagnetic part of the current density, $\mathbf{j}_P(\mathbf{r})$, is given by

$$\mathbf{j}_P(\mathbf{q})=\left(\frac{1}{2\pi}\right)^{\frac{3}{2}}\int d\mathbf{r}\, \mathbf{j}_P(\mathbf{r})e^{-i\mathbf{q}\cdot\mathbf{r}}. \quad (5.18)$$

Referring to (5.15), this can be written as

$$\mathbf{j}_P(\mathbf{q})=\frac{e^2\hbar^2}{m^2c}\left(\frac{1}{2\pi}\right)^3\int d\mathbf{k}\,(2\mathbf{k}+\mathbf{q})\mathbf{k}\cdot\mathfrak{a}(\mathbf{q})L(\epsilon_{\mathbf{k}},\epsilon_{\mathbf{k}+\mathbf{q}}), \quad (5.19)$$

where in the limit that $\mathbf{q}\to 0$, $\epsilon_{\mathbf{k}}=\epsilon_{\mathbf{k}+\mathbf{q}}$ and $L(\epsilon_{\mathbf{k}},\epsilon_{\mathbf{k}+\mathbf{q}})$ becomes

$$\lim_{\mathbf{q}\to 0} L(\epsilon_{\mathbf{k}},\epsilon_{\mathbf{k}+\mathbf{q}})=\frac{\beta e^{\beta E}}{(1+e^{\beta E})^2}. \quad (5.20)$$

Thus we want to evaluate

$$\lim_{\mathbf{q}\to 0}\mathbf{j}_P(\mathbf{q})=\frac{e^2\hbar^2}{m^2c}\left(\frac{1}{2\pi}\right)^3 2\beta\int d\mathbf{k}\,\mathbf{k}\,\mathbf{k}\cdot\mathfrak{a}(\mathbf{q})\frac{e^{\beta E}}{(1+e^{\beta E})^2}. \quad (5.21)$$

Choosing $\mathfrak{a}(\mathbf{q})$ as the polar axis, the angular integration can be done; then, using the relations $n=k_F^3/3\pi^2$ and $\mathcal{E}_F=\hbar^2k_F^2/2m$, we get

$$\lim_{\mathbf{q}\to 0}\mathbf{j}_P(\mathbf{q})=\frac{ne^2}{mc}\left(1-\frac{\Lambda}{\Lambda_T}\right)\mathfrak{a}(\mathbf{q}), \quad (5.22)$$

where

$$1-\frac{\Lambda}{\Lambda_T}\equiv\frac{2\beta\mathcal{E}_F}{k_F^5}\int_0^\infty k^4dk e^{\beta E}(1+e^{\beta E})^{-2}. \quad (5.23)$$

Letting $y=\epsilon/kT$, and using the sharp maximum of the integrand at $\epsilon=0$, we find

$$1-\frac{\Lambda}{\Lambda_T}=2\int_0^\infty dy\frac{\exp(y^2+\beta^2\epsilon_0^2)^{\frac{1}{2}}}{[1+\exp(y^2+\beta^2\epsilon_0^2)^{\frac{1}{2}}]^2}. \quad (5.24)$$

The London constant Λ_T can also be expressed in terms of derivatives of ϵ_0, as is done in Appendix C where the following result is derived

$$\Lambda_T=\Lambda\left(1+\frac{\beta}{\epsilon_0}\frac{d\epsilon_0}{d\beta}\right). \quad (5.25)$$

The total induced current thus becomes:

$$\lim_{\mathbf{q}\to 0}\mathbf{j}(\mathbf{q})=\mathbf{j}_P+\mathbf{j}_D=-\frac{\Lambda}{\Lambda_T}\frac{ne^2}{mc}, \quad \mathfrak{a}(\mathbf{q})=-\frac{1}{\Lambda_T c}\mathfrak{a}(\mathbf{q}). \quad (5.26)$$

In the two limiting situations $T\to 0$ and $T\to T_c$, Λ/Λ_T can very easily be evaluated. As $T\to 0$, β becomes infinite, so that Λ_T becomes equal to Λ, the London value

$$\lim_{T\to 0}(\Lambda/\Lambda_T)=1-2\int_0^\infty dy\exp[-(y^2+\beta^2\epsilon_0^2)^{\frac{1}{2}}]=1. \quad (5.27a)$$

This could have been seen immediately from (5.19) since in the $T\to 0$ limit $L(\epsilon_{\mathbf{k}},\epsilon_{\mathbf{k}+\mathbf{q}})=0$ and the paramagnetic part of the current density is zero. This limit just gives the equation obtained by London assuming complete rigidity of the unperturbed system wave function in the presence of an electromagnetic field. When $T\to T_c$, $\epsilon_0\beta$ goes to zero and Λ/Λ_T also goes to zero, since

$$\lim_{T\to T_c}\left(\frac{\Lambda}{\Lambda_T}\right)=1-2\int_0^\infty\frac{e^y dy}{(1+e^y)^2}=0. \quad (5.27b)$$

The small Landau diamagnetism would appear only in a higher order. We thus find that when the system is in the superconducting phase, the current density in the London limit has the form (5.1), with Λ_T varying from Λ to ∞ as T goes from 0 to T_c.

It is interesting to note that the Meissner effect occurs for any value of $\epsilon_0 \neq 0$. If $\epsilon_0=0$ (for example if we let $V=0$), then, for $T>0$, $\beta\epsilon_0=0$ and from (5.27b) we see that $\Lambda/\Lambda_T=0$. The paramagnetic part of the current density is then

$$\lim_{\mathbf{q}\to 0}\mathbf{j}_P(\mathbf{q})=\frac{ne^2}{mc}\mathfrak{a}(\mathbf{q}), \quad (5.28)$$

and the total current in the $\mathbf{q}\to 0$ limit becomes

$$\lim_{\mathbf{q}\to 0}\mathbf{j}(\mathbf{q})=0. \quad (5.29)$$

This is also true when $T=0$, though in this case one must be careful of the order in which the various limits are taken.

Current Density

We now evaluate the spacial distribution of the current density and exhibit it in a form similar to that proposed by Pippard. The method we use follows that of one of the authors[7] who carried out a similar calculation for an energy-gap model.

Beginning with (5.15), setting $\mathbf{q}=\mathbf{k}'-\mathbf{k}$, and using the fact that $\mathbf{q}\cdot\mathbf{a}(\mathbf{q})=0$ and that we can write terms like $\mathbf{k}e^{i\mathbf{k}\cdot\mathbf{r}}$ as

$$\mathbf{k}e^{i\mathbf{k}\cdot\mathbf{r}}=-i\nabla e^{i\mathbf{k}\cdot\mathbf{r}}, \tag{5.30}$$

we can reduce all of the angular integrations to integrals of the form

$$\int_0^\pi \sin\theta d\theta e^{\pm i\mathbf{k}\cdot(\mathbf{r}-\mathbf{r}')}=\frac{2}{kR}\sin kR, \tag{5.31}$$

where $R=|\mathbf{r}-\mathbf{r}'|$. Then, using the relation

$$\nabla\left(\frac{\sin kR}{kR}\right)=\frac{kR\cos kR-\sin kR}{kR^2}\nabla R, \tag{5.32}$$

we finally obtain

$$\mathbf{j}_P(\mathbf{r})=\frac{e^2\hbar^2}{2m^2c\pi^4}\int d\mathbf{r}'\frac{(\mathbf{A}(\mathbf{r}')\cdot\nabla' R)\nabla R G_s(R)}{R^4}, \tag{5.33}$$

where

$$G_s(R)=\int_0^\infty dk\int_0^\infty dk'\, kk' f(k,R)f(k',R)L(\epsilon,\epsilon') \tag{5.34}$$

and

$$f(k,R)=kR\cos kR-\sin kR. \tag{5.35}$$

These equations correspond to (21.4)–(21.6) of reference 7. The entire current density $\mathbf{j}(\mathbf{r})$ can now be written

$$\mathbf{j}(\mathbf{r})=\mathbf{j}_P(\mathbf{r})-\frac{ne^2}{mc}\mathbf{A}(\mathbf{r}). \tag{5.36}$$

It is convenient to subtract and add $G_n(R)$ inside the integral (5.33), where $G_n(R)$ is $G_s(R)$ evaluated at $\epsilon_0=0$. The integral evaluated with $G_n(R)$ gives the paramagnetic current contribution of a free electron gas and just cancels the term $-(ne^2/mc)\mathbf{A}(\mathbf{r})$, leaving the small Landau diamagnetic term in which we are not interested. We have left the interesting part of the current density:

$$\mathbf{j}(\mathbf{r})=\frac{e^2\hbar^2}{2m^2c\pi^4}\int d\mathbf{r}'\frac{[G_s(R)-G_n(R)]\mathbf{R}[\mathbf{A}(\mathbf{r}')\cdot\mathbf{R}]}{R^6}. \tag{5.37}$$

An inspection of $G_s(R)$, (5.34), reveals that the major contribution to the integral comes from the region k, $k'\simeq k_F$ (i.e., very close to the Fermi surface). Since we are interested in values of $R\gtrsim$ penetration depth which is $\simeq 10^{-5}$ cm, in the region of the major contribution, $kR\simeq k'R\gg 1$. With this in mind, the product $f(k,R)f(k'R)$ becomes

$$\begin{aligned}f(k,R)f(k',R)&\simeq kk'R^2\cos kR\cos k'R\\&=\tfrac{1}{2}kk'R^2[\cos(k+k')R+\cos(k-k')R]\\&\simeq\tfrac{1}{2}kk'R^2\cos(k-k')R,\end{aligned} \tag{5.38}$$

as $\cos(k+k')R$ is very rapidly oscillating in the region of interest.

Now, making the change of variable $dk=(dk/d\mathcal{E})d\epsilon$, approximating the slowly varying terms $k+k'\simeq 2k_F$ and $dk/d\mathcal{E}\simeq 1/\hbar v_0$, where v_0 is the velocity of an electron at the Fermi surface, and using the rapid convergence of the integral to extend large finite limits to infinite ones, we obtain

$$G_s(R)-G_n(R)=-\tfrac{1}{2}k_F{}^4R^2\left(\frac{dk}{d\mathcal{E}}\right)_F^2\pi^2\epsilon_0(0)\frac{\Lambda}{\Lambda_T}J(R,T), \tag{5.39}$$

where

$$J(R,T)=I(R,0)-I(R,\epsilon_0) \tag{5.40}$$

and

$$I(R,\epsilon_0)=\frac{\Lambda_T}{\Lambda\pi^2\epsilon_0(0)}\int\!\!\int_{-\infty}^{+\infty} d\epsilon\, d\epsilon'\, L(\epsilon,\epsilon')\times\cos\left[(\epsilon-\epsilon')\frac{R}{\hbar v_0}\right]. \tag{5.41}$$

The current density then becomes

$$\mathbf{j}(\mathbf{r})=\frac{-3}{4\pi c\Lambda_T}\frac{\pi\epsilon_0(0)}{\hbar v_0}\int d\mathbf{r}'\frac{J(R,T)\mathbf{R}[\mathbf{A}(\mathbf{r}')\cdot\mathbf{R}]}{R^4}, \tag{5.42}$$

where Λ_T has been given by (5.25).

The current density has now been written in a form in which it is easily comparable to Eq. (1.3), proposed by Pippard for a pure superconductor, where we identify $1/\xi_0$ with the microscopic quantities:

$$\frac{1}{\xi_0}=\frac{\pi\epsilon_0(0)}{\hbar v_0}, \tag{5.43}$$

and where $J(R,T)$ is to be compared with the exponential function, $\exp[-R/\xi_0]$. We have defined $J(R,T)$ so that it has the same integral as $\exp[-R/\xi_0]$:

$$\int_0^\infty J(R,T)dR=\xi_0. \tag{5.44}$$

As evaluated in Appendix C, $J(R,T)$ is given by

$$J(R,T)=\frac{\Lambda_T}{\Lambda}\frac{\epsilon_0(T)}{\epsilon_0(0)}\left[\tanh\frac{\epsilon_0(T)}{2kT}-\frac{2\epsilon_0(T)}{\pi}\int_0^\infty d\epsilon'\sin\left(\frac{2R}{\hbar v_0}\epsilon'\right)\frac{1-2f(E')}{\epsilon' E'}\right], \tag{5.45}$$

from which (5.44) can be established. In the London limit, where $\mathbf{A}(\mathbf{r})$ varies so slowly compared to the coherence distance ξ_0 that it may be taken out of the integral sign, we obtain

$$\mathbf{j}(\mathbf{r})=-\frac{1}{c\Lambda_T}\mathbf{A}(\mathbf{r}), \tag{5.46}$$

as has been shown previously.

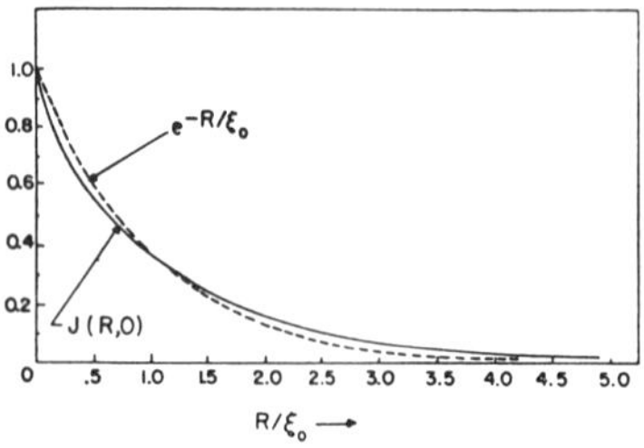

FIG. 5. The kernel $J(R,0)$ for the current density at $T=0°\text{K}$ *vs* R/ξ_0, compared with the Pippard kernel, $\exp(-R/\xi_0)$.

With this normalization of $J(R,T)$ it turns out that most of the temperature variation of the current density is contained in Λ_T, and that the integral does not produce effects that vary very much with the temperature. This will be made clear later in the calculation of the penetration depth as a function of the temperature.

At $T=0$, $J(R,0)$ has the simple form

$$J(R,0)=\frac{2}{\pi}\int_{2\epsilon_0(0)R/\hbar v_0}^{\infty} K_0(y)dy, \tag{5.47}$$

and when $R=0$

$$J(0,0)=1. \tag{5.48}$$

Thus $J(R,0)$ not only has the same integral as the exponential but also the same value at $R=0$. A comparison of the two given in Fig. 5 shows that they are quite similar.

We may express ξ_0 in a form similar to that suggested by Faber and Pippard[25]:

$$\xi_0=a(\hbar v_0/kT_c), \tag{5.49}$$

where a was adjusted empirically from observed penetration depths. From (5.43), we find

$$\xi_0=\frac{\hbar v_0}{\pi\epsilon_0(0)}=\frac{1}{\pi}\frac{kT_c}{\epsilon_0(0)}\frac{\hbar v_0}{kT_c}=0.18\frac{\hbar v_0}{kT_c}. \tag{5.50}$$

Our theoretical value of 0.18 is between the empirical estimates of 0.15 by Faber and Pippard[25] and of 0.27 by Glover and Tinkham.[27] Thus at the absolute zero our theory gives a current density very much like that proposed by Pippard. Since $J(R,T)$ varies slowly with T, most of the temperature variation is contained in the constant Λ_T.

Penetration Depths

A most important application of the equations we have derived for the current density is to the calculation of field penetration at a plane surface. The results depend to some extent on the boundary conditions for scattering of electrons at the surface. Pippard[26] has given general solutions for the limiting cases for specular reflection and for random scattering, based on corresponding expressions derived by Reuter and Sondheimer[37] for the anomalous skin effect. The penetration depth, defined by

$$\lambda=\frac{1}{H(0)}\int_0^\infty H(x)dx, \tag{5.51}$$

is given by

$$\lambda=\frac{2}{\pi}\int_0^\infty \frac{dq}{q^2+K(q)}, \tag{5.52}$$

for specular reflection and by

$$\lambda=\frac{\pi}{\int_0^\infty \ln[1+q^{-2}K(q)]}, \tag{5.53}$$

for random scattering, where $K(q)$ is defined by (5.3).

Limiting expressions have been given for ξ_0/λ large or small compared with unity. The London limit corresponds to $\xi_0\ll\lambda$, in which case $K(q)$ is a constant $=4\pi/\Lambda_T c^2$ over the important range of integration. The penetration depth is then

$$\lambda_L(T)=(4\pi/\Lambda_T c^2)^{\frac{1}{2}}. \tag{5.54}$$

For a free-electron gas, this reduces to the London value $(mc^2/4\pi ne^2)$ at $T=0°\text{K}$. Since, for most metals, $\xi_0\simeq10^{-4}$ cm and $\lambda\simeq5\times10^{-6}$ cm, it is the opposite limit, $\xi_0\gg\lambda$, which is more applicable (except possibly near T_c). In this limit $J(R,T)$ does not vary much over the penetration depth, so that it is only the value at $R=0$, $J(0,T)$, which enters. Pippard[25] has given expressions for λ in this limit, called λ_∞. For random scattering,

$$\lambda_\infty=\frac{3^{1/6}}{(2\pi)^{\frac{1}{3}}}\left(\frac{\xi_0\lambda_L^2}{J(0,T)}\right)^{\frac{1}{3}}, \tag{5.55}$$

while the value for specular reflection is smaller by a factor 8/9

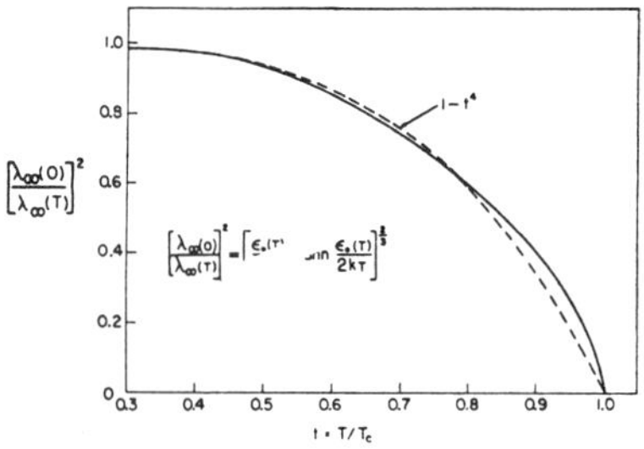

FIG. 6. The temperature variation of the penetration depth, λ_∞, in the infinite coherence distance limit, $(\xi_0/\lambda)\rightarrow\infty$, compared with the empirical law, $[\lambda(0)/\lambda(T)]^2=1-t^4$.

[37] G. E. H. Reuter and E. H. Sondheimer, Proc. Roy. Soc. (London) **A195**, 336 (1948).

TABLE IV. Penetration depth, λ, at $T=0°K$.[a]

Metal	(1) $10^6\lambda_L$ cm	(2) $10^4\hbar v_0/kT_c$ cm	(3) $10^4\xi_0$ cm	(4) ξ_0/λ_L	(5) Specular reflection λ/λ_L	(6) Specular reflection $10^6\lambda$	(7) Random scattering λ/λ_L	(8) Random scattering $10^6\lambda$	(9) Obs. $10^6\lambda$ cm
Tin	3.5	1.4	0.25	7.3	1.4	4.8	1.6	5.7	5.1
Aluminum	1.6	8.2	1.5	93.	2.8	4.4	3.2	5.2	4.9

[a] Values of λ_L and v_0 are those estimated by Faber and Pippard[25] from high-frequency skin resistance and normal electronic specific heat. The value of ξ_0 is $0.18\hbar v_0/kT_c$, as in (5.43). Ratios in columns 5 and 7 are taken from Fig. 6. Observed values are from reference 25.

The temperature dependence of λ_∞ can be obtained directly from (5.45). The second term vanishes when $R=0$, so that we have

$$\frac{\lambda_\infty^2(0)}{\lambda_\infty^2(T)}=\left[\frac{\epsilon_0(T)\tanh[\frac{1}{2}\beta\epsilon_0(T)]}{\epsilon_0(0)}\right]^{\frac{1}{2}}. \tag{5.56}$$

A plot of this quantity on a reduced temperature scale is given in Fig. 6. It is plotted in this way so that a comparison can be made with the empirical law:

$$\lambda^2(0)/\lambda^2(t)=1-t^4, \tag{5.57}$$

based on the Gorter-Casimir two-fluid model. It is seen that our theory is very close to the empirical law except for temperatures very close to T_c. It is in this region that the approximation $\xi_0 \gg \lambda$ becomes invalid as a result of λ increasing with T. The corrections are such as to reduce the theoretical values so as to bring them closer to the experimental values.

To determine $\lambda(T)$ for intermediate cases, we plot in Fig. 7 the quantity $\lambda(T)/\lambda_L(T)$ as a function of $\xi_0/\lambda_L(T)$. The calculations are based on use of the asymptotic forms for $K(q)$ near $T=0°$ and $T=T_c$ given in Appendix C, and numerical integrations of (5.52) and (5.53). The curves for the two limiting temperatures are quite close together, indicating that the effect of the variation of $J(R,T)$ with temperature is rather small. This procedure is analogous to that of Pippard[6] who gave a plot by means of which one can determine λ from known values of ξ_0 and λ_L for the case $J(R,0)=\exp(-R/\xi_0)$.

Using our theoretical expression (5.50) for ξ_0 and empirically estimated values of v_0 and $\lambda_L(0)$ for tin and aluminum,[25] we have obtained $\xi_0/\lambda_L(0)$, and, using Fig. 7, have determined $\lambda(0)/\lambda_L(0)$ for these two metals (Table IV). The agreement between theory and experiment is reasonably good, the experimental values falling between the theoretical values calculated for random scattering and specular reflection.

The theory we have presented applies to a pure metal. Pippard[6] has shown experimentally that the existence of a finite mean free path, l, due to impurity scattering, has the effect of increasing the penetration depth, and has suggested that it may be taken into account by introducing an extra factor, $\exp(-R/l)$, into the kernel for the current density. One of the authors[7] has shown why such a factor may be expected from a theory of the diamagnetic properties based on an energy-gap model. Similar considerations apply to the theory developed in this section; however, effects of coherence on the scattering matrix elements introduce complications and the proper correction factor has not yet been worked out.

To complete the electrodynamics we should give a relation corresponding to the second London equation, the one which gives the time-rate of change of current when an electric field is present. Such a theory would require a calculation, not yet completed, of transport properties with our excited-state wave functions. We expect to find something similar to a two-fluid model, in which thermally excited electrons correspond to the normal component of the fluid. For frequencies such that $h\nu \ll \epsilon_0$, one may determine $\partial\mathbf{j}/\partial t$ for the superconducting component by taking the time derivative of the integral relation (5.2) and then setting $\partial\mathbf{A}/\partial t=-c\mathbf{E}$ in the result (see reference 7).

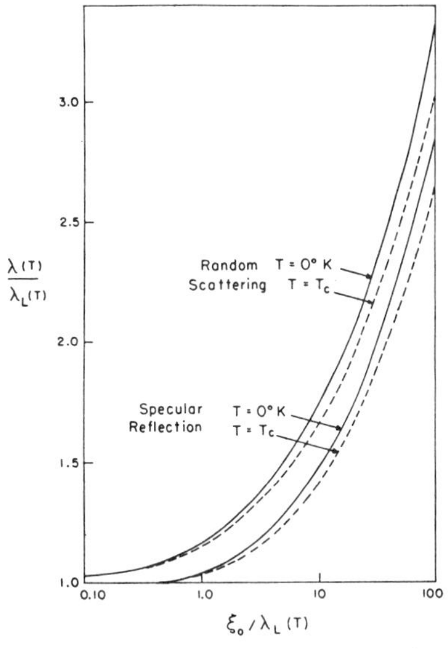

FIG. 7. The ratio $\lambda(T)/\lambda_L(T)$ vs $\xi_0/\lambda_L(T)$ for the boundary conditions of random scattering and specular reflection and for temperatures near $T=0°K$ and near $T=T_c$. The temperature variation of $\lambda_L(T)$ is given by $\lambda_L(T)=(\Lambda_T c^2/4\pi)^{\frac{1}{2}}$.

VI. CONCLUSION

Although our calculations are based on a rather idealized model, they give a reasonably good account of the equilibrium properties of superconductors. When the parameters of the theory are determined empirically, we find that we get agreement with observed specific heats and penetration depths to within the order of 10%. Only the critical temperature involves the superconducting phase; the other two parameters required (density of states and average velocity at the Fermi surface) are determined from the normal phase. This quantitative agreement, as well as the fact that we can account for the main features of superconductivity is convincing evidence that our model is essentially correct.

The basis for the theory is a net attractive interaction between electrons for transitions in which the energy difference between the electron states involved is less than the phonon energy, $\hbar\omega$. For simplicity we have assumed a constant matrix element, $-V$, for transitions within an average energy $\hbar\omega$ of the Fermi surface and have neglected the repulsive interaction outside this region. In more accurate calculations one should take an interaction region dependent on the initial states of the electron and the transition involved, and also take into account any anisotropy in the Fermi surface and in the matrix elements. The fact that there is a law of corresponding states is empirical evidence that such effects are not of great importance. Neglect of the repulsive part of the interaction is in the spirit of the Bloch approximation for normal metals, and appears to be well justified in first approximation. Our theory may be regarded as an extension of the Bloch theory to superconductors in which we introduce only those interactions responsible for the transition.

An improvement in the general formulation of the theory is desirable. We have used that of Bardeen and Pines in which screening of the Coulomb field is taken into account by the Bohm-Pines collective model, and the phonon interaction between electrons is determined only to second order. Diagonal or self-energy terms in the net interaction have been omitted with the assumption that they are included in the Bloch energies of the normal state. When the phonon interaction is so large as to give superconductivity, higher order terms than the second may well be important. One should really have used a renormalized interaction in which such higher order terms are taken into account as well as possible. Very likely the assumption of two-particle interactions is a reasonably good one, so that the only effect would be a redefinition of the interaction constant V in terms of microscopic quantities.

The discussion of the matrix elements in Sec. IV should be a good starting point for calculation of transport properties in the superconducting phase. Our excited state many-particle wave functions are not much more difficult to use in such calculations than the determinantal wave functions of the Bloch theory.

For calculation of boundary energies and related problems, one would like to introduce an order parameter which can decrease continuously from an equilibrium value for the superconducting phase to zero in the normal phase as the boundary is crossed. The Ginsburg-Landau theory and its extensions[7] appear to give a good phenomenological description of such effects. Perhaps the energy gap, $2\epsilon_0$, or, what is equivalent, the coherence distance, ξ_0, could be used for such a parameter.

Another problem, not yet solved, is the calculation of the paramagnetic susceptibility of the electrons in a superconductor, such as is required to account for Reif's data[38] on the Knight shift in the nuclear paramagnetic resonance of colloidal mercury. Our ground state is for total spin $S=0$. It is possible that there is no energy gap between this state and those for $S\neq 0$. While a finite energy is required to turn over an individual spin, it might be possible to construct states analogous to those used in spin-wave theory in which each virtual pair has a small net spin, and for which the energy varies continuously with S. The explanation of the observed electronic paramagnetism (about two-thirds that of the normal metal) would then be similar to that suggested by Reif himself.

In view of its success with equilibrium properties, it may be hoped that our theory will be able to account for these and for other so far unsolved problems.

The authors are indebted to many of their associates for discussions which have helped to clarify the problems involved. We should like to mention particularly discussions with C. P. Slichter and L. C. Hebel on calculation of matrix elements, with D. Pines on the criterion for superconductivity, and with K. A. Brueckner on the exactness of the solution for the ground state.

APPENDIX A. CORRECTIONS TO GROUND STATE ENERGY

We may estimate the accuracy of our superconducting ground state energy, W_0, measured relative to that of the normal state, by making a perturbation theory expansion, using the complete set of excited state superconducting wave functions as the basis functions of the expansion. We first consider the reduced, H_{red}, which includes only pair transitions for pair momentum $q=0$, and then the effect of the neglected portion of the Hamiltonian, $H'=H-H_{\text{red}}$.

To the second order, the ground state energy of H_{red} is:

$$\mathcal{W}_0=(\Psi_0|H_{\text{red}}|\Psi_0)+\sum_{i\neq 0}\frac{|(\Psi_i|H_{\text{red}}|\Psi_0)|^2}{W_0-W_i}+\cdots \quad \text{(A1)}$$

$$=W_0+W_0^{(2)}+\cdots.$$

[38] F. Reif, Phys. Rev. **106**, 208 (1957).

Since the smallest excitation energy is $2\epsilon_0$, the second order energy is overestimated by setting $|W_0-W_i| = 2\epsilon_0$ and performing a closure sum. We obtain the inequality

$$W_0^{(2)} \leq [(\Psi_0|H_{\text{red}}^2|\Psi_0)-(\Psi_0|H_{\text{red}}|\Psi_0)^2]/(-2\epsilon_0)$$

$$= \frac{w_a+w_b+w_c}{-2\epsilon_0}, \tag{A2}$$

where

$$w_a = (\Psi_0|H_0^2|\Psi_0)-(\Psi_0|H_0|\Psi_0)^2, \tag{A3}$$

$$w_b = (\Psi_0|H_0H_v+H_vH_0|\Psi_0) -2(\Psi_0|H_0|\Psi_0)(\Psi_0|H_v|\Psi_0), \tag{A4}$$

$$w_c = (\Psi_0|H_v^2|\Psi_0)-(\Psi_0|H_v|\Psi_0)^2, \tag{A5}$$

and

$$H_0 = \sum_{k>k_F} \epsilon_k b_k^* b_k + \sum_{k<k_F} |\epsilon_k| b_k b_k^*, \tag{A6}$$

$$H_v = -V \sum_{\mathbf{k},\mathbf{k}'} b_{\mathbf{k}'}^* b_{\mathbf{k}}, \tag{A7}$$

the sum in (A7) being carried out over the region $|\epsilon_k|$ and $|\epsilon_{k'}| < \hbar\omega$. If terms in $W_0^{(2)}$ linear in the volume of the system are evaluated, the following expressions may be derived with the aid of the decomposition (2.25):

$$w_a = 8N(0)\int_0^{\hbar\omega} d\epsilon h(\epsilon)[1-h(\epsilon)]\epsilon^2, \tag{A8}$$

$$w_b = -16N(0)V\int_0^{\hbar\omega} d\epsilon_1 \int_0^{\hbar\omega} d\epsilon_2 \{h(\epsilon_1)[1-h(\epsilon_1)] \times h(\epsilon_2)[1-h(\epsilon_2)]\}^{\frac{1}{2}}[1-2h(\epsilon_1)]\epsilon_1, \tag{A9}$$

$$w_c = 8[N(0)V]^2\left[\int_0^{\hbar\omega} d\epsilon_1 h(\epsilon_1)[1-h(\epsilon_1)]\right]^2 \times \int_0^{\hbar\omega} d\epsilon_2[1-2h(\epsilon_2)]^2. \tag{A10}$$

With the use of the relations (2.35), (2.36), and (2.37), terms linear in the volume in (A2) become

$$-2\epsilon_0 W_0^{(2)} \leq 8N(0)\int_0^{\hbar\omega} d\epsilon\, \epsilon^2 \left\{ h(\epsilon)[1-h(\epsilon)] - \frac{\epsilon_0^2}{2E^2} + \frac{\epsilon_0^2}{4E^2} \right\} = 0, \tag{A11}$$

and therefore $W_0^{(2)}/W_0$ vanishes in the limit of a large system.

It is likely that $(\Psi_0|H_{\text{red}}^n|\Psi_0)-(\Psi_0|H_{\text{red}}|\Psi_0)^n$ also vanishes in this limit for n small compared with the total number of valance electrons in the system. For Ψ_0 to be an exact eigenfunction of H_{red}, in the statistical limit, it is required that the above condition hold for all n. Since this requirement can be shown to hold in the strong coupling limit, it is possible that Ψ_0 is exact in the statistical limit for all values of the coupling constant, although no proof has been found at this time.

We may estimate the effect of the neglected portion of the Hamiltonian, $H' = H - H_{\text{red}}$ by a similar perturbation calculation. To second order in H' the correction to the superconducting ground state energy W_0 (measured relative to the normal ground state) is

$$W' = \sum_i \frac{|(i|H'|0)|^2}{W_0-W_i}, \tag{A12}$$

where

$$(i|H'|0) = -V\{[h(\mathbf{k}_0)(1-h(\mathbf{k}_0'))h(\mathbf{k})(1-h(\mathbf{k}'))]^{\frac{1}{2}} - [h(\mathbf{k}_0')(1-h(\mathbf{k}_0))h(\mathbf{k}')(1-h(\mathbf{k}))]^{\frac{1}{2}}\}. \tag{A13}$$

The first term in $(i|H'|0)$ may be identified with breaking up a pair in $\mathbf{k}_0$, the spin-down member going to $-\mathbf{k}_0'\downarrow$ and breaking up a pair in $\mathbf{k}$, the spin-up member going into $\mathbf{k}'\uparrow$. The second term arises from $\mathbf{k}_0'$ and $\mathbf{k}'$ being occupied in the ground state and arriving at the same intermediate state by $\mathbf{k}_0'\uparrow\rightarrow\mathbf{k}_0\uparrow$ and $-\mathbf{k}'\downarrow\rightarrow-\mathbf{k}\downarrow$. The transitions involving particles with parallel spin have been neglected since their contribution is reduced by exchange.

Since the distribution of particle changes differs only over an energy region several ϵ_0 wide, it is to be expected that most of W' will cancel between the normal and superconducting phases. To estimate the energy difference, we shall carry out the sums over a region $6\epsilon_0$ wide. Inserting typical values in (A12), we find

$$W' \sim -\frac{V^2[3N(0)\epsilon_0]^3}{2\epsilon_0}\left(\frac{3\epsilon_0}{E_F}\right) \sim 10^{-3}W_0, \tag{A14}$$

where the factor $(3\epsilon_0/E_F)$ comes from the average reduction in phase space as a result of conseivation of momentum of the pairs making transitions.

These estimates indicate that although the total energy associated with H' may be significant, the effect of H' on the condensation energy is very small. It should be possible to obtain any required quantitative corrections to W_0 by use of perturbation theory.

APPENDIX B. CHANGE IN ZERO-POINT ENERGY OF LATTICE VIBRATIONS

The contribution to the condensation energy from the change in zero-point energy of the lattice can be estimated on the basis of the Bardeen and Pines collective ion-electron treatment.[16] Their theory gives

$$W_{ZP}^{(s)} - W_{ZP}^{(n)} = \sum_{\mathbf{k},\boldsymbol{\kappa}} \frac{|M_\kappa|^2\{(n_{\mathbf{k}}^{(s)}-n_{\mathbf{k}}^{(n)})-(n_{\mathbf{k}+\boldsymbol{\kappa}}^{(s)}-n_{\mathbf{k}+\boldsymbol{\kappa}}^{(n)})\}}{\epsilon_{\mathbf{k}}-\epsilon_{\mathbf{k}+\boldsymbol{\kappa}}-\hbar\omega}, \tag{B1}$$

where $n_{\mathbf{k}}^{(s)}$ is the average occupation number in the superconducting state and $n_{\mathbf{k}}^{(n)}$ that for the normal state. The sum over κ ranges over the first zone and $\mathbf{k}+\kappa$ is to be interpreted as the corresponding reduced wave vector so that Umklapp processes may be taken into account. If $\mathbf{k}+\kappa$ is replaced by $\mathbf{k}$ and κ by $-\kappa$ in the terms containing $n_{\mathbf{k}+\kappa}$, (B1) reduces to

$$W_{ZP^{(s)}}-W_{ZP^{(n)}} = \sum_{\mathbf{k},\kappa} \frac{2|M_\kappa|^2(\epsilon_{\mathbf{k}}-\epsilon_{\mathbf{k}+\kappa})(n_{\mathbf{k}}^{(s)}-n_{\mathbf{k}}^{(n)})}{(\epsilon_{\mathbf{k}}-\epsilon_{\mathbf{k}+\kappa})^2-(\hbar\omega_\kappa)^2}. \quad \text{(B2)}$$

If the sum over κ is carried out, it is seen that the quantity multiplying the difference in occupation numbers is almost independent of $\mathbf{k}$. Since $n_{\mathbf{k}}^{(s)}-n_{\mathbf{k}}^{(n)}$ is antisymmetric with respect to the Fermi surface and differs from zero only in a range of the order of several ϵ_0, it follows that the change in zero-point energy will be small compared to W_0.

A rough estimate of the sum gives

$$W_{ZP^{(s)}}-W_{ZP^{(n)}} = -2\langle|M_\kappa|^2\rangle\int_0^{\hbar\omega} N(0)h(\epsilon)d\epsilon \times\int_{\mathcal{E}_F}^{\mathcal{E}_Z} N(\epsilon')\left\{\frac{\epsilon-\epsilon'}{(\epsilon-\epsilon')^2-(\hbar\omega)^2}+\frac{\epsilon+\epsilon'}{(\epsilon+\epsilon')^2-(\hbar\omega)^2}\right\}d\epsilon'$$

$$\simeq -2\langle|M_\kappa|^2\rangle\int_0^{\hbar\omega} N(0)h(\epsilon)d\epsilon\cdot 2\epsilon[N(\mathcal{E}_Z)/\mathcal{E}_Z]$$

$$\simeq -2\frac{\langle|M_\kappa|^2\rangle}{\mathcal{E}_Z}[N(0)]^2\epsilon_0^2 \simeq (\hbar\omega/\mathcal{E}_Z)W_0 \simeq 10^{-3}W_0, \quad \text{(B3)}$$

where $\mathcal{E}_Z$ is the energy at the zone boundary and typical values have been inserted for the parameters. Thus, it appears that the lattice zero-point energy should have little effect on the condensation energy.

APPENDIX C. EVALUATION OF THE KERNEL IN THE PIPPARD INTEGRAL

According to (5.41), the kernel of the Pippard integral expression for the current density at any temperature T may be written

$$J(R,T)=I(R,0)-I(R,\epsilon_0) = \frac{\Lambda_T}{\pi^2\epsilon_0(0)\Lambda}\int\int_{-\infty}^{+\infty}\left\{\frac{f(\epsilon')-f(\epsilon)}{\epsilon-\epsilon'}-L(\epsilon,\epsilon')\right\}\cos\alpha(\epsilon-\epsilon')d\epsilon d\epsilon', \quad \text{(C1)}$$

where $\alpha=R/\hbar v_0$. The limits should really be $\pm\hbar\omega$, but the convergence is sufficiently rapid in the weak coupling limit so that we may replace $\hbar\omega$ by ∞ without appreciable error. One integration can be performed if use is made of the symmetry of the integrand in ϵ and ϵ'. We shall use the second form given for $L(\epsilon,\epsilon')$ in (4.22), with the lower (+) signs. From the symmetry for positive and negative values of ϵ and ϵ', the integral may be written as the sum of two integrals with limits 0 and ∞:

$$J(R,T)=[\Lambda_T/\pi^2\epsilon_0(0)\Lambda](I_1+I_2), \quad \text{(C2)}$$

where

$$I_1=2\int_0^\infty\int_0^\infty\frac{[F(\epsilon)-F(\epsilon')]\cos\alpha\epsilon\cos\alpha\epsilon'}{\epsilon^2-\epsilon'^2}d\epsilon d\epsilon', \quad \text{(C3)}$$

$$I_2=2\int_0^\infty\int_0^\infty\frac{[G(\epsilon)-G(\epsilon')]\epsilon\epsilon'\sin\alpha\epsilon\sin\alpha\epsilon' d\epsilon d\epsilon'}{\epsilon^2-\epsilon'^2}, \quad \text{(C4)}$$

and

$$F(\epsilon)=[1-2f(\epsilon)]\epsilon-[1-2f(E)](E+\epsilon_0^2E^{-1}), \quad \text{(C5)}$$

$$G(\epsilon)=[1-2f(\epsilon)]\epsilon^{-1}-[1-2f(E)]E^{-1}. \quad \text{(C6)}$$

Unless the argument is given explicitly, $\epsilon_0\equiv\epsilon_0(T)$.

If integration over the region about $\epsilon'=\epsilon$ is made by principal parts, each of the two terms in I_1 (and each of the two terms in I_2) will give equal contributions. Care must be taken to take the same limits for ϵ and ϵ'. Thus we may write

$$I_1=4\mathcal{P}'\lim_{a\to0,b\to\infty}\int_a^b\int_a^b\frac{F(\epsilon)\cos\alpha\epsilon\cos\alpha\epsilon' d\epsilon d\epsilon'}{\epsilon^2-\epsilon'^2}. \quad \text{(C7)}$$

Here $\mathcal{P}'$ indicates the principal part of the integral of ϵ' past $\epsilon'=\epsilon$ is to be taken. Since $F(\epsilon)\to0$ as $\epsilon\to\infty$, the value of the integral does not depend on how the upper limits of the integrals over ϵ and ϵ' are approached. Thus we may set $b=\infty$ for both, and integrate over ϵ' first. This would not have been true if we had not included $I(R,0)$, the normal state contribution, which is equivalent to the term proportional to $\mathbf{A}$ in the expression for the current density. The lower limit is more critical; we must take a the same for both ϵ and ϵ' and approach the limit $a=0$ only in the final result. The integral over ϵ' may be obtained from

$$\int_a^\infty\frac{\cos\alpha\epsilon' d\epsilon'}{\epsilon^2-\epsilon'^2}=\int_0^\infty\frac{\cos\alpha\epsilon' d\epsilon'}{\epsilon^2-\epsilon'^2}-\int_0^a\frac{d\epsilon'}{\epsilon^2-\epsilon'^2}. \quad \text{(C8)}$$

In the second integral on the right we have assumed that a is sufficiently small so that $\cos\alpha\epsilon'$ may be replaced by unity. The first integral on the right may be evaluated by contour integration and the second directly to give

$$\mathcal{P}'\int_a^\infty\frac{\cos\alpha\epsilon' d\epsilon'}{\epsilon^2-\epsilon'^2}=\tfrac{1}{2}\pi\frac{\sin\alpha\epsilon}{\epsilon}-\frac{1}{2\epsilon}\ln\left(\frac{\epsilon+a}{\epsilon-a}\right). \quad \text{(C9)}$$

Similarly, to evaluate I_2, we have

$$\mathcal{P}'\int_0^\infty\frac{\epsilon'\sin\alpha\epsilon' d\epsilon'}{\epsilon^2-\epsilon'^2}=-\tfrac{1}{2}\pi\cos\alpha\epsilon. \quad \text{(C10)}$$

In this integral we may take $a=0$, since the integrand is not singular at the origin. Combining I_1 and I_2, we find

$$I_1+I_2=-2\int_a^\infty F(\epsilon)\ln\left(\frac{\epsilon+a}{\epsilon-a}\right)\frac{d\epsilon}{\epsilon}+\pi\int_0^\infty[F(\epsilon)-\epsilon^2G(\epsilon)]\frac{\sin 2\alpha\epsilon}{\epsilon}d\epsilon. \quad \text{(C11)}$$

In the limit $a\to 0$, the first integral gives a contribution only near $\epsilon=0$. We may therefore replace $F(\epsilon)$ by $F(0)$. The logarithmic integral may then be evaluated, and we find

$$I_1+I_2=\pi^2\epsilon_0[1-2f(\epsilon_0)]-2\pi\epsilon_0^2\int_0^\infty\frac{1-2f(E)}{E}\frac{\sin 2\alpha\epsilon}{\epsilon}d\epsilon. \quad \text{(C12)}$$

Note that this expression vanishes in the two limits $\epsilon_0\to 0$ and α (or R)$\to\infty$, as it should. The second term vanishes as $R\to 0$.

From (C2) and (C12), we may write

$$J(R,T)=\frac{2\Lambda_T\epsilon_0^2}{\pi\epsilon_0(0)\Lambda}\int_0^\infty\left\{\frac{1-2f(\epsilon_0)}{\epsilon_0}-\frac{1-2f(E)}{E}\right\}\frac{\sin 2\alpha\epsilon}{\epsilon}d\epsilon, \quad \text{(C13)}$$

which is equivalent to (5.45) of the text.

Since $f=0$ when $T=0$, we can write

$$J(R,O)=\frac{2\epsilon_0(0)}{\pi}\left(\frac{\pi}{2\epsilon_0(0)}-\int_0^\infty\frac{\sin 2\alpha\epsilon}{\epsilon E}d\epsilon\right). \quad \text{(C14)}$$

This can be put into a form convenient for evaluation by using

$$\frac{\partial}{\partial\alpha}\int_0^\infty\frac{\sin 2\alpha\epsilon}{\epsilon E}d\epsilon=2\int_0^\infty\frac{\cos 2\alpha\epsilon}{E}d\epsilon=2K_0(2\alpha\epsilon_0), \quad \text{(C15)}$$

where K_0 is a modified Bessel function[39] which falls off exponentially for large values of its argument. If we now observe that $\int_0^\infty K_0(y)dy=\pi/2$, we obtain

$$J(R,O)=\frac{2}{\pi}\int_{2R/\pi\xi_0}^\infty K_0(y)dy. \quad \text{(C16)}$$

We next consider the integral of $J(R,T)$ with respect to R. If we introduce a convergence factor, we may integrate under the sign in (C13):

$$\lim_{\gamma\to 0}\int_0^\infty e^{-\gamma R}\sin(2\epsilon R/\hbar v_0)dR=\tfrac{1}{2}\hbar v_0/\epsilon. \quad \text{(C17)}$$

Thus, remembering that $\xi_0=\hbar v_0/[\pi\epsilon_0(0)]$, we find

$$\frac{1}{\xi_0}\int_0^\infty J(R,T)dR=\frac{\epsilon_0^2\Lambda_T}{\Lambda}\int_0^\infty\left\{\frac{1-2f(\epsilon_0)}{\epsilon_0}-\frac{1-2f(E)}{E}\right\}\frac{d\epsilon}{\epsilon^2}. \quad \text{(C18)}$$

According to (5.44) of the text, Λ_T is defined so that the expression on the left is equal to unity. If we integrate the right hand side by parts, we find that

$$\frac{\Lambda}{\Lambda_T}=-\epsilon_0^2\int_0^\infty\frac{d}{dE}\left(\frac{1-2f(E)}{E}\right)\frac{d\epsilon}{E}. \quad \text{(C19)}$$

This integral may be expressed in terms of ϵ_0 and its temperature derivative by differentiating the defining integral for ϵ_0 with respect to $\beta=1/kT$. By a change of variable, $\beta\epsilon=x$, $\beta\epsilon_0=y$, we may write

$$\frac{1}{N(0)V}=\int_0^{\hbar\omega}\frac{1-2f(E)}{E}d\epsilon=\int_0^{\beta\hbar\omega}\frac{1-2f((x^2+y^2)^{\frac{1}{2}})}{(x^2+y^2)^{\frac{1}{2}}}dx. \quad \text{(C20)}$$

We now differentiate with respect to β, remembering that y depends on β. In the weak-coupling limit $[\epsilon_0\ll\hbar\omega,\ f(\beta\hbar\omega)\simeq 0]$, we find:

$$0=\frac{1}{\beta}\frac{d(\beta\epsilon_0)}{d\beta}\cdot\frac{\epsilon_0}{\beta}\int_0^{\hbar\omega}\frac{d}{dE}\left(\frac{1-2f(E)}{E}\right)\frac{d\epsilon}{E}. \quad \text{(C21)}$$

The integral converges sufficiently rapidly so that we may replace the upper limit by ∞. Comparing (C19) and (C21), we find that

$$\frac{\Lambda_T}{\Lambda}=\frac{1}{\epsilon_0}\frac{d(\beta\epsilon_0)}{d\beta}, \quad \text{(C22)}$$

as stated in (5.25).

Finally, we shall derive an approximate expression for the transform $K(q)$ valid when $qhv_0>\epsilon_0(0)$. First we express $K(q)$ as an integral of $J(R,T)$. The transform of $\mathbf{j}(\mathbf{r})$ may be written

$$\mathbf{j}(\mathbf{q})=-\frac{c}{4\pi}K(q)\mathfrak{a}(\mathbf{q})=-\frac{3}{4\pi c\Lambda_T\xi_0}\int\frac{\mathbf{R}[\mathbf{R}\cdot\mathfrak{a}(\mathbf{q})]e^{i\mathbf{q}\cdot\mathbf{R}}}{R^4}J(R,T)d\tau. \quad \text{(C23)}$$

[39] See Erdelyi, Magnus, Oberhettinger, and Tricomi, *Higher Transcendental Functions* (McGraw Hill Book Company, Inc., New York, 1953), Vol. 2, pp. 5 and 19.

To carry out the integration over angles, we take the polar axis in the direction of $\mathbf{q}$ and set $u=\cos\theta$. This gives

$$K(q)=\frac{3\pi}{c^2\Lambda_T\xi_0}\int_0^\infty\int_{-1}^{+1}(1-u^2)e^{iRqu}J(R,T)dudR. \quad (C24)$$

When (C13) is substituted for $J(R,T)$, we require the following integral:

$$g(b)=\int_0^\infty\int_{-1}^{+1}e^{iRqu}(1-u^2)\sin(\epsilon Rq/\epsilon_1)dudR$$

$$=(1-b^2)\ln\left(\frac{1+b}{1-b}\right)+2b, \quad (C25)$$

where $\epsilon_1=\frac{1}{2}q\hbar v_0$ and $b=\epsilon/\epsilon_1$. We thus find

$$K(q)=\frac{6\pi\epsilon_0^2}{qc^2\Lambda\hbar v_0}\int_0^\infty\left\{\frac{1-2f(\epsilon_0)}{\epsilon_0}-\frac{1-2f(E)}{E}\right\}g(b)\frac{d\epsilon}{\epsilon}. \quad (C26)$$

Expansions of $g(b)$ for b small and b large are

$$b<1:\ g(b)=4\left(b-\frac{1}{3}b^3-\frac{1}{15}b^5\cdots\right), \quad (C27)$$

$$b>1:\ g(b)=4\left(\frac{1}{3b}+\frac{1}{15b^3}+\cdots\right). \quad (C28)$$

Further,

$$\int_0^\infty g(b)\frac{db}{b}=\frac{1}{2}\pi^2. \quad (C29)$$

We may change the variable of integration in (C26) from ϵ to b, and then use different expansions for $b<1$ and $b>1$. One of the integrals required is

$$4\int_0^1\frac{1-2f(E)}{E}\left\{b-\frac{1}{3}b^3-\frac{1}{15}b^5-\cdots\right\}\frac{db}{b}$$

$$+4\int_1^\infty\frac{1-2f(E)}{E}\left\{\frac{1}{3b}+\frac{1}{15b^3}+\cdots\right\}\frac{db}{b}. \quad (C30)$$

In terms of b,

$$E=\epsilon_1(b^2+b_0^2)^{\frac{1}{2}}, \quad (C31)$$

where

$$b_0=\epsilon_0/\epsilon_1. \quad (C32)$$

An approximate expression valid in the limit $b_0\ll 1$ may be obtained by neglecting b_0 in the integral for $b>1$, and also in terms in b^3 and higher in the integral for $b<1$. The only integral in which b_0^2 is not neglected in comparison with b^2 is the first term,

$$4\int_0^1\frac{1-2f(E)}{E}db=\frac{4}{\epsilon_1}\int_0^{\epsilon_1}\frac{1-2f(E)}{E}d\epsilon, \quad (C33)$$

which can be determined from (C20), the defining integral for ϵ_0. For $\epsilon>\epsilon_1$ we may take $f(E)=0$. Thus we find

$$\frac{4}{\epsilon_1}\int_0^{\epsilon_1}\frac{1-2f(E)}{E}d\epsilon\simeq\frac{4}{\epsilon_1}\left(\frac{1}{N(0)V}-\ln\frac{\hbar\omega}{\epsilon_1}\right)$$

$$=\frac{4}{\epsilon_1}\ln\frac{q\hbar v_0}{\epsilon_0(0)}. \quad (C34)$$

The latter form follows from the expression for $\epsilon_0(0)$ in the weak-coupling limit. The remaining terms cancel when b_0^2 is neglected:

$$\int_0^1\left\{\frac{b^3}{3}-\frac{b^5}{15}\cdots\right\}\frac{db}{b^2}$$

$$+\int_1^\infty\left\{\frac{1}{3b}+\frac{1}{15b^3}+\cdots\right\}\frac{db}{b^2}=0. \quad (C35)$$

Thus we find that for $qhv_0/\epsilon_0(0)=\pi q\xi_0\gg 1$,

$$K(q)=\frac{3\pi^2\epsilon_0}{qc^2\Lambda\hbar v_0}\left\{1-2f(\epsilon_0)-\frac{16\epsilon_0}{\pi^2 q\hbar v_0}\ln(\pi q\xi_0)\right\}. \quad (C36)$$

For q very small, $K(q)$ approaches the constant value,

$$K(q)=\lambda_L^{-2}(T)=4\pi/(\Lambda_T c^2),\quad (q\to 0) \quad (C37)$$

where Λ_T is the temperature-dependent London constant defined by (C22). For intermediate values of q, one may interpolate between this constant value for $\pi q\xi_0<1$ and the asymptotic form (C36) for large q.

The two limiting cases are $T=0°$K and T close to T_c. At $T=0°$K, $\epsilon_0\to\epsilon_0(0)$ and $f(\epsilon_0)\to 0$, and we find

$$K(q)=4\pi/(\Lambda c^2)=1/\lambda_L^2(0),\quad (q\to 0); \quad (C38)$$

$$K(q)=\frac{3\pi}{4q\lambda_L^2(0)\xi_0}\left\{1-\frac{16}{\pi^3 q\xi_0}\ln(\pi q\xi_0)\right\},\ \text{(large } q). \quad (C39)$$

Near T_c, one may determine Λ_T from (C22) and (3.49):

$$\Lambda/\Lambda_T=0.23\beta_c^2\epsilon_0^2(T)=2(1-t), \quad (C40)$$

where $\beta_c=1/kT_c$ and $t=T/T_c$. When $\epsilon_0(T)$ is small,

$$1-2f(\epsilon_0)=\tanh(\tfrac{1}{2}\beta\epsilon_0)\to\tfrac{1}{2}\beta_c\epsilon_0(T). \quad (C41)$$

Thus the limiting expressions for $K(q)$ as $T\to T_c$ are

$$K(q)=\frac{1}{\lambda_L^2(T)}=\frac{2(1-t)}{\lambda_L^2(0)},\quad (q\to 0); \quad (C42)$$

$$K(q)=\frac{3.75\pi}{4q\lambda_L^2(T)\xi_0}\left\{1-\frac{18.3}{\pi^3 q\xi_0}\ln(\pi q\xi_0)\right\},\ \text{(large } q). \quad (C43)$$

The close similarity of the expressions for the two

limiting cases is evident. The reason for the similarity is that $J(R,T)$ varies slowly with T.

At temperatures very close to T_c, λ becomes larger than ξ_0, so that $\lambda(T)$ approaches $\lambda_L(T)$. A comparison of penetration depths measured near T_c to those measured at low temperatures would then be expected to give

$$\frac{\lambda(0)}{\lambda(T)}=\frac{\lambda(0)}{\lambda_L(0)}\frac{\lambda_L(0)}{\lambda(T)}=\frac{\lambda(0)}{\lambda_L(0)}\sqrt{2}(1-t)^{\frac{1}{2}}. \quad \text{(C44)}$$

This is to be compared with the ratio $2(1-t)^{\frac{1}{2}}$ which the empirical law, $(1-t^4)^{\frac{1}{2}}$, gives as $t\to 1$.

APPENDIX D. CORRELATION OF ELECTRONS OF OPPOSITE SPIN

Insight into the coherent structure of our ground state wave function may be obtained by an investigation of the correlation function for electrons of opposite spin. For the normal metal (in the Bloch approximation), electrons of antiparallel spin are entirely uncorrelated while for electrons of parallel spin there is the exchange correlation.

The correlation function $\rho_{\sigma'\sigma''}(\mathbf{r}',\mathbf{r}'')$ for an n-electron system is defined as

$$\rho_{\sigma'\sigma''}(\mathbf{r}',\mathbf{r}'')=\sum_{i,j=1}^{2n}\int\cdots\int d\tau_1\cdots d\tau_n\Psi^*(\mathbf{r}_1\cdots \mathbf{r}_n)$$
$$\times\Psi(\mathbf{r}_1\cdots\mathbf{r}_n)\delta_{\sigma'}(\mathbf{r}_i-\mathbf{r}')\delta_{\sigma''}(\mathbf{r}_j-\mathbf{r}''). \quad \text{(D1)}$$

This can be written in second quantized form, following the notation of Sec. V, as

$$\rho_{\sigma'\sigma''}(\mathbf{r}',\mathbf{r}'')=(\Psi_0|\psi_{\sigma''}^*(\mathbf{r}'')\psi_{\sigma'}^*(\mathbf{r}')\psi_{\sigma'}(\mathbf{r}')\psi_{\sigma''}(\mathbf{r}'')|\Psi_0)$$
$$=\sum_{\mathbf{k}_1,\mathbf{k}_2,\mathbf{k}_3,\mathbf{k}_4}(\Psi_0|c^*(\mathbf{k}_1,\sigma'')c^*(\mathbf{k}_2,\sigma')c(\mathbf{k}_3,\sigma')c(\mathbf{k}_4,\sigma'')|\Psi_0)$$
$$\times e^{i(\mathbf{k}_4-\mathbf{k}_1)\cdot\mathbf{r}''+i(\mathbf{k}_3-\mathbf{k}_2)\cdot\mathbf{r}'}, \quad \text{(D2)}$$

where Ψ_0 is our ground-state wave function.

Of particular interest is the correlation function for electrons of opposite spin. If we define

$$\rho_A=\rho_{\uparrow\downarrow}+\rho_{\downarrow\uparrow}, \quad \text{(D3)}$$

use the matrix elements obtained in Sec. IV, and set $\mathbf{r}=\mathbf{r}'-\mathbf{r}''$, we get at the absolute zero of temperature

$$\rho_A(r)=n[\tfrac{1}{2}n+P_A(r)],$$

where

$$P_A(r)=\frac{1}{2n}\left(\frac{1}{2\pi}\right)^6\int\int_{\mathcal{R}} d\mathbf{k}d\mathbf{k}'\left\{e^{i(\mathbf{k}'-\mathbf{k})\cdot\mathbf{r}}\frac{\epsilon_0^2(0)}{EE'}\right\}. \quad \text{(D4)}$$

The integration is over $\mathcal{R}$, the region of interaction $(|\epsilon|<\hbar\omega)$, and n is the number of electrons of both spin directions per unit volume. The terms in the bracket give the number of electrons of opposite spin per unit volume one would find a distance r away from an electron of given spin direction. The first term is just the average density, while the second gives the effects of correlation.

The integral of $P_A(r)$ gives the number of electrons of opposite spin correlated with a given electron. This is

$$\int P_A(r)d\tau=\frac{\epsilon_0^2(0)}{2n}\left(\frac{1}{2\pi}\right)^3\int_{\mathcal{R}} dk\frac{1}{\epsilon^2+\epsilon_0^2(0)}$$
$$\simeq\frac{\epsilon_0^2(0)}{2n}\frac{1}{\pi^2}\frac{dk}{d\mathcal{E}}\bigg]_F k_F^2\int_0^{\hbar\omega}\frac{d\epsilon}{\epsilon^2+\epsilon_0^2(0)}. \quad \text{(D5)}$$
$$\simeq\frac{3}{4}\frac{1}{k_F\xi_0}=\frac{\pi}{2}\frac{n_c}{n},$$

where we have given slowly varying functions their value at the Fermi surface, and have used (5.43) and $n_c=N(0)\epsilon_0(0)$. Thus the number of electrons of given spin correlated to one of opposite spin is of the order of the ratio of the number of electrons in coherent pairs to the total number of electrons in the system.

The range of the spacial correlation may be determined by investigating $P_A(r)$. We must then evaluate

$$I=\left(\frac{1}{2\pi}\right)^3\int_{\mathcal{R}} d\mathbf{k}\frac{e^{i\mathbf{k}\cdot\mathbf{r}}}{E}=\frac{1}{2\pi^2 r}\int_{k_F-\delta}^{k_F+\delta}\frac{k\sin kr dk}{[\epsilon^2+\epsilon_0^2(0)]^{\frac{1}{2}}}. \quad \text{(D6)}$$

To evaluate this, we make use of the sharp maximum of the integral near $\epsilon=0$ and the fact that the sine function is rapidly oscillating in this region for values of r of interest, $k_F r\gg 1$. We set (as in Sec. V)

$$k\simeq k_F+\frac{dk}{d\mathcal{E}}\bigg]_F\epsilon=k_F+\epsilon/(\hbar v_0), \quad \text{(D7)}$$

and

$$\sin\left(k_F+\frac{\epsilon}{\hbar v_0}\right)r\simeq\sin k_F r\cos\left(\frac{\epsilon}{\hbar v_0}r\right), \quad \text{(D8)}$$

give slowly varying functions their value at the Fermi surface, and drop terms antisymmetric in ϵ. Then, setting $x=\epsilon/\epsilon_0$, we get

$$I=\frac{k_F}{\pi^2 r}\frac{1}{\hbar v_0}\sin k_F r\int_0^a\frac{\cos(rx/\pi\xi_0)dx}{(x^2+1)^{\frac{1}{2}}}, \quad \text{(D9)}$$

where $a=\hbar\omega/\epsilon_0\gg 1$. The integral may be expressed as

$$\int_0^a\frac{\cos(r/\pi\xi_0)x}{(x^2+1)^{\frac{1}{2}}}dx=K_0\left(\frac{r}{\pi\xi_0}\right)$$
$$-\int_a^\infty\frac{\cos(rx/\pi\xi_0)}{(x^2+1)^{\frac{1}{2}}}dx, \quad \text{(D10)}$$

where K_0 is a modified Bessel function.[39] Since $a\gg 1$ except for very small values of r, $(x^2+1)^{\frac{1}{2}}\simeq x$ in the last integral. Then, setting $t=rx/(\pi\xi_0)$, we have

$$I=\frac{k_F}{\pi^2 r}\frac{1}{hv_0}\sin k_F r\left[K_0\left(\frac{r}{\pi\xi_0}\right)-\int_{ar/\pi\xi_0}^{\infty}\frac{\cos t}{t}dt\right]. \quad \text{(D11)}$$

Asymptotically

$$-\int_x^{\infty}\frac{\cos t}{t}dt\sim\frac{\sin x}{x}, \quad x\gg 1, \quad \text{(D12)}$$

while $K_0(x)$ falls off exponentially. Thus the asymptotic behavior of $P_A(r)$ is given by

$$P_A(r)\sim\frac{9n\epsilon_0^2(0)}{2(\hbar\omega)^2}\frac{\sin^2 k_F r\sin^2\left(\frac{\epsilon_0}{\hbar\omega}\frac{r}{\pi\xi_0}\right)}{(k_F r)^4}. \quad \text{(D13)}$$

Since $a=h\omega/\epsilon_0\gg 1$, the range of correlation is determined by the K_0 function which drops off rapidly when $r/(\pi\xi_0)\gtrsim 1$. Thus correlation distance is the order of $\pi\xi_0\simeq 10^{-4}$ cm.

SOVIET PHYSICS JETP VOLUME 34 (7), NUMBER 1 JULY, 1958

A NEW METHOD IN THE THEORY OF SUPERCONDUCTIVITY. I*

N. N. BOGOLIUBOV

Mathematics Institute, Academy of Sciences, U.S.S.R.

Submitted to JETP editor October 10, 1957

J. Exptl. Theoret. Phys. (U.S.S.R.) 34, 58-65 (January, 1958)

The canonical transformation method previously developed by the author for the theory of superfluids is generalized in the present paper. By application of this method and the principle of compensation of "dangerous" diagrams, it is shown that a superconducting state is inherent in the Fröhlich model. Computation of the principal parameters of this state leads to formulas that confirm those of the theory of Bardeen, Cooper, and Shrieffer.

AFTER the discovery of the isotopic effect, the idea became universal that the interaction between the electrons and the lattice must play a fundamental role in the phenomenon of superconductivity. A series of very interesting syntheses,[1-4] which treat a system of electrons interacting with the phonon field have already been carried out. In the present paper, we shall show that, by further development of a method which we have put forward previously for the theory of superfluidity, it is possible to give a systematic theory of the superconducting state. In this case, in particular, the results of the theory of Bardeen, Cooper, and Schrieffer[3] are confirmed.

For simplicity, we shall start out from the model proposed by Fröhlich,[1] in which the Coulomb interaction is not introduced explicitly and the dynamic system is characterized by the Hamiltonian†

$$H_{\mathrm{Fr}} = \sum_{k,s} E(k)\, a_{ks}^{+} a_{ks} + \sum_{q} \omega(q)\, b_q^{+} b_q + H',$$

$$H' = \sum_{\substack{k,q,s\\ k'-k=q}} g \left\{\frac{\omega(q)}{2V}\right\}^{1/2} a_{ks}^{+} a_{k's} b_q^{+} + \text{conjugate}. \qquad (1)$$

where $E(k)$ is the energy of the electron, $\omega(q)$ is the energy of the phonon, g is the coupling constant, and V the volume.

As is now well known, ordinary perturbation theory in terms of powers of the coupling constant is inapplicable, since the electron-phonon interaction, in spite of its smallness, is seen to be very considerable when close to the Fermi surface. Therefore, we shall first carry out a certain canonical transformation, starting out from the following considerations.

First, we note that the matrix elements, corresponding to virtual creation of "particles" from the vacuum, are always associated with the energy denominators

$$\{\varepsilon(k_1) + \ldots + \varepsilon(k_{2s}) + \omega(q_1) + \ldots + \omega(q_r)\}^{-1},$$

in which $\epsilon(k) \sim |E(k) - E_F|$ is the energy of the particle [electron — $E(k) > E_F$ or hole — $E(k) < E_F$] which becomes small at the Fermi surface.

Such denominators are not generally "dangerous," and do not lead to singularities in integration over the momenta $k_1, \ldots, k_{2s}, q_1, \ldots, q_r$, except when we deal with a virtual process of the creation of a single pair, without phonons. Then, because of the conservation law, the particles of this pair will have oppositely directed momenta $\pm k$, and the energy denominator $\frac{1}{2}\epsilon(k)$ will become "dangerous" for integration. We note further that their spins will also be opposite.

Thus, in the choice of the canonical transformation, it must be kept in mind that it is necessary to guarantee the mutual compensation of the diagrams which lead to virtual creation from the vacuum of pairs of particles with opposite momenta and spins.

We now underscore the analogy with the situation holding in our theory of superfluids, namely, in the consideration of a non-ideal Bose gas, where virtual creation of a pair of particles (with mo-

*Note added in proof (Dec. 20, 1957). Very recently, when our papers had already gone to press, the manuscript of the detailed work of Bardeen, Cooper, and Schrieffer became known to us. In particular, a derivation is given therein for formulas which we have cited from the brief letter by the same authors. We must note that there are several places that are not clear in the formulation of the method of Bardeen, Cooper, and Schrieffer. A detailed comparison of the two methods will be given in a paper which is currently being prepared for press.

†A system of units is employed here in which h = 1.

menta ± k) from the condensate plays the same role. We then made use of a linear transformation of the Bose amplitudes, "mixing" b_q with b^+_{-q}.

Generalizing this transformation, we now bring into consideration the case of new Fermi amplitudes

$$\alpha_{k0} = u_k a_{k,\,1/2} - v_k a^+_{-k,\,-1/2},$$
$$\alpha_{k1} = u_k a_{-k,\,-1/2} + v_k a^+_{k,\,1/2}$$

or

$$a_{k,\,1/2} = u_k \alpha_{k0} + v_k \alpha^+_{k1},$$
$$a_{-k,\,-1/2} = u_k \alpha_{k1} - v_k \alpha^+_{k0},$$

where u_k, v_k are real numbers, connected by the relation

$$u_k^2 + v_k^2 = 1.$$

It is not difficult to check that such a transformation preserves all the commutation properties of the Fermi operators and is therefore canonical.

We further note that it represents a generalization of the ordinary transformation, with the help of which the creation and annihilation operators of holes inside the Fermi surface and of electrons outside this surface are introduced. Actually, if we set

$$u_k = 1,\ v_k = 0 \quad \text{for} \quad E(k) > E_F$$
$$u_k = 0,\ v_k = 1 \quad \text{for} \quad E(k) < E_F,$$

then we obtain

$$\alpha_{k0} = a_{k,\,1/2},\ \alpha_{k1} = a_{-k,\,-1/2} \quad \text{for} \quad E(k) > E_F,$$
$$\alpha_{k0} = -a^+_{-k,\,-1/2},\ \alpha_{k1} = a^+_{k,\,1/2} \quad \text{for} \quad E(k) < E_F,$$

so that, for example, α_{k0} will be an annihilation operator for an electron of momentum k and spin $\frac{1}{2}$ outisde the Fermi sphere, and will be an annihilation operator for a hole with momentum $-k$ and spin $-\frac{1}{2}$ inside.

In the general case, when $(u_k, v_k) \neq (0, 1)$, we have to deal with the superposition of a hole and an electron.

Generalizing to the consideration of the Fröhlich Hamiltonian, we note that it will be more advantageous to us not to restrict ourselves to the relation

$$\sum_{k,\,s} a^+_{ks} a_{ks} = N_0,$$

where N_0 is the total number of electrons; we therefore make use of the ordinary procedure in such a situation and introduce a parameter λ which plays the role of the chemical potential. Then, instead of H_{Fr}, we have the Hamiltonian

$$H = H_{Fr} - \lambda N. \qquad (2)$$

We determine the parameter λ in turn from the condition that in the state under examination

$$\bar{N} = N_0. \qquad (3)$$

Transforming H to the new Fermi amplitudes, we get

$$H = U + H_0 + H_{int},\quad H_{int} = H_1 + H_2 + H_3,$$

where U is the constant

$$U = 2\sum E(k)\, v_k^2 - 2\lambda \sum v_k^2,$$

and where

$$H_1 = \sum_{\substack{k,\,k' \\ k'-k=q}} g \sqrt{\frac{\omega(q)}{2V}} \{u_k v_{k'} \alpha^+_{k0} \alpha^+_{k'1} + u_{-k} v_{-k'} \alpha^+_{-k'0} \alpha^+_{-k1}$$
$$+ u_{k'} v_k \alpha_{k1} \alpha_{k'0} + u_{-k'} v_{-k} \alpha_{-k'1} \alpha_{-k0}\}\, b^+_q + \text{conjugate}$$

$$H_2 = \sum_{\substack{k,\,k' \\ k'-k=q}} g \sqrt{\frac{\omega(q)}{2V}} \{u_k u_{k'} \alpha^+_{k0} \alpha_{k'0} + u_{-k} u_{-k'} \alpha^+_{-k1'} \alpha_{-k'1}$$
$$- v_k v_{k'} \alpha^+_{k'1} \alpha_{k1} - v_{-k} v_{-k'} \alpha^+_{-k'0} \alpha_{-k0}\}\, b^+_q + \text{conjugate}$$

$$H_3 = \sum 2(E(k) - \lambda)\, u_k v_k (\alpha^+_{k0} \alpha^+_{k1} + \alpha_{k1} \alpha_{k0}),$$

$$H_0 = \sum (E(k) - \lambda)(u_k^2 - v_k^2)(\alpha^+_{k0} \alpha_{k0} + \alpha^+_{k1} \alpha_{k1})$$
$$+ \sum \omega(q)\, b^+_q b_q.$$

We introduce the filling number

$$\nu_{k0} = \alpha^+_{k0} \alpha_{k0},\quad \nu_{k1} = \alpha^+_{k1} \alpha_{k1}$$

of new quasi-particles, which are generated by the operator α^+. Then the "interaction-free vacuum," i.e., the state C_V in which

$$H_0 C_V = 0,$$

will evidently be

$$C_V = \prod_k \delta(\nu_{k0})\, \delta(\nu_{k1})$$

with zero values of the numbers ν.

We further note that λ ought to be close to E_F, since $\lambda = E_F$ in the absence of interaction, and that consequently, the expression

$$\varepsilon(k) = (E(k) - \lambda)(u_k^2 - v_k^2)$$

ought to vanish on a surface close to the Fermi surface.

We now see that the virtual process of creation from the vacuum of a pair of quasi-particles ν_{k0} and ν_{k1} without phonons will be "dangerous" in the sense of the criterion given earlier, since the corresponding energy denominator will be

$$1 / 2\varepsilon(k).$$

The Hamiltonian H_3, which, being applied to a

vacuum, gives the diagram of Fig. 1,* leads directly to such a process. This same process is also obtained, moreover, as a result of the combined action of H_1, H_2 Thus, for example, in the second order in the coupling constant g, we have the diagrams shown in Fig. 2a.

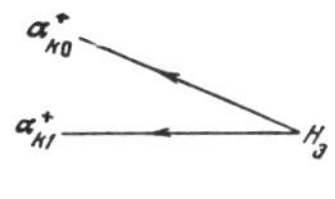

FIG. 1

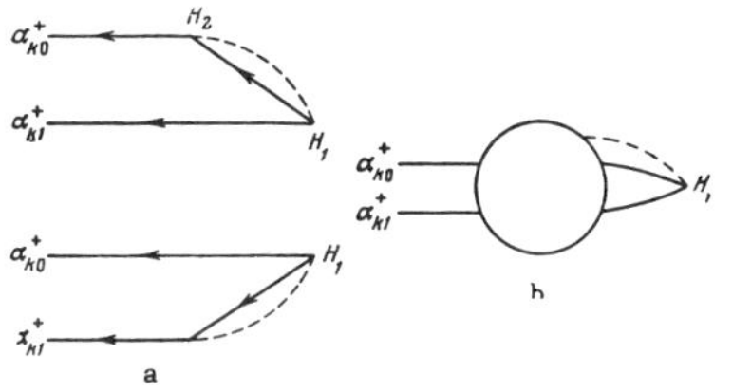

FIG. 2

In higher orders, we get diagrams of the type of 2b, where the circle denotes the coupled part, which cannot be divided into two coupled parts, and which is bound only by the two lines of one pair under consideration.

Making use of the principle of compensation of dangerous diagrams introduced above, we must equate to zero the sum of contributions from the diagrams of Fig. 1 and Fig. 2. We thus obtain an equation for u_k and v_k.

Now we no longer need to take into consideration the diagrams of Figs. 1 and 2 (and their conjugates), and therefore, in the expansions of perturbation theory, no expression will appear with a dangerous energy denominator.

We now construct a second-order equation for u_k and v_k. In this approximation, we must compensate the diagram of Fig. 1 by the diagrams of Fig. 2a. We get

$$2(E(k)-\lambda)u_k v_k + \Omega_k = 0,$$

where Ω_k is the coefficient for $\overset{+}{\alpha}_{k0}\overset{+}{\alpha}_{k1}C_V$ in the expression

$$-H_2 H_0^{-1} H_1 C_V.$$

*An account of the details of diagrams for the many body problem can be found in the detailed paper of Gugenholtz.[5]

Expanding it, we finally obtain

$$\{\widetilde{E}(k)-\lambda\}u_k v_k$$

$$= (u_k^2 - v_k^2)\frac{1}{2V}\sum_{k'} g^2 \frac{\omega(k-k')}{\omega(k-k')+\varepsilon(k)+\varepsilon(k')} u_{k'}v_{k'}, \quad (4)$$

where

$$\widetilde{E}(k) = E(k) - \frac{1}{2V}\sum_{k'} g^2 \frac{\omega(k-k')}{\omega(k-k')+\varepsilon(k)+\varepsilon(k')}(u_{k'}^2 - v_{k'}^2). \quad (5)$$

Staying within the limits of the approximation, we replace

$$\varepsilon(k) = \{E(k)-\lambda\}(u_k^2 - v_k^2)$$

in the denominator of the right side by

$$\widetilde{\varepsilon}(k) = \{\widetilde{E}(k)-\lambda\}(u_k^2 - v_k^2).$$

Then, setting

$$\widetilde{E}(k) - \lambda = \xi(k),$$

we write down the equation in the form

$$\xi(k)u_k v_k = (u_k^2 - v_k^2)$$

$$\times \frac{1}{2(2\pi)^3}\int g^2 \frac{\omega(k-k')}{\omega(k-k')+\widetilde{\varepsilon}(k)+\widetilde{\varepsilon}(k')} u_{k'}v_{k'}d\mathbf{k}'. \quad (6)$$

This equation obviously contains the trivial solution

$$uv = 0, \quad (u, v) = (0,1),$$

which corresponds to the "normal state." It possesses, however, an additional solution of another type which goes over into the trivial one at great distances from the Fermi surface.

Denoting

$$C(k) = \frac{1}{(2\pi)^3}\int g^2 \frac{\omega(k-k')}{\omega(k-k')+\widetilde{\varepsilon}(k)+\widetilde{\varepsilon}(k')} u_{k'}v_{k'}d\mathbf{k}',$$

we find from (6)

$$u_k^2 = \frac{1}{2}\left\{1 + \frac{\xi(k)}{\sqrt{C^2(k)+\xi^2(k)}}\right\};$$

$$v_k^2 = \frac{1}{2}\left\{1 - \frac{\xi(k)}{\sqrt{C^2(k)+\xi^2(k)}}\right\}, \quad (7)$$

whence

$$u_k v_k = \frac{C(k)}{2\sqrt{C^2(k)+\xi^2(k)}}; \quad \widetilde{\varepsilon} = \frac{\xi^2(k)}{\sqrt{C^2(k)+\xi^2(k)}}.$$

Thus, our equation reduces to the following form:

$$C(k) = \frac{1}{2(2\pi)^3}\int g^2 \frac{\omega(k-k')}{\omega(k-k')+\widetilde{\varepsilon}(k)+\widetilde{\varepsilon}(k')}\frac{C(k')}{\sqrt{C^2(k')+\xi^2(k')}}d\mathbf{k}'. \quad (8)$$

We note that this equation has a singular property: as $g^2 \to 0$, the solution C tends to zero as

$$\exp\{-A/g^2\}, \quad A = \text{const} > 0,$$

in view of the fact that the integral on the right side of (8) remains logarithmically divergent in the vicinity of the surface $\xi(\mathbf{k}) = 0$ if we set $C = 0$.

In such a situation, it is not difficult to obtain an asymptotic form of the solution for small g:

$$C(k) = \tilde{\omega} e^{-1/\rho} \frac{1}{2} \int_{-1}^{+1} \frac{\omega\{k_0 \sqrt{2(1-t)}\}}{\omega\{k_0 \sqrt{2(1-t)}\} + |\xi(k)|} dt, \tag{9}$$

where

$$\rho = g^2 \frac{1}{2\pi^2} \left(\frac{k^2}{d\tilde{E}(k)/dk} \right)_{k=k_0}; \qquad \tilde{E}(k_0) = \lambda,$$

$$\ln \tilde{\omega} = \int_0^\infty \ln \frac{1}{2\xi} \frac{d}{d\xi} \left\{ \frac{1}{2} \int_{-1}^{+1} \frac{\omega\{k_0 \sqrt{2(1-t)}\}}{\omega\{k_0 \sqrt{2(1-t)}\} + \xi} dt \right\}^2 d\xi. \tag{10}$$

Taking into account the additional condition (3) and the expressions (7), (9) for u, v, it can be noted that

$$k_0 = k_F.$$

Furthermore, it is clear that the corrections to Eq. (5), obtained by replacing u_k, v_k in (5) by their "normal" values

$$u_k = \theta_G(k) = \begin{cases} 1, & |k| > k_F \\ 0, & |k| < k_F \end{cases}$$
$$v_k = \theta_F(k) = \begin{cases} 0, & |k| > k_F \\ 1, & |k| < k_F \end{cases} \tag{11}$$

will be exponentially small.

We can therefore replaced $\tilde{E}(\mathbf{k})$ by the corresponding formula for the normal state in Eqs. (10) without loss of accuracy, and integrate the factor

$$\frac{1}{2\pi^2} \left(\frac{k^2}{d\tilde{E}/dk} \right)_{k=k_F} = \frac{1}{V} \left\{ \frac{V}{(2\pi)^3} \frac{4\pi k^2 dk}{dE} \right\}_0$$

as the relative density dn/dE of the number of electron levels in an infinitely narrow energy band close to the Fermi surface. Then

$$\rho = g^2 dn/dE. \tag{12}$$

We now proceed to the calculation of the energy of the ground state in the second approximation.

Of the entire H_{int}, we shall now take into account only H_1. Consequently, we get for the eigenvalue of H in the ground state

$$U - \langle C_V^* H_1 H_0^{-1} H_1 C_V \rangle =$$

$$= 2 \sum_k \{E(k) - \lambda\} v_k^2 - \frac{1}{V} \sum_{k \neq k'} g^2 \frac{\omega(k'-k)\{u_k^2 v_{k'}^2 + u_k v_k u_{k'} v_{k'}\}}{\omega(k'-k) + \varepsilon(k) + \varepsilon(k')}. \tag{13}$$

Then, substituting the expressions we have obtained for u_k, v_k, we compute the difference ΔE between the energy of the ground state and the energy of the normal state. We have

$$\frac{\Delta E}{V} = -\frac{dn}{dE} \frac{\tilde{\omega}^2}{2} \exp\left(-\frac{2}{\rho}\right). \tag{14}$$

It is interesting to observe that this result coincides with the result of Bardeen and his coworkers.[3] To see this we need only write the Bardeen parameters ω, V in the following fashion:

$$2\omega = \tilde{\omega}; \; V = g^2. \tag{15}$$

We now construct, in the given approximation, the formula for the energy of the elementary excitation. For this purpose, we choose an excited state

$$C_1 = \overset{+}{\alpha}_{k0} C_V$$

and apply perturbation theory to it in the usual fashion. We obtain the following expression for the energy of an elementary excitation with momentum k:

$$E_e(k) = \varepsilon(k) - \langle C_1^* H_{int} (H_0 - \varepsilon(k))^{-1} H_{int} C_1 \rangle'_{coup},$$

for which we find

$$E_e(k) = \tilde{\varepsilon}(k) \left\{ 1 - \frac{g^2}{V} \sum_{k'} \omega(k-k') \frac{u_k^2 u_{k'}^2 + v_k^2 v_{k'}^2}{[\omega(k-k') + \varepsilon(k')]^2 - \varepsilon^2(k)} \right\}$$
$$+ \frac{g^2}{V} 2u_k v_k \sum_{k'} \frac{\omega(k-k')(\omega(k-k') + \varepsilon(k'))}{[\omega(k-k') + \varepsilon(k')]^2 - \varepsilon^2(k)} u_{k'} v_{k'}. \tag{16}$$

Here the first term, proportional to $\tilde{\epsilon}(\mathbf{k})$ has no singular properties whatever and vanishes on the Fermi surface. The second term is equal to

$$\frac{g^2}{V} 2u_k v_k \sum_{k'} \frac{\omega(k-k')}{\omega(k-k') + \varepsilon(k')} u_{k'} v_{k'}$$
$$= 2u_k v_k C(k) = C(k_F) = \tilde{\omega} e^{-1/\rho}.$$

on the Fermi surface.

Thus the energies of the excited states are different from the energy of the ground state by the gap*

$$\Delta = \tilde{\omega} e^{-1/\rho}. \tag{17}$$

We note that in the work of Bardeen and his co-

*Here we have the fermion part of the spectrum in mind; there is, in addition, the boson part which, it appears, is not separated by a gap from the ground state. We shall not cite the corresponding formulas here since they have no relation to the theory of the superconducting state at zero temperature.

workers,[3] there is an expression of the type $2\bar{\omega} \times e^{-1/\rho}$ which is interpreted there as the energy necessary to break up a "pair."

We now consider the "ground current state," i.e., the state with the lowest energy among all the possible states with given momentum **p**.

We therefore need to find the eigenvalue of H for the additional condition

$$\sum_{k,s} \mathbf{k} a_{ks}^{+} a_{ks} = \mathbf{p}.$$

Instead, making use of the usual method, we introduce (in addition to the scalar parameter λ) an additional vector parameter **u** which plays the role of an average velocity, and choose the total Hamiltonian in the form

$$H = H_{\mathrm{Fr}} - \lambda \sum_{k,s} a_{ks}^{+} a_{ks} - \sum_{k,s} (\mathbf{uk})\, a_{ks}^{+} a_{ks} =$$

$$= \sum_{k,s} \{E(k) - (\mathbf{uk}) - \lambda\}\, a_{ks}^{+} a_{ks} + \sum_{q} \omega(q)\, b_q^{+} b_q + H_{\mathrm{int}}. \tag{18}$$

The value of **u** is determined from the condition

$$\sum_{k,s} \mathbf{k} \overline{a_{ks}^{+} a_{ks}} = \mathbf{p}.$$

Since we are always dealing in our discussions only with a small region on the Fermi surface, we can for simplicity write the following

$$E(k) = \frac{k^2}{2m} + D;\; D = E_{\mathrm{F}} - \frac{k_{\mathrm{F}}^2}{2m},$$

and then, in the final equations, take

$$m = \left(\frac{k}{dE/dk}\right)_{k=k_{\mathrm{F}}}.$$

Then

$$E(k) - (\mathbf{uk}) = E(\mathbf{k} - m\mathbf{u}) - {}^1/_2\, m\mathbf{u}^2,$$

and therefore, making the following translation in the momentum space of **k**:

$$\mathbf{k} \to \mathbf{k} + \mathbf{u}m,\quad a_{\mathbf{k}s} \to a_{\mathbf{k}+m\mathbf{u},\, s} \tag{19}$$

and putting

$$\lambda \to \lambda + m\mathbf{u}^2/2,$$

then the Hamiltonian (18) takes the form (2) and the vector **u** drops out. We again arrive at the case of a ground state with zero momentum. No new investigation is thus needed for the current state, but it will suffice only to make a transformation inverse to (19) in the formulas obtained earlier.

Thus, we can verify, for example, that the energy of the ground current state, with mean velocity **u**, differs from the energy of the ground state without current by an amount $Nmu^2/2$.

The excitations are separated from the energy of the ground current state by a gap

$$\Delta_u = \Delta - \mathbf{k}_{\mathrm{F}}\mathbf{u} > \Delta - k_{\mathrm{F}}\,|\mathbf{u}|.$$

Consequently, if

$$k_{\mathrm{F}}\,|\mathbf{u}| < \Delta,$$

the current state, although it possesses an energy greater than the non-current (so long as the action of the magnetic field is not taken into account), is shown to be stable relative to the excitations.

Thus, we see that our model contains the property of superconductivity.

Let us make a few other observations.

In our method of investigation, we must take the parameter ρ to be small in order to make it possible to restrict ourselves to asymptotic approximations. However, as was shown by Tolmachev and Tiablikov[6] (with the aid of a method which does not make use of any assumption on the smallness of ρ) the sound velocity becomes imaginary for $\rho > \frac{1}{2}$, i.e., the lattice becomes unstable. In cases in which the lattice is so rigid that the electron-phonon interaction has no essential effect on the energy of the phonon, the parameter ρ must be small. Even for $\rho = {}^1/_4$, the value of $e^{-1/\rho}$ is ${}^1/_{55}$. This, in our opinion, explains the smallness of the value of the energy gap and therefore of the critical temperature.

We further note that if we explicitly add a term of Coulomb interaction to the Fröhlich Hamiltonian, then it will be necessary to carry out a summation of the diagrams of electron-holes of the Gelman-Bruckner type, in order to guarantee the phenomenon of screening.

By way of a preliminary estimate, we could first introduce into the Hamiltonian the Coulomb interaction in the screened form, and also treat it by means of perturbation theory. Then we would have obtained substantially the same formulas as earlier, but we would need to replace g^2 by

$$g^2 - \frac{1}{2}\int_{-1}^{1} \nu_e\left(k_{\mathrm{F}}\sqrt{2(1-t)}\right) dt,$$

where $\nu_e(\mathbf{k})$ is the Fourier transform of the screened function.

Thus it could be immediately established that the Coulomb interaction contradicts the phenomenon of superconductivity.

In conclusion I consider it my pleasant duty to thank D. N. Zubarev, V. V. Tolmachev, S. V. Tiablikov and Iu. A. Tserkovinskov for their valuable discussions.

[1] H. Fröhlich, Phys. Rev. 79, 845 (1950); Proc. Roy. Soc. (London) 215A, 291 (1952).

[2] J. Bardeen, Revs. Mod. Phys. 23, 261 (1951); Handb. d. Physik (Springer, Berlin) 15, 274 (1956). J. Bardeen and D. Pines, Phys. Rev. 99, 1140 (1955).

[3] Bardeen, Cooper, and Schrieffer, Phys. Rev. 106, 162 (1957).

[4] D. Pines, Phys. Rev. (in press).

[5] N. M. Gugenholtz, Physica 23, 481 (1957).

[6] V. V. Tolmachev and S. V. Tiablikov, J. Exptl. Theoret. Phys. (U.S.S.R.) 34, 66 (1958), Soviet Phys. JETP 7, 46 (1958) (this issue).

Translated by R. T. Beyer
10

IL NUOVO CIMENTO Vol. VII, N. 6 16 Marzo 1958

Comments on the Theory of Superconductivity.

J. G. Valatin

Department of Mathematical Physics - University of Birmingham

(ricevuto il 23 Dicembre 1957)

Summary. — Some ideas of the new theory of Bardeen, Cooper and Schrieffer are expressed in a more transparent form. New collective fermion variables are introduced which are linear combinations of creation and annihilation operators of electrons, and describe elementary excitations. They lead to a simple classification of excited states and a great simplification in the calculations. The structure of the excitation spectrum is investigated without equating the matrix element of the interaction potential to a constant at an early stage, and new relationships and equations are derived. The temperature dependent problem is described by means of a statistical operator, and its relationship to that of the grand canonical ensemble is established. Simple new relationships are obtained for the correlation function.

1. - Introduction.

In a recent paper Bardeen, Cooper and Schrieffer (1) have developed a new approach to the theory of superconductivity. Their model assumes a simple attractive two body interaction between the electrons which expresses the effect of the interactions through the phonon field and of screened Coulomb interactions. It gives a good quantitative account of the thermodynamic and electromagnetic properties of superconductors.

The aim of the present paper is to simplify the conceptual background of the new theory and to obtain a clearer insight into the ideas involved. Improved mathematical tools are introduced which are related to simple physical concepts, and some simple new relationships are brought out.

(1) J. Bardeen, L. N. Cooper and J. R. Schrieffer: *Phys. Rev.*, **108**, 1175 (1957).

The important problem of obtaining the two body forces from the phonon interaction will not be touched here, so that the starting point will be a theory with a Hamiltonian containing two body interactions. It will be pointed out that the trial ground state vector introduced by BARDEEN, COOPER and SCHRIEFFER which expresses long range correlations between particles of opposite spin, is related in a natural way to new collective fermion variables which lead to a simple description of the excited states. The calculations are much simplified by the introduction of the new variables, and it can be hoped that they will be of help also in making further progress in the theory.

The structure of the excitation spectrum will then be investigated by expressing the Hamiltonian in terms of the new variables. The error involved by neglecting a non-diagonal part of the Hamiltonian will not be discussed, but the new expressions obtained also make the mathematical formulation of this problem easier. Various relationships will be established without replacing the matrix element of the interaction by an average value at an early stage. This brings out more clearly those features of the theory which are independent of the assumption of a constant matrix element. Equations are obtained for the investigation of more general interactions.

To describe temperature dependent phenomena, BARDEEN, COOPER and SCHRIEFFER formulate a variational problem with two independent functions to be varied. One describes the elementary excitations; the other, a statistically independent distribution of these excitations. This problem will be expressed here in terms of a statistical operator and it will be pointed out that this can be considered as the trial approximation for the statistical operator of the grand ensemble.

Finally, simple expressions will be obtained for the correlation functions which make the interpretation of the relationships of the theory more transparent.

2. – Collective fermion variables.

The Hamiltonian

$$H = T + V \tag{1a}$$

of the theory is the sum of the Bloch energy

$$T = E_0 + \sum_{k,\sigma} \varepsilon_k a_{k\sigma} a^*_{k\sigma} , \tag{1b}$$

$$E_0 = -2 \sum_{k<k_F} \varepsilon_k , \tag{1c}$$

and of a non-local two-body interaction energy

$$V = \frac{1}{2}\sum_{\substack{k,k',q\\ \sigma,\sigma'}} V_{kk'} a_{k\sigma} a_{q-k,\sigma'} a^*_{q-k',\sigma'} a^*_{k'\sigma} \,. \tag{1d}$$

The energy ε_k of a Bloch electron with momentum k is counted from the Fermi level. The index σ stands for a definite $\uparrow$ or $\downarrow$ spin direction. The index $\varkappa$ will be used to indicate both momentum k and spin σ, whenever this simplifies the notation. In this case, $-\varkappa$ will stand for opposite momentum and opposite spin. Creation operators are denoted by $a_\varkappa$ and annihilation operators by $a^*_\varkappa$. (This is in accordance with Dirac's notation [2] and with the geometric interpretation [3] of these operators, unstarred quantities representing the unit vectors and the stars being reserved for the basis elements of the dual vectors). It will be found convenient to write $a_{\varkappa_1} \dots a_{\varkappa_j}$ for the state vector with electrons in the one particle states $\varkappa_1, \dots, \varkappa_j$, replacing by unity the state $|0\rangle$ in which no particles are present.

The trial ground state vector of Bardeen, Cooper and Schrieffer can be written as a product of commuting factors in the form

$$\Phi_0 = \prod_\varkappa \frac{1 + g_\varkappa a_\varkappa a_{-\varkappa}}{(1 + |g_\varkappa|^2)^{\frac{1}{2}}} \,, \tag{2a}$$

with

$$g_{k\uparrow} = g_{-k\uparrow} = g_k = -g_{k\downarrow} = -g_{-k\downarrow} \,. \tag{2b}$$

This corresponds to an indeterminate number of particles, but with a sharp peak in the probability distribution of the particle number for the actual values of $g_\varkappa$. The choice of such a state vector exploits the advantages of the fact that the Hamiltonian ($1a$, b, d) is valid for any number of particles and the quantum field equations of the problem are independent of the special value of the number operator $N = \sum_\varkappa a_\varkappa a^*_\varkappa$. The projection operator of the state ($2a$) can be considèred as a trial approximation for the statistical operator of the grand ensemble in the limit of temperature $T = 0$.

Separating the state vectors and operators by a stroke, whenever this seems convenient, equations can be written like $a^*_\varkappa | a_\varkappa a_{\varkappa'} = a_{\varkappa'}$, $a^*_\varkappa | 1 = 0$ or $a_\varkappa | 1 = a_\varkappa$. One obtains the simple relations

$$a^*_\varkappa \; |(1 + g_\varkappa a_\varkappa a_{-\varkappa}) = g_\varkappa a_{-\varkappa} \,, \tag{2c}$$

$$a_{-\varkappa} |(1 + g_\varkappa a_\varkappa a_{-\varkappa}) = a_{-\varkappa} \,. \tag{2d}$$

---

[2] P. A. M. Dirac: *The Principles of Quantum Mechanics*, 3rd ed. (Oxford, 1947), p. 249.

[3] J. G. Valatin: *Journ. Phys.*, **12**, 131 (1951).

Accordingly, it is rather natural to introduce the new variables

$$(3a)\qquad \xi_\varkappa^* = \frac{a_\varkappa^* - g_\varkappa a_{-\varkappa}}{(1+|g_\varkappa|^2)^{\frac{1}{2}}}, \qquad \xi_\varkappa = \frac{a_\varkappa - g_\varkappa^* a_{-\varkappa}^*}{(1+|g_\varkappa|^2)^{\frac{1}{2}}},$$

since from (2a, c, d) one sees immediately that

$$(3b)\qquad \xi_\varkappa^* \,|\, \Phi_0 = 0 \qquad \text{for all } \varkappa.$$

The anticommutation relations of $a_\varkappa$, $a_{\varkappa'}^*$ give with (3a)

$$(3c)\qquad \xi_{\varkappa'}^*\xi_\varkappa + \xi_\varkappa\xi_{\varkappa'}^* = \delta_{\varkappa\varkappa'}, \quad \xi_\varkappa\xi_{\varkappa'} + \xi_{\varkappa'}\xi_\varkappa = 0, \quad \xi_{\varkappa'}^*\xi_\varkappa^* + \xi_\varkappa^*\xi_{\varkappa'}^* = 0.$$

The equations (3a) represent a canonical transformation, and the products

$$(3d)\qquad \Phi_{\varkappa_1 \dots \varkappa_j} = \xi_{\varkappa_1} \dots \xi_{\varkappa_j} \Phi_0$$

form a complete orthonormal set of state vectors.

From

$$(4a)\qquad \xi_\varkappa \,|\, \frac{1 + g_\varkappa a_\varkappa a_{-\varkappa}}{(1+|g_\varkappa|^2)^{\frac{1}{2}}} = a_\varkappa,$$

$$(4b)\qquad \xi_\varkappa\xi_{-\varkappa} \,|\, \frac{1 + g_\varkappa a_\varkappa a_{-\varkappa}}{(1+|g_\varkappa|^2)^{\frac{1}{2}}} = \frac{a_\varkappa a_{-\varkappa} - g_\varkappa^*}{(1+|g_\varkappa|^2)^{\frac{1}{2}}},$$

one can see that the states

$$\xi_\varkappa \Phi_0 \qquad \text{and} \qquad \xi_\varkappa \xi_{-\varkappa} \Phi_0$$

are the excited states of Bardeen, Cooper and Schrieffer with a « single particle » and a « real pair » excited. They appear here on the same footing, and the set of products $\xi_{\varkappa_1} \dots \xi_{\varkappa_j} \Phi_0$ gives a natural description of their system of excited states. As seen from (4a), though the state $\xi_\varkappa \Phi_0$ contains an explicit particle creation operator factor $a_\varkappa$, at the same time a « virtual pair » factor $(1+|g_\varkappa|^2)^{-\frac{1}{2}}(1 + g_\varkappa a_\varkappa a_{-\varkappa})$ of Φ_0 is missing. The operation of $\xi_\varkappa$ on Φ_0 corresponds to a collective excitation of all the particles; $\xi_\varkappa$ creates an elementary excitation, or a « quasi-particle », with the properties of a fermion (+).

(+) Analogous phonon variables for a boson system were introduced by Bogolubov (4) in 1947. At the time of writing this paper a preprint of a recent work by Bogolubov arrived in which essentially the same fermion variables as given here are used independently in investigating the interacting electron-phonon system.

(4) N. Bogolubov: *Journ. Phys. U.S.S.R.*, **11**, 23 (1947).

From (3*a*) one can write

$$(5) \qquad a_\varkappa = \frac{\xi_\varkappa + g_\varkappa^* \xi_{-\varkappa}^*}{(1+|g_\varkappa|^2)^{\frac{1}{2}}}, \qquad a_\varkappa^* = \frac{\xi_\varkappa^* + g_\varkappa \xi_{-\varkappa}}{(1+|g_\varkappa|^2)^{\frac{1}{2}}},$$

and any operator expressed in terms of $a_\varkappa$, $a_\varkappa^*$ can be transformed into an expression in the collective variables $\xi_\varkappa$, $\xi_\varkappa^*$. Ordering the factors in each term in such a way that creation operators $\xi_\varkappa$ stand at the left of the annihilation operators $\xi_\varkappa^*$, the constant term of the expression represents the expectation value of the operator in the ground state Φ_0.

For the number of particles in state $\varkappa$,

$$(6a) \qquad n_\varkappa = a_\varkappa a_\varkappa^*$$

one obtains

$$(6b) \qquad n_\varkappa = \frac{(\xi_\varkappa + g_\varkappa^* \xi_{-\varkappa}^*)(\xi_\varkappa^* + g_\varkappa \xi_{-\varkappa})}{1+|g_\varkappa|^2}.$$

The ground state expectation value

$$(6c) \qquad \bar{n}_\varkappa = h_\varkappa$$

results from ordering the term with $\xi_{-\varkappa}^* \xi_{-\varkappa}$, and one obtains

$$(6d) \qquad h_\varkappa = \frac{|g_\varkappa|^2}{1+|g_\varkappa|^2}.$$

From (2*b*) one concludes that

$$(6e) \qquad h_{k\uparrow} = h_{k\downarrow} = h_{-k\uparrow} = h_{-k\downarrow} = h_k.$$

For real g_k, this gives

$$(6f) \qquad (h_k)^{\frac{1}{2}} = \frac{g_k}{(1+|g_k|^2)^{\frac{1}{2}}}, \quad (1-h_k)^{\frac{1}{2}} = \frac{1}{(1+|g_k|^2)^{\frac{1}{2}}}, \quad (h_k(1-h_k))^{\frac{1}{2}} = \frac{g_k}{1+|g_k|^2},$$

which establishes the connection with the expressions of Bardeen, Cooper and Schrieffer.

The expression (6*b*) of $n_\varkappa$ contains a term with $\xi_\varkappa \xi_{-\varkappa}$ and one with $\xi_{-\varkappa}^* \xi_\varkappa^*$ which have vanishing expectation values for states $\xi_{\varkappa_1} \ldots \xi_{\varkappa_j} \Phi_0$. The remaining terms can be written in the form

$$(n_\varkappa)_0 = (1 - h_\varkappa)\mathcal{N}_\varkappa + h_\varkappa(1 - \mathcal{N}_{-\varkappa}) , \tag{6g}$$

where

$$\mathcal{N}_\varkappa = \xi_\varkappa \xi^*_\varkappa \tag{6h}$$

represents the number of elementary excitations in state $\varkappa$.

In a similar way, the substitution (5) gives for the « diagonal part » of the pair creation and annihilation operators $a_\varkappa a_{-\varkappa}$ and $a^*_{-\varkappa'} a^*_{\varkappa'}$,

$$(a_\varkappa a_{-\varkappa})_0 = \chi^*_\varkappa (1 - \mathcal{N}_\varkappa - \mathcal{N}_{-\varkappa}) , \tag{7a}$$

$$(a^*_{-\varkappa'} a^*_{\varkappa'})_0 = \chi_{\varkappa'} (1 - \mathcal{N}_{\varkappa'} - \mathcal{N}_{-\varkappa'}) , \tag{7b}$$

with

$$\chi_\varkappa = \frac{g_\varkappa}{1 + |g_\varkappa|^2} . \tag{7c}$$

One particle operators $B = \sum_{\varkappa,\varkappa'} B_{\varkappa\varkappa'} a_\varkappa a^*_{\varkappa'}$ are transformed with (5) into a form, from which the matrix elements between states $\xi_{\varkappa_1} \dots \xi_{\varkappa_j} \Phi_0$ which are tabulated in a section of the paper by BARDEEN, COOPER and SCHRIEFFER can be immediately obtained.

3. – Structure of the excitation spectrum.

The Hamiltonian (1a, b, c, d) is separated through (5) into two parts,

$$H = H_0 + H_1 , \tag{8a}$$

where H_1 has a vanishing expectation value for states of the form $\xi_{\varkappa_1} \dots \xi_{\varkappa_j} \Phi_0$ which are the eigenstates of H_0. The diagonal operator H_0 can be written as

$$\begin{aligned} H_0 = E_0 + \sum_{k,\sigma} \varepsilon_k (n_{k\sigma})_0 + \tfrac{1}{2} \sum_{k'\sigma} (n_{k'\sigma'})_0 \sum_{k\sigma} V_{kk}(n_{k\sigma})_0 - \\ - \tfrac{1}{2} \sum_{kk'\sigma} V_{kk'}(n_{k\sigma})_0 (n_{k'\sigma})_0 + \tfrac{1}{2} \sum_{kk'\sigma} V_{kk'} (a_{k,\sigma} a_{-k,-\sigma})_0 (a^*_{-k',-\sigma} a^*_{k',\sigma})_0 , \end{aligned} \tag{8b}$$

where $(n_\varkappa)_0$ and $(a_\varkappa a_{-\varkappa})_0$, $(a^*_{-\varkappa} a^*_\varkappa)_0$ are given by (6g) and (7a, b). Accordingly, H_0 is of the form

$$H_0 = W_0 + \sum_\varkappa \widetilde{E}_\varkappa \mathcal{N}_\varkappa + \sum_{\varkappa \neq \varkappa'} \mathcal{V}_{\varkappa\varkappa'} \mathcal{N}_\varkappa \mathcal{N}_{\varkappa'} , \tag{8c}$$

in which the ground state expectation value W_0 is given by

$$(8d)\qquad W_0 = E_0 + 2\sum_k \varepsilon_k h_k + 2\sum_{k'} h_{k'} \sum_k V_{kk} h_k - \sum_{kk'} V_{kk'} h_k h_{k'} + \sum_{kk'} V_{kk'} \chi_k^* \chi_{k'} = \\ = E_0 + \sum_k (\varepsilon_k h_k + \nu_k h_k - \mu_k^* \chi_k)$$

and the excitation energy $\tilde{E}_{k\uparrow} = \tilde{E}_{k\downarrow} = \tilde{E}_k$ of states $\xi_\varkappa \Phi_0$ is

$$(8e)\qquad \tilde{E}_k = \nu_k(1 - 2h_k) + (\mu_k^* \chi_k + \chi_k^* \mu_k)\,.$$

The quantities ν_k, μ_k occurring in (8d, e) are defined by

$$(8f)\qquad \nu_k = \varepsilon_k - \sum_{k'} \overline{V}_{kk'} h_{k'}\,,$$

$$(8g)\qquad \overline{V}_{kk'} = \tfrac{1}{2}(V_{kk'} + V_{k'k}) - (V_{kk} + V_{k'k'})\,,$$

$$(8h)\qquad \mu_k = -\sum_{k'} V_{kk'} \chi_{k'}\,.$$

Considering W_0 as a function of g_k^*, g_k given by (8d), (6d), (7c) and minimizing it with respect to g_k^*, the equation $\partial W^0/\partial g_k^* = 0$ reads

$$(9a)\qquad \mu_k^* g_k^2 + 2\nu_k g_k - \mu_k = 0\,.$$

In referring directly to $\partial W_0/\partial g_k^* = 0$ it is assumed that the density of states is symmetric about the Fermi level and ε_k is given with respect to it. Otherwise one would have to add a Lagrangian multiplier to ν_k in (9a), corresponding to the supplementary condition of a constant average number of particles.

The solution of the second order algebraic equation (9a) gives

$$(9b)\qquad g_k = \frac{1}{\mu_k^*}\left(-\nu_k + (\nu_k^2 + |\mu_k|^2)^{\frac{1}{2}}\right)\,,$$

or with

$$(9c)\qquad E_k = +(\nu_k^2 + |\mu_k|^2)^{\frac{1}{2}}\,,$$

$$(9d)\qquad g_k = \frac{1}{\mu_k^*}(E_k - \nu_k)\,.$$

The root with the plus sign is chosen in (9b) in order to minimize the energy W_0.

From (9*c*, *d*), (6*d*), (7*c*) one obtains

$$(10a)\qquad |g_k|^2 = \frac{(E_k - \nu_k)^2}{|\mu_k|^2},$$

$$(10b)\qquad 1 + |g_k|^2 = \frac{2E_k(E_k - \nu_k)}{|\mu_k|^2} = \frac{2E_k}{E_k + \nu_k},$$

$$(10c)\qquad h_k = \frac{1}{2}\left(1 - \frac{\nu_k}{E_k}\right),$$

$$(10d)\qquad \chi_k = \frac{\mu_k}{2E_k}.$$

With these values of h_k, χ_k the expression (8*d*) of W_0 gives

$$(10e)\qquad W_0 = E_0 + \sum_k \left\{\varepsilon_k h_k + \frac{1}{2}\left(\nu_k - \frac{\nu_k^2}{E_k} - \frac{|\mu_k|^2}{E_k}\right)\right\} =$$

$$= E_0 + \sum_k \left\{\varepsilon_k h_k + \frac{1}{2}(\nu_k - E_k)\right\},$$

and from (8*e*) one obtains

$$(10f)\qquad \tilde{E}_k = \frac{\nu_k^2}{E_k} + \frac{|\mu_k|^2}{E_k} = E_k;$$

that is, (9*c*) gives the energy of an elementary excitation (*).

The quantities ν_k, μ_k are to be determined from the non-linear equations obtained from (8*f*, *g*, *h*) with (10*c*, *d*) and (9*c*),

$$(11a)\qquad \nu_k = \bar{\varepsilon}_k + \frac{1}{2}\sum_{k'} \bar{V}_{kk'} \frac{\nu_{k'}}{(\nu_{k'}^2 + |\mu_{k'}|^2)^{\frac{1}{2}}},$$

$$(11b)\qquad \bar{\varepsilon}_k = \varepsilon_k - \frac{1}{2}\sum_{k'} \bar{V}_{kk'},$$

$$(11c)\qquad \mu_k = -\frac{1}{2}\sum_{k'} \bar{V}_{kk'} \frac{\mu_{k'}}{(\nu_{k'}^2 + |\mu_{k'}|^2)^{\frac{1}{2}}}.$$

(*) Formally very similar expressions to those obtained here, though with a rather different physical content, can be derived for a boson system. They include those of Bogolubov's 1947 method (4) as their low density limit. An investigation of this problem in collaboration with D. Butler is still in progress.

In the case of a factorisable potential $V_{kk'} = v_k v_{k'}$ these equations can be reduced to integrations and algebraic equations, by introducing quantities of the type

$$\lambda = \tfrac{1}{2} \sum_{k'} \frac{v_{k'} \mu_{k'}}{(\nu_{k'}^2 + |\mu_{k'}|^2)^{\frac{1}{2}}}, \quad \mu_k = -\lambda v_k .$$

The approximations involved in the form of the trial state vector Φ_0 are such that better approximations are obtained by excluding the occupation (or emptiness) of electron states outside a definite energy region $|\varepsilon_k| \leqslant \hbar\omega$ about the Fermi level. This can be done on the ground of physical considerations, where $\hbar\omega$ is a characteristic phonon energy. The trial state vector is chosen accordingly with $h_k = 0$ for $\varepsilon_k > \hbar\omega$ and $h_k = 1$ for $\varepsilon_k < -\hbar\omega$. The equations obtained by minimizing W_0 are unchanged, though the summation extends only over a restricted energy region. Replacing in this region $V_{kk'}$ by an average value $V_{kk'} = -V$, and replacing by their average a part of the terms in H_0 which depend only on the total number N of particles, one obtains $\nu_k \simeq \varepsilon_k$, and $\mu_k = \text{constant} = \varepsilon_0$. The relationships (9$c$), (10$c$), (10$d$) then reduce to those given by BARDEEN, COOPER and SCHRIEFFER, and the equation for the energy gap ε_0 can be solved explicitly.

The equations (11a, b, c) for determining ν_k, μ_k can be put into a simpler linearized form. They can be obtained by minimizing the expression (8d) of W_0 which is a quadratic form in h_k, χ_k with respect to the independent variables h_k, χ_k under the supplementary conditions $|\chi_k|^2 + h_k^2 = h_k$, or $|2\chi_k|^2 + (1-2h_k)^2 = 1$.

4. – Statistical operator and grand ensemble.

With the simplified classification of states, the ensemble of states considered by BARDEEN, COOPER and SCHRIEFFER corresponds to a statistical operator of the form

$$U_0 = C_0^{-1} \sum_{j=0}^{\infty} \sum_{\varkappa_1 < \ldots < \varkappa_j} w_{\varkappa_1} \ldots w_{\varkappa_j} P_{\varkappa_1 \ldots \varkappa_j}, \tag{12a}$$

where $P_{\varkappa_1 \ldots \varkappa_j}$ is the projection operator on the state $\xi_{\varkappa_1} \ldots \xi_{\varkappa_j} \Phi_0$ and C_0 is the trace of the operator sum,

$$C_0 = \operatorname{tr} \sum_{j=0}^{\infty} \sum_{\varkappa_1 < \ldots < \varkappa_j} w_{\varkappa_1} \ldots w_{\varkappa_j} P_{\varkappa_1 \ldots \varkappa_j} = \prod_{\varkappa} (1 + w_{\varkappa}), \tag{12b}$$

so that $\operatorname{tr} U_0 = 1$. The average number of elementary excitations in the state $\varkappa$ is

$$(12c)\qquad \langle \mathfrak{N}_\varkappa \rangle = \operatorname{tr} \mathfrak{N}_\varkappa U_0 = \frac{w_\varkappa}{1 + w_\varkappa} = f_\varkappa \,,$$

and it is assumed that

$$(12d)\qquad f_{k\uparrow} = f_{k\downarrow} = f_{-k\uparrow} = f_{-k\downarrow} = f_k \,.$$

This form of the statistical operator implies that the elementary excitations are statistically independent because, for $\varkappa \neq \varkappa'$ one has

$$\langle \mathfrak{N}_\varkappa \mathfrak{N}_{\varkappa'} \rangle = \langle \mathfrak{N}_\varkappa \rangle \langle \mathfrak{N}_{\varkappa'} \rangle = f_\varkappa f_{\varkappa'} \,.$$

U_0 is determined by the two independent functions $g_\varkappa$ and $f_\varkappa$ both of which will be obtained as temperature dependent. From (6g) and (7b) one has the average values

$$(13a)\qquad \langle n_\varkappa \rangle = (1 - h_\varkappa) f_\varkappa + h_\varkappa (1 - f_\varkappa) = h_\varkappa^{(T)} \,,$$

$$(13b)\qquad \langle a_{-\varkappa}^* a_\varkappa^* \rangle = \chi_\varkappa (1 - 2 f_\varkappa) = \chi_\varkappa^{(T)}$$

The average value of the energy $\langle H \rangle = \langle H_0 \rangle = W_0^{(T)}$ results from (8b) as

$$(13c)\qquad W_0^{(T)} = E_0 + 2 \sum_k \varepsilon_k h_k^{(T)} + 2 \sum_{k'} h_{k'}^{(T)} \sum_k V_{kk} h_k^{(T)} - \\ - \sum_{kk'} V_{kk'} h_k^{(T)} h_{k'}^{(T)} + \sum_{kk'} V_{kk'} \chi_k^{*(T)} \chi_{k'}^{(T)} \,,$$

which is of the same form as (8d) with h_k, χ_k replaced by $h_k^{(T)}$, $\chi_k^{(T)}$. In connection with the quantities (13a, b) the relationship

$$(13d)\qquad |\chi_\varkappa^{(T)}|^2 = h_\varkappa^{(T)} (1 - h_\varkappa^{(T)}) - f_\varkappa (1 - f_\varkappa)$$

might be mentioned.

With the standard entropy expression for a system of independent fermions given by

$$(14a)\qquad TS_0 = -2\beta \sum_k \{ f_k \log f_k + (1 - f_k) \log (1 - f_k) \} \,,$$

one can form an approximate free energy expression

$$(14b)\qquad W_0^{(T)} - TS_0$$

and minimize it independently with respect to the functions g_k and f_k.

The minimization with respect to f_k for a given g_k leads to the fermion distribution

$$f_k = \frac{1}{\exp[\beta \widetilde{E}_k^{(T)}]+1}, \tag{15a}$$

$$w_k = \exp[-\beta \widetilde{E}_k^{(T)}], \tag{15b}$$

with

$$\widetilde{E}_k^{(T)} = \frac{1}{2}\frac{\partial W_0^{(T)}}{\partial f_k} = \nu_k^{(T)}(1-2h_k) + \mu_k^{(T)*}\chi_k + \chi_k^* \mu_k^{(T)}, \tag{15c}$$

where in analogy to (8f, h)

$$\nu_k^{(T)} = \varepsilon_k - \sum_{k'} \overline{V}_{kk'} h_{k'}^{(T)}, \tag{15d}$$

$$\mu_k^{(T)} = -\sum_{k'} V_{kk'} \chi_{k'}^{(T)}. \tag{15e}$$

With (15b), the statistical operator (12a) can be, therefore, written in the form

$$U_0 = \exp[\beta \Lambda_0] \exp[-\beta \mathcal{H}_0], \tag{16a}$$

$$\exp[-\beta \Lambda_0] = \operatorname{tr} \exp[-\beta \mathcal{H}_0]. \tag{16b}$$

with

$$\mathcal{H}_0 = \overline{W}^{(T)} + \sum_{\varkappa} \widetilde{E}_{\varkappa}^{(T)} \mathcal{N}_{\varkappa}, \tag{16c}$$

where $\widetilde{E}_{k\uparrow}^{(T)} = \widetilde{E}_{k\downarrow}^{(T)} = \widetilde{E}_k^{(T)}$, and $\overline{W}^{(T)}$ is defined by

$$W_0^{(T)} = \overline{W}^{(T)} + \sum_{\varkappa} \widetilde{E}_{\varkappa}^{(T)} f_{\varkappa}, \tag{16d}$$

so that the additive Hamiltonian $\mathcal{H}_0$ has the same average value in U_0 as H_0 or H.

In the limit of $T=0$, that is $\beta \to \infty$, (15b) gives $w_k = 0$, and the statistical operator (12a) reduces to the projection operator P_0 of the state Φ_0 given by (2a). The consideration of the mixture of states (12a) includes, therefore, this special case.

The operator (12a) can be considered as the trial approximation of the theory for the statistical operator of the grand ensemble

$$U = \exp[\beta \Lambda] \exp[-\beta(H-\mu N)], \tag{17a}$$

$$\exp[-\beta \Lambda] = \operatorname{tr} \exp[-\beta(H-\mu N)]. \tag{17b}$$

A theorem due to PEIERLS [5], which as shown by SCHULTZ can be extended to include the case of the grand ensemble, shows that, with the diagonal operators H_0 and $N_0 = \sum_{\varkappa} (n_{\varkappa})_0$, the grand potential $\bar{\Lambda}_0$ defined by

$$\exp[-\beta\bar{\Lambda}_0] = \operatorname{tr}\exp[-\beta(H_0 - \mu N_0)]\,, \tag{17c}$$

is an upper bound for the grand potential Λ,

$$\exp[-\beta\Lambda] \geqslant \exp[-\beta\bar{\Lambda}_0]\,. \tag{17d}$$

The effect of the chemical potential μ in (17a) is to replace ε_k by $\varepsilon_k - \mu$ in the Hamiltonian, as can be seen from the expressions $T \simeq \sum_{\varkappa} \varepsilon_{\varkappa} n_{\varkappa}$ and $\mu N = \sum_{\varkappa} \mu n_{\varkappa}$. Assuming a density of states for the electrons which is an even function of ε_k about the Fermi level, and a constant average number of electrons, one obtains that μ is equal to the Fermi energy. The counting of ε_k from the Fermi level means, therefore, $\mu = 0$. In this case, the grand potential Λ reduces to the free energy. For any other choice of the zero of the energy scale, however, the role of the chemical potential μ becomes essential.

The grand potential $\bar{\Lambda}_0$ defined by (17c) still differs from the expression Λ_0 given by (16b) which with (15a) and (14a) is equal to (14b). By an argument due to SCHULTZ, one can, however, show that Λ_0 is an upper bound for $\bar{\Lambda}_0$ and consequently for Λ,

$$\Lambda_0 \geqslant \Lambda\,. \tag{17e}$$

As the expression of the entropy given by (14a) does not depend explicitly on g_k, the minimization of the free energy (14b) with respect to g_k, for fixed f_k, is equivalent to the minimization of $W_0^{(T)}$. This leads to an equation analogous to (9a) and to the relationships

$$h_k = \frac{1}{2}\left(1 - \frac{\nu_k^{(T)}}{E_k^{(T)}}\right), \tag{18a}$$

$$\chi_k = \frac{\mu_k^{(T)}}{2E_k^{(T)}}\,. \tag{18b}$$

with

$$E_k^{(T)} = (\nu_k^{(T)2} + |\mu_k^{(T)}|^2)^{\frac{1}{2}}\,. \tag{18c}$$

---

[5] R. E. PEIERLS: *Phys. Rev.*, **54**, 918 (1938)

With (18*a*, *b*) one obtains from (15*c*)

$$\tilde{E}_k^{(T)} = E_k^{(T)} . \tag{18d}$$

From (18*a*, *b*, *c*) and (15*d*, *e*) one can deduce equations analogous to (11*a*,*c*). For constant $V_{kk'}$, with $|\varepsilon_k| \leqslant \hbar\omega$, these equations are still of the factorizable type and can be solved exactly. One obtains the temperature dependent energy gap $|\mu_k^{(T)}| \sim \varepsilon_0$ which vanishes at the critical temperature T_c. The free energy calculated as a function of temperature gives the specific heat curve of BARDEEN, COOPER and SCHRIEFFER which explains correctly a great number of experimental facts.

For a system with a large number of particles, the statistical operator U_0 corresponds to a probability distribution for the number N of particles with a sharp peak. This can be seen by comparing the average values $\langle N^2 \rangle$ and $\langle N \rangle^2$ in U_0. A simple calculation gives

$$\langle N^2 \rangle - \langle N \rangle^2 = 2 \sum_\varkappa h_\varkappa^{(T)}(1 - h_\varkappa^{(T)}) - \sum_\varkappa f_\varkappa(1 - f_\varkappa) . \tag{19a}$$

Since one has $0 \leqslant h_\varkappa^{(T)}(1 - h_\varkappa^{(T)}) \leqslant \frac{1}{4}$ and $0 \leqslant f_\varkappa(1 - f_\varkappa) \leqslant \frac{1}{4}$, and both expressions vanish outside an energy region $|\varepsilon_k| \leqslant \hbar\omega$, (19*a*) gives

$$\langle N^2 \rangle - \langle N \rangle^2 < \sum_\varkappa \tfrac{1}{2} \simeq 2N(0)\hbar\omega, \tag{19b}$$

where $N(0)$ is the density of states near the Fermi level. With $2N(0)\hbar\omega \sim \sim 10^{-3}\langle N \rangle$, this leads to an estimate

$$\frac{\langle N^2 \rangle - \langle N \rangle^2}{\langle N \rangle^2} < 10^{-3} \frac{1}{\langle N \rangle} . \tag{19c}$$

5. – Long range correlations.

Some of the previous expressions obtain a simple physical interpretation by considering the correlation function of the particles, which is the most important quantity from the point of view of collective behaviour. The two-particle correlations are given by the expectation values of the operator

$$\varrho_{\sigma\sigma'}(x, x') = \Psi_\sigma^*(x)\Psi_{\sigma'}^*(x')\Psi_{\sigma'}(x')\Psi_\sigma(x) . \tag{20a}$$

Expressing the quantized field operators $\Psi_\sigma^*(x)$, $\Psi_{\sigma'}(x')$ by means of creation an annihilation operators of plane wave states $\psi_k(x) = \Omega^{-\frac{1}{2}} \exp[-ikx]$ in the form

$$\Psi_\sigma^*(x) = \sum_k \psi_k^*(x) a_{k\sigma} , \qquad \Psi_{\sigma'}(x') = \sum_{k'} \psi_{k'}(x') a_{k'\sigma'}^* , \tag{20b}$$

the expectation value of $\varrho_{\sigma\sigma'}(x, x')$ in the ground state Φ_0 can be obtained with the help of the relations (6*a*, *c*), (7*a*, *b*, *c*). For the correlation functions of particles with parallel and anti-parallel spin

$$\varrho_p = \varrho_{\uparrow\uparrow} + \varrho_{\downarrow\downarrow}\,, \qquad \varrho_a = \varrho_{\uparrow\downarrow} + \varrho_{\downarrow\uparrow}\,, \tag{20c}$$

a simple calculation gives

$$\bar{\varrho}_p(r) = \tfrac{1}{2}\varrho_0^2 - 2h^*(r)\,h(r)\,, \tag{21a}$$

$$\bar{\varrho}_a(r) = \tfrac{1}{2}\varrho_0^2 + 2\chi^*(r)\,\chi(r)\,, \tag{21b}$$

where $r = x - x'$, $\varrho_0 = \overline{N}/\Omega$ is the average density, $h(r)$ is the Fourier transform of the number distribution h_k given by (6*c*, *d*, *e*) and $\chi(r)$ is the Fourier transform of the quantity χ_k defined by (7*c*). This throws new light on the role of the quantities h_k, χ_k in the equations. The last two interaction terms in the expression (8*d*) of W_0 for instance correspond to the contributions from parallel and anti-parallel spin correlations.

At temperature T, the average values of ϱ_p and ϱ_a in the mixture of states (12*a*), (16*a*) are in an analogous way

$$\langle\varrho_p(r)\rangle = \tfrac{1}{2}\varrho_0^2 - 2h^{(T)*}(r)\,h^{(T)}(r)\,, \tag{22a}$$

$$\langle\varrho_a(r)\rangle = \tfrac{1}{2}\varrho_0^2 + 2\chi^{(T)*}(r)\,\chi^{(T)}(r)\,, \tag{22b}$$

where $h^{(T)}(r)$ and $\chi^{(T)}(r)$ are the Fourier transforms of the quantities $h_k^{(T)}$, $\chi_k^{(T)}$ defined by (13*a*), (13*b*).

The long range correlations are between particles with antiparallel spin and are related to the function χ_k. For $V_{kk'}$ = constant, with $|\varepsilon_k| < \hbar\omega$, they have been calculated by BARDEEN, COOPER and SCHRIEFFER. In the same model, (22*b*) gives the temperature dependence of the long range correlations, the amplitude of which tends to zero with the square of the energy gap near the critical temperature. The integrations carried out explicitly for small values of $\varepsilon_0 \sim \mu_k^{(T)}$, that is near the critical temperature, show that the form of the correlation function is practically the same as at $T = 0$, with a correlation length of the order of 10^{-4} cm.

* * *

This paper is the outcome of numerous discussions with members of the Department of Mathematical Physics at the University of Birmingham, and thanks are due to many of them. It is a pleasure to express my thanks especially to Dr. J. R. SCHRIEFFER and Dr. T. D. SCHULTZ, for stimulating discussions, for suggestions and contributions. Thanks are due to D. BUTLER for helpful co-operation.

SOVIET PHYSICS JETP VOLUME 34(7), NUMBER 3 SEPTEMBER, 1958

ON THE ENERGY SPECTRUM OF SUPERCONDUCTORS

L. P. GOR'KOV

Institute for Physical Problems, Academy of Sciences, U.S.S.R.

Submitted to JETP editor November 18, 1957

J. Exptl. Theoret. Phys. (U.S.S.R.) **34**, 735-739 (March, 1958)

A method is proposed, based on the mathematical apparatus of quantum field theory, for the calculation of the properties of a system of Fermi particles with attractive interaction.

It was shown in the work of Cooper[1] that if the interaction of electrons in a metal leads to an effective mutual attraction for two electrons close to the Fermi surface, then the pair of particles which possess mutually opposite momenta and spins can have bound states with negative coupling energies. In the works of Bardeen, Cooper and Schreiffer[2,3] and of Bogoliubov[4] a systematic theory of superconductivity has been erected on this principle. It was shown that the ground state of a system of interacting Fermi particles is located below the normal state with a filled Fermi sphere and, in consequence, is separated from the excited states by a gap in order of magnitude equal to the energy of coupling of the individual pair.

In the present work, a method is proposed, based on the physical idea of Cooper, which permits us, with the help of the apparatus of quantum field theory, to obtain all the results by a short and simple method.

We shall start out from a Hamiltonian in the form[2] which is written in the case of second quantization:

$$\hat{H} = \int\left\{-\left(\psi^+ \frac{\Delta}{2m}\psi\right) + \frac{g}{2}\left(\psi^+(\psi^+\psi)\psi\right)\right\} d^3x, \tag{1}$$

where

$$\psi_\alpha(\mathbf{x}) = V^{-1/2}\sum_{k\sigma} a_{k\sigma} s_{\sigma\alpha} e^{i\mathbf{kx}}; \quad \psi^+_\beta(\mathbf{x}') = V^{-1/2}\sum_{k\sigma} a^+_{k\sigma} s^*_{\beta\sigma} e^{-i\mathbf{kx}'}$$

satisfy the usual commutation relations:

$$\{\psi_\alpha(\mathbf{x}),\ \psi^+_\beta(\mathbf{x}')\} = \delta_{\alpha\beta}\,\delta(\mathbf{x}-\mathbf{x}'),$$
$$\{\psi_\alpha(\mathbf{x}),\ \psi_\beta(\mathbf{x}')\} = \{\psi^+_\alpha(\mathbf{x}),\ \psi^+_\beta(\mathbf{x}')\} = 0. \tag{2}$$

We shall consider the interaction to be equal to zero everywhere except in a region of energy of the particles 2κ around the Fermi surface, from $\epsilon_F - \kappa$ to $\epsilon_F + \kappa$.

We transform to the Heisenberg representation, in which the operators ψ and ψ^+ depend on the time and satisfy the following equations:

$$\{i\partial/\partial t + \Delta/2m\}\,\psi(x) - g\,(\psi^+(x)\,\psi(x))\,\psi(x) = 0,$$
$$\{i\partial/\partial t - \Delta/2m\}\,\psi^+(x) + g\psi^+(x)\,(\psi^+(x)\,\psi(x)) = 0. \tag{3}$$

We determine the Green's function $G_{\alpha\beta}(x-x')$ as an average over the ground state of the system:

$$G_{\alpha\beta}(x-x') = -i\,\langle T\,(\psi_\alpha(x),\ \psi^+_\beta(x'))\rangle, \tag{4}$$

where T is time-ordering operator.

For the derivation of the equation for the function $G(x-x')$ we take it into consideration that the ground state of the system differs from the usual state with a filled Fermi sphere by the presence of bound pairs of electrons. In the ground state, all the pairs are at rest as a whole. (This means that the interaction between particles is considered only insofar as it enters into the formation of the bound pairs. We neglect scattering effects.) A sort of "Bose condensation" of pairs takes place in the case in which the momentum of their motion as a whole is equal to zero, just as in a Bose gas such a condensation takes place by virtue of the statistics for the particles themselves. This circumstance permits us to write down in a definite way the mean form $\langle T(\psi(x_1)\,\psi(x_2)\times \psi^+(x_3)\,\psi^+(x_4))\rangle$, which appears in the equations for $G(x-x')$ by virtue of (3).

For example, we have

$$\begin{aligned}\langle T\,(\psi_\alpha(x_1)\,\psi_\beta(x_2)\,\psi^+_\gamma(x_3)\,\psi^+_\delta(x_4))\rangle = \\ -\langle T\,(\psi_\alpha(x_1)\,\psi^+_\gamma(x_3))\rangle\,\langle T\,(\psi_\beta(x_2)\,\psi^+_\delta(x_4))\rangle \\ +\langle T\,(\psi_\alpha(x_1))\,\psi^+_\delta(x_4))\rangle\,\langle T\,(\psi_\beta(x_2)\,\psi^+_\gamma(x_3))\rangle \\ +\langle N\,|\,T\,(\psi_\alpha(x_1)\,\psi_\beta(x_2))\,|\,N \\ +2\rangle\,\langle N+2\,|\,T\,(\psi^+_\gamma(x_3)\,\psi^+_\delta(x_4))\,|\,N\rangle,\end{aligned} \tag{5}$$

where $|N\rangle$ and $|N+2\rangle$ are the ground states of the system with numbers of particles N and N + 2. The quantity

$$\langle N\,|\,T\,(\psi\psi)\,|\,N+2\rangle\,\langle N+2\,|\,T\,(\psi^+\psi^+)\,|\,N\rangle, \tag{5a}$$

evidently has the order of the density of the number of pairs, while $\langle T(\psi\psi^+)\rangle$ is the particle number density.

It is easy to show that the quantities thus introduced can be written in the form

$$\langle N | T(\psi_\alpha(x)\psi_\beta(x')) | N+2\rangle = e^{-2i\mu t} F_{\alpha\beta}(x-x'),$$

$$\langle N+2 | T(\psi_\alpha^+(x)\psi_\beta^+(x')) | N\rangle = e^{2i\mu t} F_{\alpha\beta}^+(x-x'). \qquad (6)$$

The function $G(x-x')$ depends only on the difference $x-x'$, because of the homogeneity of the problem. So far as the additional dependence on t in Eq. (6) is concerned, its origin is seen from the general quantum mechanical formula for the time derivative of an arbitrary operator $\hat{A}(t)$:

$$\frac{\partial}{\partial t}\langle N | \hat{A}(t) | N+2\rangle = i(E_N - E_{N+2})\langle N | \hat{A}(t) | N+2\rangle.$$

The value of the energy difference $E_{N+2} - E_N$ is obviously equal to $2\mu\,(\partial E/\partial N = \mu)$.

Making use of Eq. (3), we obtain equations for the functions $\hat{G}(x-x')$ and $\hat{F}(x-x')$:

$$\{i\partial/\partial t + \Delta/2m\}\hat{G}(x-x')$$
$$- ig\hat{F}(0+)\hat{F}^+(x-x') = \delta(x-x'),$$
$$\{i\partial/\partial t - \Delta/2m - 2\mu\}\hat{F}^+(x-x') \qquad (7)$$
$$+ ig\hat{F}^+(0+)\hat{G}(x-x') = 0.$$

Here terms are omitted which correspond to the first two terms in Eq. (5), inasmuch as they only change μ, which one can neglect, and the notations

$$F_{\alpha\beta}^+(0+) = e^{-2i\mu t}\langle\psi_\alpha^+(x)\psi_\beta^+(x)\rangle \equiv \lim_{x\to x'(t>t')} F_{\alpha\beta}^+(x-x')$$

and introduced, and correspondingly,

$$F_{\alpha\beta}(0+) = e^{2i\mu t}\langle\psi_\alpha(x)\psi_\beta(x)\rangle.$$

The complex conjugate yields

$$(F_{\alpha\beta}^+(0+))^* = -F_{\alpha\beta}(0+). \qquad (8)$$

We transform in Eqs. (7) to the Fourier components of all functions, for example,

$$G_{\alpha\beta}(x-x') = (2\pi)^{-4}\int G_{\alpha\beta}(p\omega)\exp\{i\mathbf{p}(\mathbf{x}-\mathbf{x}')$$
$$- i\omega(t-t')\}\,d\omega d^3p.$$

Denoting $\omega - \mu = \omega'$, we find

$$(\omega' - \xi_p)\hat{G}(p\omega) - ig\hat{F}(0+)\hat{F}^+(p\omega) = 1,$$
$$(\omega' + \xi_p)\hat{F}^+(p\omega) + ig\hat{F}^+(0+)\hat{G}(p\omega) = 0, \qquad (9)$$

where

$$\xi_p = p^2/2m - \mu \approx v_F(p - p_F),$$

and p_F is the Fermi momentum. In what follows, we shall only have ω' in the formulas, hence we shall omit the prime.

It follows from Eq. (8) that $\hat{F}(0+)$ and $\hat{F}^+(0+)$ have the following matrix form:

$$\hat{F}^+(0+) = J\begin{pmatrix} 0 & 1 \\ -1 & 0\end{pmatrix} \equiv J\hat{I};\ \hat{F}(0+) = -J\hat{I},$$
$$\hat{I}^2 = -\hat{E}; \qquad (10)$$

equal both to $\hat{F}^+(p\omega)$ and $\hat{F}(p\omega)$, and the Green's function, in accord with (9), is proportional to the unit matrix. Substituting (10) in (9), we obtain

$$(\omega^2 - \xi_p^2 - g^2J^2)F^+(p\omega) = -igJ,$$
$$(\omega - \xi_p)G(p\omega) = 1 + igJF^+(p\omega). \qquad (11)$$

Hence

$$F^+(p\omega) = -ig\frac{J}{\omega^2 - \xi_p^2 - \Delta^2};$$

$$G(p\omega) = \frac{\omega + \xi_p}{\omega^2 - \xi_p^2 - \Delta^2}, \qquad (12)$$

where

$$\Delta^2 = g^2J^2.$$

As is evident from the first equation of (11), $F^+(p\omega)$ is determined with accuracy up to a solution of the homogeneous equation of the form $A(p)\delta(\omega^2 - \xi_p^2 - \Delta^2)$. It is not difficult to verify the fact that this term reduces in the expression for $G(p\omega)$ to an arbitrary imaginary part. In other words, Eqs. (11) determine only the real part of the Green's function.

We determine the rule for circling the poles in (12), making use of a theorem given by Landau,[5] according to which the imaginary part of the Green's function of a Fermi system is positive for $\omega < 0$ and changes sign in the transition from negative to positive frequencies.

As a result, we obtain

$$F^+(p\omega) = -igJ/(\omega - \varepsilon_p + i\delta)(\omega + \varepsilon_p - i\delta), \qquad (13)$$

$$G(p\omega) = u_p^2(\omega - \varepsilon_p + i\delta)^{-1} + v_p^2(\omega + \varepsilon_p - i\delta)^{-1}, \qquad (14)$$

where $\epsilon_p = \sqrt{\xi_p^2 + \Delta^2}$ and the functions μ_p^2 and v_p^2 are equal to

$$u_p^2 = \frac{1}{2}\left(1 + \frac{\xi_p}{\varepsilon_p}\right); \quad v_p^2 = \frac{1}{2}\left(1 - \frac{\xi_p}{\varepsilon_p}\right) \qquad (14')$$

For the determination of the quantity Δ, we make use of the fact that

$$J = (2\pi)^{-4}\int F^+(p\omega)\,d\omega\,d^3k. \qquad (15)$$

Substituting (13), we obtain the equation

$$1 = -\frac{g}{2(2\pi)^3}\int\frac{d^3k}{\sqrt{\xi_k^2 + \Delta^2}}\ (|\xi| < \varkappa). \qquad (16)$$

For small $g < 0$ (attraction), this equation has a solution of the form

$$\Delta = 2\varkappa e^{-1/2\rho},$$

where

$$\rho = p_F |g| m / 2\pi^2.$$

The positive pole in (14) determines the excitation spectrum which, as is shown, has a gap of magnitude Δ. These results coincide with those obtained in Refs. 2 – 4.

The chemical potential μ is connected with the particle number density by the relation

$$N/V = \langle \psi^+(x)\psi(x)\rangle = -i(2\pi)^{-4}\int G_{\alpha\alpha}(p\omega)e^{i\omega\delta}d\omega d^3p, \quad (17)$$

with accuracy up to small exponential terms: $\mu = \epsilon_F$.

The method laid out also permits us to make use of it for temperatures differing from absolute zero. In this case we consider the mean Green's function (thermodynamically averaged)

$$G_{\alpha\beta}(x - x') = -i\sum_n \exp\left\{\frac{\Omega + \mu N - E_n}{T}\right\}\langle n | T(\psi_\alpha(x)\psi_\beta^+(x')| n\rangle,$$

where Ω is the thermodynamic potential in the variables T, V, μ. As is well known, the result of averaging does not depend on whether it is carried out with the use of a Gibbs distribution or over the stationary state with a given energy. This corresponds to the choice of the quantity $\overline{E}$ as a thermodynamic variable, in place of the temperature T. Taking the averaging in such a fashion we get the earlier equations (11) for the quantities G, F^+ and F, with this difference, that the corresponding averaging of T products are taken not over the ground state of the system but over a state with total energy E equal to the energy of the system at a given temperature. Equations (11), as we have already noted above, determine uniquely only the real part of the Green's function $G(p\omega)$, which is evidently equal to the real part of Eq. (14). We write down the general solution for the function $F^+(p\omega)$:

$$F^+(p\omega) = -igJ/(\omega - \varepsilon_p + i\delta)(\omega + \varepsilon_p - i\delta) + A_1(p, T)\delta(\omega - \varepsilon_p) + A_2(p, T)\delta(\omega + \varepsilon_p). \quad (18)$$

We have made use of the fact that $1/(x \mp i\delta) = 1/x \pm \pi i\delta(x)$. The value of the quantities $A_1(p,T)$ and $A_2(p,T)$ can be obtained from the relations between the real and imaginary parts of the Green's function,[5] which has the following form at temperatures different from zero:

$$\operatorname{Re} G(\omega) = -\frac{1}{\pi}\int_{-\infty}^{+\infty}\coth\frac{x}{2T}\,\frac{\operatorname{Im} G(x)}{\omega - x}\,dx.$$

We obtain

$$A_1(p, T) = A_2(p, T) = -(\pi\Delta/\varepsilon_p)\,n(\varepsilon_p)$$

and for the Green's function,

$$G(p\omega) = u_p^2(\omega - \varepsilon_p + i\delta)^{-1} + v_p^2(\omega + \varepsilon_p - i\delta)^{-1} + 2\pi i n(\varepsilon_p)[u_p^2\delta(\omega - \varepsilon_p) - v_p^2\delta(\omega + \varepsilon_p)], \quad (19)$$

where $n(\epsilon_p)$ has the form of the Fermi distribution of excitations at the given temperature:

$$n(\varepsilon_p) = [\exp(\varepsilon_p/T) + 1]^{-1}.$$

The excitation spectrum is then

$$\varepsilon_p = \sqrt{\xi_p^2 + \Delta^2}, \quad (20)$$

where Δ is a function of the temperature. Condition (15), upon substitution of $F^+(p\omega)$ in the form (18) in it, gives a relation which determines the magnitude of the gap in its temperature dependence:

$$1 = \frac{|g|}{2(2\pi)^3}\int\frac{d^3k\,(1 - 2n(\varepsilon_k))}{\sqrt{\xi_k^2 + \Delta^2(T)}}, \quad (|\xi| < \varkappa). \quad (21)$$

Equation (21) was obtained by Bardeen, Cooper and Schrieffer[3]; it was found that the magnitude of the gap $\Delta(T)$ vanishes at $T = T_c \sim \Delta(0)$. We shall show briefly how calculation of thermodynamical quantities is carried out by our method.

The heat capacity per unit volume is equal to

$$Vc_V = (\partial\overline{E}/\partial T)_V.$$

The quantity

$$\overline{E}/V = \left\langle\left\{-\left(\psi^+\frac{\Delta}{2m}\psi\right) + \frac{g}{2}(\psi^+(\psi^+\psi)\psi)\right\}\right\rangle$$

is expressed by the function of G, F^+ and F with accuracy up to inconsequential (constant in temperature) terms in the following fashion:

$$\overline{E} = 2V(2\pi)^{-3}\int\xi_p[v_p^2(1 - n(\varepsilon_p)) + u_p^2 n(\varepsilon_p)]\,d^3p + gVJ^2.$$

The total contribution to the heat capacity at such temperatures gives the interval $|\xi_p| \ll \kappa$. Substituting here for the functions u_p^2 and v_p^2 their expressions (14′), and making use of Eq. (21), we get for the heat capacity

$$c_V = 2(2\pi)^{-3}\int\varepsilon_p\frac{\partial n(\varepsilon_p)}{\partial T}\,d^3p,$$

i.e., the ordinary formula for heat capacity of a gas of Fermi excitations with a spectrum (20). Computation of the heat capacity and the value of the gap in its temperature dependence was given in Ref. 3.

The author expresses his gratitude to Academician L. D. Landau for his valued advice and interest in the work. The author also acknowledges A. A. Abrikosov, I. M. Khalatnikov, I. E. Dzialoshinskii and L. P. Pitaev for discussion of the results of this work.

[1] L. Cooper, Phys. Rev. **104**, 1189 (1956).

[2] Bardeen, Cooper and Schrieffer, Phys. Rev. **106**, 162 (1957).

[3] Bardeen, Cooper and Schrieffer, Phys. Rev. **108**, 5 (1957).

[4] N. N. Bogoliubov, J. Exptl. Theoret. Phys. (U.S.S.R.) **34**, 58 (1958); Soviet Phys. JETP **7**, 41 (1958).

[5] L. D. Landau, J. Exptl. Theoret. Phys. (U.S.S.R.) **34**, 262 (1958); Soviet Phys. JETP **7**, 182 (1958).

Translated by R. T. Beyer
138

Random-Phase Approximation in the Theory of Superconductivity*

P. W. Anderson
Bell Telephone Laboratories, Murray Hill, New Jersey
(Received July 28, 1958)

A generalization of the random-phase approximation of the theory of Coulomb correlation energy is applied to the theory of superconductivity. With no further approximations it is shown that most of the elementary excitations have the Bardeen-Cooper-Schrieffer energy gap spectrum, but that there are collective excitations also. The most important of these are the longitudinal waves which have a velocity $v_F\{\frac{1}{3}[1-4N(0)|V|]\}^{\frac{1}{2}}$ in the neutral Fermi gas, and are essentially unperturbed plasma oscillations in the charged case. Other collective excitations resembling higher bound pair states may or may not exist but do not seriously affect the energy gap. The theory obeys the sum rules and is gauge invariant to an adequate degree throughout.

I. INTRODUCTION

RECENTLY Bardeen, Cooper, and Schrieffer proposed a theory of superconductivity[1] which has been successful in explaining experimental results of many kinds. The theory is founded on the idea that in superconductors there is a net attraction between electrons caused by the phonons. The ground-state wave function used is a product function designed to have the maximum number of pairs of electrons of zero total momentum taking advantage of this attraction. This ground-state function is

$$\Psi_g=\prod_{\mathbf{k}}[(1-h_k)^{\frac{1}{2}}+h_k^{\frac{1}{2}}c_{\mathbf{k}\uparrow}^*c_{-\mathbf{k}\downarrow}^*]\Psi_v, \tag{1}$$

where Ψ_v is the vacuum, and $c_{\mathbf{k}\sigma}^*$ is the creation operator for electrons of momentum $\mathbf{k}$ and spin σ. h_k is a number determined so as to minimize the energy. The approximations most necessary to the theory are the use of screened Coulomb and screened second-order phonon interactions according to the scheme of Bardeen and Pines[2]; and the neglect of all interactions except those between pairs with zero total momentum and spin.

Bogoliubov[3] arrived at practically the same result by an apparently different method, using from B.C.S. only the zero-momentum pairing idea. He formed pairs by introducing a new set of fermions, composed partly of an electron with ($\mathbf{k}$, spin up) and partly of a hole with ($-\mathbf{k}$, spin down):

$$\begin{aligned}\alpha_{\mathbf{k}0}&=u_kc_{\mathbf{k}\uparrow}-v_kc_{-\mathbf{k}\downarrow}^*,\\ \alpha_{\mathbf{k}1}&=u_kc_{-\mathbf{k}\downarrow}+v_kc_{\mathbf{k}\uparrow}^*,\end{aligned} \tag{2}$$

and defined the ground state as the "vacuum" with

* The final stages of this work, and all of the manuscript preparation, were done at the University of California, Berkeley, California, during a very pleasant stay made possible in part by a grant from the National Carbon Corporation.

[1] Bardeen, Cooper, and Schrieffer, Phys. Rev. **108**, 1175 (1957). We abbreviate this reference B.C.S. hereafter.

[2] J. Bardeen and D. Pines, Phys. Rev. **99**, 1140 (1955).

[3] N. N. Bogoliubov, J. Exptl. Theoret. Phys. U.S.S.R. **34**, 65 (1958)[translation: Soviet Phys. JETP **34**(7), 41 (1958)]; J. G. Valatin, Nuovo cimento **7**, 843 (1958).

respect to the new set of "particles":

$$\alpha_{\mathbf{k}0}{}^*\alpha_{\mathbf{k}0}\Psi_g=\alpha_{\mathbf{k}1}{}^*\alpha_{\mathbf{k}1}\Psi_g=0. \quad (3)$$

By this trick he was able to eliminate certain terms in perturbation theory which diverged, essentially because of the presence of the attractive interaction which is responsible for the binding of pairs.

It is easy to show that the B.C.S. state (1) and the Bogoliubov state (3) are identical if

$$v_k=h_k^{\frac{1}{2}}, \quad u_k=(1-h_k)^{\frac{1}{2}}. \quad (4)$$

Less obviously, the basic assumptions of the two theories are actually very close, as are the "energy gap" spectra of elementary excitations.[4] Bogoliubov treats the phonons only in second order, and his limitations to the lowest-order energy and to the "most divergent terms" (meaning those which cause binding of Cooper pairs[5]) are equivalent to the B.C.S. reduced-Hamiltonian assumption.

It was demonstrated[6] that the most serious question in regard to these theories is that the sum rules and gauge invariance are not obeyed, so that a consistent explanation of the Meissner effect requires, at the very least, that the whole interaction Hamiltonian be taken into account. If this is done and the long-range Coulomb forces are neglected, a new set of states of nonzero momentum must be present in the energy gap, which not only would have experimental effects but might well lead to large perturbations in the ground state. When long-range Coulomb forces are included, the gap again becomes empty, and one is led to the conclusion that the success of the zero-momentum pair theories may only be a consequence of plasma effects.

Both for this reason, and because it seems optimistic to assume that the collective and screening effects (which are vital even in determining the phonon spectrum) will be necessarily unaffected by the radical changes in the Fermi sea embodied in (1) or (3) (and vice versa), it is desirable to have a theory of the ground state of a superconductor which can simultaneously handle these collective effects in the best available approximation, that of Gell-Mann, Brueckner, Sawada, and Brout,[7,8] and yet lead to (1) or (3) and the Bardeen energy-gap excitation spectrum. Such a theory is the subject of this paper.

This theory also has a few by-products which commend it as an alternative to the earlier ones. First, it shows in a natural way why the restriction to a fixed number of electrons must be relaxed and how to handle the projection back onto $N=\text{const}$; second, it is capable of computing the correlation corrections to superconductivity—or vice versa—and showing that they are small; third, it gives a good account of other collective effects such as phonons and higher bound pair states; and finally, it seems to give a simpler and more physical picture of the nature of the superconducting state. The method we use may also have more general interest as an approach to the many-body problem, in particular in reconciling collective and individual-particle behavior.

The basic, and almost the only, assumption is a generalized form of the random-phase approximation.[9] Sawada, Sawada *et al.*, and Brout[8] have shown that the R.P.A. of Bohm and Pines and the diagram-summing method of Brueckner and Gell-Mann[7] both lead to a certain set of linear eigenvalue equations or "equations of motion" for the elementary excitations in terms of the quantities

$$\rho_{\mathbf{k},\sigma}{}^{\mathbf{Q}}=c_{\mathbf{k}+\mathbf{Q},\sigma}{}^*c_{\mathbf{k},\sigma}. \quad (5)$$

Sawada and Brout[8] showed that the Gell-Mann and Brueckner "high-density" assumption, that excited particles interact with the unperturbed Fermi sea only, is equivalent to a certain effective Hamiltonian together with altered commutation relations for the ρ's of (5), which then lead to the equations of motion; they then showed how to derive the energy and other results from these equations. The same equations were arrived at—without realizing how nearly full a solution to the correlation problem they were—by Bohm and Pines,[9] first by physical reasoning and then by a method of direct linearization of the full equations of motion. The method of this paper is a natural generalization of this last way of arriving at the "equations of motion" to the case in which the unperturbed state is not the Fermi sea but the B.C.S. state (1).

The high-density ("weak coupling") limit is the only proven domain of validity of the R.P.A., and even in that limit a consistent theory of superconductivity is interesting. However, the R.P.A. gives results in the correlation problem which appear to be satisfactory even in the intermediate range,[10] and other entirely different domains.[11] A full discussion of the domains of validity of these methods is, however, beyond the scope of this paper; in any case they represent the only approach known to be effective in studying collective effects in the many-body problem.

Briefly, the Bohm-Pines technique linearizes the full equations of motion by observing that in the Fermi sea the quantities $n_{\mathbf{k}}=c_{\mathbf{k}}{}^*c_{\mathbf{k}}$ and $1-n_{\mathbf{k}}=c_{\mathbf{k}}c_{\mathbf{k}}{}^*$ may have finite "c-number" average values. These average values

[4] N. N. Bogoliubov, J. Exptl. Theoret. Phys. U.S.S.R. **34**, 73 (1958) [translation: Soviet Phys. JETP **34(7)**, 51 (1958)].

[5] L. N. Cooper, Phys. Rev. **104**, 1189 (1956).

[6] P. W. Anderson, Phys. Rev. **110**, 827 (1958).

[7] K. A. Brueckner and M. Gell-Mann, Phys. Rev. **106**, 364 (1957).

[8] K. Sawada, Phys. Rev. **106**, 372 (1957); Sawada, Brueckner, Fukuda, and Brout, Phys. Rev. **108**, 507 (1957); see also J. Hubbard, Proc. Roy. Soc. (London) **A243**, 336 (1958).

[9] D. Bohm and D. Pines, Phys. Rev. **92**, 609 (1953). We shall abbreviate "random-phase approximation" as R.P.A. hereafter.

[10] When supplemented by a treatment of exchange terms which falls rather naturally out of the more general method of the present paper. See P. Nozières and D. Pines, Phys. Rev. **109**, 1009 (1958), and Nuovo cimento (to be published); and especially Hubbard, reference 8.

[11] E. Montroll and J. C. Ward, J. Phys. Fluids **1**, 55 (1958).

are taken as zeroth order, while the $\rho_{\mathbf{k}}{}^{\mathbf{Q}}=c_{\mathbf{k}+\mathbf{Q}}{}^{*}c_{\mathbf{k}}$ act like first-order infinitesimals. In the B.C.S. state (1) not only $n_{\mathbf{k}}$ but $b_{\mathbf{k}}{}^{*}=c_{\mathbf{k}}{}^{*}c_{-\mathbf{k}}{}^{*}$ and $b_{\mathbf{k}}$ have c-number averages, so that correspondingly we must derive equations of motion for first-order quantities $c_{\mathbf{k}+\mathbf{Q}}{}^{*}c_{-\mathbf{k}}{}^{*}=\bar{b}_{\mathbf{k}}{}^{\mathbf{Q}}$, etc. The unperturbed B.C.S. state is itself determined, in our method, by finding equations of motion for the $b_{\mathbf{k}}{}^{*}$'s and $n_{\mathbf{k}}$'s themselves and taking their stable solution; this is then equivalent to both the B.C.S. and Bogoliubov definitions. Physically it means that we construct the zeroth-order state so as to be stable against the formation of any more zero-momentum bound electron pairs.

The most direct results of the theory are the solutions of the equations of motion. These are the elementary excitations, and they fall into two groups: individual-particle-like excitations, the spectrum of which is practically the same as the B.C.S. energy-gap spectrum, but which include many of the effects of scattering; and collective solutions.

The collective solutions are calculated for two cases. The usual models implicitly ignore the long-range Coulomb forces; we may call this the "neutral" case. In this case there is, strictly speaking, no gap: we find a collective excitation of longitudinal type, which has resemblances to a longitudinal wave as well as to a bound pair of electrons of nonzero momentum, with a velocity

$$v=v_F\{\tfrac{1}{3}[1-4N(0)|V|]\}^{\frac{1}{2}}.$$

The velocity $3^{-\frac{1}{2}}v_F$ was obtained by Bogoliubov[12] and is a kinematical effect; the $N(0)|V|$ term represents the effect of interaction in this weak-coupling limit. Other collective excitations, describable physically as other bound-pair excitations orthogonal to the Cooper bound pairs, may or may not exist depending on the form of the interaction, but lie near or above the top of the gap in any case.

In the physical case of the charged Fermi gas, the longitudinal excitations have a spectrum identical with the plasmons of the normal Fermi gas, and thus the gap really exists; we see that the gap in a strict sense is enforced only by the long-range Coulomb effects. Fundamentally, also, in this case the R.P.A., with its automatic separation of the equations of excitations of different total momentum, may be expected to be more accurate, because these long-range forces single out momentum zero as having special properties. In an appendix we go on to calculate the effect of superconductivity on the long-wave phonons, and show that their spectrum is changed only to the order of the ratio of electron to ion mass.

[12] Bogoliubov, Tolmachev, and Shirkov, ***New Method in the Theory of Superconductivity***, (Academy of Sciences of the U.S.S.R., Moscow, 1958). See also V. M. Galitskii, J. Exptl. Theoret. Phys. U.S.S.R. **34**, 1011 (1958) [translation: Soviet Phys. JETP **34** (7), 698 (1958)].

II. DISCUSSION OF THE INTERACTION HAMILTONIAN

If the interaction between electrons is given, the only approximation of the method is the R.P.A. The interaction responsible for superconductivity is, however, the rather complicated interaction through the lattice phonons, and fundamentally the calculation of this interaction is outside the scope of the R.P.A. In this paper we shall assume the essential correctness of the results of Bardeen and Pines for this interaction.[13]

In a later paper this assumption will be justified by means of the renormalization methods of Hubbard.[14] Although this justification is not a part of the present paper, in view of its importance a brief qualitative description will be given here. Hubbard shows, in the free electron gas problem, that the effect of certain apparently divergent terms of perturbation theory is to replace each Coulomb interaction between electrons by a new, "effective" interaction, which may be thought of as having been modified and screened by the frequency- and wave-number-dependent dielectric constant of the electron gas.[15]

The effect of superconductivity, like that of the Coulomb correlations, can be expected to be merely a "smearing" of the properties of the surface of the Fermi sea, while (as the present paper will show) there is no serious effect on the collective modes, which primarily determine the dielectric constant in the long-wave region where the smearing might be important. Thus in interactions of secondary importance to the R.P.A., like exchange and phonon exchange, it is a good approximation to screen using a dielectric constant computed according to the unperturbed R.P.A. (including of course the phonon contribution[16]). The direct Coulomb interactions, which have the major effect on the collective modes and thus on the dielectric constant, must on the other hand be left in explicitly in deriving the excitation modes.

The procedure is thus a kind of successive approximation method, checked by demonstrating its self-consistency. In stage (1), we imagine that we have calculated the dielectric constant of the free electron gas with phonons, using only the direct interactions in the R.P.A. In stage (2), we recalculate the properties of the free electron gas including the phonon and exchange terms. These must be screened by the stage (1) dielectric constant because the corrections in stage (1) are large (in fact formally divergent). Stage (2) is the present theory of superconductivity, but also includes Hubbard's[10] method for the second-order exchange correction. In stage (3) we might recalculate the dielectric constant from stage (2) and insert it in the

[13] See Bardeen and Pines, reference 2. See also D. Pines, Phys. Rev. **109**, 280 (1958), for a probably more accurate expression.

[14] J. Hubbard, Proc. Roy. Soc. (London) **A240**, 539 (1957).

[15] See also Nozières and Pines, reference 10.

[16] See G. Wentzel, Phys. Rev. **108**, 1593 (1957) for the calculation of phonons by the method of Sawada and Brout.

screened terms; however, this correction is small and stage (3) is unnecessary. The resulting interaction is similar to that of reference 2, although it must be slightly complex to allow for real inelastic scattering; the screening is not by a cutoff as in Bardeen and Pines but smoothly decreasing, and the subsidiary condition is irrelevant to the present scheme.

We shall, then, write down as our Hamiltonian the Bardeen-Pines result. This already implies the relatively minor (at this stage) assumptions of the free-electron gas model for Bloch electrons and of the neglect of transverse phonons, as well as of the residual, real scattering interaction of electrons and phonons. It is

$$\mathcal{H}=\mathcal{H}_K+\mathcal{H}_V+\mathcal{H}_C. \tag{6}$$

Here ($\hbar=1$)

$$\mathcal{H}_K=\sum_{\mathbf{k},\sigma}\epsilon_k c_{\mathbf{k},\sigma}{}^* c_{\mathbf{k},\sigma}, \tag{7}$$

$$\mathcal{H}_V=-\tfrac{1}{2}\sum_{\mathbf{k}\neq\mathbf{k}',\mathbf{q}}\sum_{\sigma,\sigma'}\frac{\omega_{\mathbf{k}-\mathbf{k}'}M_{\mathbf{k}-\mathbf{k}'}{}^2}{(\omega_{\mathbf{k}-\mathbf{k}'})^2-(\epsilon_k-\epsilon_{k'})^2}\times c_{\mathbf{k}',\sigma'}{}^* c_{-\mathbf{k}'+\mathbf{q},\sigma}{}^* c_{-\mathbf{k}+\mathbf{q},\sigma} c_{\mathbf{k},\sigma'}. \tag{8}$$

$\mathcal{H}_C$ is a somewhat more complicated thing, because (according to the above discussion) we must understand its terms differently depending on whether we use them as direct or exchange terms. We shall write down the full Hamiltonian and simply understand that the exchange terms are to be screened:

$$\mathcal{H}_C=\sum_{\mathbf{k}\neq\mathbf{k}',\mathbf{q}}\sum_{\sigma,\sigma'}2\pi e^2(\mathbf{k}-\mathbf{k}')^{-2}\times c_{\mathbf{k}',\sigma'}{}^* c_{-\mathbf{k}'+\mathbf{q},\sigma}{}^* c_{-\mathbf{k}'+\mathbf{q},\sigma} c_{\mathbf{k},\sigma'}. \tag{9}$$

In (8), Bardeen and Pines give

$$M_k=v_k{}^i(1+6\pi ne^2/k^2\epsilon_F)^{-1}, \tag{10}$$

where $v_k{}^i$ is the true electron-phonon interaction without screening, approximately

$$v_k{}^i\cong-2\pi Ze^2ik^{-1}(n/M)^{\frac{1}{2}}. \tag{11}$$

The actual values of these constants will be of little further importance to us except in Appendix I.

In principle the description of our method implies that the direct *phonon* interactions should also be included, so that we calculate correctly both collective modes, the phonon and the plasmon. To avoid complication we do not do this in the main paper, but only in an appendix. In that case for direct interactions one must include the following two terms in the Hamiltonian:

$$\mathcal{H}_{\text{ph}}=\sum_{\mathbf{s}}\tfrac{1}{2}(p_{\mathbf{s}}p_{-\mathbf{s}}+f_s{}^2q_{\mathbf{s}}q_{-\mathbf{s}}), \tag{12}$$

and

$$\mathcal{H}_i=-\sum_{\mathbf{s}}(q_{\mathbf{s}}\rho^{-\mathbf{s}}v_s{}^i+q_{-\mathbf{s}}\rho^{\mathbf{s}}v_s{}^{i*}), \tag{13}$$

where p and q are the phonon coordinates, $\rho^{\mathbf{s}}$ is the $\mathbf{s}$ Fourier component of electron density fluctuation, and

$$f_s{}^2=4\pi ne^2Z^2M^{-1}+\omega_s{}^2,\qquad \omega_s\to 0\quad\text{as}\quad s\to 0. \tag{14}$$

(ω_s includes any interactions between the ion cores.)

One further point must be made with regard to (8). An instantaneous space interaction is a function of $\mathbf{k}-\mathbf{k}'$ only; but the phonon (and actually even the screened Coulomb) interaction is not instantaneous, and must in some sense depend on the time difference between the creation and destruction of the longitudinal photon or phonon, or in Fourier-analyzed form on a frequency variable.

Our approach is to understand the particles as being embedded in a medium with a certain frequency- and wavelength-dependent dielectric constant. The frequency which enters is the energy difference between the initial state and the intermediate state of the particle system "after" the phonon-photon is emitted. This is usually well enough approximated by $\epsilon_k-\epsilon_{k'}$ in (8). As we shall see in Sec. IV, the important thing for gauge invariance and the sum rules is that there be no q dependence in (8), which is ensured by the fact that the initial and final states always have the same energy, so that the difference from the intermediate state, whatever it be, is fixed. We shall take advantage of this fact by discarding any apparent q dependence wherever it may appear.

Thus our equations will always automatically satisfy the basic sum rule $[\mathcal{H}_V,\rho^{\mathbf{Q}}]=0$, which ensures that the usual perturbation theory will give gauge-invariant results.[6]

III. IMPROVED TREATMENT OF THE B.C.S. REDUCED HAMILTONIAN

The R.P.A. theory which we use might be thought of as a generalization of the Sawada-Brout[8] theory to the superconductor. It also follows naturally as a generalization of a certain slightly improved description of the B.C.S. theory of the "reduced" Hamiltonian. There exist in the literature elegant treatments of the Sawada-Brout theory[8,17]; since the alternative calculation of B.C.S. is new, and is of some interest in itself, we present it here in full.[18]

The first step of the B.C.S. theory is to neglect in (8) and (9) all terms involving σ and σ' parallel, as well as all terms with $q\neq 0$. This step may be justified in terms of the worst instability of the Fermi sea being caused by binding of zero-momentum pairs, and the considerations of reference 6 reinforce this point; but since we intend to relax this assumption later no extensive discussion is necessary. Then the Hamiltonian

---

[17] See Wentzel, reference 16.

[18] A similar treatment has recently appeared in N. N. Bogoliubov, J. Exptl. Theoret. Phys. U.S.S.R. 34, 73 (1958) [translation: Soviet Phys. JETP 34(7), 51 (1958)], although different in detail and interpretation.

becomes

$$\mathcal{H}_{\text{RED}}=\sum_{\mathbf{k}} \epsilon_k(n_{\mathbf{k}}+n_{-\mathbf{k}})-\sum_{\mathbf{k}\neq\mathbf{k}'} V_{\mathbf{k}\mathbf{k}'}c_{\mathbf{k}'}^*c_{-\mathbf{k}'}^*c_{-\mathbf{k}}c_{\mathbf{k}}, \quad (15)$$

where we have used the convention that an explicitly negative k has a down spin and vice versa. (This convention will be used hereafter in this paper.) $V_{\mathbf{k}\mathbf{k}'}$ is the resultant interaction obtained by subtracting the screened Coulomb exchange matrix element from the phonon exchange:

$$V_{\mathbf{k}\mathbf{k}'}=\frac{\omega_{\mathbf{k}-\mathbf{k}'}M_{\mathbf{k}-\mathbf{k}'}{}^2}{(\omega_{\mathbf{k}-\mathbf{k}'})^2-(\epsilon_k-\epsilon_{k'})^2}-\frac{4\pi e^2}{(\mathbf{k}-\mathbf{k}')^2}\times(\text{screening factor}). \quad (16)$$

Basic to the whole theory of superconductivity is the idea that, for small enough $(\epsilon_k-\epsilon_{k'})^2$, $V_{\mathbf{k}\mathbf{k}'}$ is positive on the average. Pines[19] has discussed qualitatively whether and when this may be true.

It is convenient to use as the zero of energy that of a particular Fermi sea of N^0 electrons, Fermi level $\epsilon_F{}^0$; and to sum only over k's with $\epsilon_k<2\epsilon_F{}^0$. Then (15) becomes

$$\mathcal{H}_{\text{RED}}=-\sum_{\mathbf{k}} (\epsilon_k-\epsilon_F{}^0)(1-n_{\mathbf{k}}-n_{-\mathbf{k}}) -\sum_{\mathbf{k}\neq\mathbf{k}'} V_{\mathbf{k}\mathbf{k}'}c_{\mathbf{k}'}^*c_{-\mathbf{k}'}^*c_{-\mathbf{k}}c_{\mathbf{k}}+\epsilon_F{}^0(N-N^0), \quad (17)$$

where

$$N=\sum_{\mathbf{k}} (n_{\mathbf{k}}+n_{-\mathbf{k}}). \quad (18)$$

B.C.S. pointed out that the appropriate algebra for dealing with (17) involved the operators

$$1,\quad b_{\mathbf{k}}=c_{-\mathbf{k}}c_{\mathbf{k}},\quad b_{\mathbf{k}}^*=c_{\mathbf{k}}^*c_{-\mathbf{k}}^*,\quad \text{and}\quad 1-(n_{\mathbf{k}}+n_{-\mathbf{k}}). \quad (19)$$

Their importance lies in the fact that in the subspace defined by

$$n_{\mathbf{k}}-n_{-\mathbf{k}}\equiv 0 \quad (20)$$

they are a complete set, while it is easy to show that the lowest eigenstate of (17) is in this subspace.

The properties of these operators become clear by writing them [in the subspace (20)] in the representation in which the basis functions are ($\mathbf{k}$ and $-\mathbf{k}$ empty) and ($\mathbf{k}$ and $-\mathbf{k}$ full). The operators (19) commute for different k, so only the $\mathbf{k}$, $-\mathbf{k}$ subspace need be written down:

$$1-n_{\mathbf{k}}-n_{-\mathbf{k}}=\begin{matrix} & \text{empty} & \text{full} \\ \text{empty} & 1 & 0 \\ \text{full} & 0 & -1\end{matrix},$$

$$1=\begin{pmatrix}1 & 0\\ 0 & 1\end{pmatrix},\quad b_{\mathbf{k}}=\begin{pmatrix}0 & 1\\ 0 & 0\end{pmatrix},\quad b_{\mathbf{k}}^*=\begin{pmatrix}0 & 0\\ 1 & 0\end{pmatrix}. \quad (21)$$

The further manipulation will be much clearer using the following set of Pauli spin matrices which are fully equivalent to the b's:

$$2s_{z\mathbf{k}}=1-n_{\mathbf{k}}-n_{-\mathbf{k}},$$
$$s_{x\mathbf{k}}+is_{y\mathbf{k}}=b_{\mathbf{k}}^*,\quad s_{x\mathbf{k}}-is_{y\mathbf{k}}=b_{\mathbf{k}}. \quad (22)$$

These s operators are not to be confused with real physical spin-operators; they act in an imaginary space, where z component of spin up means "empty," spin down means "full," and spin sidewise simply implies a certain phased linear combination of up and down.

In terms of the pseudospins (22), the Hamiltonian is

$$\mathcal{H}_{\text{RED}}=-2\sum_{\mathbf{k}} (\epsilon_k-\epsilon_F{}^0)s_{z\mathbf{k}}-2\epsilon_F{}^0\sum_{\mathbf{k}} s_{z\mathbf{k}} -\sum_{\mathbf{k}\neq\mathbf{k}'} V_{\mathbf{k}\mathbf{k}'}(s_{x\mathbf{k}}s_{x\mathbf{k}'}+s_{y\mathbf{k}}s_{y\mathbf{k}'}). \quad (23)$$

The second term on the right changes only when the total number of electrons changes, and may be ignored.

In the theory of magnetism such a spin problem is attacked by the so-called "semiclassical method,"[20] which is actually a perfectly well defined quantum-mechanical approximation scheme. The similarity of this scheme to the "intermediate coupling" methods of field theory is not widely appreciated but has been mentioned by Gross.[21] Later in this section we shall discuss the scheme from a fully quantum-mechanical point of view, but in the meantime we shall describe it more or less from the semiclassical viewpoint. The first approximation is to take the spin vectors $\mathbf{s}_{\mathbf{k}}$ and rotate them into the best possible classical arrangement, i.e., parallel to the field acting upon them. This, we shall see, is the same as taking the optimum product wave function. In the next approximation one finds the small oscillations about this classical equilibrium and quantizes them; then the ground-state energy and wave function are corrected for the zero-point motion of the small oscillations, the basic assumption being that these are actually small.

The field $\mathbf{H}_{\mathbf{k}}$ which $\mathbf{s}_{\mathbf{k}}$ sees is, from (23),

$$\mathbf{H}_{\mathbf{k}}=2(\epsilon_k-\epsilon_F{}^0)\hat{z}+2\sum_{\mathbf{k}'}V_{\mathbf{k}\mathbf{k}'}\mathbf{s}_{\perp\mathbf{k}'}, \quad (24)$$

where $\mathbf{s}_{\perp\mathbf{k}}$ is that portion of s perpendicular to $\hat{z}$.

In the unperturbed Fermi sea, only the z component is present, and each spin is either up or down, with a sharp break at ϵ_F [see Fig. 1(a)]. It is easy to see that because of the small fields at ϵ_F, only a small V is necessary to turn a few spins sidewise and make the configuration of Fig. 1(b) more stable: a "domain wall" in k space with states rotating smoothly from "full" to "empty."

This configuration is determined in terms of the angle θ_k between the new direction of the spin $\mathbf{k}$ and the z axis by $\mathbf{H}_{\mathbf{k}}\|\mathbf{s}_{\mathbf{k}}$:

$$s_{x\mathbf{k}}/s_{z\mathbf{k}}=\tan\theta_k=\tfrac{1}{2}(\epsilon_k-\epsilon_F{}^0)^{-1}\cdot\sum_{\mathbf{k}'} V_{\mathbf{k}\mathbf{k}'}\sin\theta_{k'}. \quad (25)$$

[19] D. Pines, reference 13.

[20] G. Heller and H. A. Kramers, Proc. Acad. Sci. Amsterdam **37**, 378 (1934); M. J. Klein and R. S. Smith, Phys. Rev. **80**, 1111 (1951); P. W. Anderson, Phys. Rev. **86**, 694 (1952).

[21] E. P. Gross, Phys. Rev. **100**, 1571 (1955).

This may easily be shown to be the same as the B.C.S. integral equation; the correspondence is

$$\sin\theta_k = 2[h_k(1-h_k)]^{\frac{1}{2}}. \tag{26}$$

It is also obvious that the product wave function of B.C.S. is of the same form as our wave function: a product of rotated factors, one for each k vector.

It is also interesting to make contact with the Bogoliubov theory. To do this we consider the spin components in the new direction:

$$\begin{aligned} s_{z'\mathbf{k}} &= \cos\theta_k s_{z\mathbf{k}} + \sin\theta_k s_{x\mathbf{k}}, \\ s_{x'\mathbf{k}} &= -\sin\theta_k s_{z\mathbf{k}} + \cos\theta_k s_{x\mathbf{k}}. \end{aligned} \tag{27}$$

On the other hand, if we set

$$2s_{z'\mathbf{k}} = 1 - (\alpha_{\mathbf{k}0}{}^*\alpha_{\mathbf{k}0} + \alpha_{\mathbf{k}1}{}^*\alpha_{\mathbf{k}1}), \tag{28}$$

in correspondence with the definition of $s_{z\mathbf{k}}$, and use (2), we get

$$2s_{z'\mathbf{k}} = (u_k{}^2 - v_k{}^2)(1 - n_{\mathbf{k}} - n_{-\mathbf{k}}) + 2u_k v_k (b_{\mathbf{k}}{}^* + b_{\mathbf{k}}).$$

This is the same as (27) if

$$\cos\theta_k = u_k{}^2 - v_k{}^2, \quad \sin\theta_k = 2u_k v_k. \tag{29}$$

Thus the first approximation gives exactly the B.C.S.-Bogoliubov results. The first approximation to the excitation spectrum is obtained by taking the energy to turn over the "spins" in the effective fields H_k:

$$2E_k = |H_k| = 2[(\epsilon_k - \epsilon_F{}^0)^2 + \tfrac{1}{4}(\textstyle\sum_{k'} V_{\mathbf{k}\mathbf{k}'} \sin\theta_{k'})^2]^{\frac{1}{2}}, \tag{30}$$

which is the energy of excitation of "real pairs" in the B.C.S. theory.[22]

An improvement on the B.C.S. theory comes when we study the true excitation spectrum modified by the interaction between "spins." First, however, we observe that this improvement can be expected to change the result *for* $\mathcal{H}_{\mathrm{RED}}$ *only* to order only $1/N$. The reason is that the field $\mathbf{H}_{\mathbf{k}}$, insofar as it involves the other spins, is a sum over a number of the order of N other spins. Thus we expect the quantum fluctuations to average out, and the semiclassical theory to be nearly valid, in contrast with the theories of reference 20.

In order to study the modified excitations we must write down the equations of motion. This is most simply done by observing that since the $\mathbf{s}_{\mathbf{k}}$'s obey the usual spin commutation relations,

$$\mathbf{s} \times \mathbf{s} = i\mathbf{s}, \tag{31}$$

the usual spin equations of motion are valid:

$$[\mathcal{H}, \mathbf{s}_{\mathbf{k}}] = i(d\mathbf{s}_{\mathbf{k}}/dt) = i[\mathbf{H}_{\mathbf{k}} \times \mathbf{s}_{\mathbf{k}}]. \tag{32}$$

Now we allow each spin $\mathbf{s}_{\mathbf{k}}$ to have, besides its static component $\mathbf{s}_{\mathbf{k}}{}^0$, an increment $\delta\mathbf{s}_{\mathbf{k}}$. Then (32) is

$$\delta(d\mathbf{s}_{\mathbf{k}}/dt) = [\mathbf{H}_k{}^0 \times \delta\mathbf{s}_{\mathbf{k}}] + [\delta\mathbf{H}_{\mathbf{k}} \times \mathbf{s}_k{}^0] + [\delta\mathbf{H}_{\mathbf{k}} \times \delta\mathbf{s}_{\mathbf{k}}], \tag{33}$$

and the last term is neglected as nonlinear. Here

$$\delta\mathbf{H}_{\mathbf{k}} = 2\textstyle\sum_{\mathbf{k}'} V_{\mathbf{k}\mathbf{k}'} \delta\mathbf{s}_{\perp\mathbf{k}'}. \tag{34}$$

Let $\mathbf{s}_k{}^0$ and $\mathbf{H}_k{}^0$ be in the x-z plane; the components of $\delta\mathbf{s}$ will be δs_y and $\delta s_{||}$, the latter meaning the component perpendicular to $\mathbf{s}^0$ in the x-z plane. Then (33) is

$$\begin{aligned} \delta(ds_{\mathbf{k}y}/dt) &= H_k{}^0 \delta s_{\mathbf{k}||} - \tfrac{1}{2}\delta H_{\mathbf{k}x} \cos\theta_k, \\ \delta(ds_{\mathbf{k}||}/dt) &= -H_k{}^0 \delta s_{\mathbf{k}y} + \tfrac{1}{2}\delta H_{\mathbf{k}y}. \end{aligned} \tag{35}$$

Equations (35) are easily solved only if we make the B.C.S. assumption that $V_{\mathbf{k}\mathbf{k}'}$ is a constant over a region in k space, and otherwise zero. That case has been worked out in detail by Suhl.[23] Using Suhl's solution as a guide, one can see some rather general properties of (33) or (35).

The unperturbed, "individual particle" solutions result from neglecting the $\delta\mathbf{H}_{\mathbf{k}}$ term. Then $\delta\mathbf{s}_{\mathbf{k}}$ simply precesses about $\mathbf{H}_k{}^0$ at the frequency $H_k{}^0$, which is the energy of the "real pair" excitation. The great majority of excitations are of approximately this form, because, from (34), if only a few $\delta\mathbf{s}_{\mathbf{k}}$ are large ($V_{\mathbf{k}\mathbf{k}'}$ is of order N^{-1}), $\delta\mathbf{H}_{\mathbf{k}}$ is indeed of order N^{-1} and may be neglected. Actually, there is a solution of (33) between every pair of unperturbed solutions, for which $\delta\mathbf{s}$ for some few particular k's is of order N larger than all other $\delta\mathbf{s}$'s.

Collective solutions may be defined as solutions for which the sum (34) is replaced by an integral, which is understood in a principal part sense; they will usually lie outside the unperturbed spectrum. A collective solution always comes at the frequency $\nu = 0$. Such a

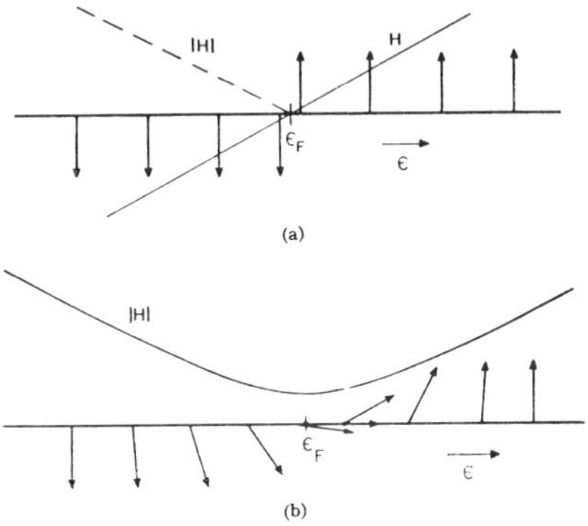

FIG. 1. Configurations in the pseudospin analogy to the electron gas. (a) Normal Fermi sea. Spin arrows up represent empty states. Also given are the effective field H_k acting on the spins, and its absolute value. (b) The superconducting ground state, showing the gradual rotation, like a domain wall, of the pseudospin vectors.

[22] Bogoliubov's fermions give a more complete excitation spectrum for practical calculations. It is easily shown that in this approximation $[\mathcal{H}_{\mathrm{RED}}, \alpha_{k0}{}^*] = \frac{1}{2}H_k\alpha_{k0}{}^*$, etc. The α's are the "singles" of the B.C.S. theory, and all thermal etc. effects may be calculated using them alone.

[23] H. Suhl (private communication).

solution has, from (35),

$$2\delta s_{\mathbf{k}y}=\delta H_{\mathbf{k}y}(H_k^0)^{-1},\quad \delta s_{\mathbf{k}||}=0. \tag{36}$$

To check this we must be sure (36) is compatible with (34). To do this we try

$$2\delta s_{\mathbf{k}y}=\epsilon \sin\theta_k. \tag{37}$$

Then (34) becomes

$$\delta H_{\mathbf{k}y}=2\epsilon \sum_{\mathbf{k}'} V_{\mathbf{k}\mathbf{k}'} \sin\theta_{k'}, \tag{38}$$

or

$$2H_k^0\epsilon \sin\theta_k=4\epsilon(\epsilon_k-\epsilon_F^0)\tan\theta_k,$$

by (25). Use of the value of H_k^0 checks (36).

With the B.C.S. assumption that $V_{\mathbf{k}\mathbf{k}'}$ is a constant, there are no other collective solutions of any interest. More realistic V's may have other solutions, but these will not lie low in the gap, at least in the weak-coupling case in which V varies much only over an energy range large compared with the gap, and thus with the energy range of variation of the wave function. The demonstration of these statements follows.

It is a good approximation to assume the problem symmetrical about ϵ_F, and then $\delta s_{\mathbf{k}y}$ is either an even or an odd function of $\epsilon_k-\epsilon_F$. If $\delta s_{\mathbf{k}y}$ is even, $\delta H_{\mathbf{k}y}$ is finite and even, the second equation of (35) makes $\delta s_{\mathbf{k}||}$ even, thus $\delta s_{\mathbf{k}z}$ odd and $\delta H_{\mathbf{k}z}=0$. Similarly the odd solutions have $\delta H_{\mathbf{k}y}=0$.

Two types of solutions may occur: angle-dependent solutions, physically like bound pairs of p, d, etc., symmetry; or s-like solutions orthogonal to the $\nu=0$ one. The lowest of the latter is necessarily the lowest odd solution; because of (35) and (34) these obey the equation

$$\delta s_{\mathbf{k}y}=[(H_k^0)^2-\nu^2]^{-1}\cos\theta_k \sum_{\mathbf{k}'} V_{\mathbf{k}\mathbf{k}'}H_{k'z}^0\delta s_{\mathbf{k}'y}.$$

When V is roughly constant this is equivalent to

$$1=\sum_{\mathbf{k}} \frac{V_{\mathbf{k}\mathbf{k}'}H_k^0}{(H_k^0)^2-\nu^2}\cos^2\theta_k,$$

which may be quickly verified to have its lowest solution precisely at the top of the gap.

We may expect the lowest angle-dependent solution to be an even one. The fundamental equation for the even solutions is

$$[(H_k^0)^2-\nu^2]\delta s_{\mathbf{k}y}=\sum_{\mathbf{k}'} V_{\mathbf{k}\mathbf{k}'}\delta s_{\mathbf{k}'y}.$$

If V does not vary rapidly with angle the sum on the right will be quite small, and ν must approach H_k^0 closely to allow a solution.

It is physically obvious (also from the discussion of reference 6) that any such solutions which actually occur in the gap are simply bound pairs of Cooper type in excited states. We shall discuss the corresponding $Q\neq 0$ excitations in a later section.

The $\nu=0$ solution could have been expected from the first, and serves a very useful and important purpose. Equations (36) and (37) show that it represents a uniform rotation of the whole "wall" about the pseudo-z axis. Now the original $\mathcal{H}_{\mathrm{RED}}$ [Eq. (23)] was axially symmetric about z. The reason for this symmetry is of course that

$$S_z^{\mathrm{tot}}=N^0-N, \tag{39}$$

must be a constant of the motion, as particles are not really created or destroyed.

On the other hand, the product solution in the semiclassical, "naive" approximation does not have a definite value of S_z^{tot}, because s_{zk} does not commute with the other components of $\mathbf{s}_k$, which are assumed constants of the motion. Correspondingly, the product solution is not axially symmetric but picks a particular direction in the x-y plane, which must be unphysical. The $\nu=0$ mode is the free rotation of this solution about the z axis. We can expect that proper inclusion of the zero-point motion of the $\nu=0$ mode will repair this situation by projecting our solution onto the space of $S_z^{\mathrm{tot}}=$ constant. In fact this can be verified using the B.C.S. representation. An eigenfunction of the $\nu=0$ rotation about the z axis is just

$$\int d\varphi\, \Psi_{\mathrm{B.C.S.}}(\varphi)\exp(in\varphi), \tag{40}$$

where $\Psi_{\mathrm{B.C.S.}}(\varphi)$ is the B.C.S. solution rotatea to a new direction φ in the x-y plane. It can be shown that (40) is

$$\Psi_n=\int d\varphi\, e^{in\varphi}\prod_{\mathbf{k}}[(1-h_k)^{\frac{1}{2}}e^{-i\varphi/2}+h_k^{\frac{1}{2}}e^{i\varphi/2}b_{\mathbf{k}}^*]\Psi_v, \tag{41}$$

which quite clearly is just such a projection.

A few final remarks will close this section. First, the correction to the energy could be calculated,

$$\Delta E=\tfrac{1}{2}\sum \nu-\tfrac{1}{2}\sum H_k^0, \tag{42}$$

just the difference of the perturbed and unperturbed zero-point energies (a special case of a relationship we will prove later). The largest part of (42) will be an amount $-H_k^0$ from the mode $\nu=0$; but the whole correction is only of order N^{-1} relative to the total energy so need not be calculated.

Second, note the existence of two formal solutions of the equations of motion:

$$[\mathcal{H}_{\mathrm{RED}},\, n_{\mathbf{k}}-n_{-\mathbf{k}}]\equiv 0,\quad [\mathcal{H}_{\mathrm{RED}},\delta s_{z'\mathbf{k}}]=0, \tag{43}$$

where z' is the direction along which $\mathbf{s}_k^0$ points. The assumed ground state satisfies

$$\begin{gathered}(n_{\mathbf{k}}-n_{-\mathbf{k}})\Psi_0\equiv 0,\\ \delta s_{z'\mathbf{k}}\Psi_0\equiv 0\quad (\text{i.e., } \delta\mathbf{s}_{\mathbf{k}}\perp \mathbf{s}_k^0),\end{gathered} \tag{44}$$

and (43) assures that these conditions remain satisfied throughout the zero-point motion.

These conditions are the conditions that essentially

nonphysical excited states do not enter the problem. Our calculation can be thought of in the following way[17]: starting from the exact ground state Ψ_0^e, we make all possible zero-momentum excitations of pairs of electrons, such as $\delta b_{\mathbf{k}}\Psi_0^e$. Certain linear combinations

$$x^\nu=\sum_{\mathbf{k}}(\alpha_{\mathbf{k}}\delta n_{\mathbf{k}}+\beta_{\mathbf{k}}\delta n_{-\mathbf{k}}+\gamma_{\mathbf{k}}\delta b_{\mathbf{k}}+\delta_{\mathbf{k}}\delta b_{\mathbf{k}}^*),$$

satisfy commutation rules

$$[\mathcal{H},x^\nu]=\nu x^\nu,$$

so that

$$\mathcal{H}(x^\nu\Psi_0^e)=(\nu+E_0)(x^\nu\Psi_0^e),$$

and our attempt is to calculate the ν and the x^ν approximately by assuming the commutators to have the values appropriate to some zeroth order Ψ_0^0. There are, however, certain pair excitations which are not possible from our unperturbed Ψ_0^0: we cannot destroy electrons in states which are empty, or create them in full states. The procedure is consistent only if the equations of motion are compatible with these restrictions, i.e., if the conditions (44) are indeed solutions of the equations, as we see from (43). We will show in the next section that this is more generally true.

This point of view may be clearer if we see that (44) may be expressed in Bogoliubov notation as

$$\begin{aligned}(\alpha_{\mathbf{k}0}^*\alpha_{\mathbf{k}0}-\alpha_{\mathbf{k}1}^*\alpha_{\mathbf{k}1})\Psi_0&=0,\\ (\alpha_{\mathbf{k}0}^*\alpha_{\mathbf{k}0}+\alpha_{\mathbf{k}1}^*\alpha_{\mathbf{k}1})\Psi_0&=0.\end{aligned}\qquad(43')$$

A little algebra with (28) and (2) shows that this rather convenient expression is the same as (43).

IV. RANDOM-PHASE APPROXIMATION TREATMENT OF THE FULL HAMILTONIAN: EQUATIONS OF MOTION

What we shall now show is that there is a single approximation scheme, starting from the full Hamiltonian and not paying undue attention to the $q=0$ part, which leads to the equations of the last section as an integral part, while reducing to the usual R.P.A. treatment of correlation energy and of plasmons in appropriate limits. In fact, in the absence of phonon attractive forces it is the same in principle as Hubbard's[10] inclusion of Coulomb exchange.

In this scheme, as in the second approximation of the last section, we calculate equations of motion of quantities which are bilinear in the original fermion operators. We generalize in two ways: we calculate the full equations of motion, by commuting with the full Hamiltonian; and we calculate equations of motion for quantities with momentum $\mathbf{Q}$ as well as momentum zero:

$$\begin{aligned}b_{\mathbf{k}}^{\mathbf{Q}}&=c_{-\mathbf{k}-\mathbf{Q}}c_{\mathbf{k}}, & \bar{b}_{\mathbf{k}}^{\mathbf{Q}}&=c_{\mathbf{k}+\mathbf{Q}}^*c_{-\mathbf{k}}^*,\\ \rho_{\mathbf{k}}^{\mathbf{Q}}&=c_{\mathbf{k}+\mathbf{Q}}^*c_{\mathbf{k}}, & \bar{\rho}_{\mathbf{k}}^{\mathbf{Q}}&=c_{-\mathbf{k}}^*c_{-\mathbf{k}-\mathbf{Q}},\end{aligned}\qquad(45)$$

as well as $b_{\mathbf{k}}$, $b_{\mathbf{k}}^*$, $n_{\mathbf{k}}$, and $n_{-\mathbf{k}}$. Note that

$$(b_{\mathbf{k}}^{\mathbf{Q}})^*=\bar{b}_{\mathbf{k}+\mathbf{Q}}^{-\mathbf{Q}},\quad(\rho_{\mathbf{k}}^{\mathbf{Q}})^*=\bar{\rho}_{-\mathbf{k}}^{-\mathbf{Q}};$$

these Hermitian conjugates of our quantities have momentum $-\mathbf{Q}$.

These full equations of motion are of course useless. However, what can be done is to linearize them, as we have done in the preceding section for $\mathcal{H}_{\mathrm{RED}}$, and as Bohm and Pines,[9] and Sawada and Brout,[8] have done for the Coulomb problem. The linearization used in the usual R.P.A. ignores the exchange terms, which are of the form of the terms we discussed in Sec. III; but there is no need to do so, as Hubbard has shown in a slightly different way, so long as the exchange terms are correctly screened. The method of linearization we use is straightforward in the extreme: each term in the interaction part of the equations of motion is a product of four fermions, and thus a product of two bilinear combinations b or ρ (or n) in a number of ways. We assume that the state about which we linearize is a B.C.S.-Bogoliubov product state, so that it may have finite zero-order values of $b_{\mathbf{k}}$, $b_{\mathbf{k}}^*$, or $[1-(n_{\mathbf{k}}+n_{-\mathbf{k}})]$. Then we keep only terms which contain one of these quantities.

When b and b^* are zero, the resulting equations are those of Bohm and Pines with exchange added.[24] The equations of motion for b and n themselves are just (32) itself. We shall go on to discuss the solutions of various kinds and to show that the individual-particle solutions give just the B.C.S. spectrum, while the longitudinal collective solutions are such as to insure the validity of the sum rules and of gauge invariance, as suggested in reference 6.

The full nonlinear equations of motion are of almost no interest in themselves. We shall write down the one for $\rho_{\mathbf{K}}^{\mathbf{Q}}$ and then show briefly how the various terms of the linear approximation follow from it. Then we shall give the linearized equations of motion for all of the quantities (45).

Let us lump together the Coulomb interaction (9) and the phonon term (8) with a common matrix element $V(\mathbf{k},\mathbf{k}')$ (which is a function of $\mathbf{k}$ and $\mathbf{k}'$ only, for the reasons put forth at the end of Sec. II). Then for the Hamiltonian we have

$$\mathcal{H}=\mathcal{H}_K+\tfrac{1}{2}\sum_{\mathbf{k}\neq\mathbf{k}',\mathbf{q}}\sum_{\sigma,\sigma'}V(\mathbf{k},\mathbf{k}')\times c_{\mathbf{k}',\sigma'}^*c_{-\mathbf{k}'+\mathbf{q},\sigma}^*c_{-\mathbf{k}+\mathbf{q},\sigma}c_{\mathbf{k},\sigma'}.\qquad(46)$$

The full equation of motion of $\rho_{\mathbf{K}}^{\mathbf{Q}}$ is

$$\begin{aligned}[\mathcal{H},\rho_{\mathbf{K}}^{\mathbf{Q}}]=(\epsilon_{\mathbf{K}+\mathbf{Q}}-\epsilon_K)\rho_{\mathbf{K}}^{\mathbf{Q}}&+\sum_{\mathbf{k},\sigma,\mathbf{q}}[V(\mathbf{K}+\mathbf{Q},\mathbf{k})c_{-\mathbf{k}+\mathbf{q},\sigma}^*c_{-\mathbf{K}-\mathbf{Q}+\mathbf{q},\sigma}c_{\mathbf{k}}^*c_{\mathbf{K}}\\ &-V(\mathbf{K},\mathbf{k})c_{-\mathbf{K}+\mathbf{q},\sigma}^*c_{-\mathbf{k}+\mathbf{q},\sigma}c_{\mathbf{K}+\mathbf{Q}}^*c_{\mathbf{k}}],\end{aligned}\qquad(47)$$

or

$$[\mathcal{H},\rho_{\mathbf{K}}^{\mathbf{Q}}]=(\epsilon_{\mathbf{K}+\mathbf{Q}}-\epsilon_K)\rho_{\mathbf{K}}^{\mathbf{Q}}+\sum_{\mathbf{k}}[V(\mathbf{K}+\mathbf{Q},\mathbf{k})\times\rho^{\mathbf{K}-\mathbf{k}+\mathbf{Q}}c_{\mathbf{k}}^*c_{\mathbf{K}}-V(\mathbf{K},\mathbf{k})\rho^{\mathbf{k}-\mathbf{K}}c_{\mathbf{K}+\mathbf{Q}}^*c_{\mathbf{k}}].$$

[24] The equations for $\rho_{\mathbf{k}}^{\mathbf{Q}}$ are the Bohm-Pines equations; those for b and $\bar{b}$, on the other hand, are the Bethe-Goldstone equations [H. A. Bethe and J. Goldstone, Proc. Roy. Soc. (London) A238, 551 (1957)].

We have used ρ^{Q} for the **Q**th Fourier component of density:

$$\rho^{Q}=\sum_{\mathbf{k},\sigma}\rho_{\mathbf{k},\sigma}{}^{Q}.$$

(The second form above, though brief, is actually quite inconvenient for calculations of all but the "direct" term.)

If we sum (47) over **K**, and in the first interaction term replace **k** by **K**+**Q** and **K** by **k** (both are now dummy variables), we see that the interaction terms in $[\mathcal{H},\rho^{Q}]$ vanish. Thus our equations of motion are such as to automatically satisfy the sum rules (this is maintained throughout) and the most important consequences of gauge invariance. The equation of motion (47) contains five types of linear terms:

$$[\mathcal{H},\rho_{\mathbf{K}}{}^{Q}]=\text{kinetic}+\text{direct}+\text{exchange self-energy} +\text{exchange scattering}+\text{superconductivity}. \quad (48)$$

The kinetic-energy term is simply the first term:

$$\text{Kinetic:}\quad [\mathcal{H}_K,\rho_{\mathbf{K}}{}^{Q}]=(\epsilon_{\mathbf{K}+\mathbf{Q}}-\epsilon_K)\rho_{\mathbf{K}}{}^{Q}. \quad (49)$$

The direct terms are those terms which result directly from the interaction with the components of fluctuation of electron density of wave number **Q**, i.e., from the term $V(\mathbf{Q})\rho^{Q}\rho^{-Q}$+cc. The linear part is, from the second half of (47):

$$\text{Direct:}\quad [\mathcal{H}_D,\rho_{\mathbf{K}}{}^{Q}]=V_D(\mathbf{Q})\rho^{Q}(n_{\mathbf{K}}-n_{\mathbf{K}+\mathbf{Q}}). \quad (50)$$

(We put the subscript D on V to indicate that only in this term should the "direct," unscreened interaction enter; this is primarily a function of $\mathbf{k}-\mathbf{k}'=\mathbf{Q}$ alone.)

The exchange self-energy terms come from the usual exchange terms,

$$\mathcal{H}_{\text{ex}}{}^{\text{self}}=\tfrac{1}{2}\sum_{\mathbf{k},\mathbf{q},\sigma}V(\mathbf{k},-\mathbf{k}+\mathbf{q})n_{\mathbf{k},\sigma}n_{-\mathbf{k}+\mathbf{q},\sigma}, \quad (51)$$

in (46). There is a phonon contribution here to $V(q)$, and it was this contribution which was the basis of the old superconductivity theories of Fröhlich and Bardeen[25]; the assumption now is, however, that this coupling is far too weak to have a visible effect. The appropriate terms are as follows:

$$\text{Self-energy:}\quad [\mathcal{H}_{\text{ex}}{}^{\text{self}},\rho_{\mathbf{K}}{}^{Q}] =-\rho_{\mathbf{K}}{}^{Q}\sum_{\mathbf{q}}[n_{\mathbf{q}-\mathbf{K}-\mathbf{Q}}V(\mathbf{K}+\mathbf{Q},\mathbf{q}-\mathbf{K}-\mathbf{Q}) -n_{\mathbf{q}-\mathbf{K}}V(\mathbf{K},\mathbf{q}-\mathbf{K})]. \quad (52)$$

The "exchange scattering" terms are the exchange terms which Hubbard includes by a device similar to ours, and follow from those exchange terms (with parallel spin) which contain ρ^{Q}'s.

$$\text{Exchange scattering:}\quad [\mathcal{H}_{\text{ex}},\rho_{\mathbf{K}}{}^{Q}] =-(n_{\mathbf{K}}-n_{\mathbf{K}+\mathbf{Q}})\sum_{\mathbf{q}}\rho_{\mathbf{q}-\mathbf{K}-\mathbf{Q}}{}^{Q}V(\mathbf{K}+\mathbf{Q},\mathbf{q}-\mathbf{K}). \quad (53)$$

Note the formal similarity to the direct terms (50), just as the exchange self-energy is similar to the kinetic term (49). For small Q, (53) is negligible compared to (50), and we shall neglect it because it has no relevance to the superconductivity problem. However, for consistency we must then also neglect the exchange self-energy terms (52). That is justifiable as a "weak-coupling" approximation.

Finally we come to the superconductivity terms. These are of two types: "individual-particle" terms, rather like self-energy terms, which come from the B.C.S. reduced Hamiltonian [$q=0$, $\sigma'=-\sigma$ in (46)]; and "collective" terms which follow from $\mathbf{q}=-\mathbf{Q}$, $\sigma'=-\sigma$ in (46). The terms are:

$$\text{Supercond.:}\quad [\mathcal{H}_S,\rho_{\mathbf{K}}{}^{Q}]=\sum_{\mathbf{k}}V(\mathbf{k},\mathbf{K})(b_{\mathbf{k}+\mathbf{Q}}{}^{*}b_{\mathbf{K}}{}^{Q} -b_{\mathbf{k}}\bar{b}_{\mathbf{K}}{}^{Q}+b_{\mathbf{K}}\bar{b}_{\mathbf{k}}{}^{Q}-b_{\mathbf{K}+\mathbf{Q}}{}^{*}b_{\mathbf{k}}{}^{Q}). \quad (54)$$

[We shall normally take Q small, so we neglect the small difference of $V(\mathbf{K}+\mathbf{Q},\mathbf{k}+\mathbf{Q})$ and $V(\mathbf{K},\mathbf{k})$.]

This method of presentation shows why the "superconductivity" terms have not appeared in previous types of theories: they come in only when b and b^* are treated as number operators are in usual theories. The relationship—or lack thereof—of superconductivity and ferromagnetism is also rather clear in this scheme. Exchange interactions involving a *repulsive* interelectronic potential act like an *attraction* between parallel spins—compare (53) and (50)—and the resulting self-energy-like terms (52) are responsible for ferromagnetism, if they are big enough to outweigh the kinetic energy. The corresponding "superconductivity" terms are repulsive, whether between parallel or antiparallel spins (as could be verified by writing down equations of motion of $c_{-\mathbf{k}-\mathbf{Q}\uparrow}c_{\mathbf{k}\uparrow}$). The interelectronic *attraction* caused by the phonons is thus detrimental to ferromagnetism, and the interactions responsible for ferromagnetism are correspondingly detrimental to superconductivity.

Now we shall write down without further explanation the linearized equations of motion for the remainder of the pair quantities (45), classifying the terms as before.

$$\text{Kinetic:}\quad \begin{aligned}[\mathcal{H}_K,\bar{\rho}_{\mathbf{K}}{}^{Q}]&=-(\epsilon_{\mathbf{K}+\mathbf{Q}}-\epsilon_K)\bar{\rho}_{\mathbf{K}}{}^{Q},\\ [\mathcal{H}_K,b_{\mathbf{K}}{}^{Q}]&=-(\epsilon_K+\epsilon_{\mathbf{K}+\mathbf{Q}})b_{\mathbf{K}}{}^{Q},\\ [\mathcal{H}_K,\bar{b}_{\mathbf{K}}{}^{Q}]&=(\epsilon_K+\epsilon_{\mathbf{K}+\mathbf{Q}})\bar{b}_{\mathbf{K}}{}^{Q}.\end{aligned} \quad (55)$$

$$\text{Exchange self-energy:}\quad [\mathcal{H}_{\text{ex}}{}^{\text{self}},\bar{\rho}_{\mathbf{K}}{}^{Q}] =\bar{\rho}_{\mathbf{K}}{}^{Q}[\sum_{\mathbf{q}}V(\mathbf{K}+\mathbf{Q},\mathbf{q}-\mathbf{K}-\mathbf{Q})n_{-\mathbf{q}+\mathbf{K}+\mathbf{Q}} -V(\mathbf{K},\mathbf{q}-\mathbf{K})n_{-\mathbf{q}+\mathbf{K}}]. \quad (56)$$

These terms may be simply taken into account by inserting into (55) an exchange self-energy

$$\delta\epsilon_{\mathbf{K},\sigma}=-\sum_{\mathbf{q}}V(\mathbf{K},\mathbf{q}-\mathbf{K})n_{\mathbf{K}-\mathbf{q},\sigma}. \quad (57)$$

$$\text{Direct:}\quad \begin{aligned}[\mathcal{H}_D,\bar{\rho}_{\mathbf{K}}{}^{Q}]&=V_D(Q)\rho^{Q}(n_{-\mathbf{K}-\mathbf{Q}}-n_{-\mathbf{K}}),\\ [\mathcal{H}_D,b_{\mathbf{K}}{}^{Q}]&=-V_D(Q)\rho^{Q}(b_{\mathbf{K}}+b_{\mathbf{K}+\mathbf{Q}}),\\ [\mathcal{H}_D,\bar{b}_{\mathbf{K}}{}^{Q}]&=V_D(Q)\rho^{Q}(b_{\mathbf{K}}{}^{*}+b_{\mathbf{K}+\mathbf{Q}}{}^{*}).\end{aligned} \quad (58)$$

[25] H. Fröhlich, Phys. Rev. **79**, 845 (1950); and J. Bardeen, Phys. Rev. **80**, 567 (1950).

Exchange scattering:

$$
\begin{aligned}
[\mathcal{H}_{\text{ex}},\bar{\rho}_{\mathbf{K}}{}^{\mathbf{Q}}]&=-(n_{-\mathbf{K}-\mathbf{Q}}-n_{-\mathbf{K}})\\
&\quad\times\textstyle\sum_{\mathbf{q}} V(\mathbf{K}+\mathbf{Q},\mathbf{q}-\mathbf{K})\bar{\rho}_{\mathbf{q}-\mathbf{K}-\mathbf{Q}}{}^{\mathbf{Q}},\\
[\mathcal{H}_{\text{ex}},b_{\mathbf{K}}{}^{\mathbf{Q}}]&=b_{\mathbf{K}}\textstyle\sum_{\mathbf{q}}\bar{\rho}_{-\mathbf{K}-\mathbf{Q}-\mathbf{q}}{}^{\mathbf{Q}}V(\mathbf{K},-\mathbf{q}-\mathbf{K}-\mathbf{Q})\\
&\quad+b_{\mathbf{K}+\mathbf{Q}}\textstyle\sum_{\mathbf{q}} V(\mathbf{K}+\mathbf{Q},-\mathbf{q}-\mathbf{K})\\
&\quad\times\rho_{-\mathbf{K}-\mathbf{Q}-\mathbf{q}}{}^{\mathbf{Q}},\\
[\mathcal{H}_{\text{ex}},\bar{b}_{\mathbf{K}}{}^{\mathbf{Q}}]&=-b_{\mathbf{K}}{}^*\textstyle\sum_{\mathbf{q}} V(\mathbf{K},-\mathbf{q}-\mathbf{K}-\mathbf{Q})\rho_{-\mathbf{K}-\mathbf{Q}-\mathbf{q}}{}^{\mathbf{Q}}\\
&\quad-b_{\mathbf{K}+\mathbf{Q}}{}^*\textstyle\sum_{\mathbf{q}} V(\mathbf{K}+\mathbf{Q},-\mathbf{q}-\mathbf{K})\bar{\rho}_{-\mathbf{K}-\mathbf{Q}-\mathbf{q}}{}^{\mathbf{Q}}.
\end{aligned}
\tag{59}
$$

Superconductivity:

$$
\begin{aligned}
[\mathcal{H}_S,\bar{\rho}_{\mathbf{K}}{}^{\mathbf{Q}}]&=\textstyle\sum_{\mathbf{k}} V(\mathbf{K},\mathbf{k})(b_{\mathbf{k}}{}^*b_{\mathbf{K}}{}^{\mathbf{Q}}-b_{\mathbf{k}+\mathbf{Q}}\bar{b}_{\mathbf{K}}{}^{\mathbf{Q}}\\
&\quad+b_{\mathbf{K}+\mathbf{Q}}\bar{b}_{\mathbf{k}}{}^{\mathbf{Q}}-b_{K}{}^*b_{\mathbf{k}}{}^{\mathbf{Q}}),\\
[\mathcal{H}_S,b_{\mathbf{K}}{}^{\mathbf{Q}}]&=\textstyle\sum_{\mathbf{k}} V(\mathbf{K},\mathbf{k})[\bar{\rho}_{\mathbf{K}}{}^{\mathbf{Q}}b_{\mathbf{k}}+\rho_{\mathbf{K}}{}^{\mathbf{Q}}b_{\mathbf{k}+\mathbf{Q}}\\
&\quad-(1-n_{\mathbf{K}}-n_{-\mathbf{K}-\mathbf{Q}})b_{\mathbf{k}}{}^{\mathbf{Q}}],\\
[\mathcal{H}_S,\bar{b}_{\mathbf{K}}{}^{\mathbf{Q}}]&=-\textstyle\sum_{\mathbf{k}} V(\mathbf{K},\mathbf{k})[b_{\mathbf{k}+\mathbf{Q}}{}^*\bar{\rho}_{\mathbf{K}}{}^{\mathbf{Q}}+b_{\mathbf{k}}{}^*\rho_{\mathbf{K}}{}^{\mathbf{Q}}\\
&\quad-\bar{b}_{\mathbf{k}}{}^{\mathbf{Q}}(1-n_{-\mathbf{K}}-n_{\mathbf{K}+\mathbf{Q}})].
\end{aligned}
\tag{60}
$$

It is clear why this method can be called a "random-phase" method: it has the effect of decoupling excitations of different momentum **Q**, because the only zeroth-order quantities have zero momentum. For this reason the $Q=0$ equations are decoupled from the rest, and are in fact almost exactly those of Sec. III. We can thus write down the $Q=0$ equations without distinguishing zeroth- and first-order quantities.

$$
\begin{aligned}
&[\mathcal{H}_K,n_{\mathbf{k}}]=[\mathcal{H}_K,n_{-\mathbf{k}}]=0,\\
&[\mathcal{H}_K,b_{\mathbf{k}}]=-2\epsilon_k b_{\mathbf{k}};\quad [\mathcal{H}_K,b_{\mathbf{k}}{}^*]=2\epsilon_k b_{\mathbf{k}}{}^*,
\end{aligned}
\tag{61}
$$

$$
\begin{aligned}
&[\mathcal{H}_{\text{ex}}{}^{\text{self}},n_{\mathbf{k}}]=[\mathcal{H}_{\text{ex}}{}^{\text{self}},n_{-\mathbf{k}}]=0,\\
&[\mathcal{H}_{\text{ex}}{}^{\text{self}},b_{\mathbf{k}}]=-2\delta\epsilon_k b_{\mathbf{k}};\quad [\mathcal{H}_{\text{ex}}{}^{\text{self}},b_{\mathbf{k}}{}^*]=2\delta\epsilon_k b_{\mathbf{k}}{}^*.
\end{aligned}
\tag{62}
$$

The "direct" terms all vanish because $V_D(0)=0$. The exchange scattering terms have coalesced with the exchange self-energy (62); they would reappear, if we separated out first-order effects, as terms involving $\sigma_{zk}{}^0\delta(\delta\epsilon_{\mathbf{k}})$ (this is why a consistent treatment must include both or neither exchange terms). Finally, the two parts of the superconducting terms coalesce, also to reappear upon applying δ's as in Sec. III:

$$
\begin{aligned}
&[\mathcal{H}_S,n_{\mathbf{K}}]=\textstyle\sum_{\mathbf{k}} V(\mathbf{k},\mathbf{K})(b_{\mathbf{k}}{}^*b_{\mathbf{K}}-b_{\mathbf{k}}b_{\mathbf{K}}{}^*)=[\mathcal{H}_S,n_{-\mathbf{K}}],\\
&[\mathcal{H}_S,b_{\mathbf{K}}]=-\textstyle\sum_{\mathbf{k}} V(\mathbf{k},\mathbf{K})(1-n_{\mathbf{K}}-n_{-\mathbf{K}})b_{\mathbf{k}},\\
&[\mathcal{H}_S,b_{\mathbf{K}}{}^*]=\textstyle\sum_{\mathbf{k}} V(\mathbf{k},\mathbf{K})(1-n_{\mathbf{K}}-n_{-\mathbf{K}})b_{\mathbf{k}}{}^*.
\end{aligned}
\tag{63}
$$

Using the definitions (22) we find, as we expect, that (63) may be written

$$
\begin{aligned}
&[\mathcal{H}_S,s_{z\mathbf{K}}]=-i(H_{y\mathbf{K}}s_{x\mathbf{K}}-H_{x\mathbf{K}}s_{y\mathbf{K}}),\\
&[\mathcal{H}_S,\,s_{x\mathbf{K}}+is_{y\mathbf{K}}]=s_{z\mathbf{K}}(H_{x\mathbf{K}}+iH_{y\mathbf{K}}).
\end{aligned}
\tag{63'}
$$

These equations may be thought of as determining the zero-order values of the quantities b, b^*, and n:

$$|b_k{}^0|=\tfrac{1}{2}\sin\theta_k,\quad (1-n_k-n_{-k})^0=\cos\theta_k.$$

Henceforth we write b_k for $b_k{}^0$, etc., so that the b's and n's appearing in the equations without superscripts are just numbers.

Thus, except for the exchange self-energy terms, which we explicitly neglected in Sec. III, (61)–(63) are identical with the equations of the spin model of the B.C.S. theory.

Equations (61)–(63) demonstrate the first feature of our method: that the equations of the B.C.S. theory separate out automatically as the zero-momentum component of a random-phase approximation. The B.C.S. equations have also the effect of determining the stablest ground state about which to linearize the rest of the theory.

This connection with the B.C.S. theory of the last section shows us the meaning of the equations we have called "equations of motion." Our attempt is to find a complete set of "elementary excitations" $x_\nu{}^{\mathbf{Q}}$ involving pairs of particles, and having momentum **Q**, analogously to the zero-momentum excitations of Sec. III:

$$x_\nu{}^{\mathbf{Q}}=\textstyle\sum_{\mathbf{k}}(\alpha_{\mathbf{k}}\rho_{\mathbf{k}}{}^{\mathbf{Q}}+\beta_{\mathbf{k}}\bar{\rho}_{\mathbf{k}}{}^{\mathbf{Q}}+\gamma_{\mathbf{k}}b_{\mathbf{k}}{}^{\mathbf{Q}}+\delta_{\mathbf{k}}\bar{b}_{\mathbf{k}}{}^{\mathbf{Q}}),\tag{64}$$

such that

$$[\mathcal{H},x_\nu{}^{\mathbf{Q}}]=\nu x_\nu{}^{\mathbf{Q}},\tag{65}$$

because then

$$\Psi_\nu{}^{\mathbf{Q}}=x_\nu{}^{\mathbf{Q}}\Psi_0,\tag{66}$$

will be an excited eigenstate of energy $E_0+\nu$. To find the ν and $x_\nu{}^{\mathbf{Q}}$ it is a valid scheme to replace the commutators by time derivatives, and assume all quantities have frequency ν.

The ν give immediately the most important observable phenomenon, the excitation spectrum. The properties of the ground state must be found more indirectly. Since the system is one with time-reversal symmetry, we can expect—and do find—that the secular equation is a function only of ν^2, so there are eigenfrequencies $\pm\nu$. If Ψ_0 is the true ground state,

$$x_{-\nu}{}^{\mathbf{Q}}\Psi_0\equiv 0\tag{67}$$

[it will be useful later to note that $x_{-\nu}{}^{\mathbf{Q}}=(x_\nu{}^{-\mathbf{Q}})^*$], and this is the simplest expression of the modification of our assumed ground state $\Psi_0{}^0$, the product wave function, by the zero-point motion. In the absence of scattering terms (67) is trivially satisfied, being simply the condition that particles cannot be destroyed in the vacuum, or created within the Fermi sea; but the coupling terms make (67) a definite modification of the product function.

In Appendix II we will show how to compute the energy using the $x_\nu{}^{\mathbf{Q}}$'s, and particularly the condition (67). In the simple cases of the pure plasmon theory, or the theory of the B.C.S. $\mathcal{H}_{\text{RED}}$, the energy is corrected simply by the sum of the zero-point shifts of the frequencies ν; but that requires that the Hamiltonian as well as the equations of motion be separable into parts identifiable with the separate momenta Q, which is not in general so. The more general expression for the energy correction which we give in the Appendix is very complicated and usually can be only approximately computed.

In the next section we will discuss the actual solutions of Eqs. (48)–(60); here we are concerned with various generalities about them, and particularly their connection with other theories. For this purpose we might examine their structure more closely.

Let $f_{\mathbf{K}}^{\mathbf{Q}}$, $(f')_{\mathbf{K}}^{\mathbf{Q}}$ be any of the quantities $\rho_{\mathbf{K}}^{\mathbf{Q}}$, $\bar{\rho}_{\mathbf{K}}^{\mathbf{Q}}$, $b_{\mathbf{K}}^{\mathbf{Q}}$, or $\bar{b}_{\mathbf{K}}^{\mathbf{Q}}$, and f^0 be any of the zeroth-order, zero-momentum quantities. The equation for $f_{\mathbf{K}}^{\mathbf{Q}}$ contains two general types of terms: self-energy-like terms like (55), (56), and the first two terms on the right in (60), which have the form

$$(f')_{\mathbf{K}}^{\mathbf{Q}} \sum_{\mathbf{k}} V f_k^0; \qquad (68)$$

and scattering (or collective) terms such as (58), (59), and the last terms on the right of (60), which have the form

$$f_K^0 \sum_{\mathbf{k}} V (f')_{\mathbf{k}}^{\mathbf{Q}}. \qquad (69)$$

These two types of terms come from two types of terms in the Hamiltonian: the self-energy terms (68) come from terms

$$\sum_{\mathbf{k},\mathbf{k}'} f_k^0 f_{k'}^0, \qquad (70)$$

which are (aside from kinetic energy) the exchange-type self-energy and the B.C.S. reduced Hamiltonian. That is, the term $(f')_{\mathbf{K}}^{\mathbf{Q}}$ in (68) comes from commuting $f_{\mathbf{K}}^{\mathbf{Q}}$ with one of the two f's in (70) having $\mathbf{k}$ or $\mathbf{k}'=\pm\mathbf{K}$ or $\pm(\mathbf{K}+\mathbf{Q})$, the remaining sum giving the self-energy sum of (68). Thus the reduced Hamiltonian gives all terms (68) correctly (aside from exchange self-energy, which was neglected in B.C.S. as well as Bogoliubov, as a weak-coupling assumption). On the other hand, the scattering terms (69) come from terms in the Hamiltonian

$$\sum_{\mathbf{k},\mathbf{k}'} f_{\mathbf{k}}^{\mathbf{Q}} (f_{\mathbf{k}'}^{\mathbf{Q}})^*, \qquad (71)$$

and result from commuting $f_{\mathbf{k}'}^{\mathbf{Q}*}$ ($\mathbf{k}'=\pm\mathbf{K}$ or $\pm(\mathbf{K}+\mathbf{Q})$) with $f_{\mathbf{K}}^{\mathbf{Q}}$ to give the zero-order f_K^0, the remaining $\mathbf{k}$-sum being the sum in (69).

Just as in the last section, there are two possible types of solutions: "individual-particle" and "collective." Individual-particle (i.p.) solutions have only one or a few $f_{\mathbf{k}}^{\mathbf{Q}}$ finite, so that terms of type (69) are of order $1/N$ relative to terms (68). Thus the frequencies of all individual-particle modes are correctly given by $\mathcal{H}_{\mathrm{RED}}$, which is the basic reason behind the success of the B.C.S.-Bogoliubov theories. As Bogoliubov has shown, energies of type (70) may be rewritten in terms of self-energies of the transformed Fermions $\alpha_{\mathbf{K}0}$, $\alpha_{\mathbf{K}1}$, and in that scheme the α's act like independent particles. Thus also for independent-particle modes with $Q\neq 0$ our scheme is fully equivalent to B.C.S.-Bogoliubov. On the other hand, the scattering as well as any collective modes require the inclusion (at least) of all terms of type (71).

There is one last point to be made about the formal structure of the theory. To understand the difficulty, let us take the simplest case of the unperturbed Fermi sea, and for definiteness take $K<k_F$, $|\mathbf{K}+\mathbf{Q}|>k_F$. We have equations for four excitations (we give also the corresponding excitation for the general case in terms of Bogoliubov operators):

$$\rho_{\mathbf{K}}^{\mathbf{Q}}:\quad \nu=\epsilon_{\mathbf{K}+\mathbf{Q}}-\epsilon_K>0 \leftrightarrow \alpha_{\mathbf{K}+\mathbf{Q}0}{}^*\alpha_{\mathbf{K}1}{}^*,$$
$$b_{\mathbf{K}}^{\mathbf{Q}}:\quad \nu=-(\epsilon_{\mathbf{K}+\mathbf{Q}}+\epsilon_K)<0 \leftrightarrow \alpha_{\mathbf{K}1}{}^*\alpha_{\mathbf{K}+\mathbf{Q}1},$$
$$\bar{\rho}_{\mathbf{K}}^{\mathbf{Q}}:\quad \nu=\epsilon_K-\epsilon_{\mathbf{K}+\mathbf{Q}}<0 \leftrightarrow \alpha_{\mathbf{K}0}\alpha_{\mathbf{K}+\mathbf{Q}1},$$
$$\bar{b}_{\mathbf{K}}^{\mathbf{Q}}:\quad \nu=\epsilon_{\mathbf{K}+\mathbf{Q}}+\epsilon_K>0 \leftrightarrow \alpha_{\mathbf{K}+\mathbf{Q}0}{}^*\alpha_{\mathbf{K}0}.$$

The ν's of course come in $\pm$ pairs, and obviously (67) is satisfied for b and $\bar{\rho}$. In the coupled theory, there will still be two conditions (67) on the ground state eliminating the new coupled version of b and $\bar{\rho}$.

There is as yet no condition built into the theory to eliminate the nonphysical *positive*-frequency quantity $\bar{b}$. The product wave function, of course, automatically satisfied

$$\bar{b}_{\mathbf{K}}^{\mathbf{Q}}\Psi_0^0=0,$$

but we have no guarantee that th s condition is maintained throughout the motion, once coupling is introduced. This means that we have to prove the theorem that even in the presence of coupling the "unphysical modes"

$$(x_\nu^{\mathbf{Q}})_u=\alpha_{\mathbf{K}+\mathbf{Q}0}{}^*\alpha_{\mathbf{K}0} \qquad (72)$$

are eigensolutions. We know this to be identically so as far as the self-energy terms (68) are concerned, but must prove the theorem in regard to the coupling terms.

This is easily done. In terms of the α's, the only finite zero-order quantities are

$$\langle\alpha_{\mathbf{K}0}\alpha_{\mathbf{K}0}{}^*\rangle^0=\langle\alpha_{\mathbf{K}1}\alpha_{\mathbf{K}1}{}^*\rangle^0=1. \qquad (73)$$

The collective terms come [see (71)] from commutators

$$\langle[(f_{\mathbf{K}}^{\mathbf{Q}})^*,\alpha_{\mathbf{K}+\mathbf{Q}0}{}^*\alpha_{\mathbf{K}0}]\rangle^0; \qquad (74)$$

but by simple enumeration we find that none of these commutators can result in quantities (73), so that $(74)\equiv 0$. Then there are no coupling terms in the equations of motion of $\alpha_{\mathbf{K}+\mathbf{Q}0}{}^*\alpha_{\mathbf{K}0}$, or for that matter the corresponding $-\nu$ quantity $\alpha_{\mathbf{K}+\mathbf{Q}1}{}^*\alpha_{\mathbf{K}1}$. This proves the theorem: the equations of motion are compatible with the requirement

$$(x_{\pm\nu}^{\mathbf{Q}})_u\Psi_0=\left\{\begin{matrix}\alpha_{\mathbf{K}+\mathbf{Q}0}{}^*\alpha_{\mathbf{K}0}\\ \alpha_{\mathbf{K}+\mathbf{Q}1}{}^*\alpha_{\mathbf{K}1}\end{matrix}\right\}\Psi_0\equiv 0, \qquad (75)$$

because

$$[\mathcal{H},(x_{\pm\nu}^{\mathbf{Q}})_u]=\pm(\nu_{\mathbf{K}}^{\mathbf{Q}})_u(x_{\pm\nu}^{\mathbf{Q}})_u. \qquad (76)$$

The analogous requirement in the spin theory was of course (43′), and that analogy shows us the nature of (75) as opposed to the similar (67). (75) is a linearized version of a *kinematical* condition, automatically satisfied by the equations of motion; we do not know the nonlinear equivalent, although it must exist. On the other hand, (67) is simply a defining condition for the ground state, which is not satisfied even in infinitesimally excited states, and also differs in that it defines a true change in the wave function from the simple product result. Equation (75) says kinematically that the motion involves *only* the simultaneous excitation of pairs; while (67) expresses the *dynamical* extent to which this occurs in the ground state. (75) will be a useful result in finding the solution of our equations.

Now let us briefly note the second limiting case: that (48)–(60) reduce to the R.P.A. theory of Coulomb correlation in the normal case in which $b_k^0=b_k^{0*}=0$. The Sawada-Brout theory corresponds to keeping only (49) and (50) [(55) and (58)], and to including the effect of exchange only by perturbation theory. As Sawada shows, in terms of diagrams this amounts to summing, besides exchange diagrams to second order, all diagrams of the form of Fig. 2(a).

Our theory in its complete form, including the screened exchange terms (51)–(52) [(56) and (59)], is an equation-of-motion equivalent of the diagram method of Hubbard, which is more accurate than Sawada-Brout.

This method sums automatically also all the diagrams of the form of Fig. 2(b), in which the interaction lines for the exchange scatterings themselves imply complete sums of terms like Fig. 2(a). A more complete discussion of the relationship of this method to other treatments of Coulomb correlation and other many-body problems[26] will be given in a later publication.

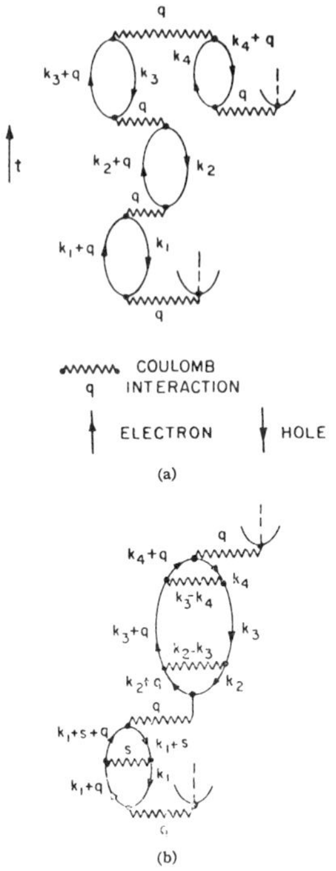

FIG. 2. Perturbation-theoretic diagrams summed by the random-phase method. Plain arcs represent electrons and holes, wavy lines Coulomb interactions ("longitudinal photons"); the momentum and momentum transfer, respectively, are given next to the lines. (a) Typical chain-type diagrams summed by the Sawada *et al.*-Brout equations. (b) More general diagrams summed by full method. Use of dielectric constant screening causes exchange interactions (like that labeled s) to imply full sums of diagrams like 2(a).

V. PARTIAL SOLUTION OF EQUATIONS OF MOTION

In this section we shall attempt a discussion of the solutions, and particularly the collective ones, of the equations of motion (49)–(60). As we pointed out in the last section, the individual-particle modes will automatically agree with the B.C.S.-Bogoliubov theory; we shall, however, verify that also. Let us first write down the equations, making use of the following abbreviations, and neglecting the exchange terms (which have the effect fundamentally of simply altering slightly the kinetic energy and the direct scattering terms, without changing their character) throughout:

$$\omega_{KQ}=\epsilon_{K+Q}-\epsilon_K,\quad \Omega_{KQ}=\epsilon_{K+Q}+\epsilon_K,$$
$$-I_K=\sum_k V(\mathbf{K},\mathbf{k})b_k=\sum_k V(\mathbf{K},\mathbf{k})b_k^*, \tag{77}$$
$$n_{KQ}=n_{K+Q}-n_K,\quad z_{KQ}=1-n_K-n_{K+Q}.$$

Here we have chosen, with no loss in generality, $b_k=b_k^*$ (the domain wall in the $+x$ direction). I_K is then half of the x component of H_K, and is defined to be positive (V is negative). With (77), we get

$$[\mathcal{H},\rho_K^Q]=\omega_{KQ}\rho_K^Q-V_D\rho^Q n_{KQ}-I_{K+Q}b_K^Q+I_K\bar{b}_K^Q$$
$$+b_K\sum_k V\bar{b}_k^Q-b_{K+Q}\sum_k Vb_k^Q, \tag{78a}$$

$$[\mathcal{H},\bar{\rho}_K^Q]=-\omega_{KQ}\bar{\rho}_K^Q+V_D\rho^Q n_{KQ}-I_K b_K^Q+I_{K+Q}\bar{b}_K^Q$$
$$+b_{K+Q}\sum_k V\bar{b}_k^Q-b_K\sum_k Vb_k^Q, \tag{78b}$$

$$[\mathcal{H},b_K^Q]=-\Omega_{KQ}b_K^Q-V_D\rho^Q(b_K+b_{K+Q})-I_K\bar{\rho}_K^Q$$
$$-I_{K+Q}\rho_K^Q-z_{KQ}\sum_k Vb_k^Q, \tag{78c}$$

$$[\mathcal{H},\bar{b}_K^Q]=\Omega_{KQ}\bar{b}_K^Q+V_D\rho^Q(b_K+b_{K+Q})$$
$$+(I_{K+Q}\bar{\rho}_K^Q+I_K\rho_K^Q)+z_{KQ}\sum_k V\bar{b}_k^Q. \tag{78d}$$

Here we have $V=V(\mathbf{K},\mathbf{k})$ for brevity, and $V_D=V_D(Q)$.

The time-reversal feature of (78) may be seen by noticing that if we take $\bar{\rho}_k^Q=(\rho_k^Q)^*$, $\bar{b}=(b)^*$, (78b) is (78a) and (78d) is (78c). Interpreting $[\mathcal{H},f]$ as $i\dot{f}$, this means that (78) connects the real parts with time

[26] Note the presence in Eqs. (60) of terms like

$$(1-n_K-n_{-K-Q})b_k^Q,$$

by means of which the scattering of excited pairs of electrons or holes in the presence of the Fermi sea may be calculated. Thus these equations reduce to the Bethe-Goldstone ones for the case of the normal Fermi sea.

derivatives of imaginary parts and vice versa; thus it is basically an equation in $(d/dt)^2$. Our method of solution will be to derive the two second-order equations for the (say) imaginary parts. For instance, we write down

$$[\mathcal{H}, \rho_K^Q+\bar{\rho}_K^Q]=\omega_{KQ}(\rho-\bar{\rho})_K^Q-(I_K+I_{K+Q})\times(b-\bar{b})_K^Q-(b_K+b_{K+Q})\textstyle\sum_k V(b-\bar{b})_k^Q, \quad (79)$$

and

$$[\mathcal{H},[\mathcal{H},(\rho-\bar{\rho})_K^Q]]=\omega_{KQ}[\mathcal{H},(\rho+\bar{\rho})_K^Q]-2V_D n_{KQ}[\mathcal{H},\rho^Q]-(I_{K+Q}-I_K)[\mathcal{H},(b+\bar{b})_K^Q]+(b_K-b_{K+Q})[\mathcal{H},\textstyle\sum_k V(b+\bar{b})_k^Q]. \quad (80)$$

In (80) we substitute from (79) and the corresponding equation for $b+\bar{b}$, except in the last sum, which is most conveniently left in its present form:

$$[\mathcal{H},[\mathcal{H},w_K^Q]]=\nu^2 w_K^Q=[\omega_{KQ}^2+(I_{K+Q}-I_K)^2]w_K^Q-2(\epsilon_{K+Q}I_K-\epsilon_K I_{K+Q})y_K^Q-2V_D n_{KQ}\Pi^Q-[\omega_{KQ}(b_K+b_{K+Q})+z_{KQ}(I_K-I_{K+Q})]\textstyle\sum_k V y_k^Q+(b_K-b_{K+Q})B_K^Q. \quad (81)$$

Here we have made some further abbreviations:

$$y_k^Q=b_k^Q-\bar{b}_k^Q;\quad w_k^Q=\rho_k^Q-\bar{\rho}_k^Q;$$
$$B_K^Q=[\mathcal{H},\textstyle\sum_k V(b_k^Q+\bar{b}_k^Q)]; \quad (82)$$
$$\Pi^Q=[\mathcal{H},\rho^Q]=\textstyle\sum_k \omega_{kQ}w_k^Q.$$

This last equality is the all-important sum rule.[6]

The equation for y_K^Q is obtained similarly:

$$\nu^2 y_K^Q=[\Omega_{KQ}^2+(I_K+I_{K+Q})^2]y_K^Q-2(I_K\epsilon_{K+Q}-I_{K+Q}\epsilon_K)w_K^Q-2V_D(b_K+b_{K+Q})\Pi^Q+[\Omega_{KQ}z_{KQ}+(I_K+I_{K+Q})(b_K+b_{K+Q})]\times\textstyle\sum_k V y_k^Q-z_{KQ}B_K^Q. \quad (83)$$

First note that, because

$$-\frac{n_{KQ}}{b_K+b_{K+Q}}=\frac{b_K-b_{K+Q}}{z_{KQ}}=\frac{\omega_{KQ}(b_K+b_{K+Q})-z_{KQ}(I_{K+Q}-I_K)}{\Omega_{KQ}z_{KQ}+(I_K+I_{K+Q})(b_K+b_{K+Q})}, \quad (84)$$

the collective parts of the two Eqs. (81) and (83) are simply proportional. This is the result of the condition (74)–(75): one solution factors from the secular equation.

Second, we derive the spectrum of the individual-particle solutions. This too we know must come out right, but we shall do it as a check on our reasoning. Leaving out the collective terms, the secular equation is

$$0=\begin{vmatrix}\Omega_{KQ}^2+(I_K+I_{K+Q})^2-\nu^2 & -2(I_K\epsilon_{K+Q}-I_{K+Q}\epsilon_K)\\ -2(I_K\epsilon_{K+Q}-I_{K+Q}\epsilon_K) & \omega_{KQ}^2+(I_{K+Q}-I_K)^2-\nu^2\end{vmatrix},$$

which indeed has the solutions

$$(\nu_K^Q)^2=[(\epsilon_K^2+I_K^2)^{\frac{1}{2}}\pm(\epsilon_{K+Q}^2+I_{K+Q}^2)^{\frac{1}{2}}]^2, \quad (85)$$

of which those with the − sign are exact solutions because of (76), but must be discarded because of (75).

Now we shall go on to discuss collective solutions. To avoid complications we make some assumptions which are valid in the weak-coupling limit or in the "cutoff" theory of B.C.S. in which V is a constant. In some cases we assume small Q, the case we study primarily (for large Q the overwhelming majority of pair excitations of momentum Q are obviously unaffected by superconductivity).

We assume $I_K=I$ independent of **K**, as it will nearly be if $\epsilon_0\ll\omega_D$. y_K^Q will be small except within ϵ_0 of the Fermi surface; then we can assume

$$\textstyle\sum_k V(\mathbf{K},\mathbf{k})y_k^Q=A^Q \text{ independent of } \mathbf{K}. \quad (86)$$

We shall find that the most important collective modes have y even; in that case

$$[\mathcal{H},\textstyle\sum_K V(b+\bar{b})_K^Q]\cong-\sum_K V\Omega_{KQ}y_K^Q=B^Q \quad (87)$$

is also zero. However, for the sake of completeness we shall retain B^Q, also assuming it to be a constant with varying **K**. Then the Eqs. (81) and (83) become

$$(\nu^2-\omega_{KQ}^2)w_K^Q=-2\omega_{KQ}Iy_K^Q-2V_D n_{KQ}\Pi^Q-\omega_{KQ}(b_K+b_{K+Q})A^Q+(b_K-b_{K+Q})B^Q, \quad (81')$$

and

$$[\nu^2-(\Omega_{KQ}^2+4I^2)]y_K^Q=-2\omega_{KQ}Iw_K^Q-2V_D(b_K+b_{K+Q})\Pi^Q+[\Omega_{KQ}z_{KQ}+2I(b_K+b_{K+Q})]A^Q-z_{KQ}B^Q. \quad (83')$$

To take advantage of (84), let us define

$$\Phi_K^Q=2V_D\Pi^Q+\frac{\omega_{KQ}}{n_{KQ}}(b_K+b_{K+Q})A^Q-\frac{(b_K-b_{K+Q})}{n_{KQ}}B^Q. \quad (88)$$

Then (81′) and (83′) may be rewritten

$$(\nu^2-\omega_{KQ}^2)w_K^Q+2\omega_{KQ}Iy_K^Q=-n_{KQ}\Phi_K^Q,$$
$$[\nu^2-(\Omega_{KQ}^2+4I^2)]y_K^Q+2\omega_{KQ}Iw_K^Q=-(b_K+b_{K+Q})\Phi_K^Q. \quad (89)$$

We may solve by multiplying the first by $2\omega_{KQ}I$, the second by $(\nu^2-\omega_{KQ}^2)$, and subtracting, or vice versa. In that case we get

$$\{[\nu^2-(\Omega_{KQ}^2+4I^2)][\nu^2-\omega_{KQ}^2]-4\omega_{KQ}^2I^2\}y_K^Q=-[(\nu^2-\omega_{KQ}^2)-2n_{KQ}\omega_{KQ}I(b_K+b_{K+Q})^{-1}]\times(b_K+b_{K+Q})\Phi_K^Q.$$

With the collective terms (the right-hand side) left out, this equation must have the solutions (85), and so the left-hand side must simply be

$$[\nu^2-(\nu_K^Q)^2][\nu^2-(\nu_K^Q)_u^2]y_K^Q,$$

where ν_K^Q is the physical (positive-sign) root of (85) and $(\nu_K^Q)_u$ the unphysical, negative-sign one. The only way, then, to satisfy the requirement (76) is for $[\nu^2-(\nu_K^Q)_u^2]$ to be a factor on *both* sides; canceling this factor out, we obtain

$$[\nu^2-(\nu_{\mathbf{K}}^{\mathbf{Q}})^2]y_{\mathbf{K}}^{\mathbf{Q}}=-(b_K+b_{\mathbf{K}+\mathbf{Q}})\Phi_{\mathbf{K}}^{\mathbf{Q}}. \quad (90)$$

By a similar procedure we can also arrive at

$$[\nu^2-(\nu_{\mathbf{K}}^{\mathbf{Q}})^2]w_{\mathbf{K}}^{\mathbf{Q}}=-n_{\mathbf{KQ}}\Phi_{\mathbf{K}}^{\mathbf{Q}}. \quad (91)$$

[So far we have not used our simplifying assumptions in any essential way, and (90) and (91) could be used to obtain complete spectra of collective modes if the resulting integral equations could be solved.]

From (90), (91), and the definition (88) of Φ, we can form the sums Π, A, and B, and arrive at the integral equations which determine the frequencies:

$$\begin{aligned}\Pi^{\mathbf{Q}}=&\{\textstyle\sum_{\mathbf{k}}\omega_{\mathbf{kQ}}n_{\mathbf{kQ}}[(\nu_{\mathbf{k}}^{\mathbf{Q}})^2-\nu^2]^{-1}\}2V_D\Pi^{\mathbf{Q}}\\&+\{\textstyle\sum_{\mathbf{k}}\omega_{\mathbf{kQ}}^2(b_{\mathbf{k}}+b_{\mathbf{k+Q}})[(\nu_{\mathbf{k}}^{\mathbf{Q}})^2-\nu^2]^{-1}\}A^{\mathbf{Q}}\\&-\{\textstyle\sum_{\mathbf{k}}\omega_{\mathbf{kQ}}(b_{\mathbf{k}}-b_{\mathbf{k+Q}})[(\nu_{\mathbf{k}}^{\mathbf{Q}})^2-\nu^2]^{-1}\}B^{\mathbf{Q}}\end{aligned} \quad (92)$$

$$\begin{aligned}A^{\mathbf{Q}}=&\{\textstyle\sum_{\mathbf{k}}V(b_{\mathbf{k}}+b_{\mathbf{k+Q}})[(\nu_{\mathbf{k}}^{\mathbf{Q}})^2-\nu^2]^{-1}\}2V_D\Pi^{\mathbf{Q}}\\&+\{\textstyle\sum_{\mathbf{k}}V\omega_{\mathbf{kQ}}(b_{\mathbf{k}}+b_{\mathbf{k+Q}})^2n_{\mathbf{kQ}}^{-1}[(\nu_{\mathbf{k}}^{\mathbf{Q}})^2-\nu^2]^{-1}\}A^{\mathbf{Q}}\\&-\{\textstyle\sum_{\mathbf{k}}V(b_{\mathbf{k}}^2-b_{\mathbf{k+Q}}^2)n_{\mathbf{kQ}}^{-1}[(\nu_{\mathbf{k}}^{\mathbf{Q}})^2-\nu^2]^{-1}\}B^{\mathbf{Q}},\end{aligned} \quad (93)$$

$$\begin{aligned}B^{\mathbf{Q}}=&-\{\textstyle\sum_{\mathbf{k}}V\Omega_{\mathbf{kQ}}(b_{\mathbf{k}}+b_{\mathbf{k+Q}})[(\nu_{\mathbf{k}}^{\mathbf{Q}})^2-\nu^2]^{-1}\}2V_D\Pi^{\mathbf{Q}}\\&-\{\textstyle\sum_{\mathbf{k}}V\omega_{\mathbf{kQ}}\Omega_{\mathbf{kQ}}(b_{\mathbf{k}}+b_{\mathbf{k+Q}})^2n_{\mathbf{kQ}}^{-1}\\&\times[(\nu_{\mathbf{k}}^{\mathbf{Q}})^2-\nu^2]^{-1}\}A^{\mathbf{Q}}+\{\textstyle\sum_{\mathbf{k}}V\Omega_{\mathbf{kQ}}\\&\quad\times(b_{\mathbf{k}}^2-b_{\mathbf{k+Q}}^2)n_{\mathbf{kQ}}^{-1}[(\nu_{\mathbf{k}}^{\mathbf{Q}})^2-\nu^2]^{-1}\}B^{\mathbf{Q}}.\end{aligned} \quad (94)$$

(92)–(94) are a set of three simultaneous linear equations in Π, A, and B. Symmetry about the sphere $|\mathbf{k}+(\mathbf{Q}/2)|=k_F$ makes the cross-terms coupling B to A and Π vanish. Thus there are, as in the $Q=0$ case, two independent types of collective solutions: the odd ones involving $B^{\mathbf{Q}}$, corresponding to δH_x in (33) finite ($\lim_{Q\to0}B^{\mathbf{Q}}\propto\delta H_x$), and the even ones, with finite $A^{\mathbf{Q}}$ and $\Pi^{\mathbf{Q}}$ (like the δH_y solutions). The odd solutions obey (94), which by use of (84) is

$$1=\textstyle\sum_{\mathbf{k}}(-V)\Omega_{\mathbf{kQ}}z_{\mathbf{kQ}}[(\nu_{\mathbf{k}}^{\mathbf{Q}})^2-\nu^2]^{-1}. \quad (95)$$

As in the $Q=0$ case, all solutions lie in or close to the continuum, and approach as $Q\to0$ the corresponding solution of (34)–(35).

As for the "even" solutions, our assumptions have been equivalent to the $V=$constant assumption which eliminates any higher bound states for $Q=0$. These higher states may also exist for $Q\neq0$, for physical V's, but we shall ignore them, simply observing that p, d, etc. states, or if you like, transverse and more complicated collective waves, may exist but will be near the continuum, and their energies will approach the corresponding $Q=0$ energies as $Q\to0$.

This leaves the coupled Eqs. (92)–(93) to determine the physically important longitudinal collective modes. The dispersion equation is a determinant

$$\begin{vmatrix}1-2V_Df & l\\ 2V_Dh & 1-g\end{vmatrix}=0, \quad (96)$$

where after some algebra we define

$$\begin{aligned}f&=\textstyle\sum_{\mathbf{k}}\omega_{\mathbf{kQ}}n_{\mathbf{kQ}}[(\nu_{\mathbf{k}}^{\mathbf{Q}})^2-\nu^2]^{-1},\\g&=\textstyle\sum_{\mathbf{k}}(-V)\nu_{\mathbf{k}}^{\mathbf{Q}}\cos^2[\frac{1}{2}(\theta_{\mathbf{k}}-\theta_{\mathbf{k+Q}})][(\nu_{\mathbf{k}}^{\mathbf{Q}})^2-\nu^2]^{-1},\\h&=\textstyle\sum_{\mathbf{k}}(-V)(b_{\mathbf{k}}+b_{\mathbf{k+Q}})[(\nu_{\mathbf{k}}^{\mathbf{Q}})^2-\nu^2]^{-1},\\l&=\textstyle\sum_{\mathbf{k}}\omega_{\mathbf{kQ}}^2(b_{\mathbf{k}}+b_{\mathbf{k+Q}})[\nu^2-(\nu_{\mathbf{k}}^{\mathbf{Q}})^2]^{-1}.\end{aligned} \quad (97)$$

The collective modes have entirely different behavior depending on whether we consider the charged or neutral cases. In the charged case, V_D is singular and large, Π is the important variable, and f determines the frequencies; in the neutral case $A^{\mathbf{Q}}$ (a variable which closely resembles the "rotation" of Sec. III) is the important variable and g mostly determines the frequency. We shall analyze the two cases separately.

Case I. Neutral Fermi Gas

This case is defined by $V_D=$const as $Q\to0$ (we might as well let $V_D=V$, the ordinary interaction). Since $f\propto Q^2$, $1-2V_Df\cong1$ in the long-wave limit except very near a $\nu_{\mathbf{k}}^{\mathbf{Q}}$; in a principal-value sense, f is small everywhere. This limit of small Q is the interesting region, since in reference 6 we proved that in this case states with $Q\to0$ lie at the bottom of the energy gap; thus we use a perturbation procedure suited to this case. Since also $l\propto Q^2$, in all terms except $1-g$ we can make approximations; in particular, we neglect f and get

$$1-g=2Vlh. \quad (98)$$

In the limit $Q=0$, this has, aside from the individual-particle solutions, only the $\nu=0$ collective mode which we discussed in Sec. III; $g=1$, $\nu=0$ leads to precisely the equilibrium condition for the ground state. Our task here is to get the dispersion of this mode to lowest order in Q. For this purpose it is adequate to expand h and l to lowest nonvanishing order in Q^2 and ν^2, and g to first order:

$$\begin{aligned}h&\cong\textstyle\sum_{\mathbf{k}}(-V)\sin\theta_{\mathbf{k}}(\nu_{\mathbf{k}}^0)^{-2}=2I\sum_{\mathbf{k}}(-V)(\nu_{\mathbf{k}}^0)^{-3};\\l&\cong-\textstyle\sum_{\mathbf{k}}\omega_{\mathbf{kQ}}^2\sin\theta_{\mathbf{k}}(\nu_{\mathbf{k}}^0)^{-2}\\&\qquad=-\tfrac{1}{3}k_F^2Q^2m^{-2}\textstyle\sum_{\mathbf{k}}\sin\theta_{\mathbf{k}}(\nu_{\mathbf{k}}^0)^{-2};\\g&\cong1+\nu^2\textstyle\sum_{\mathbf{k}}(-V)(\nu_{\mathbf{k}}^0)^{-3}\\&\quad+\{\textstyle\sum_{\mathbf{k}}(-V)(\nu_{\mathbf{k}}^{\mathbf{Q}})^{-1}\cos^2[\frac{1}{2}(\theta_{\mathbf{k+Q}}-\theta_{\mathbf{k}})]\\&\quad-\tfrac{1}{2}[\textstyle\sum_{\mathbf{k}}(-V)(\nu_{\mathbf{k}}^0)^{-1}+\sum_{\mathbf{k}}(-V)(\nu_{\mathbf{k+Q}}^0)^{-1}]\}.\end{aligned} \quad (99)$$

The calculation of this last difference follows:

$$\begin{aligned}\{\ \}&=\textstyle\sum_{\mathbf{k}}(-V)I^{-1}(\sin\theta_{\mathbf{k}}+\sin\theta_{\mathbf{k+Q}})^{-1}[\cos^2(\frac{1}{2}(\theta_{\mathbf{k+Q}}-\theta_{\mathbf{k}}))\\&\qquad\times\sin\theta_{\mathbf{k}}\sin\theta_{\mathbf{k+Q}}-\tfrac{1}{4}(\sin\theta_{\mathbf{k}}+\sin\theta_{\mathbf{k+Q}})^2]\\&=-\textstyle\sum_{\mathbf{k}}(-V)(4I)^{-1}(\sin\theta_{\mathbf{k}}+\sin\theta_{\mathbf{k+Q}})^{-1}\\&\qquad\times(\sin\theta_{\mathbf{k}}\cos\theta_{\mathbf{k+Q}}-\sin\theta_{\mathbf{k+Q}}\cos\theta_{\mathbf{k}})^2\\&\cong-\textstyle\sum_{\mathbf{k}}(-V)\omega_{\mathbf{kQ}}^2(\nu_{\mathbf{k}}^0)^{-3}.\end{aligned}$$

Thus the dispersion of this mode is given by

$$\begin{aligned}1-[1&+\nu^2\textstyle\sum_{\mathbf{k}}(-V)(\nu_{\mathbf{k}}^0)^{-3}-\sum_{\mathbf{k}}(-V)(\nu_{\mathbf{k}}^0)^{-3}\omega_{\mathbf{kQ}}^2]\\&=2V[-2I\textstyle\sum_{\mathbf{k}}(-V)(\nu_{\mathbf{k}}^0)^{-3}][\sum_{\mathbf{k}}\omega_{\mathbf{kQ}}^2\sin\theta_{\mathbf{k}}(\nu_{\mathbf{k}}^0)^{-2}],\end{aligned}$$

or

$$\nu^2 = \tfrac{1}{3} v_F^2 Q^2 \left[1 + 4IV \textstyle\sum_{\mathbf{k}} \sin\theta_{\mathbf{k}} (\nu_{\mathbf{k}}^0)^{-2}\right]$$
$$= \tfrac{1}{3} v_F^2 Q^2 \left[1 - 4N(0)|V|\right]. \qquad (100)$$

Apparently the "phonon" velocity $3^{-\frac{1}{2}} v_F$ is strictly a kinematical "ideal gas" effect, the $N(0)V$ correction being a result of the coupling to the direct interaction in the present weak coupling case. The kinematical term has been obtained in a different way by Bogoliubov.[12]

Case II. Charged Fermi Gas

In this case

$$V_D(Q) = 2\pi e^2 Q^{-2}, \qquad (101)$$

and in the important $Q \to 0$ limit, the cross-coupling $2V_D lh$ as well as the term involving f are rather large constants unless ν is large. First let us satisfy ourselves that there are no low-lying collective modes. For such modes ν could be neglected in estimating l and h; we calculated in Case I that $lh(\nu=0) \cong I^{-2} N^2(0) V \langle \omega_{\mathbf{kQ}}^2 \rangle$ so we find

$$2V_D lh(\nu=0) \cong \omega_p^2 I^{-2} N(0) V \gg 1.$$

Thus the collective mode with $\nu=0$ near $1-g=0$ disappears without trace; we can only hope for a solution at very large ν. We expect such a solution to lie near ω_p; we shall find there that now l, h, and g are small and in them we assume $\nu = \omega_p$. What we then seek are the corrections to ω_p of lowest order in the energy gap $\epsilon_0 = 2I$.

Near ω_p, g is small, if we assume that the exchange-phonon interaction, when averaged over attractive and repulsive regions, is small compared to the direct interaction. In any case $1-g$ is only coupled in by small terms, so we can neglect g and write

$$1 - 2V_D f = 2V_D lh. \qquad (102)$$

Near $\nu = \omega_p$, $Q=0$, we have

$$2V_D lh = -2V_D \left(\textstyle\sum_{\mathbf{k}} \omega_{\mathbf{kQ}}^2 \omega_p^{-2} \sin\theta_{\mathbf{k}}\right) \times \left(\textstyle\sum_{\mathbf{k}} |V| \omega_p^{-2} \sin\theta_{\mathbf{k}}\right)$$
$$= -2V_D \langle \omega_{\mathbf{kQ}}^2 \rangle_{\mathrm{Av}} \omega_p^{-4} 4I^2 |V|^{-1}. \qquad (103)$$

The correction to f may be calculated as follows:

$$f \cong -\nu^{-2} \textstyle\sum_{\mathbf{k}} \omega_{\mathbf{kQ}} n_{\mathbf{kQ}} - \omega_p^{-4} \textstyle\sum_{\mathbf{k}} \omega_{\mathbf{kQ}} n_{\mathbf{kQ}} (\nu_{\mathbf{k}}^{\mathrm{Q}})^2,$$

so that

$$2V_D f = \omega_p^2 \nu^{-2} - \langle \omega_{\mathbf{kQ}}^2 \rangle_{\mathrm{Av}} 2V_D \omega_p^{-4} \times \textstyle\sum_{\mathbf{k}} n_{\mathbf{kQ}} \omega_{\mathbf{kQ}}^{-1} (\nu_{\mathbf{k}}^{\mathrm{Q}})^2$$
$$= \omega_p^2 \nu^{-2} + 2V_D \langle \omega_{\mathbf{kQ}}^2 \rangle_{\mathrm{Av}} \omega_p^{-4} 4I^2 |V|^{-1}. \qquad (104)$$

We see that the corrections in (103) and (104) cancel, leaving the plasma frequency as $Q \to 0$ unchanged, even to the very small terms which we are calculating. It seems likely that the dispersion of the plasma mode is also unchanged, since ω_p must be the same also for large Q.

Thus, in conclusion, the predictions of reference 6 are in large part borne out: that the charged Fermi gas has no low-lying collective modes because of the strong plasma effect, while the neutral gas has a low-lying branch. The present weak-coupling theory gives no correction to the plasma mode, and derives a phonon-like mode for the neutral case with a small interaction correction. The presence of the interaction correction represents a definite improvement of the present method.

ACKNOWLEDGMENTS

I should like to thank Dr. K. Yosida for a critical reading of this paper and for the resulting useful suggestions; and Dr. H. Suhl for his help at several points.

APPENDIX I. INCLUSION OF DIRECT INTERACTIONS WITH PHONONS

In the actual physical superconductor there always remain a group of excitations in the energy gap: namely, the lattice phonons. According to our discussion of Sec. II, it should be feasible to study the effect of superconductivity on the phonons by introducing explicitly only the direct electron-phonon interaction; if the phonons are not seriously perturbed by the difference between normal and superconducting states, then their contribution to the dielectric constant, and thus to the superconducting interaction V, can be studied as though the electron wave function were normal.

Clearly it is only important to include phonons in the physical, charged case; the neutral case is of possible physical interest only in such problems as He^3 or the nucleus, where there is no lattice.

The inclusion of phonons simply involves including among the excitations of momentum $\mathbf{Q}$ the coordinate and momentum q_{Q} and p_{Q}, and the terms of (12) and (13) involving these:

$$(\mathcal{H}_{\mathrm{ph}} + \mathcal{H}_i)^{\mathrm{Q}} = (p_{\mathrm{Q}} p_{-\mathrm{Q}} + f_Q^2 q_{\mathrm{Q}} q_{-\mathrm{Q}}) - (q_{\mathrm{Q}} \rho^{-\mathrm{Q}} v_Q^i + q_{-\mathrm{Q}} \rho^{\mathrm{Q}} v_Q^{i*}). \qquad (\mathrm{A1})$$

The equations of motion for the new variables lead rather simply to

$$[\mathcal{H}, [\mathcal{H}, p_{\mathrm{Q}}]] = f_Q^2 p_{\mathrm{Q}} + 2i v_Q^{i*} \Pi^{\mathrm{Q}}. \qquad (\mathrm{A2})$$

Now we must calculate the effect of $\mathcal{H}_i$ in (A1) on the various electron coordinates. The results are

$$\begin{aligned}
[\mathcal{H}_i, \rho_{\mathbf{K}}^{\mathrm{Q}}] &= -q_{\mathrm{Q}} v_Q^i (n_{\mathbf{K}} - n_{\mathbf{K+Q}}), \\
[\mathcal{H}_i, \bar{\rho}_{\mathbf{K}}^{\mathrm{Q}}] &= q_{\mathrm{Q}} v_Q^i (n_{\mathbf{K}} - n_{\mathbf{K+Q}}), \\
[\mathcal{H}_i, b_{\mathbf{K}}^{\mathrm{Q}}] &= q_{\mathrm{Q}} v_Q^i (b_{\mathbf{K}} + b_{\mathbf{K+Q}}), \\
[\mathcal{H}_i, \bar{b}_{\mathbf{K}}^{\mathrm{Q}}] &= -q_{\mathrm{Q}} v_Q^i (b_{\mathbf{K}} + b_{\mathbf{K+Q}}).
\end{aligned} \qquad (\mathrm{A3})$$

The extra terms (A3) lead to an extra term on the right of Eqs. (81) and (82) for w and y:

$$\begin{aligned}
[\mathcal{H}_i + \mathcal{H}_{\mathrm{ph}}, [\mathcal{H}_i, w_{\mathbf{K}}^{\mathrm{Q}}]] &= 2i p_{\mathrm{Q}} v_Q^i n_{\mathbf{KQ}}, \\
[\mathcal{H}_i + \mathcal{H}_{\mathrm{ph}}, [\mathcal{H}_i, y_{\mathbf{K}}^{\mathrm{Q}}]] &= 2i p_{\mathrm{Q}} v_Q^i (b_{\mathbf{K}} + b_{\mathbf{K+Q}}).
\end{aligned} \qquad (\mathrm{A4})$$

Thus the effect of including the phonons appears (appropriately, since the phonons are collective variables) entirely as an extra term in the collective part $\Phi_K{}^Q$ of Eqs. (90) and (91); we must define a new $\Phi_K{}^Q$ which is given by

$$\Phi_K{}^Q=2V_D\Pi^Q+\frac{\omega_{KQ}}{n_{KQ}}(b_K+b_{K+Q})A^Q-\frac{(b_K-b_{K+Q})}{n_{KQ}}B^Q -2iv_Q{}^ip_Q, \quad (A5)$$

while p_Q is related to Π^Q by (A2).

Again we find that symmetry about the Fermi surface allows us to decouple B^Q from the rest of the equations, so we may assume B^Q zero. Then we are left with a three-by-three set of equations for Π^Q, A^Q, and p^Q instead of the two-by-two determinant (96). The calculation is so similar that one needs only to write down the final determinant:

$$\begin{vmatrix} f_Q{}^2-\nu^2 & 2iv_Q{}^{i*} & 0 \\ 2iv_Q{}^if & 1-2V_Df & l \\ -2iv_Q{}^ih & 2V_Dh & 1-g \end{vmatrix}=0, \quad (A6)$$

where the symbols f, g, h, and l are the same as they were in (97).

The result for the normal metal would be the limiting case of (A6) in which h and l are zero, decoupling A^Q, and $\nu_k{}^Q\rightarrow\omega_{kQ}$ in f. Then the secular equation is

$$\nu_{\text{norm}}{}^2=f_Q{}^2+\frac{4|v_Q{}^i|^2f}{1-2V_Df}. \quad (A7)$$

This is exactly the same as the secular equation [his (7a)] of Wentzel,[16] if we note that our f is his f_q; our $v_Q{}^i$, his μ_q; our V_D, his λ_q. Using (14) and (11), we find

$$\nu^2=\omega_Q{}^2+(4\pi ne^2Z^2M^{-1})\frac{1}{1-2V_Df} \cong\omega_Q{}^2+\tfrac{1}{3}(m/M)Z^2v_f{}^2Q^2, \quad (A8)$$

using the fact that $f\cong-2N(0)$. This is a well-known result in this approximation.[2]

Now we shall study the effect of superconductivity on the phonons. Adding $iv_Q{}^iV_D{}^{-1}$ times the second column to the first simplifies the determinant to

$$\begin{vmatrix} f_Q{}^2-\nu^2-2|v_Q{}^i|^2V_D{}^{-1} & 2iv_Q{}^{i*} & 0 \\ iv_Q{}^i/V_D & 1-2V_Df & l \\ 0 & 2V_Dh & 1-g \end{vmatrix}=0,$$

which, expanded, is

$$\nu^2-\omega_Q{}^2=2|v_Q{}^i|^2V_D{}^{-1}\{1-2V_D[f+lh/(1-g)]\}^{-1}. \quad (A9)$$

In this form the analogy with (A8) is clear: we have simply replaced the damping factor $(1-2V_Df)^{-1}$ by $\{1-2V_D[f+lh/(1-g)]\}^{-1}$. In the normal case, f is a constant and the singularity in V_D makes the second term the large one; now $f\propto Q^2$ and if the phonon frequency is to remain similar $lh/(1-g)$ must have the correct value. Fortunately, in the "neutral case" calculations of Sec. V we have the values of l, h, and $(1-g)$ in the appropriate limit, and we get

$$\frac{lh}{1-g}=-2N(0)\left(1+\frac{\nu^2}{\langle\omega_{kQ}{}^2\rangle_{\text{Av}}}\right). \quad (A10)$$

Since phonon frequencies are quite small relative to ω_{kQ}, this agrees well with the normal value (A8). The velocity change caused by superconductivity, given by the last term, is of order m/M or $\sim10^{-4}$. Whether this is physical depends on whether our calculation is really accurate for these very small terms.

Note added in proof.—Dr. A. W. Overhauser has pointed out to me that an effect agreeing with this correction in order of magnitude and sign was measured on Sn and Pb by B. Welber and S. L. Quimby, Acta Metallurgica **6**, 351 (1958).

APPENDIX II. TOTAL ENERGY CALCULATIONS

As Wentzel has pointed out, in the simple Sawada-Brout method the energy change caused by the interactions is given simply by the change in zero-point energy summed over all the excitations. Unfortunately, even the inclusion of phonon and photon exchange effects complicates this simple prescription very much. We shall show briefly here how the energy might be calculated on our method, but since it is relatively unimportant physically—representing only a small change from the B.C.S. result—we shall not make any attempt at evaluating it. This appendix is included primarily just to show that the equation of motion method is a complete and satisfactory substitute for the diagram method, for the energy as well as for the excitation spectrum which it exhibits so naturally. Since this is the purpose, we confine the calculation to the case of an ordinary space potential:

$$V=\sum_Q V(Q)\rho^{Q*}\rho^Q, \quad (A11)$$

and then the potential energy (the kinetic can be obtained by the trick of integrating with respect to e^2)[8] is

$$\text{P.E.}=\sum_Q V(Q)(\rho^Q\Psi_0,\rho^Q\Psi_0). \quad (A12)$$

The essential point of the method is to expand ρ^Q in terms of the eigenexcitations x_ν (64) (we shall work entirely with momentum $\mathbf{Q}$ alone, so we omit the $\mathbf{Q}$ index where possible hereafter) which we expand [instead of (64)] for convenience in Bogoliubov form:

$$\begin{aligned} x_\nu&=\sum_k(\lambda_{\nu k}\alpha_{k+Q0}{}^*\alpha_{k1}{}^*+\mu_{\nu k}\alpha_{k+Q1}\alpha_{k0}),\\ x_{-\nu}&=\sum_k(\mu_{\nu k}\alpha_{k+Q0}{}^*\alpha_{k1}{}^*-\lambda_{\nu k}\alpha_{k+Q1}\alpha_{k0}). \end{aligned} \quad (A13)$$

We also shall need the inverse transformation

$$\begin{aligned} \alpha_{k+Q0}{}^*\alpha_{k1}{}^*&=\sum_\nu(l_{k\nu}x_\nu+m_{k\nu}x_{-\nu}),\\ \alpha_{k+Q1}\alpha_{k0}&=\sum_\nu(m_{k\nu}x_\nu-l_{k\nu}x_{-\nu}). \end{aligned} \quad (A14)$$

Simply by substitution of (A14) in (A13) we get

$$\sum_{\mathbf{k}} (\lambda_{\nu\mathbf{k}} l_{\mathbf{k}\nu'} + \mu_{\nu\mathbf{k}} m_{\mathbf{k}\nu'}) = \delta_{\nu\nu'}. \tag{A15}$$

Let us define two quantities which relate ρ and x_ν, and which could if necessary be written out in terms of the expansion coefficients:

$$U_\nu = (x_\nu\Psi_0, \rho\Psi_0) = (\Psi_0, (x_\nu^*\rho - \rho x_\nu^*)\Psi_0); \qquad \sum_\nu Y_\nu x_\nu = \rho. \tag{A16}$$

Then clearly the term in the energy (A12) for momentum Q is

$$V(Q)\sum_{\nu>0} U_\nu Y_\nu. \tag{A17}$$

[U_ν is zero for $\nu<0$ by (67).]

We find that the equations of motion, together with (A15), give us a relationship for $U_\nu Y_\nu$. Suppose, for example, that we assume that (90) and (91) are the correct equations of motion. A little manipulation leads us to the following equations for the α's:

$$\nu\alpha_{\mathbf{k}+\mathbf{Q}0}^*\alpha_{\mathbf{k}1}^* = \nu_{\mathbf{k}}^{\mathbf{Q}}\alpha_{\mathbf{k}+\mathbf{Q}0}^*\alpha_{\mathbf{k}1}^* + (1/2\nu)f_1(\mathbf{k})\Phi_{\mathbf{k}}^{\mathbf{Q}}, \tag{A18a}$$

$$\nu\alpha_{\mathbf{k}+\mathbf{Q}1}\alpha_{\mathbf{k}0} = -\nu_{\mathbf{k}}^{\mathbf{Q}}\alpha_{\mathbf{k}+\mathbf{Q}1}\alpha_{\mathbf{k}0} - (1/2\nu)f_1(\mathbf{k})\Phi_{\mathbf{k}}^{\mathbf{Q}}, \tag{A18b}$$

where $f_1(\mathbf{k})$ is the complicated but known function

$$f_1(\mathbf{k}) = \cos[\tfrac{1}{2}(\theta_{\mathbf{k}+\mathbf{Q}} - \theta_{\mathbf{k}})][b_{\mathbf{k}} + b_{\mathbf{k}+\mathbf{Q}}] + \sin[\tfrac{1}{2}(\theta_{\mathbf{k}+\mathbf{Q}} - \theta_{\mathbf{k}})]n_{\mathbf{k}\mathbf{Q}}. \tag{A19}$$

Note added in proof.—G. Rickaysen has pointed out to me that Eqs. (A18) are in error, and that only their consequence

$$[\nu^2 - (\nu_{\mathbf{k}}^{\mathbf{Q}})^2](\alpha_{\mathbf{k}+\mathbf{Q}0}^*\alpha_{\mathbf{k}1}^* - \alpha_{\mathbf{k}+\mathbf{Q}1}\alpha_{\mathbf{k}0}) = f_1(\mathbf{k})\Phi_{\mathbf{k}}^{\mathbf{Q}}$$

is valid. The principle of the energy calculation is not affected by this error, and also (A24) is still correct.

Now we can derive equations both for the λ's and the l's from (A18). We take the average value in the ground state of the commutator of x_ν^* first with (A18a) and then (A18b). This gives

$$\lambda_{\nu\mathbf{k}}(\nu - \nu_{\mathbf{k}}^{\mathbf{Q}}) = (2\nu)^{-1}f_1(\mathbf{k})(x_\nu\Psi_0, \Phi_{\mathbf{k}}^{\mathbf{Q}}\Psi_0), \qquad \mu_{\nu\mathbf{k}}(\nu + \nu_{\mathbf{k}}^{\mathbf{Q}}) = (2\nu)^{-1}f_1(\mathbf{k})(x_\nu\Psi_0, \Phi_{\mathbf{k}}^{\mathbf{Q}}\Psi_0). \tag{A20}$$

Second, we expand (A18) by means of (A14). Then if

$$\Phi_{\mathbf{k}}^{\mathbf{Q}} = \sum_\nu \varphi_{\mathbf{k}\nu} x_\nu, \tag{A21}$$

we have

$$(\nu - \nu_{\mathbf{k}}^{\mathbf{Q}})l_{\mathbf{k}\nu} = (2\nu)^{-1}f_1(\mathbf{k})\varphi_{\mathbf{k}\nu} = -(\nu + \nu_{\mathbf{k}}^{\mathbf{Q}})m_{\mathbf{k}\nu}. \tag{A22}$$

The only remaining task is to study $\Phi_{\mathbf{k}}^{\mathbf{Q}}$. To show what might be done with this let us take the simple case of Sawada-Brout, where

$$\Phi_{\mathbf{k}}^{\mathbf{Q}} = 2V(Q)\Pi^{\mathbf{Q}}.$$

Since

$$(x_\nu\Psi_0, \Pi^{\mathbf{Q}}\Psi_0) = \nu U_\nu, \tag{A23}$$

(A20) becomes

$$\lambda_{\nu\mathbf{k}} = f_1(\mathbf{k})(\nu - \nu_{\mathbf{k}}^{\mathbf{Q}})^{-1}U_\nu V(Q),$$

$$\mu_{\nu\mathbf{k}} = -f_1(\mathbf{k})(\nu + \nu_{\mathbf{k}}^{\mathbf{Q}})^{-1}U_\nu V(Q),$$

while, using (A16) and the definition of $\Pi^{\mathbf{Q}}$,

$$\varphi_{\mathbf{k}\nu} = 2V\nu Y_\nu,$$

so that (A22) becomes

$$l_{\mathbf{k}\nu} = VY_\nu(\nu - \nu_{\mathbf{k}}^{\mathbf{Q}})^{-1},$$

$$m_{\mathbf{k}\nu} = -VY_\nu(\nu + \nu_{\mathbf{k}}^{\mathbf{Q}})^{-1}.$$

Then (A15) gives us our desired relation for $U_\nu Y_\nu$:

$$U_\nu Y_\nu \sum_{\mathbf{k}} V^2(Q)f_1^2(\mathbf{k})[(\nu - \nu_{\mathbf{k}}^{\mathbf{Q}})^{-2} - (\nu + \nu_{\mathbf{k}}^{\mathbf{Q}})^{-2}] = 1. \tag{A24}$$

Eventually this leads to exactly the simple result of Sawada, when we insert the correct value of $f_1(\mathbf{k})$.

In the more general case $\Phi_{\mathbf{k}}^{\mathbf{Q}}$ does not consist of $\Pi^{\mathbf{Q}}$ alone. In any soluble case, however, it consists of a finite number of sums like $A^{\mathbf{Q}}$ and $B^{\mathbf{Q}}$ which we must perforce consider as constants, and the solution of the problem involves finding a linear equation set like (92)–(94) connecting the parts of Φ. If this is so we can solve (92)–(94) for the $A^{\mathbf{Q}}$, $B^{\mathbf{Q}}$, and any further such sums in terms of $\Pi^{\mathbf{Q}}$; then the only change in the theory is a redefinition of the $\mathbf{k}$- and ν-dependent quantity $f_1(\mathbf{k})$, which will have to contain factors coming from the solution for Φ in terms of Π.

Collective Excitations in the Theory of Superconductivity*

G. Rickayzen†
Department of Physics, University of Illinois, Urbana, Illinois
(Received March 5, 1959)

The complex dielectric constant of a superconductor and the Meissner effect are derived in a manner which is gauge invariant, from the theory of superconductivity due to Bardeen, Cooper, and Schrieffer. The collective excitations are important in maintaining gauge invariance; the longitudinal collective excitations ensure that a static vector potential produces no longitudinal current and the transverse collective excitations contribute to the Meissner current an amount which depends on the angular properties of the two-body interaction. This contribution is estimated to be small. An earlier calculation of ultrasonic absorption in superconductors is justified. The whole investigation is based upon the generalized random-phase approximation introduced by Anderson and applies whether or not the Coulomb interaction between the electrons is taken into account. The equations of motion are linearized in such a way that the exchange terms are automatically screened if the Coulomb interaction is, in fact, taken into account. The region of applicability of most of the results is limited by the approximations to temperatures at or near absolute zero.

1. INTRODUCTION

MANY properties of a superconductor, notably the thermodynamic properties, can be understood in terms of independent quasi-particle excitations of the system. However, the force between one electron and another which brings about the superconducting transition also ensures that the system possess certain collective excitations. These collective excitations are essential for a complete understanding of certain properties of the superconducting system, particularly its interaction with external electric and magnetic fields. In this paper we wish to stress the collective aspects of the theory of superconductivity of Bardeen, Cooper, and Schrieffer.[1] The main contributions of the present paper are a completely gauge-covariant calculation of the Meissner effect and of the complex dielectric constant of a superconductor at absolute zero. These calculations take into account both the longitudinal and transverse collective excitations of the system.

Using the random-phase approximation, Anderson[2] and Bogoliubov, Tolmachov, and Shirkov[3] studied the existence and frequencies of the longitudinal collective excitations. They have found that the plasma frequency and the collective coordinates are practically unchanged in the transition to the superconducting state, in the long-wavelength limit. They have pointed out the existence of transverse and more complicated oscillations and BTS have attempted to calculate their frequencies. Anderson has shown that when the longitudinal collective modes are taken into account then, at least to order $(\epsilon_0/\hbar\omega)^2$, the longitudinal sum rules are satisfied.

Already in a normal metal the collective aspects of the interacting electron system are important. As is well known, if an external charge interacts with the electrons of the metal, all the electrons are perturbed in such a way that each electron is acted upon by the field of the external charge together with the perturbed fields of all the other electrons. The result is that every electron is perturbed by a screened field. (The screened field can be calculated by a self-consistent Hartree method[4] or equivalently by a canonical transformation.[5,6]) If the external charge density is a wave of long wavelength the screening is practically complete.

Another way of looking at the screening is to note that a part of the charge density of the electrons is a plasmon variable and that a low-frequency external field will not excite the plasmon states. We should like to develop this viewpoint using some of the ideas of Lipkin.[7] The system of electrons possesses longitudinal collective modes of wave vector $\mathbf{k}$ with coordinates Q_k, conjugate momenta P_k, and frequency ω_k. In the long-wavelength limit, Q_k is $(4\pi e^2)^{\frac{1}{2}}\sum_i \exp(i\mathbf{k}\cdot\mathbf{x}_i/2)(\mathbf{k}\cdot\mathbf{p}_i/nk)\exp(i\mathbf{k}\cdot\mathbf{x}_i/2)$ and P_k is $i(4\pi e^2/k^2)^{\frac{1}{2}}\sum_i \exp(i\mathbf{k}\cdot\mathbf{x}_i)$, where the $\mathbf{x}_i$ are the coordinates of the electrons and $\mathbf{p}_i$ are their momenta. There will be other operators, functions of $\mathbf{x}_i$ and $\mathbf{p}_i$, which, added to P_k, Q_k, will form a complete set. What these other variables are we leave aside for the moment. Now if an external field (say that associated with an incoming phonon) of wave vector $\mathbf{k}$, frequency Ω, and amplitude r_k acts on the system, there is an extra interaction term in the Hamiltonian,

$$M_k^2 r_k \rho_{-k}\exp(i\Omega t)+\text{c.c.},\quad \rho_{-k}=\sum_i \exp(-i\mathbf{k}\cdot\mathbf{x}_i),$$

where M_k^2 is the strength of the interaction. The component of the charge density ρ_{-k} can be expanded in terms of P_k, Q_k, and the other variables of the complete

* This work was supported in part by the Office of Ordnance Research, U. S. Army.

† Present address: Nuclear Physics Research Laboratory, University of Liverpool, England.

[1] Bardeen, Cooper, and Schrieffer, Phys. Rev. **108**, 1175 (1957). This paper is referred to as BCS.

[2] P. W. Anderson, Phys. Rev. **114**, 1002 (1959).

[3] Bogoliubov, Tolmachov, and Shirkov, *A New Method in the Theory of Superconductivity* (Academy of Sciences of USSR Press, Moscow, 1958, translated by Consultants Bureau, Inc., New York, 1959), Chaps. IV and VII. This book is referred to as BTS.

[4] J. Bardeen, Phys. Rev. **52**, 688 (1937).

[5] S. Nakajima, *Proceedings of the International Conference on Theoretical Physics, Kyoto and Tokyo, September, 1953* (Science Council of Japan, Tokyo, 1954).

[6] J. Bardeen and D. Pines, Phys. Rev. **99**, 1140 (1955).

[7] H. J. Lipkin, Phys. Rev. Letters **2**, 159 (1959).

set; it then has the structure

$$\rho_{-k}=\alpha_k P_k+\rho_{-k,s},$$

where

$$-i\hbar\alpha_k=[\rho_{-k},Q_k]_-, \quad =-\hbar(4\pi e^2/k^2)^{-\frac{1}{2}},$$

and $\rho_{-k,s}$ is the residual screened charge density which vanishes in the long-wavelength limit. ($\rho_{-k,s}$ is of order g_k^2 relative to P_k, where g_k is the electron-plasmon coupling constant.) The first term of ρ_{-k} can lead to real transitions only if energies $\hbar\omega_k$ are involved. In general, however, $\omega_k\gg\Omega$, so that the only part of ρ_{-k} that can cause real transitions is $\rho_{-k,s}$, that is, the screened part.

One can eliminate the collective part of ρ_{-k} from the Hamiltonian by performing a canonical transformation with the unitary operator

$$\exp[ir_k e^{i\Omega t}M_k^2\alpha_k Q_{-k}/\hbar-\text{c.c.}];$$

as $k\rightarrow 0$ this approaches

$$\exp[ir_k e^{i\Omega t}\textstyle\sum_i \exp(-\frac{1}{2}i\mathbf{k}\cdot\mathbf{x}_i) \times\mathbf{p}_i \exp(-\frac{1}{2}i\mathbf{k}\cdot\mathbf{x}_i)\cdot\mathbf{k}/nk^2-\text{c.c.}]$$

which is of the form found by Pines and Schrieffer[8] and can be interpreted as giving rise to a "dipolar backflow." The backflow is calculated in more detail at the end of Sec. 4.

In a metal the screening and collective coordinates are also affected by the extra phonon interaction between electrons which leads to the superconducting state. However, the contribution to the screening comes from all electrons within the Fermi sphere, whereas in the superconducting transition only those electrons within a small energy range kT_c of the Fermi surface are involved. Thus it is to be expected that in the superconducting state the screening will be practically the same as in the normal state. This is confirmed by the calculations of Anderson and BTS and by that of Sec. IV which is also based upon the generalized random phase approximation.

One problem that requires the introduction (explicit or implicit) of the collective modes for its solution is that of the Meissner effect. BCS have calculated by perturbation theory the current density produced by a static transverse vector potential. They made no attempt to derive the current density produced by a static longitudinal vector potential although it was earlier pointed out by Bardeen[9] that it would be necessary to take account of the longitudinal collective modes to do this. A static longitudinal vector potential contributes neither to the electric field nor to the magnetic field; it should, therefore, have no physical effects and should not give rise to a current. Because the BCS method, if applied without modification to the longitudinal vector potential, would give a nonzero (not even small) current, doubt has been cast on their calculation of the transverse current.[10] Anderson's verification[11] of the sum rules shows that the longitudinal current density is of order $(\epsilon_0/\hbar\omega)^2$, but as he does not introduce the wavefunctions explicitly his proof is not a test of the method of BCS for calculating the transverse current.

Pines and Schrieffer,[8] by exploiting the smallness of the electron-plasmon coupling constant, g, have shown that the longitudinal current density is small. Their proof introduces the wave functions for the collective states and depends on the fact that the collective coordinates are practically unchanged in the superconducting transition. As we have seen, in the normal state the long-wavelength components of charge and longitudinal paramagnetic current densities are just collective coordinates (P_k and Q_k, respectively) to order g^2 and this is still true in the superconducting state. As Pines and Schrieffer show, the collective part of the longitudinal paramagnetic current density just cancels the diamagnetic current density and the total longitudinal current density is at the most of order g^4. An objection[10] has been raised against the argument of Pines and Schrieffer on the grounds that they use the controversial subsidiary condition of Bohm and Pines,[12] but in the way we have put the argument the subsidiary condition does not arise. In Sec. VI of this paper the corrections to the collective coordinates are taken into account and from the equations of motion obtained by Anderson within the generalized random-phase approximation, the transverse and longitudinal currents are calculated simultaneously. Because the longitudinal current is found to be zero it is believed that the calculation of the transverse current is to be trusted.

Because of the properties of the interaction which leads to the superconducting transition there exist, in a superconductor, transverse collective modes. (The effect of these excitations on the Meissner current has already been reported in a preliminary letter.[13]) In the presence of a static transverse vector potential they do not contribute to the current in the London limit (where the penetration depth is much greater than the coherence distance); only the single-particle excitations have to be considered and the calculation is exactly that of BCS. In the Pippard limit (where the penetration depth is much less than coherence distance) there is a contribution from the transverse modes and this depends on the angular properties of the two-particle interaction. We have assumed a simple angular dependence for the interaction to estimate the order of magnitude of the correction and we conclude that it is, in fact, small.

A simple model for a superconductor that is commonly used[3,14,15] is a gas of Fermi particles which do not interact with each other through the Coulomb interaction, but only through the phonons. For this model the

[8] D. Pines and J. R. Schrieffer, Nuovo cimento **10**, 496 (1958).
[9] J. Bardeen, Nuovo cimento **5**, 1766 (1957).
[10] G. Wentzel, Phys. Rev. Letters **2**, 33 (1959).
[11] P. W. Anderson, Phys. Rev. **110**, 827 (1958).
[12] D. Bohm and D. Pines, Phys. Rev. **92**, 609 (1953).
[13] G. Rickayzen, Phys. Rev. Letters **2**, 90 (1959).
[14] G. Wentzel, Phys. Rev. **111**, 1488 (1958).
[15] G. Rickayzen, Phys. Rev. **111**, 817 (1958).

long-wavelength longitudinal excitations are not charge or current density fluctuations and it is not so easy to see their part in the Meissner effect and the screening. The method given below applies to both this case and a model which is more realistic and the results are formally the same. Of course if the parameters involved in the two models were calculated from first principles they would be different. If the treatment of Bogoliubov is used and the Meissner effect calculated according to the method used previously by the author,[15] we believe that to obtain the correct result for the effect of the longitudinal vector potential, it is necessary to sum an infinite set of graphs. This same set of graphs should lead to a correction to the transverse current of the kind calculated in Sec. VI.

If one wishes to compute quantities of order g^2, for example the screening of a time-dependent external field, then the single-particle excitations of the system must be examined more closely and one must calculate $\rho_{-k,s}$. In the first calculations of the interaction of a time-dependent longitudinal external field with the superconducting electrons[1,16] it was assumed that one can use the normal form for the interaction with the same screening as in the normal state. For instance, it was implicitly assumed in BCS that the interaction between an acoustic wave and the superconducting electrons is of the form

$$\sum_Q (4\pi e^2/Q^2) r_{-Q} e^{-i\Omega t} \rho_Q f(Q) + \text{c.c.}, \qquad (1.1)$$

where $r_{-Q}\exp(i\Omega t)$ is the charge fluctuation associated with the acoustic wave and $f(Q)$ is the same screening factor as in the normal state. However, if the theory is gauge invariant one should be able to describe the interaction by either a vector potential or a scalar potential. Thus an equally good choice for the interaction would appear to be

$$\sum_Q (4\pi e^2/Q^2) r_{-Q} e^{-i\Omega t} [\mathbf{Q}\cdot\mathbf{j}(Q)] f(Q)/\Omega + \text{c.c.}, \qquad (1.2)$$

which is also the interaction in the normal state. If this interaction is actually used, then a result for the attenuation is obtained very different from that of BCS. This discrepancy arises because the equivalence of Eqs. (1.1) and (1.2) depends on the equation for the conservation of charge,

$$(E_n - E_m)\langle n|\rho|m\rangle = \hbar\langle n|\mathbf{Q}\cdot\mathbf{j}|m\rangle,$$

and this equation is not satisfied by the wave functions of BCS [not even to within $(\epsilon_0/\hbar\omega)^2$]. There is no general principle to show which result is correct. Therefore in order to ensure that the result be independent of gauge, it is necessary that the screening be calculated from improved wave functions, in a way that is gauge invariant. This calculation is performed in Sec. III, and it is concluded that for cases of practical importance $\langle n|\rho_Q|m\rangle$ can be replaced by $\langle n|\rho_Q|m\rangle_{\text{BCS}} f(Q)$ so that one can use the scalar potential screened as in the normal state. The interaction with a vector potential has to be modified so as to produce agreement.

The basis of this paper is the set of linear equations of motion determined by Anderson[2]; he used the generalized random-phase approximation and neglected the exchange terms. We solve these equations in Sec. III and obtain wave functions for which the equation governing the conservation of charge is satisfied. We then find a simple approximation to the matrix elements. In Sec. IV we add the effect of an external time-dependent charge fluctuation to the equations of motion and from the solution determine the generalized dielectric constant which depends on the frequency and wave vector of the external field. A form for the interaction with the external field is given, which can be used with the wave functions of BCS. From this interaction it is shown in Sec. V that corrections to the calculation of ultrasonic attenuation by BCS are negligible. It is also shown that the exact wave functions incorporate a backflow around an external charge. In Sec. VI, the effect of an external static vector potential is added to the equations of motion and from the solution the Meissner effect is calculated in an arbitrary gauge.

As the results of this paper are based upon the generalized random-phase approximation (RPA), their region of applicability is limited. The essence of the RPA is that products of a pair of single-particle operators are treated as bosons. The approximation takes account of the transition of a pair of quasi-particles out of the Fermi sea or into it but neglects the scattering of a particle outside the sea. This approximation should be valid as long as the number of single-particle excitations outside the sea is small, a condition which restricts the discussion to the region of temperature near absolute zero. The result for the real part of the dielectric constant is as accurate as all but the most recent calculations of the dielectric constant of a free-electron gas. If we were to include the exchange terms we should even obtain Hubbard's result[17] but we shall not enlarge on this here.

One defect of the present analysis is that the two-body interaction is arbitrarily chosen so that the Hamiltonian is gauge invariant and still leads to a superconducting transition. (The two-body interaction used by BCS is such that their Hamiltonian is not strictly gauge invariant.) Most of the results are not sensitive to the potential used and can be applied directly to the model of BCS. The physical principles underlying the mathematics are generally so clear that it is easy to see how the results can be applied. Those corrections which are sensitive to the form of the interaction cannot be trusted quantitatively except perhaps so far as the order of magnitude. In any case our model provides an example against which calculations of the electromagnetic properties of superconductors can be tested. To

[16] Bardeen, Tewordt, and Rickayzen, Phys. Rev. **113**, 982 (1959).

[17] J. Hubbard, Proc. Roy. Soc. (London) **A240**, 539 (1957).

improve upon our method it would be necessary to go back to the original gauge-covariant Hamiltonian from which the electron-phonon interaction has not been eliminated.

An unsatisfactory feature of Anderson's approach is the assumption that in the terms of the equations of motion that lead to the superconductlng transition, the two-body interactions are screened. This point has been discussed by Anderson, who suggests that these equations are the second step in a self-consistent calculation. It is clear from the work of BCS and BTS that the two-body interaction in this term is in fact screened, but it would be more satisfactory to have a justification of Anderson's equations of motion which is basic. In Appendix A we attempt to provide this justification. We are led to a system of nonlinear equations which can be approximated by Anderson's equations in just the way he suggests. Although we do not take into account the exchange terms, it is possible to do this by the method of Appendix A and to obtain a set of linear equations with the exchange terms properly screened.

2. NOTATION

In this section the notation to be used is summarized. The operators which create electrons in the states of momentum $\mathbf{k}$ and spin $\boldsymbol{\sigma}$ are denoted by $c_{k,\sigma}$. The Bloch energies of the electrons in the normal state are ϵ_k, measured from the Fermi level, ϵ_F. In general ϵ_k will be assumed to be $(\hbar^2k^2/2m)-\epsilon_F$. The velocity at the Fermi surface is v_0. We define

$$\rho_k{}^Q=c_{k+Q\uparrow}{}^*c_{k\uparrow},\quad \bar{\rho}_k{}^Q=c_{-k\downarrow}{}^*c_{-k-Q\downarrow},$$

$$b_k{}^Q=c_{-k-Q\downarrow}c_{k\uparrow},\quad \bar{b}_k{}^Q=c_{k+Q\uparrow}{}^*c_{-k\downarrow}{}^*.$$

The operators b_k, ρ_k corresponding to $\mathbf{Q}=0$ are always taken as c-numbers, their expectation values in the BCS ground state,

$$b_k=\langle c_{-k\downarrow}c_{k\uparrow}\rangle_0=b_k{}^*,$$

$$\rho_k=\langle c_{k\uparrow}{}^*c_{k\uparrow}\rangle_0=\bar{\rho}_k=n_k.$$

The wave functions are normalized to unit volume. The Hamiltonian is

$$H=H_K+H_V+H_C,$$

where

$$H_K=\sum_{k,\sigma}\epsilon_k c_{k,\sigma}{}^*c_{k,\sigma},$$

$$H_V=\tfrac{1}{2}\sum_{k,k',q}\sum_{\sigma,\sigma'}v(\mathbf{k},\mathbf{k}')c_{k'\sigma'}{}^*c_{-k'+q\sigma}{}^*c_{-k+q\sigma}c_{k\sigma'},$$

$$H_C=\sum_{k,k',q}\sum_{\sigma,\sigma'}2\pi e^2|\mathbf{k}-\mathbf{k}'|^{-2}c_{k'\sigma'}{}^*c_{-k'+q\sigma}{}^*c_{-k+q\sigma}c_{k\sigma'}.$$

For the theory to be gauge invariant, $v(\mathbf{k},\mathbf{k}')$ is taken to be a function of $(\mathbf{k}-\mathbf{k}')$ only, $v(\mathbf{k}-\mathbf{k}')$. The direct interaction is denoted by

$$V_D(\mathbf{Q})=4\pi e^2Q^{-2}+v(\mathbf{Q}),$$

and the interaction which leads to the superconducting transition is denoted by

$$V(\mathbf{k},\mathbf{k}')=[v(\mathbf{k},\mathbf{k}')+4\pi e^2|\mathbf{k}-\mathbf{k}'|^{-2}]\times(\text{screening factor}).$$

The potential $V(\mathbf{k},\mathbf{k}')$ is predominantly negative when $\mathbf{k}$ and $\mathbf{k}'$ are near the Fermi surface. If the method of Appendix A is followed, the screening factor is zero for $|\mathbf{k}-\mathbf{k}'|<Q_{\max}$ and unity otherwise. The wave number $Q_{\max}$ is the cutoff of Sawada *et al.*[18]

The energy gap parameter, analogous to ϵ_0 of BCS, is I_K where

$$I_K=-\textstyle\sum_k V(\mathbf{K},\mathbf{k})b_k=-\sum_k V(\mathbf{K},\mathbf{k})b_k{}^*.$$

Near the Fermi surface it is a constant which we shall denote by ϵ_0. The energy of a single quasi-particle excitation is

$$E_k=(\epsilon_k{}^2+I_k{}^2)^{\frac{1}{2}}.$$

The energy of a pair of particles with momenta $-\mathbf{k}$ and $\mathbf{k}+\mathbf{Q}$ is

$$\nu_k(\mathbf{Q})=E_k+E_{k+Q}.$$

The coherence distance is $\xi_0=\hbar v_0/\pi\epsilon_0$. The ground state is such that

$$b_k=b_k{}^*=u_kv_k>0,\quad n_k=v_k{}^2,$$

where

$$u_k{}^2=\tfrac{1}{2}(1+\epsilon_k/E_k),\quad v_k{}^2=\tfrac{1}{2}(1-\epsilon_k/E_k).$$

We shall use the operators,

$$\gamma_{k0}=u_kc_{k\uparrow}-v_kc_{-k\downarrow}{}^*,$$

$$\gamma_{k1}=u_kc_{-k\downarrow}+v_kc_{k\uparrow}{}^*,$$

the four coherence factors,

$$l(\mathbf{k},\mathbf{Q})=u_ku_{k+Q}+v_kv_{k+Q},$$

$$m(\mathbf{k},\mathbf{Q})=u_kv_{k+Q}+v_ku_{k+Q},$$

$$n(\mathbf{k},\mathbf{Q})=u_ku_{k+Q}-v_kv_{k+Q},$$

$$p(\mathbf{k},\mathbf{Q})=u_kv_{k+Q}-v_ku_{k+Q},$$

and the three collective variables

$$\begin{aligned}\rho(\mathbf{Q})&=\textstyle\sum_k(\rho_k{}^Q+\bar{\rho}_k{}^Q)\\&=\textstyle\sum_k[m(\mathbf{k},\mathbf{Q})(\gamma_{k+Q0}{}^*\gamma_{k1}{}^*+\gamma_{k+Q1}\gamma_{k0})\\&\qquad+n(\mathbf{k},\mathbf{Q})(\gamma_{k1}{}^*\gamma_{k+Q1}+\gamma_{k+Q0}{}^*\gamma_{k0})],\end{aligned}$$

$$\begin{aligned}B_K(\mathbf{Q})&=\textstyle\sum_k V(\mathbf{K},\mathbf{k})(b_k{}^Q+\bar{b}_k{}^Q)\\&=\textstyle\sum_k V(\mathbf{K},\mathbf{k})[n(\mathbf{k},\mathbf{Q})(\gamma_{k+Q0}{}^*\gamma_{k1}{}^*+\gamma_{k+Q1}\gamma_{k0})\\&\qquad-m(\mathbf{k},\mathbf{Q})(\gamma_{k+Q0}{}^*\gamma_{k0}+\gamma_{k1}{}^*\gamma_{k+Q1})],\end{aligned}$$

$$\begin{aligned}A_K(\mathbf{Q})&=\textstyle\sum_k V(\mathbf{K},\mathbf{k})(b_k{}^Q-\bar{b}_k{}^Q)\\&=-\textstyle\sum_k V(\mathbf{K},\mathbf{k})[l(\mathbf{k},\mathbf{Q})(\gamma_{k+Q0}{}^*\gamma_{k1}{}^*-\gamma_{k+Q1}\gamma_{k0})\\&\qquad+p(\mathbf{k},\mathbf{Q})(\gamma_{k+Q0}{}^*\gamma_{k0}-\gamma_{k1}{}^*\gamma_{k+Q1})].\end{aligned}$$

[18] Sawada, Brueckner, Fukuda, and Brout, Phys. Rev. **108**, 507 (1957).

3. SOLUTION OF THE EQUATIONS OF MOTION

In this section we are concerned with the solution of the equations of motion derived from the Hamiltonian H, and establishing the equation of conservation of charge,

$$(E_n-E_0)(\Psi_n,\rho(\mathbf{Q})\Psi_0)=\hbar(\Psi_n,\mathbf{Q}\cdot\mathbf{j}(\mathbf{Q})\Psi_0). \quad (3.1)$$

Having obtained matrix elements for which the equation of conservation holds, we find approximate expressions for them which are valid in the usual physical situations for which they are required.

The equations of motion that are the basis for this work are the linear equations derived by Anderson. They were obtained from the full equations of motion for $\rho_k{}^Q$, $\bar{\rho}_k{}^Q$, $b_k{}^Q$, $\bar{b}_k{}^Q$, by replacing those products of pairs of operators which have nonzero expectation values in the ground state by those expectation values, i.e., $c_{k\sigma}{}^*c_{k\sigma}$, $c_{-k\downarrow}c_{k\uparrow}$ are replaced by the c-numbers n_k, b_k, respectively. All remaining terms containing the product of four operators are neglected. In those terms in which b_k appears the potential is screened. Anderson gave reasons for this screening but since this procedure is not in the spirit of his approximations we provide a justification for his equations in Appendix A. The resultant linearized equations of motion are

$$\begin{aligned}[H,\rho_K{}^Q]=&(\epsilon_{K+Q}-\epsilon_K)\rho_K{}^Q-V_D(\mathbf{Q})\rho(\mathbf{Q})(n_{K+Q}-n_K)\\&-I_{K+Q}b_K{}^Q+I_K\bar{b}_K{}^Q+b_K\textstyle\sum_k V\bar{b}_k{}^Q\\&-b_{K+Q}\textstyle\sum_k Vb_k{}^Q,\end{aligned} \quad (3.2a)$$

$$\begin{aligned}[H,\bar{\rho}_K{}^Q]=&(\epsilon_K-\epsilon_{K+Q})\bar{\rho}_K{}^Q+V_D(\mathbf{Q})\rho(\mathbf{Q})(n_{K+Q}-n_K)\\&-I_Kb_K{}^Q+I_{K+Q}\bar{b}_K{}^Q+b_{K+Q}\textstyle\sum_k V\bar{b}_k{}^Q\\&-b_K\textstyle\sum_k Vb_k{}^Q,\end{aligned} \quad (3.2b)$$

$$\begin{aligned}[H,b_K{}^Q]=&-(\epsilon_K+\epsilon_{K+Q})b_K{}^Q-V_D(\mathbf{Q})\rho(\mathbf{Q})\\&\times(b_K+b_{K+Q})-I_K\bar{\rho}_K{}^Q-I_{K+Q}\rho_K{}^Q\\&-(1-n_K-n_{K+Q})\textstyle\sum_k Vb_k{}^Q,\end{aligned} \quad (3.2c)$$

$$\begin{aligned}[H,\bar{b}_K{}^Q]=&(\epsilon_K+\epsilon_{K+Q})\bar{b}_K{}^Q+V_D(\mathbf{Q})\rho(\mathbf{Q})\\&\times(b_K+b_{K+Q})+I_{K+Q}\bar{\rho}_K{}^Q+I_K\rho_K{}^Q\\&+(1-n_K-n_{K+Q})\textstyle\sum_k V\bar{b}_k{}^Q.\end{aligned} \quad (3.2d)$$

As the mathematical detail may obscure the essentially simple steps involved, it will be useful to outline the procedure beforehand. We first find those linear combinations $\mu_k{}^*(\mathbf{Q})$, $\mu_k(\mathbf{Q})$, of the operators ρ, $\bar{\rho}$, b, $\bar{b}$, which are the normalized normal modes of the equations of motion (3.2), and then we define the ground state Ψ_0 of the problem by

$$\mu_k(\mathbf{Q})\Psi_0=0,$$

where the $\mu_k(\mathbf{Q})$ are the destruction operators. Then the unscreened charge density and current density are written in terms of the normal modes, and the matrix elements

$$(\Psi_0\mu_k{}^*(\mathbf{Q}),\rho(\mathbf{Q})\Psi_0),\quad(\Psi_0\mu_k{}^*(\mathbf{Q}),\mathbf{Q}\cdot\mathbf{j}(\mathbf{Q})\Psi_0)$$

are calculated. It is shown explicitly that Eq. (3.1) governing the conservation of charge is satisfied. As already pointed out, Eq. (3.1) implies that the calculation of transition probabilities is independent of gauge. In the succeeding sections the effects of external electromagnetic fields are added to the equations of motion. These produce changes in the operators, $\mu_k(\mathbf{Q})$, from which the complex dielectric constant and the Meissner effect can be calculated.

It is actually easier to work with the operators γ_k of Bogoliubov[19] and Valatin.[20] In terms of these operators it is found (after considerable algebraic manipulation) that

$$\begin{aligned}[H,\gamma_{k+Q0}{}^*\gamma_{k1}{}^*]=&(E_k+E_{k+Q})\gamma_{k+Q0}{}^*\gamma_{k1}{}^*\\&+V_D(\mathbf{Q})m(\mathbf{k},\mathbf{Q})\rho(\mathbf{Q})+\tfrac{1}{2}n(\mathbf{k},\mathbf{Q})\\&\times B_k(\mathbf{Q})-\tfrac{1}{2}l(\mathbf{k},\mathbf{Q})A_k(\mathbf{Q}),\end{aligned} \quad (3.3a)$$

$$\begin{aligned}[H,\gamma_{k+Q1}\gamma_{k0}]=&-(E_k+E_{k+Q})\gamma_{k+Q1}\gamma_{k0}\\&-V_D(\mathbf{Q})m(\mathbf{k},\mathbf{Q})\rho(\mathbf{Q})-\tfrac{1}{2}n(\mathbf{k},\mathbf{Q})\\&\times B_k(\mathbf{Q})-\tfrac{1}{2}l(\mathbf{k},\mathbf{Q})A_k(\mathbf{Q}),\end{aligned} \quad (3.3b)$$

$$[H,\gamma_{k+Q0}{}^*\gamma_{k0}]=(E_{k+Q}-E_k)\gamma_{k+Q0}{}^*\gamma_{k0},$$

$$[H,\gamma_{k1}{}^*\gamma_{k+Q1}]=-(E_{k+Q}-E_k)\gamma_{k1}{}^*\gamma_{k+Q1}.$$

Evidently, half of the normal modes are given by the operators $\gamma_{k+Q0}{}^*\gamma_{k0}$, $\gamma_{k1}{}^*\gamma_{k+Q1}$, which have eigenvalues $(E_{k+Q}-E_k)$ and (E_k-E_{k+Q}), respectively. These are just the modes called unphysical by Anderson. Since all the physical states Ψ satisfy

$$\gamma_{k+Q0}{}^*\gamma_{k0}\Psi=\gamma_{k1}{}^*\gamma_{k+Q1}\Psi=0,$$

these operators may be taken to be zero throughout the remainder of this section.

In the absence of the collective coordinates $\rho(\mathbf{Q})$, $A_K(\mathbf{Q})$, $B_K(\mathbf{Q})$, the other normal modes are $\gamma_{k+Q0}{}^*\gamma_{k1}{}^*$ and $\gamma_{k+Q1}\gamma_{k0}$ which oscillate with frequencies (E_k+E_{k+Q}) and $-(E_k+E_{k+Q})$, respectively. In the presence of the collective coordinates there will be modes with frequencies $\pm(E_k+E_{k+Q})$ and also collective modes with frequencies outside the range of the $\pm(E_k+E_{k+Q})$. We shall let $\mu_i(\mathbf{Q})$ denote any collective coordinates. We shall use $\mu_k(\mathbf{Q})^*$ to denote a mode of frequency (E_k+E_{k+Q}) and $\mu_{-k}(-\mathbf{Q})$ to denote the mode of frequency $-(E_k+E_{k+Q})$. The operator $\mu_k{}^*(\mathbf{Q})$ adds energy to the system and is, therefore, a creation operator; $\mu_k(-\mathbf{Q})$ subtracts energy from the system and is, therefore, a destruction operator which must satisfy $\mu\Psi_0=0$. The notation is consistent because the equations of motion are invariant under time reversal. The products $\gamma_{k+Q0}{}^*\gamma_{k1}{}^*$, $\gamma_{k+Q1}\gamma_{k1}$, and the collective variables $\rho(\mathbf{Q})$, $A_k(\mathbf{Q})$, $B_k(\mathbf{Q})$, can be expanded in terms of the μ's and

[19] N. N. Bogoliubov, Nuovo cimento 7, 794 (1958).
[20] J. G. Valatin, Nuovo cimento 7, 843 (1958).

written as

$$\gamma_{k+Q0}^*\gamma_{k1}^*=\sum_{k'}\alpha_1(\mathbf{k},\mathbf{k}',\mathbf{Q})\mu_{k'}^*(\mathbf{Q})+\sum_{k'}\alpha_2(\mathbf{k},\mathbf{k}',\mathbf{Q})\mu_{-k'}(-\mathbf{Q})+\sum_i\alpha_i(\mathbf{k},\mathbf{Q})\mu_i(\mathbf{Q}),\quad(3.4a)$$

$$\gamma_{k+Q1}\gamma_{k0}=\sum_{k'}\beta_1(\mathbf{k},\mathbf{k}',\mathbf{Q})\mu_{k'}^*(\mathbf{Q})+\sum_{k'}\beta_2(\mathbf{k},\mathbf{k}',\mathbf{Q})\mu_{-k'}(-\mathbf{Q})+\sum_i\beta_i(\mathbf{k},\mathbf{Q})\mu_i(\mathbf{Q}),\quad(3.4b)$$

$$\rho(\mathbf{Q})=\sum_{k'}M_{k'}\mu_{k'}^*(\mathbf{Q})+\text{terms in }\mu_{-k'}(-\mathbf{Q})\text{ and }\mu_i,$$

$$A_k(\mathbf{Q})=\sum_{k'}L_{kk'}\mu_{k'}^*(\mathbf{Q})+\text{terms in }\mu_{-k'}(-\mathbf{Q})\text{ and }\mu_i,$$

$$B_k(\mathbf{Q})=\sum_{k'}N_{kk'}\mu_{k'}^*(\mathbf{Q})+\text{terms in }\mu_{-k'}(-\mathbf{Q})\text{ and }\mu_i.$$

If these expressions are substituted into the equations of motion, the α's and β's must be chosen so that the coefficients of all the $\mu_k^*(\mathbf{Q})$ are separately zero. Hence

$$-[\nu_k(\mathbf{Q})-\nu_{k'}(\mathbf{Q})]\alpha_1(\mathbf{k},\mathbf{k}',\mathbf{Q})=\Phi_{k,k'}+\tfrac{1}{2}l(\mathbf{k},\mathbf{Q})L_{kk'},$$

$$[\nu_k(\mathbf{Q})+\nu_{k'}(\mathbf{Q})]\beta_1(\mathbf{k},\mathbf{k}',\mathbf{Q})=-\Phi_{k,k'}+\tfrac{1}{2}l(\mathbf{k},\mathbf{Q})L_{kk'},$$

where

$$\Phi_{kk'}=V_Dm(\mathbf{k},\mathbf{Q})M_{k'}+\tfrac{1}{2}n(\mathbf{k},\mathbf{Q})N_{kk'}.$$

A set of orthogonal solutions of the equations are given by

$$\alpha_1(\mathbf{k},\mathbf{k}',\mathbf{Q})=\delta_{k,k'}+\frac{\Phi_{kk'}+\frac{1}{2}l(\mathbf{k},\mathbf{Q})L_{kk'}}{\nu_{k'}-\nu_k+i\epsilon},\quad(3.5a)$$

$$\beta_1(\mathbf{k},\mathbf{k}',\mathbf{Q})=-\frac{\Phi_{kk'}-\frac{1}{2}l(\mathbf{k},\mathbf{Q})L_{kk'}}{\nu_k+\nu_{k'}}.\quad(3.5b)$$

In the Appendix it is shown that these solutions are normalized. From the definitions of ρ, A, and B and Eqs. (3.3) for $\gamma^*\gamma^*$ and $\gamma\gamma$ it is found that

$$M_{k'}=\sum_k m(\mathbf{k},\mathbf{Q})[\alpha_1(\mathbf{k},\mathbf{k}',\mathbf{Q})+\beta_1(\mathbf{k},\mathbf{k}',\mathbf{Q})],\quad(3.6a)$$

$$L_{kk'}=\sum_{k''}V(\mathbf{k},\mathbf{k}'')l(\mathbf{k}'',\mathbf{Q})\times[\alpha_1(\mathbf{k}'',\mathbf{k}',\mathbf{Q})-\beta_1(\mathbf{k}'',\mathbf{k}',\mathbf{Q})],\quad(3.6b)$$

$$N_{kk'}=\sum_{k''}V(\mathbf{k},\mathbf{k}'')n(\mathbf{k}'',\mathbf{Q})\times[\alpha_1(\mathbf{k}'',\mathbf{k}',\mathbf{Q})+\beta_1(\mathbf{k}'',\mathbf{k}',\mathbf{Q})].\quad(3.6c)$$

When the α's and β's, as given by Eqs. (3.5), are substituted in these equations we have a set of three linear simultaneous integral equations for the functions $M_{k'}$, $L_{kk'}$, $N_{kk'}$. The situation is not as bad as it seems, for in most cases of interest $M_{k'}$ is much greater than $L_{kk'}$ and $N_{kk'}$, and it is necessary to solve only a simple algebraic equation.

It is at once apparent from the definition of M_k and the fact that the operators are normalized, that

$$(\Psi_0\mu_k^*(\mathbf{Q}),\rho(\mathbf{Q})\Psi_0)=M_k(\mathbf{Q}).$$

In the normal state $M_k(\mathbf{Q})$ is the function $[\dot{\phi}_+(\omega_Q)]^{-1}$ of Brout.[21] For the continuity equation it is necessary to calculate

$$(\Psi_0\mu_k^*(\mathbf{Q}),\ \mathbf{Q}\cdot\mathbf{j}(\mathbf{Q})\Psi_0),$$

where

$$\mathbf{j}(\mathbf{Q})=(\hbar/2m)\sum_k(2\mathbf{k}+\mathbf{Q})(\rho_k{}^Q-\bar{\rho}_k{}^Q).$$

[Note that

$$\mathbf{j}(\mathbf{r})=\sum_Q\mathbf{j}(\mathbf{Q})\exp(i\mathbf{Q}\cdot\mathbf{r})].$$

Now

$$\hbar\mathbf{Q}\cdot\mathbf{j}(Q)=\sum_k(\epsilon_{k+Q}-\epsilon_k)(\rho_k{}^Q-\bar{\rho}_k{}^Q)$$
$$=\sum_k(\epsilon_{k+Q}-\epsilon_k)[l(\mathbf{k},\mathbf{Q})(\gamma_{k+Q0}^*\gamma_{k0}-\gamma_{k1}^*\gamma_{k+Q1})-p(\mathbf{k},\mathbf{Q})(\gamma_{k+Q0}^*\gamma_{k1}^*-\gamma_{k+Q1}\gamma_{k0})].$$

Hence

$$\hbar(\Psi_0\mu_k^*(Q),\ \mathbf{Q}\cdot\mathbf{j}(Q)\Psi_0)=-\sum_{k'}(\epsilon_{k'+Q}-\epsilon_{k'})p(\mathbf{k}',\mathbf{Q})[\alpha_1(\mathbf{k}',\mathbf{k},\mathbf{Q})-\beta_1(\mathbf{k}',\mathbf{k},\mathbf{Q})].$$

We now use the identity

$$-p(\mathbf{k},\mathbf{Q})(\epsilon_{k+Q}-\epsilon_k)\equiv\nu_k(\mathbf{Q})m(\mathbf{k},\mathbf{Q})-(I_k+I_{k+Q})l(\mathbf{k},\mathbf{Q})\quad(3.7)$$

to obtain

$$\hbar(\Psi_0\mu_k^*(\mathbf{Q}),\ \mathbf{Q}\cdot\mathbf{j}(\mathbf{Q})\Psi_0)=\sum_{k'}\nu_{k'}(\mathbf{Q})m(\mathbf{k}',\mathbf{Q})\left[\delta_{k'',k}+\frac{\Phi_{k'k}2\nu_k(\mathbf{Q})}{\nu_k{}^2-\nu_{k'}{}^2}+\frac{l(k',Q)L_{k'k}\nu_{k'}}{\nu_k{}^2-\nu_{k'}{}^2}\right]-\sum_{k'}(I_{k'}+I_{k'+Q})l(\mathbf{k}',\mathbf{Q})\times[\alpha_1(\mathbf{k}',\mathbf{k},\mathbf{Q})-\beta_1(\mathbf{k}',\mathbf{k},\mathbf{Q})].$$

Now

$$\sum_{k'}\frac{\nu_{k'}(Q)^2m(\mathbf{k}',\mathbf{Q})l(\mathbf{k}',\mathbf{Q})L_{k'k}}{\nu_k{}^2-\nu_{k'}{}^2}=\sum_{k'}\frac{\nu_k{}^2m(\mathbf{k}',\mathbf{Q})l(\mathbf{k}',\mathbf{Q})L_{k'k}}{\nu_k{}^2-\nu_{k'}{}^2}-\sum_{k'}m(\mathbf{k}',\mathbf{Q})l(\mathbf{k}',\mathbf{Q})L_{kk'},$$

and from Eq. (3.6b)

$$\sum_{k'}m(\mathbf{k}',\mathbf{Q})l(\mathbf{k}',\mathbf{Q})L_{k'k}=\sum_{k',k''}(u_{k'}v_{k'}+u_{k'+Q}v_{k'+Q})V(\mathbf{k}',\mathbf{k}'')l(\mathbf{k}'',\mathbf{Q})\times[\alpha_1(\mathbf{k}'',\mathbf{k},\mathbf{Q})-\beta_1(\mathbf{k}'',\mathbf{k},\mathbf{Q})]$$
$$=-\sum_{k''}(I_{k''}+I_{k''+Q})l(\mathbf{k}'',\mathbf{Q})\times[\alpha_1(\mathbf{k}'',\mathbf{k},\mathbf{Q})-\beta_1(\mathbf{k}'',\mathbf{k},\mathbf{Q})].$$

Therefore

$$\hbar(\Psi_0\mu_k^*(Q),\ \mathbf{Q}\cdot\mathbf{j}(Q)\Psi_0)=\nu_k(\mathbf{Q})\sum_{k'}m(\mathbf{k}',\mathbf{Q})\left[\delta_{k,k'}+\frac{2\Phi_{k'k}\nu_{k'}}{\nu_k{}^2-\nu_{k'}{}^2}+\frac{l(\mathbf{k}',\mathbf{Q})L_{k'k}\nu_k}{\nu_k{}^2-\nu_{k'}{}^2}\right]$$
$$=\nu_k(\mathbf{Q})\sum m(\mathbf{k}',\mathbf{Q})[\alpha_1(\mathbf{k}',\mathbf{k},\mathbf{Q})+\beta_1(\mathbf{k}',\mathbf{k},\mathbf{Q})]$$
$$=\nu_k(\mathbf{Q})M_k(\mathbf{Q}),$$

[21] R. Brout, Phys. Rev. **108**, 515 (1957).

and

$$\hbar(\Psi_0\mu_k^*(\mathbf{Q}),\ \mathbf{Q}\cdot\mathbf{j}(\mathbf{Q})\Psi_0)$$
$$=\nu_k(\mathbf{Q})(\Psi_0\mu_k(\mathbf{Q})^*,\rho(\mathbf{Q})\Psi_0), \quad (3.1)$$

and the wave functions are such that our results will be gauge invariant.

Now that the matrix elements satisfy the equation of continuity, it is possible to approximate them in a consistent manner. We shall compute M_k. In general the states $\mathbf{k}$ and $\mathbf{k}+\mathbf{Q}$ are close to the Fermi surface, much closer than $\hbar\omega$, the average spread in energy allowed by $V(\mathbf{k},\mathbf{k}')$, which in turn is much less than the Fermi energy. This means that $A_k(\mathbf{Q})$, $B_k(\mathbf{Q})$, $L_{kk'}$, $N_{kk'}$ are all approximately independent of $\mathbf{k}$ and that the states $\mathbf{k}'$, $\mathbf{k}'+\mathbf{Q}$ in the integrals involving L and N will lie within an energy $\hbar\omega$ of the Fermi surface. As the interaction with a *longitudinal* field is being calculated, all these variables will be independent of angle, too.[3] In this case the set of three integral equations reduces to a set of three linear algebraic equations which are

$$M_k(\mathbf{Q})=m(\mathbf{k},\mathbf{Q})+\sum_{k'} m(\mathbf{k}',\mathbf{Q})\Phi_{k'k}\frac{2\nu_{k'}}{\nu_k^2-\nu_{k'}^2}+\sum_{k'} m(\mathbf{k}',\mathbf{Q})l(\mathbf{k}',\mathbf{Q})\frac{\nu_k L_k}{\nu_k^2-\nu_{k'}^2},$$

$$L_k(\mathbf{Q})=Vl(\mathbf{k},\mathbf{Q})+\sum_{k'} Vl(\mathbf{k}',\mathbf{Q})\Phi_{k'k}\frac{2\nu_k}{\nu_k^2-\nu_{k'}^2}+\sum_{k'} Vl^2(\mathbf{k}',\mathbf{Q})\frac{\nu_{k'}L_k}{\nu_k^2-\nu_{k'}^2},$$

$$N_k(\mathbf{Q})=Vn(\mathbf{k},\mathbf{Q})+\sum_{k'} Vn(\mathbf{k}',\mathbf{Q})\Phi_{k'k}\frac{2\nu_{k'}}{\nu_k^2-\nu_{k'}^2}+\sum_{k'} Vn(\mathbf{k}',\mathbf{Q})l(\mathbf{k}',\mathbf{Q})\frac{\nu_k L_k}{\nu_k^2-\nu_{k'}^2}.$$

Normally $M_k(\mathbf{Q})$ is required to obtain the probability of absorption of a wave of energy $\nu_k(\mathbf{Q})$, for which $\hbar v_0 Q\gg\nu_k$ (e.g., acoustic wave). Since the formalism above applies only for absorption when the system is initially in the ground state, the condition $\nu_k\geq 2\epsilon_0$ must hold. Therefore $\hbar v_0 Q\gg 2\epsilon_0$. In this case it is easy to show that to order $(\nu_k/\hbar v_0 Q)^2$ and $(\epsilon_0/\hbar\omega)^2$, L_k and N_k can be neglected. Then

$$M_k(\mathbf{Q})=m(\mathbf{k},\mathbf{Q})\Big/\left(2V_D(\mathbf{Q})\sum_{k'}\frac{m^2(\mathbf{k}',\mathbf{Q})\nu_{k'}}{\nu_{k'}^2-\nu_k^2}+1\right)$$
$$=\langle 0|\rho(\mathbf{Q})|\mathbf{k},\mathbf{k}+\mathbf{Q}\rangle_{\mathrm{BCS}}\Big/\left(2V_D(\mathbf{Q})\sum_{k'}\frac{m^2(\mathbf{k}',\mathbf{Q})\nu_{k'}}{\nu_{k'}^2-\nu_k^2}+1\right).$$

To the same order of accuracy the denominator of this expression can be replaced by its value in the normal state. This shows that to calculate the probability of absorption of the system at the absolute zero of temperature one can use the interaction term in the Hamiltonian,

$$H_1=\sum_{k,Q} V_D(\mathbf{Q})r_Q e^{i\Omega t}c_{k,\sigma}{}^*c_{k-Q,\sigma}+\text{complex conj.},$$

where, if we choose r_Q to be the external charge fluctuation screened as in the normal state, we need consider only the single-particle wave functions of BCS. In the next section we approach the problem from a slightly different point of view and see how to generalize this result so that it applies at all temperatures and takes into account the corrections to the screening from the superconducting transition.

4. DIELECTRIC CONSTANT OF A SUPERCONDUCTOR

The procedure developed by Nozières and Pines[22] for the normal metal is adopted; a time-dependent longitudinal external field (unscreened) is allowed to act on the system and the polarization it induces is calculated and related to the complex dielectric constant. Let us suppose there is an oscillating test charge of wave vector $\mathbf{Q}$ and frequency Ω acting on the system. Its charge density is

$$r_Q \exp[-i(\Omega t-\mathbf{Q}\cdot\mathbf{r})]+\text{c.c.}$$

The interaction of this test charge with the system adds the term H_1 to the Hamiltonian where

$$H_1=V_D(\mathbf{Q})[\rho(-\mathbf{Q})r_Q e^{-i\Omega t}+\text{c.c.}]e^{\eta t}.$$

[In this section the contribution of the phonons to $V_D(\mathbf{Q})$ is neglected so that $V_D(\mathbf{Q})$ is just the Coulomb interaction. The term neglected is only of the order of the electron mass divided by the mass of an ion.] The infinitesimally small quantity η is introduced to ensure that the test charge is switched on adiabatically; in the mathematics, η indicates which contour to choose for the integrals that arise.

The interaction H_1 leads to extra terms in the equations of motion so that these equations become

$$[H,\gamma_{k+Q0}{}^*\gamma_{k1}{}^*]=\nu_k(\mathbf{Q})\gamma_{k+Q0}{}^*\gamma_{k1}{}^*+V_D m(\mathbf{k},\mathbf{Q})\times[\rho(\mathbf{Q})+r_Q e^{-i\Omega t+\eta t}+\text{c.c.}]+\tfrac{1}{2}n(\mathbf{k},\mathbf{Q})\times B_k(\mathbf{Q})-\tfrac{1}{2}l(\mathbf{k},\mathbf{Q})A_k(\mathbf{Q}), \quad (4.1\text{a})$$

$$[H,\gamma_{k+Q1}\gamma_{k0}]=-\nu_k(\mathbf{Q})\gamma_{k+Q1}\gamma_{k0}-V_D m(\mathbf{k},\mathbf{Q})\times[\rho(\mathbf{Q})+r_Q e^{-i\Omega t+\eta t}+\text{c.c.}]-\tfrac{1}{2}n(\mathbf{k},\mathbf{Q})\times B_k(\mathbf{Q})-\tfrac{1}{2}l(\mathbf{k},\mathbf{Q})A_k(\mathbf{Q}), \quad (4.1\text{b})$$

$$[H,\gamma_{k+Q0}{}^*\gamma_{k0}]=(E_{k+Q}-E_k)\gamma_{k+Q0}{}^*\gamma_{k0},$$

$$[H,\gamma_{k1}{}^*\gamma_{k+Q1}]=(E_k-E_{k+Q})\gamma_{k1}{}^*\gamma_{k+Q1}.$$

[22] P. Nozières and D. Pines, Nuovo cimento 9, 470 (1958).

As shown by Nozières and Pines,[22] only the part of $\langle\rho(\mathbf{Q})\rangle$ that varies as $\exp(-i\Omega t)$ is required. Then $[\epsilon(\mathbf{Q},\Omega)^{-1}-1]$ is the ratio of this to $r_Q\exp(-i\Omega t)$. Now the effect of the extra terms in the equations of motion is to change the normal coordinates by adding c-numbers, and since $\rho(\mathbf{Q})$ is linear in the old normal coordinates it too is increased only by a c-number. The expectation value of $\rho(\mathbf{Q})$ in the ground state will be just this c-number. Hence it is necessary to solve the equations of motion treating the operators as c-numbers, remembering that the commutators on the left-hand sides are to be replaced by time derivatives. Thus as it is necessary to treat only the part of the test charge that varies as $e^{-i\Omega t+\eta t}$, we find

$$\begin{aligned}\rho(\mathbf{Q})=\sum_k[&V_D(\rho(\mathbf{Q})+r_Qe^{-i\Omega t+\eta t})m(\mathbf{k},\mathbf{Q})\\&+\tfrac{1}{2}n(\mathbf{k},\mathbf{Q})B_k(\mathbf{Q})]m(\mathbf{k},\mathbf{Q})\\&\times\left[\frac{1}{-\hbar\Omega-i\eta-\nu_k(Q)}-\frac{1}{-\hbar\Omega-i\eta+\nu_k(Q)}\right]\\&-\tfrac{1}{2}\sum_k l(\mathbf{k},\mathbf{Q})A_k(Q)m(\mathbf{k},\mathbf{Q})\\&\times\left[\frac{1}{-\hbar\Omega-i\eta-\nu_k}+\frac{1}{-\hbar\Omega-i\eta+\nu_k}\right],\end{aligned}\quad(4.2\text{a})$$

$$\begin{aligned}B_K(\mathbf{Q})=\sum_k[&V_D(\rho(\mathbf{Q})+r_Qe^{-i\Omega t})m(\mathbf{k},\mathbf{Q})\\&+\tfrac{1}{2}n(\mathbf{k},\mathbf{Q})B_k(Q)]V(\mathbf{K},\mathbf{k})n(\mathbf{k},\mathbf{Q})\\&\times\left[\frac{1}{-\hbar\Omega-i\eta-\nu_k}-\frac{1}{-\hbar\Omega-i\eta+\nu_k}\right]\\&-\tfrac{1}{2}\sum_k l(\mathbf{k},\mathbf{Q})A_k(\mathbf{Q})V(\mathbf{K},\mathbf{k})n(\mathbf{k},\mathbf{Q})\\&\times\left[\frac{1}{-\hbar\Omega-i\eta-\nu_k}+\frac{1}{-\hbar\Omega-i\eta+\nu_k}\right],\end{aligned}\quad(4.2\text{b})$$

$$\begin{aligned}A_K(\mathbf{Q})=-\sum_k[&V_D(\rho(\mathbf{Q})+r_Qe^{-i\Omega t})m(\mathbf{k},\mathbf{Q})\\&+\tfrac{1}{2}n(\mathbf{k},\mathbf{Q})B_k(\mathbf{Q})]l(\mathbf{k},\mathbf{Q})V(\mathbf{K},\mathbf{k})\\&\times\left[\frac{1}{-\hbar\Omega-i\eta-\nu_k}+\frac{1}{-\hbar\Omega-i\eta+\nu_k}\right]\\&+\tfrac{1}{2}\sum_k l(\mathbf{k},\mathbf{Q})A_k(\mathbf{Q})V(\mathbf{K},\mathbf{k})l(\mathbf{k},\mathbf{Q})\\&\times\left[\frac{1}{-\hbar\Omega-i\eta-\nu_k}-\frac{1}{-\hbar\Omega-i\eta+\nu_k}\right].\end{aligned}\quad(4.2\text{c})$$

From these three integral equations, $\rho(\mathbf{Q})$ and hence the complex dielectric constant can be determined. It is not difficult to show that the imaginary part of the dielectric constant leads to a result for the absorption which is in agreement with that calculated from the matrix elements of the previous section.

It is apparent from the structure of the equations of motion (4.1) that if one wishes to calculate the probability of a transition caused by an external charge fluctuation $\rho(Q)e^{-i\Omega t}$ $(\Omega\ll v_0Q)$ one can take as the interaction term in the Hamiltonian

$$\begin{aligned}H'=V_D(\rho(\mathbf{Q})+r_Qe^{-i\Omega t})\rho_{0p}(-\mathbf{Q})\\+\tfrac{1}{2}\textstyle\sum_k B_k(\mathbf{Q})(b_k^{-Q}+\bar{b}_k^{-Q})\\-\tfrac{1}{2}\textstyle\sum_k A_k(\mathbf{Q})(b_k^{-Q}-\bar{b}_k^{-Q}),\end{aligned}\quad(4.3)$$

where $\rho(\mathbf{Q})$, $B_k(\mathbf{Q})$, and $A_k(\mathbf{Q})$ are the solutions of Eqs. (4.2) and are c-numbers. ρ, B, and A are all proportional to $r_Qe^{-i\Omega t}$ and the constants of proportionality need be determined once for all interactions. The result (4.3) has been proved only for transitions into and out of the ground state. We guess that it is correct for all transitions and that $\rho(\mathbf{Q})$ is hardly altered by a change of temperature. For temperatures $kT\ll\epsilon_0$, $A(\mathbf{Q})$ and $B(\mathbf{Q})$ will also be unchanged. For many problems $A(\mathbf{Q})$ and $B(\mathbf{Q})$ can be neglected; for these problems the interaction is H_1 even up to T_c. As an example of the use of Eq. (4.3), we shall show in Sec. V that the corrections to ultrasonic attenuation as calculated by BCS are of order $(u/v_0)^2$, where u is the phase velocity of the sound wave. Notice that if the corrections are important then, because the terms involving A and B are not single-particle operators, the interaction cannot be described as a screened charge acting on each excitation.

As a check on the formula for $\epsilon(\mathbf{Q},\Omega)$ we shall investigate its behavior as $\Omega\to\infty$. According to Nozières and Pines[22] we should obtain

$$\epsilon(Q,\Omega)-1\to-\omega_p^2/\Omega^2\quad\text{as}\quad\Omega\to\infty.\quad(4.4)$$

If we write

$$A_k(\mathbf{Q})=V_D(\mathbf{Q})\alpha_k(\mathbf{Q})[\rho(\mathbf{Q})+r_Qe^{-i\Omega t}],$$
$$B_k(\mathbf{Q})=V_D(\mathbf{Q})\beta_k(\mathbf{Q})[\rho(\mathbf{Q})+r_Qe^{-i\Omega t}],$$

then

$$\begin{aligned}\epsilon-1&=\frac{-\rho(\mathbf{Q})}{\rho(\mathbf{Q})+r_Qe^{-i\Omega t}}\\&=-V_D\sum_k\left\{[2m(\mathbf{k},\mathbf{Q})+n(\mathbf{k},\mathbf{Q})\beta_k(\mathbf{Q})]\times\frac{m(\mathbf{k},\mathbf{Q})\nu_k}{\hbar^2\Omega^2-\nu_k^2}+\frac{l(\mathbf{k},\mathbf{Q})m(\mathbf{k},\mathbf{Q})\alpha_k(\mathbf{Q})\hbar\Omega}{\hbar^2\Omega^2-\nu_k^2}\right\}\to\\&-(V_D/\Omega^2)\sum_k\{[2m(\mathbf{k},\mathbf{Q})+n(\mathbf{k},\mathbf{Q})\beta_k(\mathbf{Q})]\nu_k\\&\qquad+l(\mathbf{k},\mathbf{Q})\alpha(\mathbf{k},\mathbf{Q})\hbar\Omega\}m(\mathbf{k},\mathbf{Q}).\end{aligned}$$

As $\Omega\to\infty$ both β and α are proportional to Ω^{-1}. Therefore β can be neglected and

$$\alpha_K(\mathbf{Q})=(1/\hbar\Omega^2)\sum_k V(\mathbf{K},\mathbf{k})2m(\mathbf{k},\mathbf{Q})l(\mathbf{k},\mathbf{Q})\Omega$$
$$=-(2/\hbar\Omega)(I_K+I_{K+Q}).$$

Therefore

$$\epsilon-1=-[V_D(\mathbf{Q})/\hbar^2\Omega^2]\sum_k m(\mathbf{k},\mathbf{Q})$$
$$\times[2m(\mathbf{k},\mathbf{Q})\nu_k(\mathbf{Q})-2(I_k+I_{k+Q})l(\mathbf{k},\mathbf{Q})]$$
$$=[2V_D(\mathbf{Q})/\hbar^2\Omega^2]\sum_k m(\mathbf{k},\mathbf{Q})p(\mathbf{k},\mathbf{Q})(\epsilon_{k+Q}-\epsilon_k)$$
$$=[2V_D(\mathbf{Q})/\hbar^2\Omega^2]\sum(u_k^2v_{k+Q}^2-v_k^2u_{k+Q}^2)(\epsilon_{k+Q}-\epsilon_k)$$
$$=[2V_D(\mathbf{Q})/\hbar^2\Omega^2]\sum(v_{k+Q}^2-v_k^2)(\epsilon_{k+Q}-\epsilon_k)$$
$$=-[4V_D(\mathbf{Q})/\hbar^2\Omega^2]\sum v_k^2(\hbar^2Q^2/2m)$$
$$=-[V_D(\mathbf{Q})/\Omega^2](Q^2/m)N$$
$$=-(\omega_p^2/\Omega^2).$$

Had we neglected α we should have found that $(\epsilon-1)$ behaves like Q^{-2}. The proof of Eq. (4.4) is implicitly a proof of the sum rules

$$\sum_k \nu_k(\mathbf{Q})M_k^2(\mathbf{Q})=N\hbar^2Q^2/2m. \quad (4.5)$$

Notice that as Q tends to zero the energy $\nu_k(\mathbf{Q})$ remains finite but the matrix element $M_k(\mathbf{Q})$ is proportional to Q. In the normal metal it is the matrix element that remains finite while the energy is proportional to Q^2. One can see from this discussion that $\epsilon(\mathbf{Q},\Omega)$ is significantly different from its value in the normal metal only if $\hbar\Omega\sim\hbar v_0Q\lesssim\epsilon_0$.

Before leaving the subject of the dielectric constant we shall consider the connection of this work with the ideas of backflow.[8] An external charge fluctuation $r_Qe^{-i\Omega t}$ causes a charge fluctuation $\rho(\mathbf{Q})$ in the superconducting system. Therefore there is a current flow given by the density $\Omega\mathbf{Q}\rho(\mathbf{Q})Q^{-2}$. If an external point charge moving with velocity $\mathbf{V}$ interacts with the system the external charge density is

$$\delta(\mathbf{r}-\mathbf{V}t)=\sum_Q \exp[i\mathbf{Q}\cdot(\mathbf{r}-\mathbf{V}t)],$$

and the induced current flow is

$$\mathbf{j}(\mathbf{r})=\sum_Q(\mathbf{Q}\cdot\mathbf{V})\mathbf{Q}\rho(\mathbf{Q})Q^{-2}\exp[i\mathbf{Q}\cdot(\mathbf{r}-\mathbf{V}t)]/r_Q\exp(-i\Omega t)$$
$$=\sum_Q(\mathbf{Q}\cdot\mathbf{V})\mathbf{Q}\left(\frac{1}{\epsilon(\mathbf{Q},\mathbf{Q}\cdot\mathbf{V})}-1\right)Q^{-2}\exp[i\mathbf{Q}\cdot(\mathbf{r}-\mathbf{V}t)]$$
$$=-\nabla(\mathbf{V}\cdot\nabla)\sum_Q(1/\epsilon-1)Q^{-2}\exp[i\mathbf{Q}\cdot(\mathbf{r}-\mathbf{V}t)].$$

When $Q\to 0$, $\epsilon(\mathbf{Q},\mathbf{Q}\cdot\mathbf{V})^{-1}\to 0$. Therefore, at large distances from the moving charge the flow is just that due to a dipole of strength[23] $(-\mathbf{V}/4\pi)$ as it is in the normal metal. If $V\ll v_0$ or $V\gg v_0$, the flow everywhere is as in the normal metal. However, if $V\sim v_0$ then the flow in the superconductor is different from that in the normal metal at distances less than the coherence distance from the external charge.

[23] This backflow is analogous to the backflow around a foreign atom in liquid helium. The strength is the same, see R. P. Feynman and M. Cohen, Phys. Rev. **102**, 1189 (1956).

5. ULTRASONIC ABSORPTION

The interaction (4.3) will be used to show that the correction to the ultrasonic absorption calculated by BCS is (at least at the lowest temperatures) of order $(u/v_0)^2$, where u is the velocity of sound in the metal. Near absolute zero, $kT<\epsilon_0(T)$, $\hbar\Omega\ll\hbar v_0Q\ll\epsilon_0$ for the acoustic waves of interest. If the integrands of Eqs. (4.2) are expanded in powers of (Ω/v_0Q) and $(\hbar v_0Q/\epsilon_0)$, it is found that

$$A=-(40/17)V_D[r_Qe^{-i\Omega t}+\rho(\mathbf{Q})]\epsilon_0\Omega(\hbar v_0Q)^{-2}$$
$$=\alpha_QV_D[r_Qe^{-i\Omega t}+\rho(Q)]=\alpha_QV_Dr_Qe^{-i\Omega t}\epsilon^{-1},$$
$$B=\tfrac{4}{5}V_D[r_Qe^{-i\Omega t}+\rho(\mathbf{Q})]\hbar v_0Q/\epsilon_0$$
$$=\beta_QV_D[r_Qe^{-i\Omega t}+\rho(Q)]=\beta_QV_Dr_Qe^{-i\Omega t}\epsilon^{-1}.$$

The correction to $\rho(\mathbf{Q})$ is of order $(\Omega/\hbar v_0Q)^2$ and $(\hbar v_0Q/\epsilon_0)^2$. As only the order of magnitude of the correction is being estimated, these corrections to $\rho(\mathbf{Q})$ can be ignored. (BCS have already ignored corrections of this order of magnitude.) Hence, the absorption is proportional to

$$\int d^3k\,\delta(E_k-E_{k+Q}+\hbar\Omega_Q)(f_{k+Q}-f_k)\{[n(\mathbf{k},\mathbf{Q})+\beta_Qm(\mathbf{k},\mathbf{Q})$$
$$+\alpha_Qp(\mathbf{k},\mathbf{Q})]^2+[n(\mathbf{k},\mathbf{Q})+\beta_Qm(\mathbf{k},\mathbf{Q})-\alpha_Qp(\mathbf{k},\mathbf{Q})]^2\}$$
$$=2\int d^3k\,\delta(E_k-E_{k+Q}+\hbar\Omega_Q)(f_{k+Q}-f_k)$$
$$\times\left\{\frac{\epsilon^2}{E^2}+\frac{4}{5}\frac{\epsilon(\hbar v_0Q)}{E^2}+\frac{16}{25}\left(\frac{\hbar v_0Q}{E}\right)^2+\frac{5}{4}\frac{\Omega^2}{(\hbar v_0Q)^2}\frac{\epsilon_0^4}{E^4}\right\}.$$

The first term in the curly brackets gives the result of BCS. The corrections are evidently no bigger than terms already neglected. Since the term involving $\rho(\mathbf{Q})$ is the only important one, this result will be valid for temperatures up to T_c.

6. MEISSNER EFFECT

We shall calculate the current within the superconductor due to an external static magnetic field. The connection between this current and the existence of a Meissner effect and the calculation of the penetration depth has been discussed sufficiently elsewhere[24] for it to be omitted here. If the static magnetic field is described by the vector potential $\mathbf{a}(\mathbf{Q})\exp(i\mathbf{Q}\cdot\mathbf{r})$, the extra perturbing term in the Hamiltonian is H_1, given

[24] J. Bardeen, *Handbuch der Physik* (Springer-Verlag, Berlin, 1956), Vol. 19. This review article contains further references on this topic.

by

$$H_1=-\alpha \sum_k \mathbf{a}(\mathbf{Q})\cdot(2\mathbf{k}+\mathbf{Q})(\rho_k{}^Q-\bar{\rho}_k{}^Q)^*$$
$$=-\alpha\sum_k \mathbf{a}(\mathbf{Q})\cdot(2\mathbf{k}+\mathbf{Q})[l(\mathbf{k},\mathbf{Q})(\gamma_{k+Q0}{}^*\gamma_{k0}-\gamma_{k1}{}^*\gamma_{k+Q1}) - p(\mathbf{k},\mathbf{Q})(\gamma_{k+Q0}{}^*\gamma_{k1}{}^*-\gamma_{k+Q1}\gamma_{k0})]^*, \quad (6.1)$$

$\alpha=e\hbar/2mc.$

A straightforward way of calculating the paramagnetic part of the current to first order in $\mathbf{a}(\mathbf{Q})$ would be to use first-order perturbation theory and obtain

$$\mathbf{j}_p(\mathbf{Q})=\sum_n \frac{\langle 0|j_{p,0p}|n\rangle\langle n|H_1|0\rangle}{E_0-E_n}+\text{c.c.},$$

where the states $|n\rangle$ are the states $\mu^*|0\rangle$ of Sec. III. To obtain the result explicitly (i.e., without invoking sum rules) it would be necessary to obtain all the solutions μ^*. In order to avoid the excessive computation involved we use the following quicker method which involves no extra assumptions.

The equations of motion when H_1 is added to the Hamiltonian are

$$[H,\gamma_{k+Q0}{}^*\gamma_{k1}{}^*]$$
$$=\nu_k(\mathbf{Q})\gamma'_{+Q0}{}^*\gamma_{k1}{}^*+V_D(\mathbf{Q})\rho(\mathbf{Q})m(\mathbf{k},\mathbf{Q}) +\tfrac{1}{2}n(\mathbf{k},\mathbf{Q})B_k(\mathbf{Q})-\tfrac{1}{2}l(\mathbf{k},\mathbf{Q})A_k(\mathbf{Q}) +\alpha p(\mathbf{k},\mathbf{Q})\mathbf{a}(\mathbf{Q})\cdot(2\mathbf{k}+\mathbf{Q}), \quad (6.2a)$$

$$[H,\gamma_{k+Q1}\gamma_{k0}]$$
$$=-\nu_k(\mathbf{Q})\gamma_{k+Q1}\gamma_{k0}-V_D(\mathbf{Q})\rho(\mathbf{Q})m(\mathbf{k},\mathbf{Q}) -\tfrac{1}{2}n(\mathbf{k},\mathbf{Q})B_k(\mathbf{Q})-\tfrac{1}{2}l(\mathbf{k},\mathbf{Q})A_k(\mathbf{Q}) +\alpha p(\mathbf{k},\mathbf{Q})\mathbf{a}(\mathbf{Q})\cdot(2\mathbf{k}+\mathbf{Q}). \quad (6.2b)$$

[These and the following equations still apply if the Coulomb term is omitted. We then have to omit the Coulomb contributions to $V_D(\mathbf{Q})$ and $V(\mathbf{k}',\mathbf{k})$.] As in Sec. IV, we look only for the steady-state solution of these equations. Because the external field is static the left-hand side of the equations is zero. Subtracting the two equations, one quickly finds

$$\rho(\mathbf{Q})=B_k(\mathbf{Q})=0.$$

From the sum of the two equations

$$-\nu_k(\mathbf{Q})(\gamma_{k+Q0}{}^*\gamma_{k1}{}^*-\gamma_{k+Q1}\gamma_{k0}) =-l(\mathbf{k},\mathbf{Q})A_k(\mathbf{Q})+2\alpha p(\mathbf{k},\mathbf{Q})\mathbf{a}\cdot(2\mathbf{k}+\mathbf{Q}).$$

It follows from this equation and the definition of $A_K(\mathbf{Q})$,

$$A_K(\mathbf{Q})=-\sum_k V(\mathbf{K},\mathbf{k})l(\mathbf{k},\mathbf{Q})(\gamma_{k+Q0}{}^*\gamma_{k1}{}^*-\gamma_{k+Q1}\gamma_{k0}),$$

that

$$A_K(\mathbf{Q})=-\sum_k V(\mathbf{K},\mathbf{k})[l(\mathbf{k},\mathbf{Q})A_k(\mathbf{Q}) -2\alpha p(\mathbf{k},\mathbf{Q})\mathbf{a}\cdot(2\mathbf{k}+\mathbf{Q})]\nu_k{}^{-1}l(\mathbf{k},\mathbf{Q}). \quad (6.3)$$

The paramagnetic part of the current density $\mathbf{j}(\mathbf{Q})$ is given by

$$(2m/e\hbar)\mathbf{j}_p(\mathbf{Q})$$
$$=\sum_k(2\mathbf{k}+\mathbf{Q})(\rho_k{}^Q-\bar{\rho}_k{}^Q)$$
$$=\sum_k(2\mathbf{k}+\mathbf{Q})[-l(\mathbf{k},\mathbf{Q})A_k(\mathbf{Q}) +2\alpha p(\mathbf{k},\mathbf{Q})\mathbf{a}\cdot(2\mathbf{k}+\mathbf{Q})]p(\mathbf{k},\mathbf{Q})\nu_k{}^{-1}. \quad (6.4)$$

The second term on the right is the one that is calculated from the BCS wave functions alone. As the result is linear in $\mathbf{a}(\mathbf{Q})$, one can calculate the effects of longitudinal and transverse fields separately. This we proceed to do.

(1) Longitudinal Field

As pointed out in the Introduction, a longitudinal static vector potential cannot give rise to any current. It will now be shown that the effect of including the collective term $A_k(\mathbf{Q})$ is to ensure that this result is satisfied. If $\mathbf{a}(\mathbf{Q})$ is a longitudinal potential, one can write

$$\mathbf{a}(\mathbf{Q})=\mathbf{Q}(\hbar^2/2m)\phi(\mathbf{Q}).$$

Then the solution of the integral equation for $A_k(\mathbf{Q})$ is

$$A_k(\mathbf{Q})=2(I_k+I_{k+Q})\alpha\phi(\mathbf{Q}). \quad (6.5)$$

This can be checked directly. If the formula (6.5) is substituted into the right-hand side of Eq. (6.3), that side becomes [when Eq. (3.7) is used]

$$-2\alpha\phi\sum_k V(\mathbf{K},\mathbf{k})[l(\mathbf{k},\mathbf{Q})(I_k+I_{k+Q}) -(\epsilon_{k+Q}-\epsilon_k)p(\mathbf{k},\mathbf{Q})]l(\mathbf{k},\mathbf{Q})\nu_k{}^{-1}$$
$$=-2\alpha\phi\sum_k V(\mathbf{K},\mathbf{k})m(\mathbf{k},\mathbf{Q})l(\mathbf{k},\mathbf{Q})$$
$$=2\alpha\phi(I_K+I_{K+Q})$$
$$=A_K(\mathbf{Q}).$$

Therefore the two sides of Eq. (6.3) are equal. If one substitutes for $A_k(\mathbf{Q})$ in the current density it is found that

$$(2m/e\hbar)\mathbf{j}_p(\mathbf{Q})=+2\alpha\phi\sum_k(2\mathbf{k}+\mathbf{Q})[(\epsilon_{k+Q}-\epsilon_k)p(\mathbf{k},\mathbf{Q}) -(I_k+I_{k+Q})l(\mathbf{k},\mathbf{Q})]p(\mathbf{k},\mathbf{Q})\nu_k{}^{-1}$$
$$=-2\alpha\phi\sum_k(2\mathbf{k}+\mathbf{Q})m(\mathbf{k},\mathbf{Q})p(\mathbf{k},\mathbf{Q})$$
$$=-2\alpha\phi\sum_k(2\mathbf{k}+\mathbf{Q})(u_k{}^2v_{k+Q}{}^2-v_k{}^2u_{k+Q}{}^2)$$
$$=-2\alpha\phi\sum(2\mathbf{k}+\mathbf{Q})(v_{k+Q}{}^2-v_k{}^2)$$
$$=2\alpha\phi 2\sum(2\mathbf{k}+\mathbf{Q})v_k{}^2$$
$$=2\alpha\phi\mathbf{Q}N$$
$$=(4m/\hbar^2)\alpha N\mathbf{a}(\mathbf{Q}),$$

$$\mathbf{j}_p(\mathbf{Q})=(ne^2/mc)\mathbf{a}(\mathbf{Q}).$$

Hence the paramagnetic current density just cancels the diamagnetic current density, the total longitudinal current being zero.

(2) Transverse Field

If $\mathbf{a}$ is transverse and $V(\mathbf{K},\mathbf{k})$ is independent of angle, then

$$\mathbf{a}\cdot\sum_k V(\mathbf{K},\mathbf{k})(2\mathbf{k}+\mathbf{Q})l(\mathbf{k},\mathbf{Q})p(\mathbf{k},\mathbf{Q})\nu_k^{-1}$$

is zero because the sum must be proportional to $\mathbf{Q}$, and $\mathbf{a}\cdot\mathbf{Q}=0$. In this case $A_k(\mathbf{Q})$ is zero and there are no corrections to the result of BCS. In general, however, $V(\mathbf{K},\mathbf{k})$ is a function of angle and there will be a contribution from the transverse excitations. In the London limit, $Q\rightarrow 0$, $p(\mathbf{k},\mathbf{Q})\rightarrow 0$ and as there is no singularity in the solution one finds no contribution to $\mathbf{j}_p$ in the limit. Thus the London equation is obtained as $Q\rightarrow 0$. We have tried to estimate the order of magnitude of the correction in the Pippard limit, $Q\xi_0\gg 1$, by treating a specific example. The dependence of V on $|k|$ and $|K|$ is not important provided we cut off the integrals appropriately. Because the phonon interaction tends to zero as the angle, θ, between $\mathbf{k}$ and $\mathbf{K}$ tends to zero we have chosen a $V(\mathbf{K},\mathbf{k})$ that possesses this property. The simplest potential that gives a nonzero contribution is

$$V(\mathbf{K},\mathbf{k})=-\tfrac{3}{4}V(1-\cos\theta)^2. \tag{6.6}$$

Then

$$-\langle V(\mathbf{K},\mathbf{k})\rangle_{\text{Av}}=V$$

is the same parameter as used by BCS. With this form for $V(\mathbf{K},\mathbf{k})$, it is found that

$$\sum_k V(\mathbf{K},\mathbf{k})\mathbf{a}\cdot(2\mathbf{k}+\mathbf{Q})\nu_k^{-1}p(\mathbf{k},\mathbf{Q})l(\mathbf{k},\mathbf{Q})$$
$$=-\tfrac{3}{4}V\sum_k(\mathbf{a}\cdot 2\mathbf{k})(1-\cos\theta)^2\nu_k^{-1}\epsilon_0(\epsilon-\epsilon')/2EE',$$
$$\epsilon'=\epsilon_{k+Q}.$$

If the x axis is chosen along $\mathbf{Q}$ and the z axis along $\mathbf{a}$, the sum is (neglecting terms of order Q/k_0)

$$\frac{3Va}{2}\sum_k k_z\left[-\frac{k_zK_z}{kK}+\frac{k_zK_z}{k^2K^2}(k_xK_x+k_yK_y)\right]\times\frac{\epsilon_0\hbar^2k_xQ_x/m}{E_kE_{k+Q}(E+E')}.$$

If we substitute $(-k_x-Q_x)$ for k_x we find that the integrand, apart from the term in square brackets, is an odd function of k_x (to order Q/k_0). Hence the sum is

$$\frac{3Va}{2}\frac{K_xK_z}{K^2}Q\frac{\hbar^2}{m}\sum_k\frac{k_x^2k_z^2}{k^2}\frac{\epsilon_0}{E_kE_{k+Q}(E_k+E_{k+Q})}$$
$$=\frac{3}{2}\frac{V(\mathbf{a}\cdot\mathbf{K})(\mathbf{Q}\cdot\mathbf{K})}{K^2Q}\sum_k\frac{k_z\frac{1}{2}(k^2-k_z^2)}{k^2}\frac{\epsilon_0(\epsilon'-\epsilon)}{EE'(E+E')}$$
$$=\tfrac{3}{4}N(0)V\frac{(\mathbf{a}\cdot\mathbf{K})(\mathbf{Q}\cdot\mathbf{K})}{K^2Q}k_0\int d\epsilon\times\int_{-1}^{1}\frac{d\mu}{2}\mu(1-\mu^2)\frac{\epsilon_0(\epsilon'-\epsilon)}{EE'(E+E')}$$
$$=\tfrac{3}{8}N(0)Vk_0\frac{(\mathbf{a}\cdot\mathbf{K})(\mathbf{Q}\cdot\mathbf{K})}{K^2Q}J(Q)\quad\text{(say)}.$$

This result suggests that we try

$$A_K(Q)=\frac{(\mathbf{a}\cdot\mathbf{K})(\mathbf{Q}\cdot\mathbf{K})}{K^2}\bar{A}(Q).$$

In that case

$$\sum_k V(\mathbf{K},\mathbf{k})A_k(Q)l(\mathbf{k},\mathbf{Q})^2\nu_k(\mathbf{Q})^{-1}$$
$$=-\tfrac{3}{4}V\bar{A}(Q)\sum_k\left(1-\frac{\mathbf{k}\cdot\mathbf{K}}{kK}\right)^2\times\frac{(\mathbf{a}\cdot\mathbf{k})(\mathbf{Q}\cdot\mathbf{k})}{k^2}\frac{1}{2}\left(1+\frac{\epsilon\epsilon'+\epsilon_0^2}{EE'}\right)\nu_k(\mathbf{Q})^{-1}$$
$$=-\tfrac{3}{4}V\bar{A}(Q)\sum_k\frac{2k_x^2K_xk_z^2K_z}{k^4K^2}aQ\tfrac{1}{2}\left(1+\frac{\epsilon\epsilon'+\epsilon_0^2}{EE'}\right)\nu_k(Q)^{-1}$$
$$=-\tfrac{3}{4}N(0)V\bar{A}(Q)\frac{(a\cdot K)(Q\cdot K)}{K^2}\times\int_0^{\hbar\omega}d\epsilon\int\frac{d\mu}{2}\frac{\mu^2(1-\mu^2)}{E+E'}\frac{1}{2}\left(1+\frac{\epsilon\epsilon'+\epsilon_0^2}{EE'}\right).$$

After some calculation it is found that if $\hbar v_0Q<\hbar\omega$ and $\epsilon_0\ll\hbar v_0Q$, the sum is

$$-(3/20)N(0)VA_K(\mathbf{Q})\ln(2\hbar\omega/\hbar v_0Q)$$
$$\cong-(3/20)A_K(\mathbf{Q})[1-N(0)V\ln(\hbar v_0Q/\epsilon_0)].$$

Then the equation for $A_k(\mathbf{Q})$ is

$$\left[1-\frac{3}{20}+\frac{3}{20}N(0)V\ln(\hbar v_0Q/\epsilon_0)\right]A_K(Q)=\tfrac{3}{4}\alpha N(0)Vk_0\frac{(\mathbf{a}\cdot\mathbf{K})(\mathbf{Q}\cdot\mathbf{K})}{K^2Q}J(Q).$$

As we are only estimating the order of magnitude of the correction, we shall keep only the first term of the square brackets. Hence

$$A(Q)=\tfrac{3}{4}(\alpha/q)N(0)Vk_0J(Q).$$

The correction to the current density is given by

$$\left(\frac{2m}{e\hbar}\right)j_z(Q)=-\frac{3}{4}\frac{\alpha}{q}N(0)Vk_0J(Q)\times\sum_k k_z\frac{\epsilon_0(\epsilon-\epsilon')}{EE'(E+E')}\frac{aQk_xk_z}{k^2}$$
$$=\tfrac{3}{4}\alpha N(0)^2Vk_0J(Q)a\times\int d\epsilon\int_{-1}^{1}\frac{d\mu}{2}\mu(1-\mu^2)\frac{\epsilon_0(\epsilon'-\epsilon)}{2EE'(E+E')}$$
$$=\tfrac{3}{16}a\alpha N(0)^2Vk_0^2J(Q)^2.$$

$J(Q)$ has been evaluated and is

$$-\frac{16\epsilon_0}{3\hbar v_0 Q}[\ln(\hbar v_0 Q/\epsilon_0)-\tfrac{4}{3}].$$

Therefore, the total transverse current density is

$$j_\perp(Q)=\frac{-3\epsilon}{16Q\xi_0\lambda_L{}^2(0)}\left\{1-\frac{16}{\pi^3 Q\xi_0}\ln(\pi Q\xi_0)-\frac{8N(0)V}{3\pi^3 Q\xi_0}[\ln(\pi Q\xi_0)-\tfrac{4}{3}]^2\right\}a_\perp(Q).$$

The first two terms are those given by BCS while the third is the new correction. As the formula is valid only for $Q\xi_0>1$, it is reasonable to test the correction using $\pi Q\xi_0=10$. If we also choose $N(0)V=0.3$, the ratio of the third term to the second is 0.02, which suggests that the correction is small. It *is* possible to choose a potential that makes the correction large by making the potential vary considerably with angle and change sign. For example, if one chooses

$$V(\mathbf{K},\mathbf{k})=V_1-\tfrac{3}{4}V_2(1-\cos\theta)^2,\quad V_1-V_2<0,$$

the correction is enhanced by the factor $V_2/(V_1-V_2)$ which can be made as large as one pleases by making (V_1-V_2) sufficiently small. But, although $V(\mathbf{K},\mathbf{k})$ may oscillate widely over small angles because of the contribution of the umklapp processes, we expect that on the average it will not vary widely enough over 180° to make the correction large. As the correction is sensitive to the dependence of the potential on angle the argument is not conclusive.

One can see the connection with the work of Pines and Schrieffer[8] in the following way. The operator which creates a plasmon is

$$\mu_{\rm pl}{}^*(\mathbf{Q})=\sum_k[\alpha(\mathbf{k},\mathbf{Q})\gamma_{k+Q1}{}^*\gamma_{k0}{}^*-\beta(\mathbf{k},\mathbf{Q})\gamma_{k+Q0}\gamma_{k1}],$$

where

$$\alpha(\mathbf{k},\mathbf{Q})=[\Phi_k+\tfrac{1}{2}l(\mathbf{k},\mathbf{Q})L_k]/(\hbar\omega_{\rm pl}-\nu_k),$$

$$\beta(\mathbf{k},\mathbf{Q})=-[\Phi_k-\tfrac{1}{2}l(\mathbf{k},\mathbf{Q})L_k]/(\hbar\omega_{\rm pl}+\nu_k),$$

and $\omega_{\rm pl}$ is the plasma frequency. Φ_k and L_k satisfy Eqs. (3.6) without the inhomogeneous terms and with $\nu_{k'}$ replaced by $\hbar\omega_{\rm pl}$. Hence

$$\mu_{\rm pl}{}^*(\mathbf{Q})-\mu_{\rm pl}(-\mathbf{Q})=\sum_k[\alpha(\mathbf{k},\mathbf{Q})+\beta(\mathbf{k},\mathbf{Q})]\times[\gamma_{k+Q1}{}^*\gamma_{k0}{}^*-\gamma_{k+Q0}\gamma_{k1}].$$

If terms of second order in the electron-plasmon coupling constant are neglected,

$$\begin{aligned}\alpha(\mathbf{k},\mathbf{Q})+\beta(\mathbf{k},\mathbf{Q})&=[2\Phi_k\nu_k+\hbar\omega_{\rm pl}l(\mathbf{k},\mathbf{Q})L_k](\hbar\omega_{\rm pl})^{-2}\\&=2M_{\rm pl}V_D(\mathbf{Q})[m_k\nu_k-(I_k+I_{k+Q})l(\mathbf{k},\mathbf{Q})](\hbar\omega_{\rm pl})^{-2}\\&=-2M_{\rm pl}V_D(\mathbf{Q})(\epsilon_{k+Q}-\epsilon_k)p(\mathbf{k},\mathbf{Q})(\hbar\omega_{\rm pl})^{-2}.\end{aligned}$$

Hence

$$\mu_{\rm pl}{}^*(\mathbf{Q})-\mu_{\rm pl}(-\mathbf{Q})=-\frac{2M_{\rm pl}V_D(\mathbf{Q})}{(\hbar\omega_{\rm pl})^2}\sum_k(\epsilon_{k+Q}-\epsilon_k)(\gamma_{k+Q1}{}^*\gamma_{k0}{}^*-\gamma_{k+Q0}\gamma_{k1}),$$

which is proportional to $\mathbf{j}_p(\mathbf{Q})$. $M_{\rm pl}$ has to be chosen so that the creation operators are properly normalized. Then the analysis follows that of Pines and Schrieffer.

Note added in proof.—Since this paper was submitted a number of papers and preprints have appeared on the theory of the Meissner effect. The reader is referred to K. Yosida [Prog. Theoret. Phys. (Kyoto) **21**, 731 (1959)], Blatt, Matsubara, and May [Prog. Theoret. Phys. (Kyoto) **21**, 745 (1959)], N. N. Bogoliubov (preprint) and Y. Nambu (preprint).

ACKNOWLEDGMENTS

It is a pleasure to acknowledge the advice of Professor J. Bardeen at all stages of this work and several stimulating discussions with Professor H. J. Lipkin and Professor D. Pines. The author is also grateful to Dr. P. W. Anderson for letting him see a copy of his paper prior to its publication.

APPENDIX A

In this Appendix, Anderson's linear equations of motion are derived and the screening of the exchange terms justified. It is apparent from a comparison of the treatments of Nakajima[5] and of Bardeen and Pines[6] of the electron-phonon interaction, that in order to obtain the interaction properly screened it is necessary to separate out the plasma degrees of freedom. Accordingly, let us try to separate out these degrees of freedom. We will suppose at first that the operators $\mu_Q{}^*$ which create plasma oscillations of wave vector $\mathbf{Q}$ in the superconductor are known. For these modes the RPA is certainly a good approximation; one can write

$$\mu_Q{}^*=\sum_k[\alpha(\mathbf{k},\mathbf{Q})\rho_k{}^Q+\beta(\mathbf{k},\mathbf{Q})\bar\rho_k{}^Q+\gamma(\mathbf{k},\mathbf{Q})b_k{}^Q+\phi(\mathbf{k},\mathbf{Q})\bar b_k{}^Q].$$

Ultimately the coefficients α, β, γ, ϕ will have to be determined.

The "intrinsic" Hamiltonian, $H_{\rm int}$, is introduced by

$$H_{\rm int}=H-\sum_{Q<Q_{\rm max}}\hbar\omega_Q\mu_Q{}^*\mu_Q;$$

$H_{\rm int}$ is a function of operators $\mu_k(\mathbf{Q})$ and $\mu_k(\mathbf{Q})^*$ which commute with μ_Q and $\mu_Q{}^*$ and which destroy and create states in which a pair of particles are excited. We can find these operators from the equations of motion derived from $H_{\rm int}$. Let us consider just one of these equations, the one for $\rho_K{}^Q$. One finds

$$\begin{aligned}
&[H_{\text{int}},\rho_K{}^Q]\\
&=[H-\sum \hbar\omega_{Q'}\mu_{Q'}{}^*\mu_{Q'},\ \rho_K{}^Q]\\
&=[H,\rho_K{}^Q]-\sum_{Q'}\hbar\omega_{Q'}\mu_{Q'}{}^*[\mu_{Q'},\rho_K{}^Q]\\
&\qquad\qquad -\sum_{Q'}\hbar\omega_{Q'}[\mu_{Q'}{}^*,\rho_K{}^Q]\mu_{Q'}\\
&=[H,\rho_K{}^Q]-\sum_{Q'}\hbar\omega_{Q'}\mu_{Q'}{}^*\sum_{k'}\{\alpha(\mathbf{k}',\mathbf{Q}')\\
&\quad\times[c_{k'\uparrow}{}^*c_{k\uparrow}\delta_{k'+Q',\,k+Q}-c_{k+Q\uparrow}{}^*c_{k'+Q'\uparrow}\delta_{k',\,k}]\\
&\quad-\gamma(\mathbf{k}',\mathbf{Q}')[c_{k+Q\uparrow}{}^*c_{-k'-Q'\downarrow}{}^*\delta_{k',\,k}]+\phi(\mathbf{k}',\mathbf{Q}')\\
&\quad\times[c_{-k'\downarrow}c_{k\uparrow}\delta_{k'+Q',\,k+Q}]\}-\sum_{Q'}\hbar\omega_{Q'}\sum_{k'}\{\alpha(\mathbf{k}',\mathbf{Q}')\\
&\quad\times[c_{k'+Q'\uparrow}{}^*c_{k\uparrow}\delta_{k',\,k+Q}-c_{k+Q\uparrow}{}^*c_{k'\uparrow}\delta_{k'+Q',\,k}]\\
&\quad+\gamma(\mathbf{k}',\mathbf{Q}')c_{-k'-Q'\downarrow}c_{k\uparrow}\delta_{k',\,k+Q}\\
&\qquad\qquad-\phi(\mathbf{k}',\mathbf{Q}')c_{k+Q\uparrow}{}^*c_{-k'\downarrow}{}^*\delta_{k'+Q',\,k}]\}\mu_{Q'}\\
&=[H,\rho_K{}^Q]-\sum_{Q'}\hbar\omega_{Q'}\mu_{Q'}{}^*\{\alpha(\mathbf{k}+\mathbf{Q}-\mathbf{Q}',\ \mathbf{Q}')\\
&\quad\times[c_{k+Q-Q'\uparrow}{}^*c_{k\uparrow}-\alpha(\mathbf{k},\mathbf{Q}')c_{k+Q\uparrow}{}^*c_{k+Q'\uparrow}]\\
&\quad-\gamma(\mathbf{k},\mathbf{Q}')c_{k+Q\uparrow}{}^*c_{-k-Q'\downarrow}{}^*\\
&\quad+\phi(\mathbf{k}+\mathbf{Q}-\mathbf{Q}',\ \mathbf{Q}')c_{-k-Q+Q'\downarrow}c_{k\uparrow}\}\\
&\quad-\sum_{Q'}\hbar\omega_{Q'}\{\alpha(\mathbf{k}+\mathbf{Q},\ \mathbf{Q}')c_{k+Q+Q'\uparrow}{}^*c_{k\uparrow}\\
&\quad-\alpha(\mathbf{k}-\mathbf{Q}',\ \mathbf{Q}')c_{k+Q\uparrow}{}^*c_{k-Q'\uparrow}\\
&\quad+\gamma(\mathbf{k}+\mathbf{Q},\ \mathbf{Q}')c_{-k-Q-Q'\downarrow}c_{k\uparrow}\\
&\qquad\qquad-\phi(\mathbf{k}-\mathbf{Q}',\ \mathbf{Q}')c_{k+Q\uparrow}{}^*c_{-k+Q'\downarrow}{}^*\}\mu_{Q'}.
\end{aligned}$$

To linearize this equation, products of pairs are replaced by their expectation values in the ground state. The first term when linearized looks just like the right-hand side of Eq. (3.1a) but the potential appearing in it is unscreened. This term will be written as $[H,\rho_K{}^Q]_L$. Then (omitting the exchange terms to save space)

$$\begin{aligned}
&[H_{\text{int}},\rho_K{}^Q]\\
&=[H,\rho_K{}^Q]_L+\sum_{Q'}\hbar\omega_{Q'}\{-\mu_{Q'}{}^*\langle[\mu_{Q'},\rho_K{}^Q]\rangle\\
&\quad+\langle[\mu_{Q'}{}^*,\rho_K{}^Q]\rangle\mu_{Q'}-\beta(\mathbf{k}',\mathbf{Q}')\alpha(\mathbf{k}+\mathbf{Q}-\mathbf{Q}',\ \mathbf{Q}')\\
&\quad\times[b_{k'}\delta_{k',\,k+Q-Q'}c_{-k'-Q'\downarrow}c_{k\uparrow}\\
&\quad-c_{-k'\downarrow}{}^*c_{k+Q-Q'\uparrow}{}^*\delta_{k'+Q',\,k}b_k]+\beta(\mathbf{k}',\mathbf{Q}')\alpha(\mathbf{k},\mathbf{Q}')\\
&\quad\times[b_{k'}\delta_{k',\,k+Q}c_{-k'-Q'\downarrow}c_{k+Q'\uparrow}-b_{k'+Q'}c_{-k'\downarrow}{}^*c_{k+Q\uparrow}{}^*\delta_{k,\,k'}]\\
&\quad-\alpha(\mathbf{k}+\mathbf{Q},\ \mathbf{Q}')\beta(\mathbf{k}',\mathbf{Q}')[-b_{k'+Q'}c_{k\uparrow}c_{-k'\downarrow}\delta_{k+Q,\,k'}\\
&\quad+b_kc_{k+Q+Q'\uparrow}{}^*c_{-k'-Q'\downarrow}{}^*\delta_{k,\,k'}]+\alpha(\mathbf{k}-\mathbf{Q}',\ \mathbf{Q}')\beta(\mathbf{k}',\mathbf{Q}')\\
&\quad\times[-b_{k+Q}\delta_{k+Q,\,k'+Q'}c_{k-Q'\uparrow}c_{-k'\downarrow}\\
&\qquad+b_{k-Q'}\delta_{k-Q',\,k'}c_{k+Q\uparrow}{}^*c_{-k'-Q'\downarrow}{}^*]+\text{terms in }\gamma\text{ and }\delta,
\end{aligned}$$

where $\langle[\mu_{Q'},\rho_K{}^Q]\rangle$ is the expectation value of the commutator in the ground state, i.e.,

$$\begin{aligned}
\langle[\mu_{Q'},\rho_K{}^Q]\rangle&=\delta_{Q,\,Q'}\{\alpha(\mathbf{k},\mathbf{Q}')[n_k-n_{k+Q}]\\
&\qquad-\gamma(\mathbf{k},\mathbf{Q}')b_{k+Q}+\phi(\mathbf{k},\mathbf{Q}')b_k\},
\end{aligned}$$

and

$$\begin{aligned}
\langle[\mu_{Q'}{}^*,\rho_K{}^Q]\rangle&=\delta_{Q,-Q'}\{\alpha(\mathbf{k}+\mathbf{Q},\ \mathbf{Q}')(n_k-n_{k+Q})\\
&\qquad+\gamma(\mathbf{k}+\mathbf{Q},\ \mathbf{Q}')b_k-\phi(\mathbf{k}+\mathbf{Q},\ \mathbf{Q}')b_{k+Q}\}.
\end{aligned}$$

One now has a set of equations (not all independent) from which to determine the single-particle excitations. Since the coefficients α, β, γ, and ϕ are also unknown, these equations together with

$$[H_{\text{int}},\mu_Q{}^*]=[H_{\text{int}},\mu_Q]=0$$

determine these coefficients and also the cutoff on Q'. The equations are not linear in the coefficients.

In order to see the connection with the Eqs. (3.1) we first make what appears to be a reasonable approximation and replace α, β, γ, ϕ by the values we should obtain in the normal state (neglecting second-order terms in the electron-plasmon coupling constant), that is, we take

$$\gamma=\phi=0,\quad \beta(\mathbf{k},\mathbf{Q})=\alpha(\mathbf{k},\mathbf{Q})=[V(\mathbf{Q})/2\hbar\omega_Q]^{\frac{1}{2}},\qquad \text{(A1)}$$

where $V(Q)$ is the sum of the Coulomb and phonon interactions. $V(Q)$ is unscreened, the phonon part corresponds to the interaction obtained by Nakajima, not that of Bardeen and Pines. This approximation can be made the first step of a self-consistent calculation. It follows that

$$\begin{aligned}
&[H_{\text{int}},\rho_k{}^Q]\\
&=[H,\rho_k{}^Q]_L-\sum_{Q'}V(\mathbf{Q}')\\
&\quad\times[b_{k+Q-Q'}b_k{}^Q+b_k\bar{b}_{k-Q'}{}^Q-b_{k+Q}b_{k+Q'}{}^Q-b_{k+Q'}\bar{b}_k{}^Q]\\
&\qquad-\hbar\omega_Q\{\mu_Q{}^*\langle[\mu_Q,\rho_k{}^Q]\rangle+\langle[\mu_{-Q}{}^*,\rho_k{}^Q]\rangle\mu_Q\}\\
&=[H,\rho_k{}^Q]_A-\hbar\omega_Q\{\mu_Q{}^*\langle[\mu_Q,\rho_k{}^Q]\rangle\\
&\qquad\qquad+\langle[\mu_{-Q}{}^*,\rho_k{}^Q]\rangle\mu_{-Q}\},\qquad \text{(A2)}
\end{aligned}$$

where $[H,\rho_k{}^Q]_A$ stands symbolically for the commutator written down by Anderson. This equation and the corresponding equations for $\bar{\rho}$, b, and $\bar{b}$ are together equivalent to those of Anderson. This can be seen in the following way. The equations for the coefficients $\alpha(\mathbf{k},\mathbf{Q})$, β, γ, and ϕ are found from the equation for $\mu_Q{}^*$. From Eqs. (A2) and the corresponding equations for b, $\bar{b}$, and $\bar{\rho}$ one obtains

$$\begin{aligned}
[H_{\text{int}},\mu_Q{}^*]=[H,\mu_Q{}^*]_A-\hbar\omega_Q\{\mu_Q{}^*\langle[\mu_Q,\mu_Q{}^*]\rangle&\\
+\langle[\mu_{-Q}{}^*,\mu_Q{}^*]\rangle\mu_Q\}&.
\end{aligned}$$

Since $\mu_Q{}^*$ commutes with H_{int},

$$[H,\mu_Q{}^*]_A=\hbar\omega_Q\mu_Q{}^*,$$

which is the equation one obtains from Anderson. For the single-particle excitations, $\mu_k{}^*(\mathbf{Q})$, one finds in the same way

$$[H,\mu_k{}^*(\mathbf{Q})]_A=[H_{\text{int}},\mu_k{}^*(\mathbf{Q})]=[H,\mu_k{}^*(\mathbf{Q})],$$

since $\mu_k{}^*(\mathbf{Q})$ commutes with μ_Q and $\mu_Q{}^*$. This proves that Eqs. (A2) are equivalent to Eqs. (3.1). The reason the terms that lead to the superconducting transition appear screened is that as far as these terms are concerned H_{int} is

$$H-\sum_{Q'<Q_{\max}}\hbar\omega_{Q'}\mu_{Q'}{}^*\mu_{Q'}\approx H-\tfrac{1}{2}\sum_{Q'<Q_{\max}}V(\mathbf{Q}')\rho_{-Q'}\rho_{Q'}.$$

In this Hamiltonian the two-body interaction is screened. For the same reason, had the exchange terms been kept we should have found that these, too, are screened.

It would seem possible to generalize the equations by supposing all the operators μ_i^* to be known linear combinations of ρ_k^Q, $\bar{\rho}_k^Q$, b_k^Q, and $\bar{b}_k^Q$. Then one would obtain a set of nonlinear integral equations for the coefficients by making the equations

$$[H-\sum_{j\neq i}\hbar\omega_j\mu_j^*\mu_j,\,\mu_i^*]=\hbar\omega_i\mu_i^*$$

linear in the operators. Of course, Eqs. (A1) would only be a first approximation to the plasmon operators.

APPENDIX B

In this appendix it will be shown that the operators, μ_k^* defined by Eqs. (3.4) and (3.5) form an orthonormal set. The plan is to show that the operators $X_k^*(\mathbf{Q})$, defined by

$$X_k^*(\mathbf{Q})=\sum_{k'}[\alpha_1^*(\mathbf{k}',\mathbf{k},\mathbf{Q})\gamma_{k'+Q1}^*\gamma_{k'0}^* -\beta_1^*(\mathbf{k}',\mathbf{k},\mathbf{Q})\gamma_{k'+Q0}\gamma_{k'1}],$$

form an orthonormal set. It then will follow that the coefficients of $X_k^*(\mathbf{Q})$ in the expansions of $\gamma_{k+Q1}^*\gamma_{k0}^*$ and $\gamma_{k+Q0}\gamma_{k1}$ in terms of the X's are, respectively, $\alpha_1(\mathbf{k},\mathbf{k}',\mathbf{Q})$ and $\beta_1(\mathbf{k},\mathbf{k}',\mathbf{Q})$. By direct substitution it can be seen that $X_k^*(\mathbf{Q})$ satisfies the equations of motion with eigenvalue $\nu_k(\mathbf{Q})$. Hence $X_k(\mathbf{Q})$ can be identified with $\mu_k(\mathbf{Q})$ and the result will be proved. Now

$$[X_{k''}(\mathbf{Q}),X_{k'}(\mathbf{Q})^*]$$

$$=\sum_k[\alpha_1(\mathbf{k},\mathbf{k}'',\mathbf{Q})\alpha_1(\mathbf{k},\mathbf{k}',\mathbf{Q})^*-\beta_1(\mathbf{k},\mathbf{k}'',\mathbf{Q})\beta_1(\mathbf{k},\mathbf{k}',\mathbf{Q})^*]$$

$$=\delta_{k'k''}+\frac{\Phi_{k''k'}^*+\frac{1}{2}l(\mathbf{k}'',\mathbf{Q})L_{k''k'}^*}{\nu_{k'}-\nu_{k''}-i\epsilon}+\frac{\Phi_{k'k''}+\frac{1}{2}l(\mathbf{k}'',\mathbf{Q})L_{k''k'}}{\nu_{k''}-\nu_{k'}+i\epsilon}$$
$$+\sum_k\frac{(\Phi_{kk'}^*+\frac{1}{2}l(\mathbf{k},\mathbf{Q})L_{kk'}^*)(\Phi_{kk''}+\frac{1}{2}l(\mathbf{k},\mathbf{Q})L_{kk''})}{(\nu_k-\nu_{k''}-i\epsilon)(\nu_k-\nu_{k'}+i\epsilon)}$$
$$-\sum_k\frac{(\Phi_{kk'}^*-\frac{1}{2}l(\mathbf{k},\mathbf{Q})L_{kk'}^*)(\Phi_{kk''}-\frac{1}{2}l(\mathbf{k},\mathbf{Q})L_{kk''})}{(\nu_k+\nu_{k''})(\nu_k+\nu_{k'})}$$

$$=\Bigg\{\tfrac{1}{2}\delta_{k'k''}+\frac{1}{\nu_{k'}-\nu_{k''}-i\epsilon}\Bigg[\Phi_{k''k'}^*+\tfrac{1}{2}l(\mathbf{k}'',\mathbf{Q})L_{k''k'}^*$$
$$-\sum_k\Bigg(\frac{[\Phi_{kk'}^*+\frac{1}{2}l(\mathbf{k},\mathbf{Q})L_{kk'}^*][\Phi_{kk''}+\frac{1}{2}l(\mathbf{k},\mathbf{Q})L_{kk''}]}{(\nu_k-\nu_{k'}+i\epsilon)}$$
$$+\frac{[\Phi_{kk'}^*-\frac{1}{2}l(\mathbf{k},\mathbf{Q})L_{kk'}^*][\Phi_{kk''}-\frac{1}{2}l(\mathbf{k},\mathbf{Q})L_{kk''}]}{(\nu_k+\nu_{k'})}\Bigg)\Bigg]\Bigg\}+\{\mathbf{k}'\leftrightarrow\mathbf{k}''\}^*$$

$$=\Bigg\{\tfrac{1}{2}\delta_{k'k''}+\frac{1}{\nu_{k'}-\nu_{k''}-i\epsilon}[\Phi_{k''k'}^*+\tfrac{1}{2}l(\mathbf{k}'',\mathbf{Q})L_{k''k'}^*$$
$$-\sum_k\Phi_{kk''}\Bigg[\Phi_{kk'}^*\Bigg(\frac{1}{\nu_k-\nu_{k'}+i\epsilon}-\frac{1}{\nu_k+\nu_{k'}}\Bigg)+\tfrac{1}{2}l(\mathbf{k},\mathbf{Q})L_{kk'}^*\Bigg(\frac{1}{\nu_k-\nu_{k'}+i\epsilon}+\frac{1}{\nu_k+\nu_{k'}}\Bigg)\Bigg]$$
$$-\sum_k\tfrac{1}{2}l(\mathbf{k},\mathbf{Q})L_{kk''}\Bigg[\Phi_{kk'}^*\Bigg(\frac{1}{\nu_k-\nu_{k'}+i\epsilon}+\frac{1}{\nu_k+\nu_{k'}}\Bigg)+\tfrac{1}{2}l(\mathbf{k},\mathbf{Q})L_{k,k'}^*\Bigg(\frac{1}{\nu_k-\nu_{k'}+i\epsilon}-\frac{1}{\nu_k+\nu_{k'}}\Bigg)\Bigg]\Bigg\}$$
$$+\{\mathbf{k}'\leftrightarrow\mathbf{k}''\}^*$$

$$=\{\tfrac{1}{2}\delta_{k'k''}+\frac{1}{\nu_{k'}-\nu_{k''}-i\epsilon}[\Phi_{k''k'}^*+\tfrac{1}{2}l(\mathbf{k}'',\mathbf{Q})L_{k''k'}^*$$
$$-\sum_k\Phi_{kk''}[\alpha_1(\mathbf{k},\mathbf{k}',\mathbf{Q})^*+\beta_1(\mathbf{k},\mathbf{k}',\mathbf{Q})^*-\delta_{kk'}]$$
$$-\sum_k\tfrac{1}{2}l(\mathbf{k},\mathbf{Q})L_{kk''}[\alpha_1(\mathbf{k},\mathbf{k}',\mathbf{Q})^*-\beta_1(\mathbf{k},\mathbf{k}',\mathbf{Q})^*-\delta_{kk'}]\}+\{\mathbf{k}'\leftrightarrow\mathbf{k}''\}^*.$$

Now

$$\sum_k\tfrac{1}{2}l(\mathbf{k},\mathbf{Q})L_{kk''}[\alpha_1(\mathbf{k},\mathbf{k}',\mathbf{Q})^*-\beta_1(\mathbf{k},\mathbf{k}',\mathbf{Q})^*]$$
$$=\sum_{k,k'''}\tfrac{1}{2}l(\mathbf{k},\mathbf{Q})V(\mathbf{k},\mathbf{k}''')l(\mathbf{k}''',\mathbf{Q})\times[\alpha_1(\mathbf{k}''',\mathbf{k}'',\mathbf{Q})-\beta_1(\mathbf{k}''',\mathbf{k}'',\mathbf{Q})]\times[\alpha_1(\mathbf{k},\mathbf{k}',\mathbf{Q})^*-\beta_1(\mathbf{k},\mathbf{k}',\mathbf{Q})^*].$$

This remains unchanged when $\mathbf{k}'$ and $\mathbf{k}''$ are interchanged and the complex conjugate is taken $[V(k,k')=V(k',k)^*]$. Hence this sum disappears from the final result. Similarly, the term

$$\sum_k\Phi_{kk''}[\alpha_1(\mathbf{k},\mathbf{k}',\mathbf{Q})^*+\beta_1(\mathbf{k},\mathbf{k}',\mathbf{Q})^*]$$

does not contribute. Hence

$$[X_{k''}(\mathbf{Q}),X_{k'}(\mathbf{Q})^*]$$
$$=\Bigg\{\tfrac{1}{2}\delta_{k'k''}+\frac{1}{\nu_{k'}-\nu_{k''}-i\epsilon}[\Phi_{k''k'}^*+\tfrac{1}{2}l(\mathbf{k}'',\mathbf{Q})L_{k''k'}^*+\Phi_{k'k''}+\tfrac{1}{2}l(\mathbf{k}',\mathbf{Q})L_{k'k''}]\Bigg\}+\{\mathbf{k}'\leftrightarrow\mathbf{k}''\}^*=\delta_{k'k''},$$

as was to be proved.

Ground-State Energy and Green's Function for Reduced Hamiltonian for Superconductivity*

J. BARDEEN AND G. RICKAYZEN†
Department of Physics, University of Illinois, Urbana, Illinois

(Received December 14, 1959)

In their theory of superconductivity, Bardeen, Cooper, and Schrieffer made use of a reduced Hamiltonian which included only scattering of pairs of particles of opposite momentum and spin. It is shown that the solution they obtained by a variational method is correct to $O(1/n)$ for a large system. The single particle Green's function is derived and used to calculate the interaction energy.

IN their theory of superconductivity, Cooper, Schrieffer, and one of the authors[1] made use of a reduced Hamiltonian which included only scattering of pairs of particles of opposite momentum and spin. They suggested that the solution obtained by a variational method might well be correct to $O(1/n)$ in the number of particles, n. They showed in particular that $\langle H_{\text{red}}^p\rangle$ differs from $\langle H_{\text{red}}\rangle^p$ by $O(1/n)$ for $p=2,3$ and very likely for $p\ll n$, where the averages are taken with respect to the variational wave function for the ground state. If this were true for all p, the energy would be exact. Anderson[2] showed that the reduced problem is analogous to one of a system of interacting spins and gave a physical argument, based on the correspondence principle, which also indicated that the solution is correct to $O(1/n)$. Particularly because the validity of the solution has been questioned, we thought it desirable to give a direct proof based on the structure of the wave functions. We also derive the single particle Green's function for the system and use it to calculate the energy of H_{red} from an expression given by Galitskii and Migdal.[3]

The reduced Hamiltonian may be written in the form:

$$H_{\text{red}}=\sum_{\mathbf{k},\sigma}(\epsilon_k+\mu)c_{\mathbf{k}\sigma}^*c_{\mathbf{k}\sigma}+\sum V_{\mathbf{k}\mathbf{k}'}b_{\mathbf{k}'}^*b_{\mathbf{k}}, \quad (1)$$

where $c_{\mathbf{k}\sigma}^*$ is a creation operator for a particle in a state of wave vector $\mathbf{k}$ and spin σ, $b_{\mathbf{k}}=c_{-\mathbf{k}\downarrow}c_{\mathbf{k}\uparrow}$ destroys the pair $\mathbf{k}\equiv(\mathbf{k}\uparrow,-\mathbf{k}\downarrow)$ and ϵ_k is the Bloch energy measured from the Fermi energy, μ. For simplicity we assume that the interaction $V_{\mathbf{k}\mathbf{k}'}=V_{\mathbf{k}'\mathbf{k}}$ is real. The ground-state wave function Ψ_0 of H_{red} is a linear combination of configurations in which the single particle states are occupied in pairs of opposite momentum and spin.

Following BCS it is convenient to decompose Ψ_0 into parts with definite occupancy of a particular pair, $\mathbf{k}$:

$$\Psi_0=u_{\mathbf{k}}b_{\mathbf{k}}^*\varphi_{0\mathbf{k}}(n-2)+v_{\mathbf{k}}\varphi_{0\mathbf{k}}(n), \quad (2)$$

where $u_{\mathbf{k}}^2=1-v_{\mathbf{k}}^2=h(\mathbf{k})$ is the probability of occupancy of $\mathbf{k}$. Here $\varphi_{0\mathbf{k}}(n)$ is a function of n particles in which the pair $\mathbf{k}$ is not occupied.

The functions $\varphi_{0\mathbf{k}}$ may be further decomposed into parts with definite occupancy of some other pair $\mathbf{k}'$:

$$\varphi_{0\mathbf{k}}(n)=u_{\mathbf{k}\mathbf{k}'}b_{\mathbf{k}}^*\varphi_{0\mathbf{k}\mathbf{k}'}(n-2)+v_{\mathbf{k}\mathbf{k}'}\varphi_{0\mathbf{k}\mathbf{k}'}(n). \quad (3)$$

This function is similar to Ψ_0 except that it contains no configurations in which the pair $\mathbf{k}$ is occupied. To terms of $O(1/n)$, the coefficients of this second decomposition must be independent of $\mathbf{k}$ and the same as those in (2), i.e.,

$$u_{\mathbf{k}\mathbf{k}'}=u_{\mathbf{k}'}+O(1/n), \quad (4a)$$

$$v_{\mathbf{k}\mathbf{k}'}=v_{\mathbf{k}'}+O(1/n). \quad (4b)$$

This is true because the occupancy of $\mathbf{k}'$ depends on the interaction of pairs in $\mathbf{k}'$ with all other pairs, and can be changed only to $O(1/n)$ if configurations with one pair state, e.g., $\mathbf{k}$, are omitted. In the BCS solution, there is no correlation between the coefficients of the first and second decomposition, which corresponds to omitting the terms of $O(1/n)$ in (4). The interaction energy to this order is:

$$U=\sum(0|V_{\mathbf{k}\mathbf{k}'}b_{\mathbf{k}'}^*b_{\mathbf{k}}|0)$$
$$=\sum V_{\mathbf{k}\mathbf{k}'}[u_{\mathbf{k}}v_{\mathbf{k}}u_{\mathbf{k}'}v_{\mathbf{k}'}+O(1/n)]. \quad (5)$$

As the size of the system is increased, keeping the particle density constant, the matrix element $V_{\mathbf{k}\mathbf{k}'}$ is inversely proportional to volume and thus to n, so that the total interaction energy is proportional to n. The terms of $O(1/n)$ would contribute only a constant energy, independent of volume.[4]

To calculate the Green's function,[5] we need the

* This research was supported in part by the Office of Ordnance Research, U. S. Army.

† Present address is Department of Physics, University of Liverpool, England.

[1] J. Bardeen, L. N. Cooper, and J. R. Schrieffer, Phys. Rev. **108**, 1175 (1957). We refer to this paper as BCS.

[2] P. W. Anderson, Phys. Rev. **110**, 827 (1958).

[3] V. M. Galitskii and A. B. Migdal, J. Exptl. Theoret. Phys. (U.S.S.R.) **34**, 139 (1958) [translation: Soviet Phys.—JETP **34**(7), 96 (1958)]. V. M. Galitskii, J. Exptl. Theoret. Phys. (U.S.S.R.) **34**, 145 (1958) [translation: Soviet Phys.—JETP **7**, 104 (1958)].

[4] It should be noted that a similar argument cannot be applied to the complete Hamiltonian. There are then n^2 interaction terms and errors of $O(1/n)$ can pile up to give a correction proportional to n. In fact, almost the entire normal-superconducting transition energy comes from the interaction terms in H_{red}, and these contain only $O(1/n)$ of the total interaction.

[5] The Green's function for H_{red} has been derived by others, e.g., L. P. Gorkov, J. Exptl. Theoret. Phys. (U.S.S.R.) **34**, 735 (1958) [translation: Soviet Phys.—JETP **34**(7), 505 (1958)], by use of wave functions with variable numbers of particles. It is believed that the present derivation is new.

matrix elements of $c_{k\sigma}^*$ and $c_{k\sigma}$ which connect the ground-state Ψ_0 of n particles with excited states of $n+1$ and $n-1$ particles, respectively. The excited states of interest are what are called single particle excitations in BCS and consist of linear combinations in which one state of a given pair is occupied in all configurations and the other is not. Thus, we shall denote by $\Psi(\mathbf{k}\sigma, n+1)$ an excited state of a system of $n+1$ particles in which $\mathbf{k}\sigma$ is certainly occupied and the remaining particles are in ground pair configurations. It should be noted that there are no terms in H_{red} which scatter an electron in a singly occupied state. It is this fact which makes possible the determination of the excitation energies and matrix elements to $O(1/n)$ from the structure of the wave functions without use of a perturbation or diagram expansion.

We shall use Lehmann's expression for the momentum-energy representation of the single-particle Green's function in a form given by Galitskii[3]:

$$G(k,\epsilon)=\int_0^\infty dE\left(\frac{\rho^+(\mathbf{k},E)}{E-\epsilon-i\delta}-\frac{\rho^-(\mathbf{k},E)}{E+\epsilon-i\delta}\right), \tag{6}$$

where $\rho^+(\mathbf{k},E)$ and $\rho^-(\mathbf{k},E)$ are defined by

$$\rho^+(\mathbf{k},E)dE=\sum_{\mathbf{k}}|(\mathbf{k}, n+1|c_{\mathbf{k}}^*|0)|^2,\quad E<E_{\mathbf{k}}<E+dE; \tag{7a}$$

$$\rho^-(\mathbf{k},E)dE=\sum_{\mathbf{k}}|(\mathbf{k}, n-1|c_{\mathbf{k}}|0)|^2,\quad E<E_{\mathbf{k}}<E+dE. \tag{7b}$$

The single particle energies, $E_{\mathbf{k}}$ are defined by[6]

$$E_{\mathbf{k}}=W(\mathbf{k}, n+1)-W_0(n)-\mu, \tag{8a}$$

$$E_{\mathbf{k}}=W(\mathbf{k}, n-1)-W_0(n)+\mu, \tag{8b}$$

in which $W(\mathbf{k}, n\pm1)$ are energies of $\Psi(\mathbf{k}, n\pm1)$ and W_0 is the ground-state energy of H_{red}.

To terms of $O(1/n)$,

$$\Psi(\mathbf{k}\uparrow, n+1)=c_{\mathbf{k}\uparrow}^*\varphi_{0\mathbf{k}}(n),$$
$$\Psi(\mathbf{k}\uparrow, n-1)=c_{-\mathbf{k}\downarrow}\varphi_{0\mathbf{k}}(n), \tag{9}$$

so that the matrix elements are

$$(\mathbf{k}\uparrow, n+1|c_{\mathbf{k}\uparrow}^*|0)=u_{\mathbf{k}}=[1-h(\mathbf{k})]^{\frac{1}{2}}, \tag{10a}$$

$$(\mathbf{k}\uparrow, n-1|c_{-\mathbf{k}\uparrow}|0)=v_{\mathbf{k}}=h(\mathbf{k})^{\frac{1}{2}}. \tag{10b}$$

Thus

$$G(\mathbf{k},\epsilon)=\frac{1-h(\mathbf{k})}{E_{\mathbf{k}}-\epsilon-i\delta}-\frac{h(\mathbf{k})}{E_{\mathbf{k}}+\epsilon-i\delta}. \tag{11}$$

The energies $W_0(n)$ and $W(\mathbf{k}, n+1)$ may be expressed in the form:

$$W_0(n)=W_{0\mathbf{k}}(n)+2\epsilon_{\mathbf{k}}h(k)+U_{\mathbf{k}}, \tag{12}$$

$$W(\mathbf{k}, n+1)=W_{0\mathbf{k}}(n)+\epsilon_{\mathbf{k}}+\mu. \tag{13}$$

Here $W_{0\mathbf{k}}(n)$ is the energy corresponding to $\varphi_{0\mathbf{k}}(n)$ and is what the ground-state energy of the system would be if the state $\mathbf{k}$ were omitted from the Hamiltonian. The last two terms of (12) represent the contributions of the kinetic energy and of the interaction energy $U_{\mathbf{k}}$ from the pair state $\mathbf{k}$ to W_0. The expression (13) follows because the singly occupied state $\mathbf{k}$ cannot contribute to the pair interaction energy of H_{red} and the Fermi energy μ is unchanged to $O(1/n)$ if one particle or one state is added or subtracted from the system. From (8) we find

$$E_{\mathbf{k}}-\epsilon_{\mathbf{k}}=-2\epsilon_{\mathbf{k}}h(k)-U_{\mathbf{k}}. \tag{14}$$

The explicit expression for $U_{\mathbf{k}}$ is

$$\begin{aligned}U_{\mathbf{k}}&=(0|\Sigma_{\mathbf{k}'}V_{\mathbf{k}\mathbf{k}'}b_{\mathbf{k}'}^*b_{\mathbf{k}}+\Sigma_{\mathbf{k}'}V_{\mathbf{k}'\mathbf{k}}b_{\mathbf{k}}^*b_{\mathbf{k}'}|0)\\&=2\Sigma_{\mathbf{k}'}V_{\mathbf{k}\mathbf{k}'}[h(1-h)h'(1-h')]^{\frac{1}{2}}+O(1/n).\end{aligned} \tag{15}$$

Another expression for $U_{\mathbf{k}}$ can be obtained from the equations of motion for G, or, perhaps more directly, for $c_{\mathbf{k}}$. For a pair interaction,

$$\begin{aligned}U_{\mathbf{k}}&=[0|(H,c_{\mathbf{k}}^*)c_{\mathbf{k}}-c_{\mathbf{k}}^*(H,c_{\mathbf{k}})-2(\epsilon_{\mathbf{k}}+\mu)c_{\mathbf{k}}^*c_{\mathbf{k}}|0]\\&=[0|Hc_{\mathbf{k}}^*c_{\mathbf{k}}-2c_{\mathbf{k}}^*Hc_{\mathbf{k}}+c_{\mathbf{k}}^*c_{\mathbf{k}}H\\&\qquad-2(\epsilon_{\mathbf{k}}+\mu)c_{\mathbf{k}}^*c_{\mathbf{k}}|0].\end{aligned} \tag{16}$$

For our case, this reduces to

$$\begin{aligned}U_{\mathbf{k}}&=h(\mathbf{k})[2W_0-2(E_{\mathbf{k}}-\mu+W_0)-2(\epsilon_{\mathbf{k}}+\mu)]\\&=-2(\epsilon_{\mathbf{k}}+E_{\mathbf{k}})h(\mathbf{k}).\end{aligned} \tag{17}$$

The same result may be obtained by an expression derived by Galitskii and Migdal[3] for pair interactions in terms of the Green's function:

$$U_{\mathbf{k}}=\frac{i}{2\pi}\int_C(\epsilon-\epsilon_{\mathbf{k}})G(\mathbf{k},\epsilon)d\epsilon. \tag{18}$$

The integral is over a contour C in the complex ϵ plane which consists of the real axis and a semicircle over the upper half-plane.

By combining (14) and (17), we get the same relation between $E_{\mathbf{k}}$ and $h(\mathbf{k})$ that was found by BCS from their variational method:

$$h(\mathbf{k})=\tfrac{1}{2}[1-(\epsilon_{\mathbf{k}}/E_{\mathbf{k}})]. \tag{19}$$

Finally, from (14), (15), and (17), we obtain an integral equation for $h(\mathbf{k})$ which is the same as Eq. (2.33) of BCS.

ACKNOWLEDGMENT

The authors wish to thank Dr. J. R. Schrieffer for helpful discussions.

[6] Our $E_{\mathbf{k}}$ is identical with $E_{\mathbf{s}}$ of Galitskii (reference 3).

COMMENTS AND CORRECTIONS

Professors Bardeen, Cooper, and Schrieffer have requested that the following corrections to their paper be noted:

1186 1st col., 3rd line from bottom: For "(2.36) and (3.28)" read "(2.40) and (3.29)"
2nd col., 2nd line under Eq. (3.35): For "(3.34)" read "(3.35)"

1187 1st col., 1st line: For "(3.31)" read "(3.33)"
2nd col., Eq. (3.49): For "10.2" read "9.4"
2nd col., Eq. (3.50): For "1.52" read "1.43"
2nd col., Eq. (3.52): For "19.4" read "18.0"

1188 1st col., 2nd line under Eq. (3.54): Add "The last equality holds if $5 > T_c/T > 1.5$."

1189 Table II:
Line 3, col. 7: For "$E + E$" read "$E + E'$"
Line 5, col. 9: For "$(hh')^{1/2}$" read "$-(hh')^{1/2}$"
Line 5, col. 10: For "$-[1 - h)(1 - h')]^{1/2}$" read "$[(1 - h)(1 - h')]^{1/2}$"
Line 8, col. 9: For "$[(1 - h)h')]^{1/2}$" read "$-[(1 - h)h']^{1/2}$"
Line 8, col. 10: For "$[h(1 - h')]^{1/2}$" read "$-[h(1 - h')]^{1/2}$"

1193 2nd col., Eq. (5.15): For "$Zm^2 c$" read "$m^2 c$"

1202 2nd col., Eq. (C36): For "$3\pi^2 \epsilon_0$" read "$3\pi^3 \epsilon_0$"
2nd col., Eq. (C43): For "3.75π" read "4.0π"